P9-BXX-387

MICROELECTRONIC CIRCUITS

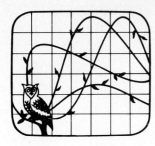

HRW
Series in
Electrical and
Computer Engineering

M. E. Van Valkenburg, Series Editor

MICROELECTRONIC CIRCUITS

ADEL S. SEDRA
Department of Electrical Engineering
University of Toronto

KENNETH C. SMITH
Department of Electrical Engineering and Computer Science
University of Toronto

Holt, Rinehart and Winston

New York   Chicago   San Francisco   Philadelphia
Montreal   Toronto   London   Sydney   Tokyo
Mexico City   Rio de Janeiro   Madrid

TO OUR PARENTS

Library of Congress Cataloging in Publication Data

Sedra, Adel S.
　Microelectronic circuits.

　(HRW series in electrical and computer engineering)
　Bibliography: p.
　Includes index.
　1. Electronic circuits.　I. Smith, Kenneth Carless.
II. Title.　III. Series.
TK7867.S39　　　621.3815′3　　　82-1037
ISBN 0-03-056729-7　　　　　　AACR2

Printed in the United States of America
4 5 6　016　10 9 8 7 6

CBS COLLEGE PUBLISHING
Holt, Rinehart and Winston
The Dryden Press
Saunders College Publishing

PREFACE

Microelectronic Circuits is intended as a text for the core courses in electronic circuits taught to majors in electrical engineering. It should also prove useful for engineers wishing to update their knowledge through self-study.

The objective of the book is to develop in the reader the ability to analyze and design electronic circuits both analog and digital, discrete and integrated. While devoting a good part of the book to the application of integrated circuits (such as op amps, digital logic, and memories), traditional topics such as transistor amplifiers are dealt with in detail. This is done because of our belief that even if discrete circuit design were to disappear entirely, knowledge of what is inside the IC package would enable intelligent and innovative application of these chips.

Although traditional topics are treated, the approach taken is modern. For example, we do not consider the effect of I_{CBO} on the BJT bias point nor do we analyze transistor circuits using *h*-parameters. Rather our treatment enables the student to deal with circuits that contain a large number of transistors as is the case in today's IC packages. This is achieved by focusing attention on circuit configurations and by instilling in the student the ability to perform rapid manual analysis. Design is taught not by the use of "neat" procedures for calculating component values of simple circuits but through consistently pointing out and analyzing the trade-offs available in the selection of circuit configuration and in the selection of component values once the configuration has been decided. We have also attempted to avoid the usual approach of treating simple circuits in great detail with the extension to complex real-life circuits left to the reader.

In addition to providing topics which we consider to be relevant and practical, the book is organized in a modern and flexible manner. Except for an appendix dealing with IC fabrication, device physics is kept to a minimum. We strongly feel that it is no longer necessary to study physical electronics prior to electronic circuits. In fact we have found the reverse sequence, that is, circuits and then device physics, to be more effective. We have, of course, included an elementary description of the physical operation of each device, leaving the more rigorous treatment of device physics to other texts and courses. This physical description is adequate to provide an underlying model which serves to bind together and interrelate higher level circuit concepts.

The prerequisite for studying the material in this book is a first course on circuit analysis. The *s*-domain concept and techniques are not needed until Chapter 11. Thus a first course in electronics based on Chapters 1 to 9 can be taught at the sophomore level, perhaps concurrently with a second course on circuit analysis.

To aid in the learning process, a large number of solved examples are included. We urge the reader to attempt an independent solution of these examples. In addition, almost every section includes one or more exercise problems with answers (about 220 exercises in total). Solving these problems will help the reader gauge his or her grasp of the subject matter of the particular section, as well as, of course, helping her or him in the learning process. Finally, a large number of problems (about 530) keyed to the individual chapters are given in Appendix B. These problems range from simple to very challenging. Answers to a selected number of problems are given in Appendix C. Also, a complete set of problem and exercise solutions is available in the Instructor's Manual which can be obtained by contacting the publisher.

AN OUTLINE FOR THE READER

We shall now present an outline of the material covered. The book starts with an overview of electronic systems and signal processing in Chapter 1. This chapter discusses some electronic system examples and introduces the idea of signal as information container. Its purpose is motivational as well as placing the study of electronic circuits in proper perspective.

Chapter 2 serves as a bridge between the material on circuit analysis normally taught in a first course on this subject and the electronic circuits topics treated in this book. Other objectives of Chapter 2 include: establishing notation and conventions, introducing new concepts such as that of biasing, small-signal operation, amplifier frequency response, etc.

Chapter 3 deals with operational amplifiers, their terminal characteristics, simple applications, and limitations. We have chosen to discuss the op amp as a circuit building block at this early stage simply because it is easy to deal with and because the student can experiment with op-amp circuits that perform nontrivial tasks with relative ease and with a sense of accomplishment. We have found this approach to be highly motivating to the student.

Chapter 4 introduces the ideal diode and the real *pn* junction diode. Here we concentrate on providing an understanding of the diode terminal characteristics and its hierarchy of models. Assuming that the reader has had no prior exposure to physical electronics, we provide a qualitative description of the physical operation of the *pn* junction.

Drawing on the knowledge of op amps and diodes acquired in Chapters 3 and 4, Chapter 5 presents many interesting and practical nonlinear circuit applications. These include various types of rectifiers, limiters and comparators, the bistable, waveform generation, etc. Though it is customary in texts and courses to place this material at an advanced stage, we believe that it fits nicely here since no advanced mathematical methods or concepts are required. These circuits are exciting and have "lots of action" in them, and the circuits perform useful and significant tasks. Of course, the reader may skip some sections of this chapter and return to them at a later stage in his or her study.

Chapter 6 serves as an introduction to digital concepts. In addition to providing traditional material on Boolean algebra, our purpose here is to enable the reader to understand the data sheets of digital ICs. We also treat topics such as flip-flops, registers, counters, multivibrators, and A/D and D/A conversion, and introduce microprocessors.

Each of the following three chapters deals with a single type of transistor: the JFET in Chapter 7, the MOSFET in Chapter 8, and the BJT in Chapter 9. For each device we present a qualitative description of its physical operation. Emphasis is placed on various modes of operation and on using simple models (mostly implicitly) for the rapid analysis of dc, linear, and digital circuits. Our hope is that by the end of each of these chapters the reader becomes thoroughly familiar and intimately comfortable with the device treated. In each of these chapters we consider both discrete and integrated-circuit design.

By the end of Chapter 9 the reader will have learned about the basic building blocks of electronic circuits (op amps, diodes, logic gates, JFETs, MOSFETs, and BJTs), will have become familiar with their characteristics, and will have applied them in a variety of circuits. He or she will be ready to consider the more advanced topics covered in the remainder of the book.

Chapter 10 is the first of a sequence of four chapters devoted to the study of analog circuits emphasizing amplifiers. In Chapter 10 we deal with a number of amplifier configurations, including the emitter follower and the differential pair. We study biasing methods and focus on design trade-offs. In Chapter 11 we study amplifier frequency response. At this stage the reader is expected to know *s*-domain analysis and related topics. We emphasize the choice of configuration to obtain wideband operation. A variety of techniques for manual circuit analysis is studied.

Chapter 12 deals with the important topic of feedback. Practical circuit applications of negative feedback are presented. We also discuss the stability problem in feedback amplifiers and treat frequency compensation in detail.

Chapter 13 presents an introduction to the topic of analog ICs. We discuss circuit design techniques in the context of studying the complete circuit of a popular general-purpose op amp, the 741 type. This chapter should serve to tie together the ideas and methods presented in the previous chapters.

Chapter 13 is concerned with the design of filters and oscillators. Emphasis is placed on circuits utilizing op amps and RC networks. In addition, the recently developed (1977) and highly promising switched-capacitor filters are treated.

The last two chapters of the book, 15 and 16, are concerned with digital circuits. With a few obvious exceptions, these two chapters do not rely on the material in Chapters 10 to 14 and, if desired, can be studied earlier.

In Chapter 15 we present a detailed comparative study of logic circuit families with emphasis placed on TTL, ECL, CMOS, and I^2L. Here we should note that NMOS is treated in Chapter 8 but, if desired, its study can be delayed and performed in conjunction with the material in Chapter 15.

Through a detailed study of a variety of memory types and technologies, Chapter 16 introduces the reader to the subject of VLSI circuits. Included are the latest in dynamic and static RAMs, ROMs, PROMs and EPROMs, and CCDs. In addition, PLAs and gate arrays are introduced. The latter represent the current trend toward semicustom IC design.

TO THE INSTRUCTOR

The book contains sufficient material for a sequence of two one-semester courses (each of about 40 lecture hours). The first of these can be taught at the sophomore (second year) level. The only prerequisite for the first course is a course in elementary circuit analysis. A first course in electronics may be based on Chapters 1 to 9 as follows:

Chapter 1   light treatment and/or reading assignment
Chapter 2   pertinent topics depending on student background

Chapter 3   all, except perhaps for Sections 3.9, 3.10, and 3.11 which may be post-poned to the second course

Chapter 4   all

Chapter 5   all or a selection of topics; some sections may be postponed for the second course

Chapter 6   pertinent topics depending on student background and curriculum structure; e.g., to interact with a separate course dealing with digital systems

Chapter 7   all or Sections 7.1 to 7.7; 7.8 to 7.13 may be postponed for the second course

Chapter 8   all or the digital part may be postponed and studied in conjunction with Chapter 15

Chapter 9   all

The second course requires knowledge of the *s*-plane and related concepts. Also, though not essential, it may be beneficial if a course in physical electronics is given concurrently with or preceding the second course. The second course may be based on Chapters 10 to 16 in addition to sections of the preceding chapters which were not covered in the first course. Essential ingredients of the second course are Chapters 10, 11, 12, and 15. A selection of Chapters 13, 14, and 16 can be included depending on the desired course orientation (e.g., predominantly analog or predominantly digital), curriculum structure, availability of time, etc.

ACKNOWLEDGEMENTS

We wish to express our appreciation for the help given us by a number of individuals: Mrs. Amelia Ma cheerfully and skillfully typed the entire manuscript a number of times. Professor F. E. Holmes supplied us with Appendix A on IC technology. Professors A. J. Cousin and F. E. Holmes and Mr. K. Pelonis used earlier versions of the manuscript in a number of courses and provided us with valuable feedback. Mrs. Laura Jermyn provided expert assistance in preparing the index. Cobb/Dunlop ably guided the manuscript through the various production phases.

It has been a real pleasure to work with our editor, Paul Becker. He provided encouragement at times when it was badly needed. Last but not least, we wish to thank our families for their moral support and understanding.

Adel S. Sedra
Kenneth C. Smith

CONTENTS

ELECTRONIC SYSTEMS

1

The subject of this book is modern electronics, a field that has come to be known as *microelectronics*. Microelectronics refers to the integrated-circuit (IC) technology that at the time of this writing (1981) is capable of producing circuits that contain more than 100,000 components in a small piece of silicon (known as a *silicon chip*) whose area is of the order of 25 mm². One such microelectronic circuit, for example, is a complete digital computer which, accordingly, is known as a *microcomputer* or, more loosely, a *microprocessor*.

In this book we shall study electronic devices that can be used singly (in the design of *discrete* circuits) or as components of an integrated-circuit chip. We will study the design and analysis of interconnections of these devices which form discrete and integrated circuits of varying complexity and which perform a wide variety of functions. We will also learn about available IC chips and their application in the design of electronic systems.

The purpose of this introductory chapter is to set the stage on which electronics will be seen to play a major part as interpreter of the vast body of work known broadly as signal and information processing. While there are other interpreters of this art—mechanics, fluidics, and optics being examples—none has played center stage nor is likely to without the strong support that electronics can provide. Our challenge then has a broad range: to equip the reader with some appreciation of the underlying themes of signal and information processing, and to provide some of the ambience in which (and through which) the play unfolds.

1.1 INFORMATION AND SIGNALS

In the broadest sense, *information* is simply knowledge of a situation, of an event, or of a trend. Likewise, *signals* are the means by which such knowledge is communicated. Thus the status of a mechanical system such as an automobile can be described by many pieces of information, the number of which depends on the need and the context of the situation. Thus while it is usually important for the observer of a parked car to know only its location, for a moving car information on velocity and even acceleration may be necessary. Such information is communicated to the observer by signals, com-

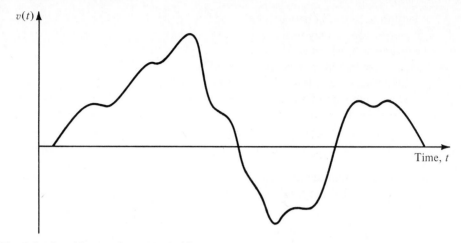

Fig. 1.1   An arbitrary voltage signal v(t).

ponents of the situation which the observer notices. Of course such signals are often not explicit: While the driver of an automobile can observe his speedometer directly, the pedestrian must derive a car's velocity in other ways; although the driver is given a speedometer to read, he must sense acceleration by other means. In such cases an observer receives a collection of signals in which the information he wishes is *encoded:* The information itself is obtained only after some reflection, an activity which is really *signal processing.*

Electronic systems are employed to aid in the process of extracting the required information from a received set of signals. However, in order for a signal to be processed by an electronic system, it must first be converted into an electrical signal, that is, a voltage or a current. This process is accomplished by devices known as *transducers.* A variety of transducers exist, each suitable for one of the various forms of physical signals. For instance, the sound waves generated by a human can be converted to electrical signals using a microphone which is in effect a pressure transducer. It is not our purpose here to study transducers; rather, we shall assume that the signals of interest already exist in the electrical domain. Our concern will be with the processing of such signals using electronic circuits.

From the discussion above it should be apparent that a signal is a time-varying quantity that can be represented by a graph such as that shown in Fig. 1.1. In fact, the information content of the signal is represented by the changes in its magnitude as time progresses; that is, the information is contained in the "wiggles" in the signal waveform. In general, such waveforms are difficult to characterize mathematically. In other words, it is not easy to succinctly describe an arbitrary looking waveform such as that of Fig. 1.1. Of course, such a description is of great importance for the purpose of designing appropriate signal-processing circuits that perform desired functions on the given signal.

1.2   FREQUENCY SPECTRUM OF SIGNALS

An extremely useful characterization of a signal, and for that matter of any arbitrary function of time, is in terms of its *frequency spectrum.* Such a description of signals is obtained through the mathematical tools of *Fourier series* and *Fourier transform.*[1] The Fourier series applies only to signal waveforms that are periodic functions of time. The Fourier transform, which is a limiting case of the Fourier series, can be used to obtain the frequency spectrum of nonperiodic functions.

The Fourier series allows us to express a given periodic function of time as the sum of an infinite number of sinusoids whose frequencies are harmonically related. For instance, the symmetrical square-wave signal in Fig. 1.2 can be expressed as

$$v(t) = \frac{4V}{\pi} (\sin \omega_0 t + \tfrac{1}{3} \sin 3\omega_0 t + \tfrac{1}{5} \sin 5\omega_0 t + \cdots) \tag{1.1}$$

where V is the amplitude of the square wave and $\omega_0 = 2/T$ (T is the period of the square wave) is called the *fundamental frequency*. Note that because the amplitudes of

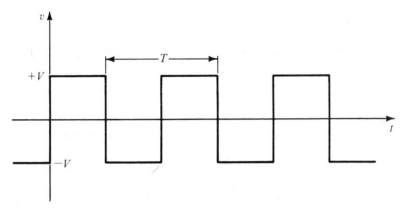

Fig. 1.2   *A symmetrical square-wave signal of amplitude V.*

the harmonics progressively decrease, the infinite series can be truncated, with the truncated series providing an approximation to the square waveform.

The sinusoidal components in the series of Eq. (1.1) constitute the frequency spectrum of the square-wave signal. Such a spectrum can be graphically represented as in Fig. 1.3, where the horizontal axis represents the angular frequency ω in rad/s.

The Fourier transform can be applied to a nonperiodic function of time, such as that depicted in Fig. 1.1, and provides its frequency spectrum as a *continuous* function of frequency, as indicated in Fig. 1.4. Unlike the case of periodic signals where the spectrum consists of discrete frequencies (at ω_0 and its harmonics), the spectrum of a nonperiodic signal contains in general all possible frequencies. Nevertheless, the essential parts of the spectra of practical signals are usually confined to relatively short segments of the frequency (ω) axis—an observation that is very useful in the processing of such signals. For instance, the spectrum of audible sounds such as speech and music extends from about 20 Hz to about 20 kHz—a frequency range known as the *audio*

---

[1] The reader who has not yet studied these topics should not be alarmed. No detailed application of this material will be made until Chapter 11. Nevertheless, the reading of Section 1.2 would be very helpful in earlier parts of the book.

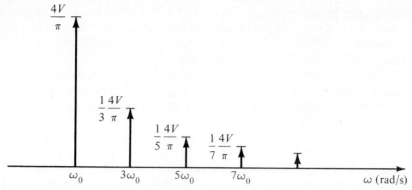

Fig. 1.3 Frequency spectrum (also known as line spectrum) of the periodic square wave of Fig. 1.2.

band. Here we should note that although some music tones have frequencies above 20 kHz, the human ear is incapable of hearing frequencies that are much above 20 kHz.

The concept of frequency spectrum of signals is very important in the design of communication systems such as radio, TV, and telephone systems.

EXERCISE

1.1 When the square-wave signal of Fig. 1.2, whose Fourier series is given in Eq. (1.1), is applied to a resistor, the total power dissipated may be calculated directly using the relationship $P = 1/T \int_0^T (v^2/R) \, dt$, or indirectly by summing the contribution of each of the harmonic components, that is, $P = P_1 + P_3 + P_5 + \dots$ which may be found directly from RMS values. (Hint: $V_{rms} = V_{peak}/\sqrt{2}$ for a sine wave). Verify that the two approaches are equivalent. What fraction of the energy of a square wave is in its fundamental? In its first five harmonics? In its first seven? First nine? In what number of harmonics is 90% of the energy?

Ans. 0.81; 0.96; 0.97; 0.977; 2

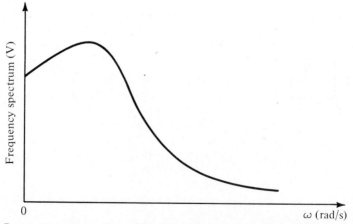

Fig. 1.4 Frequency spectrum of an arbitrary waveform such as that in Fig. 1.1.

1.3 ANALOG AND DIGITAL SIGNALS

The voltage signal depicted in Fig. 1.1 is called an *analog signal*. The name derives from the fact that such a signal is analogous to the physical signal that it represents. The magnitude of an analog signal can take on any value; that is, the amplitude of an analog signal exhibits a continuous variation over its range of activity. The vast majority of signals in the world around us are analog. Electronic circuits that process such signals are known as *analog circuits*. A variety of analog circuits will be studied in this book.

An alternative form of signal representation is that of a sequence of numbers, each number representing the signal magnitude at an instant of time. The resulting signal is called *digital signal*. To see how a signal can be represented in this form—that is, how signals can be converted from analog to digital form—consider Fig. 1.5a. Here the curve represents a voltage signal, identical to that in Fig. 1.1. At equal intervals along the time axis we have marked the time instants t_0, t_1, t_2, and so on. At each of these

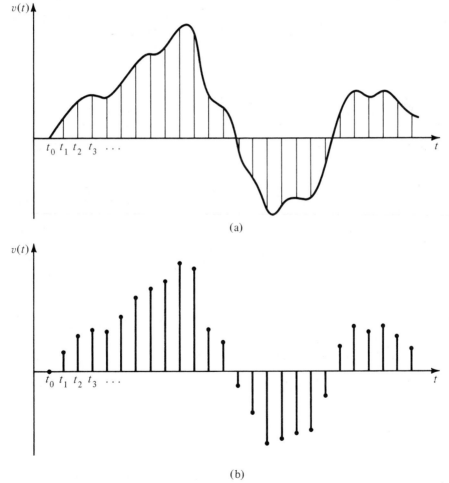

(a)

(b)

Fig. 1.5 *Sampling the continuous-time analog signal in (a) results in the discrete-time signal in (b).*

time instants the magnitude of the signal is measured, a process known as *sampling*. Figure 1.5b shows a representation of the signal of Fig. 1.5a in terms of its samples. The signal of Fig. 1.5b is defined only at the sampling instants; it no longer is a continuous function of time, rather it is a *discrete-time* signal. However, since the magnitude of each sample can take any value in a continuous range, the signal in Fig. 1.5b is still an analog signal.

Now if we represent the magnitude of each of the signal samples in Fig. 1.5b by a number having a finite number of digits, then the signal amplitude will no longer be continuous, rather it is said to be *quantized, discretized,* or *digitized*. The resulting digital signal then is simply a sequence of numbers that represent the magnitudes of the successive signal samples. We will discuss the process of signal sampling and that of converting the samples into numbers (known as analog-to-digital or A/D conversion) in Chapter 6.

Electronic circuits that process digital signals are called *digital circuits*. The digital computer is a system that is constructed of digital circuits. All the internal signals in a digital computer are digital signals. Digital processing of signals has become quite popular primarily because of the tremendous advances made in the design and fabrication of digital circuits. Another reason for the popularity of digital signal processing is that one generally prefers to deal with numbers. For instance, there is little doubt that the majority of us find the digital display of time (as in a digital watch) much more convenient than the analog display (hands moving relative to a graduated dial). While the latter form of display calls for interpretation on the part of the observer, the former is explicit, eliminating any subjective judgment. This is an important point that perhaps is better appreciated in the context of an instrumentation system, such as that for monitoring the status of a nuclear reactor. In such a system human interpretation of instrument readings and the associated inevitable lack of consistency could be hazardous. Furthermore, in such an instrumentation system the measurement results usually have to be fed to a digital computer for further analysis. It would be convenient therefore if the signals obtained by the measuring instruments were already in digital form.

Digital processing of signals is economical and reliable. Furthermore, it allows a wide variety of processing functions to be performed—functions that are either impossible or impractical to implement by analog means. Nevertheless, as already mentioned, most of the signals in the physical world are analog. Also, there remain many signal-processing tasks that are best performed by analog circuits. It follows that a good electronics engineer must be proficient in both forms of signal processing. Such is the philosophy adopted in this text.

Before leaving this discussion of signals we should point out that not all the signals with which electronic systems deal originate in the physical world. For instance, the electronic calculator and the digital computer perform mathematical and logical operations to solve problems. The internal digital signals represent the variables and parameters of these problems and are obviously not supplied directly from external physical signals.

EXERCISE

1.2   Consider the process of communicating information by samples which are too infrequent. For example, list the sequence of output samples of a system which samples at 5-ms intervals starting at time $t = 0$, when supplied with a 100 Hz, ± 1 V symmetrical square wave which goes positive 2.5 ms earlier. What if the input is a sine wave of ± 1 V amplitude at 100 Hz which has the same phase as the fundamental of the square wave? What if a square wave of 200 Hz starting at -1.25 ms is applied? What if a square wave of 400 Hz starting at -0.625 ms is applied?

Ans. $+1, -1, +1, -1, \ldots; +1, -1, +1, -1, \ldots; +1, +1, +1 \ldots; +1, +1, +1,$
$\ldots .$

Note: To convey useful information, sampling must be done at a frequency at least twice that of the highest-frequency component in the signal to be communicated. This is called *Shannon's sampling theorem.*

1.4 AMPLIFICATION AND FILTERING

In this section we shall introduce two fundamental signal-processing functions that are employed in some form in almost every electronic system. These functions are amplification and filtering.

Signal Amplification

From a conceptual point of view the simplest signal processing task is that of *signal amplification.* The need for amplification arises because transducers provide signals that are said to be "weak," that is, in the microvolt (μV) or millivolt (mV) range and possessing little energy. Such signals are too small for reliable processing and processing is much easier if the signal magnitude is made larger. The functional block which accomplishes this task is the *signal amplifier.*

It is appropriate at this point to discuss the need for *linearity* in amplifiers. When amplifying a signal, care must be exercised so that the information contained in the signal is not changed and no new information is introduced. Thus when feeding the signal shown in Fig. 1.1 to an amplifier we want the output signal of the amplifier to be an exact replica of that at the input, except of course for having larger magnitude. In other words, the "wiggles" in the output waveform must be identical to those in the input waveform. Any change in waveform is considered to be *distortion* and is obviously undesirable.

An amplifier that preserves the details of the signal waveform is characterized by the relationship

$$v_o(t) = Av_i(t) \tag{1.2}$$

where v_i and v_o are the input and output signals and A is a constant representing the magnitude of amplification which is known as *amplifier gain.* Equation (1.2) is a linear relationship, hence the amplifier it describes is a *linear amplifier.* It should be easy to see that if the relationship between v_o and v_i contains higher powers of v_i then the waveform of v_o will no longer be identical to that of v_i. The amplifier is then said to exhibit *nonlinear distortion.*

The amplifiers discussed so far are primarily intended to operate on very small input signals. Their purpose is to make the signal magnitude larger and therefore are thought of as *voltage amplifiers.* The *preamplifier* in the home stereo system is an example of a voltage amplifier. However, it usually does more than just amplify the signal; specifically it performs some shaping of the frequency spectrum of the input signal. This topic, however, is beyond our need at this moment.

At this time we wish to mention another type of amplifier, namely, the power amplifier. Such an amplifier provides little or no voltage gain but substantial current gain. Thus while absorbing little power from the input signal source to which it is connected—often a preamplifier, it delivers large amounts of power to its load. An example is found in the power amplifier of the home stereo system, whose purpose is to provide sufficient power to drive the loudspeaker. Here we should note that the loudspeaker is the output transducer of the stereo system; it converts the electrical output signal of the system into an acoustic signal. A further appreciation of the need for linearity can be acquired by reflecting on the power amplifier. A linear power amplifier causes both soft and loud music passages to be reproduced without distortion.

Signal Filtering

Another common signal processing task is that of signal filtering. Filtering is the process by which the essential and useful part of a signal is separated from extraneous and undesirable components that are generally referred to as noise and that somehow get mixed with the signal.

Thus a general view of an electrical filter is that it separates the signal from noise. Here noise is used to refer to two different things. The first is the extraneous signals (also known as *interference*) that might be picked up by the electronic system under consideration. An example of this is found in the interference caused to a radio or a TV set by the switching action in the mechanism of a home appliance. Such an interfering signal is usually coupled to the radio or TV set via transmission over the power supply wiring in the home. In other cases, however, the interference is coupled directly via electromagnetic radiation.

The second item included in the general concept of noise refers to signals generated by the electronic components themselves. Such noise signals are due to the continuous movement of charged particles in the materials of the electronic components. A detailed study of noise in electronic components will not be carried out in this text; the interested reader may consult reference 1.2.[2]

A specific filter example will be discussed in Section 1.5 and the analysis and design of filters will be studied more completely in subsequent chapters. Finally, it should be noted that filtering functions can be implemented by either analog or digital circuits.

EXERCISE

1.3   An amplifier with a gain A of 1,000 remains linear for outputs up to ± 10 V peak-to-peak. For larger outputs (potentially beyond ± 10 V) the output remains at (is limited to) ± 10 V. What are the outputs which result for a square-wave signal peak-to-peak inputs of 1 μV, 100 μV, 1 mV, 10 mV, 100 mV, and 1 V?
   Ans. ± 0.5 mV; ± 50 mV; ± 0.5 V; ± 5 V; ± 10 V; ± 10 V

[2]The references are listed in Appendix D.

1.5 COMMUNICATIONS

Historically, the most important application for electronics has been in the communication of information. It was the goal of information dissemination, exchange, manipulation, and storage that fostered developments in telegraphy, telephony, radio, television, and computers. Accordingly a brief overview of communication systems may serve to introduce concepts that will be found useful in the study of modern electronics.

A Radio Broadcast System

We shall first consider a simple radio broadcast system. Here the information is speech and music, originating from a human source, a record player, or a tape recorder. After transduction and amplification we obtain electrical signals of reasonable strength that are ready for transmission to remotely located listeners. However, a problem arises if one attempts to transmit these signals directly. The problem is twofold: First, since the signals are in the audio-frequency range their wavelengths are quite long and hence the transmitting antenna required to convert the signals to electromagnetic radiation would have to be very long. Second, if all radio stations were to transmit the same frequency band (audio), the remote listener would find it impossible to distinguish between the broadcasts received from the various stations.

A solution to this problem is provided by the process of *modulation* which allows one to shift the frequency band to be transmitted from the audio range to a location at a much higher frequency [in a range known as the *radio frequency* (RF) range]. In this way the required antenna is of moderate length. Furthermore, since each of the radio stations in a given geographical area is assigned a different location in the radio-frequency band for its transmission, the listener can easily select a desired station by simply *tuning* the receiver to the frequency location at which this particular station is broadcasting.

Amplitude Modulation

To be specific let us consider the simplest form of modulation, namely, *amplitude modulation* (AM). In amplitude-modulation systems each station is assigned a separate frequency, known as the carrier frequency f_c, in the range 535 to 1605 kHz. The carrier wave is a sinusoid that can be described by

$$v_c(t) = \hat{V}_c \cos \omega_c t \tag{1.3}$$

where $\hat{V}_c$ is the carrier amplitude and ω_c is its frequency. Amplitude modulation consists of causing the carrier amplitude to vary in correspondence with the instantaneous magnitude of the audio-frequency signal (called the *modulating signal*). To illustrate, assume that the audio signal consists of a single sinusoidal tone given by

$$v_s(t) = \hat{V}_s \cos \omega_s t \tag{1.4}$$

We have to arrange that the instantaneous carrier amplitude becomes $\hat{V}_c + k\hat{V}_s \cos \times \omega_s t$, where k is a constant. This results in an amplitude-modulated wave given by

$$v_{AM}(t) = (\hat{V}_c + k\hat{V}_s \cos \omega_s t) \cos \omega_c t$$
$$= \hat{V}_c(1 + m \cos \omega_s t) \cos \omega_c t \qquad (1.5)$$

where $m \equiv k\hat{V}_s/\hat{V}_c$ is called the *modulation depth* or *modulation index*. Figure 1.6 illustrates the process. It shows that the information to be transmitted is contained in the *envelope* of the AM wave. Note that to avoid distortion, the modulation depth m must be kept less than or equal to unity.

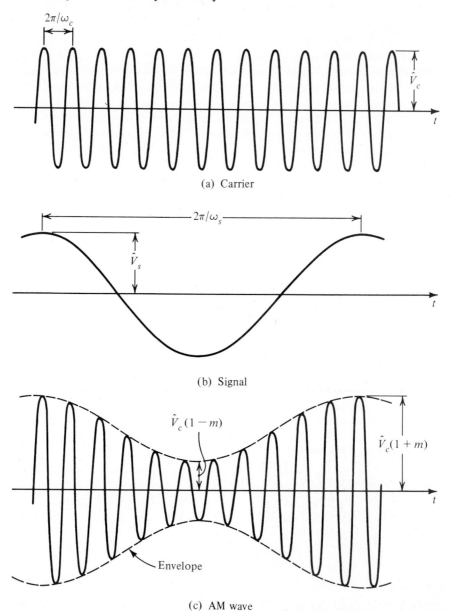

(a) Carrier

(b) Signal

(c) AM wave

Fig. 1.6   *Amplitude-modulating the carrier wave in (a) with the signal waveform in (b) results in the AM wave in (c).*

The AM wave in Eq. (1.5) can be expressed as

$$v_{AM}(t) = \hat{V}_c \cos \omega_c t + \tfrac{1}{2}m\hat{V}_c \cos (\omega_c + \omega_s)t + \tfrac{1}{2}m\hat{V}_c \cos (\omega_c - \omega_s)t \qquad (1.6)$$

Thus the spectrum of the AM wave consists of the carrier at ω_c, a component at $\omega_c + \omega_s$, and another component at $\omega_c - \omega_s$.

An actual audio-frequency signal has a spectrum that occupies a band of frequencies, say from ω_{s1} to ω_{s2}. Correspondingly, the spectrum of the resulting AM wave will consist of the carrier and two bands: the *upper sideband* occupying the range $\omega_c + \omega_{s1}$ to $\omega_c + \omega_{s2}$ and the *lower sideband* occupying the range $\omega_c - \omega_{s2}$ to $\omega_c - \omega_{s1}$. The spectrum therefore extends over the band from $\omega_c - \omega_{s2}$ to $\omega_c + \omega_{s2}$ and is centered around the carrier frequency ω_c, as illustrated in Fig. 1.7. This is the type of spectrum that an AM radio station transmits.

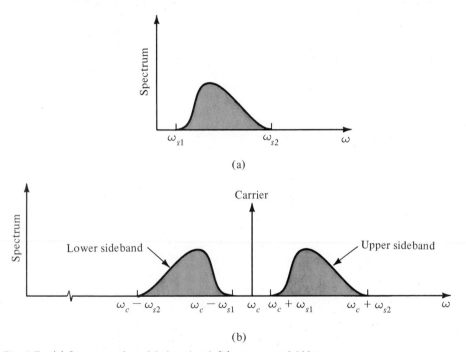

Fig. 1.7 (a) Spectrum of modulating signal; (b) spectrum of AM wave.

Since all the radio stations in a given geographical location have to share the AM broadcast band (535 to 1605 kHz), the band allotted to each station is limited to 10 kHz. This means that the highest audio-frequency transmitted must be only 5 kHz! This in turn implies that the higher-frequency notes present in music are suppressed, which explains the modest quality observed in music received on AM radio.

The AM Radio Receiver

We shall now discuss the AM radio receiver. The receiver antenna (which can be just a piece of wire), located in the field of the transmitting antenna, will have small induced voltage signals corresponding to the various radio transmissions in the geographical area. The antenna delivers these signals to the first stage of the receiver, whose function is twofold: first to select the signal of the desired station, and second to amplify the received signal to a level suitable for further processing. The first stage therefore consists of an amplifier and a *tunable bandpass filter*. A bandpass filter is a filter that passes signals whose spectrum occupies a specified band and rejects (stops or attenuates) other signals. To receive a particular station, the bandpass filter at the receiver input must be tuned so as to make its *passband* coincide with the spectrum of the AM wave emanating from the selected station.

In order to recover the audio signal, the received AM wave is fed to a *demodulator,* also known as a *detector*. A circuit for AM demodulation will be studied in Chapter 5. The audio output of the demodulator is then amplified and fed to the loudspeaker.

This completes our oversimplified description of one of the simplest and most common communication systems. The reader will note the variety of signal-processing functions embodied in such a system.

Telephone Communications

Telephone communication systems have undergone tremendous development in the last few years. Many of these developments have been a direct result of advances in microelectronics. Currently, the telephone system is used to communicate not only speech but also television programs and digital data. Such *data communications* take place between computer terminals located at widely separated locations and a central computer, as well as between two or more computers. Familiar examples of data communication systems include systems used in banks and for airline reservations. Perhaps less glamorous but extremely useful is the ability to dial up a computer from one's home, connect the telephone via an *acoustic coupler* to a computer terminal, and then proceed to enter one's programs on the terminal. This arrangement was employed for typing and editing a good part of the manuscript for this book.

Advances in microelectronics are also allowing the addition of "intelligence" capabilities to the telephone set. As an example, "smart" telephones that "remember" frequently dialed numbers and that allow the user to reach any one of such numbers by pushing a single button are currently available.

Let us briefly examine the telephone communication system. For the sake of simplicity, let us restrict outselves to speech communication. In order to make it possible to use a single telephone cable for the transmission of many separate conversations (separated in what is called *telephone channels*) the bandwidth allotted for each channel is limited to about 4 kHz (actually, from 300 to 3400 Hz). It has been established that this relatively narrow bandwidth is sufficient for the transmission of speech signals and provides speech of reasonable quality. After transduction, the speech signal is fed to a *low-pass filter* whose passband extends to about 4 kHz. This low-pass filter eliminates the higher-frequency components, thus avoiding the possibility of the various channels carried on the same cable "running into one another."

To carry a number of telephone channels on a single cable, modulation is used to shift the spectrum of each channel to a different frequency band. Such a system is known as a *frequency division multiplexing* (FDM) system. At the receiving end the channels are separated using filtering and demodulation.

Another more recent approach to multiplexing a number of telephone channels on a single cable is the use of *time division multiplexing* (TDM). In TDM the signals from the channels to be multiplexed (say 12 channels) are sampled in sequence. That is, a sample is taken from Channel 1, then a sample taken from Channel 2, and so on to the last channel, Channel 12. Then a new round of sampling is started and the process is periodically repeated. The samples are transmitted on the same cable in the same order in which they were obtained. Thus samples 1, 13, 25, etc. belong to the signal of Channel 1, and so on.

At the receiving end, the samples have to be separated, a process known as *demultiplexing*. The samples of each particular channel are then used to reconstruct a good approximation to the audio signal being transmitted.

TDM systems usually convert the signal samples to digital form and the signals transmitted are actually digits. Due to the advances made in the digital circuits area and to the fact that the telephone network is carrying more and more digital data, the current trend is to perform a great deal of the processing digitally. It is probable therefore that in the not too distant future each telephone will contain its own analog-to-digital (A/D) and digital-to-analog (D/A) converters. Nevertheless, analog circuits will still be needed for processing the speech signals prior to converting to digital form and for reconstructing an analog signal from the received digital information.

EXERCISE

1.4 To preserve adequate information in a TDM telephone system, signal sampling must be done at least at twice the upper frequency of the 300- to 3400-Hz channel. Available technology allows a sample amplitude to be transmitted in 10 μs. What is the greatest number of channels that can be multiplexed on a single line in such a system?
Ans. 14

1.6 COMPUTERS

The digital computer has already been mentioned on a number of occasions in this introductory chapter, especially in connection with data communication systems which is an area that is growing and advancing at a rapid pace. The digital computer will also be mentioned in Section 1.7 because of the key role it is playing in instrumentation and control systems. In fact, computers from microprocessors to large machines are playing a role in almost every aspect of our daily lives. It is therefore futile to attempt to list the variety of tasks in which computers are employed. Rather, we wish first to point out the close relationship between developments in microelectronics and those in computers. These two areas have greatly influenced each other and will likely continue to do so. Although the original digital computers used vacuum tubes, it was the invention of the transistor about 30 years ago and the silicon integrated circuit a decade later that made possible the digital computers that we know today. In turn, the demand for digital computers that are cheaper, smaller, faster, more powerful, and so on, has accelerated the development of microelectronics.

The development of the microprocessor some eight years ago has to be regarded as the beginning of a new era in the application of microelectronics. For the first time the concept of distributed computing or *distributed intelligence* has become feasible. By including microprocessors in a variety of devices, such as instruments and computer terminals, such devices acquire a degree of intelligence or "smartness" that enables them to do considerable processing. If desired, these local processors can be made to communicate among themselves as well as with a more powerful central computer for further processing and higher-level decision making.

We conclude this section with a discussion of the oversimplified block diagram of a digital computer, shown in Fig. 1.8. The digital computer consists of three basic

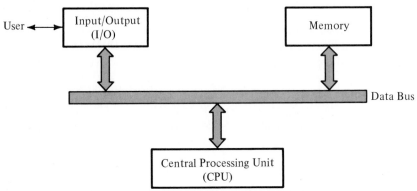

Fig. 1.8 *An oversimplified block diagram of a digital computer.*

blocks: The *input/output* (I/O) devices that enable the user to interact with the machine, to enter programs and data, and to obtain results; the *memory* in which the program and data are stored; and the *central processing unit* (CPU) which is the heart of the machine where arithmetic and logical operations are performed and the various control signals are generated.

Under the control of a program whose instructions reside in memory, the CPU operates on the input data, which is also stored in memory. As a result of these arithmetic and logical operations, decisions are made regarding, for instance, the sequence of executing program instructions. Ultimately, output data are generated.

The required I/O devices such as video terminals, magnetic tape units, printers, plotters, and so on, as well as digital computers themselves and other digital systems are constructed of two digital circuit types: logic gates and memory circuits. We shall study both logic and memory circuits in detail in subsequent chapters.

1.7 INSTRUMENTATION AND CONTROL

Another area in which microelectronics is playing a key role is that of measurement and control. The measurement of a variety of phenomena is the key to the automatic control of variables and processes ranging from the temperature in our homes to the operation of industrial plants.

An instrumentation system can be as simple as just one instrument measuring a variable of interest and displaying the result as the reading of an analog or a digital meter. In such a system it is left to the human observer to react to the measurement result and make adjustments to the process being monitored, if such adjustments are called for.

A further level of sophistication is achieved by having the instrument compare the result of its measurement with a preset value or range of values. If the measured value is found to be outside the preset range, a "flag" is automatically raised, for instance an alarm lamp is lit. Then again it is up to the human observer to take corrective action.

An even greater level of sophistication is obtained by having the instrumentation system determine what kind of corrective action is required, initiate such action, and supervise its completion. A simple example of such a system is found in the home thermostat, a device no doubt familiar to most readers. Such an *automatic control system* is based on a *negative-feedback loop*. The loop consists essentially of a *sensor* (transducer), a *processor* that today is most often electronic, and an *actuator* which is a device that responds to the commands of the processor and causes the corrective action to take place physically.

Traditionally, the processor in an instrumentation system has been analog with a growing use of separate digital logic and memory circuits. Only the most sophisticated systems employ a central computer to do the bulk of the processing. However, the advent of microprocessors and other advances in analog and digital microelectronic circuits is causing a major change to occur in instrumentation systems. Each sensor can now be equipped with sophisticated analog circuits as well as a microprocessor, thus enabling the sensor to do considerable processing locally. Such a "smart" sensor would also be capable of making local decisions. It would then need to communicate only infrequently with a central computer. The central computer would therefore be relieved of routine processing and be able to implement more extensive programs of automation. Some such automatic control systems are already available; witness the use of robots in the manufacture of automobiles.

Advances in microelectronics are also influencing such routine laboratory instruments as the oscilloscope, the function generator, the spectrum analyzer, the digital multimeter, and so on. Newer versions of these instruments feature such facilities as automatic adjustment, improved display, self-calibration, and others. The result is an instrument that is much easier to use and that relies less on the skill and attentiveness of the human operator. In addition, such features permit many more measurements to be made in a given length of time.

Finally, there is the area of *automatic testing,* an essential aspect of an assembly line. This area also has benefited greatly from advances in microelectronics and specifically those in digital computer technology. For further reading on the influence of microelectronics on instrumentation the reader is urged to consult reference 1.4.

In conclusion, the area of instrumentation and control is a fertile one for the application of the latest developments in analog and digital microelectronic circuits.

1.8 CONCLUDING REMARKS

The focus of this chapter has been on signals and signal processing. The electronic system examples discussed are intended to illustrate the variety of signal-processing tasks with which electronics engineers deal. We have attempted also to illustrate the role played by microelectronics in at least three broad-ranging areas: communications, computers, and instrumentation and control.

It is not the purpose of this book to study the design of any particular kind of electronic system (such as a radio, a TV, a computer, etc.). Rather, we shall study the principles of design and analysis of electronic circuits that realize many of the signal-processing functions needed in a variety of electronic systems. The circuits treated are both analog and digital, discrete and integrated. While our approach is not to provide recipes for designing specific circuits, it is very down to earth. It is our goal that the reader develop the abilities and sharpen the skills required to innovate and create in this exciting field.

If the reader has not already done so, he is encouraged to read the overview of the text given in the Preface. Such an effort will prove useful before delving further into this book.

LINEAR CIRCUITS

2

Electronic circuits can be broadly divided into two categories: *linear circuits* and *nonlinear and digital circuits.* Filters and amplifiers are examples of linear circuits, whereas logic and memory elements belong to the digital circuits category.

Introduction

The present chapter is devoted to the study of fundamentals of linear circuit analysis. A good part of the material presented is a review of circuit analysis techniques, but our objective here is to demonstrate and illustrate specific applications of these techniques in electronics. As will become apparent throughout this book, an attribute of a "good" circuit designer is the ability to perform rapid circuit analysis.

In addition to reviewing important topics from linear circuit theory, this chapter introduces a number of new concepts and establishes terminology and conventions used throughout the text.

2.1 RESISTIVE ONE-PORT NETWORKS

Figure 2.1a shows a two-terminal, or one-port, network labeled A. The *total instantaneous voltage* between the two terminals is denoted $v_A(t)$, and the total instantaneous current through the terminals is denoted $i_A(t)$. The static *i-v* characteristic of this one-port network is shown in Fig. 2.1b. Such a characteristic can be measured point by point by first applying a constant (dc) voltage V_A and measuring the corresponding current I_A; then the value of V_A is changed and a new value of I_A is measured, and so on. The characteristic curve shown in Fig. 2.1b is clearly nonlinear; it is not simply a straight line. Nonlinear networks play a very important role in electronic systems. For example, the design of logic circuits makes effective use of the nonlinearity of transistors; however, there are a large number of signal-processing tasks where linear operation is a requirement, such as in the design of linear amplifiers, which constitutes a large part of this text. Since all active devices, such as transistors and vacuum tubes, are basically nonlinear, a technique had to be found for using these devices in a linear or an almost linear mode. The technique, which we illustrate below, is very important and will be employed on numerous occasions throughout this book.

A method for using a nonlinear element to obtain linear operation is illustrated in Fig. 2.1c. A dc voltage V_A is applied to *bias* the device at a point Q, called the *quiescent* or *bias point*. The device is conducting a dc current I_A corresponding to the dc bias voltage V_A. Superimposed on V_A is the time-varying signal $v_a(t)$. For illustration, we

17

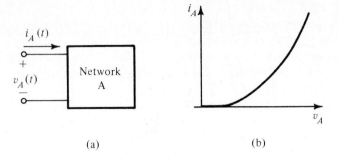

(a)   (b)

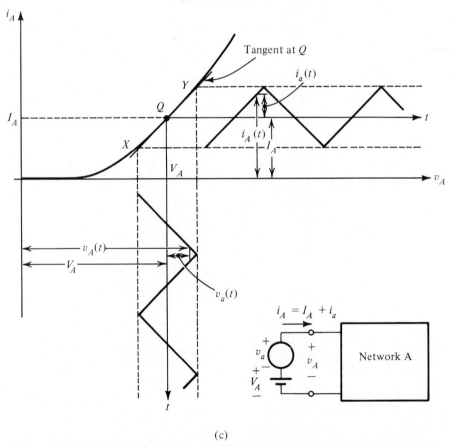

(c)

Fig. 2.1   (a) General one-port network A. (b) The i-v characteristic of network A. (c) Network A biased by a dc voltage V_A at a point Q; a small voltage signal $v_a(t)$ is superimposed on V_A.

are assuming in Fig. 2.1c that $v_a(t)$ has a triangular waveform. If the amplitude of $v_a(t)$ is small enough so that operation is restricted to an almost linear segment (the segment labeled *XY*) of the device's *i-v* characteristic, then the signal current $i_a(t)$ will have the same waveform as $v_a(t)$. The signal current $i_a(t)$ will be related to $v_a(t)$ through the slope of this almost linear segment of the characteristic curve. This slope is equal to the slope of the tangent to the *i-v* characteristic at the bias point *Q*. If this slope is denoted *g*,

$$ g = \left. \frac{\partial i_A}{\partial v_A} \right|_{v_A = V_A} $$

then

$$ i_a(t) = g v_a(t) $$

Note that *g* will have the dimension of conductance. Since *g* relates the small-signal voltage $v_a(t)$ to the corresponding current $i_a(t)$, it is referred to as the *small-signal* or *incremental conductance*. The value of the small-signal conductance obviously is dependent on the bias point *Q*.

It should be evident from Fig. 2.1c that if the amplitude of the voltage signal is increased, operation will no longer be restricted to a linear segment of the *i-v* characteristic, and the waveform of the current signal $i_a(t)$ will be different from that of $v_a(t)$. This change in waveform is undesirable and constitutes a distortion. Nonlinear distortion introduces new and spurious "information" into the signal and is thus undesirable.

Once the circuit is considered linear, analysis is greatly simplified. The remainder of this chapter is devoted to the study of linear networks.

Before leaving this section we wish to draw the reader's attention to the terminology just introduced. Total instantaneous quantities are denoted by a lowercase symbol with an uppercase subscript, for example, $i_A(t)$, $v_C(t)$. Direct-current (dc) quantities will be denoted by an uppercase symbol with an uppercase subscript, for example, I_A, V_C. Finally, incremental signal quantities will be denoted by a lowercase symbol with a lowercase subscript, for example, $i_a(t)$, $v_c(t)$.

EXERCISE

2.1 Consider a one-port network D having the *i-v* characteristic

$$ i_D = I_S e^{v_D / V_T} $$

where $I_S = 10^{-15}$ A and $V_T = 25$ mV. Find the value of the incremental resistance r_D at a bias current $I_D = 1$ mA. Also find the value of the dc bias voltage V_D.
Ans. 25 Ω, 690.8 mV

2.2 TWO-PORT NETWORKS

Figure 2.2a shows a black-box representation of a two-port network. In this book we are concerned with the analysis and design of many two-port networks. Let us assume that the two-port network of Fig. 2.2a is a linear one, and let us as an example assume that its purpose is to amplify voltage signals. Figure 2.2b shows the two-port network fed from a signal source whose voltage is $v_S(t)$ and whose output resistance is R_S. The two-port amplifier delivers its output signal to a load resistance R_L. Port 1 is the input port of the amplifier, and port 2 is its output port. Since in general the input of the amplifier will draw a current $i_1(t)$, the voltage that appears between the two input terminals of the amplifier—that is, $v_1(t)$—will not be equal to the source voltage $v_S(t)$. The difference between the two voltages is equal to the voltage drop $R_S i_1(t)$. The gain

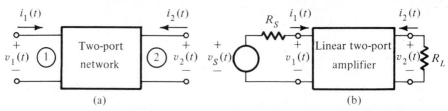

Fig. 2.2 (a) Black-box representation of a two-port network. (b) A linear network used as an amplifier. The amplifier is fed from a signal source $v_S(t)$ whose resistance is R_S; it delivers its output into a load resistance R_L.

of the amplifier is equal to the ratio of the voltage across the load resistance R_L—that is, $v_2(t)$—to the source voltage $v_S(t)$. We shall return to the analysis of two-port networks—specifically, amplifying two-port networks—very shortly.

At this point we wish to introduce a number of important concepts and terminology. The word *amplifier* deserves some clarification. An amplifier increases the signal power, an important feature that distinguishes an amplifier from a transformer. In the case of a transformer, although the voltage delivered to the load could be greater than the voltage feeding the input side (primary), the power delivered to the load is equal to or smaller than the power supplied by the signal source. On the other hand, a two-port network operating as an amplifier provides the load with power greater than that obtained from the signal source. The question then arises as to where the additional power comes from. The answer is found by observing that amplifiers need dc supplies for their operation. These dc sources supply the extra power delivered to the load as well as any power that might be dissipated in the two-port network itself.

Figure 2.3a shows the circuit symbol commonly used for a two-port network operating as an amplifier. Note that the symbol clearly distinguishes the input and output ports. For generality we have shown the amplifier to have two input terminals that are distinct from the two output terminals. A more common situation is illustrated in Fig. 2.3b, where a common terminal exists between the input and the output ports of the amplifier. This common terminal is used as a reference point and is called the *circuit ground*.

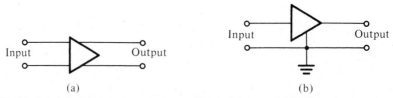

Fig. 2.3 (a) Circuit symbol for a two-port amplifier. (b) An amplifier with a common (ground) terminal between the input and output ports.

As already mentioned, amplifiers require dc sources for their operation. In Fig. 2.3 we have not explicitly shown these dc sources. Figure 2.4a shows an amplifier that requires two dc sources, one positive, of value V_1, and one negative, of value V_2. The amplifier has two terminals, marked V^+ and V^- for connection to the dc sources. For the amplifier to operate, the terminal labeled V^+ has to be connected to the positive side

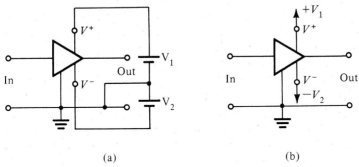

(a)   (b)

Fig. 2.4   *Illustrating an amplifier that requires two dc power supplies for operation.*

of a dc source whose voltage is V_1 and whose negative side is connected to the circuit ground. Also, the terminal labeled V^- has to be connected to the negative side of a dc source whose voltage is V_2 and whose positive side is connected to the circuit ground.

In order to simplify circuit diagrams, we shall adopt the convention illustrated in Fig. 2.4b. Here the V^+ terminal is shown connected to an arrowhead pointing upward and the V^- terminal to an arrowhead pointing downward. Next to each arrowhead the corresponding voltage is indicated. Note that in some cases we will not explicitly show the connections of the amplifier to the dc power sources. Finally, we note that some amplifiers require only one power supply.

2.3   VOLTAGE AMPLIFIERS

The Ideal Amplifier

Figure 2.5a shows the equivalent circuit model of an *ideal* voltage amplifier. The ideal voltage amplifier is essentially a voltage-controlled voltage source. It accepts voltage signals between its two input terminals, terminal 1 and ground, and provides a magni-

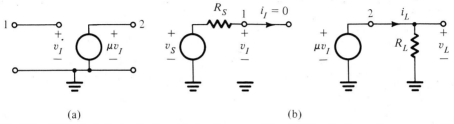

(a)   (b)

Fig. 2.5   *(a) Equivalent circuit of an ideal voltage amplifier. (b) The ideal amplifier connected to signal source and load.*

fied replica of v_I as an ideal voltage source between the output terminals, terminal 2 and ground. As the model indicates, the input port of the amplifier looks like an open circuit. Thus if the amplifier is fed by a voltage source v_S having a source resistance R_S, as shown in Fig. 2.5b, no current will flow into the amplifier input terminals. The result of the infinite input resistance of the ideal amplifier is that the entire source voltage v_S will appear between the amplifier input terminals,

$$v_I = v_S$$

Thus v_S will be amplified by the amplifier voltage gain μ.

At the output side, since the controlled source is ideal, it will supply any required current to the load resistance R_L without changing the value of the load voltage. The result of the zero output resistance of the ideal voltage amplifier is that the voltage (μv_I) will appear directly across the load resistance R_L,

$$v_L = \mu v_I$$

Thus the overall voltage gain A_v, defined

$$A_v \equiv \frac{v_L}{v_S}$$

will be given by

$$A_v = \mu$$

In the amplifier circuit of Fig. 2.5b the power supplied by the source is zero, while that delivered to the load is

$$P_L = \frac{v_L^2}{R_L}$$

Thus the power gain is infinite, with the load power being supplied by the amplifier dc supplies, which are not explicitly shown.

EXERCISE

2.2   Consider an amplifier that requires one dc supply of 15 V for its operation. The amplifier is fed by a 2-V signal and draws negligible current from the signal source. The amplifier delivers a 10-V signal to a 1-kΩ load resistance. It is known that the power dissipation in the amplifier is 20 mW. Calculate the dc current that must be supplied by the 15-V power supply.
   Ans.   8 mA

Effect of Input and Output Resistances

Real voltage amplifiers have a finite (that is, not infinite) input resistance and a finite (that is, nonzero) output resistance. Figure 2.6a shows the equivalent circuit of a voltage amplifier having an input resistance R_i and an output resistance R_o. To appreciate the effect of the input and output resistances, consider the circuit in Fig. 2.6b. The amplifier input voltage v_I can be found using the voltage divider rule,

$$v_I = v_S \frac{R_i}{R_i + R_S} \tag{2.1}$$

It follows that in order for us not to lose a significant part of the input signal in coupling the signal source to the amplifier input, the input resistance R_i should be high in comparison to the source resistance R_S. How high R_i should be will depend on the particular application. For example, some transducers have output resistances on the order of few megohms, in which case an appropriate amplifier should have an input resistance on the order of 10^7 Ω.

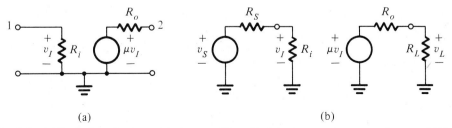

(a)   (b)

Fig. 2.6   (a) Equivalent circuit model of an amplifier with finite input and output resistance. (b) The real amplifier in part (a) connected to signal source and load.

In the circuit of Fig. 2.6b the load voltage v_L can be found by applying the voltage-divider rule,

$$v_L = \mu v_I \frac{R_L}{R_L + R_o} \tag{2.2}$$

It follows that in order for us not to lose gain in coupling the amplifier output to a load, the output resistance R_o should be much smaller than the load resistance R_L. Substituting v_I from Eq. (2.1) into Eq. (2.2) results in the overall voltage gain A_v:

$$A_v \equiv \frac{v_L}{v_S} = \mu \frac{R_i}{R_i + R_S} \frac{R_L}{R_L + R_o}$$

To emphasize the effects of R_i and R_o in lowering the voltage gain we rewrite this equation in the form

$$A_v = \mu \frac{1}{1 + R_S/R_i} \frac{1}{1 + R_o/R_L} \tag{2.3}$$

Obviously, for $R_i \gg R_S$ and $R_o \ll R_L$ the voltage gain becomes

$$A_v \simeq \mu$$

In the case of a finite input resistance, the power gain of the amplifier is no longer infinite; however, it should still be high, or there would be no need for the amplifier. In fact there are situations in which one is interested not in the voltage gain but in a significant power gain. For instance, the source signal can be of a respectable voltage but the source resistance can be much greater than the load resistance. Connecting the source directly to the load would result in significant signal attenuation. In such a case one requires an amplifier with a high input resistance (much greater than the source resistance) and a low output resistance (much smaller than the load resistance) but with a modest voltage gain (or even unity gain). Such an amplifier is referred to as a *buffer amplifier*. We will encounter buffer amplifiers often throughout this text.

Expressing Gain in Decibels

For a number of reasons, some of them historic, electronics engineers express amplifier gain with a logarithmic measure. Specifically the voltage gain A_v can be expressed as

$$\text{Voltage gain in decibels} = 20 \log|A_v| \quad \text{dB}$$

Since power is related to voltage squared, the power gain A_p can be expressed in decibels:

$$\text{Power gain in decibels} = 10 \log A_p \quad \text{dB}$$

If in a specific application the current gain A_i is of interest, it too can be expressed in decibels:

$$\text{Current gain in decibels} = 20 \log|A_i| \quad \text{dB}$$

The absolute values of the voltage and current gains are used because in many cases A_v or A_i may be negative numbers. A negative gain A_v simply means that there is a 180° phase difference between input and output signals; it does not imply that the amplifier is *attenuating* the signal. On the other hand, an amplifier whose voltage gain is -20 dB is in fact attenuating the input signal by a factor of 10 (that is, $A_v = 0.1$).

Example 2.1

Figure 2.7 depicts an amplifier composed of a cascade of three stages. The amplifier is fed by a signal source with a source resistance of 100 kΩ and delivers its output into a

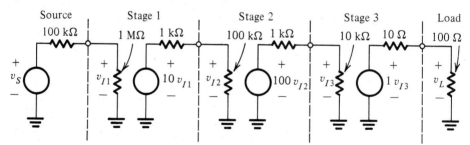

Fig. 2.7 *Three-stage amplifier for Example 2.1.*

load resistance of 100 Ω. The first stage has a relatively high input resistance and a modest gain of 10. The second stage has a higher gain but lower input resistance. Finally, the last, or output, stage has unity gain but a low output resistance. We wish to evaluate the voltage and power gains.

Solution

The voltage gain of the first stage, taking the source resistance and the input resistance of the second stage into account, is

$$A_{v1} \equiv \frac{v_{I2}}{v_S} = \frac{1000}{1000 + 100} \, 10 \, \frac{100}{100 + 1} = 9$$

The voltage gain of the second stage, taking the input resistance of the third stage into account, is

$$A_{v2} \equiv \frac{v_{I3}}{v_{I2}} = 100 \frac{10}{10 + 1} = 90.9$$

Finally, the voltage gain of the output stage is given by

$$A_{v3} \equiv \frac{v_L}{v_{I3}} = 1 \frac{100}{100 + 10} = 0.909$$

The overall voltage gain can be obtained by multiplying the gains of the three stages:

$$A_v \equiv \frac{v_L}{v_S} = A_{v1}A_{v2}A_{v3}$$

Thus

$$A_v = 743.7$$

or, in decibels,

$$\text{Voltage gain} = 57.4 \text{ dB}$$

To obtain the power gain we first find the load power,

$$P_L = \frac{v_L^2}{100}$$

and the source power,

$$P_S = \frac{v_S^2}{10^5 + 10^6}$$

Thus the power gain is

$$A_p \equiv \frac{P_L}{P_S} = \left(\frac{v_L}{v_S}\right)^2 (1.1 \times 10^4)$$

Thus

$$A_p = A_v^2 \times 1.1 \times 10^4$$

In decibels we have

$$\text{Power gain} = 20 \log A_v + 40.4 = 97.8 \text{ dB}$$

$$\bullet \quad \bullet \quad \bullet$$

EXERCISE

2.3 Consider a unity-gain buffer amplifier with 1-MΩ input resistance and 100-Ω output resistance. It is fed by a source having a resistance of 200 kΩ and connected to a load of 1-kΩ resistance. Calculate the voltage, current, and power gains in decibels.
Ans. -2.4 dB; 59.2 dB; 28.4 dB

2.4 OTHER AMPLIFIER TYPES

In the design of an electronic system the signal of interest, whether at the system input, at an intermediate stage, or at the output, can be either a voltage or a current. For instance, some transducers have very high output resistances and can be more appropriately modeled as current sources. Similarly, there are applications in which the output current rather than the voltage is of interest. Thus, although it is the most popular, the voltage amplifier considered in the previous section is just one of four possible amplifier types. The other three are current amplifier, transconductance amplifier, and transresistance amplifier.

Current Amplifier

The ideal current amplifier can be modeled by a current-controlled current source, as shown in Fig. 2.8a. Figure 2.8b shows a practical current amplifier fed by a current signal source and connected to a load resistance R_L. In order for the practical current

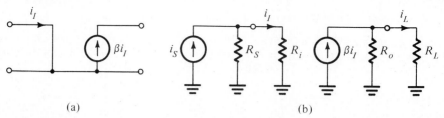

(a)   (b)

Fig. 2.8   (a) Ideal current amplifier. (b) Practical current amplifier connected to source and load.

amplifier to approach the ideal behavior, its input resistance R_i should be much smaller than the source resistance R_S and its output resistance R_o should be much larger than the load resistance R_L. In this case the current gain

$$A_i \equiv \frac{i_L}{i_S} = \frac{R_S}{R_S + R_i} \beta \frac{R_o}{R_o + R_L}$$

approaches the ideal value β,

$$A_i \simeq \beta$$

Transconductance Amplifier

The ideal transconductance amplifier can be modeled by a voltage-controlled current source, as shown in Fig. 2.9a. Figure 2.9b shows a practical transconductance amplifier

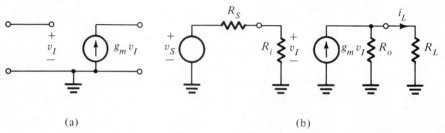

(a)   (b)

Fig. 2.9   (a) Ideal transconductance amplifier. (b) Practical transconductance amplifier connected to source and load.

fed by a voltage signal source and connected to a load resistance R_L. In order for the practical transconductance amplifier to approach the ideal behavior, its input resistance R_i should be much larger than the source resistance R_S and its output resistance R_o should be much larger than the load resistance R_L. In this case the transconductance gain A_g, which has the dimension of mhos or A/V,

$$A_g \equiv \frac{i_L}{v_S} = \frac{R_i}{R_S + R_i} g_m \frac{R_o}{R_o + R_L}$$

approaches the ideal value g_m,

$$A_g \simeq g_m$$

Transresistance Amplifier

The ideal transresistance amplifier can be modeled by a current-controlled voltage source, as shown in Fig. 2.10a. Figure 2.10b shows a practical transresistance amplifier

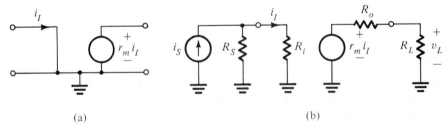

(a)   (b)

Fig. 2.10 (a) Ideal transresistance amplifier. (b) Practical transresistance amplifier connected to source and load.

fed by a current signal source and connected to a load resistance R_L. In order for the practical transresistance amplifier to approach the ideal behavior, its input resistance R_i should be much smaller than the source resistance R_S and its output resistance R_o should be much smaller than the load resistance R_L. In this case the transresistance gain A_r, which has the dimension of ohms,

$$A_r \equiv \frac{v_L}{i_S} = \frac{R_S}{R_S + R_i} r_m \frac{R_L}{R_L + R_o}$$

approaches the ideal value r_m,

$$A_r \simeq r_m$$

Example 2.2

The *bipolar junction transistor (BJT)* is a three-terminal device having the circuit symbol shown in Fig. 2.11a with the terminals called *emitter* (E), *base* (B), and *collector* (C). The device is basically nonlinear; thus the relationships between the total instantaneous terminal currents and voltages are generally nonlinear. Nevertheless, the linearization technique discussed in Section 2.1 can be used to provide linear operation for *small signals,* that is, incremental variations in currents and voltages around a *bias or*

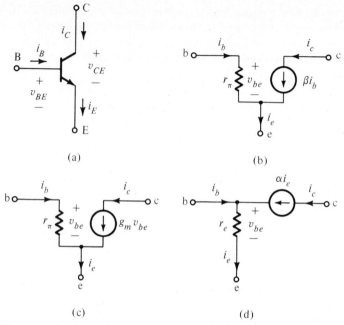

Fig. 2.11 *(a) Circuit symbol for the bipolar junction transistor (BJT), illustrating the definition of the total instantaneous terminal quantities. (b, c, d) Equivalent circuit models for the small-signal linear operation of the BJT. (See Example 2.2.)*

quiescent point. Thus the total instantaneous quantities can be expressed as the sums of dc or bias quantities and signal quantities as

$$v_{BE} = V_{BE} + v_{be} \qquad i_B = I_B + i_b$$
$$v_{CE} = V_{CE} + v_{ce} \qquad i_C = I_C + i_c$$
$$i_E = I_E + i_e$$

It will be shown in Chapter 9 that under the so-called small-signal approximation the relationships between the signal quantities are linear and that the three-terminal transistor can be represented by one of the equivalent circuit models of Fig. 2.11b, c, and d. If these models are equivalent, we wish to find the relationships between their parameters.

Solution
For the model in Fig. 2.11b we have

$$i_b = \frac{v_{be}}{r_\pi} \tag{2.4}$$

$$i_c = \beta i_b \tag{2.5}$$

$$i_e = (\beta + 1)i_b \tag{2.6}$$

For the model in Fig. 2.11c we have

$$i_c = g_m v_{be} \qquad (2.7)$$

Use of Eqs. (2.5), (2.4), and (2.7) gives

$$g_m = \frac{\beta}{r_\pi}$$

For the model in Fig. 2.11d we have

$$i_b = (1 - \alpha)i_e \qquad (2.8)$$
$$i_c = \alpha i_e \qquad (2.9)$$
$$i_e = \frac{v_{be}}{r_e} \qquad (2.10)$$

Use of Eqs. (2.5) and (2.6) gives

$$\frac{i_c}{i_e} = \frac{\beta}{\beta + 1}$$

Comparing this result with Eq. (2.9), we get

$$\alpha = \frac{\beta}{\beta + 1}$$

Equations (2.4) and (2.6) can be combined to yield

$$i_e = \frac{\beta + 1}{r_\pi} v_{be}$$

Comparison of this result with Eq. (2.10) provides

$$r_e = \frac{r_\pi}{\beta + 1}$$

We have thus obtained expressions for the parameters of the models in Fig. 2.11c and d in terms of those for the model in Fig. 2.11b.

· · ·

Example 2.3

A *gyrator* is a two-port network that can be realized by connecting in parallel and back to back two voltage-controlled current sources of opposite polarities. Such an arrangement is shown in Fig. 2.12a (we assume ideal voltage-controlled current sources). It is required to show that if port 2 of the gyrator is terminated in a capacitance, as shown in Fig. 2.12b, then input port 1 will have the *i-v* relationship of an inductance.

Solution

We must show that in the circuit of Fig. 2.12b

$$v_1 = L \frac{di_1}{dt}$$

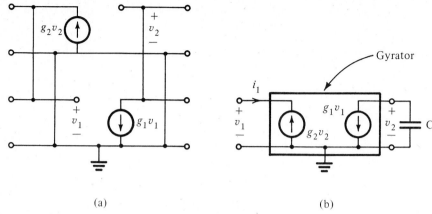

(a)   (b)

Fig. 2.12   (a) Two voltage-controlled current sources of opposite polarities connected back-to-back in parallel to form a gyrator. (b) The gyrator of (a) terminated at port 2 in a capacitance C. (See Example 2.3.)

where L is a constant representing an inductance. To obtain the relationship between i_1 and v_1, consider first port 2, where we have

$$v_2 = -\frac{1}{C}\int_0^t g_1 v_1 \, dt$$

where it has been assumed that C was originally (at $t = 0$) uncharged. The current i_1 is given by

$$i_1 = -g_2 v_2 = \frac{g_1 g_2}{C}\int_0^t v_1 \, dt$$

which can be written in differential form as

$$v_1 = \frac{C}{g_1 g_2}\frac{di_1}{dt}$$

Thus the i_1-v_1 relationship is of the form found in an inductance, with the inductance L given by

$$L = \frac{C}{g_1 g_2}$$

Thus the gyrator can be used to realize an inductance using active elements (controlled sources) and a capacitance. This is a significant and useful result because physical inductors are almost impossible to realize using integrated circuit (IC) technology.

· · ·

EXERCISES

2.4   Consider a transistor modeled as in Fig. 2.11b to be connected between a voltage source with a 1-kΩ source resistance and a load resistance of 1 kΩ. Let $r_\pi = 1$ kΩ and $\beta = 100$. Calculate the voltage gain, current gain, and power gain, all in decibels.
Ans.   34 dB; 40 dB; 37 dB

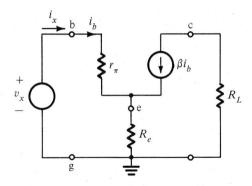

Fig. E2.5

2.5 Find the input resistance between terminals b and g in the circuit shown in Fig. E2.5. The voltage v_x is a test voltage with the input resistance R_{in} defined as $R_{in} \equiv v_x/i_x$.

Ans. $R_{in} = r_\pi + (\beta + 1)R_e$

2.5 FREQUENCY RESPONSE OF AMPLIFIERS

The Sine-Wave Signal

The concept of the frequency spectrum of a signal was introduced in Chapter 1. To repeat, a signal that is an arbitrary function of time can be represented as the sum of sine waves through the mathematical devices of Fourier series and Fourier transform.[1]

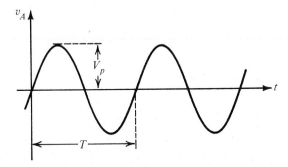

Fig. 2.13 Sine-wave voltage signal of amplitude V_p and frequency $f = 1/T$ Hz.

This fact has made the sine wave, or sinusoid, a standard signal waveform useful in the characterization of linear networks such as filters and amplifiers. Figure 2.13 shows a sine-wave voltage signal $v_A(t)$,

$$v_A(t) = V_p \sin(\omega t) \tag{2.11}$$

---

[1] If the reader has not encountered these topics in the study of circuit theory, there is no need for concern; no significant use will be made of them until Chapter 11.

where V_p denotes the peak value or amplitude and ω denotes the angular frequency in radians per second, that is,

$$\omega = 2\pi f \quad \text{rad/s}$$

where f is the frequency in hertz,

$$f = \frac{1}{T} \quad \text{Hz}$$

and T is the period in seconds.

The sine-wave signal is completely characterized by its peak value V_p, its frequency ω, and its phase with respect to an arbitrary reference time. In this case the time origin has been chosen so that the phase angle is 0. It should be mentioned that it is common to express the amplitude of a sine-wave signal in terms of its root-mean-square value V_{rms} rather than the peak value V_p. The two quantities are related by

$$V_{rms} = V_p / \sqrt{2}$$

For instance, when we speak of the wall power supply in our homes as being 110 V, we mean that it has a sine wave of $110\sqrt{2}$ peak value.

Because of the special role played by sinusoids in the characterization of linear networks we will use a special symbol to refer to them. The symbol $V_a(\omega)$ will refer to the sine-wave signal of Eq. (2.11), with the frequency ω indicated in the functional description for emphasis.

Measuring the Frequency Response

If the sinusoid $V_a(\omega)$ is applied at the input of an amplifier, the output will be a sinusoid of the same frequency. In fact the sine wave is the only signal that does not change shape in passing through a linear network. The output sinusoid $V_b(\omega)$, however, could have a different amplitude and phase than the input $V_a(\omega)$. Thus an amplifier can be characterized in terms of the change it causes in the amplitude and phase of sinusoids of various frequencies applied at the input. To be more specific, the *frequency response* of an amplifier is measured as follows: A sinusoid of a given frequency and amplitude is applied at the input, and the amplitude of the output sinusoid and its phase angle relative to the input sinusoid are measured. Thus at this particular frequency we know the magnitude of the amplifier gain or transmission, which is the ratio of the output amplitude to the input amplitude, and we know the phase angle of the amplifier gain. Then the frequency of the input sinusoid is changed and the experiment is repeated, and so on. The end result will be a table or graph of gain magnitude versus frequency and a table or graph of phase angle versus frequency. Let us for the time being concentrate on the first, the graph of gain magnitude versus frequency. This is called the *magnitude response* or, loosely, the frequency response[2] of the amplifier.

---

[2] Frequency response refers to the combination of magnitude response and phase response.

Amplifier Bandwidth

Figure 2.14 shows the magnitude response of an amplifier. It indicates that the gain is almost constant over a wide frequency range, roughly between ω_1 and ω_2. Signals whose frequencies are below ω_1 or above ω_2 will experience lower gain, with the gain decreasing as we move farther away from ω_1 and ω_2. The band of frequencies over which the gain of the amplifier is almost constant, to within a certain number of decibels (usually 3

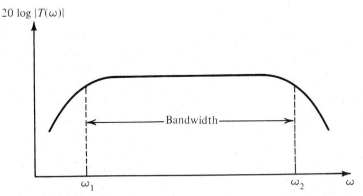

Fig. 2.14 Typical magnitude response of an amplifier. T(ω) is the amplifier transfer function — that is, the ratio of the output V₀(ω) to the input Vᵢ(ω).

dB), is called the *amplifier bandwidth*. Normally the amplifier is designed so that its bandwidth coincides with the spectrum of the signals it is required to amplify. If this were not the case, the amplifier would *distort* the frequency spectrum of the input signal, with different components of the input signal amplified by different amounts.

Evaluating the Frequency Response of Amplifiers

Above we described the method used to measure the frequency response of an amplifier. We now briefly discuss the method for analytically obtaining an expression for the frequency response. What we are about to say is just a preview of this important subject whose detailed study starts in Chapter 11.

To evaluate the frequency response of an amplifier one has to analyze the amplifier equivalent circuit model, taking into account all reactive components.[3] Circuit analysis proceeds in the usual fashion but with inductances and capacitances represented by their reactances. An inductance L has a reactance or impedance $j\omega L$, and a capacitance C has a reactance or impedance $1/j\omega C$, or equivalently a susceptance or admittance $j\omega C$. Thus in a *frequency-domain* analysis we deal with impedances and/or admittances. The result of the analysis is the amplifier transfer function $T(\omega)$,

$$T(\omega) = \frac{V_o(\omega)}{V_i(\omega)}$$

[3] Note that in the models considered in previous sections no reactive components were included.

where $V_i(\omega)$ and $V_o(\omega)$ denote the input and output sinusoids, respectively. $T(\omega)$ is generally a complex function whose magnitude $|T(\omega)|$ gives the magnitude of transmission or the magnitude response of the amplifier. The phase of $T(\omega)$ gives the phase response of the amplifier.

Example 2.4

Derive an expression for the transfer function of the two-port RC network shown in Fig. 2.15.

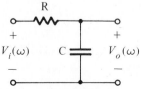

Fig. 2.15 *Two-port RC network whose transfer function T(ω) = V$_o$(ω)/V$_i$(ω) is derived in Example 2.4.*

Solution

Replacing the capacitor C by its impedance $1/j\omega C$, we can use the voltage-divider rule to obtain the transfer function $T(\omega)$,

$$T(\omega) \equiv \frac{V_o(\omega)}{V_i(\omega)}$$

$$= \frac{1/j\omega C}{R + 1/j\omega C}$$

Thus

$$T(\omega) = \frac{1}{1 + j\omega CR}$$

Thus the magnitude response $|T(\omega)|$ is given by

$$|T(\omega)| = \frac{1}{\sqrt{1 + (\omega CR)^2}}$$

and the phase response $\phi(\omega)$ is

$$\phi(\omega) = -\tan^{-1}(\omega CR)$$

Figure 2.16 shows sketches of the magnitude and phase response using a logarithmic frequency axis with the frequency normalized with respect to $\omega_0 = 1/CR$ and using a decibel scale for the magnitude axis.[4] Note that this network *passes low frequencies* and attenuates high-frequency signals. The magnitude of transmission decreases as the frequency increases. Thus this network is called a *low-pass filter*. At any specific frequency the phase response gives the phase angle by which the output sinusoid leads the input sinusoid. The significance of the phase response will be explained at a later stage in the text.

---

[4] Such diagrams are known as *Bode plots*.

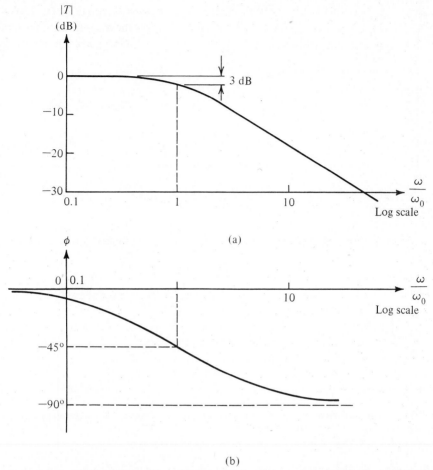

(a)

(b)

Fig. 2.16   The magnitude and phase response of the two-port network in Fig. 2.15. Note that frequencies have been normalized to $\omega_0 = 1/RC$. At this frequency the transmission drops by 3 dB below the transmission at dc, and the phase is $-45°$.

Finally, it should be noted that the reason for the falloff in magnitude of transmission of the network in Fig. 2.15 with frequency is that as the frequency increases the impedance of the capacitor decreases; at very high frequencies the capacitor acts almost as a short circuit. Thus while it acts as an open circuit for dc, a capacitor acts as a short circuit for sinusoids with infinite frequency.

· · ·

Classification of Amplifiers Based on Frequency Response

Amplifiers can be classified based on the shape of their magnitude-response curve. Figure 2.17 shows typical frequency response curves for various amplifier types. In Fig. 2.17a the gain remains constant over a wide frequency range but falls off at low and

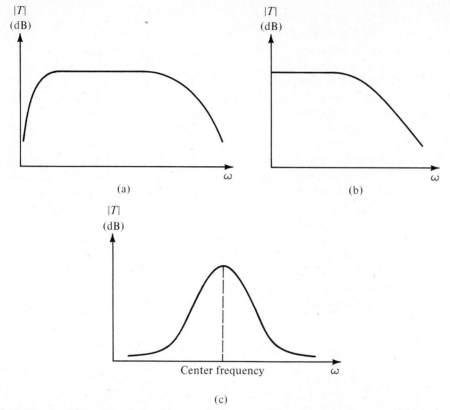

Fig. 2.17 *Frequency response for (a) a capacitively coupled amplifier, (b) a direct-coupled ampli-*
fier, and (c) a tuned or bandpass amplifier.

high frequencies. This is a common type of frequency response found in audio and video
amplifiers. For video amplifiers, however, the bandwidth has to be quite wide.

As will be shown in Chapter 11 *internal capacitances* in the device (transistor)
cause the falloff of gain at high frequencies. On the other hand, the falloff of gain at
low frequencies is usually caused by *coupling capacitors* used to connect one amplifier
stage to another, as indicated in Fig. 2.18. This practice is usually adopted to simplify
the design process of the different stages. The coupling capacitors are usually chosen

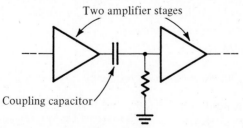

Fig. 2.18 *Use of capacitors to couple amplifier stages.*

quite large (a fraction of a microfarad to a few tens of microfarads) so that their reactance (impedance) is small at the frequencies of interest. Nevertheless, at sufficiently low frequencies the reactance of a coupling capacitor will become large enough to cause part of the signal being coupled to appear as a voltage drop across the coupling capacitor and thus not reach the subsequent stage. Coupling capacitors will thus cause loss of gain at low frequencies and cause the gain to be zero at dc.

There are many applications in which it is important that the amplifier maintain its gain at low frequencies down to dc. Furthermore, monolithic integrated circuit (IC) technology does not allow the fabrication of large coupling capacitors. Thus IC amplifiers are usually designed as *directly coupled* or *dc amplifiers* (as opposed to *capacitively coupled* or *ac amplifiers*). Figure 2.17b shows the frequency response of a dc amplifier. Such a frequency response characterizes what is referred to as a *low-pass amplifier*. Although it is not very appropriate, the term low-pass amplifier is also usually used to refer to the amplifier whose response is shown in Fig. 2.17a.

In the discussion of a radio receiver in Chapter 1, the need for a tuned circuit became apparent. Amplifiers whose frequency response peaks around a certain frequency (called the *center frequency*) and falls off on both sides of this frequency, as shown in Fig. 2.17c, are called *tuned amplifiers, bandpass amplifiers,* or *bandpass filters*. The design of bandpass amplifiers is studied in Chapter 14.

EXERCISES

2.6   Find the frequencies f and ω of a sine-wave signal with a period of 1 ms.
   Ans. $f = 1000$ Hz; $\omega = 2\pi \times 10^3$ rad/s
2.7   Consider the RC circuit analyzed in Example 2.4. Find the output $v_o(t)$ if the input is a sinusoid with frequency $\omega = 10^3$ rad/s and 10-V amplitude. The component values are $R = 1$ kΩ, $C = 1$ μF.
   Ans. $v_o(t) = 7.07 \sin(10^3 t - \tfrac{1}{4}\pi)$
2.8   Derive an expression for the magnitude of transmission of the two-port RC network shown in Fig. E2.8.
   Ans. $\left| \dfrac{V_o(\omega)}{V_i(\omega)} \right| = \dfrac{\omega CR}{\sqrt{1 + (\omega CR)^2}}$

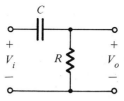

Fig. E2.8

2.6   SOME USEFUL NETWORK THEOREMS

Thévenin's Theorem

Thévenin's theorem is used to represent a part of a network by a voltage source V_t and a series impedance Z_t, as shown in Fig. 2.19. Figure 2.19a shows a network divided into two parts, A and B. In Fig. 2.19b part A of the network has been replaced by its Thév-

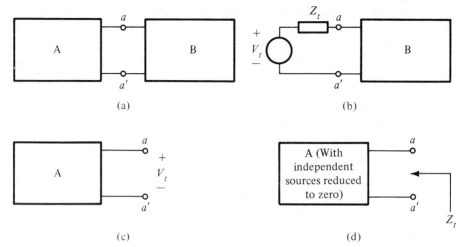

Fig. 2.19   Thévenin's theorem.

enin's equivalent: a voltage source V_t and a series impedance Z_t. Figure 2.19c illustrates how V_t is to be determined: simply open-circuit the two terminals of network A and measure (or calculate) the voltage that appears between these two terminals. To determine Z_t we reduce all external (i.e., independent) sources in network A to zero by short-circuiting voltage sources and open-circuiting current sources. The impedance Z_t will be equal to the input impedance of network A after this reduction has been performed, as illustrated in Fig. 2.19d.

Norton's Theorem

Norton's theorem is the *dual* of Thévenin's theorem. It is used to represent a part of a network by a current source I_n and a parallel impedance Z_n, as shown in Fig. 2.20.

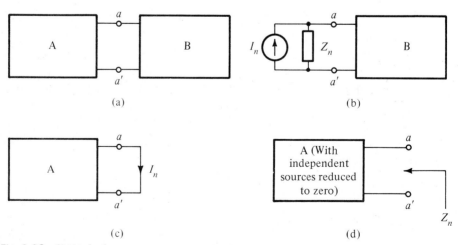

Fig. 2.20   Norton's theorem.

Figure 2.20a shows a network divided into two parts, A and B. In Fig. 2.20b part A has been replaced by its Norton's equivalent: a current source I_n and a parallel impedance Z_n. The Norton's current source I_n can be measured (or calculated) as shown in Fig. 2.20c. The terminals of the network being reduced (network A) are shorted, and the current I_n will be equal simply to the short-circuit current. To determine the impedance Z_n we first reduce the external excitation in network A to zero, that is, short-circuit independent voltage sources and open-circuit independent current sources. The imped-ance Z_n will be equal to the input impedance of network A after this source elimination process has taken place. Thus the Norton impedance Z_n is equal to the Thévenin imped-ance Z_t. Finally, note that $I_n = V_t/Z$, where $Z = Z_n = Z_t$.

Example 2.5
Figure 2.21a shows a bipolar junction transistor circuit. The transistor is a three-ter-minal device with the terminals labeled E (emitter), B (base), and C (collector). As

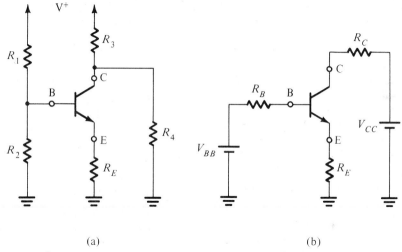

(a)   (b)

Fig. 2.21 *Thévenin's theorem applied to simplify the circuit of (a) to that in (b). (See Example 2.5.)*

shown, the base is connected to the dc power supply V^+ via the voltage divider composed of R_1 and R_2. The collector is connected to the dc supply V^+ through R_3 and to ground through R_4. To simplify the analysis we wish to reduce the circuit through application of Thévenin's theorem.

Solution
Thévenin's theorem can be used at the base side to reduce the network composed of V^+, R_1, and R_2 to a dc voltage source V_{BB},

$$V_{BB} = V^+ \frac{R_2}{R_1 + R_2}$$

and a resistance R_B,

$$R_B = R_1 \| R_2$$

where $\parallel$ denotes "in parallel with." At the collector side Thévenin's theorem can be applied to reduce the network composed of V^+, R_3, and R_4 to a dc voltage source V_{CC},

$$V_{CC} = V^+ \frac{R_4}{R_3 + R_4}$$

and a resistance R_C,

$$R_C = R_3 \| R_4$$

The reduced circuit is shown in Fig. 2.21b.

$\cdot \quad \cdot \quad \cdot$

Source-Absorption Theorem

Consider the situation shown in Fig. 2.22. In the course of analyzing a network we find a controlled current source I_x appearing between two nodes whose voltage difference is

Fig. 2.22   The source absorption theorem.

the controlling voltage V_x. That is, $I_x = g_m V_x$ where g_m is a conductance. We can replace this controlled source by an impedance $Z_x = V_x/I_x = 1/g_m$, as shown in Fig. 2.22, because the current drawn by this impedance will be equal to the current of the controlled source that we have replaced.

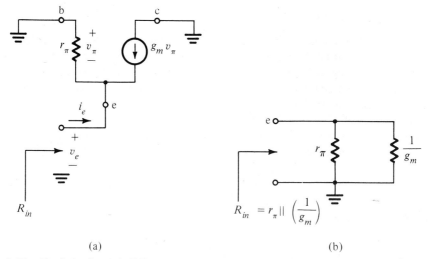

(a)   (b)

Fig. 2.23   Circuit for Example 2.6.

Example 2.6

Figure 2.23a shows the small-signal equivalent circuit model of a transistor. We want to find the resistance R_{in} "looking into" the emitter terminal e, that is, between the emitter and ground, with the base b and collector c grounded.

Solution

From Fig. 2.23a we see that the voltage v_π will be equal to $-v_e$. Thus looking between e and ground we see a resistance r_π in parallel with a current source drawing a current $g_m v_e$ away from terminal e. This latter source can be replaced by a resistance $(1/g_m)$, resulting in the input resistance R_{in} given by

$$R_{in} = r_\pi \| (1/g_m)$$

as illustrated in Fig. 2.23b.

. . .

Miller's Theorem

Consider the situation shown in Fig. 2.24a. We have identified two nodes 1 and 2 together with the reference or ground terminal of a particular network. As shown, an admittance Y is connected between nodes 1 and 2. In addition, nodes 1 and 2 may be

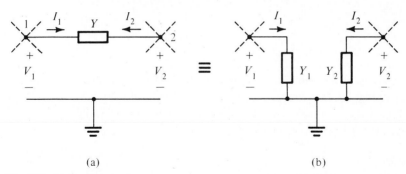

(a)   (b)

Fig. 2.24 Miller's theorem.

connected by other components to other nodes in the network, which is signified by the broken lines emanating from nodes 1 and 2. Miller's theorem provides the means for replacing the "bridging" admittance Y by two admittances; Y_1 between node 1 and ground and Y_2 between node 2 and ground. This replacement is usually quite helpful in simplifying circuit analysis.

The values of Y_1 and Y_2 can be determined as follows: It may be seen from Fig. 2.24a that the only way that node 1 "knows of the existence" of the admittance Y is through the current drawn by Y away from node 1. This current, I_1, is given by

$$I_1 = Y(V_1 - V_2)$$
$$= YV_1(1 - V_2/V_1)$$

Now, let the ratio of V_2 to V_1 be known and denoted by K:

$$\frac{V_2}{V_1} = K$$

Thus

$$I_1 = Y(1 - K)V_1$$

For the circuit of Fig. 2.24b to be equivalent to that of Fig. 2.24a it is essential that the admittance Y_1 be of such value that the current it draws from node 1 is equal to I_1:

$$Y_1 V_1 = I_1$$

which leads to

$$Y_1 = Y(1 - K) \qquad (2.12)$$

Similarly, node 2 "feels" the existence of admittance Y only through the current I_2 drawn by Y away from node 2 (note that $I_2 = -I_1$):

$$I_2 = Y(V_2 - V_1)$$
$$= YV_2(1 - V_1/V_2)$$

Thus

$$I_2 = Y(1 - 1/K)V_2$$

For the circuit of Fig. 2.24b to be equivalent to that of Fig. 2.24a it is essential that the value of Y_2 be such that the current it draws from node 2 is equal to I_2:

$$Y_2 V_2 = I_2$$

which leads to

$$Y_2 = Y(1 - 1/K) \qquad (2.13)$$

Equations (2.12) and (2.13) are the two necessary and sufficient conditions for the network in Fig. 2.24b to be equivalent to that of Fig. 2.24a. Note that in both networks the voltage gain from node 1 to node 2 is equal to K. We shall apply Miller's theorem quite frequently throughout this book.

Example 2.7
We want to find the input impedance of the circuit in Fig. 2.25a, where the amplifier can be assumed an ideal voltage amplifier with a gain of -100.

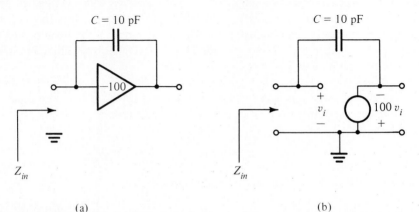

(a)   (b)

Fig. 2.25   See Example 2.7.

Solution

Figure 2.25b shows the circuit with the amplifier replaced with an ideal voltage-controlled voltage source. Application of Miller's theorem at the input results in an input admittance given by

$$Y_{in} = Y(1 - K)$$
$$= j\omega C[1 - (-100)]$$

Thus the input impedance is simply composed of a capacitance C_{in},

$$C_{in} = 101C = 1010 \text{ pF}$$

$\bullet \quad \bullet \quad \bullet$

EXERCISES

2.9   In the circuit shown in Fig. E2.9 the diode has a voltage drop $V_D \simeq 0.7$ V. Use Thévenin's theorem to simplify the circuit and hence calculate the diode current I_D.

Ans.   1 mA

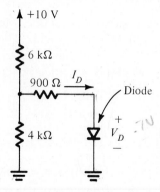

Fig. E2.9

2.10   The two-terminal device M in the circuit of Fig. E2.10 has a current $I_M \simeq 1$ mA independent of the voltage V_M across it. Use Norton's theorem to simplify the circuit and hence calculate the voltage V_M.

Ans.   5 V

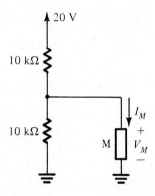

Fig. E2.10

2.11   Consider an ideal voltage amplifier with gain $\mu = 0.95$ and a resistance $R = 100$ kΩ connected in the feedback path, that is, between the output and input terminals. Use Miller's theorem to find the input resistance of this circuit.
An.   2 MΩ

2.7   CHARACTERIZATION OF LINEAR TWO-PORT NETWORKS[5]

A two-port network (Fig. 2.26) has four port variables: V_1, I_1, V_2, and I_2. If the two-port network is linear, we can use two of the variables as excitation and the other two

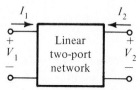

Fig. 2.26   *The reference directions of the four port variables in a linear two-port network.*

as response variables. For instance, the network can be excited by a voltage V_1 at port 1 and a voltage V_2 at port 2 and the two currents, I_1 and I_2, measured to represent the network response. In this case V_1 and V_2 are independent variables and I_1 and I_2 are dependent variables, and the network operation can be described by the two equations

$$I_1 = y_{11}V_1 + y_{12}V_2$$
$$I_2 = y_{21}V_1 + y_{22}V_2$$

Here the four parameters y_{11}, y_{12}, y_{21}, and y_{22} are admittances, and their values completely characterize the linear two-port network.

Depending on which two of the four port variables are used to represent the network excitation, a different set of equations (and a correspondingly different set of parameters) is obtained for characterizing the network. In the following we shall present the four parameter sets commonly used in electronics.

y Parameters

The short-circuit admittance or y-parameter characterization is based on exciting the network by V_1 and V_2, as shown in Fig. 2.27a. The describing equations are

$$I_1 = y_{11}V_1 + y_{12}V_2 \tag{2.14}$$
$$I_2 = y_{21}V_1 + y_{22}V_2 \tag{2.15}$$

The four admittance parameters can be defined according to their roles in Eqs. (2.14) and (2.15).

Specifically, from Eq. (2.14) we see that y_{11} is defined

$$y_{11} = \left.\frac{I_1}{V_1}\right|_{V_2=0}$$

[5] The material of this section will not be applied until Chapter 11.

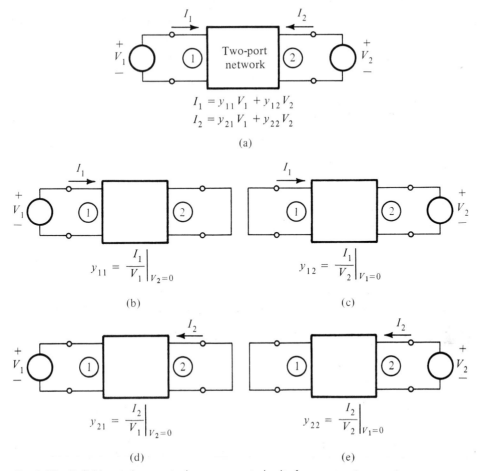

$$I_1 = y_{11} V_1 + y_{12} V_2$$
$$I_2 = y_{21} V_1 + y_{22} V_2$$

(a)

$$y_{11} = \left. \frac{I_1}{V_1} \right|_{V_2=0}$$

(b)

$$y_{12} = \left. \frac{I_1}{V_2} \right|_{V_1=0}$$

(c)

$$y_{21} = \left. \frac{I_2}{V_1} \right|_{V_2=0}$$

(d)

$$y_{22} = \left. \frac{I_2}{V_2} \right|_{V_1=0}$$

(e)

Fig. 2.27 Definition and conceptual measurement circuits for y parameters.

Thus y_{11} is the input admittance at port 1 with port 2 short-circuited. This definition is illustrated in Fig. 2.27b, which also provides a conceptual method for measuring the input short-circuit admittance y_{11}.

The definition of y_{12} can be obtained from Eq. (2.14) as

$$y_{12} = \left. \frac{I_1}{V_2} \right|_{V_1=0}$$

Thus y_{12} represents transmission from port 2 to port 1. Since in amplifiers port 1 represents the input port and port 2 the output port, y_{12} represents internal *feedback* in the network. Figure 2.27c illustrates the definition and the method for measuring y_{12}.

The definition of y_{21} can be obtained from Eq. (2.15) as

$$y_{21} = \left. \frac{I_2}{V_1} \right|_{V_2=0}$$

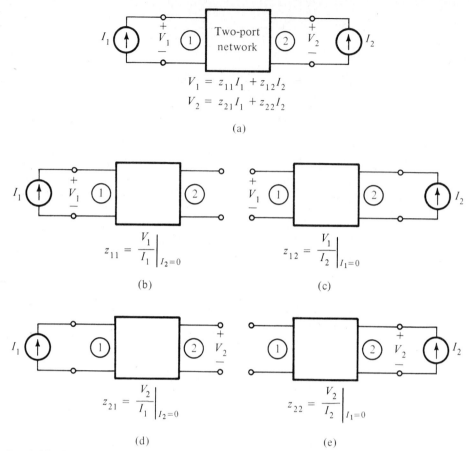

$$V_1 = z_{11}I_1 + z_{12}I_2$$
$$V_2 = z_{21}I_1 + z_{22}I_2$$

(a)

$$z_{11} = \frac{V_1}{I_1}\bigg|_{I_2=0}$$

(b)

$$z_{12} = \frac{V_1}{I_2}\bigg|_{I_1=0}$$

(c)

$$z_{21} = \frac{V_2}{I_1}\bigg|_{I_2=0}$$

(d)

$$z_{22} = \frac{V_2}{I_2}\bigg|_{I_1=0}$$

(e)

Fig. 2.28   Definition and conceptual measurement circuits for z parameters.

Thus y_{21} represents transmission from port 1 to port 2. If port 1 is the input port and port 2 the output port of an amplifier, then y_{21} provides a measure of the forward gain or transmission. Figure 2.27d illustrates the definition and the method for measuring y_{21}.

The parameter y_{22} can be defined, based on Eq. (2.15), as

$$y_{22} = \frac{I_2}{V_2}\bigg|_{V_1=0}$$

Thus y_{22} is the admittance looking into port 2 while port 1 is short-circuited. For amplifiers y_{22} is the output short-circuit admittance. Figure 2.27e illustrates the definition and the method for measuring y_{22}.

z Parameters

The open-circuit impedance (or z-parameter) characterization of two-port networks is based on exciting the network by I_1 and I_2, as shown in Fig. 2.28a. The describing equations are

$$V_1 = z_{11}I_1 + z_{12}I_2 \qquad (2.16)$$
$$V_2 = z_{21}I_1 + z_{22}I_2 \qquad (2.17)$$

Owing to the duality between the z- and y-parameter characterizations we shall not give a detailed discussion of z parameters. The definition and method of measuring each of the four z parameters is given in Fig. 2.28.

h Parameters

The hybrid (or h-parameter) characterization of two-port networks is based on exciting the network by I_1 and V_2, as shown in Fig. 2.29a (note the reason behind the name hybrid). The describing equations are

$$V_1 = h_{11}I_1 + h_{12}V_2 \qquad (2.18)$$
$$I_2 = h_{21}I_1 + h_{22}V_2 \qquad (2.19)$$

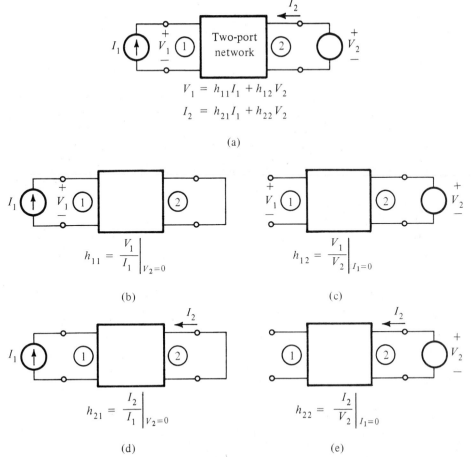

Fig. 2.29 Definition and conceptual measurement circuits for h parameters.

from which the definition of the four h parameters can be obtained as

$$h_{11} = \frac{V_1}{I_1}\bigg|_{V_2=0} \qquad h_{21} = \frac{I_2}{I_1}\bigg|_{V_2=0}$$

$$h_{12} = \frac{V_1}{V_2}\bigg|_{I_1=0} \qquad h_{22} = \frac{I_2}{V_2}\bigg|_{I_1=0}$$

Thus h_{11} is the input impedance at port 1 with port 2 short-circuited. The parameter h_{12} represents the reverse or feedback voltage ratio of the network, measured with the input port open-circuited. The forward-transmission parameter h_{21} represents the current gain of the network with the output port short-circuited; for this reason h_{21} is called the *short-circuit current gain*. Finally, h_{22} is the output admittance with the input port open-circuited.

The definitions and conceptual measuring setups of the h parameters are given in Fig. 2.29.

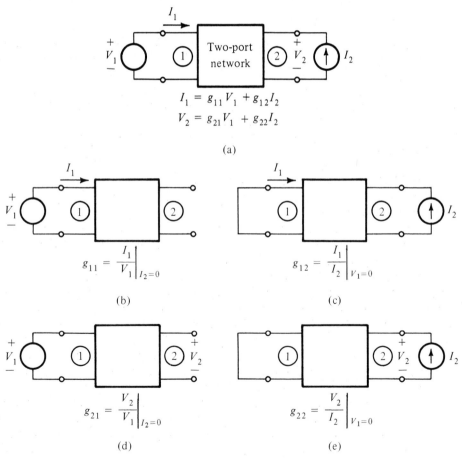

Fig. 2.30 *Definition and conceptual measurement circuits for g parameters.*

g Parameters

The inverse-hybrid (or g-parameter) characterization of two-port networks is based on excitation of the network by V_1 and I_2, as shown in Fig. 2.30a. The describing equations are:

$$I_1 = g_{11}V_1 + g_{12}I_2 \tag{2.20}$$
$$V_2 = g_{21}V_1 + g_{22}I_2 \tag{2.21}$$

The definitions and conceptual measuring setups are given in Fig. 2.30.

Equivalent Circuit Representation

A two-port network can be represented by an equivalent circuit based on the set of parameters used for its characterization. Figure 2.31 shows four possible equivalent circuits corresponding to the four parameter types discussed above. Each of these equivalent circuits is a direct pictorial representation of the corresponding two equations describing the network in terms of the particular parameter set.

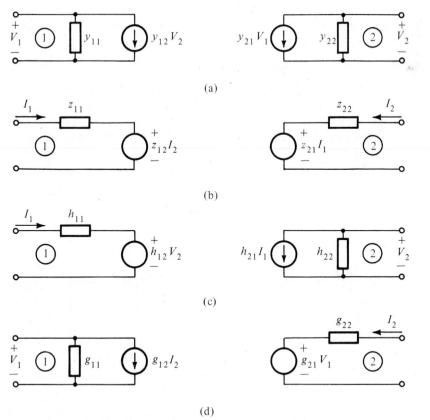

Fig. 2.31 Equivalent circuits for (a) y, (b) z, (c) h, and (d) g parameters.

Finally, it should be mentioned that other parameter sets exist for characterizing two-port networks but these will not be discussed or used in this book.

EXERCISE

2.12   Figure E2.12 shows the small-signal equivalent circuit model of a transistor. Calculate the values of the *h* parameters.

Ans. $h_{11} \simeq 2.6 \text{ k}\Omega$, $h_{12} \simeq 2.5 \times 10^{-4}$, $h_{21} = 100$, $h_{22} = 10^{-5} \text{ } \mho$

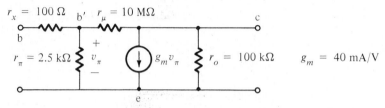

Fig. E2.12

2.8   SINGLE-TIME-CONSTANT NETWORKS

Single-time-constant (STC) networks are those networks that are composed of, or can be reduced to, one reactive component (inductance or capacitance) and one resistance. An STC network formed of an inductance *L* and a resistance *R* has a time constant $\tau = L/R$. The time constant τ of an STC network composed of a capacitance *C* and a resistance *R* is given by $\tau = CR$.

Although STC networks are quite simple, they play an important role in the design and analysis of linear and digital circuits. For instance, it will be shown in later chapters that the analysis of an amplifier circuit can usually be reduced to the analysis of one or more STC networks. For this reason we will review the process of evaluating the response of STC networks to sinusoidal and other input signals such as step and pulse waveforms. These latter signal waveforms are encountered in some amplifier applications but are more important in switching circuits, including digital circuits.

Example 2.8

Reduce the network in Fig. 2.32a to an STC network, and find its time constant.

Solution

The reduction process is illustrated in Fig. 2.32 and consists of repeated applications of Thévenin's theorem. The final circuit is shown in Fig. 2.32c, from which we obtain the time constant as

$$\tau = C\{R_4 \| [R_3 + (R_2 \| R_1)]\}$$

$$\cdot \quad \cdot \quad \cdot$$

Rapid Evaluation of τ

In many instances it will be important to be able to rapidly evaluate the time constant τ of a given STC network. A simple method for accomplishing this goal consists of first

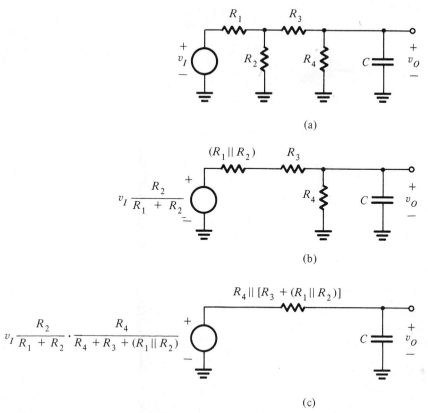

Fig. 2.32 *The reduction of the network in (a) to the STC network in (c) through the repeated application of Thévenin's theorem. (See Example 2.8.)*

reducing the excitation to zero—that is, if the excitation is by a voltage source, short it, and if by a current source, open it. Then if the network has one reactive component and a number of resistances, "grab hold" of the two terminals of the reactive component (capacitance or inductance) and find the equivalent resistance R_{eq} seen by the component. The time constant is then either L/R_{eq} or CR_{eq}. As an example, in the circuit of Fig. 2.32a we find that the capacitor C "sees" a resistance R_4 in parallel with the series combination of R_3 and (R_2 in parallel with R_1). Thus

$$R_{eq} = R_4 \| [R_3 + (R_2 \| R_1)]$$

and the time constant is CR_{eq}.

In some cases it may be found that the network has one resistance and a number of capacitances or inductances. In such a case the procedure should be inverted; that is, "grab hold" of the resistance terminals and find the equivalent capacitance C_{eq}, or equivalent inductance L_{eq}, seen by this resistance. The time constant is then found as $C_{eq}R$ or L_{eq}/R. This is illustrated in Example 2.9.

Example 2.9
Find the time constant of the circuit in Fig. 2.33.

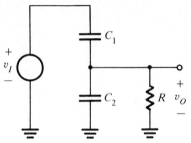

Fig. 2.33   Circuit for Example 2.9.

Solution
After reducing the excitation to zero by short-circuiting the voltage source, we see that the resistance R "sees" an equivalent capacitance $C_1 + C_2$. Thus the time constant τ is given by

$$\tau = (C_1 + C_2)R$$

. . .

Finally, there are cases where an STC network has more than one resistance and more than one capacitance (or more than one inductance). In such a case some initial work must be performed to simplify the network, as illustrated by Example 2.10.

Example 2.10
Here we show that the response of the network in Fig. 2.34a can be obtained using the method of analysis of STC networks.

Solution
The analysis steps are illustrated in Fig. 2.34. In Fig. 2.34b we show the network excited by two separate but equal voltage sources. The reader should convince himself or herself of the equivalence of the networks in Fig. 2.34a and Fig. 2.34b. The "trick" employed to obtain the arrangement in Fig. 2.34b is a very useful one.

Application of Thévenin's theorem to the circuit to the left of the line XX' and then to the circuit to the right of that line results in the circuit of Fig. 2.34c. Since this is a linear circuit, the response may be obtained using the principle of superposition. Specifically, the output voltage v_O will be the sum of the two components v_{O1} and v_{O2}. The first component, v_{O1}, is the output due to the left-hand-side voltage source with the other voltage source reduced to zero. The circuit for calculating v_{O1} is shown in Fig. 2.34d. It is an STC network with time constant given by

$$\tau = (C_1 + C_2)(R_1 \| R_2)$$

Similarly, the second component v_{O2} is the output obtained with the left-hand-side voltage source reduced to zero. It can be calculated from the circuit of Fig. 2.34e, which is an STC network with a time constant equal to that given above.

. . .

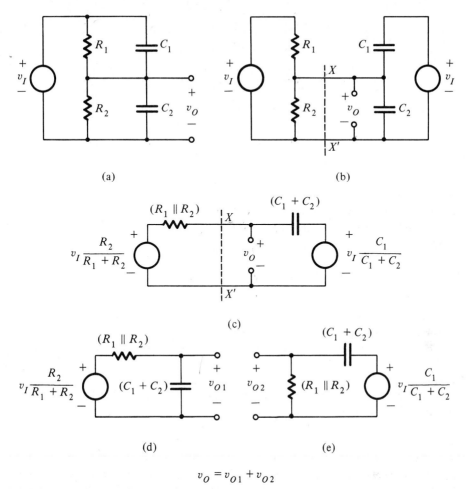

$$v_O = v_{O1} + v_{O2}$$

Fig. 2.34 Details for Example 2.10.

Classification of STC Networks

STC networks can be classified into two categories, *low-pass* (LP) and *high-pass* (HP) types, with each of the two categories displaying distinctly different signal responses. The task of finding whether an STC network is of LP or HP type may be accomplished in a number of ways, the simplest of which uses the frequency-domain response. Specifically, low-pass networks pass dc (that is, signals with zero frequency) and attenuate high frequencies, with the transmission being zero at $\omega = \infty$. Thus we can test for the network identity either at $\omega = 0$ or at $\omega = \infty$. At $\omega = 0$ capacitors should be replaced by open circuits ($1/j\omega C = \infty$) and inductors should be replaced by short circuits ($j\omega L = 0$). Then if the output is zero, the circuit is of the high-pass type, while if the output is finite, the circuit is of the low-pass type. Alternatively, we may test at $\omega = \infty$ by replacing capacitors by short circuits ($1/j\omega C = 0$) and inductors by open circuits

TABLE 2.1   *Rules for Finding the Type of STC Network*

Test at	Replace	Network is LP if	Network is HP if
$\omega = 0$	C by o.c. L by s.c.	Output is finite	Output is zero
$\omega = \infty$	C by s.c. L by o.c.	Output is zero	Output is finite

($j\omega L = \infty$). Then if the output is finite, the circuit is of the HP type, while if the output is zero, the circuit is of the LP type.

Table 2.1 provides a summary of these rules (s.c., short circuit; o.c., open circuit).

Figure 2.35 shows examples of low-pass STC networks, and Fig. 2.36 shows examples of high-pass STC networks. For each circuit we have indicated the input and output variables of interest. Note that a given network can be of either category, depending on the input and output variables. The reader is urged to verify, using the rules of Table 2.1, that the circuits of Figs. 2.35 and 2.36 are correctly classified.

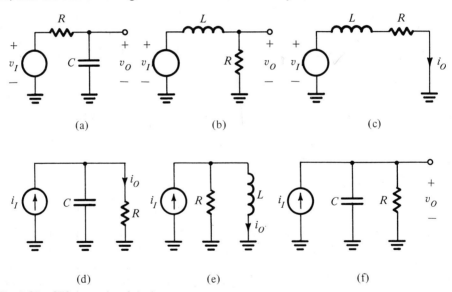

Fig. 2.35   *STC networks of the low-pass type.*

2.9   *FREQUENCY RESPONSE OF STC NETWORKS*

Low-Pass Networks

The transfer function $T(\omega)$ of an STC low-pass network always can be written in the form

$$T(\omega) = \frac{K}{1 + j(\omega/\omega_0)} \qquad (2.22)$$

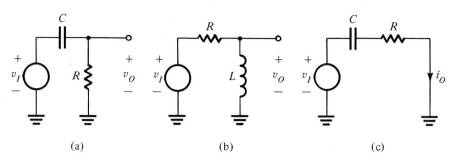

(a)   (b)   (c)

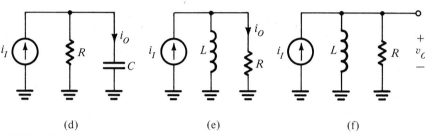

(d)   (e)   (f)

Fig. 2.36 STC networks of the high-pass type.

where K is the magnitude of the transfer function at $\omega = 0$ (dc) and ω_0 is defined

$$\omega_0 = \frac{1}{\tau}$$

with τ being the time constant. Thus the magnitude response is given by

$$|T(\omega)| = \frac{K}{\sqrt{1 + (\omega/\omega_0)^2}} \tag{2.23}$$

and the phase response is given by

$$\phi(\omega) = -\tan^{-1}(\omega/\omega_0) \tag{2.24}$$

Figure 2.37 shows sketches of the magnitude and phase responses for an STC low-pass network. The magnitude response shown in Fig. 2.37a is simply a graph of the function in Eq. (2.23). The magnitude is normalized with respect to the dc gain K and is expressed in dB, that is, the plot is for $20 \log |T(\omega)/K|$, with a logarithmic scale used for the frequency axis. Furthermore, the frequency variable has been normalized with respect to ω_0. As shown, the magnitude curve is closely defined by two straight-line asymptotes. The low-frequency asymptote is a horizontal straight line at 0 dB. To find the slope of the high-frequency asymptote consider Eq. (2.23) and let $\omega/\omega_0 \gg 1$ resulting in

$$|T(\omega)| \simeq K\frac{\omega_0}{\omega}$$

It follows that if ω doubles in value, the magnitude is halved. On a logarithmic frequency axis, doublings of ω represent equally spaced points, with each interval called

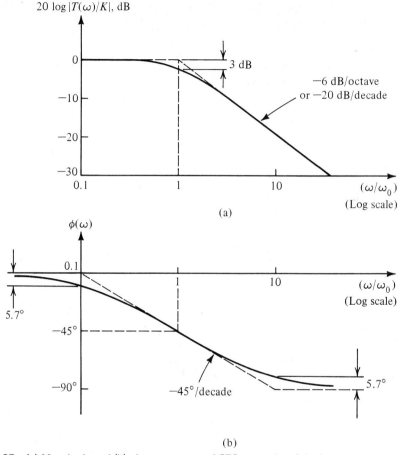

Fig. 2.37   (a) Magnitude and (b) phase response of STC networks of the low-pass type.

an *octave*. Halving the magnitude function corresponds to a 6-dB reduction in transmission (20 log 0.5 = −6 dB). Thus the slope of the high-frequency asymptote is −6 dB/octave. This can be equivalently expressed as −20 dB/decade, where a decade refers to an increase in frequency by a factor of 10.

The two straight-line asymptotes of the magnitude-response curve meet at the "corner frequency" or "break frequency" ω_0. The difference between the actual magnitude-response curve and the asymptotic response is largest at the corner frequency, where its value is 3 dB. To verify that this value is correct, simply substitute $\omega = \omega_0$ in Eq. (2.23) to obtain

$$|T(\omega_0)| = \frac{K}{\sqrt{2}}$$

Thus at $\omega = \omega_0$ the gain drops by a factor of $\sqrt{2}$ relative to the dc gain, which corresponds to a 3-dB reduction in gain. The corner frequency ω_0 is appropriately referred to as the 3-dB frequency.

Similar to the magnitude response, the phase-response curve, shown in Fig. 2.37b, is closely defined by straight-line asymptotes. Note that at the corner frequency the phase is $-45°$, and that for $\omega \gg \omega_0$ the phase approaches $-90°$.

Example 2.11

Consider the circuit shown in Fig. 2.38a, where an ideal voltage amplifier of gain $\mu = -100$ has a small (10 pF) capacitance connected in its feedback path. The amplifier is

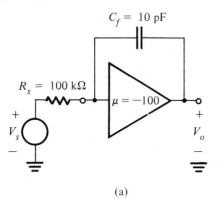

(a)

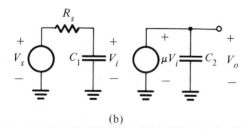

(b)

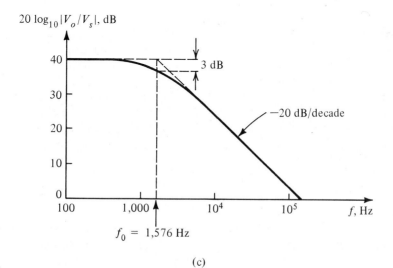

(c)

Fig. 2.38 See Example 2.11.

fed by a voltage source having a source resistance of 100 kΩ. Show that the frequency response V_o/V_s of this amplifier is equivalent to that of an STC network, and sketch the magnitude response.

Solution
According to Miller's theorem the feedback capacitance C_f can be replaced by a capacitance C_1 at the input,

$$C_1 = C_f(1 - \mu) = 1010 \text{ pF},$$

and a capacitance C_2 at the output,

$$C_2 = C_f(1 + 1/\mu) \simeq C_f = 10 \text{ pF}$$

Thus the equivalent circuit shown in Fig. 2.38b is obtained, from which we immediately recognize the low-pass RC network at the input. Thus the amplifier transfer function is given by

$$\frac{V_o}{V_s} = \frac{V_i}{V_s}\frac{V_o}{V_i} = \frac{V_i}{V_s}\mu$$

where V_i/V_s is the transfer function of the RC network formed by R_s and C_1. Note that since the output of the amplifier is an ideal voltage source, C_2 has no effect on the transfer function V_o/V_s. Now the input RC network has a dc transmission of unity; thus

$$\frac{V_o}{V_s} = \frac{\mu}{1 + j\omega/\omega_0}$$

where $\omega_0 = 1/C_1R_s = 0.99 \times 10^4$ rad/s. It follows that the amplifier response is that of a low-pass STC network; thus its magnitude can be readily sketched, as shown in Fig. 2.38c.

$$\cdot \quad \cdot \quad \cdot$$

High-Pass Networks

The transfer function $T(\omega)$ of an STC high-pass network always can be expressed in the form

$$T(\omega) = \frac{K}{1 - j\omega_0/\omega} \tag{2.25}$$

where K denotes the gain as ω approaches infinity and ω_0 is the inverse of the time constant τ,

$$\omega_0 = 1/\tau$$

The magnitude response

$$|T(\omega)| = \frac{K}{\sqrt{1 + (\omega_0/\omega)^2}} \tag{2.26}$$

and the phase response

$$\phi(\omega) = \tan^{-1}(\omega_0/\omega) \qquad (2.27)$$

are sketched in Fig. 2.39. As in the low-pass case, the magnitude and phase curves are well defined by straight-line asymptotes. Because of the similarity (or more appropriately duality) with the low-pass case, no further explanation will be given.

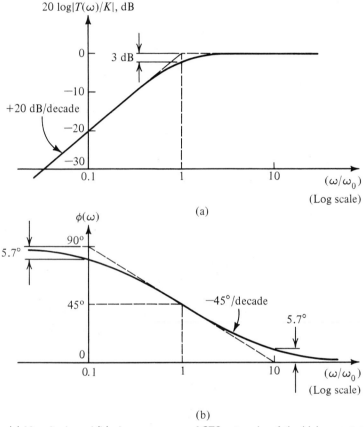

(a)

(b)

Fig. 2.39   (a) Magnitude and (b) phase response of STC networks of the high-pass type.

EXERCISES

2.13   Find the dc transmission, the corner frequency f_0, and the transmission at $f = 2$ MHz for the low-pass STC network shown in Fig. E2.13.
Ans. -6 dB; 318 kHz; -22 dB

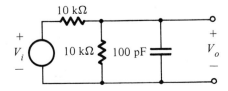

Fig. E2.13

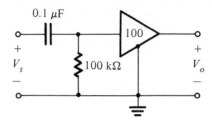

Fig. E2.14

2.14   Find the high-frequency gain, the 3-dB frequency f_0, and the gain at $f = 1$ Hz of the capacitively coupled amplifier shown in Fig. E2.14. Assume the voltage amplifier to be ideal.

Ans.   40 dB; 15.9 Hz; 16 dB

2.10   STEP RESPONSE OF STC NETWORKS

In this section we consider the response of STC networks to the step-function signal shown in Fig. 2.40. Knowledge of the step response enables rapid evaluation of the response to other switching signal waveforms, such as pulses and square waves.

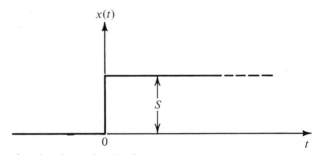

Fig. 2.40   A step-function signal of height S.

Low-Pass Networks

In response to an input step signal of height S, a low-pass STC network produces the waveform shown in Fig. 2.41. Note that while the input rises from 0 to S at $t = 0$, the output does not respond to this transient and simply begins to rise exponentially toward the *final* dc value of the input, S. In the long term—that is, for $t \gg \tau$—the output approaches the dc value S, a manifestation of the fact that low-pass networks faithfully pass dc.

The equation of the output waveform can be obtained from the expression

$$y(t) = Y_\infty - (Y_\infty - Y_{0+})e^{-t/\tau} \tag{2.28}$$

where Y_∞ denotes the *final* value or the value toward which the output is heading and Y_{0+} denotes the value of the output immediately after $t = 0$. This equation states that the output at any time t is equal to the difference between the final value Y_∞ and a gap

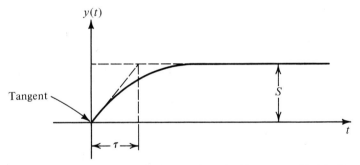

Fig. 2.41   *The output y(t) of a low-pass STC network excited by a step of height S.*

whose initial value is $Y_\infty - Y_{0+}$ and which is "shrinking" exponentially. In our case $Y_\infty = S$ and $Y_{0+} = 0$; thus

$$y(t) = S(1 - e^{-t/\tau}) \qquad (2.29)$$

The reader's attention is drawn to the slope of the tangent to $y(t)$ at $t = 0$, which is indicated in Fig. 2.41.

High-Pass Networks

The response of an STC high-pass network to an input step of height S is shown in Fig. 2.42. The high-pass network faithfully transmits the transient of the input signal (the step change) but blocks the dc. Thus the output at $t = 0$ follows the input,

$$Y_{0+} = S$$

and then it decays toward zero,

$$Y_\infty = 0$$

Substituting for Y_{0+} and Y_∞ in Eq. (2.28) results in the output $y(t)$,

$$y(t) = Se^{-t/\tau} \qquad (2.30)$$

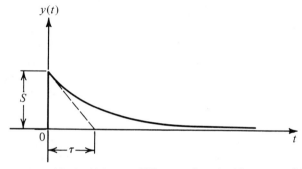

Fig. 2.42   *The output y(t) of a high-pass STC network excited by a step of height S.*

The reader's attention is drawn to the slope of the tangent to $y(t)$ at $t = 0$, indicated in Fig. 2.42.

Example 2.12
This example is a continuation of the problem considered in Example 2.10. For an input v_I that is a 10-V step, find the condition under which the output v_O is a perfect step.

Solution
Following the analysis in Example 2.10, which is illustrated in Fig. 2.34, we have

$$v_{O1} = k_r[10(1 - e^{-t/\tau})]$$

where

$$k_r \equiv \frac{R_2}{R_1 + R_2}$$

and

$$v_{O2} = k_c(10e^{-t/\tau})$$

where

$$k_c \equiv \frac{C_1}{C_1 + C_2}$$

and

$$\tau = (C_1 + C_2)(R_1 \| R_2)$$

Thus

$$v_O = v_{O1} + v_{O2}$$
$$= 10k_r + 10e^{-t/\tau}(k_c - k_r)$$

It follows that the output can be made a perfect step of height $10k_r$ V if we arrange that

$$k_c = k_r$$

that is, the resistive voltage-divider ratio is equal to the capacitive voltage-divider ratio.

This example illustrates an important technique, namely, that of the "compensated attenuator." An application of this technique is found in the design of the oscilloscope probe. The oscilloscope probe problem is investigated in Problem 2.45.

$$\cdot \quad \cdot \quad \cdot$$

EXERCISES
2.15   For the circuit of Fig. 2.35f find v_O if i_I is a 3-mA step, $R = 1$ kΩ, and $C = 100$ pF.
   Ans. $3(1 - e^{-10^7 t})$
2.16   In the circuit of Fig. 2.36f find $v_O(t)$ if i_I is a 2-mA step, $R = 2$ kΩ, and $L = 10$ μH.
   Ans. $4 e^{-2 \times 10^8 t}$

2.11   PULSE RESPONSE OF STC NETWORKS

Figure 2.43 shows a pulse signal whose height is P and whose width is T. We wish to find the response of STC networks to input signals of this form. Note at the outset that a pulse can be considered as the sum of two steps: a positive one of height P occurring at $t = 0$ and a negative one of height P occurring at $t = T$. Thus the response of a network to the pulse signal can be obtained by summing the responses to the two step signals.

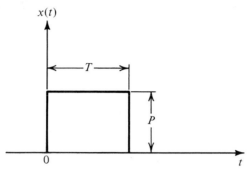

Fig. 2.43   *A pulse signal with height P and width T.*

Low-Pass Networks

Figure 2.44a shows the response of a low-pass STC network to an input pulse of the form shown in Fig. 2.43. In this case we have assumed that the time constant τ is in the same range as the pulse width T. As shown, the LP network does not respond to the step change at the leading edge of the pulse; rather the output starts to rise exponentially toward a final value of P. This exponential rise, however, will be stopped at time $t = T$, that is, at the trailing edge of the pulse when the input undergoes a negative step change. Again the output will respond by starting an exponential decay toward the final value of the input, which is zero. Finally, note that the area under the output waveform will be equal to the area under the input pulse waveform, since the LP network faithfully passes dc.

In connecting a pulse signal from one part of an electronic system to another, a low-pass effect usually occurs. The low-pass network in this case is formed by the output resistance (Thévenin's equivalent resistance) of the system part from which the signal originates and the input capacitance of the system part to which the signal is fed. This unavoidable low-pass filter will cause distortion—of the type shown in Fig. 2.44a—of the pulse signal. In a well designed system such distortion is kept to a low value by arranging that the time constant τ be much smaller than the pulse width T. In this case the result will be a slight rounding of the pulse edges, as shown in Fig. 2.44b. Note, however, that the edges are still exponential.

The distortion of a pulse signal by a parasitic (that is, unwanted) low-pass network is measured by its *rise time* and *fall time*. The rise time is the time taken by the amplitude to increase from 10% to 90% of the final value. Similarly, the fall time is the time

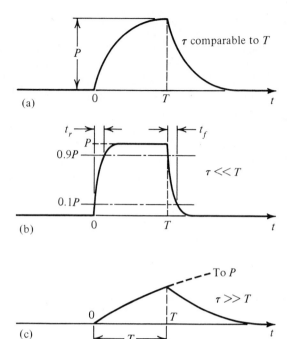

Fig. 2.44   *Pulse response of STC low-pass networks.*

during which the pulse amplitude falls from 90% to 10% of the maximum value. These definitions are illustrated in Fig. 2.44b. By use of the exponential equations of the rising and falling edges of the output waveform it can be easily shown that

$$t_r = t_f \simeq 2.2\tau$$

which can be also expressed in terms of $f_0 = \omega_0/2\pi = 1/2\pi\tau$ as

$$t_r = t_f \simeq \frac{0.35}{f_0}$$

Finally, we note that the effect of the parasitic low-pass networks that are always present in a system is to "slow down" the operation of the system: in order to keep the signal distortion within acceptable limits one has to use a relatively long pulse width (for a given time constant).

The other extreme case—namely, when τ is much larger than T—is illustrated in Fig. 2.44c. As shown, the output waveform rises exponentially toward the level P. However, since $\tau \gg T$, the value reached at $t = T$ will be much smaller than P. At $t = T$ the output waveform starts its exponential decay toward zero. Note that in this case the output waveform bears little resemblance to the input pulse. Also note that because $\tau \gg T$ the portion of the exponential curve from $t = 0$ to $t = T$ is almost linear. Since the slope of this linear curve is proportional to the height of the input pulse, we see that the output waveform approximates the time integral of the input pulse. That is, a low-pass network with a large time constant approximates the operation of an *integrator*.

High-Pass Networks

Figure 2.45a shows the output of an STC HP network excited by the input pulse of Fig. 2.43, assuming that τ and T are comparable in value. As shown, the step transition at the leading edge of the input pulse is faithfully reproduced at the output of the HP network. However, since the HP network blocks dc, the output waveform immediately starts an exponential decay toward zero. This decay process is stopped at $t = T$ when the negative step transition of the input occurs and the HP network faithfully reproduces it. Thus at $t = T$ the output waveform exhibits an *undershoot*. Then it starts an exponential decay toward zero. Finally, note that the area of the output waveform above the zero axis will be equal to that below the axis for a total average area of zero, consistent with the fact that HP networks block dc.

In many applications an STC high-pass network is used to couple a pulse from one part of a system to another part. In such an application it is necessary to keep the distortion in the pulse shape as small as possible. This can be accomplished by selecting the time constant τ to be much longer than the pulse width T. If this is indeed the case,

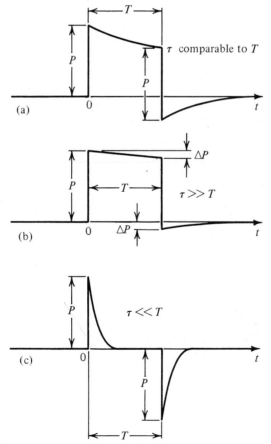

Fig. 2.45   Pulse response of STC high-pass networks.

the loss in amplitude during the pulse period T will be very small, as shown in Fig. 2.45b. Nevertheless, the output waveform still swings negatively, and the area under the negative portion will be equal to that under the positive portion.

Consider the waveform in Fig. 2.45b. Since τ is much larger than T, it follows that the portion of the exponential curve from $t = 0$ to $t = T$ will be almost linear and that its slope will be equal to the slope of the exponential curve at $t = 0$, which is P/τ. We can use this value of the slope to determine the loss in amplitude ΔP as

$$\Delta P \simeq \frac{P}{\tau} T$$

The distortion effect of the high-pass network on the input pulse is usually specified in terms of the per unit or percentage loss in pulse height. This quantity is taken as an indication of the "sag" in the output pulse,

$$\text{Percentage sag} \equiv \frac{\Delta P}{P} \times 100$$

Thus

$$\text{Percentage sag} = \frac{T}{\tau} \times 100$$

Finally, note that the magnitude of the undershoot at $t = T$ is equal to ΔP.

The other extreme case—namely, $\tau \ll T$—is illustrated in Fig. 2.45c. In this case the exponential decay is quite rapid, resulting in the output becoming almost zero shortly beyond the leading edge of the pulse. At the trailing edge of the pulse the output swings negatively by an amount almost equal to the pulse height P. Then the waveform decays rapidly to zero. As seen from Fig. 2.45c, the output waveform bears no resemblance to the input pulse. It consists of two spikes: a positive one at the leading edge and a negative one at the trailing edge. Note that the output waveform is approximately equal to the time derivative of the input pulse. That is, for $\tau \ll T$ an STC high-pass network approximates a *differentiator*. However, the resulting differentiator is not an ideal one; an ideal differentiator would produce two impulses. Nevertheless, high-pass STC networks with short time constants are employed in some applications to produce sharp pulses or spikes at the transitions of an input waveform.

EXERCISES

2.17 Find the rise and fall times of a 1-μs pulse after it passes through a low-pass RC network with a corner frequency of 10 MHz.
Ans. 35 ns

2.18 The output of an amplifier stage is connected to the input of another stage via a capacitance C. If the first stage has an output resistance of 1 kΩ and the second stage has an input resistance of 4 kΩ, find the minimum value of C such that a 10-μs pulse exhibits less than 1% sag.
Ans. 0.2 μF

2.19 A high-pass STC network with a time constant of 10 μs is excited by a pulse of 10-V height and 10-μs width. Calculate the value of the undershoot in the output waveform.
Ans. 6.32 V

2.12   CONCLUDING REMARKS

This chapter should serve as a "bridge" between the study of circuit theory and that of electronic circuits. At this stage it is expected that the reader has become familiar with the techniques used in the analysis of linear circuits. These techniques, together with the terminology and conventions introduced here, will be employed frequently throughout this text.

A good number of the ideas and methods introduced in this chapter will be put to practical use in the next chapter which is concerned with the operational amplifier. This is a special yet very versatile amplifier commercially available as a single integrated circuit (IC) chip.

OPERATIONAL 3
AMPLIFIERS

Introduction

Having reviewed the fundamentals of linear circuit theory, we are now ready to consider their application to actual circuits. This chapter is devoted to the study of a circuit building block of universal importance: the operational amplifier (op amp). Although op amps have been in use for a long time, their applications were initially in the areas of analog computation and instrumentation. Early op amps were constructed from discrete components (transistors and resistors), and their cost was prohibitively high (tens of dollars). In the mid-1960s the first integrated-circuit (IC) op amp was produced. This unit (the μA 709) was made up of a relatively large number of transistors and resistors all on the same silicon chip. Although its characteristics were poor (by today's standards) and its price was still quite high, its appearance signaled a new era in electronic circuit design. Electronics engineers started using op amps in large quantities, which caused their price to drop dramatically. They also demanded better-quality op amps. Semiconductor manufacturers responded quickly, and within the span of few years high-quality op amps became available at extremely low prices (tens of cents) from a large number of suppliers.

One of the reasons for the popularity of the op amp is its versatility. As we will shortly see, one can do almost anything with op amps! More importantly, the IC op amp has characteristics that closely approach the assumed ideal. This implies that it is quite easy to design circuits using the IC op amp. Also, op-amp circuits work at levels that are quite close to their predicted theoretical performance. It is for this reason that we are studying op amps at this early stage. It is expected that by the end of this chapter the reader should be able to successfully design nontrivial circuits using op amps.

As already implied, an IC op amp is made up of a large number of transistors, resistors, and (sometimes) one capacitor connected in a rather complex circuit. Since we have not yet studied transistor circuits, the circuit inside the op amp will not be discussed in this chapter. Rather, we will treat the op amp as a "black box" and study its terminal characteristics and its applications. This approach is quite satisfactory in many op amp applications. Nevertheless, for the more difficult and demanding applications it is quite useful to know what is inside the op-amp package. This topic will be studied in Chapter 13. Finally, it should be mentioned that more advanced applications of op amps will appear in later chapters.

3.1 THE OP-AMP TERMINALS

From a signal point of view the op amp has three terminals: two input terminals and one output terminal. Figure 3.1 shows the symbol that we shall use to represent the op amp. Terminals 1 and 2 are input terminals, and terminal 3 is the output terminal.

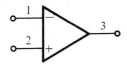

Fig. 3.1 Circuit symbol for the op amp.

Since the op amp is an active device, it requires dc power to operate. Most IC op amps require two dc power supplies, as shown in Fig. 3.2. Two terminals, 4 and 5, are brought out of the op-amp package and connected to a positive voltage V^+ and a negative voltage V^-, respectively. In Fig. 3.2b we explicitly show the two dc power supplies as batteries with a common ground. It is interesting to note that the reference grounding point in op-amp circuits is just the common terminal of the two power supplies; that is, no terminal of the op-amp package is physically connected to ground. In what follows we will not explicitly show the op-amp power supplies.

In addition to the three signal terminals and the two power supply terminals, an op amp may have other terminals for specific purposes. These other terminals can include terminals for frequency compensation and terminals for offset nulling; both functions will be explained in later sections.

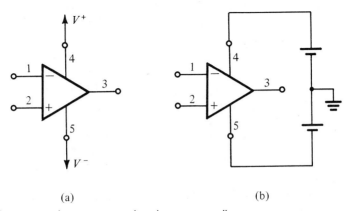

(a)   (b)

Fig. 3.2 The op amp shown connected to dc power supplies.

3.2   THE IDEAL OP AMP

We now consider an important question: What is the op amp supposed to do? The answer is simple: the op amp is supposed to sense the difference between the voltage signals applied at its two input terminals (that is, the quantity $v_2 - v_1$), multiply this by a number A, and cause the resulting voltage $A(v_2 - v_1)$ to appear at output terminal 3. Here it should be emphasized that when we talk about the voltage at a terminal we mean the voltage between that terminal and ground; thus v_1 means the voltage applied between terminal 1 and ground.

The ideal op amp is not supposed to draw any input current—that is, the signal current into terminal 1 and the signal current into terminal 2 are both zero. In other words, the input impedance of an ideal op amp is supposed to be infinite.

How about the output terminal 3? This terminal is supposed to act as the output terminal of an ideal voltage source. That is, the voltage between terminal 3 and ground

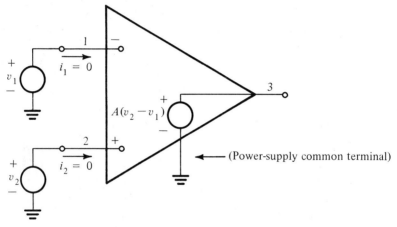

Fig. 3.3   Equivalent circuit of the ideal op amp.

will always be equal to $A(v_2 - v_1)$ independent of the current that may be drawn from terminal 3 into a load impedance. In other words, the output impedance of an ideal op amp is supposed to be zero.

Putting together all of the above we arrive at the equivalent circuit model shown in Fig. 3.3. Note that the output is in phase with (has the same sign as) v_2 and out of phase with (has the opposite sign of) v_1. For this reason input terminal 1 is called the *inverting input terminal* and is distinguished by a "$-$" sign, while input terminal 2 is called the *noninverting input terminal* and is distinguished by a "$+$" sign.

As can be seen from the above description the op amp responds only to the *difference* signal $v_2 - v_1$ and hence ignores any signal *common* to both inputs. That is, if $v_1 = v_2 = 1$ V, then the output will—ideally—be zero. We call this property *common-mode rejection*, and we conclude that an ideal op amp has infinite common-mode rejection. We will have more to say about this point later. For the time being note that the op amp is a *differential-input, single-ended-output* amplifier, with the latter term referring to the fact that the output appears between terminal 3 and ground. Furthermore, gain A is called the *differential gain,* for obvious reasons. Perhaps not so obvious is another name that we will attach to A: the *open-loop gain.* The reason for this latter name will become obvious later on when we "close the loop" around the op amp and define another gain, the closed-loop gain.

An important characteristic of op amps is that they are direct-coupled devices or dc amplifiers, where dc stands for direct coupled (it could equally well stand for direct current, since a direct-coupled amplifier is one that amplifies signals whose frequency is as low as zero). The fact that op amps are direct-coupled devices will allow us to use them in many important applications. Unfortunately, though, the direct-coupling property can cause some serious problems, as will be discussed in a later section.

How about bandwidth? The ideal op amp has a gain A that remains constant down to zero frequency and up to infinite frequency. That is, ideal op amps will amplify signals of any frequency with equal gain.

We have discussed all of the properties of the ideal op amp except for one, which in fact is the most important. This has to do with the value of A. The ideal op amp should have a gain A whose value is very large and ideally infinite. One may justifiably ask, If the gain A is infinite, how are we going to use the op amp? The answer is very simple: in almost all applications the op amp will *not* be used in an open-loop configuration. Rather, we will apply feedback to close the loop around the op amp, as will be illustrated in detail in Section 3.3.

3.3 ANALYSIS OF CIRCUITS CONTAINING IDEAL OP AMPS— THE INVERTING CONFIGURATION

Consider the circuit shown in Fig. 3.4, which consists of one op amp and two resistors R_1 and R_2. Resistor R_2 is connected from the output terminal of the op amp, terminal 3, *back* to the *inverting* or *negative* input terminal, terminal 1. We speak of R_2 as applying *negative feedback;* if R_2 were connected between terminals 3 and 2 we would have called this *positive feedback.* Note also that R_2 closes the loop around the op amp. In addition to adding R_2, we have grounded terminal 2 and connected a resistor R_1 between terminal 1 and an input signal source with a voltage v_I. The output of the overall circuit is taken at terminal 3 (that is, between terminal 3 and ground). Terminal 3 is, of course, a convenient point to take the output, since the impedance level there is ideally zero. Thus the voltage v_O will not depend on the value of the current that might be supplied to a load impedance connected between terminal 3 and ground.

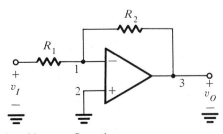

Fig. 3.4 The inverting closed-loop configuration.

We now wish to analyze the circuit in Fig. 3.4 to determine the *closed-loop gain G*,

$$G \equiv \frac{v_O}{v_I}$$

We will do so assuming the op amp to be ideal. The procedure is as follows: The gain A is very large (ideally infinite). If we assume that the circuit is "working" and producing a finite output voltage at terminal 3, then the voltage between the op amp input terminals should be negligibly small. Specifically, if we call the output voltage v_O, then, by definition,

$$v_2 - v_1 = \frac{v_O}{A} \simeq 0$$

It follows that the voltage at the inverting input terminal (v_1) is given by

$$v_1 \simeq v_2$$

That is, because the gain A approaches infinity, the voltage v_1 approaches v_2. We speak of this as the two input terminals "tracking each other in potential." We also speak of a "virtual short circuit" that exists between the two input terminals. Here the word *virtual* should be emphasized, and one should *not* make the mistake of physically shorting terminals 1 and 2 together while analyzing a circuit. A *virtual short circuit* means that whatever voltage is at 2 will automatically appear at 1 because of the infinite gain A. But terminal 2 happens to be connected to ground; thus $v_2 = 0$ and $v_1 \simeq 0$. We speak of terminal 1 as being a *virtual ground*—that is, having zero voltage but not physically connected to ground.

Now that we have determined v_1 we are in a position to apply Ohm's law and find the current i_1 through R_1,

$$i_1 = \frac{v_I - v_1}{R_1} \simeq \frac{v_I}{R_1}$$

Where will this current go? It cannot go into the op amp, since the ideal op amp has an infinite input impedance and hence draws zero current. It follows that i_1 will have to "take a detour" and go through R_2 to the low-impedance terminal 3. We can then apply Ohm's law to R_2 and determine v_O,

$$v_O = v_1 - i_1 R_2$$
$$= 0 - \frac{v_I}{R_1} R_2$$

Thus

$$\frac{v_O}{v_I} = -\frac{R_2}{R_1}$$

which is the required closed-loop gain. Figure 3.5 illustrates some of these analysis steps.

We thus see that the closed-loop gain is simply the ratio of the two resistances R_2 and R_1. The minus sign means that the closed-loop amplifier provides signal inversion. Thus if $R_2/R_1 = 10$ and we apply at the input (v_I) a sine-wave signal of 1 V peak-to-peak, then the output v_O will be a sine wave of 10 V peak-to-peak and 180° phase shift with respect to the input sine wave. Because of the minus sign associated with the closed-loop gain this configuration is called the *inverting configuration*.

The fact that the closed-loop gain depends entirely on external passive components

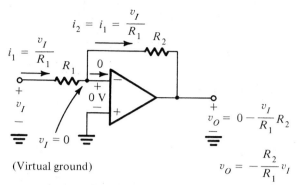

Fig. 3.5 Analysis of the inverting configuration.

(resistors R_1 and R_2) is a very interesting one. It means that we can make the closed-loop gain as accurate as we want by selecting passive components of appropriate accuracy. It also means that the closed-loop gain is (ideally) independent of the op amp gain. This is a dramatic illustration of negative feedback: we started out with an amplifier having very large gain A, and through applying negative feedback we have obtained a closed-loop gain R_2/R_1 that is much smaller than A but is stable and predictable. That is, we are trading gain for accuracy. We will illustrate this point further in Chapter 12, which deals with feedback amplifiers.

The input impedance of our closed-loop inverting amplifier is simply equal to R_1. This can be seen from Fig. 3.5, where,

$$R_{\text{in}} \equiv \frac{v_I}{i_1} = \frac{v_I}{v_I/R_1} = R_1$$

Thus to make R_{in} high we should select a high value for R_1. However, if the required gain R_2/R_1 is also high, then R_2 could become impractically large. We may conclude that the inverting configuration suffers from a low input resistance.

EXERCISES

3.1 Design an inverting amplifier having a gain of -100 and a 1-kΩ input resistance.
Ans. $R_1 = 1$ kΩ; $R_2 = 100$ kΩ

3.2 Find the closed-loop gain of the circuit shown in Fig. E3.2. Assume the op amp to be ideal.
Ans. -1020

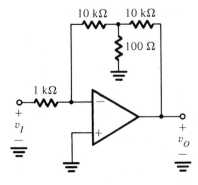

Fig. E3.2

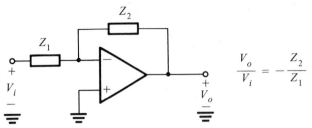

Fig. 3.6   *The inverting configuration with general impedances in the feedback and at the input.*

3.4   OTHER APPLICATIONS OF THE INVERTING CONFIGURATION

Rather than using two resistors R_1 and R_2, we may use two impedances Z_1 and Z_2, as shown in Fig. 3.6. The closed-loop gain, or more appropriately the closed-loop transfer function, is given by

$$\frac{V_o}{V_i} = - \frac{Z_2}{Z_1}$$

As a special case consider first the following:

$$Z_1 = R \quad \text{and} \quad Z_2 = \frac{1}{j\omega C}$$

$$\frac{V_o}{V_i} = - \frac{1}{j\omega CR} \tag{3.1}$$

which can be shown to correspond to integration; that is, $v_o(t)$ will be the integral of $v_I(t)$. To see this in the time domain, consider the corresponding circuit shown in Fig. 3.7. It is easy to see that the current i_1 is given by

$$i_1 = \frac{v_I(t)}{R}$$

If at time $t = 0$ the voltage across the capacitor is V, then

$$v_o(t) = V - \frac{1}{C} \int_0^t i_1(t)\, dt$$

$$= V - \frac{1}{CR} \int_0^t v_I(t)\, dt$$

Thus $v_o(t)$ is the time integral of $v_I(t)$, and voltage V is the initial condition of this integration process. The time constant CR is called the *integration time constant*. This integrator circuit is inverting because of the minus sign associated with its transfer function; it is known as the *Miller integrator*.

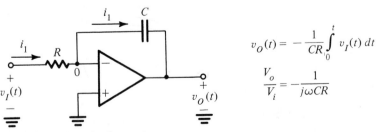

$$v_O(t) = - \frac{1}{CR} \int_0^t v_I(t)\, dt$$

$$\frac{V_o}{V_i} = - \frac{1}{j\omega CR}$$

Fig. 3.7   *The Miller or inverting integrator.*

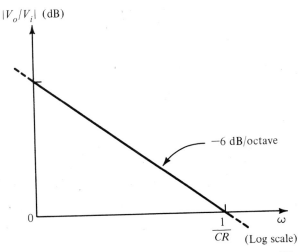

$|V_o/V_i|$ (dB)

-6 dB/octave

0

$\dfrac{1}{CR}$ (Log scale)

ω

Fig. 3.8 Frequency response of an ideal integrator with a time constant CR.

From the transfer function in Eq. (3.1) and our discussion of the frequency response of low-pass single-time-constant networks in Chapter 2, it is easy to see that the Miller integrator will have the magnitude plot shown in Fig. 3.8, which is identical to that of a low-pass network with zero break frequency. It is important to note that at zero frequency the closed-loop gain is infinite. That is, at dc the op amp is operating as an open loop, which could be easily seen when we recall that capacitors behave as open circuits with dc. When op-amp imperfections are taken into account, we will find it necessary to modify the integrator circuit to make the closed-loop gain at dc finite. This will be illustrated in a later section.

As another special case consider

$$Z_1 = \frac{1}{j\omega C} \quad \text{and} \quad Z_2 = R$$

$$\frac{V_o}{V_i} = -j\omega CR$$

which corresponds to a differentiation operation, that is,

$$v_o(t) = -CR\frac{dv_I(t)}{dt}$$

The reader can easily verify that the circuit of Fig. 3.9 indeed implements this differentiation operation. Figure 3.10 shows a plot for the magnitude of the transfer function

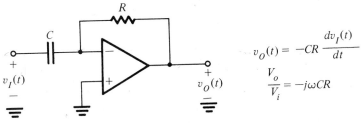

$$v_o(t) = -CR\frac{dv_I(t)}{dt}$$

$$\frac{V_o}{V_i} = -j\omega CR$$

Fig. 3.9 A differentiator.

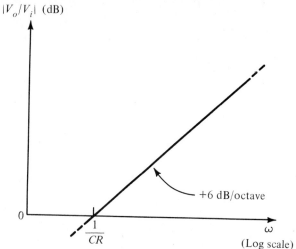

Fig. 3.10   *Frequency response of a differentiator with a time constant CR.*

of the differentiator, which is identical to that of an STC high-pass network with an infinite-frequency break point.

The very nature of a differentiator circuit causes it to be a "noise magnifier." This is due to the spikes introduced at the output every time there is a sharp change in $v_i(t)$; such a change could be a "picked up" interference. For this reason and because they suffer from stability problems (Chapter 12), differentiator circuits are generally avoided in practice.

As a final application of the inverting configuration consider the circuit shown in Fig. 3.11. Here we have a resistance R_f in the negative-feedback path (as before), but we have a number of input signals $v_1, v_2, \ldots, v_n$ each applied to corresponding resistor $R_1, R_2, \ldots, R_n$, which are connected to the inverting terminal of the op amp. From our previous discussion, the ideal op amp will have a virtual ground appearing at its negative input terminal. Ohm's law then tells us that the currents $i_1, i_2, \ldots, i_n$ are given by

$$i_1 = \frac{v_1}{R_1}, \qquad i_2 = \frac{v_2}{R_2}, \qquad \ldots, \qquad i_n = \frac{v_n}{R_n}$$

All these currents sum together to produce the current i,

$$i = i_1 + i_2 + \cdots + i_n$$

that will be forced to flow through R_f (since no current flows into the input terminals of ideal op amps). The output voltage v_O may now be determined by another application of Ohm's law,

$$v_O = 0 - iR_f = -iR_f$$

Thus

$$v_O = -\left(\frac{R_f}{R_1} v_1 + \frac{R_f}{R_2} v_2 + \cdots + \frac{R_f}{R_n} v_n \right)$$

That is, the output voltage is a weighted sum of the input signals $v_1, v_2, \ldots, v_n$. This circuit is therefore called a *weighted summer*. Note that each summing coefficient may

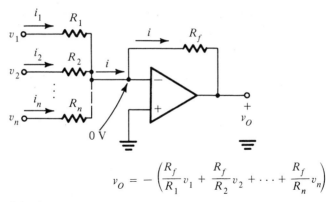

$$v_O = -\left(\frac{R_f}{R_1}v_1 + \frac{R_f}{R_2}v_2 + \cdots + \frac{R_f}{R_n}v_n\right)$$

Fig. 3.11 A weighted summer.

be independently adjusted by adjusting the corresponding "feed-in" resistor (R_1 to R_n). This nice property, which greatly simplifies circuit adjustment, is a direct consequence of the virtual ground that exists at the inverting op-amp terminal. As the reader will soon come to appreciate, virtual grounds are extremely "handy."

In the above we have seen that op amps can be used to multiply a signal by a constant, integrate it, differentiate it, and sum a number of signals with prescribed weights. These are all mathematical operations—hence the name operational amplifier. The circuits above are functional building blocks needed to perform analog computation. For this reason the op amp has been the basic element of analog computers. Op amps, however, can do much more than just perform the mathematical operations required in analog computation. In this chapter we will get a taste of this versatility, with other applications presented in later chapters.

EXERCISES

3.3 Consider a symmetrical square wave of 20-V peak-to-peak, 0 average, and 1-ms period applied to a Miller integrator. Find the value of the time constant CR such that the triangular waveform at the output has a 20-V peak-to-peak amplitude.
Ans. 250 μs

3.4 Show that the circuit shown in Fig. E3.4 has an STC low-pass transfer function. Find the dc gain and the 3-dB frequency.
Ans. -100; 10^5 rad/s

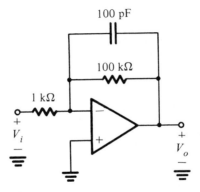

Fig. E3.4

3.5   THE NONINVERTING CONFIGURATION

The second closed-loop configuration we shall study is shown in Fig. 3.12. Here the input signal v_I is applied directly to the positive input terminal of the op amp while one terminal of R_1 is connected to ground. The reader is encouraged to investigate the difference between this configuration and the inverting configuration of Fig. 3.4. Note that one can be obtained from the other by interchanging the input terminal with ground—

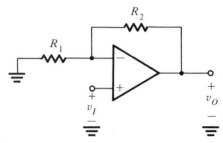

Fig. 3.12   The noninverting configuration.

that is, the terminal that was connected to v_I should be grounded, and the terminal that was connected to ground should be connected to v_I. This process is called the *complementary transformation.* When this transformation is applied to a network whose transfer function is T_1, the resulting network will have a transfer function T_2,

$$T_2 = 1 - T_1$$

This transformation is illustrated in Fig. 3.13, where we have a three-terminal network N. In Fig. 3.13a voltage source V_b is applied to terminal b, terminal c is grounded, and

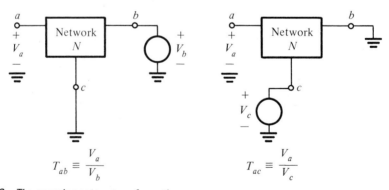

Fig. 3.13   The complementary transformation.

the voltage at a is measured to obtain the transfer function T_{ab} (from b to a). In Fig. 3.13b we have interchanged the input and ground terminals, so that now terminal b is grounded while the input signal is applied to terminal c. Again the voltage at a is measured to obtain the transfer function T_{ac} (from c to a). Network theory tells us that

$$T_{ac} = 1 - T_{ab}$$

Let us now return to the noninverting configuration of Fig. 3.12. The above network theorem enables us to directly obtain the transfer function or gain of the noninverting configuration as

$$\frac{v_O}{v_I} = 1 - \left(-\frac{R_2}{R_1} \right)$$

Thus

$$\frac{v_O}{v_I} = 1 + \frac{R_2}{R_1}$$

That this result is correct could be easily verified by analyzing the noninverting circuit directly. Details of the analysis are shown in Fig. 3.14. Note that a virtual short circuit

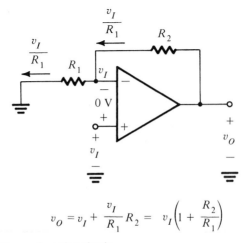

$$v_O = v_I + \frac{v_I}{R_1} R_2 = v_I \left(1 + \frac{R_2}{R_1} \right)$$

Fig. 3.14   Analysis of the noninverting circuit.

exists between the two input terminals; hence the voltage at the inverting input terminal is equal to v_I.

The gain of the noninverting configuration is positive—hence the name noninverting. The input impedance of this closed-loop amplifier is ideally infinite, since no current flows into the positive input terminal of the op amp. This property of high input impedance is a very desirable feature of the noninverting configuration. It enables using this circuit as an isolation amplifier to connect a source with a high impedance to a low impedance load. We have discussed the need for isolation amplifiers in Chapter 2. In many applications the isolation amplifier is not required to provide any voltage gain; rather it is used mainly as an impedance transformer or a power amplifier. In such cases we may make $R_2 = 0$ and $R_1 = \infty$ to obtain the unity-gain amplifier shown in Fig. 3.15. This circuit is commonly referred to as a *voltage follower*, since the output "follows" the input. In the ideal case, $v_O = v_I$, $R_{in} = \infty$, and $R_{out} = 0$.

Since the noninverting configuration has a gain greater than or equal to unity, depending on the choice of R_2/R_1, some prefer to call it "a follower with gain."

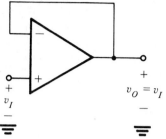

Fig. 3.15   The unity-gain or follower amplifier.

EXERCISES

3.5   Use the superposition principle to find the output voltage of the circuit shown in Fig. E3.5.
        Ans. $v_O = 6v_1 + 4v_2$

3.6   Show that application of the complementary transformation to a single-time-constant low-pass RC network results in a high-pass network.

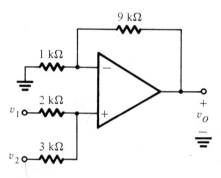

Fig. E3.5

3.6   EXAMPLES OF OP-AMP CIRCUITS

Now that we have studied the two most common closed-loop configurations of op amps, we present a number of examples. Our objective is twofold: first, to enable the reader to gain experience in analyzing circuits containing op amps; second, to introduce the reader to some of the many interesting and exciting applications of op amps—in short, to "whet his (or her) appetite."

Example 3.1

Figure 3.16 shows a circuit for an analog voltmeter of very high input impedance that uses an inexpensive moving-coil meter. As shown, the moving-coil meter is connected in the negative-feedback path of the op amp. The voltmeter measures the voltage v applied between the op-amp positive input terminal and ground. Assume that the moving coil produces full-scale deflection when the current passing through it is 100 μA; we wish to find the value of R such that the full-scale reading for v is $+10$ V.

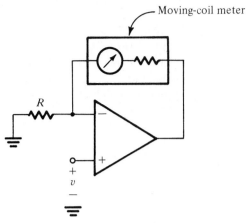

Fig. 3.16 An analog voltmeter with a high input impedance.

Solution
The current in the moving-coil meter is v/R because of the virtual short circuit at the op-amp input and the infinite input impedance of the op amp. Thus we have to choose R such that

$$\frac{10}{R} = 100 \ \mu A$$

Thus

$$R = 100 \ k\Omega$$

Note that the resulting voltmeter will produce readings directly proportional to the value of v, irrespective of the value of the internal resistance of the moving-coil meter— a very desirable property.

· · ·

Example 3.2
We need to find an expression for the output voltage v_o in terms of the input voltages v_1 and v_2 for the circuit in Fig. 3.17.

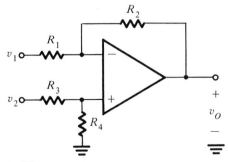

Fig. 3.17 A difference amplifier.

Solution

There are a number of ways to solve this problem; perhaps the easiest is using the principle of superposition. Obviously superposition may be employed here, since the network is linear. To apply superposition we first reduce v_2 to zero—that is, ground the terminal to which v_2 is applied—and then find the corresponding output voltage, which will be due entirely to v_1. We denote this output voltage v_{O1}. Its value may be found from the

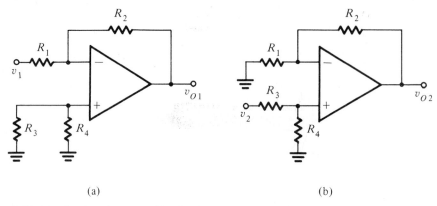

(a)   (b)

Fig. 3.18   *Application of superposition for analysis of the circuit of Fig. 3.17.*

circuit in Fig. 3.18a, which we recognize as that of the inverting configuration. The existence of R_3 and R_4 does not affect the gain expression, since no current flows through either of them. Thus

$$v_{O1} = - \frac{R_2}{R_1} v_1$$

Next, we reduce v_1 to zero and evaluate the corresponding output voltage v_{O2}. The circuit will now take the form shown in Fig. 3.18b, which we recognize as the noninverting configuration with an additional voltage divider, made up of R_3 and R_4, connected across the input v_2. The output voltage v_{O2} is therefore given by

$$v_{O2} = v_2 \frac{R_4}{R_3 + R_4} \left(1 + \frac{R_2}{R_1} \right)$$

The superposition principle tells us that the output voltage v_O is equal to the sum of v_{O1} and v_{O2}. Thus we have

$$v_O = - \frac{R_2}{R_1} v_1 + \frac{1 + R_2/R_1}{1 + R_3/R_4} v_2 \qquad (3.2)$$

This completes the analysis of the circuit in Fig. 3.17. However, because of the practical importance of this circuit we shall pursue it further. We will ask, What is the condition under which this circuit will act as a differential amplifier? In other words, we wish to make the circuit respond (produce an output) in proportion to the difference signal $v_2 - v_1$ and reject common mode (that is, produce zero output when $v_1 = v_2$).

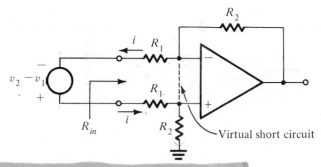

Fig. 3.19 *Finding the input resistance of the differential amplifier.*

The answer can be obtained from the expression we have just derived [Eq. (3.2)]. Let us set $v_1 = v_2$ and require that $v_O = 0$. It is easy to see that this process leads to the condition

$$\frac{R_2}{R_1} = \frac{R_4}{R_3}$$

Substituting in Eq. (3.2) results in the output voltage

$$v_O = \frac{R_2}{R_1}(v_2 - v_1)$$

which is clearly that of a differential amplifier with a gain of R_2/R_1.

We next inquire about the input resistance seen between the two input terminals. The circuit is redrawn in Fig. 3.19 with the condition $R_2/R_1 = R_4/R_3$ imposed. In fact, to simplify matters and for other practical considerations we have made $R_3 = R_1$ and $R_4 = R_2$. We wish to evaluate the input differential resistance R_{in}, defined

$$R_{in} \equiv \frac{v_2 - v_1}{i}$$

Since the two input terminals of the op amp track each other in potential, we may write a loop equation and obtain

$$v_2 - v_1 = R_1 i + 0 + R_1 i$$

Thus

$$R_{in} = 2R_1$$

Note that if the amplifier is required to have a large differential gain, then R_1, by necessity, will be relatively small and the input resistance will be correspondingly small, a drawback of this circuit.

$\cdot \quad \cdot \quad \cdot$

Example 3.3
We desire to find the input resistance R_{in} of the circuit in Fig. 3.20.

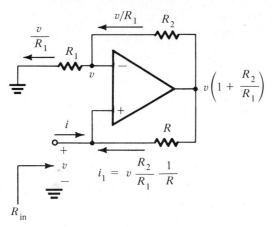

Fig. 3.20 *Circuit for Example 3.3.*

Solution

To find R_{in} we apply an input voltage v and evaluate the input current i. Then R_{in} may be found from its definition, $R_{in} \equiv v/i$.

Owing to the virtual short circuit between the op-amp input terminals, the voltage at the inverting terminal will be equal to v. The current through R_1 will therefore be v/R_1. Owing to the infinite input impedance of the op amp, the current through R_2 will also be v/R_1. Thus the voltage at the op-amp output will be:

$$v + \frac{v}{R_1} R_2 = \left(1 + \frac{R_2}{R_1}\right) v$$

We may now apply Ohm's law to R and obtain the current through it as

$$i_1 = \frac{v(1 + R_2/R_1) - v}{R} = v \frac{R_2}{R_1} \frac{1}{R}$$

Since no current flows into the positive input terminal of the op amp, we have

$$i = -i_1 = -\frac{v}{R} \frac{R_2}{R_1}$$

Thus

$$R_{in} = -R \frac{R_1}{R_2}$$

that is, the input resistance is negative with a magnitude equal to R, the resistance in the positive-feedback path, multiplied by the ratio R_1/R_2. This circuit is therefore called a *negative impedance converter* (NIC), where R may in general be replaced by an arbitrary impedance Z.

Let us investigate the application of this circuit further. Consider the case $R_1 = R_2 = r$, where r is an arbitrary value; it follows that $R_{in} = -R$. Let the input be fed with a voltage source V_s having a source resistance equal to R, as shown in Fig. 3.21a. We want to evaluate the current I_l that flows in an impedance Z_L connected as shown.

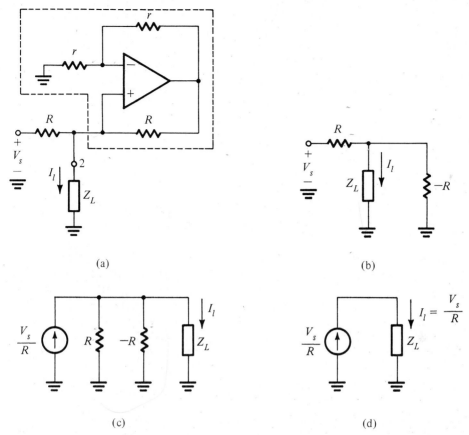

(a)

(b)

(c)

(d)

Fig. 3.21 *Illustrating the application of the negative impedance converter of Fig. 3.20 in the design of a voltage-to-current converter.*

In Fig. 3.21b we have utilized the information gained in the above and replaced the circuit in the dashed box by a resistance $-R$. Figure 3.21c illustrates the conversion of the voltage source to its Norton's equivalent. Finally, the two parallel resistances R and $-R$ are combined to produce an infinite resistance, resulting in the circuit in Fig. 3.21d, from which we see that the load current I_l is given by

$$I_l = \frac{V_s}{R}$$

independent of the value of Z_L! This is an interesting result; it tells us that the circuit of Fig. 3.21a acts as a *voltage-to-current converter*, providing a current I_l that is directly proportional to V_s ($I_l = V_s/R$) and is independent of the value of the load impedance. That is, terminal 2 acts as a current-source output, with the impedance looking back into 2 equal to infinity. Note that this infinite resistance is obtained via the cancellation of the positive source resistance R with the negative input resistance $-R$.

As it is, the circuit is quite useful and has practical applications where it is necessary to generate a current signal in proportion to a given voltage signal. A specific application is illustrated in Fig. 3.22, where a capacitor C is used as a load. From the above analysis we conclude that capacitor C will be supplied by a current $I = V_i/R$ and thus its voltage V_2 will be given by

$$V_2 = \frac{I}{j\omega C} = \frac{V_i}{j\omega CR}$$

That is,

$$\frac{V_2}{V_i} = \frac{1}{j\omega CR}$$

which is the transfer function of an integrator,

$$v_2 = \frac{1}{CR}\int_0^t v_i \, dt + V$$

where V is the voltage across the capacitor at $t = 0$. This integrator has some interesting properties. The transfer function does not have an associated negative sign, as is the case with the Miller integrator. Noninverting, or positive, integrators are required in many applications. Another useful property is the fact that one terminal of the capacitor

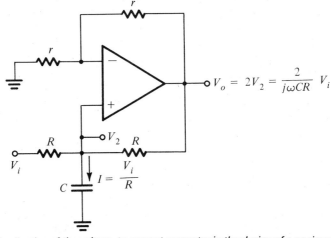

Fig. 3.22 *Application of the voltage-to-current converter in the design of a noninverting integrator.*

is grounded. Among other things, this would simplify the initial charging of the capacitor, as may be necessary to simulate an initial condition in the solution of a differential equation on an analog computer.

The integrator circuit of Fig. 3.22 has a serious problem, though. We cannot "take the output" at terminal 2 as indicated, since terminal 2 is a high-impedance point, which means that connecting any load resistance there will change the transfer function V_2/V_i. Fortunately, however, a low-impedance point exists where the signal is propor-

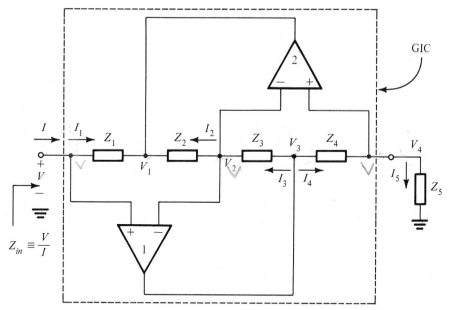

Fig. 3.23 Circuit for Example 3.4.

tional to V_2. We are speaking of the op-amp output terminal, where, as the reader can easily verify,

$$V_o = 2V_2$$

Thus

$$\frac{V_o}{V_i} = \frac{2}{j\omega CR}$$

. . .

Example 3.4
We wish to analyze the circuit shown in Fig. 3.23 in order to derive an expression for the input impedance Z_{in}.

Solution
To find Z_{in} we apply a voltage V at the input and attempt to find an expression for the input current I. Then, the input impedance can be determined from

$$Z_{in} \equiv \frac{V}{I}$$

Because of the virtual short circuit between the input terminals of op amp 1, the voltage V_2 will be equal to V. Similarly, the virtual short circuit between the input terminals of op amp 2 will cause $V_4 = V_2 = V$. We now can evaluate the current in Z_5 as

$$I_5 = \frac{V_4}{Z_5} = \frac{V}{Z_5}$$

The current in Z_4 will be equal to I_5, since no current flows into the positive input terminal of op amp 2. Thus

$$I_4 = I_5 = \frac{V}{Z_5}$$

We are now able to determine V_3 as

$$V_3 = V_4 + I_4 Z_4 = V + \frac{V}{Z_5} Z_4 = \left(1 + \frac{Z_4}{Z_5}\right) V$$

Next the current in Z_3 can be determined as

$$I_3 = \frac{V_3 - V_2}{Z_3} = \frac{V(1 + Z_4/Z_5) - V}{Z_3} = \frac{V}{Z_3} \frac{Z_4}{Z_5}$$

Application of Ohm's law to Z_2 enables us to find V_1. The current in Z_2 is, of course, equal to I_3, since no current flows into the negative input terminals of the two op amps. Thus

$$V_1 = V_2 - I_2 Z_2 = V - I_3 Z_2 = V - \frac{V}{Z_3} \frac{Z_4}{Z_5} Z_2$$

The current I_1 in Z_1 may be determined from

$$I_1 = \frac{V - V_1}{Z_1} = \frac{V - V + (V/Z_3)(Z_4/Z_5)Z_2}{Z_1}$$

$$= \frac{V}{Z_3} \frac{Z_4}{Z_5} \frac{Z_2}{Z_1}$$

Finally, we recognize that the input current I should be equal to I_1, since no current flows into the positive input terminal of op amp 1. Thus

$$I = \frac{V}{Z_3} \frac{Z_4}{Z_5} \frac{Z_2}{Z_1}$$

from which we obtain

$$Z_{in} \equiv \frac{V}{I}$$

$$Z_{in} = \frac{Z_1 Z_3}{Z_2 Z_4} Z_5$$

This circuit is called the *generalized impedance converter (GIC)*. The name follows from the above expression by considering the GIC to be the circuit within the box in Fig. 3.23 and Z_5 to be a terminating impedance. The GIC causes the impedance Z_5 to be multiplied by a general conversion factor $Z_1 Z_3/Z_2 Z_4$ and the resulting converted impedance to appear at the input terminals.

The GIC finds applications in the area of active inductorless filter design (Chapter 14). As a specific example of its use, consider the realization of an inductance. That is, let it be required to find suitable components (resistors and capacitors) for the imped-ances Z_1 to Z_5 such that

$$Z_{in} = j\omega L$$

where L is the value of the required inductance. From the equation for Z_{in} derived above, we see that we have two choices. First, we have

$$Z_1 = R_1, \qquad Z_2 = \frac{1}{j\omega C}, \qquad Z_3 = R_3, \qquad Z_4 = R_4, \qquad Z_5 = R_5$$

resulting in

$$Z_{in} = j\omega C \frac{R_1 R_3 R_5}{R_2}$$

Thus

$$L = C \frac{R_1 R_3 R_5}{R_2}$$

Second, we have

$$Z_1 = R_1, \qquad Z_2 = R_2, \qquad Z_3 = R_3, \qquad Z_4 = \frac{1}{j\omega C}, \qquad Z_5 = R_5$$

resulting in

$$Z_{in} = j\omega C \frac{R_1 R_3 R_5}{R_4}$$

Thus

$$L = C \frac{R_1 R_3 R_5}{R_4}$$

Note that we have restricted ourselves to the use of a single capacitor, a practical constraint.

· · ·

EXERCISES

3.7 Find values for the resistances in the circuit of Fig. 3.17 such that the circuit behaves as a differential amplifier with an input resistance of 4 kΩ and a gain of 100.
  Ans. $R_1 = R_3 = 2$ kΩ; $R_2 = R_4 = 200$ kΩ

3.8 For the circuit shown in Fig. E3.8 find the magnitude and phase of the transfer function V_o/V_i at $\omega = 5000$ rad/s.
  Ans. $|V_o/V_i| = 1$; $\phi = +90°$

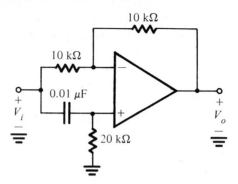

Fig. E3.8

3.7   NONIDEAL PERFORMANCE OF OP AMPS

Above we defined the ideal op amp, and we presented a number of circuit applications of op amps. The analysis of these circuits assumed the op amps to be ideal. Although in many applications such an assumption is not a bad one, a circuit designer has to be thoroughly familiar with the characteristics of real-life op amps and the effects of such characteristics on her or his circuits. Only then will he or she be able to use the op amp intelligently, especially if the application at hand is not a straightforward one. The non-ideal properties of op amps will, of course, limit the range of operation of the circuits analyzed in the previous examples.

In the following sections we consider the nonideal properties of the op amp. We do this by treating one parameter at a time.

3.8   FINITE OPEN-LOOP GAIN AND BANDWIDTH

The differential open-loop gain of an op amp is not infinite; rather it is finite and decreases with frequency. Figure 3.24 shows a plot for $|A|$, with the numbers typical of most general-purpose op amps (such as the 741-type op amp, which is available from many semiconductor manufacturers and whose internal circuit is studied in Chapter 13).

Note that although the gain is quite high at dc and low frequencies it starts to fall off at a rather low frequency (10 Hz in our example). The uniform -20 dB/decade gain rolloff shown is typical of *internally compensated* op amps. These are units that have a network (usually a single capacitor) included on the same IC chip whose function is to cause the op-amp gain to have the single-time-constant low-pass response shown. This process of modifying the open-loop gain is termed *frequency compensation,* and its purpose is to ensure that op-amp circuits will be stable (as opposed to oscillatory). The subject of stability of op-amp circuits, or, more generally, of feedback amplifiers, will be studied in Chapter 12.

By analogy to the response of low-pass STC circuits (Chapter 2), the gain $A(\omega)$ of an internally compensated op amp may be expressed

$$A(\omega) = \frac{A_0}{1 + j\omega/\omega_b} \tag{3.3}$$

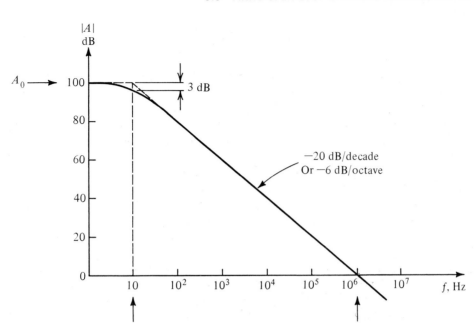

Fig. 3.24 *Open-loop gain of a typical general-purpose internally compensated op amp.*

where A_0 denotes the dc gain and ω_b is the 3-dB frequency (or "break" frequency). For the example shown in Fig. 3.24, $A_0 = 10^5$ and $\omega_b = 2\pi \times 10$ rad/s. For frequencies $\omega \gg \omega_b$ (about ten times and higher) Eq. (3.3) may be approximated by

$$A(\omega) \simeq \frac{(A_0\omega_b)}{j\omega} \tag{3.4}$$

from which it can be seen that the gain $|A|$ reaches unity (0 dB) at a frequency denoted by ω_t and given by

$$\omega_t = A_0\omega_b$$

Substituting in Eq. (3.4) gives

$$A(\omega) \simeq \frac{\omega_t}{j\omega} \tag{3.5}$$

where ω_t is called the *unity-gain bandwidth*. The unity-gain bandwidth $f_t = \omega_t/2\pi$ is usually specified on the data sheets of op amps.

The gain magnitude can be obtained from Eq. (3.5) as

$$|A(\omega)| \simeq \frac{\omega_t}{\omega} = \frac{f_t}{f} \tag{3.6}$$

Thus if f_t is known (10^6 Hz in our example), one can easily estimate the magnitude of the op-amp gain at a given frequency f. Finally, it should be mentioned that an op amp having this uniform -6 dB/octave gain rolloff is said to have a "single-pole" model. More will be said about poles and zeros in Chapter 11.

We next consider the effect of the limited op-amp gain and bandwidth on the closed-loop transfer functions of the two simple configurations shown in Fig. 3.25. For the inverting amplifier the analysis proceeds as follows: Denote the output voltage V_o. The voltage between the two input terminals will, by definition, be V_o/A.

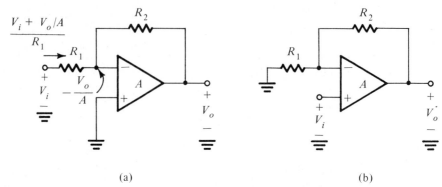

(a)   (b)

Fig. 3.25   (a) The inverting and (b) the noninverting configurations, with the op amp having finite gain A.

Since the positive input terminal is grounded, the voltage at the inverting input terminal will be $-V_o/A$. The current in R_1 can now be obtained as

$$I_1 = \frac{V_i - (-V_o/A)}{R_1} = \frac{1}{R_1}\left(V_i + \frac{V_o}{A}\right)$$

Because of the infinite input impedance of the op amp, the current in R_2 will be equal to I_1, and the output voltage V_o can be obtained directly as

$$V_o = -\frac{V_o}{A} - \frac{1}{R_1}\left(V_i + \frac{V_o}{A}\right)R_2$$

By collecting terms and with simple manipulations we obtain the transfer function

$$\frac{V_o}{V_i} = -\frac{R_2/R_1}{1 + (1 + R_2/R_1)/A} \tag{3.7}$$

which we note reduces to the ideal expression

$$\frac{V_o}{V_i} = -\frac{R_2}{R_1}$$

as A approaches ∞.

Similar analysis for the noninverting configuration provides

$$\frac{V_o}{V_i} = \frac{1 + R_2/R_1}{1 + (1 + R_2/R_1)/A} \tag{3.8}$$

Substituting for A from Eq. (3.3) into Eq. (3.7) gives

$$\frac{V_o(\omega)}{V_i(\omega)} = -\frac{R_2/R_1}{1 + \dfrac{1}{A_0}\left(1 + \dfrac{R_2}{R_1}\right) + j\dfrac{\omega}{\omega_t/(1 + R_2/R_1)}}$$

For $A_0 \gg 1 + R_2/R_1$, which is the case in normal applications,

$$\frac{V_o(\omega)}{V_i(\omega)} \simeq -\frac{R_2/R_1}{1 + j\dfrac{\omega}{\omega_t/(1 + R_2/R_1)}} \tag{3.9}$$

which says that the inverting configuration has a dc gain of magnitude equal to R_2/R_1, the ideal value. The closed-loop gain rolls off at a uniform -20 dB/decade slope with a corner frequency (3-dB frequency) given by

$$\omega_{3\ dB} = \frac{\omega_t}{1 + R_2/R_1} \tag{3.10}$$

The -20 dB/decade slope follows from the STC low-pass response given by Eq. (3.9). Similarly, the noninverting amplifier will have a dc gain of $1 + R_2/R_1$, a uniform gain rolloff with a slope of -20 dB/decade, and a 3-dB frequency given also by Eq. (3.10).

Example 3.5
Consider an op amp with $f_t = 1$ MHz. Find the 3-dB frequency of closed-loop amplifiers with nominal gains of $+1,000$, $+100$, $+10$, $+1$, -1, -10, -100, and $-1,000$. Sketch the magnitude frequency response for the amplifiers with closed-loop gains of $+10$ and -10.

Solution
Using Eq. (3.10) we obtain the results given in the following table:

Closed-Loop Gain	$\dfrac{R_2}{R_1}$	$f_{3\ dB} = f_t/(1 + R_2/R_1)$
$+1,000$	999	1 kHz
$+100$	99	10 kHz
$+10$	9	100kHz
$+1$	0	1 MHz
-1	1	0.5 MHz
-10	10	90.9 kHz
-100	100	9.9 kHz
$-1,000$	1,000	$\simeq 1$ kHz

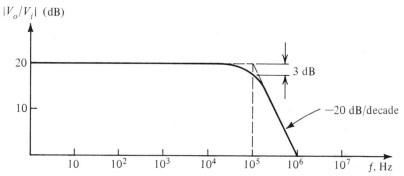

Fig. 3.26   Frequency response of an amplifier with a nominal gain of + 10.

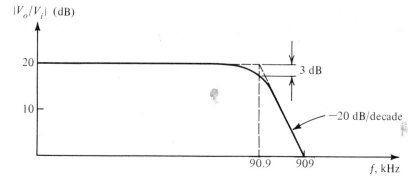

Fig. 3.27   Frequency response of an amplifier with a nominal gain of − 10.

Figure 3.26 shows the frequency response for the amplifier whose nominal dc gain is +10, and Fig. 3.27 shows the frequency response for the −10 case. An interesting observation follows from the table above: the unity-gain inverting amplifier has a 3-dB frequency of $f_t/2$ as compared to f_t for the unity-gain noninverting amplifier.

· · ·

Example 3.5 clearly illustrates the trade-off between gain and bandwidth. For instance, the noninverting configuration exhibits a constant *gain–bandwidth product*. An interpretation of these results in terms of feedback theory will be given in Chapter 12. For the time being it should be mentioned that both the inverting and the noninverting configurations have identical "feedback loops." This can be seen by eliminating the excitation (that is, short-circuiting the input voltage source), resulting in both cases in the feedback loop shown in Fig. 3.28. Since their feedback loops are identical, the two configurations have the same dependence on the finite op-amp gain and bandwidth (for example, identical expressions for $f_{3\ dB}$).

EXERCISES

3.9   Consider an op amp having 106-dB gain at dc and a single-pole frequency response with $f_t = 2$ MHz. Find the magnitude of gain at $f = 1$ kHz, 10 kHz, and 100 kHz.
   Ans.   2,000; 200; 20

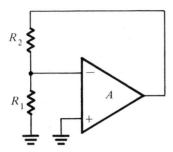

Fig. 3.28   The feedback loop of the inverting and the noninverting configurations.

3.10   If the op amp in Exercise 3.9 is used to design a noninverting amplifier with nominal dc gain of 100, find the 3-dB frequency of the closed-loop gain. Also find the rise time of the output waveform if a voltage step is applied at the input.
Ans.   20 kHz; 17.5 μs

3.9   SLEW RATE AND FULL-POWER BANDWIDTH

Although it was not specifically mentioned, the analysis of the previous section assumes small-signal operation of the op amp. In other words, if we measure the frequency response of a closed-loop amplifier having a gain of, say, +10, the predicted 3-dB frequency of $f_t/10$ would be achieved only if the output voltage is quite small (a volt or so). On the other hand, op amps are capable of providing output signal swings that approach the voltage of the power supplies used. For instance, a 741-type op amp operated from ±15-V power supplies is rated to provide output signal swings of at least ±10 V. However, if we attempt to measure the frequency response of our +10-gain amplifier with the full-power output (that is, with a sine-wave signal that produces a 20-V peak-to-peak output voltage), we encounter a phenomenon called *slew-rate limiting*. Specifically, we will find that at a frequency much lower than the expected $f_{3\ dB}$ the output sine wave will become distorted. The distortion increases as the frequency of the input signal is increased.

The distortion observed will be of the type shown in Fig. 3.29, where we note that the rate of change (slope) of the output signal is smaller than that of the sine wave.

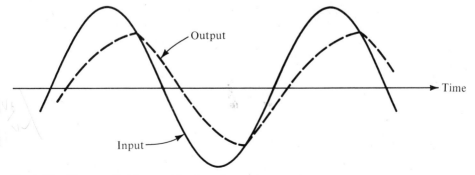

Fig. 3.29   Slew-rate limiting resulting in a distorted output sine wave.

Obviously, at low frequencies the maximum slope of the sine wave (which occurs at its zero crossings) will be small, and no distortion will be observed. As the frequency is increased, the slope of the sine wave increases, and we reach a point at which the op amp will not be able to "keep up" with the rate of change required to produce an output sine wave. This is slew-rate limiting.

The op amp's slew rate is normally specified on its data sheets as so many volts per microsecond. It can be measured by applying a square wave to, say, a voltage follower

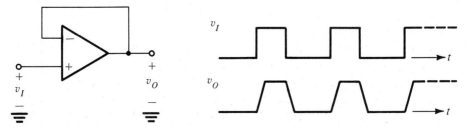

Fig. 3.30   *Measuring the slew rate of an op amp by applying a square wave to a voltage follower.*

and measuring the slope of the ramp edges of the output signal, as shown in Fig. 3.30. Slew rate is formally defined as the maximum rate of change of the output voltage with respect to time that the op amp is capable of producing.

Slew rate can be easily related to the *full-power bandwidth f_M,* which is defined as the frequency at which a sine-wave output whose amplitude is equal to the rated op-amp output voltage starts to show distortion (due to op-amp slewing). Consider the follower of Fig. 3.30 and let the input signal be a sine wave with an amplitude equal to the maximum specified output $\hat{V}$,

$$v = \hat{V}\sin(\omega t)$$

Then,

$$\frac{dv}{dt} = \omega\hat{V}\cos(\omega t)$$

$$\left.\frac{dv}{dt}\right|_{max} = \omega\hat{V}$$

Denoting the slew rate by SR, it follows that

$$SR = \omega_M\hat{V}$$

or

$$f_M = \frac{SR}{2\pi\hat{V}} \tag{3.11}$$

As an example, if the slew rate of an op amp is specified to be 2 V/μs and its maximum rated output is ± 10 V, then the full-power frequency f_M will be

$$f_M = \frac{2 \times 10^6}{2\pi \times 10} = 31.8 \text{ kHz}$$

Mention was made above of the maximum output voltage swing of the op amp. This is the voltage swing that the op amp can produce at the output without signal distortion (at frequencies lower than f_M). The rated output voltage defines the limits of the linear operating range of the op amp. As an example, consider an op amp with maximum output voltage of ± 12 V connected as a noninverting amplifier with a gain of $+10$. It follows that at low frequencies ($f < f_M$) one can apply input signals in the range ± 1.2 V with the amplifier remaining in the linear range. On the other hand, if an input signal of, say, $+1.5$ V is applied, the amplifier output will be "saturated" at a voltage close to the upper specified limit of $+12$ V. This limit on signal swing is a nuisance in linear circuit design; however, it can be put to use in nonlinear applications of the op amp, as shown in Chapter 5.

3.10   COMMON-MODE REJECTION

Real op amps have finite nonzero *common-mode gain*—that is, if the two input terminals are tied together and a signal V_{cm} is applied, the output will not be zero. The ratio of the output voltage to the input V_{cm} is called the common-mode gain A_{CM}. Figure 3.31 illustrates this definition.

The ability of an op amp to reject common-mode signals is specified in terms of the common-mode rejection ratio (CMRR), defined as

$$CMRR = \frac{|A|}{|A_{CM}|} \qquad (3.12)$$

Usually the CMRR is expressed in dB:

$$CMRR = 20 \log \left| \frac{A}{A_{CM}} \right| \qquad dB$$

The CMRR is a function of frequency, decreasing as the frequency is increased. Typical values of CMRR at low frequencies are 80 to 100 dB.

The finite CMRR of op amps is unimportant in the case of the inverting configuration, since the positive input terminal is grounded and hence the common-mode input signal is zero. In the noninverting amplifier the input signal is in effect a common-mode input signal and the finite CMRR might have to be taken into account in applications that demand high accuracy. The circuit that suffers most from the finite CMRR of the op amp is that of the difference amplifier, shown in Fig. 3.32. Recall that this circuit was analyzed in Example 3.2 assuming an ideal op amp. There we found that if the

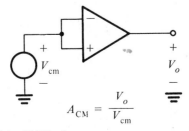

Fig. 3.31   Illustrating the finite CMRR of an op amp.

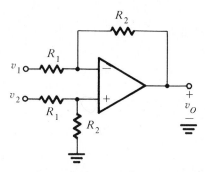

Fig. 3.32   The difference amplifier configuration suffers from the finite CMRR of the op amp.

resistors are matched as shown in Fig. 3.32, the amplifier responds only to the difference signal $v_2 - v_1$. This will no longer be true if the finite CMRR of the op amp is taken into account.

A simple method for taking into account the effect of the finite CMRR in calculating closed-loop gains is as follows: Dividing the common-mode input voltage V_{cm} by the CMRR (expressed as a ratio) we obtain the *common-mode error voltage V_{cr}*,

$$V_{cr} = \frac{V_{cm}}{|A|/|A_{cm}|}$$

From this definition it follows that if V_{cr} is multiplied by the differential gain $|A|$, the result is the output voltage component due to the common-mode input V_{cm}. Thus the effect of the finite CMRR can be taken into account by including V_{cr} as an additional difference input signal to the op amp. A simple way to accomplish this is to include a voltage source equal to V_{cr} in series with one of the op-amp input leads and assume that the op amp is ideal (that is, it has an infinite CMRR). This procedure is illustrated in Fig. 3.33 for the case of the noninverting amplifier. Note that the voltage source V_{cr} is not assigned a polarity, since this is not known.

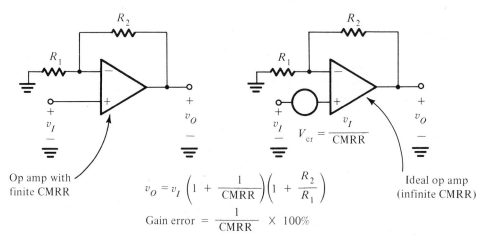

Op amp with finite CMRR

$$v_O = v_I \left(1 + \frac{1}{CMRR}\right)\left(1 + \frac{R_2}{R_1}\right)$$

$$\text{Gain error} = \frac{1}{CMRR} \times 100\%$$

$$V_{cr} = \frac{v_I}{CMRR}$$

Ideal op amp (infinite CMRR)

Fig. 3.33   Analysis of the noninverting amplifier with the finite CMRR of the op amp taken into account.

3.11 Use the method described above to calculate the common-mode gain of the difference amplifier circuit in Fig. 3.32 for the case of an op-amp CMRR $= 80$ dB and $R_2/R_1 = 1,000$.

Ans. CM gain $= 0.1$

3.11 INPUT AND OUTPUT RESISTANCES

Figure 3.34 shows an equivalent circuit of the op amp, incorporating its finite input and output resistances. As shown, the op amp has a differential input resistance R_{id} seen between the two input terminals. In addition, if the two input terminals are tied together

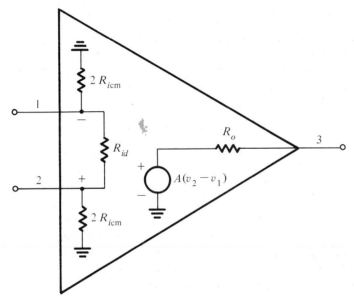

Fig. 3.34 Op-amp model with the input and output resistances shown.

and the input resistance (to ground) is measured, the result is the common-mode input resistance R_{icm}. In the equivalent circuit we have split R_{icm} into two equal parts $(2 R_{icm})$, each connected between one of the input terminals and ground.

Input Resistance

Typical values for the input resistances of general purpose op amps are $R_{id} = 1$ MΩ and $R_{icm} = 100$ MΩ. The value of the input resistance of a particular closed-loop circuit will depend on the values of R_{id} and R_{icm} as well as on the circuit configuration. For the inverting configuration the input resistance is approximately equal to R_1. Detailed analysis shows that taking R_{id} and R_{icm} into account has a negligible effect on the value of the input resistance of the inverting circuit. On the other hand, the input resistance of the noninverting configuration is strongly dependent on the values of R_{id} and R_{icm} as well as on the value of A and R_2/R_1. Straightforward analysis of the noninverting circuit using the op-amp model of Fig. 3.34 and assuming that

$$R_o = 0, \qquad R_1 \ll R_{icm}, \qquad \frac{R_2}{R_{id}} \ll A, \qquad \frac{R_2}{R_1} \ll A$$

results in the following approximate expression for the input resistance of the noninverting circuit:

$$R_{in} \simeq \{2R_{icm}\}\|\{[A/(1 + R_2/R_1)]\, R_{id}\} \tag{3.13}$$

An interpretation of this result in terms of feedback theory will be given in Chapter 12.

Example 3.6
Consider an op amp having $f_t = 1$ MHz, $R_{id} = 1$ MΩ, and $R_{icm} = 100$ MΩ. Find the input impedance of a noninverting amplifier with a nominal gain of 100.

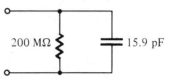

200 MΩ   15.9 pF

Fig. 3.35   *The input impedance of an amplifier with a nominal closed-loop dc gain of + 100.*

Solution
The input impedance can be obtained by substituting in Eq. (3.13) the following:

$$R_{icm} = 10^8 \ \Omega$$
$$R_{id} = 10^6 \ \Omega$$
$$1 + \frac{R_2}{R_1} = 100$$
$$A \simeq \frac{\omega_t}{j\omega} = \frac{2\pi \times 10^6}{j\omega}$$

The result is

$$Z_{in} = [2 \times 10^8 \ \Omega]\|[2\pi \times 10^6 \times 10^6/(j\omega \times 100)]$$

The second component is obviously a capacitor whose value is $100/2\pi$ pF. Figure 3.35 shows an equivalent circuit of the input impedance obtained above.

$\cdot$ $\cdot$ $\cdot$

Output Resistance

We now turn to the effect of the finite output resistance R_o shown in the model of Fig. 3.34. Typical values for the open-loop output resistance R_o are 75 to 100 Ω, although some amplifiers with much higher output resistance exist. We wish to find the output resistance of closed-loop amplifiers. To do this, we short the signal source, which makes the inverting and noninverting configurations identical. Then, the output resistance can be obtained by straightforward analysis of the circuit in Fig. 3.36 as follows.

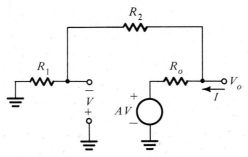

Fig. 3.36   Derivation of the closed-loop output resistance.

$$V = -V_o \frac{R_1}{R_1 + R_2} = -\beta V_o$$

$$I = \frac{V_o}{R_1 + R_2} + \frac{V_o - AV}{R_o}$$

$$= \frac{V_o}{R_1 + R_2} + \frac{(1 + A\beta)V_o}{R_o}$$

Thus

$$\frac{1}{R_{\text{out}}} \equiv \frac{I}{V_o} = \frac{1}{R_1 + R_2} + \frac{1 + A\beta}{R_o}$$

where the constant β is defined

$$\beta = \frac{R_1}{R_1 + R_2}$$

This means that the closed-loop output resistance is composed of two parallel components,

$$R_{\text{out}} = [R_1 + R_2] \| [R_o/(1 + A\beta)]$$

Normally R_o is much smaller than $R_1 + R_2$, resulting in

$$R_{\text{out}} \simeq \frac{R_o}{1 + A\beta}$$

Again, this result has a nice interpretation in terms of feedback theory, as will be explained in Chapter 12. A further simplification of the expression for R_{out} is obtained by noting that normally $A\beta \gg 1$, which gives

$$R_{\text{out}} \simeq \frac{R_o}{A\beta} \tag{3.14}$$

At very low frequencies A is real and large, resulting in a very small R_{out}. As an example, a voltage follower ($\beta = 1$) designed using an op amp with $R_o = 100 \ \Omega$ and $A_0 = 10^5$ will have

$$R_{\text{out}} = \frac{100}{10^5 \times 1} = 1 \ \text{m}\Omega$$

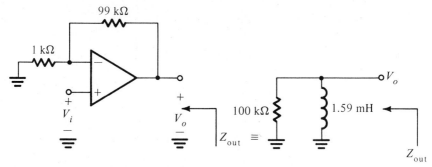

Fig. 3.37 *Illustrating the output impedance of a noninverting amplifier.*

It is interesting to inquire about the effect of the finite op-amp bandwidth on the closed-loop output impedance. Substituting $A = \omega_t/j\omega$ in Eq. (3.14) results in

$$Z_{out} = j\omega \frac{R_o}{\beta\omega_t}$$

Thus the output impedance becomes inductive with the value of the inductance equal to $R_o/\beta\omega_t$. As an example consider an amplifier with a nominal closed-loop gain of 100. Let the op amp have $R_o = 100\ \Omega$ and $f_t = 1$ MHz. The value of β will be 0.01, resulting in an output inductance of

$$L_{out} = \frac{100}{0.01 \times 2\pi \times 10^6} = 1.59\ \text{mH}$$

which is quite large and could cause serious problems in certain applications. For instance, at $f = 1$ kHz the output impedance will be

$$Z_{out} \simeq j10\ \Omega$$

Figure 3.37 illustrates this example.

3.12   DC PROBLEMS

Offset Voltage

Because op amps are direct-coupled devices with large gains at dc, they are prone to dc problems. The first such problem is the dc offset voltage. To understand this problem consider the following conceptual experiment: If the two input terminals of the op amp are tied together and connected to ground, it will be found that a finite dc voltage exists at the output. This is the *output dc offset voltage*. To take this dc offset voltage into account in the analysis of closed-loop configurations, it has been found convenient to "refer it back" to the input. Specifically, if we divide the output dc offset voltage by the gain A_0 we obtain the *input offset voltage* V_{off}. The latter may be represented by a voltage source connected in series with one of the input leads of an ideal op amp, that is, one with zero offset voltage. This arrangement is shown in Fig. 3.38, where we note that the V_{off} source has no assigned polarity because this is usually not known.

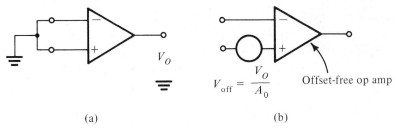

(a)   (b)

Fig. 3.38   Illustrating the definition of the input offset voltage V_{off}.

It should be obvious from Fig. 3.38b that if the two input terminals are tied together the output will be $A_0 V_{off} = V_O$, which correlates with our original definition of V_{off}. Another interpretation of V_{off} is that it is the magnitude of the voltage which if applied between the two input terminals of a real op amp, the output dc voltage would be reduced to zero. This interpretation assumes that the polarity of V_{off} is known, so that the external source could be of opposite polarity.

General-purpose op amps have offset voltages V_{off} on the order of few millivolts (2 to 5 mV). The value of V_{off} and its temperature coefficient are usually specified in the op-amp data sheets. The latter specification is very important, especially if one is contemplating nulling V_{off} to zero. Many op amps are provided with two extra terminals and a prescribed technique for reducing V_{off} to zero. The question then becomes, Will V_{off} remain zero with changes in temperature? The answer can be found if we know the specification on *offset voltage drift* (usually given in $\mu V/^\circ C$).

Let us now consider the effect of V_{off} on the performance of closed-loop amplifiers. In doing this we may simplify matters by grounding the signal source. In this manner both the inverting and the noninverting configurations will be identical, as shown in Fig. 3.39. It is easily seen that V_{off} will be amplified by the quantity $1 + R_2/R_1$, and the resulting dc voltage will appear at the output as an offset voltage. In the presence of an input signal this output dc voltage will be superimposed on the output signal. If the closed-loop gain is large, the output dc offset voltage could be high, thus reducing the maximum possible output signal swing.

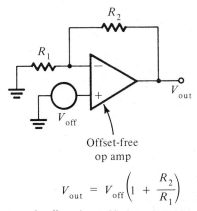

$$V_{out} = V_{off}\left(1 + \frac{R_2}{R_1}\right)$$

Fig. 3.39   Evaluating the output dc offset due to V_{off} in a closed-loop amplifier.

One way to overcome the dc offset problem is to capacitively couple the amplifier. This, however, will be possible only in applications where the closed-loop amplifier is not required to amplify dc or very-low-frequency signals. Figure 3.40 shows a capacitively coupled inverting amplifier. The coupling capacitor will cause the gain to be zero at dc. In fact, the circuit will have an STC high-pass response with break frequency $\omega_p = 1/CR_1$, and the gain will be $-R_2/R_1$ for frequencies $\omega \gg \omega_p$. The advantage of

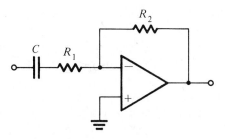

Fig. 3.40   *A capacitively coupled inverting amplifier.*

this arrangement is that V_{off} will not be amplified. Thus the output dc voltage will be equal to V_{off} rather than $V_{off}(1 + R_2/R_1)$, which is the case without the coupling capacitor. The reader should convince himself or herself that the V_{off} generator indeed sees a unity-gain follower.

Input Bias Currents

The second dc problem encountered in op amps is illustrated in Fig. 3.41. In order for the op amp to operate, its two input terminals have to be supplied by finite dc currents, termed the *input bias currents*. In Fig. 3.41 these two currents are represented by two

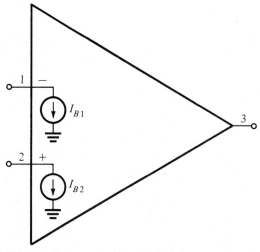

Fig. 3.41   *The input bias currents represented by two current sources I_{B1} and I_{B2}.*

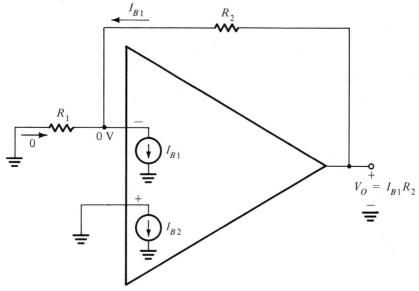

Fig. 3.42   *Analysis of the closed-loop amplifier, taking into account the input bias currents.*

current sources (or, more appropriately, current sinks), I_{B1} and I_{B2}, connected to the two input terminals. It should be emphasized that the input bias currents are independent of the fact that the op amp has finite input resistance. The op-amp manufacturer usually specifies the average value of I_{B1} and I_{B2} as well as their expected difference. The average value I_B is called the input bias current,

$$I_B = \frac{I_{B1} + I_{B2}}{2}$$

while the difference is called the *input offset current* and is given by

$$I_{\text{off}} = |I_{B1} - I_{B2}|$$

Typical values for general-purpose op amps are $I_B = 100$ nA and $I_{\text{off}} = 10$ nA.

We now wish to find the dc output voltage of the closed-loop amplifier due to the input bias currents. To do this we ground the signal source and obtain the circuit shown in Fig. 3.42 for both the inverting and noninverting configurations. As shown in Fig. 3.42, the output dc voltage is given by

$$V_O = I_{B1}R_2 \simeq I_B R_2 \tag{3.15}$$

This obviously places an upper limit on the value of R_2. Fortunately, however, a technique exists for reducing the value of the output dc voltage due to the input bias currents. The method consists of introducing a resistance R_3 in series with the noninverting input lead, as shown in Fig. 3.43. From a signal point of view, R_3 has a negligible effect. The appropriate value for R_3 can be determined by analyzing the circuit in Fig. 3.43, where analysis details are shown and the output voltage is given by

$$V_O = -I_{B2}R_3 + R_2(I_{B1} - I_{B2}R_3/R_1) \tag{3.16}$$

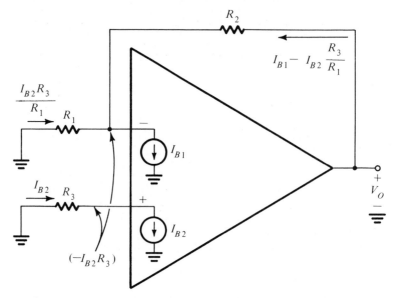

Fig. 3.43   *Reducing the effects of the input bias currents by introducing a resistor R_3.*

Consider first the case $I_{B1} = I_{B2} = I_B$, which results in

$$V_O = I_B[R_2 - R_3(1 + R_2/R_1)]$$

Thus we may reduce V_O to zero by selecting R_3 such that

$$R_3 = \frac{R_2}{1 + R_2/R_1} = \frac{R_1 R_2}{R_1 + R_2}$$

That is, R_3 should be made equal to the parallel equivalent of R_1 and R_2.

Having selected R_3 as in the above, let us evaluate the effect of a finite offset current I_{off}. Let $I_{B1} = I_B + I_{off}/2$ and $I_{B2} = I_B - I_{off}/2$, and substitute in Eq. (3.16). The result is

$$V_O = I_{off} R_2$$

which is usually about an order of magnitude smaller than the value obtained without R_3 [Eq. (3.15)]. We conclude that to minimize the effect of the input bias currents one should place in the positive lead a resistance equal to the dc resistance seen by the inverting terminal. We should emphasize the word dc in the last statement; note that if the amplifier is ac coupled, we should select $R_3 = R_2$, as shown in Fig. 3.44.

While we are on the subject of ac-coupled amplifiers we should note that one must always provide a continuous dc path between each of the input terminals of the op amp and ground. For this reason the ac-coupled noninverting amplifier of Fig. 3.45 will *not* work without the resistance R_3 to ground. Unfortunately, including R_3 lowers considerably the input resistance of the closed-loop amplifier.

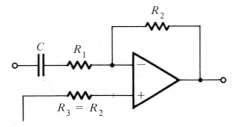

Fig. 3.44   In an ac-coupled amplifier the dc resistance seen by the inverting terminal is R_2; hence $R_3 = R_2$.

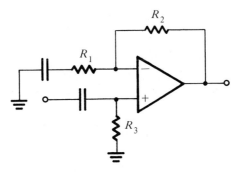

Fig. 3.45   Illustrating the need for a continuous dc path for each of the op-amp input terminals.

Example 3.7

We wish to evaluate the effects of the input offset voltage and the input bias current on the performance of the Miller integrator shown in Fig. 3.46.

Solution

Straightforward analysis shows that irrespective of the input signal there will be a dc current through the capacitor given by

$$I = I_{B1} + \frac{V_{off}}{R}$$

Correspondingly, there will be an output voltage given by

$$v_O = V_{off} + \frac{1}{C} \int I \, dt$$

$$= V_{off} + \frac{V_{off}}{CR} t + \frac{I_{B1}}{C} t$$

which means that the output voltage will increase linearly with time until the amplifier leaves its linear range and "saturates"—that is, the output remains constant at a value close to the power supply voltage. Obviously this is an unacceptable situation! To over-

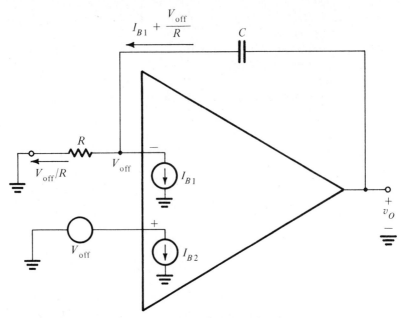

Fig. 3.46 *Evaluating the effects of the dc bias and offsets on the performance of the Miller integrator.*

come this problem we have to provide a dc path for the current I. This can be achieved by connecting a resistance in parallel with the capacitor. Unfortunately the resulting circuit will no longer be an ideal integrator. The performance of this "damped" or "leaky" integrator is investigated in Problem 3.33.

• • •

EXERCISES

3.12 Consider the differential amplifier circuit in Fig. 3.32. Let $R_1 = 10$ kΩ and $R_2 = 1$ MΩ. If the op amp has $V_{off} = 3$ mV, $I_B = 0.2$ μA, and $I_{off} = 50$ nA, find the dc offset voltage at the output.
Ans.   353 mV

3.13 Consider the capacitively coupled amplifer of Fig. 3.40. Find values of C, R_1, and R_2 such that the corner frequency f_0 is 10 Hz and that for $f \gg f_0$ the gain is approximately 100 and the input resistance is approximately 1 kΩ. Also find the magnitude and phase of the gain at $f = 100$ Hz.
Ans. $R_1 = 1$ kΩ; $R_2 = 100$ kΩ; $C = 15.9$ μF; $|G| = 99.5$; $\phi = 180° + 5.7°$

3.13   CONCLUDING REMARKS

At this stage the reader should be able to design and construct useful op-amp circuits. Because IC op amps in actuality nearly meet their assumed ideal performance, it is not difficult to experiment with them. In fact, experimenting with op amps is essential at this stage. The examples, exercises, and problems included in this chapter suggest many possible circuits for experimentation.

Other than mentioning the fact that op amps saturate if their linear operating region is exceeded, we have dealt solely with their linear operation and applications. A number of nonlinear applications of op amps are discussed in Chapter 5, and other linear and nonlinear applications are studied in later chapters.

DIODES   4

All the circuits considered in the two previous chapters are
linear; the only nonlinearity encountered was the nonlinear
one-port circuit of Section 2.1 that we used to demonstrate the
techniques of biasing and small-signal analysis, and the
saturation effect exhibited by an op amp when its output
voltage exceeds the maximum allowable signal swing. In this
chapter we begin our study of nonlinear circuits by considering
the most fundamental nonlinear two-terminal device: the
diode. Nonlinear circuits play a major role in signal processing.
Examples include signal generators, modulators, demodulators,
dc power supplies, and digital circuits.

Introduction

We begin by defining the characteristics of the *ideal
diode.* Then, to illustrate the use of diodes, we consider one of
their most dramatic applications, namely, their use in the
design of rectifiers. Rectifiers, which are utilized in dc power
supplies, will be studied in more detail in Chapter 5, which is
concerned with diode circuit applications. The main purpose of
the present chapter is to introduce the silicon junction diode,
explain its terminal characteristics, and provide techniques for
the analysis of diode circuits. This last task involves the
important subject of modeling the terminal characteristics of
the diode. Though the majority of diode applications are
deferred until the next chapter, a few are introduced here by
way of circuit analysis examples.

The latter part of this chapter is concerned with the
physical operation of the *pn junction.* In addition to being a
diode, the *pn* junction is the basis of many other solid-state
devices. Thus an understanding of the *pn* junction is essential
to the study of modern electronics. Though simplistic, the
qualitative view presented here provides sufficient background
for our purpose, the study of circuits. A more detailed study of
the *pn* junction can be found in courses and texts dealing with
physical electronics.

4.1   THE IDEAL DIODE

The ideal diode may be considered the most fundamental nonlinear element. It is a two-
terminal device having the circuit symbol of Fig. 4.1a and the *i-v* characteristic shown
in Fig. 4.1b. The terminal characteristic of the ideal diode can be interpreted as follows:
If a negative voltage (relative to the reference direction indicated in Fig. 4.1a) is applied
to the diode, no current flows and the diode behaves as an open circuit. Diodes operated

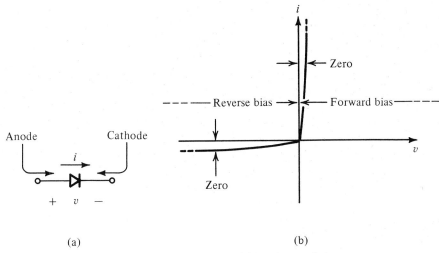

Fig. 4.1   The ideal diode: (a) diode circuit symbol; (b) i-v characteristic.

in this mode are said to be *reverse-biased,* or operated in the reverse direction. An ideal diode has zero current when operated in the reverse direction.

On the other hand, if a positive current (relative to the reference direction indicated in Fig. 4.1a) is applied to the ideal diode, zero voltage drop appears across the diode. In other words, the ideal diode behaves as a short circuit in the *forward* direction; it passes any current with zero voltage drop.

Note that in the above description we have deliberately assumed that in the reverse direction the diode is excited by a voltage source while in the forward direction the diode is excited by a current source. The reason for doing so is that otherwise the nature of the response is undefined. For instance, feeding a negative current to an ideal diode produces, ideally, an infinite voltage. Although no infinite voltages or currents occur in practice, the above description of ideal diode suggests the manner in which diodes should be used in practice. Specifically, in the forward direction diode current should be limited by the external circuit. Similarly, in the reverse direction the diode voltage should be limited by the external circuit. Examples are given in Fig. 4.2.

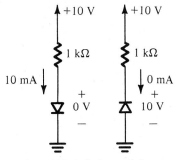

Fig. 4.2   The two modes of operation of ideal diodes and the use of an external circuit to limit the forward current and the reverse voltage.

The positive terminal of the diode is called the anode and the negative terminal the cathode, a carryover from the days of vacuum tube diodes. The *i-v* characteristic of the ideal diode should explain the choice of its circuit symbol.

As should be evident from the above, the *i-v* characteristic of the ideal diode is highly nonlinear; it consists of two straight-line segments at 90° to one another. A nonlinear curve that consists of straight-line segments is said to be *piecewise linear*. If a device having a piecewise linear characteristic is used in a particular application in such a way that the signal across its terminals swings only along one of the linear segments, then the device can be considered a linear circuit element as far as that particular circuit application is concerned. On the other hand, if signals swing past one or more of the break points in the characteristics, linear analysis is no longer possible. A fundamental application of the diode nonlinearity is presented in Section 4.2.

EXERCISES
4.1   Find the values of *I* and *V* in the circuits shown in Fig. E4.1.
   Ans.   (a) 2 mA, 0 V; (b) 0 mA, 5 V; (c) 0 mA, 5 V; (d) 2 mA, 0 V

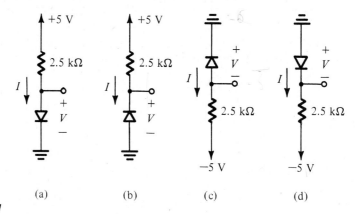

(a)   (b)   (c)   (d)

Fig. E4.1

4.2   A FUNDAMENTAL APPLICATION OF DIODES— SIGNAL RECTIFICATION

One of the most dramatic applications of the gross nonlinearity exhibited by ideal diodes is found in their use in signal rectification. A signal rectifier is a circuit which converts an alternating-current (ac) signal into a unidirectional one. Figure 4.3 shows a diode rectifier fed by a sine-wave voltage source v_S. Also shown is the waveform of the output voltage v_O that appears across the resistance R_L. The operation of the circuit is quite straightforward. When v_S is positive the diode conducts and acts as a short circuit and the voltage v_S appears directly at the output—that is, $v_O = v_S$, and the diode forward current is equal to v_S/R_L. On the other hand, when v_S is negative the diode *cuts off*— that is, no current will flow, the output voltage v_O will be zero, and the diode becomes reverse-biased by the value of the input voltage v_S. It follows that the output voltage waveform will consist of the positive half cycles of the input sinusoid. Thus the signal is said to have been rectified; because only half cycles are utilized, the circuit is called a *half-wave rectifier*.

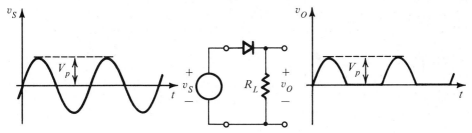

Fig. 4.3   *Half-wave rectification.*

It should be noted that while the input sinusoid has a zero average value, the output waveform has a finite average value or dc component. Therefore rectifiers are used to generate dc voltages from ac voltages. More details on rectifiers and their use in the design of dc power supplies will be given in Chapter 5.

Before leaving this section we wish to introduce two related ideas. The first is the concept of the *transfer characteristic,* used to describe a nonlinear two-port network. Figure 4.4 shows the half-wave rectifier previously discussed together with its transfer characteristic. The transfer characteristic is simply a plot of v_O versus v_I for the two-port network. As can be seen the half-wave rectifier produces an output voltage equal to the input voltage when the input voltage is positive and produces zero output voltage when the input voltage is negative.

The second idea relates to the circuit of Fig. 4.5. It may be seen that this circuit is identical to the half-wave rectifier discussed above except that here the output is taken across the diode. This observation enables us to easily obtain the transfer characteristic shown in Fig. 4.5b. As the characteristics indicate, if the input v_I goes positive, the diode conducts and causes the output voltage v_O to be zero. On the other hand, if v_I goes negative, the diode is reverse-biased and no current flows, causing the voltage drop across the resistance to be zero and v_O to equal v_I. Figure 4.5c shows the output waveform obtained when a triangular wave signal is applied to the circuit. Note that while the negative halves of the input waveform are reproduced at the output, the positive peaks are "clipped off." This circuit is therefore called a *clipper.* Another name commonly used is a *diode limiter,* since the amplitude of the output waveform is limited to a certain value (zero, in the positive direction). We will return to the study of rectifiers and other diode circuit applications in Chapter 5.

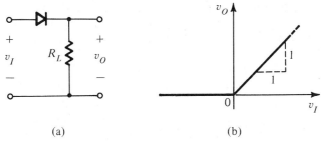

(a)   (b)

Fig. 4.4   *Transfer characteristic of a half-wave rectifier.*

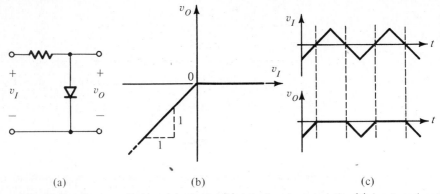

Fig. 4.5 Diode clipper or limiter: (a) circuit; (b) transfer characteristic; (c) input and output waveforms.

EXERCISES

4.2 Figure E4.2 shows a circuit for charging a battery. If v_S is a sinusoid with 24 V peak amplitude, find the fraction of a cycle during which the diode conducts. Also find the peak value of the diode current.

Ans. 1/3; 120 mA

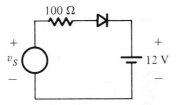

Fig. E4.2

4.3 Figure E4.3 shows a circuit for an ac voltmeter. It utilizes a moving-coil meter that gives a full-scale reading when the *average* current flowing through it is 1 mA. The moving-coil meter has a 50-Ω resistance. Find the value of R that results in the meter indicating a full-scale reading when the input sine-wave voltage v_I is 20-V peak-to-peak. (*Hint:* The average value of half sine waves is V_p/π.)

Ans. 3.133 kΩ

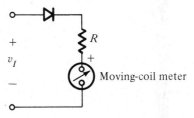

Fig. E4.3

4.3   TERMINAL CHARACTERISTICS OF REAL JUNCTION DIODES

Before proceeding further with the study of diode applications we will consider the characteristics of real diodes—specifically semiconductor junction diodes made of silicon.

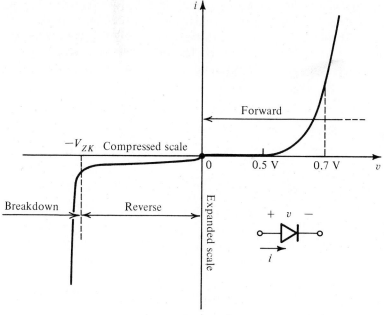

Fig. 4.6 *The i-v relationship of real silicon junction diodes.*

The physical processes that give rise to the diode terminal characteristics, and to the name "junction diode," will be studied in the latter part of this chapter.

Figure 4.6 shows the *i-v* characteristic of a silicon junction diode. As indicated, the characteristic curve consists of three distinct regions:

1. The forward-bias region, determined by $v > 0$
2. The reverse-bias region, determined by $v < 0$
3. The breakdown region, determined by $v < -V_{ZK}$

These three regions of operation are described in the following three sections. It should be noted that in order to show the details of the reverse region different scales are used for the negative parts of the i and v axes, resulting in an apparent discontinuity at the origin.

4.4 THE FORWARD-BIAS REGION

The forward-bias—or, simply, forward—region of operation is entered when the terminal voltage v is positive. In the forward region the *i-v* relationship is closely approximated by

$$i = I_S (e^{v/nV_T} - 1) \tag{4.1}$$

In this equation I_S is a constant for a given diode at a given temperature. The current I_S is usually called the *saturation current* (for reasons that will become apparent in the next section). A better name, one that we will use for I_S, is the *scale current*. This name arises from the fact that I_S is directly proportional to the cross-sectional area of the diode. Thus doubling of the junction area results in a diode with double the value of I_S and, as the diode equation indicates, double the value of current i for a given forward voltage v. For "small-signal" diodes, which are small-size diodes intended for low-power applications, I_S is of the order of 10^{-15} A. The value of I_S is, however, a very strong function of temperature. As a rule of thumb I_S doubles in value for every $10°$C rise in temperature.

The voltage V_T in Eq. (4.1) is a constant called the thermal voltage, given by

$$V_T = \frac{KT}{q}$$

where K = Boltzmann constant
    T = the absolute temperature in degrees Kelvin
    q = the magnitude of electronic charge
At room temperature ($22°$C) the value of V_T is 25 mV, a number that is easy to remember.

In the diode equation the constant n has a value between 1 and 2, depending on the material and the physical structure of the diode.

For appreciable current i in the forward direction, specifically for $i \gg I_S$, Eq. (4.1) can be approximated by the exponential relationship

$$i \simeq I_S e^{v/nV_T} \tag{4.2}$$

This relationship can be expressed alternatively in the logarithmic form

$$v = nV_T \ln \frac{i}{I_S} \tag{4.3}$$

where ln denotes the natural (base e) logarithm.

The exponential relationship of the current i to the voltage v holds over many decades of current (a span of as many as seven decades—that is, a factor of 10^7—has been reported). This is quite a remarkable property of junction diodes, one that is also found in bipolar junction transistors and one that has been exploited in many interesting applications.

Let us consider the forward i-v relationship in Eq. (4.2) and evaluate the current I_1 corresponding to a diode voltage V_1:

$$I_1 = I_S e^{V_1/nV_T}$$

Similarly, if the voltage is V_2 the diode current I_2 will be

$$I_2 = I_S e^{V_2/nV_T}$$

These two equations can be combined to produce

$$\frac{I_1}{I_2} = e^{(V_1-V_2)/nV_T}$$

which can be rewritten

$$V_1 - V_2 = nV_T \ln \frac{I_1}{I_2}$$

or, in terms of base 10 logarithms,

$$V_1 - V_2 = 2.3nV_T \log \frac{I_1}{I_2} \tag{4.4}$$

This equation simply states that for a decade (factor of 10) change in current the diode voltage drop changes by $2.3nV_T$, which is 60 mV for $n = 1$ and 120 mV for $n = 2$. This also suggests that the diode i-v relationship is most conveniently plotted on a semilog paper. Using the vertical, linear axis for v and the horizontal, log axis for i, one obtains a straight line with a slope of $2.3nV_T$ per decade of current. Finally, it should be mentioned that not knowing the exact value of n (which can be obtained from a simple experiment), circuit designers use the convenient approximate number of 0.1 V/decade for the slope of the diode logarithmic characteristic.

A glance at the i-v characteristic in the forward region (Fig. 4.6) reveals that the current is negligibly small for v smaller than about 0.5 V. This value is usually referred to as the *cut-in voltage*. It should be emphasized, however, that this apparent threshold in the characteristic is simply a consequence of the exponential relationship. Another consequence of this relationship is the rapid increase of i. Thus for a "fully conducting" diode the voltage drop lies in a narrow range, approximately 0.6 to 0.8 V. This gives rise to a simple "model" for the diode where it is assumed that a conducting diode has approximately a 0.7-V drop across it. Diodes with different current ratings (that is, different areas and correspondingly different I_S) will exhibit the 0.7-V drop at different currents. For instance, a small-signal diode may be considered to have a 0.7-V drop at $i = 1$ mA, while a higher-power diode probably has a 0.7-V drop at $i = 1$ A. We will return to the topics of diode circuit analysis and diode models shortly.

Example 4.1
A silicon diode said to be a 1-mA device displays a forward voltage of 0.7 V at a current of 1 mA. Evaluate the junction scaling constant I_S in the event that n is either 1 or 2. What scaling constants would apply for a 1-A diode of the same manufacture that conducts 1 A at 0.7 V?

Solution
Since

$$i = I_S e^{v/nV_T}$$

then

$$I_S = i e^{-v/nV_T}$$

For the 1-mA diode:

If $n = 1$: $I_S = 10^{-3} e^{-700/25} = 6.9 \times 10^{-16}$ A,   or about 10^{-15} A

If $n = 2$: $I_S = 10^{-3} e^{-700/50} = 8.3 \times 10^{-10}$ A,   or about 10^{-9} A

The diode conducting 1 A at 0.7 V corresponds to 1,000 1-mA diodes in parallel with a total junction area 1,000 times greater. Thus I_S is also 1,000 times greater, being 1 pA and 1 μA, respectively, for $n = 1$ and $n = 2$.

From this example it should be apparent that the value of n used is quite important.

· · ·

Temperature Dependence

The temperature dependence of the forward i-v characteristic is illustrated in Fig. 4.7. At a given constant diode current the voltage drop across the diode decreases by approximately 2 mV for every 1 °C increase in temperature, owing to the dependence of both I_S and V_T on temperature. The change in diode voltage with temperature has been exploited in the design of electronic thermometers.

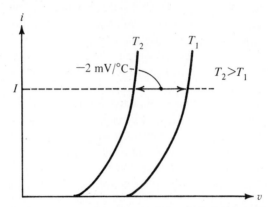

Fig. 4.7 *Illustrating the temperature dependence of the diode forward characteristics. At a constant current I the voltage drop decreases by approximately 2 mV for every degree C increase in temperature.*

EXERCISES

4.4 Consider a silicon junction diode with $n = 1.5$. Find the change in voltage if the current changes from 0.1 mA to 10 mA.
   Ans. 172.5 mV

4.5 A silicon junction diode with $n = 2$ has $v = 0.7$ V at $i = 1$ mA. Find the voltage drop at $i = 0.1$ mA and $i = 10$ mA.
   Ans. 0.58 V; 0.82 V

4.6 Using the fact that a silicon diode has $I_S = 10^{-14}$ A at room temperature (TEMP = 22 °C) and that it doubles for every 10 °C rise in temperature, express I_S as a function of TEMP.
   Ans. $I_S = 2.2 \times 10^{-15} \times 2^{0.1 \times \text{TEMP}}$

4.5   THE REVERSE-BIAS REGION

The reverse-bias region of operation is entered when the diode voltage v is made negative. Equation (4.1) predicts that if v is negative and few times larger than V_T (25 mV) in magnitude, the exponential term becomes negligibly small compared to unity and the diode current becomes

$$i \simeq -I_S$$

that is, the current in the reverse direction is constant and equal to I_S. This is the reason behind the term *saturation current*.

Real diodes exhibit reverse currents that although quite small are much larger than I_S. For instance, a small-signal or a 1-mA diode whose I_S is of the order of 10^{-14} to 10^{-15} A could show a reverse current of the order of 1 nA. The reverse current also increases somewhat with the increase in magnitude of the reverse voltage. Note that because of the very small magnitude of current these details are not clearly evident on the diode i-v characteristic of Fig. 4.6.

A good part of the reverse current is due to leakage effects. These leakage currents are proportional to the junction area, just as I_S is. Finally, it should be mentioned that the reverse current is a strong function of temperature, with the rule of thumb being that it doubles for every $10°C$ rise in temperature.

4.6   THE BREAKDOWN REGION AND ZENER DIODES

The third distinct region of diode operation is the breakdown region, which can be easily identified on the diode i-v characteristic in Fig. 4.6. For convenience the i-v character-istic in the reverse and breakdown regions are redrawn in Fig. 4.8. The breakdown region is entered when the magnitude of the reverse voltage exceeds a threshold value specific to the particular diode and called the *breakdown voltage*. This is the voltage at the "knee" of the i-v curve in Fig. 4.8 and is denoted V_{ZK}, where the subscript Z stands for zener (to be explained shortly) and K denotes knee.

As can be seen from Fig. 4.8, in the breakdown region the reverse current increases rapidly, with the associated increase in voltage drop being very small. Diode breakdown is normally not destructive provided that the power dissipated in the diode is limited by external circuitry to a "safe" level. This safe value is normally specified on the device data sheets. It therefore is necessary to limit the reverse current in the breakdown region to a value consistent with the permissible power dissipation.

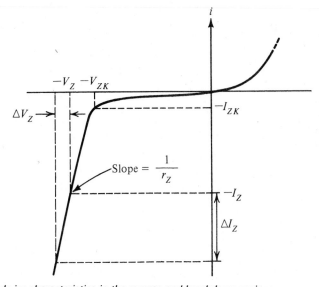

Fig. 4.8   Diode i-v characteristics in the reverse and breakdown regions.

Voltage Regulators

The fact that the diode *i-v* characteristic in breakdown is almost a vertical line enables it to be used in voltage regulation. A voltage regulator is a circuit whose purpose is to provide a constant dc voltage between its output terminals. This output voltage must remain as constant as possible in spite of changes in the load current drawn from the regulator and possible changes in the dc power supply that feeds the regulator circuit.

Fig. 4.9 *Circuit symbol for a zener diode.*

The diode in the breakdown region exhibits a voltage drop that is almost constant and independent of the current through the diode. Special diodes have been manufactured to operate specifically in the breakdown region. Such diodes are called *breakdown diodes* or, more commonly, *zener diodes* for reasons that will be explained shortly.

Figure 4.9 shows the circuit symbol of the zener diode. In normal application of zener diodes, current flows into the cathode, and the cathode is positive with respect to the anode; thus *I* and *V* in Fig. 4.9 have positive values. To illustrate the application of zener diodes, consider the circuit shown in Fig. 4.10. Here a zener diode, specified to

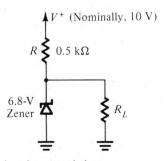

Fig. 4.10 *Use of zener diodes in voltage regulation.*

have a voltage of 6.8 V, is shown being fed from a dc supply of $+10$ V through a resistance R of 0.5 kΩ. When no load resistance is connected the current through the zener will be

$$I = \frac{10 - 6.8}{0.5} = 6.4 \text{ mA}$$

Now let a resistance $R_L = 2$ kΩ be connected across the zener diode. Assuming that the voltage across the zener will remain approximately constant, an assumption whose validity we will have to check shortly, we see that the load current I_L is given by

$$I_L = \frac{6.8}{2} = 3.4 \text{ mA}$$

Since the current through R is still approximately 6.4 mA, it follows that the current through the zener is now 3 mA. Thus the zener current has changed from 6.4 mA to 3 mA. Nevertheless, because of the sharp i-v curve the decrease in the voltage drop across the zener will be negligibly small. That is, the voltage remains almost constant in spite of changes in load current. Note, however, that the load current cannot exceed the current being fed to the zener through the resistance R. In fact, if the current in the zener decreases below a certain specified value (equal to the *knee current* I_{ZK} in Fig. 4.8), the voltage across the zener starts to decrease, and eventually the diode leaves the breakdown region and stops functioning as a voltage regulator. For instance, if R_L is 0.5 kΩ, the zener in Fig. 4.10 will be off and the load voltage will be equal to the value determined by the voltage divider composed of R and R_L,

$$V_L = 10 \frac{0.5}{0.5 + 0.5} = 5 \text{ V}$$

which is less than the 6.8 V required for the zener to operate in the breakdown region.

Zener Resistance

Every zener diode has a specified value for its breakdown voltage, also called the *zener voltage* V_Z. This is the voltage drop across the zener when a specified nominal current I_Z is flowing through it (see Fig. 4.8). As the current through the zener deviates from the nominal value I_Z by a small amount ΔI_Z, the voltage across it will deviate from V_Z by an amount ΔV_Z given by

$$\Delta V_Z = r_Z \Delta I_Z$$

where r_Z is the incremental resistance of the zener diode at the operating point specified by V_Z and I_Z. (The concept of incremental resistance was introduced in Section 2.1.) The value of r_Z is equal to the inverse of the slope of the i-v curve at the operating point. Typically, r_Z is a few tens of ohms in value, but it increases as the current decreases, reaching a very high value at the knee of the i-v curve. Obviously, as r_Z increases the zener diode becomes less effective as a voltage regulator. The current at the knee, I_{ZK}, is also normally specified on the zener data sheets.

To illustrate the use of r_Z let us return to the voltage-regulator circuit of Fig. 4.10. Consider the no-load case. We have already found that when $V^+ = 10$ V the zener has a current of 6.4 mA. Let us assume that at this current the zener resistance r_Z is 50 Ω. Now let V^+ change by 10%, to 11 V. The corresponding change in zener voltage can be calculated using the voltage-divider rule with the divider composed of R and r_Z,

$$\Delta V_Z = \Delta V^+ \frac{r_Z}{r_Z + R}$$

Thus

$$\Delta V_Z = 1 \times \frac{50}{50 + 500} \simeq 91 \text{ mV}$$

Hence the regulated voltage changes by 1.3%, corresponding to the 10% change in supply voltage; the 1.3% figure is usually called *voltage regulation*.

Avalanche Versus Zener Diodes

As will be explained in Section 4.13 there are two different mechanisms for diode break-down: zener breakdown and avalanche breakdown. The zener effect, named in honor of an early worker in the area, is responsible for breakdown if it occurs at low voltages (2 to 5 V). If breakdown occurs at voltages about 7 V or greater, the mechanism respon-sible is avalanche breakdown. Avalanche breakdown results in diodes with sharper knees; these therefore make better voltage regulators. For diodes whose breakdown occurs between 5 and 7 V either or both mechanisms may be responsible. Irrespective of the breakdown mechanism, breakdown diodes are commonly called zener diodes.

Zener diodes are manufactured in a variety of standard voltages. The lower-voltage units, however, have rather weak knees. For this reason one may consider using a string of forward-biased diodes to provide low-voltage regulators. This application is based on the observation that a forward-conducting diode has a voltage drop of about 0.7 V and that this voltage drop changes very little with changes in current (0.1 V per decade of current change). Thus, for example, one can obtain a constant voltage drop of approx-imately 2 V by connecting three diodes in series. We will have more to say about this shortly.

Curves illustrating the sharpness of the breakdown knee are shown in Fig. 4.11. Here the attempt (although difficult to appreciate on a linear plot) is to show the very

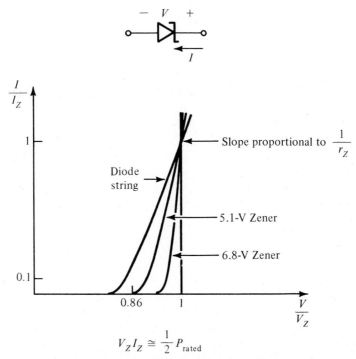

Fig. 4.11   Sharpness of the i-v characteristics of zener diodes of different nominal voltages and of a diode string. Note that I_Z and V_Z denote the nominal values of current and voltage.

sharp knee of an avalanche zener of 6.8 V. A nominal 20-mA unit (normally rated at 400 mW) will hold its voltage within 0.3 V or so down to tens of microamperes, exhibiting a slope resistance of a few tens of ohms or less at high currents. Here it should be mentioned that the reversed base–emitter junction of bipolar transistors (Chapter 9) suffers avalanche breakdown at about 7 V and may be used as a zener diode. Breakdown diodes of higher and lower voltages exhibit higher slope resistances and weaker knees, as shown by the characteristic of a 5.1-V unit, sketched in Fig. 4.11. Also indicated in Fig. 4.11, for comparison, is the forward characteristic of a silicon junction diode.

Example 4.2

In this example we wish to contrast the use of a 1N5235 zener diode as a regulator in two applications. In each the supply voltage is nominally 9 V. The first application is one in which the supply of regulator current is no problem but the source, the "raw" supply, changes ± 1 V (called *ripple* for reasons explained in Chapter 5). The second application is for a portable instrument where power is obtained from a rechargeable battery whose operating range extends from 10 V down to 8 V. For the purpose of this initial design, assume no load.

The 1N5235 unit is rated at $V_z = 6.8$ V for $I_z = 20$ mA with a typical zener resistance r_z of 5 Ω. It is also specified that r_z will be no greater than 750 Ω at 0.25 mA (nearer the knee). At other current values r_z may be considered inversely proportional to the current.

Solution

In both cases, to obtain the security of using specified data and thus to avoid relying unduly on extrapolated assumptions, we will use the high and low values of current at which the zener resistance has been specified.

For the first case we will design for a zener current of 20 mA. Establishment of this current in the 6.8-V zener from a nominal 9-V supply in a circuit such as that in Fig. 4.10 requires a series resistance R of

$$R = \frac{9 - 6.8}{20} = 110 \ \Omega$$

The resulting regulator having $r_z = 5 \ \Omega$ will reduce the ± 1-V ripple in the source to an output voltage ripple of

$$\Delta V_z = \pm \frac{5}{110 + 5} \times 1 = \pm 43 \ \text{mV} \qquad \text{for 0.63\% regulation}$$

Because of the limited current capacity of the battery in the second application we will design for a zener current of 0.25 mA. To establish this current we need a resistance R of

$$R = \frac{9 - 6.8}{0.25} = 8.8 \ \text{k}\Omega$$

The resulting regulator having $r_z = 750\ \Omega$ will reduce the ± 1-V change in the source to an output voltage change of

$$\Delta V_z = \pm \frac{750}{750 + 8800} \times 1 = \pm 79\ \text{mV} \qquad \text{for 1.2\% regulation}$$

. . .

Example 4.2 indicates that while the equivalent reduction of disturbances is somewhat less good at low currents, regulation is not dramatically different in the two designs. In fact, because the value of r_z is approximately inversely proportional to the operating current regulation will be almost independent of bias design for all other conditions being equal (for example, no load).

Temperature Effects

Zener diodes exhibit a temperature coefficient that depends on both voltage and current in a rather complex way, as indicated in Fig. 4.12. In particular, zeners with a nominal breakdown of 5.1 V exhibit a point of zero temperature coefficient at modest current levels. Such zeners, while exhibiting a relatively high zener impedance, can be used in selected constant-current designs to establish temperature-independent voltages (particularly with amplifier assistance). On the other hand, zeners of about 6.8 V exhibit a temperature coefficient of about $+2\ \text{mV}/^\circ\text{C}$, almost exactly the complement of that of

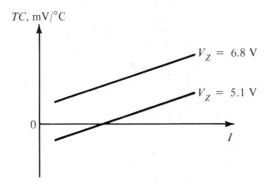

Fig. 4.12 *The zener diode temperature coefficient (TC) versus the current level I for a 5.1-V and a 6.8-V unit.*

a forward-conducting diode. Thus a series combination of a 6.8-V zener and a single forward junction exhibits a drop of about 7.5 V with a greatly reduced temperature coefficient. Other low-temperature-coefficient diodes (said also to have *low temco* or *low TC*) exist at somewhat higher voltages, utilizing other combinations of forward-conducting and breakdown diodes.

EXERCISE
4.7 Consider the regulator designed in Example 4.2. For both cases find the change in zener voltage when a 2-kΩ load resistance is connected.
Ans. -17 mV; -4.95 V

4.7   ANALYSIS OF DIODE CIRCUITS

Consider the circuit shown in Fig. 4.13 and consisting of a dc source V_{DD}, a resistance R, and a diode. We wish to analyze this circuit to determine the diode current I and voltage V.

Since the diode is biased in the forward direction it will obviously be conducting (assuming that $V_{DD} > 0.5$ V), and we can write the two equations

$$I \simeq I_S e^{V/nV_T} \tag{4.5}$$

$$V = V_{DD} - RI \tag{4.6}$$

These are two equations in the two unknowns I and V, assuming that the diode parameters I_S and n are given. Although it is possible to solve these two equations to determine

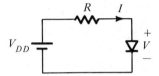

Fig. 4.13   A simple diode circuit.

the values of I and V, the amount of work involved is usually not justified. In other words, one is usually interested in finding quick approximate answers.

Before presenting techniques of approximate analysis we will discuss a method for graphical analysis. However, it should be emphasized at the outset that one rarely uses graphical techniques to solve simple diode circuits. Nevertheless, it is important to understand the technique of graphical analysis, since similar techniques will be used in conjunction with transistor circuits to obtain insights not otherwise available.

Graphical Analysis

Graphical analysis consists simply of plotting the relationships of Eqs. (4.5) and (4.6) on the i-v plane. The solution can then be obtained as the coordinates of the point of intersection of the two graphs. Equation (4.5) represents the diode equation, and a sketch of it is given in Fig. 4.14. The other equation, Eq. (4.6), represents a straight line, called the *load line,* that intersects the voltage axis at $v = V_{DD}$ and has a slope of $-1/R$. As shown in Fig. 4.14, the load line intersects the diode curve at point Q, which represents the operating point of the diode and whose coordinates give the values of I and V.

Approximate Analysis

The method of approximate analysis is based on our earlier observation regarding the diode voltage drop. To repeat, a forward-conducting diode has a voltage drop that lies in a narrow range, approximately 0.6 to 0.8 V. Therefore as a gross approximation (one that in fact is not bad in many applications) we may assume that a 1-mA diode con-

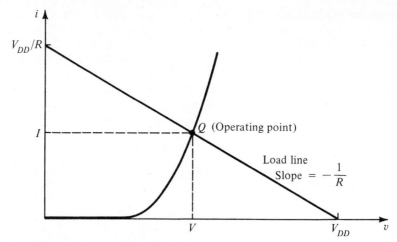

Fig. 4.14   Graphical analysis of the circuit in Fig. 4.13.

ducting current in the millampere range will have a voltage drop of approximately 0.7 V.

Our approach to the problem depicted in Fig. 4.13 would then be as follows: First, we assume that the diode is conducting; thus its voltage drop is approximately 0.7 V. Then we proceed to use Kirchhoff's loop equation to determine the current I,

$$I = \frac{V_{DD} - 0.7}{R}$$

If we find that I is in the range assumed for the diode (that is, in the mA range for a 1-mA diode), then our original assumption is not a bad one and our work is completed. If, on the other hand, we find that the calculated value of I differs considerably from the assumed order of magnitude, then clearly our result needs further refinement, as will be explained shortly.

Example 4.3
Find I and V in the circuit of Fig. 4.13; assume that the diode is a 1-mA diode, with $V_{DD} = 10$ V, and R $= 10$ kΩ.

Solution
Assuming $V \simeq 0.7$ V, then

$$I = \frac{10 - 0.7}{10} = 0.93 \text{ mA}$$

which is in the normal operating range of the diode. Thus our assumption of $V \simeq 0.7$ V is not a bad one.

· · ·

Refining the Results

It is possible to use a simple iterative method to refine somewhat the results obtained by the approximate-analysis scheme above. The method consists of using the value of current obtained to calculate a new value for the diode voltage drop. Then we use this new value to calculate a better approximation for the value of the current. This completes one iteration. If better accuracy is desired, we can proceed with one or more further iterations. This process is best illustrated by numerical examples.

Example 4.4
Find better approximations for the values of *I* and *V* in the circuit considered in Example 4.3. Assume that the diode is specified for 0.7 V at 1 mA and that $n = 1.8$.

Solution
The results of iteration 0, performed in Example 4.3, are $V = 0.7$ V, $I = 0.93$ mA. Since the diode is characterized by

$$I = I_S e^{V/nV_T}$$

then

$$10^{-3} = I_S e^{700/(1.8 \times 25)}$$

which results in

$$I_S = 10^{-3} e^{-700/(1.8 \times 25)}$$

Thus

$$I = 10^{-3} e^{(V-700)/(1.8 \times 25)}$$

Substituting $I = 0.93$ mA gives

$$V = 696.7 \text{ mV}$$

Using Kirchhoff's voltage equation

$$I = \frac{V_{DD} - V}{R}$$

we obtain

$$I = \frac{10 - 0.6967}{10} = 0.9303 \text{ mA}$$

which is very close to the originally calculated value of 0.93 mA. Thus no further iterations are justified. In fact, because the result of our iteration 0 (0.93 mA) is so close to the current of 1 mA (corresponding to the 0.7-V drop) even the first iteration was not quite justified.

The following example will better illustrate the need for iterating.

Finally, note that for this diode the value of *n* was given. If *n* is not known, we will have to assume a value (a good assumption is a value of 2 for silicon diodes, otherwise 0.1 V per decade of current is often used).

· · ·

Example 4.5
Calculate the values of V_D and I_D in the circuit of Fig. 4.15. Let the diode have $V_D = 0.7$ V at $I_D = 1$ mA, and let $n = 2$.

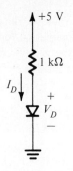

Fig. 4.15   Circuit for Example 4.5

Solution
From the diode equation

$$i = I_s e^{v/nV_T}$$

it follows that corresponding to a voltage V_1 the current I_1 is

$$I_1 = I_s e^{V_1/nV_T}$$

and corresponding to a voltage V_2 the current I_2 is

$$I_2 = I_s e^{V_2/nV_T}$$

Thus

$$\frac{I_2}{I_1} = e^{(V_2 - V_1)/nV_T}$$

or, alternatively,

$$V_2 - V_1 = nV_T \ln \frac{I_2}{I_1} \qquad (4.7)$$

These two relationships are useful in performing our iterative analysis.
Iteration 0

$$V_D = 0.7 \text{ V} \qquad \text{and} \qquad I_D = \frac{5 - 0.7}{1} = 4.3 \text{ mA}$$

Iteration 1   Use of Eq. (4.7) gives

$$V_2 - V_1 = 2 \times 25 \ln \frac{4.3}{1} = 72.9 \text{ mV}$$

Thus

$$V_D = 0.7729 \text{ V} \qquad \text{and} \qquad I_D = \frac{5 - 0.7729}{1} = 4.227 \text{ mA}$$

Iteration 2

$$V_2 - V_1 = 2 \times 25 \ln \frac{4.227}{4.3} = -0.85 \text{ mV}$$

Thus

$$V_D = 0.772 \text{ V} \quad \text{and} \quad I_D = \frac{5 - 0.772}{1} = 4.23 \text{ mA}$$

Obviously no further iterations are called for; in fact, even iteration 2 is hardly justified.

· · ·

EXERCISES

4.8 Calculate the values of V and I in the circuits shown in Fig. E4.8. Assume that the diode has 0.7 V drop at a current of 1 mA, and let $n = 2$.
Ans. (a) 647.6 mV, 0.352 mA; (b) 810.9 mV, 9.189 mA; (c) 696.4 mV, 0.930 mA

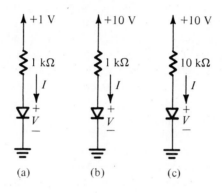

(a)　　　　(b)　　　　(c)

Fig. E4.8

4.9 In the circuit shown in Fig. E4.9 assume that the three diodes are identical, with drops of 0.7 V at currents of 1 mA and $n = 2$. Calculate the value of V_1. Also calculate the change in the value of V_1 when the dc supply voltage changes to $+15$ V and then to $+5$ V.
Ans. 2.4 V; 72 mV; -153 mV

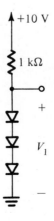

Fig. E4.9

4.8   MODELING THE DIODE FORWARD CHARACTERISTIC

Philosophy of Device Modeling

For the purpose of analyzing and designing circuits, each device has to be represented by a model. Generally speaking, such a model takes the form of one or more relationship between the terminal voltages and currents. For instance, the model for a resistor is given by Ohm's law,

$$v = Ri$$

where R is the value of the resistance. Also, the various two-port network representations discussed in Chapter 2 are models. If one represents the equations of the model by a circuit, one obtains an *equivalent circuit model*. We did this in Chapter 2, where equivalent circuit representations of linear two-port networks were given (Fig. 2.31).

Each device model has its limitations: the model will represent the actual physical operation of the device over a limited range of variables and only to a certain degree of accuracy. As an example, the $v = Ri$ model of a resistor assumes that the resistance remains constant as current is varied, an assumption that is valid only for a limited range of current values. Also, this resistor model does not account for the dependence of the resistance value on temperature, humidity, aging, etc.

Although it is generally possible to find device models that are as accurate as possible, this is usually neither required nor desirable. As the model becomes more accurate it also becomes more complex, thus resulting in more effort being required in using the model for analyzing and designing circuits. More seriously, using complex models does not enable the circuit designer to obtain insight into the physical operation of his or her circuit. It is this insight that allows one to improve on the design. Thus one should use as simple a device model as the application at hand permits. After the initial design is obtained one may analyze the circuit using a computer circuit-analysis program and employing more accurate device models. As a result, the design may be further refined.

The Exponential Model

The exponential i-v relationship of the junction diode is itself a model of the physical processes that take place inside the pn junction that forms the diode. These physical processes will be described later in this chapter. The exponential relationship is a very good model in the forward region of diode operation. Use of this elaborate model, however, requires the solving of nonlinear equations even if the circuit being analyzed is a very simple one. To cope with this problem we introduced the iterative scheme of Section 4.7. Nevertheless, there are many diode circuit applications for which a much simpler diode model would be sufficient.

The "Ideal-Diode" Model

The simplest diode model is the ideal-diode model depicted in Fig. 4.16. Such a model assumes that the voltage drop across the diode in the forward direction is negligibly small. This model is quite appropriate for applications that make use of the gross non-

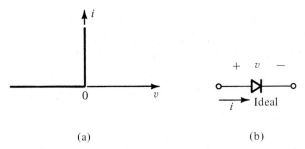

(a)   (b)

Fig. 4.16   (a) The i-v characteristic for the ideal-diode model. (b) Circuit symbol for ideal diode.

linearity exhibited by diodes, such as rectifiers of large voltages (much greater than 0.7 V). If the voltages involved are much larger than the diode voltage drop, we can safely neglect the latter and use the ideal diode as a model for the real junction diode.

Even in applications in which the diode voltage drop cannot be neglected, insights into circuit operation can be obtained by initially assuming that the diode is ideal. Subsequently the results can be refined with a more accurate diode model.

The Constant-Voltage-Drop Model

In the hierarchy of diode models the constant-voltage-drop model is the one next to the ideal-diode model. We have already alluded to the constant-voltage-drop model in the course of analyzing simple diode circuits. This model, depicted in Fig. 4.17, assumes that a conducting diode has a voltage drop that is almost constant and independent of the current flowing through the diode. Typically a value of 0.7 V is assumed for the constant voltage drop.

The "Battery-Plus-Resistance" Model

To account for the fact that the diode forward voltage drop is not constant but increases with the current flowing through the diode, a constant resistance may be included with the battery (or constant voltage) in the diode model. The result is the battery-plus-resistance model illustrated in Fig. 4.18. Obviously there is latitude in selecting values for the battery voltage V_B and for the series resistance R_D. In doing this one should

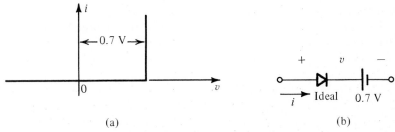

(a)   (b)

Fig. 4.17   (a) The i-v characteristic for the constant-voltage-drop model. (b) Circuit representation for the constant-voltage-drop model in terms of ideal diode and battery.

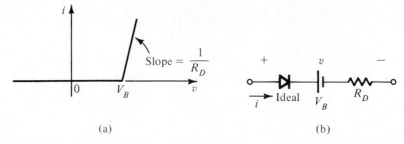

Fig. 4.18   (a) The i-v characteristic for the battery-plus-resistance model. (b) Circuit representation for the model.

attempt to select values that best fit the i-v characteristic over the expected range of operation. However, if the range of operation is not known, a simple scheme to be used is as follows: Since the diode conducts negligible currents when the forward voltage is less than about 0.5 V, we may take this voltage to be the battery voltage V_B. Then for a 1-mA diode—that is, a diode that conducts 1 mA when the voltage drop is about 0.7 V—we need a series resistance of

$$R_D = \frac{0.2 \text{ V}}{1 \text{ mA}} = 200 \ \Omega$$

Example 4.6
Find the current I_D in the circuit of Fig. 4.19 for the two cases (a) $V^+ = 10$ V and (b) $V^+ = 1$ V. In each case use (i) the ideal-diode model, (ii) the constant-voltage-drop

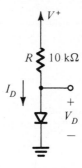

Fig. 4.19   Circuit for Example 4.6.

model, (iii) the battery-plus-resistance model, and (iv) the exponential model, assuming the diode to be a 1-mA diode with $n = 2$.

Solution
Case a: $V^+ = 10$ V
(i) Use of the ideal-diode model gives

$$V_D = 0 \text{ V} \quad \text{and} \quad I_D = \frac{V^+}{R} = \frac{10}{10} = 1 \text{ mA}$$

(ii) Use of the constant-voltage-drop model gives

$$V_D = 0.7 \text{ V} \quad \text{and} \quad I_D = \frac{V^+ - V_D}{R} = \frac{10 - 0.7}{10} = 0.93 \text{ mA}$$

(iii) Use of the battery-plus-resistance model with a battery voltage of 0.5 V and a series resistance R_D of 200 Ω gives

$$I_D = \frac{V^+ - 0.5}{R + R_D} = \frac{10 - 0.5}{10 + 0.2} = 0.931 \text{ mA} \quad \text{and} \quad V_D = 0.5 + I_D R_D = 0.69 \text{ V}$$

(iv) Use of the exponential model with $I_D = 1$ mA at $V_D = 0.7$ V and $n = 2$ together with the iterative scheme described in Section 4.7 gives

$$I_D = 0.930 \text{ mA} \quad \text{and} \quad V_D = 0.70 \text{ V}$$

Case b: $V^+ = 1$ V
(i) Use of the ideal-diode model gives

$$V_D = 0 \text{ V} \quad \text{and} \quad I_D = \frac{V^+}{R} = \frac{1}{10} = 0.1 \text{ mA}$$

(ii) Use of the constant-voltage-drop model gives

$$V_D = 0.7 \text{ V} \quad \text{and} \quad I_D = \frac{1 - 0.7}{10} = 0.03 \text{ mA}$$

(iii) Use of the battery-plus-resistance model gives

$$I_D = 0.049 \text{ mA} \quad \text{and} \quad V_D = 0.51 \text{ V}$$

(iv) Use of the exponential model with two iterations gives

$$I_D = 0.045 \text{ mA} \quad \text{and} \quad V_D = 0.55 \text{ V}$$

. . .

Example 4.6 shows that the constant-voltage-drop model provides good results when the voltages involved (V^+ in this case) are much greater than the diode voltage drop. When this is not the case, as in case b above, the slightly more elaborate battery-plus-resistance model provides a result much closer to the value obtained with the more elaborate exponential model. The choice of an appropriate device model is indeed an attribute of a "good" circuit designer.

The Small-Signal Model

There are applications in which a diode is biased to operate at a point on the forward i-v characteristic and a small ac signal is superimposed on the dc quantities. For this situation the diode is best modeled by a resistance equal to the inverse of the slope of the tangent to the i-v characteristic at the bias point. This arrangement was explained in general terms in Section 2.1 and will be developed for the junction diode in Section 4.9, where circuit examples will also be given.

EXERCISE

4.10 Consider a voltage source v_S, $v_S = t$ volts, where t is the time in seconds, connected through a resistance of 10 kΩ to a 1-mA diode. Use the battery-plus-resistance diode model to find an expression for the current i flowing through the diode.

$$\textit{Ans.} \quad i = 0, \qquad\qquad t \leq 0.5 \text{ s}$$
$$\qquad = \frac{t - 0.5}{10.2} \text{ mA}, \qquad t \geq 0.5 \text{ s}$$

4.9   THE SMALL-SIGNAL MODEL AND ITS APPLICATION

Consider the conceptual circuit in Fig. 4.20a and the corresponding graphical representation in Fig. 4.20b. A dc voltage V_D, represented by a battery, is applied to the diode

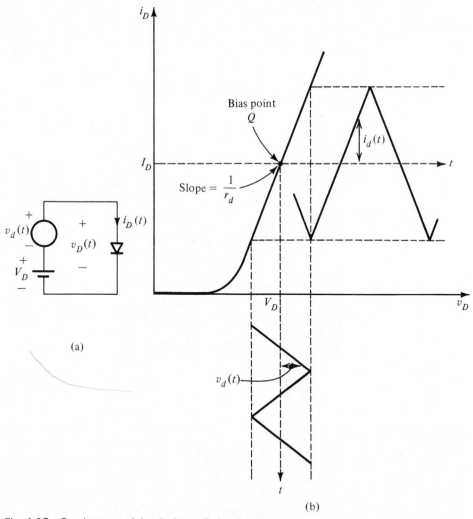

Fig. 4.20   Development of the diode small-signal model.

and a time-varying signal $v_d(t)$, assumed (arbitrarily) to have a triangular waveform, is superimposed on the dc voltage V_D. In the absence of the signal $v_d(t)$ the diode voltage is equal to V_D, and correspondingly the diode will conduct a dc current I_D given by

$$I_D = I_S e^{V_D/nV_T} \tag{4.8}$$

When the signal $v_d(t)$ is applied, the total instantaneous diode voltage $v_D(t)$ will be given by

$$v_D(t) = V_D + v_d(t) \tag{4.9}$$

Correspondingly, the total instantaneous diode current $i_D(t)$ will be

$$i_D(t) = I_S e^{v_D/nV_T} \tag{4.10}$$

Substituting for v_D from Eq. (4.9) gives

$$i_D(t) = I_S e^{(V_D+v_d)/nV_T}$$

which can be rewritten

$$i_D(t) = I_S e^{V_D/nV_T} e^{v_d/nV_T}$$

Using Eq. (4.8) we obtain

$$i_D(t) = I_D e^{v_d/nV_T} \tag{4.11}$$

Now if the amplitude of the signal $v_d(t)$ is kept sufficiently small such that

$$\frac{v_d}{nV_T} \ll 1 \tag{4.12}$$

then we may expand the exponential of Eq. (4.11) in a series and truncate the series after the first two terms to obtain the approximate expression

$$i_D(t) \simeq I_D\left(1 + \frac{v_d}{nV_T}\right) \tag{4.13}$$

This is the *small-signal approximation.* It is valid for signals whose amplitudes are smaller than about 10 mV [see Eq. (4.12) and recall that $V_T = 25$ mV].

From Eq. (4.13) we have

$$i_D(t) = I_D + \frac{I_D}{nV_T} v_d \tag{4.14}$$

Thus superimposed on the dc current I_D we have a signal current component directly proportional to the signal voltage v_d. That is,

$$i_D = I_D + i_d \tag{4.15}$$

where

$$i_d = \frac{I_D}{nV_T} v_d \tag{4.16}$$

The quantity relating the signal current i_d to the signal voltage v_d has the dimensions of conductance (℧) and is called the *diode small-signal conductance*. The inverse of this parameter is the *diode small-signal resistance*, or *incremental resistance*, r_d,

$$r_d = \frac{nV_T}{I_D} \tag{4.17}$$

Note that the value of r_d is inversely proportional to the bias current I_D.

Let us return to the graphical representation in Fig. 4.20b. It is easy to see that using the small-signal approximation is equivalent to assuming that the signal amplitude is sufficiently small such that the excursion along the *i-v* curve is limited to a short, almost linear segment. The slope of this segment, which is equal to the slope of the *i-v* curve at the operating point Q, is equal to the small-signal conductance. The reader is encouraged to prove that the slope of the *i-v* curve at $i = I_D$ is equal to I_D/nV_T, which is $1/r_d$.

Example 4.7
Consider the circuit shown in Fig. 4.21. Let the nominal value of the power supply voltage V_{DD} be 10 V and let $R = 10$ kΩ. Calculate the value of the diode voltage,

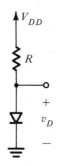

Fig. 4.21   Circuit for Example 4.7.

assuming that $n = 2$ and the diode is a 1-mA diode. If V_{DD} changes by ± 1 V, what are the corresponding changes in the diode voltage?

Solution
When V_{DD} is at its nominal value of 10 V one may assume that the diode voltage is

$$V_D \simeq 0.7 \text{ V}$$

Thus the diode current will be

$$I_D = \frac{10 - 0.7}{10} = 0.93 \text{ mA}$$

At this operating point the diode incremental resistance r_d is

$$r_d = \frac{nV_T}{I_D} = \frac{2 \times 25}{0.93} = 53.8 \text{ Ω}$$

The ± 1-V change in V_{DD} may be considered as a signal of 2 V peak-to-peak. Assuming the corresponding change in the diode voltage v_d to be small, we may find the peak-to-peak value of v_d using the ratio of the voltage divider composed of r_d and R, that is,

$$v_d \text{ (peak-to-peak)} = 2 \times \frac{r_d}{R + r_d}$$

$$= 2 \times \frac{0.0538}{10 + 0.0538} = 10.7 \text{ mV}$$

Thus the change in diode voltage is ± 5.35 mV. Since this value is quite small, our use of the small-signal model of the diode is justified.

$$\cdot \quad \cdot \quad \cdot$$

Use of the Diode Forward Drop in Voltage Regulation

Example 4.7 suggests an important application of the diode forward characteristics, namely, the use of the diode in low-voltage (smaller than 3 or 4 V) regulators. Since the diode voltage drop remains almost constant at approximately 0.7 V independent of changes in the supply voltage, a string of forward-conducting diodes can be used to provide a constant voltage. For instance, the use of three diodes in series provides a voltage of about 2 V and would be equivalent to a zener diode. Since low-voltage zeners do not have sharp knees, diode strings are quite competitive in generating constant (regulated) dc voltages lower than about 3 or 4 V. In such applications the small-signal model of the diode can be used to calculate the changes in the regulated voltage due to changes in power supply voltage and/or changes in load current. This process is illustrated in Example 4.8.

Example 4.8
Consider the circuit shown in Fig. 4.22. A string of three diodes is used to provide a constant voltage of about 2.1 V. We want to calculate the percentage change in this regulated voltage caused by (a) $\pm 10\%$ change in the power supply voltage and (b) connection of a 1-kΩ load resistance.

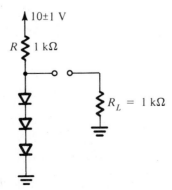

Fig. 4.22 *Circuit for Example 4.8.*

Solution

With no load the nominal value of the current in the diode string is given by

$$I = \frac{10 - 2.1}{1} = 7.9 \text{ mA}$$

Thus each diode will have an incremental resistance of

$$r_d = \frac{nV_T}{I}$$

Using $n = 2$ gives

$$r_d = \frac{2 \times 25}{7.9} = 6.3 \ \Omega$$

The three diodes in series will have a total incremental resistance of

$$r = 3r_d = 18.9 \ \Omega$$

This resistance along with the resistance R forms a voltage divider whose ratio can be used to calculate the change in output voltage due to a $\pm 10\%$ (that is, ± 1 V) change in supply voltage. Thus the peak-to-peak change in output voltage will be

$$\Delta v_o = 2 \frac{r}{r + R} = 2 \frac{0.0189}{0.0189 + 1} = 37.1 \text{ mV}$$

that is, corresponding to the ± 1-V ($\pm 10\%$) change in supply voltage the output voltage will change by ± 18.5 mV or $\pm 0.9\%$. Since this implies a change of about ± 6.2 mV per diode, our use of the small-signal model is justified.

When a load resistance of 1 kΩ is connected across the diode string it draws a current of approximately 2.1 mA. Thus the current in the diodes decreases by 2.1 mA, resulting in a decrease in voltage across the diode string given by

$$\Delta v_o = -2.1 \times r = -2.1 \times 18.9 = -39.7 \text{ mV}$$

Since this implies that the voltage across each diode decreases by about 13.2 mV, our use of the small-signal model is not entirely justified. Nevertheless, a detailed calculation of the voltage change using the exponential model results in $\Delta v_o = -35.5$ mV, which is not too different from the approximate value obtained using the incremental model.

· · ·

4.10   PHYSICAL OPERATION OF DIODES—BASIC SEMICONDUCTOR CONCEPTS

Having studied the terminal characteristics of junction diodes, we will now briefly consider the physical processes that give rise to these characteristics. The following treatment of device physics is qualitative; nevertheless, it should provide sufficient background for the design of diode and other semiconductor circuits.

The pn Junction

The semiconductor diode is basically a *pn* junction, as shown schematically in Fig. 4.23. As indicated, the *pn* junction consists of *p*-type semiconductor material (such as silicon) brought into close contact with *n*-type semiconductor material (silicon). In actual practice both the *p* and *n* regions are part of the same silicon crystal; that is, the *pn* junction is formed within a single silicon crystal by creating regions of different "dopings" (*p*

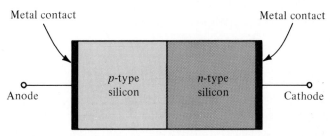

Fig. 4.23   *Physical structure of the junction diode. (Actual geometries are given in Appendix A.)*

and *n* regions). Appendix A provides a brief description of the process employed in the fabrication of *pn* junctions. As indicated in Fig. 4.23, external wire connections to the *p* and *n* regions (that is, diode terminals) are made through metal (aluminum) contacts.

In addition to being essentially a diode, the *pn* junction is the basic component of field-effect transistors (FETs) and bipolar-junction transistors (BJTs). Thus an understanding of the physical operation of *pn* junctions is essential to the understanding of the operation and terminal characteristics of diodes and transistors.

Intrinsic Silicon

Although both silicon and germanium can be used to manufacture semiconductor devices—indeed earlier diodes and transistors were made of germanium—today's integrated-circuit technology is based entirely on silicon. For this reason we will deal exclusively with silicon devices throughout this book.

A crystal of pure or intrinsic silicon has a regular lattice structure where the atoms are held in their positions by bonds, called *covalent bonds,* formed by the four valence electrons associated with each silicon atom. At sufficiently low temperatures all covalent bonds are intact and no (or very few) *free electrons* are available to conduct electric current. However, at room temperature some of the bonds are broken by thermal ionization and some electrons are freed. When a covalent bond is broken an electron leaves its parent atom; thus a positive charge, equal to the magnitude of the electron charge, is left with the parent atom. An electron from a neighboring atom may be attracted to this positive charge, leaving its parent atom. This action fills up the *hole* that existed in the ionized atom but creates a new *hole* in the other atom. This process may repeat itself with the result that we effectively have a positively charged carrier, called a hole, moving through the silicon crystal structure and being available to conduct electric current. The charge of a hole is equal in magnitude to the charge of an electron.

Thermal ionization results in free electrons and holes in equal numbers and hence equal concentrations. These free electrons and holes move randomly through the silicon crystal structure, and in the process some electrons may fill some of the holes. This process, called *recombination,* results in the disappearance of free electrons and holes. The recombination rate is proportional to the number of free electrons and holes, which is in turn determined by the ionization rate. The ionization rate is a strong function of temperature. In thermal equilibrium the recombination rate is equal to the ionization or thermal generation rate, and one can calculate the concentration of free electrons n, which is equal to the concentration of holes p,

$$n = p = n_i$$

where n_i denotes the concentration of free electrons or holes in intrinsic silicon at a given temperature.

Finally, it should be mentioned that the reason that silicon is called a semiconductor is that its conductivity, which is determined by the number of charge carriers available to conduct electric current, is between that of conductors (such as metals) and that of insulators (such as glass).

Diffusion and Drift

There are two mechanisms by which holes and electrons move through a silicon crystal—*diffusion* and *drift*. Diffusion is associated with random motion due to thermal agitation. In a piece of silicon with uniform concentrations of free electrons and holes this random motion does not result in a net flow of charge (that is, current). On the other hand, if by some mechanism the concentration of, say, free electrons is made higher in one part of the piece of silicon than in another, then electrons will diffuse from the region of high concentration to the region of low concentration. This diffusion process gives rise to a net flow of charge, or *diffusion current*. As an example, consider the bar of silicon shown in Fig. 4.24a, in which the hole *concentration profile* shown in Fig. 4.24b has been created along the x axis by some unspecified mechanism. The existence of such a concentration profile results in a hole diffusion current in the x direction, with the magnitude of the current at any point being proportional to the slope of the concentration curve, or the concentration gradient, at that point.

The other mechanism for carrier motion in semiconductors is drift. Carrier drift occurs when an electric field is applied across a piece of silicon. Free electrons and holes are accelerated by the electric field and acquire a velocity component (superimposed on the velocity of their thermal motion) called *drift velocity*. The resulting hole and electron current components are called *drift currents*. The relationship between the drift current and the applied electric field represents one form of Ohm's law.

Doped Semiconductors

The intrinsic silicon crystal described above has equal concentrations of free electrons and holes generated by thermal ionization. These concentrations, denoted n_i, are strongly dependent on temperature. Doped semiconductors are materials in which carriers of one kind (electrons or holes) predominate. Doped silicon in which the majority

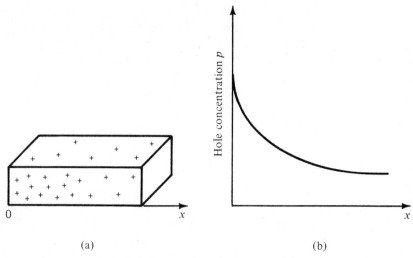

Fig. 4.24 A bar of intrinsic silicon (a) in which the hole concentration profile shown in (b) has been created along the x axis by some unspecified mechanism.

of charge carriers are the *negatively* charged electrons is called *n type,* while silicon doped so that the majority of charge carriers are the *positively* charged holes is called *p type.*

Doping of a silicon crystal to turn it into *n* type or *p* type is achieved by introducing a small number of impurity atoms. For instance, introducing impurity atoms of a pentavalent element such as phosphorus results in *n*-type silicon because the phosphorus atoms that replace some of the silicon atoms in the crystal structure have five valence electrons, four of which form bonds with the neighboring silicon atoms while the fifth becomes a free electron. Thus each phosphorus atom *donates* a free electron to the silicon crystal, and the phosphorus impurity is called a *donor.* It should be clear, though, that no holes are generated by this process; hence the majority of charge carriers in the phosphorus-doped silicon will be electrons. In fact, if the concentration of donor atoms (phosphorus) is N_D, in thermal equilibrium the concentration of free electrons in the *n*-type silicon, n_{n0}, will be

$$n_{n0} \simeq N_D$$

where the additional subscript 0 denotes thermal equilibrium. In this *n*-type silicon the concentration of holes, p_{n0}, that are generated by thermal ionization will be

$$p_{n0} \simeq \frac{n_i^2}{N_D}$$

Since n_i is a function of temperature, it follows that the concentration of the *minority* holes will be a function of temperature, while that of the *majority* electrons is independent of temperature.

To produce a *p*-type semiconductor, silicon has to be doped with a trivalent impurity such as boron. Each of the impurity boron atoms *accepts* one electron from the silicon crystal, so that they may form covalent bonds in the lattice structure. Thus each boron atom gives rise to a hole, and the concentration of the majority holes in p-type silicon, under thermal equilibrium, is approximately equal to the concentration N_A of the *acceptor* (boron) impurity,

$$p_{p0} \simeq N_A$$

In this p-type silicon the concentration of the minority electrons, which are generated by thermal ionization, will be

$$n_{p0} \simeq \frac{n_i^2}{N_A}$$

It should be mentioned that a piece of n-type or p-type silicon is electrically neutral; the majority free carriers (electrons in n-type silicon and holes in p-type silicon) are neutralized by *bound charges* associated with the impurity atoms.

4.11   THE pn JUNCTION UNDER OPEN-CIRCUIT CONDITIONS

Figure 4.25 shows a *pn* junction under open-circuit conditions—that is, the external terminals are left open. The "+" signs in the p-type material denote the majority holes. The charge of these holes is neutralized by an equal amount of bound negative charge associated with the acceptor atoms. For simplicity these bound charges are not shown in the diagram. Also not shown are the minority electrons generated in the p-type material by thermal ionization.

In the n-type material the majority electrons are indicated by "−" signs. Here also the bound positive charge, which neutralizes the charge of the majority electrons, is not shown in order to keep the diagram simple. The n-type material also contains minority holes generated by thermal ionization and not shown on the diagram.

The Diffusion Current I_D

Because the concentration of holes is high in the p region and low in the n region, holes diffuse across the junction from the p side to the n side; similarly, electrons diffuse across the junction from the n side to the p side. These two current components add together to form the diffusion current I_D, whose direction is from the p side to the n side, as indicated in Fig. 4.25.

The Depletion Region

The holes that diffuse across the junction into the n region quickly recombine with some of the majority electrons present there and thus disappear from the scene. This recombination process results in the disappearance of some free electrons from the n-type material. Thus some of the bound positive charge will no longer be neutralized by free electrons, and this charge is said to have been *uncovered*. Since recombination takes place close to the junction, there will be a region close to the junction that is depleted of free electrons and contains uncovered bound positive charge, as indicated in Fig. 4.25.

The electrons that diffuse across the junction into the *p* region quickly recombine with some of the majority holes present there and thus disappear from the scene. This results also in the disappearance of some majority holes, causing some of the bound negative charge to be uncovered (that is, no longer neutralized by holes). Thus in the *p* material close to the junction there will be a region depleted of holes and containing uncovered bound negative charge, as indicated in Fig. 4.25.

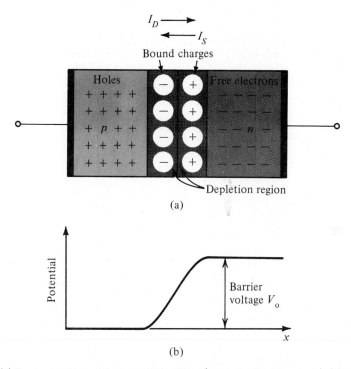

(a)

(b)

Fig. 4.25   (a) The pn junction with no applied voltage (open-circuited terminals). (b) The potential distribution along an axis perpendicular to the junction.

From the above it follows that a *carrier-depletion region* will exist on both sides of the junction, with the *n* side of this region positively charged and the *p* side negatively charged. This carrier-depletion region—or, simply, *depletion region*—is also called the *space-charge region*. The charge on both sides of the depletion region cause an electric field to be established across the region; hence a potential difference results across the depletion region, with the *n* side at a positive voltage relative to the *p* side, as shown in Fig. 4.25b. Thus the resulting electric field opposes the diffusion of holes into the *n* region and electrons into the *p* region. In fact, the voltage drop across the depletion region acts as a *barrier* that has to be overcome for holes to diffuse into the *n* region and electrons to diffuse into the *p* region. The larger the barrier voltage, the smaller the number of carriers that will be able to overcome the barrier, and hence the lower the magnitude of diffusion current. Thus the diffusion current I_D depends strongly on the voltage drop V_0 across the depletion region.

The Drift Current I_S and Equilibrium

In addition to the current component I_D due to majority carrier diffusion, a component due to minority carrier drift exists across the junction. Specifically, some of the thermally generated holes in the n material diffuse through the n material to the edge of the depletion region. There they experience the electric field in the depletion region, which sweeps them across that region into the p side. Similarly, some of the minority thermally generated electrons in the p material diffuse to the edge of the depletion region and get swept by the electric field in the depletion region across that region into the n side. These two current components—electrons moved by drift from p to n and holes moved by drift from n to p—add together to form the drift current I_S, whose direction is from the n side to the p side of the junction, as indicated in Fig. 4.25. Since the current I_S is carried by thermally generated minority carriers, its value is strongly dependent on temperature; however, it is independent of the value of the depletion layer voltage V_0.

Under open-circuit conditions (Fig. 4.25) no external current exists; thus the two opposite currents across the junction should be equal in magnitude:

$$I_D = I_S$$

This equilibrium condition is maintained by the barrier voltage V_0. Thus if for some reason I_D exceeds I_S, then more bound charge will be uncovered on both sides of the junction, the depletion layer will widen, and the voltage across it (V_0) will increase. This in turn causes I_D to decrease until equilibrium is achieved with $I_D = I_S$. On the other hand, if I_S exceeds I_D, then the amount of uncovered charge will decrease, the depletion layer will narrow, and the voltage across it will decrease. This causes I_D to increase until equilibrium is achieved with $I_D = I_S$.

The Terminal Voltage

When the pn junction terminals are left open-circuited the voltage measured between them will be zero. That is, the voltage V_0 across the depletion region *does not* appear between the diode terminals. This is because of the contact voltages existing at the metal–semiconductor junctions at the diode terminals, which counter and exactly balance the barrier voltage. If this were not the case, we would have been able to draw energy from the isolated pn junction, which would clearly violate the principle of conservation of energy.

Width of the Depletion Region

From the above it should be apparent that the depletion region exists in both the p and n materials and that equal amounts of charge exist on both sides. However, since usually the doping levels are not equal in the p and n materials, one can reason that the width of the depletion region will not be the same on the two sides. Rather, in order to uncover the same amount of charge the depletion layer will extend deeper into the more lightly doped material. In actual practice it is usual that one side of the junction is much more lightly doped than the other, with the result that the depletion region exists almost entirely in one of the two semiconductor materials.

4.12 THE pn JUNCTION UNDER REVERSE BIAS CONDITIONS

The behavior of the *pn* junction in the reverse direction is more easily explained on a microscopic scale if we consider exciting the junction with a constant-current source (rather than with a voltage source), as shown in Fig. 4.26. The current source I is obviously in the reverse direction. For the time being let the magnitude of I be less than I_S; if I is greater than I_S, breakdown will occur, as explained in Section 4.13.

The current I will be carried by electrons flowing in the external circuit from the *n* material to the *p* material (that is, in the direction opposite to that of I). This will cause electrons to leave the *n* material and holes to leave the *p* material. The free electrons leaving the *n* material cause the uncovered positive bound charge to increase. Similarly, the holes leaving the *p* material result in an increase in the uncovered negative bound charge. Thus the reverse current I will result in an increase in the width of, and the charge stored in, the depletion layer. This, in turn, will result in a higher voltage across the depletion region—that is, a greater barrier voltage V_0—which causes the diffusion current I_D to decrease. The drift current I_S, being independent of the barrier voltage, will remain constant. Finally, equilibrium (steady state) will be reached when

$$I_S - I_D = I$$

In equilibrium, the increase in depletion-layer voltage will appear as an external voltage between the diode terminals, with *n* being positive with respect to *p*.

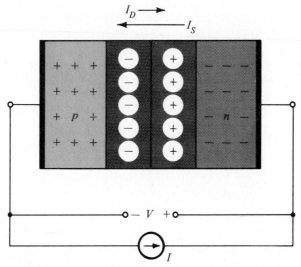

Fig. 4.26 *The pn junction excited by a constant-current source I in the reverse direction. To avoid breakdown I is kept smaller than I_S. Note that the depletion layer widens and the barrier voltage increases by V volts, which appears between the terminals as a reverse voltage.*

We can now consider exciting the *pn* junction by a reverse voltage V, where V is less than the breakdown voltage V_{ZK}. When the voltage V is first applied a reverse current flows in the external circuit from p to n. This current causes the increase in width and charge of the depletion layer. Eventually the voltage across the depletion layer will increase by the magnitude of the external voltage V, at which time an equilibrium is

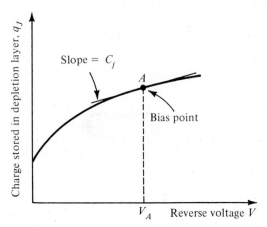

Fig. 4.27   *The charge stored on either side of the depletion layer as a function of the reverse voltage V.*

reached with the external reverse current I equal to $(I_S - I_D)$. Note, however, that initially the external current can be much greater than I_S. The purpose of this initial transient is to *charge* the depletion layer and increase the voltage across it by V volts.

From the above we observe the analogy between the depletion layer of a *pn* junction and a capacitor. As the voltage across the *pn* junction changes, the charge stored in the depletion layer changes accordingly. Figure 4.27 shows a sketch of typical charge-versus-external-voltage characteristic of a *pn* junction. Since this *q-v* characteristic is nonlinear, one has to be careful in speaking of the "depletion capacitance." As we did in Section 4.9 with the nonlinear *i-v* characteristic, we may consider operation around a bias point on the *q-v* curve, such as point *A* in Fig. 4.27, and define the small-signal or incremental *depletion capacitance* C_j as the slope of the *q-v* curve at the operating point,

$$C_j = \frac{dq_J}{dv} \bigg|_{v=V_A}$$

It can be shown that

$$C_j = \frac{k}{(V_0 - V_D)^m}$$

where $V_0 =$ the depletion layer voltage with zero external voltage,

$V_D =$ the voltage between the diode terminals (V_D is negative in the reverse direction),

> k = a constant depending on the area of the junction and impurity concentrations, and
>
> m = a constant depending on distribution of impurity near the junction. The value of m ranges between $\frac{1}{3}$ and 3 or 4 for junctions of various types

To recap, as a reverse bias voltage is applied to a *pn* junction a transient occurs during which the depletion capacitance is charged to the new bias voltage. After the transient dies the steady-state reverse current is simply equal to $I_S - I_D$. Usually I_D is very small when the diode is reverse-biased and the reverse current is approximately equal to I_S. This, however, is only a theoretical model that does not apply very well. In actual fact currents as high as few nanoamps (10^{-9} A) flow in the reverse direction, although I_S is usually of the order of 10^{-15} A. This large difference is due to leakage and other effects. Furthermore, the reverse current is dependent to a certain extent on the magnitude of the reverse voltage, contrary to the theoretical model, which states that $I \simeq I_S$ independent of the value of the reverse voltage applied. Nevertheless, because of the very low current involved one is usually not interested in the details of the diode *i-v* characteristic in the reverse direction.

4.13   THE pn JUNCTION IN THE BREAKDOWN REGION

In considering the diode operation in the reverse-bias region in Section 4.12 it was assumed that the reverse current source I (Fig. 4.26) is smaller than I_S or, equivalently, that the reverse voltage V is smaller than the breakdown voltage V_{ZK}. We now wish to consider the breakdown mechanisms in *pn* junctions and explain the reasons behind the

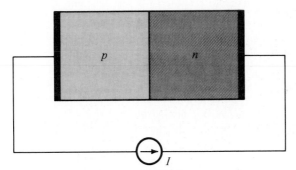

Fig. 4.28   The pn junction excited by a reverse-current source I, where $I > I_S$.

almost vertical line representing the *i-v* relationship in the breakdown region. For this purpose, let the *pn* junction be excited by a current source that causes a constant current I greater than I_S to flow in the reverse direction, as shown in Fig. 4.28. This current source will move holes from the *p* material through the external circuit[1] into the *n* material and electrons from the *n* material through the external circuit into the *p* material. This action results in more and more of the bound charge being uncovered; hence the depletion layer widens and the barrier voltage rises. This latter effect causes

---

[1]The current in the external circuit will, of course, be carried entirely by electrons.

the diffusion current to decrease; eventually it will be reduced to almost zero. Nevertheless, this is not sufficient to reach a steady state, since I is greater than I_S. Therefore the process leading to the widening of the depletion layer continues until a sufficiently high junction voltage develops, at which point a new mechanism sets in to supply the charge carriers needed to support the current I. As will be now explained, this mechanism for supplying reverse currents in excess of I_S can take one of two forms depending on the pn junction material, structure, and so on.

The two possible breakdown mechanisms are the *zener effect* and the *avalanche effect*. If a pn junction breaks down with $V_Z < 5$ V, the breakdown mechanism is usually the zener effect. Avalanche breakdown occurs when V_Z is greater than about 7 V. For junctions that breakdown between 5 and 7 V, the breakdown mechanism can be either the zener or the avalanche effect or a combination of the two.

Zener breakdown occurs when the electric field in the depletion layer increases to the point where it can break covalent bonds and generate electron–hole pairs. The electrons generated this way will be swept by the electric field into the n side and the holes into the p side. Thus these electrons and holes constitute a reverse current across the junction that helps support the external current I. Once the zener effect starts, a large number of carriers can be generated with a negligible increase in the junction voltage. Thus the reverse current in the breakdown region will be determined by the external circuit, while the reverse voltage appearing between the diode terminals will remain close to the rated breakdown voltage V_Z.

The other breakdown mechanism is avalanche breakdown, which occurs when the minority carriers that cross the depletion region under the influence of the electric field gain sufficient kinetic energy to be able to break covalent bonds in atoms with which they collide. The carriers liberated by this process may have sufficiently high energy to be able to cause other carriers to be liberated in another ionizing collision. This process occurs in the fashion of an avalanche, with the result that many carriers are created that are able to support any value of reverse current, as determined by the external circuit, with a negligible change in the voltage drop across the junction.

As mentioned before, pn junction breakdown is not a destructive process, provided that the maximum specified power dissipation is not exceeded. This maximum power dissipation rating, in turn, implies a maximum value for the reverse current.

4.14   THE pn JUNCTION UNDER FORWARD-BIAS CONDITIONS

We next consider operation of the pn junction in the forward-bias region. Again it is easier to explain physical operation if we excite the junction by a constant-current source supplying a current I in the forward direction, as shown in Fig. 4.29. The electrons carrying the current I in the external circuit from the p material to the n material cause holes to be extracted from the n region and electrons to be extracted from the p region. This results in majority carriers being supplied to both sides of the junction by the external circuit: holes to the p material and electrons to the n material. These majority carriers will neutralize some of the uncovered bound charge, causing less charge to be stored in the depletion layer. Thus the depletion layer narrows and the depletion barrier voltage reduces. The reduction in barrier voltage enables more holes to cross the barrier from the p material into the n material, and more electrons from

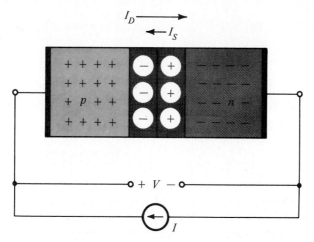

Fig. 4.29 *The pn junction excited by a constant-current source supplying a current I in the forward direction. The depletion layer narrows and the barrier voltage decreases by V volts, which appears as an external voltage in the forward direction.*

the n side to cross into the p side. Thus the diffusion current I_D increases until equilibrium is achieved with $I_D - I_S = I$, the externally supplied forward current.

Let us now examine closely the current flow across the forward-biased pn junction in the steady state. The barrier voltage is now lower than V_0 by an amount V that appears between the diode terminals as a forward voltage drop (that is, the anode of the diode will be more positive than the cathode by V volts). Owing to the decrease in the barrier voltage or, alternatively, because of the forward voltage drop V, holes are *injected* across the junction into the n region and electrons are injected across the junction into the p region. The holes injected into the n region will cause the minority carrier concentration there, p_n, to exceed the thermal equilibrium value, p_{n0}. The *excess* concentration $p_n - p_{n0}$ will be highest near the edge of the depletion layer and will decrease (exponentially) as one moves away from the junction, eventually reaching zero. Figure 4.30 shows such a minority carrier distribution.

In the steady state the concentration profile of *excess minority carriers* remains constant, and indeed it is such a distribution that gives rise to the increase of diffusion current I_D above the value I_S. This is because the distribution shown causes injected minority holes to diffuse away from the junction into the n region and disappear by recombination. To maintain equilibrium, an equal number of electrons will have to be supplied by the external circuit, thus replenishing the electron supply in the n material.

Similar statements can be made about the minority electrons in the p material. The diffusion current I_D is, of course, the sum of the electron and hole components.

Thus in the steady state excess minority carrier distributions as shown in Fig. 4.30 exist in both the n and p materials. It can be shown that the hole current crossing the junction is proportional to the total excess hole charge stored in the n material. This charge is proportional to the area under the hole concentration curve, which is shown hatched in Fig. 4.30 and which, in turn, is proportional to the excess hole concentration at the edge of the junction. Similarly, it can be shown that the current component car-

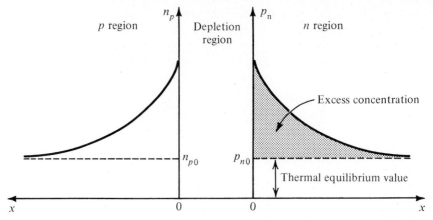

Fig. 4.30 *Minority carrier distribution as a function of the distance from the edges of the depletion layer.*

ried by electrons is proportional to the excess stored electron charge in the *p* material. Thus the electron-current component is proportional to the area under the minority distribution curve and hence is proportional to the excess concentration of electrons at the edge of the depletion layer.

Finally, we note that if we change the value of the external current *I* or, alternatively, change the forward voltage drop *V*, the minority carrier charge stored in the *p* and *n* materials will have to be changed for a new steady state to be established (this, of course, is in addition to the change in charge stored in the depletion region). Thus the *pn* junction exhibits a capacitive effect—in addition to the depletion capacitance— related to the storage of minority carrier charges. Since these charges q_M are proportional to the current flowing across the junction *I*, it follows from the diode equation [Eq. (4.1)] that q_M is related to the forward voltage drop by a relationship of the form

$$q_M = q_0(e^{v/nV_T} - 1)$$

where q_0 is a constant charge proportional to the current I_S. Thus the q-v curve of this capacitive effect is clearly a nonlinear one. We may model this capacitive behavior, however, by a small-signal capacitance C_d,

$$C_d = \frac{dq_M}{dv}\bigg|_{v=V_A}$$

where V_A is the dc diode voltage at the operating point *A* around which the small-signal model is valid. The capacitance C_d is called the *diffusion capacitance*. From the above relationships one can easily establish that C_d is proportional to the value of $q_M + q_0$. Whereas in the reverse-bias region C_d is zero, its value in the forward-bias region is approximately proportional to the bias current I_A at the operating point.

4.15 THE COMPLETE SMALL-SIGNAL MODEL

From the above we conclude that an appropriate small-signal model for the *pn* junction consists of the diode resistance r_d in parallel with the depletion layer capacitance C_j, in parallel with the diffusion capacitance C_d. This model is depicted in Fig. 4.31. If the diode is biased in the forward region at a point A on the i-v curve, then

$$r_d = \frac{nV_T}{I_A}$$
$$C_d = k_c I_A$$
$$C_j = \frac{k}{(V_0 - V_A)^m}$$

where k_c is a constant and V_A is a positive number. On the other hand, if the diode is reverse-biased with a reverse-bias voltage $v = -V_A$ (V_A is a positive number), then

$$r_d \simeq 0$$
$$C_d \simeq 0$$
$$C_j = \frac{k}{(V_0 + V_A)^m}$$

The values of C_d and C_j are, of course, dependent on the size of the junction, being directly proportional to the cross-sectional area. For a small-signal diode reverse-biased by few volts, C_j is typically of the order of 1 pF. The same diode forward-biased with a current of few milliamps has a diffusion capacitance of the order of 10 pF.

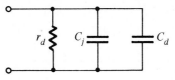

Fig. 4.31   Complete small-signal model of pn junction.

4.16   CONCLUDING REMARKS

The object of this chapter is to present the concept of the ideal diode and familiarize the reader with the characteristics of real silicon junction diodes. Of particular importance is the topic of modeling the diode terminal characteristics. These modeling ideas and philosophies will be employed in conjunction with field effect transistors (FETs) and bipolar junction transistors (BJTs). Equally important are the approximate methods presented for the rapid analysis of diode circuits. Again, similar techniques will be extensively used in the analysis of transistor circuits.

The brief qualitative view of the physical operation of *pn* junctions will be relied upon in understanding the physical operation of FETs and BJTs.

In addition to the few diode circuit applications presented in this chapter, many more are studied in Chapter 5. In fact, now that the reader is familiar with op amps and diodes, we can put them together to design many interesting nonlinear circuits. Some such circuits are presented in Chapter 5.

NONLINEAR CIRCUIT APPLICATIONS 5

Having studied the op amp in Chapter 3 and the diode in Chapter 4, we are now in a position to consider combining them in the design of nonlinear circuits. The circuits considered in this chapter find a variety of applications, including dc power generation, measurement and instrumentation, generation of signals having a variety of waveforms, analog computation, and modulation and demodulation.

Introduction

It will be seen that connecting the diode in the negative-feedback path of an op amp provides idealized behavior, with all the nonidealities of the diode being masked by the high gain of the op amp. This allows for the design of circuits having precise characteristics, as is usually needed in instrumentation and computation applications. This is another illustration of the utility of negative feedback, a concept that we encountered in the linear circuits of Chapter 3 and that we will study formally in Chapter 12.

In the current chapter we will also study some applications of *positive feedback*. As the name implies, positive feedback is exactly the opposite of negative feedback; rather than providing gain stability, as negative feedback does, positive feedback allows for the design of circuits that *oscillate* (that is, provide an output with no input) in a controlled manner. Such circuits can be made to generate signals having square, triangular, pulse, and other waveforms. Furthermore, we will show how circuits composed of diodes and resistors can be used for the purpose of shaping the waveform of a signal in a predetermined manner. The "sine shaper" discussed in Section 5.10 is one such circuit. It can be used to convert triangular waveforms into sinusoids.

This chapter has many objectives. One of these is to introduce the reader to interesting, challenging, and practical circuits that can be designed, assembled, and experimented with. Another objective of this chapter is to expose the reader

to methods of analysis of nonlinear circuits. In keeping with
the spirit of this book, our methods are usually approximate,
but they "do the job."

5.1   HALF-WAVE RECTIFICATION

In Section 4.2 we considered the simple diode circuit shown in Fig. 5.1a. With an ideal-
diode model the transfer characteristic shown in Fig. 5.1b is obtained. For example, an
input signal with triangular waveform (Fig. 5.1c) provides the output waveform shown

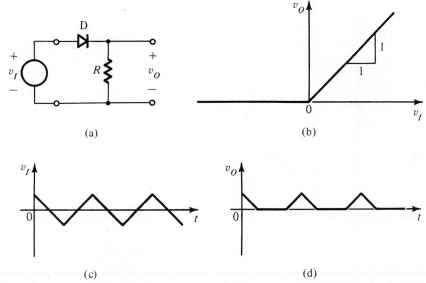

Fig. 5.1   *Half-wave rectification. (a) Circuit. (b) Transfer characteristics using the ideal-diode
model. (c) An input signal having a triangular waveform. (d) The corresponding output waveform.*

in Fig. 5.1d. Since the output is unidirectional, the alternating-current (ac) signal at
the input is said to have been *rectified*. Also, because only half-cycles of the input signal
are utilized the circuit is called a *half-wave rectifier*.

Effect of the Diode Nonideal Characteristics

We will now consider the effect of the diode finite forward-voltage drop on the operation
of the half-wave rectifier. In order for our discussion to be concrete we will consider the
circuit operation with a specific input signal, namely, a triangular waveform with 10 V
peak amplitude. Figure 5.2a shows the positive half-cycle of this input waveform. To
find the corresponding output waveform we first observe that the diode will not conduct
appreciably until the input voltage exceeds about 0.5 V. Thus until this point is reached
the output voltage will remain approximately zero. Then, as the diode conducts, the
output voltage v_O will be less than that of the input by the diode drop. This diode drop,
however, will change as the input voltage changes.

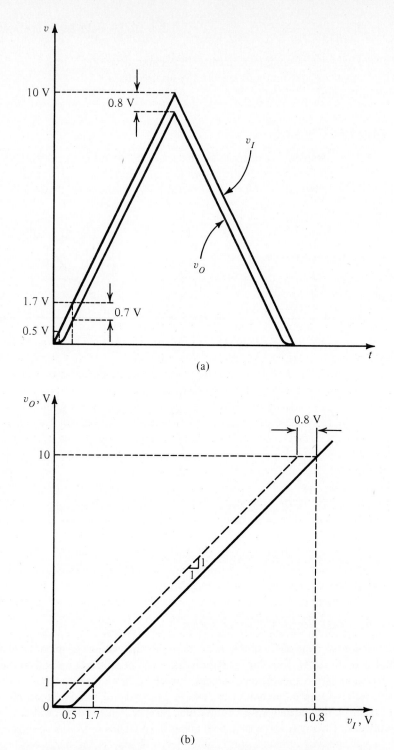

Fig. 5.2 *The effect of the diode finite forward voltage drop on the operation of the half-wave rectifier circuit of Fig. 5.1a. (a) The positive half cycle of an input triangular waveform and the corresponding output waveform. (b) The transfer characteristic.*

To obtain quantitative results, let $R = 1$ kΩ and assume that the diode is a 1-mA diode; that is, it displays a 0.7-V drop at a current of 1 mA. Furthermore, let us assume that we do not know the exact value of the constant n for this particular diode and therefore will use the 0.1 V/decade approximation. Under this approximation the 1-mA diode will exhibit a voltage drop of 0.5 V at 0.01 mA, 0.6 V at 0.1 mA, 0.8 V at 10 mA, and so on.

Now at the point where the output is 0.1 V the current is 0.1 V/1 kΩ = 0.1 mA and the input will be 0.7 V. When the output reaches 1 V the current becomes 1 mA and the corresponding input voltage is 1.7 V. Finally, as the input reaches its peak value of 10 V the output will lag by about 0.8 V, leading to a current of 9.2 mA, which indeed corresponds to a diode drop of almost 0.8 V.

Using the numbers obtained above we can sketch the output waveform shown in Fig. 5.2a and the transfer characteristic shown in Fig. 5.2b. Note that because of the continuous nature of the diode i-v characteristics the corners of the resulting transfer characteristic and output waveform will be rounded, as shown in Fig. 5.2.

Although the process of finding the exact output waveform by taking into account the actual diode characteristics might be an interesting exercise, the amount of work involved is usually too great to be justified in practice. Depending on the particular application, one might just ignore the effects of the finite diode drop completely, investigate such effects qualitatively, or, if the application demands it, use a more elaborate circuit that "masks" these effects. For instance, if we are rectifying large input signals it will most probably make little difference to the system operation whether the diode drop is 0.6 V or 0.8 V, and in some cases the assumption of an ideal diode is probably not a bad one. On the other hand, if the signals involved are reasonably small and if the application demands accuracy, we should either expend effort in accurately analyzing the circuit or, better yet, expend money in designing a more elaborate circuit, one in which the exact diode characteristics do not matter. As an illustration of the need for this latter alternative, consider the case of rectifying a signal only 100 mV in amplitude. Obviously our simple rectifier circuit will not do. As usual, the ubiquitous op amp will come to the rescue.

5.2 PRECISION HALF-WAVE RECTIFIER—THE "SUPERDIODE"

Figure 5.3a shows a precision half-wave-rectifier circuit consisting of a diode placed in the negative-feedback path of an op amp, with R being the rectifier load resistance. The circuit works as follows: If v_I goes positive, the output voltage v_A of the op amp will go positive and the diode will conduct, thus establishing a closed feedback path between the op-amp output terminal and the negative input terminal. This negative-feedback path will cause the op amp to operate "normally," in the manner we studied in Chapter 3, and a virtual short circuit appears between the two input terminals. Thus the voltage at the negative input terminal, which is also the output voltage v_O, will equal (to within a few millivolts) that at the positive input terminal, which is the input voltage v_I,

$$v_O = v_I, \qquad v_I \geq 0$$

Note that the offset voltage ($\simeq 0.5$ V) exhibited in the simple circuit of Fig. 5.1 is no longer present. For the op-amp circuit to start operation v_I has to exceed only a negli-

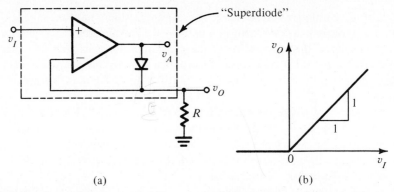

Fig. 5.3   The "superdiode" precision half-wave rectifier and its almost ideal transfer characteristic. Note that when $v_I > 0$ and the diode conducts, the op amp supplies the load current and the source is conveniently buffered, an added advantage.

gibly small voltage equal to the diode drop divided by the op-amp open-loop gain. In other words, the straight-line transfer characteristic $v_O = v_I$ almost passes through the origin. This makes this circuit suitable for applications involving very small signals.

Consider now the case when v_I goes negative. The op-amp output voltage v_A will tend to follow and go negative. This will reverse-bias the diode, and no current will flow through resistance R, causing v_O to remain equal to 0 V. Thus for $v_I < 0$, $v_O = 0$. The transfer characteristic of this circuit will be that shown in Fig. 5.3b, which is almost identical to the ideal characteristic of a half-wave rectifier. The nonideal diode characteristics have been almost completely masked by placing it in the negative-feedback path of an op amp. This is another dramatic application of negative feedback. The combination of diode and op amp, shown in the dotted box in Fig. 5.3a, is appropriately referred to as a "superdiode."

As usual, though, not all is well. The circuit of Fig. 5.3 has one disadvantage. When v_I goes negative and $v_O = 0$, the entire magnitude of v_I appears between the two input terminals of the op amp. If this magnitude is greater than few volts, the op amp may be damaged unless it is equipped with what is called "overvoltage protection" (a feature that most modern IC op amps have). Another disadvantage is that when v_I is negative the op amp will be saturated and its output will be equal to a value near the negative power supply level. Although not harmful to the op amp, saturation should usually be avoided, since getting the op amp out of the saturation region and back into its linear region of operation requires a wait of some time. This time delay will obviously slow down circuit operation and limit the frequency range of operation of the half-wave-rectifier circuit.

An Alternative Circuit

An alternative precision rectifier circuit that does not suffer from the disadvantages mentioned above is shown in Fig. 5.4. The circuit operates in the following manner: For positive v_I, diode D_2 conducts and closes the negative-feedback loop around the op amp. A virtual ground therefore will appear at the inverting input terminal, and the op-amp

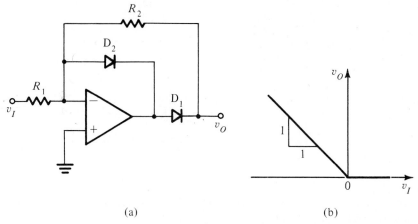

(a)   (b)

Fig. 5.4 *An improved version of the precision half-wave rectifier. Here diode D_2 is included to keep the feedback loop closed around the op amp during the off times of the rectifier diode D_1, thus preventing the op amp from saturating. For $R_2 = R_1$, the transfer characteristic is shown in (b).*

output will be *clamped* at one diode drop below ground. This negative voltage will keep diode D_1 off, and no current will flow in the feedback resistance R_2. It follows that the rectifier output voltage will be zero.

As v_I goes negative, the voltage at the inverting input terminal will tend to go negative, causing the voltage at the op-amp output terminal to go positive. This will cause D_2 to be reverse-biased and hence cut off. Diode D_1, however, will conduct through R_2, thus establishing a negative-feedback path around the op amp and forcing a virtual ground to appear at the inverting input terminal. The current through the feedback resistance R_2 will be equal to the current through the input resistance R_1. Thus for $R_1 = R_2$ the output voltage v_O will be

$$v_O = -v_I, \qquad v_I \le 0$$

The transfer characteristic of the circuit is shown in Fig. 5.4b. Note that unlike the previous circuit, here the slope of the characteristic can be set to any desired value, including unity, by selecting appropriate values for R_1 and R_2.

As mentioned before, the major advantage of this circuit is that the feedback loop around the op amp remains closed at all times. Hence the op amp remains in its linear operating region, avoiding the possibility of saturation and the associated time delay required to "get out" of saturation.

An Application: Measuring AC Voltages

As one of the many possible applications of the precision rectifier circuits discussed in this section, consider the basic ac voltmeter circuit shown in Fig. 5.5. The circuit consists of a half-wave rectifier—formed by op amp A_1, diodes D_1 and D_2, and resistors R_1 and R_2—and a first-order low-pass filter—formed by op amp A_2, resistors R_3 and R_4, and capacitor C. For an input sinusoid having a peak amplitude V_p the output v_1 of the rectifier will consist of half sine waves having peak amplitudes of $V_p R_2 / R_1$. It can be

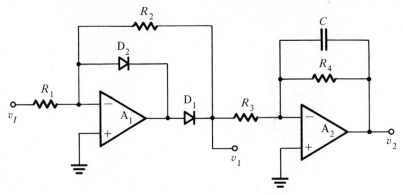

Fig. 5.5 *A simple ac voltmeter consisting of a precision half-wave rectifier followed by a first-order low-pass filter.*

shown using Fourier series analysis that the waveform of v_1 has an average value of $(V_p/\pi)(R_2/R_1)$ in addition to harmonics of the frequency ω of the input signal. To reduce the amplitudes of all of these harmonics to negligible levels, the corner frequency of the low-pass filter should be chosen much smaller than the lowest expected frequency ω_{min} of the input sine wave. This leads to

$$\frac{1}{CR_4} \ll \omega_{min}$$

Then the output voltage v_2 will be mostly dc, with a value

$$V_2 = -\frac{V_p}{\pi}\frac{R_2}{R_1}\frac{R_4}{R_3}$$

where (R_4/R_3) is the dc gain of the low-pass filter. Note that this voltmeter essentially measures the average value of the negative parts of the input signal but can be calibrated to provide root-mean-square (RMS) readings for input sinusoids.

EXERCISES

5.1 If the diode in the circuit of Fig. 5.3a is reversed, find the transfer characteristic v_O as a function of v_I.
    Ans. $v_O = 0$ for $v_I \geq 0$; $v_O = v_I$ for $v_I \leq 0$

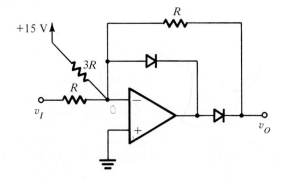

Fig. E5.3

5.2   If the diodes in the circuit of Fig. 5.4a are reversed, find the transfer characteristic v_O as a function of v_I.

  Ans. $v_O = -v_I$ for $v_I \geq 0$; $v_O = 0$ for $v_I \leq 0$

5.3   Find the transfer characteristic for the circuit in Fig. E5.3.

  Ans. $v_O = 0$ for $v_I \geq -5$ V; $v_O = -v_I - 5$ for $v_I \leq -5$ V

5.3   FULL-WAVE RECTIFICATION

There are many applications in instrumentation (and in the design of power supplies) where the information (or the energy) provided by both halves of an ac signal is either useful or necessary. This need can be met by arranging that both the original signal and its inverse (or negative version) be made available and supplying each to a half-wave rectifier joined in turn to a common load. Such an arrangement, shown conceptually in Fig. 5.6, is called a *full-wave rectifier*. Here, when the voltage at A is positive, that at B is negative, causing D_A to conduct while D_B remains cut off. When the voltage at A goes negative, that at B goes positive and the diode roles interchange: D_B conducts while D_A is cut off. The output voltage resulting appears as a simple combination of alternate half-cycles of the input waveform, the negative halves having been inverted. The resulting waveforms, assuming ideal diodes, are depicted in Fig. 5.6. Note that these waveforms are more continuous and energetic than the corresponding half-wave ones. Also note that the average voltage at the output will be twice that obtained from a half-wave rectifier.

In practice there are many possible implementations of the full-wave rectifier. In the remainder of this section we develop implementations utilizing op amps. Circuits utilizing transformers for obtaining signal inversion will be presented in Section 5.4. Yet another alternative, the diode bridge rectifier, is presented in Section 5.5.

Amplifier-Coupled Full-Wave Rectifiers

Figure 5.7a shows a full-wave rectifier in which op amp A_1 and two equal resistors labeled R_1 and R_2 are used to implement the unity-gain inverting amplifier that provides the antiphase input to the second rectifier. Since for some applications an asymmetry exists owing to second-order effects related to the amplifier, including its dynamics and its output resistance, it may be useful to introduce a unity-gain follower in the direct

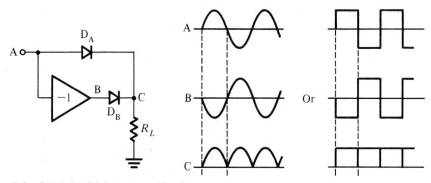

Fig. 5.6   Principle of full-wave rectification.

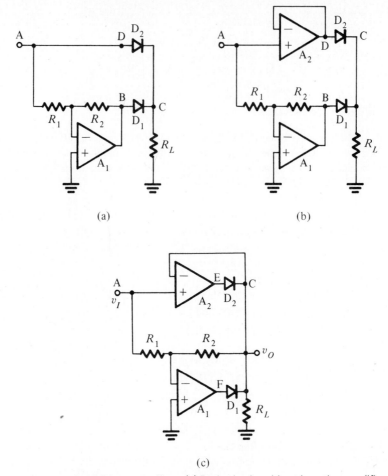

(a) (b)

(c)

Fig. 5.7 Amplifier-coupled full-wave rectifiers. (a) Basic circuit, with an inverting amplifier used to provide the required inversion. (b) Same as (a) but with a buffer included in the in-phase path. (c) Same as (b) but with the diodes placed in the feedback loops.

path, as shown in Fig. 5.7b. An incidental result that accrues and is particularly useful in instrumentation applications is that the input is not required to supply the output current directly; the output is said to be buffered. Note that if asymmetry is not a problem, then the input resistance can be made even higher by connecting the input resistor R_1 of the inverting amplifier to the output of the follower rather than directly to the input.

But there are other useful and more interesting results to be obtained by simple reconnections of the two amplifiers. Such a possibility is shown in Fig. 5.7c. An examination of this circuit reveals that it differs from its predecessor in Fig. 5.7b only in that diode D_1 has been placed in the negative-feedback loop of op amp A_1 and diode D_2 has been placed in the negative-feedback loop of op amp A_2. A further examination of this circuit reveals that it consists of two parallel circuits; the upper one, consisting of A_2

and D_2, may be recognized as the "superdiode" discussed in Section 5.2 that gives rise to the transfer characteristic shown in Fig. 5.8a. The lower circuit, composed of A_1, D_1, and the two resistors R_1 and R_2, may be recognized as the half-wave rectifier of Fig. 5.4a but without the catching diode. This lower circuit gives rise to the transfer characteristics shown in Fig. 5.8b. Connecting these two circuits in parallel implies combin-

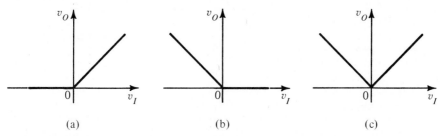

(a)   (b)   (c)

Fig. 5.8 *Development of the transfer characteristic of the circuit in Fig. 5.7c.*

ing their transfer characteristics to yield the characteristic shown in Fig. 5.8c, which may be recognized as that of a full-wave rectifier or, equivalently, an *absolute-value circuit.*

A word of caution is in order. In connecting two circuits in parallel one should be careful that nothing unusual results. For this reason it is prudent to reexamine the circuit of Fig. 5.7c to make sure that it indeed functions in the assumed manner. Toward that end, consider first the case where the input at A is positive. The output of A_2 will go positive, turning D_2 on, which will conduct through R_L and thus close the feedback loop around A_2. A virtual short circuit will thus be established between the two input terminals of A_2, and the voltage at the negative input terminal, which is the output voltage of the circuit, will become equal to the input. This will in turn cause the voltage at the inverting input of A_1 to become equal to the input and hence positive. Therefore the output terminal (F) of A_1 will go negative until A_1 saturates. This causes D_1 to be turned off.

Next consider the case when A goes negative. The tendency for a negative voltage at the negative input of A_1 causes F to rise, making D_1 conduct to supply R_L and allowing the feedback loop around A_1 to be closed. Thus a virtual ground appears at the negative input of A_1 and the two equal resistances R_1 and R_2 force the voltage at C, which is the output voltage, to be equal to the negative of the input voltage at A and thus positive. The negative voltage at A causes the output of A_2 to saturate in the negative direction, thus (because C is positive) keeping D_2 off.

The overall result is perfect full-wave rectification. This precision is, of course, a result of placing the diodes in op-amp feedback loops, thus masking their nonidealities. This circuit is one of many possible precision full-wave-rectifier or absolute-value circuits. Another related implementation of this function is examined in Exercise 5.4.

EXERCISE

5.4 The block diagram shown in Fig. E5.4a gives another possible arrangement for implementing the absolute-value or full-wave-rectifier operation depicted symbolically in Fig.

E5.4b. As shown, the block diagram consists of two boxes: a half-wave rectifier that can be implemented by the circuit in Fig. 5.4a after reversing both diodes, and a weighted inverting summer. Convince yourself that this block diagram does in fact realize the absolute-value operation. Then draw a complete circuit diagram, giving reasonable values for all resistors.

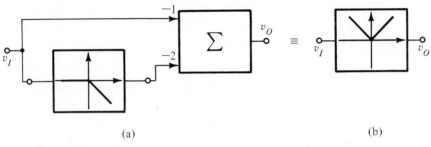

(a)   (b)

Fig. E5.4

5.4   TRANSFORMER COUPLING OF RECTIFIERS

Figure 5.9 illustrates the use of a transformer in generating the complementary signals required for full-wave rectification as well as *isolating* the input and output circuits. Polarity marks on the windings indicate the direction of coupling, that is the fact that

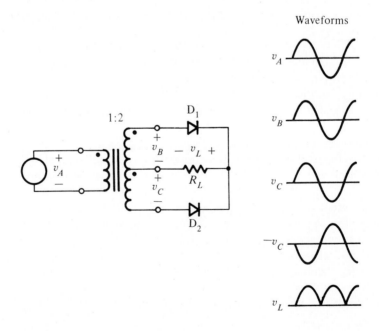

Waveforms

Fig. 5.9   *Use of a center-tapped transformer to generate the complementary signals required for full-wave rectification.*

all marked terminals will be polarized (relative to the corresponding unmarked terminals) in the same way at any moment. As Fig. 5.9 illustrates, the winding providing v_B and v_C is continuous with a connection, at its midpoint, called a *center tap*.

Waveforms accompanying Fig. 5.9 illustrate the principle directly. Taking the center tap as a conceptual reference (in view of the symmetry of the circuit around it) the anode of D_1 will be positive when the anode of D_2 is negative, and vice versa. Thus diodes D_1 and D_2 will conduct alternately to provide full-wave operation.

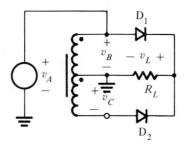

Fig. 5.10   *If isolation is not required, the transformer in the circuit of Fig. 5.9 can be replaced by a center-tapped coil.*

Note that the transformer provides complete isolation between the input v_A and the output v_L by virtue of magnetic flux coupling of otherwise separate windings. Isolation is a very important feature in power supply systems, particularly in the context of connection to the power line. It prevents undesirable current flow between circuits so separated. Thus its use in a consumer product such as a stereo receiver is very important to ensure that the user, while connecting loudspeakers, for example, be totally protected from the possibility of shock from power-line voltages.

Figure 5.10 illustrates, however, that an isolating transformer is not really essential for the full-wave function; only a center-tapped, well coupled coil is required.

EXERCISE

5.5   Consider the full-wave rectifier circuit of Fig. 5.9. If the input is a sinusoid with 100 V peak amplitude and $R_L = 1$ kΩ, calculate
   (a)   The peak current in each diode;
   (b)   The peak reverse voltage appearing on the off diode;
   (c)   The average voltage across the load;
   (d)   The peak amplitude of the sinusoidal current supplied by the source;
   (e)   The power supplied by the source. Assume ideal diodes.
   Ans: 100 mA; 200 V; 63.7 V; 100 mA; 5 W

5.5   THE BRIDGE RECTIFIER

While the center-tapped transformer provides both isolation and a relatively simple circuit implementation of a full-wave rectifier, the effort required to tap the transformer is relatively great in practice. Moreover, each half of the secondary winding is used only half the time. Thus to provide a load voltage of V volts peak, the secondary must be capable of providing $2V$ volts total.

However, there is an alternative: At the modest expense of two extra diodes (and their voltage drops) and under the condition that the source and load do not share an essential common connection, a solution exists in what is known as the *diode bridge configuration.* To emphasize the fact that the source and load should not have a common terminal, the circuit is first shown in Fig. 5.11a using a transformer. Note, however,

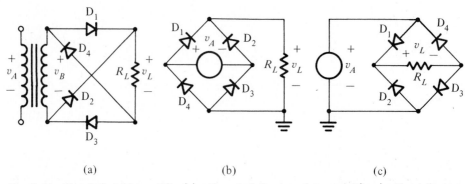

(a)   (b)   (c)

Fig. 5.11   The diode bridge rectifier (a) with an isolation transformer and (b, c) without the transformer. In (b) the source is floating and the load is grounded. In (c) the source is grounded and the load is floating.

that the secondary of this transformer is quite simple, having no tap and a voltage only slightly greater (to account for the one additional diode drop) than half that of the secondary for a simple full-wave design. It should be obvious that one cannot have a common ground between the secondary winding and the load resistance, since this would result in shorting one of the two diodes D_1 and D_3. Of course with the transformer included such a common ground can exist between the source (primary winding) and the load (R_L).

Let us examine the operation: When v_B is positive, current will flow through D_1, R_L, and D_3, while at the same moment D_2 and D_4 will be reverse-biased. When v_B reverses polarity and becomes negative, conduction will result through D_2, R_L, and D_4, with D_1 and D_3 reversed. Note that the current in R_L is unidirectional, supplied by either D_1 or D_2 and removed by either D_3 or D_4.

It is useful to reflect that there are many ways of looking at, and of drawing, the diode bridge. Two of these, emphasizing the traditional bridge shape, are shown in Figs. 5.11b and c with component labels corresponding to those in Fig. 5.11a. For simplicity, and to imply other possibilities, the transformer has been eliminated and the source and load are, of course, assumed to share no common terminals. Figure 5.11b is most often used when the source is truly floating, isolated, and symmetrical, in which case the lower end of R_L may be grounded. Figure 5.11c is used when the source is asymmetric and ground referenced while the load is floating. A classic example of the latter case is when the load is a dc meter movement and the source is an ac signal in a ground-referenced signal-processing or instrumentation system.

A Precision Bridge Rectifier for Instrumentation Applications

From Fig. 5.11c it should be apparent that the load current comes directly from, and is replicated in, the ac signal source. Consider next the circuit of Fig. 5.12a. Here the bridge load is assumed to be a moving-coil meter M, and we have placed a unity-gain follower at the input. This follower will provide the circuit with a high input resistance while causing a replica of the input signal to appear at a low impedance level at point C. Now comes the important point: In order to control the meter sensitivity we have included a resistance R in series with the bridge. We could, of course, have placed this resistance directly in series with meter M with identical results. However, placing it in series with the bridge emphasizes its role as being in effect a voltage-to-current converter. For a given value of the input voltage v_A the value of R, the resistance of the meter, and the diode voltage drops determine the average meter current and hence the meter reading.

To eliminate the effects of the usually unpredictable meter resistance and diode characteristics as well as to make the circuit work for small voltages (less than the minimum of 1 V required in the circuit of Fig. 5.12a) a simple reconnection can be made. This is illustrated in Fig. 5.12b, where all that we have done is simply to place the diode bridge inside the negative-feedback loop of the op amp. If we assume that this feedback loop will be closed for both input polarities (which it will, as will be seen shortly), a virtual short circuit appears between the two input terminals of the op amp. This causes an almost perfect replica of the input voltage v_A to appear across resistance

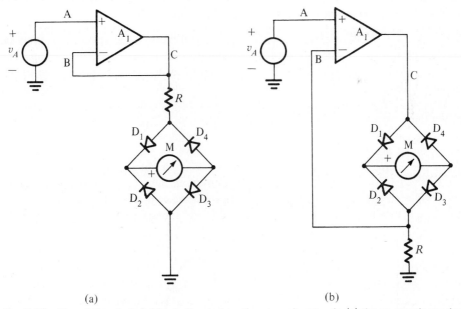

(a)   (b)

Fig. 5.12   Use of the diode bridge in the design of an ac voltmeter. In (a) the op amp is used as a buffer, and the resistance R determines the meter sensitivity. In (b) the bridge is included in the feedback loop, resulting in a precision instrument.

R. Thus the current through R will be alternating and directly proportional to the input voltage v_A (assuming an ideal op amp),

$$i_R = \frac{v_A}{R}$$

Now the positive half-cycles of this current will be coming from the op-amp output through diode D_1, meter M, and diode D_3. On the other hand, the negative half-cycles will be going into the op-amp output through diode D_4, the meter, and diode D_2. In both half-cycles the current in meter M will flow in the same direction.

The circuit in Fig. 5.12b thus provides a relatively accurate high-input-impedance ac voltmeter using an inexpensive moving-coil meter. Again through the intelligent use of negative feedback, all nonidealities have been masked.

EXERCISES

5.6   In the circuit of Fig. 5.12b find the value of R that would cause the meter to provide a full-scale reading when the input voltage is a sine wave of 5 V RMS. Let meter M have a 1-mA, 50-Ω movement (that is, its resistance is 50 Ω and it provides full-scale deflection when the average current through it is 1 mA). What are the approximate maximum and minimum voltages at the op-amp output?
Ans.   4.5 kΩ; $+8.52$ V; -8.52 V

5.7   Consider the full-wave bridge rectifier of Fig. 5.11c. Assume that the input is a sine wave with 100 V peak and that the diodes are ideal. What is the maximum magnitude of reverse bias voltage that appears across the off diodes?
Ans.   100 V

5.6   THE RECTIFIER IN POWER SUPPLIES—FILTERING

While a simple rectifier circuit, whether half or full wave, is capable of supplying dc current or voltage, it does so with the addition of variable or ac components. For instance, an ideal half-wave rectifier fed by a sinusoidal waveform produces the output waveform shown in Fig. 5.13. This waveform can be decomposed into the sum of various components—a dc one, one at the fundamental frequency of the input sine wave, one at the third harmonic of the input sinusoid, and so on.

In a measurement application using a moving-coil meter, the ac components have no net effect provided that their frequency lies above the natural resonance frequency of the mechanical meter because of the fact that when the frequency is sufficiently high the needle cannot follow the pulse-to-pulse variation, and the ac components are in effect "filtered out" by mechanical inertia. In other measurement applications one may require a constant output voltage proportional to the average or dc component. In such

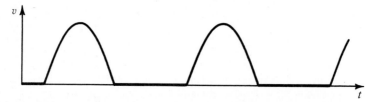

Fig. 5.13   The output waveform of an ideal half-wave rectifier fed by a sinusoid.

a case the ac components may be removed by an electronic filter, as we have done in the circuit of Fig. 5.5. Both of these filtering processes fall into the class of linear filtering. We will deal with the subject of linear signal filtering in greater detail in Chapter 14.

In power-supply applications also one desires a pure dc output from the rectifier; hence some form of filtering is needed to remove the ac components. In this section we introduce an alternative approach to filtering that is useful in power-supply and other applications. As will be seen, the scheme is quite distinct from the linear filter circuits just mentioned. The process makes use of the interaction of the basic nonlinear element (diode) and the energy-storage element (capacitor). It will also be seen that from the point of view of power dissipation this is a much more efficient filter and hence is ideally suited to power-supply applications. Since the dc output voltage is almost equal to the peak of the input waveform the circuit is called a *peak rectifier*.

The Peak Rectifier

Figure 5.14a shows the basic peak rectifier circuit. Let the input signal be the triangular waveform shown in Fig. 5.14b and let the capacitor be initially discharged. To start with, let us also assume the diode to be ideal. As the input voltage increases above zero (at time t_0) the diode conducts and the capacitor charges up to the instantaneous value of the input voltage. In fact, if the diode is ideal, the voltage across the capacitor will be equal to the input voltage. This will continue until the time the positive peak is reached, time t_1. At $t = t_1$ the capacitor voltage will be equal to the peak of the input, V_p. Now at $t > t_1$ the input decreases, thus reverse-biasing the diode. The capacitor voltage therefore remains constant at the peak value V_p unless the value of the positive

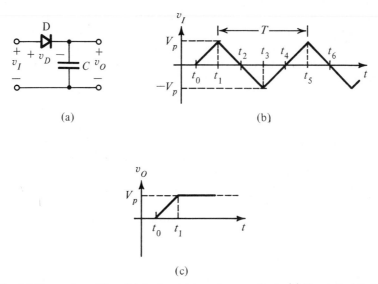

(a)

(b)

(c)

Fig. 5.14   (a) The peak rectifier. (b) An input triangular waveform. (c) The output that would be obtained if the diode were ideal.

peak of the input signal increases. If this happens the diode conducts for the time interval during which $v_I > V_p$ and the capacitor charges up to the new peak value.

From the above description we see that this circuit detects the positive peak of the input signal and provides an output dc voltage equal to the value of this peak. For this reason the circuit is called a *peak rectifier* or a *peak detector*. It now remains to consider the effect on circuit operation of the characteristics of real diodes and of connecting a load resistance R across the capacitor.

The Nonideal Case With No Load

Although it is not of immense practical interest, it is instructive to examine the operation of the peak-rectifier circuit taking nonideal diode characteristics into account. For simplicity we will still assume that the input signal is of triangular waveform as shown in Fig. 5.14b. At the outset we note that one can write equations to accurately describe the circuit operation and solve for the output voltage waveform and for the diode current (Problem 5.13). For the time being, however, we wish to qualitatively describe circuit operation.

Assume that the capacitor was initially discharged, and consider the interval t_0 to t_1. As the input rises the diode begins to conduct with a voltage drop related to current flow. After an initial transient during which the diode voltage exceeds 0.5 or 0.6 V (that is, as the diode current becomes much larger than I_S) it is easy to show that the output will be rising linearly with a slope equal to that of the input. To see that this indeed is the case, we will assume that it is true and then verify our assumption. If the output is linearly increasing with time, then the current

$$i = C \frac{dv_O}{dt}$$

will be constant. This means that the diode will have a constant voltage drop v_D. Now since $v_O = v_I - v_D$, it follows that v_O will be increasing linearly with time with the same slope as that of v_I. The picture should be now clear: after an initial transient the diode conducts a constant current and exhibits a constant voltage drop. The output is identical to the input except that it is lower by a diode drop. The value of the diode current can be easily estimated as

$$i = C \frac{dv_O}{dt} = C \frac{dv_I}{dt} = C \frac{V_p}{T/4}$$

The corresponding diode voltage can be found from the diode characteristics.

As a numerical example, consider $V_p = 10$ V, $T = 40$ ms, $C = 10$ μF. It follows that $i = 10$ mA and $v_D = 0.8$ V. Thus at the time of the peak ($t = t_1$) the output voltage will be $10 - 0.8 = 9.2$ V.

Let us now consider what happens beyond the positive peak of the input. As the input voltage reverses slope and begins to decrease, the voltage drop across the diode will begin to decrease, causing the diode current to decrease. Nevertheless, current continues to flow beyond the peak point and thus the capacitor voltage continues to increase as shown in Fig. 5.15. Shortly after the peak, however, the diode current will decrease to zero (that is, the diode cuts off), and the output voltage remains constant at a value

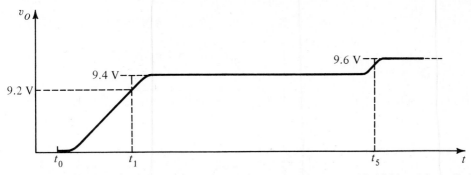

Fig. 5.15   Details of the output voltage waveform as a triangular wave with 10 V peak is applied to the peak rectifier and the diode characteristics are taken into account.

slightly greater than 9.2 V (say, 9.3 or 9.4 V). The exact value reached by the output voltage can be found by solving the circuit equations using the exponential i-v model of the diode.

Now during the interval t_1 to nearly t_5 the output remains constant at, say, 9.4 V, since the diode is off and no current whatsoever can flow. Then when the input reaches 9.4 V again the diode enters the forward-bias region, with the ultimate possibility of having as much as 0.6 V of forward bias. The result is that small currents will flow and the capacitor will charge up slightly more (say, to 9.6 V) before the diode is once again reversed. It is probably obvious that after a few cycles of this process the output will reach nearly 10 V. Thus in the steady state that is reached, the diode does not conduct and the output remains constant (pure dc) at a value equal to the peak of the input waveform.

The Peak Rectifier With Load

The no-load situation considered above is an idealization that cannot exist in real life. Practically speaking, because of leakage effects the capacitor will always discharge slightly even if no actual load resistance is connected across it. Furthermore, in actual practical situations, such as the design of a power supply, a load resistance will be inherently present. In other applications of the peak rectifier, a load resistance may be essential in order to allow the capacitor to discharge in the event that the input positive peak decreases in value. Therefore we will now consider the operation of the peak rectifier with a load resistance R connected across the capacitor, as shown in Fig. 5.16a.

At the very peak of the input, while the diode conducts, the load current, which is approximately equal to V_p/R, will be supported directly by the input. However, between peaks—for example, in the interval t_1 to t_5—charge from the capacitor must supply the load current. This load current in turn will discharge the capacitor, lowering the output voltage in an exponential fashion (with time constant CR) for the greater part of the interval t_1 to t_5. At the point where the falling capacitor voltage intersects the rising edge of the input, the diode begins to conduct once more. Again conduction continues until a point just past the peak of the input. In a steady state the diode conducts for a short interval near the peak to replenish the charge lost by the capacitor during the

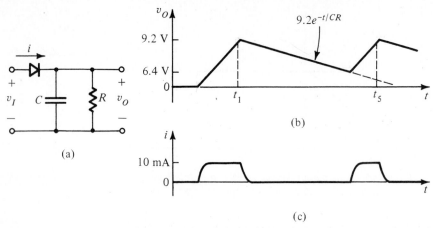

Fig. 5.16   (a) The peak rectifier with a load resistance. (b, c) The output voltage and diode current for the case $C = 10 \, \mu F$, $R = 10 \, k\Omega$, and the input a triangular wave with 10 V peak and 40 ms period.

interval in which the diode is off (most of the peak-to-peak interval). Figures 5.16b and c show the waveforms of the output voltage and diode current for the same numerical example considered before but with a load resistance $R = 10 \, k\Omega$. Note that the output is no longer a dc voltage but may be considered as composed of a dc voltage on which an ac component, called *ripple,* is superimposed. The dc voltage can be considered as the average of the two extreme values of the output voltage. In the numerical example illustrated in Fig. 5.16 the output voltage changes from 6.4 V to 9.2 V. Thus the dc component is approximately $\frac{1}{2}(6.4 + 9.2) = 7.8$ V. The ac component has a peak-to-peak amplitude of $9.2 - 6.4 = 2.8$ V and a frequency equal to that of the input signal. In conclusion, we see that the effect of adding a load resistance to the peak rectifier is to cause the output to have an ac or ripple component, with the ripple amplitude increasing as the load resistance is decreased.

Calculation of the Ripple Voltage

Consider a half-wave peak rectifier fed by a sine-wave signal with a peak voltage V_p. Let the rectifier have a capacitance C and a load resistance R. We wish to find an approximate expression for the peak-to-peak ripple voltage V_r. For this purpose we shall assume that the diode is ideal. Furthermore, we will consider only the practical situation where the time constant CR is sufficiently large relative to the period T such that the capacitor discharges almost linearly. This approximation is equivalent to assuming that the ripple voltage is much smaller than the peak voltage V_p—hence the name *low-ripple approximation.*

Under the above assumption the output waveform in the steady state takes the shape shown in Fig. 5.17. Note that the discharge interval (the diode cutoff interval) is almost equal to the period T, while the charging interval (the diode conduction interval) is a small fraction of T. In steady state the charge lost by the capacitor during the off interval is equal to the charge supplied to it during the on interval. Consider first the

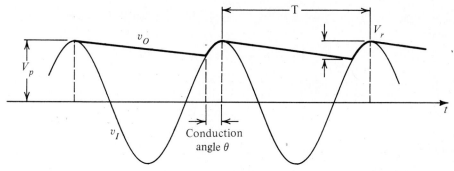

Fig. 5.17   *Input and output waveforms for the calculation of the ripple voltage V, at the output of the peak rectifier.*

off interval. The capacitor voltage remains almost constant and close to the peak value V_p, resulting in a constant discharge current equal to V_p/R. This current flows for an interval almost equal to T, with the result that the charge lost by the capacitor is given by

$$Q_{\text{lost}} \simeq \frac{V_p}{R} T$$

Next consider the charging interval. During this diode conduction time the capacitor voltage increases by V_r volts. Thus it acquires a charge of

$$Q_{\text{acquired}} = CV_r$$

Equating the two charge amounts results in the following simple expression for the peak-to-peak ripple voltage V_r:

$$V_r = \frac{V_p T}{CR} \tag{5.1}$$

which can be expressed alternatively in the form

$$V_r = \frac{V_p}{fCR} \tag{5.2}$$

where f is the frequency of the input sine wave.

Example 5.1
Consider a peak rectifier fed by a 60-Hz sinusoid having a peak value $V_p = 100$ V. Let the load resistance $R = 10$ kΩ. Find the value of the capacitance C that will result in a peak-to-peak ripple of 2 V. Also calculate the fraction of the cycle during which the diode is conducting and hence estimate the average value of diode current.

Solution
From Eq. (5.2) we obtain the value of C as

$$C = \frac{V_p}{V_r fR} = \frac{100}{2 \times 60 \times 10 \times 10^3} = 83.3 \ \mu\text{F}$$

By reference to Fig. 5.17 we see that the angle θ during which the diode conducts is defined by

$$V_p \cos \theta = V_p - V_r \tag{5.3}$$

Since θ is small, we may approximate $\cos \theta$ by

$$\cos \theta \simeq 1 - \frac{\theta^2}{2}$$

Substituting in Eq. (5.3) yields

$$1 - \frac{\theta^2}{2} \simeq 1 - \frac{V_r}{V_p}$$

Thus

$$\theta = \sqrt{\frac{2V_r}{V_p}}$$

For our case

$$\theta = \sqrt{\frac{2 \times 2}{100}} = 0.2 \text{ rad}$$

Thus the diode will be conducting for $(0.2/2\pi) \times 100 = 3.18\%$ of the cycle. Since during the off period the current supplied is $100 \text{ V}/10 \text{ k}\Omega = 10$ mA, it follows that (for charge equality) the average diode current in the conduction interval will be

$$I_D \simeq \frac{10}{0.0318} = 0.314 \text{ A}$$

. . .

From Example 5.1 we note that although increasing the filter capacitance C results in a proportional decrease in ripple, the price paid is an increase (again proportional) of diode current.

Full-Wave Peak Rectifiers

The full-wave rectifier circuits of Figs. 5.9, 5.10, and 5.11 can be converted to peak rectifiers by including a capacitor across the load resistor. As in the half-wave case the output dc voltage will be almost equal to the peak value of the input sine wave. The ripple frequency, however, will be twice that of the input. The peak-to-peak ripple voltage, for this case, can be derived using a procedure identical to that above but with the discharge period T replaced by $T/2$, resulting in

$$V_r = \frac{V_p}{2fCR}$$

Precision Peak Rectifiers

Including the diode of the peak rectifier inside the negative-feedback loop of an op amp, as shown in Fig. 5.18, results in a precision peak rectifier. The diode–op-amp combination will be recognized as the superdiode presented in Section 5.2 (Fig. 5.3). Operation of the circuit in Fig. 5.18 is quite straightforward. For v_I greater than the output voltage the op amp will drive the diode on, thus closing the negative-feedback path and

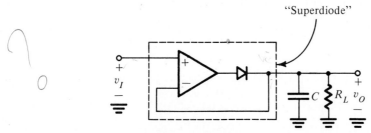

Fig. 5.18   *A precision peak rectifier obtained by placing the diode in the feedback loop of an op amp.*

causing the op amp to act as a follower. The output voltage will therefore follow that of the input, with the op amp supplying the capacitor charging current. This process continues until the input reaches its peak value. Beyond the positive peak the op amp will see a negative input voltage. Thus its output will go negative to the saturation level and the diode will turn off. Except for possible discharge through the load resistance the capacitor will retain a voltage equal to the positive peak of the input. Inclusion of a load resistance is essential if the circuit is required to detect reductions in the magnitude of the positive peak, as in the case of the amplitude-modulation (AM) detector discussed below.

A Buffered Precision Peak Detector

When the peak detector is required to hold the value of the peak for long times, the capacitor should be buffered, as shown in the circuit of Fig. 5.19. Here op amp A_2,

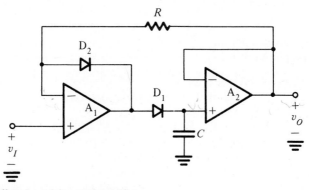

Fig. 5.19   *A buffered precision peak rectifier.*

which should have high input impedance and low input bias current, is connected as a voltage follower. The remainder of the circuit is quite similar to the half-wave-rectifier circuit of Figure 5.3. While diode D_1 is the essential diode for the peak rectification operation, diode D_2 acts as a catching diode to prevent negative saturation, and the associated delays, of op amp A_1. During the holding state, follower A_2 supplies D_2 with current through R. The output of op amp A_1 will then be clamped at one diode drop below the input voltage.

Finally, it should be noted that this circuit provides its output at a convenient low-impedance point.

EXERCISES

5.8   Consider a bridge rectifier with a capacitance C connected across the the load resistance R. If the input is a 60-Hz sine wave with 100 V peak and $R = 10$ kΩ, find the value of C such that the peak-to-peak ripple voltage is limited to 2 V.
   Ans.   41.7 μF

5.9   In the circuit of Fig. 5.18 let $C = 1$ μF and $R_L = \infty$ and assume that the op amp is specified to have a slew rate of 0.1 V/μs and a maximum output current of 10 mA. What is the maximum rate of change of output voltage?
   Ans.   10 V/ms

The Peak Detector as an AM Demodulator

The peak rectifier finds a great many applications, beyond the simple supply of power, in the areas of signal detection and processing. For example, the information contained in an amplitude-modulated (AM) signal can be extracted or *detected* using a diode peak rectifier. This is possible because of the ability of the diode peak rectifier to follow the peak value of the input signal from cycle to cycle. For this to happen, however, the time constant CR must be carefully chosen.

Figure 5.20 illustrates the process. Note that the output will consist of the modulating signal with some superimposed ripple at the carrier frequency. This high-frequency ripple can be removed by following the AM detector with a simple RC filter.

Obviously we have to select the detector time constant CR sufficiently small to be able to track every peak of the input, even at the steepest point on the envelope of the AM wave. On the other hand, selecting CR too small would result in excessive amount of ripple on the detected output.

Let us consider the question of selecting an appropriate value for the detector time constant quantitatively. The AM waveform can be described by (see Chapter 1)

$$v_{AM} = V(1 + m \sin \omega_m t) \sin \omega_c t$$

where $V(1 + m \sin \omega_m t)$ represents the envelope. This envelope has a slope of $m\omega_m V \cos \omega_m t$, which has a maximum magnitude of $m\omega_m V$. This occurs at point A in Fig. 5.20a, where the value of the AM waveform is V. Now note that this envelope slope is proportional to the modulation depth m and the modulating frequency ω_m. Thus the highest possible slope occurs when $m = 1$ and $\omega_m = \hat{\omega}_m$ (where the caret "^" denotes maximum). For detection to be perfect, as measured by the ability to follow the

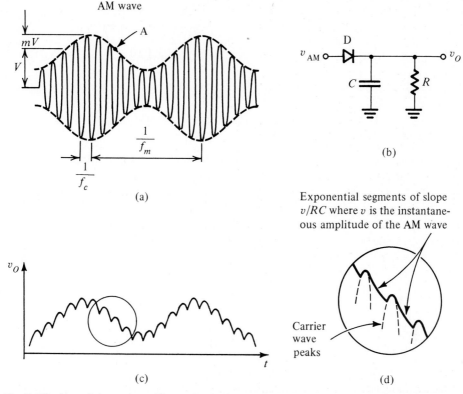

Fig. 5.20   Use of the peak rectifier to detect the envelope of an amplitude-modulated wave.

most rapid variation of the input presented, the slope of the detector decay curve must exceed this. That is,

$$\frac{V}{CR} > \hat{\omega}_m V$$

which results in

$$CR < \frac{1}{\hat{\omega}_m}$$

Example 5.2

The highest-frequency signal components transmitted on the AM broadcast band are limited to 5 kHz. Design an AM detector to supply a resistive load of 10 kΩ, assuming that an ideal diode is available.

Solution

The appropriate circuit is the one shown in Fig. 5.20b, where D is an ideal diode and $R = 10$ kΩ. The challenge is to choose a value for the capacitance C. Clearly this follows directly from the previous result: at the limit,

$$C = \frac{1}{\hat{\omega}_m R}$$

$$= \frac{1}{2\pi \times 10^4 \times 5 \times 10^3} = 3,200 \text{ pF}$$

If the receiver cost and size are important, a somewhat smaller capacitor may be used.

· · ·

5.7   THE CLAMPED CAPACITOR OR DC RESTORER

If in the basic peak rectifier circuit the output is taken across the diode rather than across the capacitor, an interesting circuit with important applications results. The circuit, called a dc restorer, is shown in Fig. 5.21 fed with a square wave. Because of the polarity in which the diode is connected, the capacitor will charge to a voltage v_C (see Fig. 5.21) equal to the magnitude of the most negative peak of the input signal. Subsequently, the diode turns off and the capacitor retains its voltage indefinitely. If, for

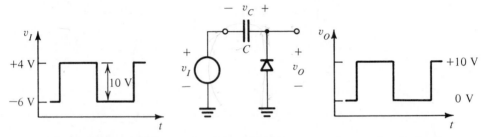

Fig. 5.21   The clamped capacitor or dc restorer with a square-wave input and no load.

instance, the input square wave has the arbitrary levels -6 V and $+4$ V, then v_C will be equal to 6 V. Now, since the output voltage v_O is given by

$$v_O = v_I + v_C$$

it follows that the output waveform will be identical to that of the input except shifted upwards by v_C volts. In our example the output will be a square wave with levels of 0 V and $+10$ V.

Another way of visualizing the operation of the circuit in Fig. 5.21 is to note that because the diode is connected across the output with the polarity shown, it prevents the output voltage from going below 0 V (by conducting and charging up the capacitor, thus causing the output to rise to 0 V), but this connection will not constrain the positive excursion of v_O. The output waveform will therefore have its lowest peak *clamped* to 0 V, which is why the circuit is called a *clamped capacitor*. It should be obvious that reversing the diode polarity will provide an output waveform whose highest peak is clamped to 0 V. In either case the output waveform will have a finite average value or dc component. This dc component is entirely unrelated to the average value of the input waveform. As an application, consider a pulse signal being transmitted through a capacitively or ac-coupled system. The capacitive coupling will cause the pulse train to lose whatever dc component it originally had. Feeding the resulting pulse waveform to a clamping circuit provides it with a well-determined dc component; a process known as *dc restoration*. This is why the circuit is also called a *dc restorer*.

Restoring dc is useful because the dc component of a pulse waveform is an effective measure of its duty cycle. The duty cycle of a pulse waveform can be modulated (pulse-width modulation) and made to carry information. In such a system, detection or demodulation could be achieved simply by feeding the received pulse waveform to a dc restorer and then using a simple RC low-pass filter to separate the average of the output waveform from the superimposed pulses.

Effect of Load Resistance

When a load resistance R is connected across the diode in a clamping circuit, as shown in Fig. 5.22, the situation changes significantly. While the output is above ground, a net dc current must flow in R. Since at this time the diode is off, this current obviously

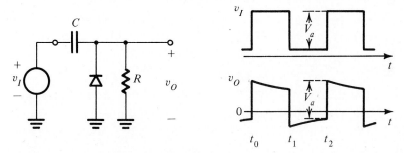

Fig. 5.22   The clamped capacitor with a load resistance R.

comes from the capacitor, thus causing the capacitor to discharge and the output voltage to fall. This is shown in Fig. 5.22 for a square-wave input. During the interval t_0 to t_1 the output voltage falls exponentially with time constant CR. At t_1 the input decreases by V_a volts and the output attempts to follow. This causes the diode to conduct heavily and to quickly charge the capacitor. At the end of the interval t_1 to t_2 the output voltage would normally be a few tenths of a volt negative (say, -0.5 V). Then as the input rises by V_a volts (at t_2) the output follows, and the cycle repeats itself. In a steady state the charge lost by the capacitor during the interval t_0 to t_1 is recovered during the interval t_1 to t_2. This charge equilibrium enables us to calculate the average diode current as well as the details of the output waveform, as illustrated in Example 5.3.

Example 5.3
Calculate the extremes of output voltage and of diode current in a clamped capacitor circuit utilizing a 10-μF capacitor, a 10-kΩ load, and a diode having a 0.7 V drop at 1 mA and representable by the logarithmic diode model. The circuit is driven by a 1-kHz train of (a) 10-V positive pulses of 10 μs duration, and (b) 10-V negative pulses of 10 μs duration.

Solution

The time constant of the capacitor and load is $RC = (10 \times 10^3)(10 \times 10^{-6})$ or 10^{-1} s, a factor of 100 greater than the pulse period. Thus the pulse "droop" or sag will be very small and the output will appear identical in shape to the input; only its most positive and negative levels will change. The lower level will lie slightly below zero, while the upper will lie slightly below $+10$ V. The problem is, by what amount?

(a)   *For Positive Pulses*   When the output is positive, at about $+10$ V, the current in the load is nearly constant at 10 V/10 kΩ or 1 mA. Thus the charge lost by the capacitor during the positive 10-μs interval is

$$Q = \text{Current} \times \text{Time interval}$$
$$Q = (1 \times 10^{-3})(10 \times 10^{-6}) = 10^{-8} \text{ coulombs}$$

producing a change in capacitor voltage of

$$\Delta V = Q/C$$
$$= 10^{-8}/(10 \times 10^{-6})$$
$$= 1 \text{ mV}$$

When the input goes negative, the charge (and this voltage change) must be recovered. Since the recovery interval is 99 times the loss interval, the average current flowing must be about 1% of the load current, or, more precisely,

$$I = \text{Charge/Time}$$
$$= 10^{-8}/(990 \times 10^{-6}) \text{ A}$$

or about 10 μA. Thus one concludes that the diode will be forced to conduct an average of only 10 μA for nearly 1 ms, with a net result of only 1 mV voltage change. This miniscule change implies (from a 0.1 V/decade model) that the diode current remains essentially fixed. Since 10 μA lies two decades below 1 mA, this further implies that the diode voltage must be $0.7 - 2(0.1)$ or 0.5 V. Thus one concludes that the output consists of a positive pulse train with amplitude ranging from -0.5 to $+9.5$ V.

(b)   *For Negative Pulses*   When the output is positive (for 990 μs) the charge lost by the capacitor, in direct correspondence to the previous calculation, will be 99×10^{-8} coulombs, corresponding to a voltage change of 99 mV. When the input goes negative this charge must be recovered in 10 μs. Thus the average current over 10 μs must be

$$I = \text{Charge/Time}$$
$$= (99 \times 10^{-8})/(10 \times 10^{-6})$$
$$= 99 \text{ mA}$$

or about 100 mA, which must result in a net voltage change of about 0.1 V. Thus the current must vary over about one decade from I down to $I/10$, with an average of 100 mA. If one approximates the exponential reality with linearity, the result is

$$\frac{I + I/10}{2} = 100$$

from which one concludes that the peak current is about 180 mA decaying to about 18 mA, and for which the 1-mA-diode voltage drop varies from $0.7 + 0.1 \log (18)$ or 0.93 V to 0.83 V. In view of the very large currents involved, resistances in the source, in connections, and in the capacitor and diode components would modify this result, causing the peak current to be reduced somewhat and the final current to be raised. If one ignores stray resistance (because of the difficulty in its specification), one concludes that the output would consist of a negative pulse train falling to -0.93 V with upward droop to -0.83 V, rising from there to $+9.17$ V and drooping to $+9.07$ V for the start of another cycle.

· · ·

A Precision Clamping Circuit

By replacement of the diode in the clamping circuit of Fig. 5.21 by a "superdiode," the precision clamp of Fig. 5.23 is obtained. Operation of this circuit should be self-explanatory.

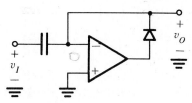

Fig. 5.23 A precision clamping circuit.

> **EXERCISE**
> **5.10** Consider a precision clamping circuit such as the one in Fig. 5.23 but with the diode reversed. Let the input be a sinusoid with zero average value, 5 V peak, and 1 kHz frequency. If the output is fed to an RC low-pass filter with a corner frequency of 10 Hz, find the value of the dc component and the amplitude of the sinusoidal component at the output of the filter.
> *Ans.* 5 V; 0.05 V

5.8 LIMITERS AND COMPARATORS

We have studied rectifiers in some detail and now we shall consider other important nonlinear functional blocks or operators. Specifically, in this section we introduce two important nonlinear functions: the limiter operator and the comparator operator. Although both operations can be realized using diodes (including zener diodes) and resistors, to improve performance, add precision, and in fact simplify the design, op amps are usually used. A variety of circuit realizations will be discussed in Sections 5.9 and 5.10. Limiters and comparators are used in a variety of signal-processing applications, including analog computation and signal generation.

The Limiter Operator

The general transfer characteristic of the limiter operator is shown in Fig. 5.24. As indicated, the limiter acts as an amplifier of gain K, which can be either positive or negative, for inputs in a certain range; $L_-/K \leq v_I \leq L_+/K$. If v_I exceeds the upper

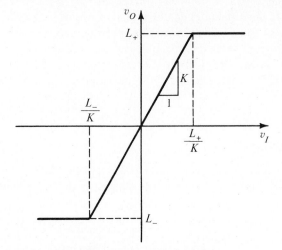

Fig. 5.24   *General transfer characteristics for the limiter operator.*

threshold (L_+/K), the output voltage is *limited* or clamped to the upper limiting level L_+. On the other hand, if v_I is reduced below the lower limiting threshold (L_-/K), the output voltage v_O is limited to the lower limiting level L_-.

The general transfer characteristic of Fig. 5.24 describes a *double limiter,* that is, a limiter that works on both the positive and negative peaks of an input waveform. *Single limiters,* of course, exist. Finally, note that if an input waveform such as that shown in Fig. 5.25 is fed to a double limiter its two peaks will be *clipped off.* Limiters therefore are sometimes referred to as *clippers.*

The limiter whose characteristics are depicted in Fig. 5.24 is described as a *hard limiter. Soft limiting* is characterized by smoother transitions between the linear region and the saturation regions and a slope greater than zero in the saturation regions, as illustrated in Fig. 5.26. Depending on the application, either hard or soft limiting may be preferred.

The Comparator Operator

Figure 5.27 shows a block-diagram representation and transfer characteristics for the comparator operator. As indicated, the comparator compares the input v_I with a ref-

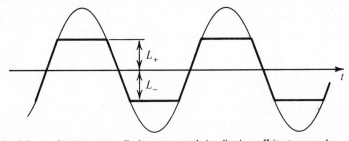

Fig. 5.25   *Applying a sine wave to a limiter can result in clipping off its two peaks.*

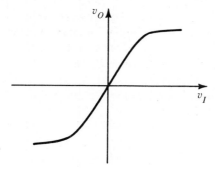

Fig. 5.26   Soft limiting.

erence level V_R that constitutes the *comparator threshold*. If v_I is greater than the reference V_R, the comparator provides a high output L_+. Alternatively, for $v_I < V_R$ the comparator output will be the low level L_-.

Although special integrated circuits are manufactured to realize the comparator function, there are many instances where the comparator operator is more conveniently implemented using a general-purpose op amp. We will discuss some such arrangements in Section 5.9.

The comparator is a very useful circuit block. It finds application in such obvious operations as detecting the zero crossings of an arbitrary signal waveform and converting sine waves into square waves, as well as in more involved applications.

As an illustration of an important point that we wish to make, consider an application where it is required to count the instances of time at which a given arbitrary waveform crosses zero. This process can be implemented using a comparator that provides a step output every time a zero crossing occurs. These step changes at the output of the comparator can then be used to *trigger* a *monostable (one-shot)* circuit. This is a circuit that produces an output pulse of specified duration and height every time it is triggered. We will discuss monostables in Chapter 6. The standardized output pulses of the monostable can be then fed to a *counter,* which provides us with the required total number of zero crossings. Counters also will be discussed in Chapter 6.

Imagine now what happens if the signal being processed has, as it usually does, interference superimposed on it, say of a frequency much higher than that of the signal. It follows that the signal might cross the zero axis a number of times around each of

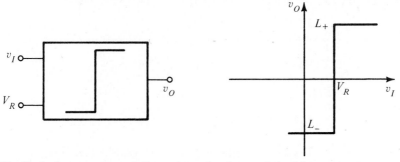

Fig. 5.27   Block-diagram representation and transfer characteristics for a comparator.

the zero-crossing points we are trying to detect, as shown in Fig. 5.28. Correspondingly the comparator would change state a number of times at each of the zero crossings and our count would obviously be in error. If we have an idea of the expected peak-to-peak amplitude of the interference, the problem can be solved by introducing *hysteresis* in the comparator characteristics. Basically, this means that instead of having one threshold the comparator will have two thresholds, a low threshold V_{TL} and a high threshold V_{TH}. The two thresholds would be placed on either side of the reference level V_R and would differ by an amount of, say, 50 or 100 mV, depending on the magnitude of interference we wish to reject. This difference, $V_{TH} - V_{TL}$, is called the magnitude of hysteresis. Now if the input signal is increasing in magnitude, the comparator with hysteresis will remain in the low state until the input level exceeds the high threshold V_{TH}. Subsequently the comparator will remain in the high state even if, owing to interference, the signal decreases below V_{TH}. The comparator will switch to the low state only if the input signal is decreased below the low threshold V_{TL}. This situation is illustrated in Fig. 5.28, from which we see that including hysteresis in the comparator characteristics provides an effective means for rejecting interference (thus providing another form of filtering).

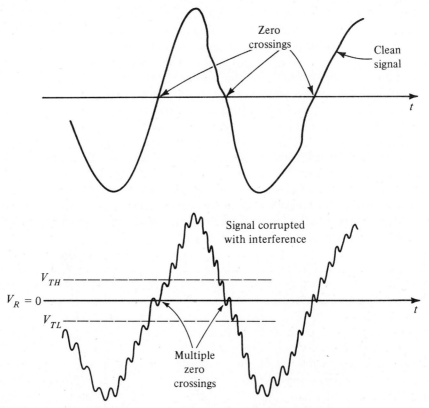

Fig. 5.28   Illustrating the need for hysteresis in the comparator characteristics as a means of rejecting interference.

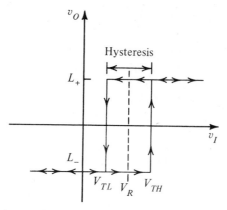

Fig. 5.29 Comparator characteristics with hysteresis.

Figure 5.29 shows the comparator characteristic with hysteresis included. The arrows on the transfer characteristic illustrate the operation, as described above. Comparators with hysteresis are very useful circuits because they provide a *memory* function, as will be explained in Section 5.10. Finally, it should be mentioned that the comparator with hysteresis is also known as a *Schmitt trigger*.

5.9 COMPARATOR AND LIMITER CIRCUITS

We will now present a number of circuit realizations of the two nonlinear operators introduced in Section 5.8.

Basic Circuits

First we note that the half-wave-rectifier circuits of Sections 5.1 and 5.2 can be considered as limiters. The diode clamp or clipper obtained by taking the output across the diode, as discussed in Section 4.2, is also a limiter. Figure 5.30 shows a number of rudimentary realizations of the limiter operator, all based on the principle of the simple diode clamp. In each part of the figure both the circuit and a sketch of its transfer characteristic are given.

The circuits of Figs. 5.30a and b are identical except for the reversal of diode polarity. That in Fig. 5.30c uses two opposite-polarity diodes in parallel to provide double limiting. Thus feeding a sinusoid to the circuit of Fig. 5.30c results in a crude approximation of a square wave, with about 1.4 V peak-to-peak amplitude. The limiting thresholds and levels can be controlled by using strings of diodes and/or by biasing the diode. The latter idea is illustrated in Fig. 5.30d. Finally, rather than strings of diodes we may use two zener diodes in series, as shown in Fig. 5.30e. In this circuit limiting occurs in the positive direction at a voltage of $V_{Z2} + 0.7$, where 0.7 V represents the voltage drop across zener diode Z_1 when conducting in the *forward* direction. For negative inputs, Z_1 acts as a zener, while Z_2 conducts in the forward direction. It should be mentioned that pairs of zener diodes connected in series are available commercially for applications of this type under the name *double-anode zener*.

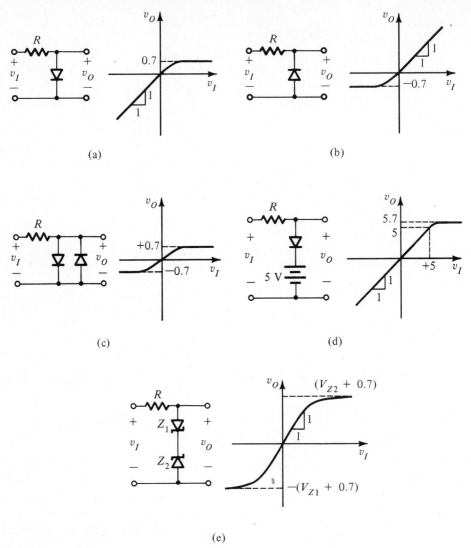

Fig. 5.30   *A variety of basic limiting circuits.*

Circuits Using Op Amps

All of the circuits in Fig. 5.30 suffer from the lack of flexibility in independently setting the linear gain, limiting thresholds, and limiting levels. In addition, all of them suffer from the essential inaccuracies due to the nonideal characteristics of the diodes involved. Moreover, we have not yet been able to implement a limiter with gain or a comparator. Better results can be obtained by combining diodes with op amps. As a first example, consider placing a zener diode in the negative-feedback path of an op amp, as shown in Fig. 5.31a. As v_I goes sufficiently positive to supply the zener with a current v_I/R greater than the knee current, the zener will enter the breakdown region and the output

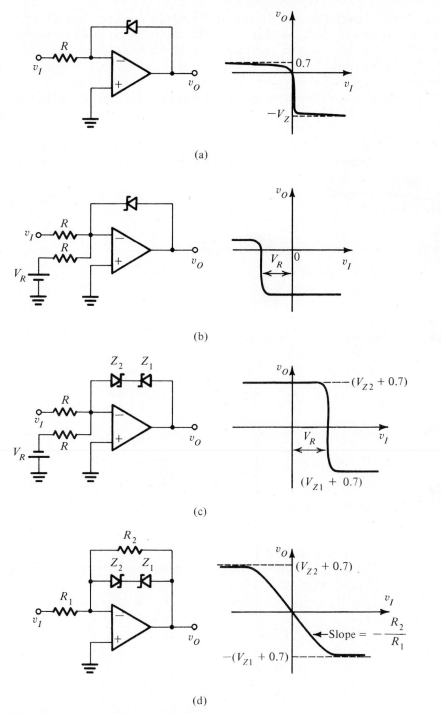

Fig. 5.31 *Comparator and limiter circuits that employ op amps.*

voltage will be limited to $-V_Z$. On the other hand, for negative v_I the zener functions as a forward-biased diode, limiting the output voltage to one diode drop or approximately $+0.7$ V. Thus it can be seen that the circuit functions as a comparator with the threshold set at approximately 0 V. This threshold, however, can be changed by biasing the circuit as shown in Fig. 5.31b.

The double-anode zener can be placed in the op-amp negative-feedback path as shown in Fig. 5.31c, thus providing a comparator with two symmetrical output levels and with a threshold controlled by the bias voltage.

Finally, consider the circuit of Fig. 5.31d. Here we have simply added a feedback resistance R_2 in parallel with the double-anode zener. The result of this simple change, though, is rather dramatic. The circuit becomes a limiter rather than a comparator. It is left to the reader to convince himself or herself that the transfer characteristic will be that shown in Fig. 5.31d.

A Comparator/Limiter Circuit With Adjustable Levels

As a last example, consider the circuit in Fig. 5.32a. This is a useful and practical circuit that functions as a comparator if the feedback resistance R_f is not included and as a limiter with R_f included. In the following we will explain the circuit operation in the comparator configuration.

Assuming ideal diodes, the transfer characteristic of the circuit in Fig. 5.32a without the resistance R_f will take the shape shown in Fig. 5.32b. The circuit operates as follows: If v_I goes positive, diode D_1 conducts and clamps point a at zero voltage (assuming, for simplicity, an ideal diode). If the current through D_1 as it starts to conduct is

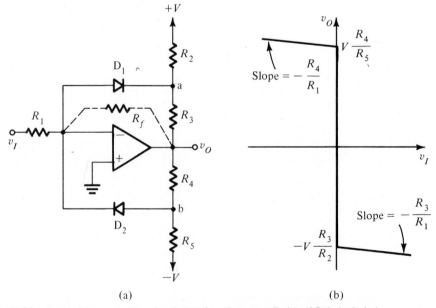

(a)   (b)

Fig. 5.32   A popular comparator circuit that functions as a limiter if R_f is included.

negligibly small, the current through R_3 will be equal to that supplied from $+V$ through R_2. Thus

$$\frac{V - 0}{R_2} = \frac{0 - v_O}{R_3}$$

$$v_O = -V\frac{R_3}{R_2}$$

Since v_O is negative, diode D_2 will be off. As v_I increases in the positive direction diode D_1 conducts more and more current. Nevertheless, the voltage at point a remains at about one diode drop below ground. Thus the current through R_2 remains almost constant, and hence R_2 will have little effect on the operation of the circuit in this region. The extra current pushed through D_1 will therefore have to flow through R_3, thus causing the output voltage v_O to decrease. It may therefore be seen that in this region of operation the op amp operates as an inverting amplifier with a feedback resistance equal to R_3 plus the resistance presented by diode D_1. The latter resistance can be usually ignored in comparison with R_3, resulting in a linear transfer characteristic with a slope (gain) of $-R_3/R_1$. Normally, R_3 is chosen much smaller than R_1, rendering the slope quite small.

Consider next the situation as the input signal v_I goes negative. Diode D_1 will turn off and diode D_2 will turn on. This will cause the voltage at point b to become approximately zero. Neglecting the initial current through D_2, we see that the currents through R_4 and R_5 will be equal,

$$\frac{v_O - 0}{R_4} = \frac{0 + V}{R_5}$$

Thus

$$v_O = V\frac{R_4}{R_5}$$

As v_I is increased in the negative direction, the op amp operates as an inverting amplifier with a feedback resistance approximately equal to R_4 (neglecting the resistance presented by diode D_2). This results in a linear transfer characteristic with a slope of $-R_4/R_1$ that can be made small by selecting $R_4 \ll R_1$.

The nonideal diode characteristics will slightly modify the output levels of the comparator as well as making the transitions at the break points less sharp than those indicated in Fig. 5.32b. Thus as a limiter this circuit provides relatively soft limiting.

Example 5.4
In the circuit of Fig. 5.32a let $V = 15$ V. Find suitable values for all resistances so that the comparator output levels are ± 5 V and that the slope of the limiting characteristics is 0.1. Also, bias the comparator to cause the threshold to occur at $v_I = +5$ V.

Solution
To cause the threshold to occur at $v_I = +5$ V the comparator can be biased by connecting an additional resistance R_B from the inverting input terminal to a negative reference voltage. Selecting $R_B = 3R_1$, we may use -15 V as the reference voltage.

To make the slope of the limiting characteristics equal to 0.1 we select

$$R_3 = R_4 = 0.1R_1$$

Now for ± 5-V output levels we have

$$5 = 15\,\frac{R_4}{R_5}, \quad \text{leading to} \quad R_5 = 3R_4$$

$$5 = 15\,\frac{R_3}{R_2}, \quad \text{leading to} \quad R_2 = 3R_3$$

We may therefore select

$$R_3 = R_4 = 3\text{ k}\Omega; \quad R_2 = R_5 = 9\text{ k}\Omega; \quad R_1 = 30\text{ k}\Omega; \quad R_B = 90\text{ k}\Omega$$

$$\bullet \quad \bullet \quad \bullet$$

EXERCISES
5.11   Assuming the diodes to be ideal, describe the transfer characteristic of the circuit shown in Fig. E5.11.

Ans. $v_O = v_I \qquad$ for $-5 \le v_I \le +5$
$\qquad\quad v_O = \tfrac{1}{2}v_I - 2.5 \qquad$ for $v_I \le -5$
$\qquad\quad v_O = \tfrac{1}{2}v_I + 2.5 \qquad$ for $v_I \ge +5$

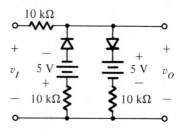

Fig. E5.11

5.12   Consider the circuit of Fig. 5.32a with R_f included and with $V = 15$ V, $R_1 = 30$ kΩ, $R_f = 60$ kΩ, $R_2 = R_5 = 9$ kΩ, $R_3 = R_4 = 3$ kΩ. First convince yourself that the circuit operates as a limiter, and then find the value of the lower and upper threshold, the slope in the linear region, the output levels just as limiting starts, and the slope in the limiting regions (assume ideal diodes).

Ans. -2.5 V; $+2.5$ V; -2; ± 5 V; -0.095

5.10   COMPARATOR WITH HYSTERESIS: THE BISTABLE CIRCUIT

The advantages obtained by adding hysteresis to the comparator transfer characteristic were discussed in Section 5.8. We will now show how this may be accomplished. It will also be shown that the resulting circuit has *memory;* its usefulness therefore extends far beyond comparator applications.

Consider first the use of a single op amp as a comparator. If one of the op amp's input terminals is connected to a reference voltage V_R, then the open-loop op amp will obviously function as a comparator with the two output-limiting levels being the saturation voltages of the op amp. In view of the delay time required for an op amp to get out of saturation it is sometimes desirable to fix, by external circuitry, the output voltages at levels lower than those of saturation. Except for this speed limitation, the op amp functions as an excellent comparator.

Next consider the circuit shown in Fig. 5.33a. Here we have connected the negative input terminal to ground while applying *positive feedback* through resistance R_2 and feeding the input v_I to the positive input terminal through another resistance R_1. We

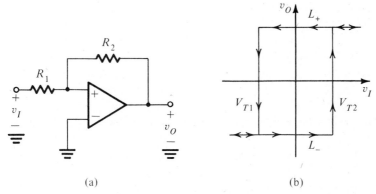

(a)   (b)

Fig. 5.33   *A comparator circuit that employs an op amp with positive feedback to obtain transfer characteristics with hysteresis (a), shown graphically in (b).*

will now show that this circuit functions as a comparator with hysteresis and that it has the transfer characteristics depicted in Fig. 5.33b, where L_+ and L_- denote the values of the op-amp saturation voltages (normally each is within about 2 V of the corresponding power-supply voltage).

By superposition, it follows that the voltage v_+ at the positive input terminal of the op amp is given by

$$v_+ = v_O \frac{R_1}{R_1 + R_2} + v_I \frac{R_2}{R_1 + R_2} \tag{5.4}$$

Consider now the case when the input v_I is at a high positive value (greater than the yet undetermined value V_{T2}). The transfer characteristic shows that the output should be L_+. Using these values in Eq. (5.4) shows that v_+ is positive which indicates that the op amp is saturated in the positive direction and $v_O = L_+$ as originally assumed.

From the circuit we see that for the output to switch to the low level L_-, v_+ has to become slightly negative—that is, switching occurs when v_+ just passes zero in the negative direction. We can now use Eq. (5.4) to determine the value of v_I at which switching occurs, V_{T1}, by setting $v_+ = 0$ and $v_O = L_+$. The result is

$$V_{T1} = -L_+ \frac{R_1}{R_2}$$

If v_I is reduced below V_{T1}, the output remains at the low level L_-.

Next consider increasing v_I. Again we see from the circuit that for v_O to change from L_- to L_+ the voltage at the noninverting input terminal should become slightly positive. Thus the input threshold occurs at the value of v_I that results in v_+ being equal to zero. Setting $v_+ = 0$ and $v_O = L_-$ in Eq. (5.4) results in the upper threshold V_{T2} given by

$$V_{T2} = -L_- \frac{R_1}{R_2}$$

Increasing v_I beyond the positive value V_{T2} keeps v_+ positive and v_O equal to L_+.

We have thus shown that the circuit of Fig. 5.33a acts as a comparator with hysteresis. The reference level is 0 V, and the hysteresis width is $V_{T2} - V_{T1}$.

The comparison level can be changed by connecting the inverting input terminal to an appropriate reference voltage.

The circuit just described has more significance than just being a comparator with hysteresis. This circuit is *bistable*—that is, it has two stable states, one with the output equal to L_+, the other with the output equal to L_-. The circuit can remain in either stable state indefinitely. To change its state we have to apply an input signal of appropriate value. Thus if the circuit is in the L_+ state we can *trigger* it to change state by applying an input $v_I < V_{T1}$. Here it is important to note that the word trigger is an excellent descriptor of the action caused by the input signal; the input signal merely starts the action of switching states. The switching process is completed by the action of the positive feedback. To see this consider the case when the output is high (at L_+) and an input signal $v_I < V_{T1}$ is applied. According to Eq. (5.4) the voltage v_+ at the positive input terminal goes negative. This causes the op-amp output to go negative, which in turn, according to Eq. (5.4), causes v_+ to become more negative. Once this *regenerative* process starts we can remove the input signal and the process will continue on its own until the bistable switches completely to the other state.

Another important property of the bistable is that it has memory. This can be seen by noting that the response of the circuit at any moment is not determined solely by the value of the input signal at that moment but rather by the value of the input signal as well as the state in which the circuit is in. For instance, applying an input signal v_I in the range $V_{T1} < v_I < V_{T2}$ will result in output equal to L_+ if the circuit *was* in the L_+ state and in output equal to L_- if the circuit was in the L_- state. Memory is a very useful property and will be exploited in Chapter 6 in the context of digital circuits.

Finally, we should note that the circuit of Fig. 5.33a is just one of many possible implementations of a bistable.

EXERCISES

5.13 In the circuit of Fig. 5.33a let $L_+ = -L_- = 10$ V, $R_1 = 10$ kΩ, and $R_2 = 20$ kΩ. Find the two threshold voltages V_{T1} and V_{T2}.
Ans. -5 V; $+5$ V

5.14 In the circuit of Fig. 5.33a let $L_+ = -L_- = 10$ V and $R_1 = 1$ kΩ. Find a value for R_2 that gives 100-mV hysteresis.
Ans. $R_2 = 200$ kΩ

5.11   WAVEFORM GENERATORS

Op amps, together with diodes, resistors, and capacitors, can be used to design circuits whose purpose is to generate signals having square, triangular, sine, and other waveforms. The subject of designing *waveform* or *function generators,* as they are normally called, is an important and challenging one. In this section we describe simple circuits for the generation of square and triangular waveforms. The resulting triangular waveform can be used as an input to a "sine-wave shaper," described in Section 5.12, to produce sinusoidal outputs. Other, more direct techniques for the generation of sinusoidal waveforms will be studied in Chapter 14. Circuits for generating and processing pulse waveforms will be presented in Chapter 6.

A Single Op-Amp Square-Wave Generator— The Astable Multivibrator

Figure 5.34 shows a simple circuit that can be designed to provide a symmetrical square waveform of arbitrary frequency (limited normally by the speed of the op amp to the audio frequency range). The circuit operates in the following manner: Assume that for some reason a slight positive voltage develops between the op amp input terminals; that is, point C becomes more positive than point B. Because of the high open-loop gain of

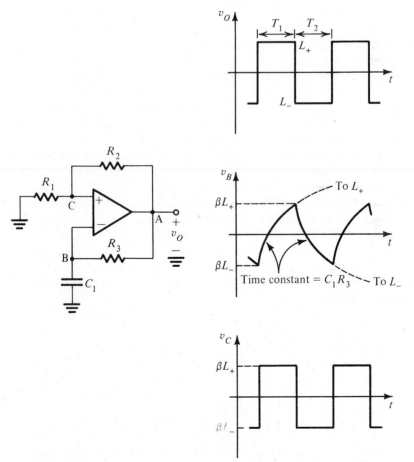

Fig. 5.34   *A simple square-wave generator and its waveforms. Note that $\beta \equiv R_1/(R_1 + R_2)$.*

the op amp its output will become positive at the upper saturation level L_+. This positive voltage L_+ at point A will, through the voltage divider composed of R_1 and R_2, cause the voltage at point C to become

$$v_C = v_A \frac{R_1}{R_1 + R_2} = \beta L_+ \tag{5.5}$$

where

$$\beta \equiv \frac{R_1}{R_1 + R_2}$$

Now B is lower than C in voltage, a situation that cannot continue because capacitor C_1 is connected to point A through R_3 and point A is at the higher positive voltage L_+. Thus C_1 will charge and its voltage will rise exponentially, with a time constant $C_1 R_3$, toward L_+. This process will be terminated when the voltage at B reaches and just exceeds the voltage at C, βL_+. At this moment the op amp will see a negative input voltage that because of the high open-loop gain will cause its output to change to the negative saturation level L_-. Now this negative voltage at A causes—through the R_1, R_2 divider—the voltage at C to change to the negative value βL_-. In turn, this negative voltage at C maintains the circuit for a time in its newly acquired state, the one with the output equal to L_-.

At this point we should note the *regenerative* action present: Once the voltage at B has risen above that at C the switching action is started. Then the action is sustained and completed by the negative voltage fed back to point C through the R_1, R_2 divider. This regenerative action, which we have encountered before in the bistable, is a result of the judicious application of positive feedback.

Continuing with the description of circuit operation, we see that with the output at the negative voltage L_-, capacitor C_1 begins to discharge toward this negative level through R_3. The voltage v_B decreases exponentially, with a time constant $C_1 R_3$, from the positive value βL_+. The discharge process will be terminated at the moment the magnitude of voltage at B just exceeds that at C (that is, βL_-), and so on. Thus the circuit oscillates generating a square waveform at A.

Figure 5.34 shows the waveforms at various circuit points, from which the period T of the square wave can be found as follows: During the charging interval T_1 the voltage v_B across the capacitor at any time t, with $t = 0$ at the beginning of T_1, is given by (see Section 2.10)

$$v_B = L_+ - (L_+ - \beta L_-)e^{-t/\tau}$$

where

$$\tau = C_1 R_3$$

Substituting $v_B = \beta L_+$ at $t = T_1$ gives

$$T_1 = \tau \ln \frac{1 - \beta(L_-/L_+)}{1 - \beta} \tag{5.6}$$

Similarly, during the discharge interval T_2 the voltage v_B at any time t, with $t = 0$ at the beginning of T_2, is given by

$$v_B = L_- - (L_- - \beta L_+)e^{-t/\tau}$$

Substituting $v_B = \beta L_-$ at $t = T_2$ gives

$$T_2 = \tau \ln \frac{1 - \beta(L_+/L_-)}{1 - \beta} \tag{5.7}$$

Equations (5.6) and (5.7) can then be combined to obtain the period $T = T_1 + T_2$. Normally, $L_+ = -L_-$, resulting in symmetrical square waves of period T given by

$$T = 2\tau \ln \frac{1 + \beta}{1 - \beta} \tag{5.8}$$

Finally, we should mention that this square-wave generator can be made to have variable frequency by switching different capacitors C_1 (usually in decades) and by continuously adjusting R_3 (to obtain continuous frequency control within each decade of frequency). Also, although the waveform at B can be made almost triangular by using a small value for the parameter β, triangular waveforms of superior linearity can be easily generated using the scheme discussed next.

*A General Scheme for Generating Triangular
and Square Waveforms*

Figure 5.35a shows a general scheme for generating triangular and square waveforms. The circuit consists of a feedback loop incorporating an inverting integrator and a bistable circuit. The latter is shown in block form, with the transfer characteristic shown inside the box. This bistable can be realized by a variety of circuits, the simplest of which is the one discussed in Section 5.10.

We now proceed to show how the feedback loop of Fig. 5.35a oscillates and generates a triangular waveform v_1 at the output of the integrator and a square waveform v_2 at the output of the bistable. Let the output of the bistable be at L_+. A current equal to L_+/R will flow into the resistor R and through capacitor C, causing the output of the integrator to *linearly* decrease with a slope of $-L_+/CR$, as shown in Fig. 5.35c. This will continue until the integrator output reaches the lower threshold V_{TL} of the bistable, at which point the bistable will switch states, its output becoming negative and equal to L_-. At this moment the current through R and C will reverse direction, and its value will become equal to $|L_-|/R$. It follows that the integrator output will start to increase linearly with a positive slope equal to $|L_-|/CR$. This will continue until the integrator output voltage reaches the positive threshold of the bistable, V_{TH}. At this point the bistable switches, its output becomes positive (L_+), the current into the integrator switches direction, and the output of the integrator starts to decrease linearly, beginning a new cycle.

From the above discussion it is relatively easy to derive an expression for the period T of the square and triangular waveforms. During the interval T_1 we have, from Figs. 5.35a and c,

$$\frac{V_{TH} - V_{TL}}{T_1} = \frac{L_+}{CR}$$

from which we obtain

$$T_1 = CR \frac{V_{TH} - V_{TL}}{L_+}$$

Similarly, during T_2 we have

$$\frac{V_{TH} - V_{TL}}{T_2} = \frac{-L_-}{CR}$$

from which we obtain

$$T_2 = CR \frac{V_{TH} - V_{TL}}{-L_-}$$

Thus to obtain symmetrical square waves we design the bistable to have $L_+ = -L_-$.

EXERCISES

5.15   Consider the circuit in Fig. 5.34a. Let the op-amp saturation voltages be ± 10 V. For $R_1 = 100$ kΩ, $R_2 = R_3 = 1$ MΩ, and $C_1 = 0.01$ µF, find the frequency of oscillation.
Ans.   63.4 Hz

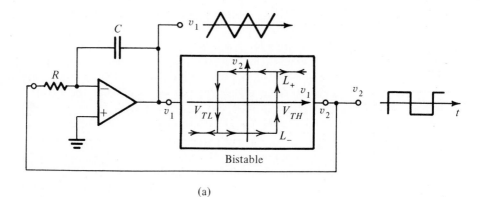

(a)

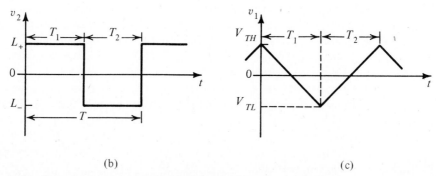

(b)   (c)

Fig. 5.35   *General scheme for generating triangular and square waveforms.*

5.16 Consider the circuit of Fig. 5.35a with the bistable realized by the circuit in Fig. 5.33a. If the op amps used have saturation voltages of ± 10 V and using a capacitor $C = 0.01$ μF and a resistor $R_1 = 10$ kΩ, find the values of R and R_2 (note that R_1 and R_2 are associated with the bistable of Fig. 5.33a) such that the frequency of oscillation is 1 kHz and the triangular waveform has a 10-V peak-to-peak amplitude.
Ans. 50 kΩ; 20 kΩ

5.12 NONLINEAR WAVE SHAPING

Diodes together with resistors can be used to synthesize two-port networks having arbitrary nonlinear transfer characteristics. Such two-port networks can be employed in *waveform shaping*—that is, changing the waveform of an input signal in a prescribed manner to produce at the output a waveform of a desired shape. In this section we shall illustrate this application by a concrete example: the *sine-wave shaper*. This is a circuit whose purpose is to change the waveform of an input triangular wave signal to a sine wave. Though simple, this sine shaper is a practical building block used extensively in function generators. There are, of course, other, more direct techniques for generating sinusoidal signals; the topic of sinusoidal oscillators will be discussed in Chapter 14. These direct techniques, however, are not very convenient at very low frequencies, and one might prefer the simple schemes discussed in this chapter, namely, generation of stable triangular waveforms and then the processing of them through the sine shaper discussed next.

Consider the circuit shown in Fig. 5.36a. It consists of a chain of resistors connected across the entire symmetric voltage supply $+V$, $-V$. The purpose of this voltage

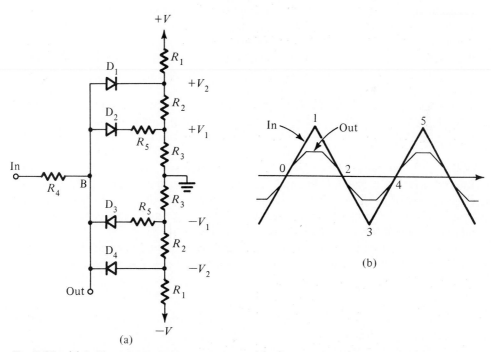

Fig. 5.36 (a) A three-segment sine-wave shaper. (b) The input, triangular waveform and the output, approximately sinusoidal waveform.

divider is to generate reference voltages that will serve to determine the breakpoints in the transfer characteristic. In our example these reference voltages are denoted $+V_2$, $+V_1$, $-V_1$, $-V_2$. Note that the entire circuit is quite symmetric, driven by a symmetric triangular wave and generating a symmetric sine wave. The circuit approximates each quarter-cycle of the sine wave by three straight-line segments, with the breakpoints between these segments determined by the reference voltages V_1 and V_2.

The circuit works as follows: Let the input be the triangular wave shown in Fig. 5.36b, and consider first the quarter-cycle defined by the two points labeled 0 and 1. When the input signal is less in magnitude than V_1, none of the diodes conduct. Thus zero current flows through R_4, and the output voltage at B will be equal to the input voltage. But as the input rises to V_1 and above, D_2 (assumed ideal) begins to conduct. Assuming that the conducting D_2 behaves as a short circuit, we see that for $v_I > V_1$

$$v_O = V_1 + (v_I - V_1)\frac{R_5}{R_4 + R_5}$$

This implies that as the input continues to rise above V_1 the output follows but with a reduced slope. This gives rise to the second segment in the output waveform, as shown in Fig. 5.36b. Note that in developing the above equation we have assumed that the resistances in the voltage divider are low valued so as to cause the voltages V_1 and V_2 to be constant independent of the current coming from the input.

Next consider what happens as the voltage at point B reaches the second break-point determined by V_2. At this point D_1 conducts, thus limiting the output v_B to V_2 (plus, of course, the voltage drop across D_1 if it is not assumed ideal). This gives rise to the third segment, which is flat, in the output waveform. The result is to "bend" the waveform and shape it into an approximation of the first quarter-cycle of a sine wave. Then, beyond the peak of the input triangular wave, as the input voltage decreases the process unfolds, the output becoming progressively more like the input. Finally, when the input goes sufficiently negative, the process begins to repeat at $-V_1$ and $-V_2$ for the negative half-cycle.

While the circuit is relatively simple, its performance is surprisingly good. A measure of goodness usually taken is to quantify the purity of the output sine wave by specifying the percentage *total harmonic distortion* (THD). This is the percentage ratio of the RMS voltage of all harmonic components above the fundamental frequency (which is the frequency of the triangular wave) to the RMS voltage of the fundamental. Interestingly, one reason for the good performance of the diode shaper is the beneficial effects produced by the nonideal *i-v* characteristics of the diodes, that is, the exponential knee of the junction diode as it goes into forward conduction. The consequence is a relatively smoothed transition from one line segment to the next.

The circuit discussed is an example of a family of circuits known as *diode function generators*. To enhance accuracy, such circuits usually employ op amps.

EXERCISE

5.17 The circuit in Fig. E5.17 is required to provide a three-segment approximation to the nonlinear *i-v* characteristic

$$i = 0.1v^2$$

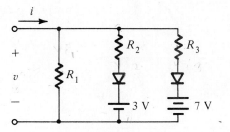

Fig. E5.17

where v is the voltage in volts and i is the current in milliamperes. Find the values of R_1,
R_2, and R_3 such that the approximation is perfect at $v = 2, 4$ and 8 volts. Calculate the
error in current value at $v = 3, 5, 7,$ and 10 volts. Assume an ideal diode.

Ans.   5 kΩ, 1.25 kΩ, 1.25 kΩ; −0.3 mA, +0.1 mA, −0.3 mA, 0.

5.13 CONCLUDING REMARKS

The object of this chapter has been to expose the reader to a variety of techniques for
the design and analysis of nonlinear circuits. Many of the circuits presented are practical and should therefore provide an excellent opportunity for experimentation.

One of the most significant ideas discussed in this chapter is the use of negative
feedback to mask the diode nonideal characteristics and thus provide precise nonlinear
functions. Equally significant is the concept of using positive feedback to provide the
regenerative action required in the synthesis of bistable or memory circuits and astable
multivibrators and other waveform-generation circuits. Both negative and positive feedback will be applied in many other occasions throughout this book, with the formal
study of feedback presented in Chapter 12.

A special yet very important and extensive class of nonlinear circuits is that of
digital circuits. These circuits are the fundamental building blocks of digital computers
and other instrumentation, control, and communication systems. An introduction to digital circuits is the topic of Chapter 6.

DIGITAL CIRCUITS

6

In a digital circuit, signals take on a limited number of values. The most common digital systems employ two values and are said to be *binary* systems. In such a system voltage signals are either "high" or "low," and symbols such as "1" and "0" are used to denote the two possible levels.

Digital circuits operate on binary-valued input signals and produce binary-valued output signals. The operation of digital circuits can be described by a special kind of algebra called *Boolean algebra*. It is not our purpose here to study Boolean algebra; such a study is normally a part of courses and texts dealing exclusively with digital systems. Nevertheless, we shall briefly introduce the fundamental elementary operations performed by digital circuits. A complex digital system such as the digital computer is made up of thousands of digital circuits performing relatively simple operations. It is such repetition that makes possible the design of complex digital systems.

Digital circuits play a very important role in today's electronic systems. They are employed in almost every facet of electronics, including communications, control, instrumentation, and, of course, computing. This widespread usage is due mainly to the availability of inexpensive integrated-circuit packages that contain powerful digital circuitry. The complexity of circuitry on an IC chip ranges from a small number of circuits performing identical elementary operations to a complete computer on one chip (a *microprocessor*).

In this chapter we begin our study of digital circuits by presenting the fundamental methods and techniques describing their operation. By the end of the chapter the reader should be able to build simple but meaningful digital systems from available standard IC packages. In the next three chapters, in conjunction with the study of various electronic devices, we shall present circuits that implement some of the functions described here. Finally, a detailed study of the circuits that exist in available digital IC packages will be presented in Chapters 15 and 16.

6.1 FUNDAMENTAL LOGIC OPERATIONS

Digital circuits operate on binary-valued signals according to the rules of logic. In this section we shall introduce the three most fundamental logic operations: OR, AND, and NOT.

Positive and Negative Logic

In digital circuits two distinct voltage levels are chosen to represent the two possible values of binary variables. Let us denote these two voltage levels by V_1 and V_2 and assume that $V_2 > V_1$. If we choose to let V_1 represent logic 0 and V_2 represent logic 1, then the system is called a positive-logic system. Alternatively, we may choose to let the lower voltage V_1 represent logic 1 and the higher voltage V_2 represent logic 0, resulting in a negative-logic system. Although positive logic is intuitively more appealing, there are situations in which negative logic is more convenient.

If a positive-logic convention is adopted, one tends to use the words high and low interchangeably with 1 and 0, respectively.

The NOT Operation

Figure 6.1a shows the symbol for the NOT logic operation, with A denoting the input variable and B denoting the output variable. The NOT operation is defined by the table in Fig. 6.1b, which lists all possible input conditions (two) and the corresponding output values. As seen, when the input is 0 the output is 1, and when the input is 1 the output

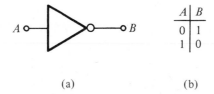

A	B
0	1
1	0

(a)   (b)

Fig. 6.1   (a) Circuit symbol and (b) truth table for the logic inverter or NOT operator.

is 0. Thus the NOT circuit *inverts* the signal and is therefore referred to as a *logic inverter*. We shall denote the inverse or complement of a logic variable by placing a bar above the symbol denoting the variable. Thus the NOT operation is described by

$$B = \overline{A}$$

The table of Fig. 6.1b completely specifies the logic operation of the inverter. It is called a *truth table*. It is quite usual and convenient to use truth tables to describe the operation of logic circuits.

As will be discussed in the following chapters, the logic inverter can be implemented by an amplifying device such as a transistor with a load resistance. Figure 6.2 shows a typical input-output transfer characteristic for a transistor logic inverter. If we

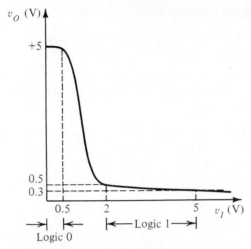

Fig. 6.2   *Typical transfer characteristic of a transistor logic inverter.*

take logic 0 to be represented by voltages between 0 V and 0.5 V and logic 1 to be represented by voltages between 2 V and 5 V, then we can easily verify that this transfer characteristic results in proper inverter operation.

The OR Operation

Figure 6.3a shows the circuit symbol of a two-input OR operator. A circuit that performs the logic OR operation is called an *OR gate*. Since the OR gate of Fig. 6.3a has two inputs and since each input can have one of two values (0 or 1), there are four

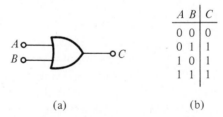

A	B	C
0	0	0
0	1	1
1	0	1
1	1	1

(a)   (b)

Fig. 6.3   *(a) Circuit symbol and (b) truth table for a two-input OR gate.*

possible input combinations. These four input combinations together with the output corresponding to each combination are listed in the truth table of Fig. 6.3b. Note that the output is 1 if A is 1 *or* if B is 1. In other words, the output is 1 if any of the inputs is 1. This operation can be symbolically described by the Boolean expression

$$C = A + B$$

where the "+" sign is used to denote the logic OR.

The addition of more inputs to the OR gate is straightforward conceptually. As an example, Fig. 6.4 shows a four-input OR. The output Y is 1 if A or B or C or D (or *any* input) is 1,

$$Y = A + B + C + D$$

The simplest implementation of the OR function uses diodes and is illustrated in Fig. 6.5. Assuming a positive-logic convention under which 0 is denoted by 0 V and 1

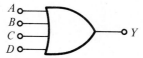

Fig. 6.4 *A four-input OR gate.*

is denoted by a positive voltage (say, $+5$ V), we can easily verify that this circuit implements the OR function. Specifically, the output v_C will be 0 V (logic 0) only in one case: when $v_A = 0$ V (logic 0) and $v_B = 0$ V (logic 0). If v_A or v_B (or both) is high, the corresponding diode (or both diodes) will conduct and the output will be high (logic 1). Thus we see that the truth table in Fig. 6.3b is verified.

However, a problem exists with the diode logic gate of Fig. 6.5. When one of the inputs is high, say, $+5$ V, the output will be high but its level will be lower than $+5$ V

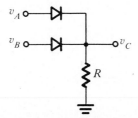

Fig. 6.5 *Implementation of the two-input OR gate of Fig. 6.3a using diode logic. This circuit uses the positive-logic convention.*

by the finite diode voltage drop ($\simeq 0.7$ V). Thus the output will only be about $+4.3$ V. Since this voltage is "close" to $+5$ V it could be correctly interpreted by the succeeding circuit as a logic 1. Unfortunately, though, if we have to cascade a number of such gates, as is normally the case in a digital system of some complexity, the diode voltage drops add and the level might degrade to the point where it could no longer be correctly interpreted as logic 1.

The problem of logic-level degradation is serious, limiting the usefulness of diode logic. The problem can be easily overcome by using an amplifying device, such as a transistor, that *restores* the logic level to its appropriate value. A typical transfer characteristic for such a restorer is shown in Fig. 6.6. Following the diode logic gate by this level-restoring circuit eliminates the level-degradation problem, as can be readily verified. Integrated-circuit logic gates include level-restoration circuitry as an integral part of the gate circuit and are therefore referred to as *level-restoring logic*.

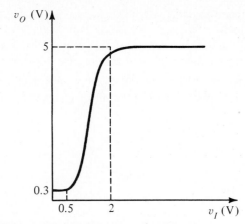

Fig. 6.6 *Typical transfer characteristic for a logic level restorer.*

The AND Operation

Figure 6.7a shows the circuit symbol for a two-input AND operator. A circuit that performs the AND operation is called an *AND gate*. The truth table for the two-input AND gate is given in Fig. 6.7b. As indicated, the output of the AND gate is 1 only in

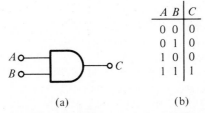

A	B	C
0	0	0
0	1	0
1	0	0
1	1	1

(a)   (b)

Fig. 6.7 *(a) Circuit symbol and (b) truth table for a two-input AND gate.*

one case: when A is 1 *and* B is 1. The AND operation is symbolically described by the Boolean expression

$$C = A \cdot B$$

where "$\cdot$" is used to denote logic AND. For simplicity the dot is usually omitted.

The addition of more inputs to the AND gate is straightforward and is illustrated in Fig. 6.8 with a four-input AND. For this gate the output is given by

$$Y = ABCD$$

The simplest implementation of the AND function uses diodes, as illustrated in Fig. 6.9. For a positive-logic convention under which 0 is represented by 0 V and 1 is

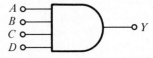

Fig. 6.8 *A four-input AND gate.*

represented by +5 V one can easily verify that the circuit of Fig. 6.9 implements the AND function. Specifically, note that if any of the two inputs is low, the corresponding diode will conduct and the output will be low. Only when both inputs are high will the two diodes be off and the output be high (equal to the power-supply voltage).

Because of the finite diode drop the output voltage in the low state will be not zero but about 0.7 V. This degradation of the logic 0 level is similar to that encountered in

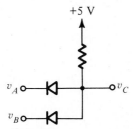

Fig. 6.9 *A two-input AND gate using diode logic with positive-logic convention.*

the AND gate (for logic 1 level) and can be eliminated by using level-restoring circuitry.

The Duality of AND and OR

A form of duality exists between the AND and OR functions. Specifically, a gate that performs the OR operation under a positive-logic convention performs the AND operation under a negative-logic convention. Similarly, a "positive" AND is a "negative" OR. That this is true can easily be verified with the diode logic gates described above or by reflecting on the truth tables. This form of duality will be encountered in a number of other occasions and is useful in the design of logic circuits.

Fig. 6.10 *An OR gate with input inversion.*

Inverted Inputs

Some commercially available gates include an inversion at the input, denoted usually by a circle placed at the end of the input line. An example is shown in Fig. 6.10, where the function implemented is

$$Y = \overline{A} + \overline{B}$$

Example 6.1
This example illustrates the application of logic systems to the solution of engineering problems. We wish to create a logic device to operate an elevator door safely under a combination of conditions. The device should close the door when one or more floor

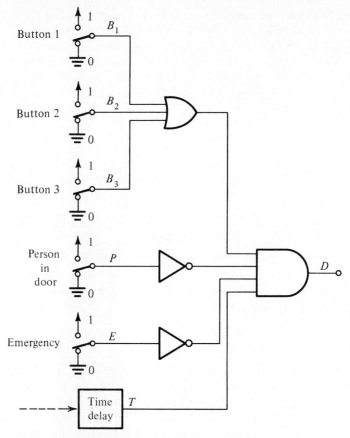

Fig. 6.11 Combinational circuit for the elevator door control of Example 6.1.

buttons have been pressed, when no one remains in the doorway, when a preset delay time has passed, and when an emergency switch has not been activated.

Solution
The first part of the solution consists of assigning symbols to the various variables involved, as follows:

1. Let D denote the signal to close the door, with $D = 1$ if the door is to be closed.
2. For the buttons let $B_1 = 1$ when button 1 has been pressed, $B_2 = 1$ when 2 has been pressed, and so on.
3. Let P denote the presence of a person in the doorway, with $P = 0$ when no person is in the doorway.
4. For the delay time we use the variable T, with $T = 1$ indicating that the delay time has elapsed.
5. Finally, we use E to indicate whether or not an emergency switch has been pressed, with $E = 1$ indicating that an emergency has been signaled.

The second part of the solution consists of an analysis of the problem statement in terms of the variables that we have assigned. The problem states that the door should be closed ($D = 1$) when

(a)   one of the buttons has been pressed, that is, when B_1 or B_2 or B_3 or ... = 1, *and*
(b)   no person is present in the doorway, that is, when $P = 0$ or $\overline{P} = 1$, *and*
(c)   the delay time has expired, that is, when $T = 1$, *and*
(d)   the emergency switch has not been pressed, that is, when $E = 0$ or $\overline{E} = 1$.

This analysis leads to the logic expression

$$D = (B_1 + B_2 + B_3 + \cdots) \cdot \overline{P} \cdot T \cdot \overline{E}$$

A possible logic implementation of the door-control circuit is shown in Fig. 6.11 for a positive-logic convention. Single-pole, double-throw switches are used to indicate the function of the pushbuttons and switches.

$\bullet \quad \bullet \quad \bullet$

EXERCISES
6.1   In the circuit of Fig. 6.9 find v_C if $v_A = +1$ V and $v_B = +2$ V.
      Ans. $+ 1.7$ V
6.2   For the digital circuit in Fig. E6.2 find Y as a function of A, B, C, and D.
      Ans. $Y = AB + CD$

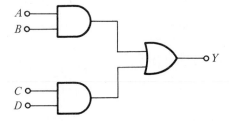

Fig. E6.2

6.2 SYNTHESIS OF LOGIC FUNCTIONS

Logic functions are most conveniently described in truth-table form. In this section we present a systematic procedure for obtaining an implementation using AND, OR, and NOT gates of a logic function given in truth-table form.

Obtaining a Logic Expression

The first step in implementing a logic function consists of finding an algebraic expression describing the function. Many such expressions are possible, with some simpler than others. We shall start by describing a method for obtaining a standard form for expressing logic functions. The method will be presented in the context of an example. Consider the truth table in Fig. 6.12, where two functions f_1 and f_2 of three variables A, B, and C are given. We wish to express f_1 as a function of the three variables

A B C	f_1	f_2
0 0 0	1	1
0 0 1	1	1
0 1 0	0	1
0 1 1	1	0
1 0 0	0	1
1 0 1	0	1
1 1 0	0	0
1 1 1	1	0

Fig. 6.12   Truth table for three-variable functions f_1 and f_2.

A, B, and C. A way to do this is to search the table for the logic 1 entries under f_1. One such entry exists in the first row. The first row, in effect, says that $f_1 = 1$ when $A = 0$ and $B = 0$ and $C = 0$. This can be restated as $f_1 = 1$ when $\overline{A} = 1$ and $\overline{B} = 1$ and $\overline{C} = 1$ or more succinctly when $(\overline{A} \cdot \overline{B} \cdot \overline{C}) = 1$. Another 1 exists in the second row, which can be stated as $f_1 = 1$ when $(\overline{A} \cdot \overline{B} \cdot C) = 1$. We continue in this manner with all other 1 entries and then express f_1 as an OR function of all these conditions. In this way we obtain

$$f_1 = \overline{A}\overline{B}\overline{C} + \overline{A}\overline{B}C + \overline{A}BC + ABC \qquad (6.1)$$

This expression is said to have a *canonic sum-of-products* form. The name derives from the practice of calling the OR operation a "sum" function (using the "+" operator) and the AND operation a product function (using the "·" operator). Note that there is a product term corresponding to each logic 1 entry under f_1. In this product term, called a *minterm*, each variable appears either in true form (if its entry in the table is 1) or complemented (if its entry is 0).

In expressing a logic function in the sum-of-products form there is an implied hierarchy among the operations AND, OR, and NOT. These operations should be performed in the following order: NOT followed by AND followed by OR. This allows us to obtain directly an implementation using AND, OR, and NOT gates. Such an implementation for the function f_1 is given in Fig. 6.13.

In some cases it is more convenient to invert the above process and use the logic 0 entries of a function. Consider as an example the function f_2 given in Fig. 6.12. We can use the above procedure and express f_2 as

$$f_2 = \overline{A}\overline{B}\overline{C} + \overline{A}\overline{B}C + \overline{A}B\overline{C} + A\overline{B}\overline{C} + A\overline{B}C$$

Alternatively we may observe that under f_2 there are only three 0 entries, and thus we may obtain a simpler expression if we express f_2 using these 0 entries. We can do this by writing $\overline{f_2}$ as a sum of the product terms corresponding to the three 0 entries to obtain

$$\overline{f_2} = \overline{A}BC + AB\overline{C} + ABC$$

If we require f_2 we can simply write

$$f_2 = \overline{\overline{A}BC + AB\overline{C} + ABC}$$

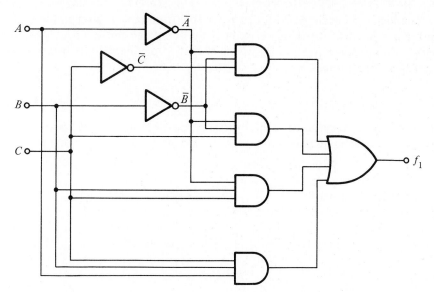

Fig. 6.13   Realization of the function f, (Fig. 6.12) using AND, OR, and NOT gates.

That this expression is equivalent to the one obtained above can be verified using methods described below.

EXERCISES

6.3   Obtain the canonic sum-of-products expression for the function f described by the truth table in Fig. E6.3. This function is called Exclusive OR, or EXOR, and is symbolized by

A	B	f
0	0	0
0	1	1
1	0	1
1	1	0

Fig. E6.3   Truth table for the Exclusive-OR function, $f = A \oplus B$.

$A \oplus B$. The name derives from the fact that the function is 1 if one *or* the other *but not both* of A and B equal 1. The Exclusive OR is an important conceptual building block in logic systems.

Ans. $f = \overline{A}B + A\overline{B}$

6.4   For three variables, what is the number of possible minterms?

Ans.   8

The Equivalence of Logic Expressions

We have shown how to find *one* possible function expression from a given truth table. In fact, there are *many* equivalent expressions and logic networks for any particular truth table. In general, two logic expressions or logic networks are equivalent if they have identical truth tables. In fact, the most straightforward (though sometimes tedious) way of proving the equivalence of two logic expressions is by means of truth tables.

Consider as an example the function f_1 whose truth table is given in Fig. 6.12 and whose sum-of-product expression is given in Eq. (6.1), and for which a circuit implementation is given in Fig. 6.13. We will show that f_1 can alternatively be described by the much simpler expression

$$f_1 = \overline{A}\overline{B} + BC \qquad (6.2)$$

The truth table corresponding to this expression can be easily constructed, as shown in Fig. 6.14. If we compare this truth table with that in Fig. 6.12, we see that the two are

$A\ B\ C$	$\overline{A}\overline{B}$	BC	f_1
0 0 0	1	0	1
0 0 1	1	0	1
0 1 0	0	0	0
0 1 1	0	1	1
1 0 0	0	0	0
1 0 1	0	0	0
1 1 0	0	0	0
1 1 1	0	1	1

Fig. 6.14   Truth table for the function f, given by Eq. (6.2).

indeed identical. Thus the expressions in Eqs. (6.1) and (6.2) are equivalent. Figure 6.15 shows a direct implementation of the expression in Eq. (6.2) that is obviously simpler and more economical than that of Fig. 6.13.

The Minimization of Logic Expressions

Now that we know how to verify the equivalence of logic expressions and networks, we will explore the creation or synthesis of alternate forms, ideally simpler and ultimately minimal. To simplify logic expressions we will perform a series of algebraic manipulations using a set of algebraic identities.

Table 6.1 lists some of the most basic and important identities. The reader should

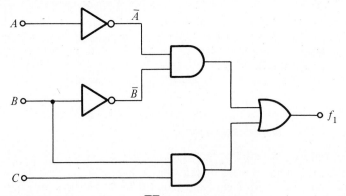

Fig. 6.15   Realization of the function $f_1 = \overline{A}\overline{B} + BC$.

TABLE 6.1   Rules of Binary Logic

Name	Algebraic Identity	
Involution	$A = \overline{\overline{A}}$	
Dominance	$1 + A = 1$	$0 \cdot A = 0$
	$0 + A = A$	$1 \cdot A = A$
Complementation	$A + \overline{A} = 1$	$A \cdot \overline{A} = 0$
Indempotence	$A + A = A$	$A \cdot A = A$
Commutation	$A + B = B + A$	$A \cdot B = B \cdot A$
Association	$(A + B) + C = A + (B + C)$	$(A \cdot B) \cdot C = A \cdot (B \cdot C)$
Distribution	$A \cdot (B + C) = A \cdot B + A \cdot C$	$A + B \cdot C = (A + B) \cdot (A + C)$
DeMorgan	$\overline{A + B} = \overline{A} \cdot \overline{B}$	$\overline{A \cdot B} = \overline{A} + \overline{B}$

satisfy herself or himself of their truth by means of truth tables. Note that except for the involution law, which involves only one variable, two *dual* identities are given in each line. The dual identities are obtained from each other by replacing the OR operator ($+$) by the AND operator ($\cdot$), and vice versa, and replacing 1 by 0, and vice versa.

The identities of Table 6.1 can be used to help simplify a given logic expression. The general strategy in performing algebraic manipulations on an expression for the purpose of simplification is as follows: First, group in pairs product terms that differ only in that some variable appears complemented ($\overline{X}$) in one and true (X) in the other. When the common subproduct consisting only of variables other than X is factored out of the pair via the distribution rule, only $\overline{X} + X$ remains, a combination which always equals 1. For example, let us apply this idea to the large expression for f_1 in Eq. (6.1),

$$f_1 = \overline{A}\overline{B}\overline{C} + \overline{A}\overline{B}C + \overline{A}BC + ABC$$

$$= \overline{A}\overline{B}(\overline{C} + C) + BC(\overline{A} + A)$$

$$= \overline{A}\overline{B} + BC$$

This is the expression in Eq. (6.2); we have already demonstrated that it is equivalent to the initial expression for f_1.

Although grouping in pairs for minimization is a potentially successful ploy, it is not as easy as the above example suggests; it is often necessary to manufacture suitable extra terms for the purpose. The most helpful fact is that

$$A = A + A$$

allowing replication of terms for multiple pairing, as will be seen in Example 6.2.

Example 6.2
Consider the function f_2 whose truth table is given in Fig. 6.12. From this table we can write the canonic sum-of-products expression for f_2 as

$$f_2 = \overline{A}\overline{B}\overline{C} + \overline{A}\overline{B}C + \overline{A}B\overline{C} + A\overline{B}\overline{C} + A\overline{B}C$$

By repeating the first product term $\overline{A}\overline{B}\overline{C}$ we obtain

$$f_2 = \overline{A}\overline{B}\overline{C} + \overline{A}\overline{B}C + \overline{A}B\overline{C} + \overline{A}\overline{B}\overline{C} + A\overline{B}\overline{C} + A\overline{B}C$$

Grouping in pairs and factoring yields

$$f_2 = \overline{A}\overline{B}(\overline{C} + C) + \overline{A}\overline{C}(B + \overline{B}) + A\overline{B}(\overline{C} + C)$$

$$= \overline{A}\overline{B} + \overline{A}\overline{C} + A\overline{B}$$

or

$$f_2 = \overline{A}\overline{B} + A\overline{B} + \overline{A}\overline{C}$$

which reduces by factoring of the first pair of terms to

$$f_2 = \overline{B} + \overline{A}\overline{C} \tag{6.3}$$

$$\cdot \quad \cdot \quad \cdot$$

EXERCISE

6.5 We have previously obtained the following alternate form for f_2:

$$\overline{f}_2 = \overline{A}BC + AB\overline{C} + ABC$$

Manipulate this expression to obtain the form in Eq. (6.3).

The above technique for minimization of logic functions relies on intuition and hence, though convenient, is not the most straightforward. There exist other, organized procedures for function minimization, the most popular of which uses the Karnaugh map. This topic, however, is beyond the scope of this book.

6.3 THE NAND AND NOR OPERATIONS

There exist two logic operations that are composites of the elementary AND, OR, and NOT operations but that have special importance because they are the most natural functions to implement in electronic circuit form. These are the NAND and NOR operations, which are defined in the following subsections. We shall also show how NAND and NOR gates can be employed in the synthesis of logic functions.

The NAND Function

The NAND function is an AND followed by a NOT. Its name is a shortened version of NOTAND. Figure 6.16 shows the circuit symbol for a two-input NAND gate. The

Fig. 6.16 Circuit symbol for a two-input NAND gate.

$$
\begin{array}{cc|c}
A & B & C \\
\hline
0 & 0 & 1 \\
0 & 1 & 1 \\
1 & 0 & 1 \\
1 & 1 & 0 \\
\end{array}
$$

Fig. 6.17   Truth table for NAND gate.

symbol is simply that of an AND with a circle representing the output inversion. The truth table for the two-input NAND is shown in Fig. 6.17. The NAND operation is described by

$$C = \overline{AB}$$

which, using DeMorgan's law, can be expressed in the alternate form

$$C = \overline{A} + \overline{B}$$

Thus the output is 1 if A or B is 0, as the truth table shows.

The NAND function for more than two variables is defined in a similar manner. For example, the NAND function for four variables is

$$Y = \overline{ABCD}$$

or

$$Y = \overline{A} + \overline{B} + \overline{C} + \overline{D}$$

Note that this form for expressing the NAND operation suggests the alternative symbol in Fig. 6.18.

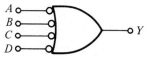

Fig. 6.18   An alternative symbol for NAND gate; $Y = \overline{A} + \overline{B} + \overline{C} + \overline{D}$.

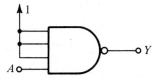

Fig. 6.19   Using a four-input NAND gate as an inverter; $Y = \overline{A}$.

A NAND gate can be used as an inverter by connecting its unused inputs to logic 1, as indicated in Fig. 6.19.

The NOR Function

The NOR function is an OR followed by a NOT. Its name is a shortened version of NOTOR. Figure 6.20 shows the circuit symbol for a two-input NOR gate. The symbol is that of an OR with an additional circle representing the output inversion. The truth table of the two-input NOR is shown in Fig. 6.21. The NOR operation is described by

$$C = \overline{A + B}$$

which can be expressed, using DeMorgan's law, in the alternative form

$$C = \overline{A}\,\overline{B}$$

Thus C will be 1 only if A and B are simultaneously 0, in correspondence with the truth table.

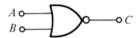

Fig. 6.20   Circuit symbol for NOR gate.

A	B	C
0	0	1
0	1	0
1	0	0
1	1	0

Fig. 6.21   Truth table for the NOR function; $C = \overline{A + B} = \overline{A}\,\overline{B}$.

The NOR definition extends to more than two variables. For example, the NOR of four variables is

$$Y = \overline{A + B + C + D}$$

or, alternatively,

$$Y = \overline{A}\,\overline{B}\,\overline{C}\,\overline{D}$$

Note that this form for expressing the NOR operation suggests the alternative symbol in Fig. 6.22, which is simply an AND with input inversions.

A NOR gate can be used as an inverter by simply connecting the unused inputs to logic 0 level, as indicated in Fig. 6.23.

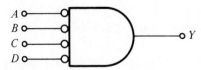

Fig. 6.22   An alternative symbol for NOR gate; $Y = \overline{\overline{A}\,\overline{B}\,\overline{C}\,\overline{D}}$.

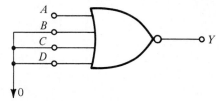

Fig. 6.23   Using a four-input NOR gate as an inverter.

The Completeness of the NAND and NOR Operators

In Section 6.2 we showed how to realize logic functions using AND, OR, and NOT gates. All three operators together, however, are not necessary. It can be shown that one needs only either AND and NOT gates, or OR and NOT gates. This suggests that we may be able to realize arbitrary logic functions using NAND gates alone (or NOR gates alone). That this is true can be proved by showing how NAND (or NOR) gates can be used to realize the three elementary logic functions. We can thus say that the NAND (NOR) operator is logically complete.

EXERCISE

6.6   Find the functions realized by the circuits in Fig. E6.6.
    Ans.   (a) $Y = A + B$; (b) $Y = AB$

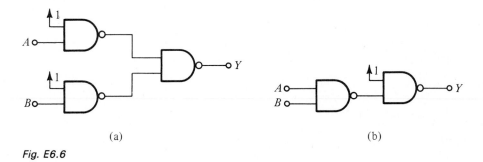

(a)   (b)

Fig. E6.6

Synthesis with NAND and NOR Gates

Synthesis of logic functions using only NAND gates can be done in the following manner: First, express the function in the sum-of-products form and sketch its realization using AND, OR, and NOT gates. As an example, consider the function

$$Y = AB + CD$$

whose AND-OR realization is shown in Fig. 6.24a. Then add inverters (circles) that cancel each other, as shown in Fig. 6.24b. In this figure there exist two cascaded inverters in each signal path, with the result that the function realized is identical to that of the original circuit in Fig. 6.24a. In the circuit of Fig. 6.24b, however, all of the gates are NAND gates. We may therefore redraw the circuit using the standard NAND symbols, as shown in Fig. 6.24c.

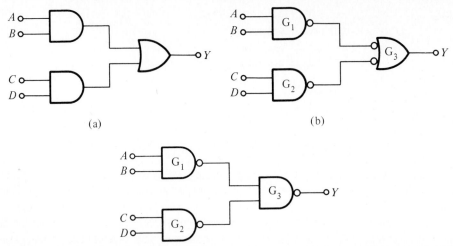

(a)   (b)

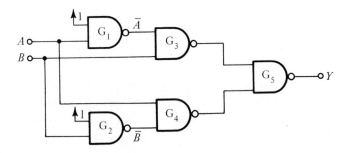

Fig. 6.24   *Illustrating the realization of the function Y = AB + CD using NAND gates. (a) AND-OR realization. (b) Addition of inverters that cancel, thus leaving the function unchanged. (c) Same as (b) but with G₃ redrawn.*

The synthesis of logic functions using NOR gates follows a parallel, but dual, procedure. Here we start with a product-of-sums form and a realization using ORs and one AND. We then add canceling inverters. In this way all gates are converted to NOR gates.

EXERCISE

6.7   Find a realization using NAND gates of the Exclusive OR function

$$Y = \overline{A}B + A\overline{B}$$

Ans.   See Fig. E6.7

Fig. E6.7

6.4   *PRACTICAL CIRCUIT CONSIDERATIONS*

Thus far we have concentrated on the synthesis of logic functions using ideal logic gates. In this section we shall discuss some of the properties and limitations of actual integrated-circuit logic packages. Our objective at this stage is to present sufficient information to enable the reader to experiment with this powerful class of circuits. The following treatment is in a sense parallel to that of the op amp in Chapter 3. It should be emphasized, however, that one's ability to work with logic circuits is tremendously enhanced by the knowledge of the circuits inside the package. Such a study is presented in Chapter 15.

Logic Families

Integrated-circuit (IC) logic packages are classified into a number of different families. Members of each family are made with the same technology, have similar circuit structure, and exhibit the same basic features. Currently the most popular families are transistor-transistor logic (TTL) and complementary metal-oxide-semiconductor (CMOS) logic.

In designing a digital system one selects an appropriate logic family and attempts to implement as much of the system as possible using packages that belong to this family. In this way interconnection of the various packages is relatively straightforward. If, on the other hand, packages from more than one family have to be used, one has to design suitable *interface circuits*. The selection of a logic family is based on such considerations as logic flexibility, speed of operation, availability of complex functions, noise immunity, operating-temperature range, power dissipation, and cost. We will discuss these and other considerations in this chapter and in Chapter 15.

Many types of logic functions are usually available within each logic family. Depending on the complexity of the circuit on the IC chip, the package can be classified as either a

1. Small-scale integrated (SSI) circuit, or
2. Medium-scale integrated (MSI) circuit, or
3. Large-scale integrated (LSI) circuit, or
4. Very-large-scale integrated (VLSI) circuit.

Although the boundaries between the different levels of integration are not very sharp, a rough guide, based on the number of "equivalent logic gates" on the chip, is as follows: SSI, 1 to 10 gates; MSI, 10 to 100 gates; LSI, 100 to 1,000 gates; and VLSI, > 1,000 gates.

The basic circuit properties of a logic family can be established by studying the basic inverter circuit of that family. In the following we shall consider the general characteristics and structure of the logic inverter circuit.

The Basic Inverter

The logic inverter is basically a voltage-controlled switch such as that represented schematically in Fig. 6.25. As shown, the switch connected between terminals 2 and 3 is controlled by the input signal v_I applied between terminals 1 and 3 (terminal 3 is connected to the reference or ground point). The inverter output voltage v_O is taken across the switch, that is, between 2 and ground. When v_I is low (around 0 V) the switch is open and the output voltage v_O is high (equal to the supply voltage V^+). When v_I is

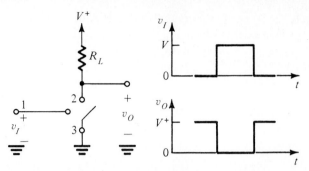

Fig. 6.25   *A conceptual representation of the logic inverter as a voltage-controlled switch.*

high (above a specified threshold voltage, as explained below) the switch is closed and the output voltage is low (0 V). It should therefore be obvious that this circuit realizes the logic inversion operation.

   Practical logic inverter circuits differ from the conceptual circuit in Fig. 6.25 in a number of ways. First, the input terminal of the inverter usually draws some current from the driving source. Second, the switch is not ideal; specifically, when the switch is closed it does not behave as a short circuit but rather has a finite closure resistance (called *on resistance*) and sometimes an additional voltage drop (called offset voltage). Figure 6.26 shows an equivalent circuit of the switch in the closed position. As a result

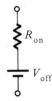

Fig. 6.26   *Equivalent circuit for a typical switch in the on position.*

of these imperfections the voltage v_O will not be zero in the on state. Third, the inverter switch does not switch instantaneously; rather there is a delay time between the application of the input change and the appearance of the output change. This point will be discussed in some detail shortly. Fourth, actual inverters do not exhibit a well defined switching threshold as implied by the discussion of the conceptual circuit in Fig. 6.25. Transfer characteristics of logic inverters are discussed next.

The Inverter Transfer Characteristic

Figure 6.27 shows the ideal transfer characteristic for a logic inverter fed with a power supply V^+. As indicated, the inverter exhibits a threshold voltage $V_t = \frac{1}{2}V^+$. Input signals below this threshold voltage are interpreted as being low, and the inverter output is equal to the supply voltage V^+. Input signals above the threshold are interpreted as high, and the inverter output is equal to 0 V. Note that this inverter is quite tolerant of

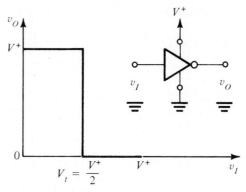

Fig. 6.27   Ideal transfer characteristic of the logic inverter.

errors in the value of the input signal. Also, with the threshold voltage at half the power supply, this tolerance is equally distributed between the two domains of input signal (low and high). Finally, the inverter characteristics are ideal in the sense that, independent of the exact value of input signal, the output is either V^+ or 0 V.

Actual logic inverters have transfer characteristics that only approximate the ideal one of Fig. 6.27. Figure 6.28 shows a typical inverter transfer characteristic. Note that the threshold voltage is no longer well defined and that there exists a *transition region*

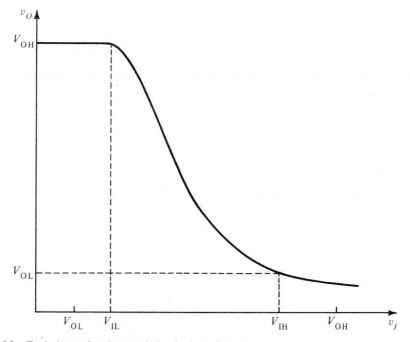

Fig. 6.28   Typical transfer characteristic of a logic inverter.

between the high and low states. Also, the high output (V_{OH}) and low output (V_{OL}) are no longer equal to V^+ and 0 V, respectively.

Examining the inverter transfer characteristic in Fig. 6.28 in more detail, we note that there are three distinct regions:

1. the low-input region; $v_I < V_{IL}$,
2. the transition region; $V_{IL} \leq v_I \leq V_{IH}$, and
3. the high-input region; $v_I > V_{IH}$

Input voltages less than V_{IL} are acknowledged by the gate as representing logic 0. Thus, V_{IL} *is the maximum allowable logic 0 value.* Similarly, input voltages greater than V_{IH} are acknowledged by the gate as representing logic 1. Thus, V_{IH} *is the minimum allowable logic 1 value.*

We should note in passing that the inverter transfer characteristic indicates that it is a grossly nonlinear device. In fact, digital circuits are extreme cases of the nonlinear circuits discussed in Chapter 5. If the inverter whose transfer characteristic is depicted in Fig. 6.28 were to be used as a linear amplifier, it would be biased at a point somewhere in the transition region and the input signal swing would be kept small to restrict operation to a short, almost linear segment in the transition region of the characteristic.

Noise Immunity

A great advantage of binary digital circuits is their tolerance to variations in value of input signals. As long as the input signal is correctly interpreted as low or high, the accuracy of operation is not affected. This tolerance to variation in signal level also can be considered as immunity to noise superimposed on the input signal. We shall now define "noise immunity."

Consider once more the inverter transfer characteristic of Fig. 6.28 where the four parameters V_{IL}, V_{IH}, V_{OL} and V_{OH} are indicated. On the horizontal axis we have also marked the points corresponding to the output levels V_{OL} and V_{OH}. In a logic system, one gate usually drives another. Thus a gate whose output is high at V_{OH} drives an identical gate whose specified minimum input logic 1 level is V_{IH}. For proper logic operation it should be obvious that V_{OH} has to be greater than V_{IH}. Furthermore, we see that the difference $V_{OH} - V_{IH}$ represents a margin of safety; for if noise were superimposed on the output signal of the driving gate (V_{OH}), the driven gate would not be bothered as long as the amplitude of the noise voltage was lower than ($V_{OH} - V_{IH}$). This difference is therefore called *logic 1 noise margin* and is denoted $\Delta 1$,

$$\Delta 1 = V_{OH} - V_{IH}$$

The *logic 0 noise margin,* $\Delta 0$, is similarly defined,

$$\Delta 0 = V_{IL} - V_{OL}$$

Due to the unavoidable variabilities in the values of circuit components and power supply voltage, the manufacturer usually specifies *worst-case* values for the four parameters V_{OH}, V_{IH}, V_{OL}, and V_{IL}. These values are defined as follows:

V_{OH} the minimum voltage that will be available at a gate output when the output is supposed to be logic 1.

V_{IH} the minimum gate input voltage that will be unambiguously recognized by the gate as corresponding to logic 1.

V_{OL} the maximum voltage that will be available at a gate output when the output is supposed to be logic 0.

V_{IL} the maximum gate input voltage which will be unambiguously recognized by the gate as corresponding to logic 0.

The manufacturer thus guarantees that for $v_I \leq V_{IL}$, $v_O \geq V_{OH}$ and that for $v_I \geq V_{IH}$, $v_O \leq V_{OL}$.

Instead of drawing the complete transfer characteristic, one usually is satisfied with the logic band diagram of Fig. 6.29. From this diagram it can be inferred that to max-

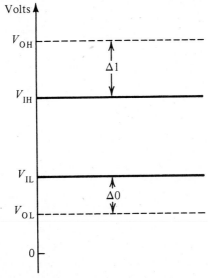

Fig. 6.29   Logic band diagram indicating the noise margins.

imize and equalize $\Delta 0$ and $\Delta 1$, one ideally desires that $V_{IL} = V_{IH} =$ a value midway in the *logic swing* V_{OL} to V_{OH}. This implies that the transfer characteristic should switch abruptly, that is, it should exhibit high gain in the transition region. It also implies that switching should occur in the middle of the logic swing. To maximize this swing, V_{OL} should be as low as possible (ideally 0 V) and V_{OH} should be as high as possible (ideally equal to the power supply voltage V^+). Such an idealized transfer characteristic is shown in Fig. 6.27. As will be seen in Chapter 15, the CMOS logic family provides an excellent approximation to the ideal performance.

Power Dissipation

Of interest to a logic-circuit designer is the amount of power consumed by his or her circuits. With this fact she or he can determine the current that the power supply for a digital system must provide. The power dissipated in a logic circuit is composed of two components: static and dynamic. The *static power* is the power dissipated while the circuit is not changing states. Thus in the idealized inverter of Fig. 6.25 we see that when the output is high the static power is zero, while when the output is low the static power is $(V^+/R_L)V^+$. If one assumes that on average a gate spends half the time in either state, then the average static power dissipation will be $(V^+)^2/2R_L$.

To visualize *dynamic-power dissipation,* consider the idealized inverter of Fig. 6.25 when driving a load capacitance C_L. This could be the input capacitance of another logic gate or the capacitance of the wire interconnection between the inverter and another part of the system. Let us assume that initially, at $t = 0-$, the input was high and the switch was closed. Thus the capacitor was initially discharged. Now let v_I change to low at $t = 0$. It follows that the switch will open, but since the capacitor voltage cannot change instantaneously, v_O will rise exponentially toward V^+. The charging current will be supplied through R_L, and thus power will be dissipated in R_L. If the instantaneous supply current is denoted by i, then the energy drawn from the supply will be $\int V^+ i \, dt = V^+ \int i \, dt = V^+ Q$, where Q is the charge supplied to the capacitor, that is, $Q = C_L V^+$. Thus the energy drawn from the supply is $C_L(V^+)^2$. Now, since the capacitor initially had zero energy and will finally have a stored energy of $\frac{1}{2}C_L(V^+)^2$, it follows that the energy dissipated in R_L is $\frac{1}{2}C_L(V^+)^2$.

If we next let v_I go high, the switch will close and C_L will discharge through the switch resistance R_{on}, again dissipating power through the switch. If we assume that V_{off} is zero, then the energy dissipated in R_{on} will be equal to the energy stored on the capacitor, $\frac{1}{2}C_L(V^+)^2$. Thus if the inverter is switched on and off f times per second, the dynamic power dissipation will be

$$\text{Dynamic power dissipation} = fC_L(V^+)^2$$

which is the sum of the power dissipations in R_L and R_{on}.

Fan-in and Fan-out

Fan-in of a gate is the number of its inputs. Thus a four-input NOR gate has a fan-in of 4. Fan-out is the maximum number of similar gates that a gate can drive while remaining within the guaranteed specifications. Much more will be said about fan-in and fan-out in Chapter 15.

Propagation Delay

Because of the dynamics involved in the action of a switching device such as a transistor, the switch in the inverter of Fig. 6.25 does not respond instantaneously to the control signal v_I. In addition, the inevitable load capacitance at the inverter output causes the waveform of v_O to depart from an ideal pulse. Figure 6.30 illustrates typical response of an inverter to an input pulse with finite rise and fall times. The output pulse, of course, exhibits finite rise and fall times. In addition, there is a finite nonzero delay time between the input and output pulses. There are various ways for expressing this *propagation delay*. One way is to specify the times between the 50% points of the input and output waveforms at both the leading and trailing edges. These times are referred to as

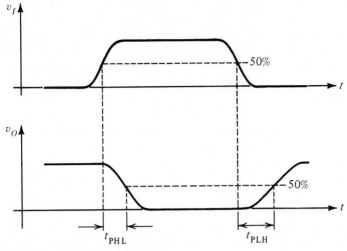

Fig. 6.30 *Propagation delay of a logic inverter.*

t_{PHL} (where HL indicates high-to-low transition of the output) and t_{PLH} (where LH indicates low-to-high transition of the output) in Fig. 6.30. The propagation time delay t_{PD} can then be defined as the average of these two times,

$$t_{PD} = \tfrac{1}{2}(t_{PHL} + t_{PLH})$$

Delay-Power Product

One is usually interested in high-speed performance (low t_{PD}) combined with low power dissipation. Unfortunately these two requirements are in conflict since, generally, if one attempts when designing a gate to reduce power dissipation by decreasing supply current, the gate delay will increase. It follows that a figure of merit for comparing logic families is the delay-power product DP, defined

$$DP = t_{PD}P_{diss}$$

where P_{diss} is the power dissipation of the gate.[1] The lower the DP figure for a family, the more effective this logic family is. We shall provide a comparison of logic families on this basis in Chapter 15.

Physical Packaging of Logic Circuits

Figure 6.31a shows the most common physical package used to house IC logic circuits. This package is made either of plastic or ceramic and is called a *dual-in-line* (DIP) package. It has 14 leads, 7 brought out to each side. Other packages with 16, 24, and 40 pins exist.

A functional diagram of a quad two-input NAND package is shown in Fig. 6.31b.

---

[1]Usually only the static power dissipation is employed in evaluating the delay-power product.

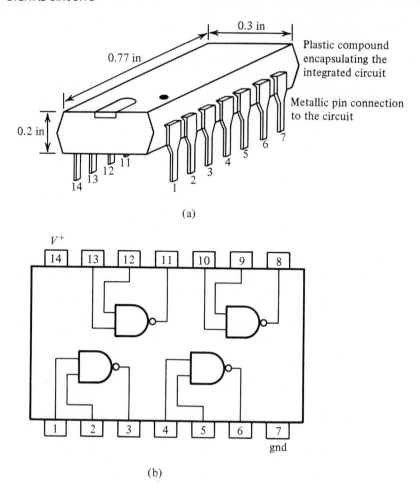

(a)

(b)

Fig. 6.31 A 14-pin integrated circuit. (a) Physical appearance. (b) Schematic of an integrated circuit providing 4 two-input NAND gates.

EXERCISE

6.8 The data sheet of the SN7400 quad 2-input NAND gates of the TTL family provides the following:

Logic 1 input voltage required at both input terminals to ensure logic 0 level at output: MIN (minimum) 2 V

Logic 0 input voltage required at either input terminal to ensure logic 1 level at output: MAX (maximum) 0.8 V

Logic 1 output voltage: MIN 2.4 V, TYP (typical) 3.3 V

Logic 0 output voltage: TYP 0.22 V, MAX 0.4 V

Logic 0 level supply current: TYP 12 mA, MAX 22 mA

Logic 1 level supply current: TYP 4 mA, MAX 8 mA

Propagation delay time to logic 0 level: TYP 7 ns, MAX 15 ns

Propagation delay time to logic 1 level: TYP 11 ns, MAX 22 ns

(a) Find the noise margin in both the 0 and 1 states.

(b) Assuming that the gate is 50% of the time in the 1 state and 50% of the time in the 0 state, find the average static power dissipated in a typical gate. The power supply voltage is $+5$ V.

(c) Assuming that the gate drives a capacitance $C_L = 45$ pF and is switched at 1 MHz frequency, find the dynamic power dissipation per gate using the typical values of the logic 1 and 0 levels at the output.

(d) Find the typical value of the gate delay–power product.

Ans.   (a) 0.4 V, 0.4 V; (b) 10 mW; (c) 0.7 mW; (d) 90 picojoules

6.5   FLIP-FLOPS

The logic circuits considered thus far are called *combinational* (or *combinatorial*). Their output depends only on the current value of the input. Thus these circuits do *not* have memory.

Memory is a very important part of digital systems. Its availability in digital computers allows for storing programs and data. Furthermore, it is important for the temporary storage of the output produced by a combinational circuit for use at a later stage in the operation of the digital system.

In this section we shall study some basic digital memory elements. Logic circuits that incorporate memory are called *sequential circuits*—their output depends not only on the present value of the input but also on the input's previous values.

The SR Flip-Flop

We shall begin by considering the simplest digital memory element, the *set–reset (SR) flip-flop* shown in Fig. 6.32a. As indicated, the flip-flop is formed by cross-coupling two NOR gates. It has two input terminals, called set (S) and reset (R), and two complementary outputs, denoted Q and $\overline{Q}$. In the *rest state* or *memory state* the inputs are both 0 ($R = 0$ and $S = 0$). Under this condition the flip-flop can have one of two output states, either $Q = 1$, $\overline{Q} = 0$ or $Q = 0$, $\overline{Q} = 1$. Furthermore, the flip-flop will remain in either of these two states as long as the rest condition $R = 0$, $S = 0$ prevails. That this is true can be easily verified with the circuit diagram in Fig. 6.32a. If $Q = 1$, the output $\overline{Q}$ of NOR gate G_2 will be 0. Thus both inputs to G_1 are 0, causing its

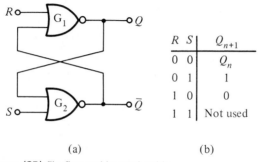

R	S	Q_{n+1}
0	0	Q_n
0	1	1
1	0	0
1	1	Not used

(a)   (b)

Fig. 6.32   *The set-reset (SR) flip-flop and its truth table.*

output Q to be 1, as initially assumed. On the other hand, if $Q = 0$, then both inputs to G_2 are 0 and thus its output $\overline{Q}$ is 1. This 1 input to G_1 causes its output Q to be 0.

Since the flip-flop can have one of two possible stable states, it is capable of *storing one bit* of information. If the flip-flop is in the state $Q = 1$, $\overline{Q} = 0$, it will be referred to as storing a 1. Conversely, in the state $Q = 0$, $\overline{Q} = 1$ the flip-flop will be considered to be storing a 0. When the flip-flop is storing a 1, it is said to be *set*; when it is storing a 0 it is said to be *reset*.

Consider the flip-flop to be in the reset state ($Q = 0$, $\overline{Q} = 1$). Let us raise the input S to logic 1 while leaving R at 0. From the circuit diagram in Fig. 6.32a we see that the 1 at the S terminal will force $\overline{Q}$ to 0. Thus the two inputs to G_1 will be 0 and its output Q will go to 1. Now even if S returns to 0 the flip-flop remains in the newly acquired set state. Obviously, if we raise S to 1 again (with R remaining at 0) no change will occur. To reset the flip-flop we need to raise R to 1 while leaving $S = 0$. We can readily show that this forces the flip-flop into the reset state and that the flip-flop remains in this state even after R returns to 0.

Finally, we inquire into what happens if both S and R are simultaneously raised to 1. The two NOR gates will cause both Q and $\overline{Q}$ to become 0 (note that in this case the complementary labeling of these two variables is incorrect). However, if R and S return to the rest state simultaneously, the state of the flip-flop will be undefined. In other words it will be impossible to predict the final state of the flip-flop. For this reason this input combination is usually disallowed (that is, not used). Note, however, that this situation arises only in the idealized case, when both R and S return to 0 precisely simultaneously. In actual practice one of the two will return to 0 first, and the final state will be determined by the input that remains high longest.

The operation of the flip-flop is summarized by the truth table in Fig. 6.32b, where Q_n denotes the value of Q at time t_n just before the application of the R and S signals and Q_{n+1} denotes the value of Q at time t_{n+1} after the application of input signals. The first row of the table corresponds to the memory or rest condition, the second row to the set condition, and the third row to the reset condition; the fourth row represents the disallowed input condition.

At this point the reader should note the similarity of this digital memory element and the bistable circuit described in Section 5.9. The SR flip-flop is a particular implementation of a bistable multivibrator.

Rather than using two NOR gates, one can implement an SR flip-flop by cross-coupling two NAND gates, as shown in Fig. 6.33. In this case, however, the rest or

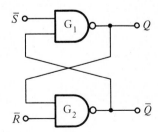

Fig. 6.33   SR flip-flop formed from NAND gates.

memory condition is obtained with the two input terminals at logic 1 level. To set the flip-flop the set line (labeled $\bar{S}$) should be lowered to the 0 level. Similarly, to reset the flip-flop the reset line (labeled $\bar{R}$) should be lowered to the 0 level. The labeling of the set and reset lines makes the truth table of Fig. 6.32b applicable in this case as well.

Finally, we show in Fig. 6.34 a black-box representation of the SR flip-flop of Fig. 6.33.

Fig. 6.34   Black-box representation of the SR flip-flop in Fig. 6.33.

The Clocked (Gated) SR Flip-Flop

There are many situations in which one requires that the flip-flop sense its inputs only at a certain time. For instance, the R and S input lines are usually connected to the output terminals of another part of the digital system. The signals on these lines could therefore be changing, and we might wish that the flip-flop not respond to these changes until such time as we are sure that "things have settled down" and the correct signals have appeared on the R and S lines. This can be achieved by *gating* the inputs to the flip-flop, using two additional logic gates, as shown in Fig. 6.35. It is easy to verify that only with $G = 1$ are S and R sensed by the flip-flop. With $G = 0$ the flip-flop remains in the memory state independent of changes on the S and R lines.

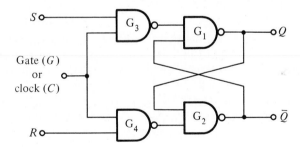

Fig. 6.35   A gated (clocked) SR flip-flop.

In many digital systems, operations (such as the setting or resetting of flip-flops) occur only at times determined by a square-wave generator called the *clock*. Connecting this generator to the G terminal of the flip-flop in Fig. 6.35 ensures that the flip-flop senses the S and R signals only when the clock signal is at the high level. In designing such a system one should ensure that the correct signals on the S and R lines should appear and stabilize before the rising edge of the clock pulse. Then, as the clock rises the flip-flop senses the S and R signals and changes state (if required). Digital systems in which operations are controlled by periodic clock pulses are called *synchronous* systems.

6.9   Figure E6.9 shows a clocked SR flip-flop in which additional logic is included to obtain the *direct set and reset* terminals ($\overline{S}_d$ and $\overline{R}_d$). Show that when input signals are applied to these terminals the flip-flop becomes completely unresponsive to the clock pulse. Specifically, show the following:

(a)   If $\overline{S}_d = 0$ and $\overline{R}_d = 1$, then $Q = 1$, and if $\overline{S}_d = 1$ and $\overline{R}_d = 0$, then $Q = 0$, each independent of the signal at C, S, or R.

(b)   The flip-flop is controlled by the C, S, and R inputs only when $\overline{S}_d = \overline{R}_d = 1$.

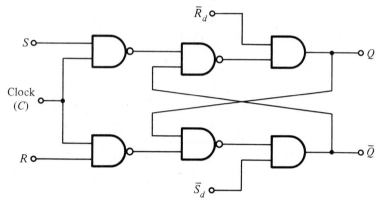

Fig. E6.9

The Latch or Data Flip-Flop

The $R = S = 1$ restriction that still applies to the gated RS flip-flop in Fig. 6.35 may be overcome with the simple addition shown in Fig. 6.36. Here an inverter is added to make $R = \overline{S}$, thus preventing the disallowed $R = S = 1$ input state. The result is a flip-flop configuration that implements two functions of considerable importance. The first of these, called the *latch mode*, creates a *latch* or *latching flip-flop* in which the gate G is normally open ($G = 1$) so that $Q = S$, the logic behaving as a wire (with delay). However, when the gate G is closed ($G = 0$) the outputs of G_3 and G_4 are raised simultaneously and the current state is maintained or latched with $Q_{n+1} = S_n$.

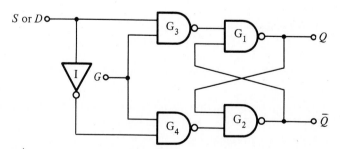

Fig. 6.36   *Including the inverter I converts the SR flip-flop to a latch or D-type flip-flop.*

The second mode of operation is the *gated data mode,* creating a special form of *data* or *D flip-flop.* Here the gate G is normally closed ($G = 0$) and is opened ($G = 1$) only briefly to capture the data at S (or D, as it is usually called in this case).

A somewhat more important form of the D flip-flop is the *edge-triggered* type, which operates on the rising edge of G (which in this case is often called C, for clock). The most popular form of this edge-triggered D flip-flop is shown in Fig. 6.37. For this

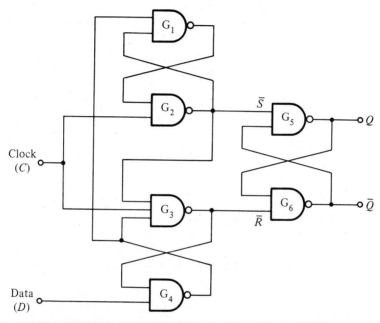

Fig. 6.37 *An edge-triggered (positive or rising edge of clock) D-type flip-flop.*

circuit it can be shown that when C makes a positive-going transition from $C = 0$ to $C = 1$, Q becomes 0 if D was 0 and Q becomes 1 if D was 1. It can be also shown that there is no other way to change Q and that changes in D when $C = 1$ do not affect Q.

EXERCISE

6.10 For the circuit in Fig 6.37 verify the following:
  (a) When $C = 0$, $\overline{S} = \overline{R} = 1$ independently of D; hence the flip-flop can be, and remain, in either of the two states.
  (b) With $D = 0$ and $C = 0$, Q goes to 0 as C goes to 1. Then, changes in D do not affect Q while $C = 1$.
  (c) With $D = 1$ and $C = 0$, Q goes to 1 as C goes to 1. Then, changes in D do not affect Q while $C = 1$.

Figure 6.38 shows the D flip-flop in black-box form. It also shows waveforms for input data, clock, and output Q. Note that the output waveform is identical to that of

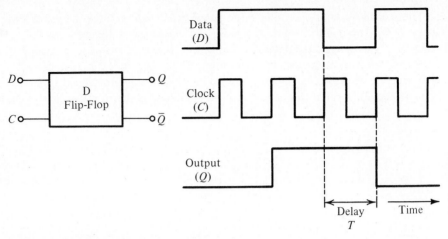

Fig. 6.38   The D-type flip-flop and its waveforms.

the input except for a delay time T equal to the period of the clock. This application of the D flip-flop in delaying an input signal gives rise to calling it a delay (or D) flip-flop.[2]

The Master-Slave Flip-Flop

The output of the clocked SR flip-flop of Fig. 6.35 changes state when the clock rises to the 1 level. In the design of digital systems where the output of one flip-flop may provide the input to another, it is important that flip-flop outputs not change at the rising edge of the clock but rather *after* the clock has returned to the 0 level. This can

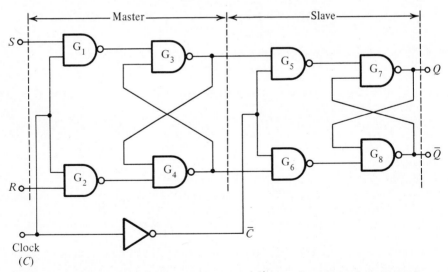

Fig. 6.39   A master-slave flip-flop. This flip-flop changes state after the clock returns to 0.

[2]It should be noted that the waveforms in Fig. 6.38 are idealized in the sense that the D signal changes at the same instant as the clock. Normally one would delay the change of the D signal slightly in order to allow for the nonzero setup time of the flip-flop.

be accomplished by using a cascade of two flip-flops, as shown in Fig. 6.39. As indicated, this clocked SR flip-flop is composed of two flip-flops called *master* and *slave*. While the input gates of the master are operated by the clock (C), those of the slave are operated by the complement of the clock ($\overline{C}$). Thus when the clock rises the slave flip-flop is isolated from the master flip-flop, and Q and $\overline{Q}$ remain unchanged. Simultaneously the master flip-flop responds to the S and R input signals and goes to the state dictated by these input signals. Then when the clock goes low the master flip-flop is isolated from the input signals and its outputs control the state of the slave flip-flop. Thus the output of the slave, which is the output of the composite flip-flop, changes when the clock goes to the 0 level and acquires the state dictated by the signals present at the S and R input terminals when the clock *was* high.

Fig. 6.40   The JK flip-flop as a black box.

The JK Flip-Flop

The type of flip-flop that we shall discuss finally is the JK flip-flop, shown in black-box form in Fig. 6.40. This is a clocked flip-flop whose output changes when the clock pulse reverts to 0. It is usually constructed using the master–slave structure with additional feedback. It has the truth table of Fig. 6.41.

J	K	Q_{n+1}
0	0	Q_n
0	1	0
1	0	1
1	1	$\overline{Q}_n$

Fig. 6.41   Truth table for the JK flip-flop.

EXERCISE

6.11   Figure E6.11 shows a JK flip-flop formed from a master-slave flip-flop (such as that in Fig. 6.39) in which each of the input NANDs has an additional input terminal connected to one of the outputs. Verify that this flip-flop follows the truth table of Fig. 6.41.

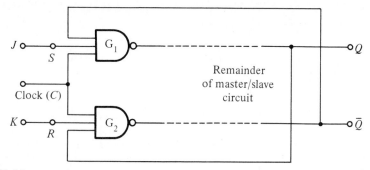

Fig. E6.11

Computer Memory Circuits

The flip-flop circuits studied in this section play an important role in the design of sequential logic circuits. An example of their application is in the design of counters, as explained in Section 6.6. These flip-flops, however, are too complicated (that is, they require too many components) to be used for data and program storage in digital computers. This application requires that large arrays of memory circuits, called *memory cells,* be fabricated on a single IC chip. Although each memory cell can be formed by cross coupling two logic inverters, there are other simpler, more efficient designs. We shall study the topic of memory circuits in some detail in Chapter 16.

6.6   REGISTERS AND COUNTERS

In this section we shall explore the use of an array of flip-flops to form two useful digital system building blocks: the shift register and the ripple counter. There are many other important digital blocks, but these two will serve to illustrate the basic principles involved.

The Shift Register

A flip-flop can be used to store or *register* a binary digit or bit. We can therefore use an array of N flip-flops to store or register N bits. If we further arrange that these bits can be *shifted* through the array, we obtain a *shift register.*

Figure 6.42 shows a 4-bit shift register formed by connecting four JK flip-flops. The input to the shift register is I. Binary data that appear on this line are shifted through the register under the control of the clock pulses. For instance, let us assume that initially the four flip-flops each have 0 stored and that the input line I is connected to logic 1. Let us further assume that the flip-flops used are of the type that change state when the clock pulse returns to 0. We can then see that, following the first clock pulse the first flip-flop, FF_0, will have an output of 1 while all other flip-flops will have 0. After the second clock pulse both FF_0 and FF_1 will have logic 1 stored in them, while FF_3 and FF_4 will have 0. The 1 stored in FF_1 is due to the 1 that was stored in FF_0 after the first clock pulse. Continuing in this manner, we see that the 1 applied at the input is shifted one flip-flop to the right every clock pulse.

The *timing diagram* in Fig. 6.43 illustrates the operation of the shift register when the arbitrary sequence (data word) 10011 is applied at the input. Here it is assumed that this sequence of input pulses is *synchronous* with the clock—that is, the changes

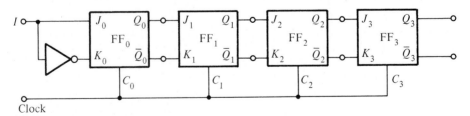

Clock

Fig. 6.42   A 4-bit shift register.

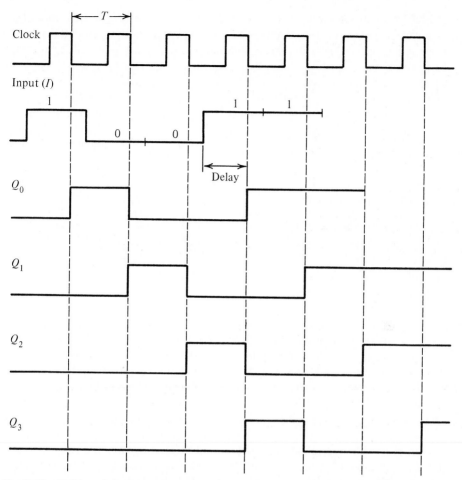

Fig. 6.43 Shifting of the data word 10011 through the shift register of Fig. 6.42. It is assumed that initially all flip-flops were reset and that flip-flops change state after the clock returns to 0.

in the pulse sequence occur at a fixed time relative to the clock pulses. Note that the waveform at the output of the first flip-flop, FF_0, of the shift register is identical to that on the input line except for a delay time. This statement applies to the waveform at the output of FF_1, which is delayed T seconds (T is the clock period) relative to that at the output of FF_0. Similarly, the waveform at the output of FF_2 is delayed T seconds relative to that at the output of FF_1, and so on.

At this point we should note that the shift register operates correctly only if each flip-flop changes state *after* the clock pulse has returned to 0 and the flip-flop input logic (that is, gates G_1 and G_2 in Fig. E6.11) has been disabled. To see that this condition is necessary, consider what would happen if a flip-flop were to change state when the clock pulse is high. FF_2, for instance, would first change state to correspond with the previous state of FF_1 (the correct situation), but then FF_1 would change state and FF_2 would follow this newly acquired state of FF_1, which obviously is not the intended operation.

Shift registers can be used to convert digital data appearing serially on a single line to parallel data appearing simultaneously on a number of lines, and vice versa. To convert N bits of serial data to parallel form, all we need is an N-bit shift register and a clock synchronized with the serial data. The serial data is then shifted through the register, and after N clock pulses it will be available at the outputs of the N flip-flops. Conversely, to convert from parallel to serial we first store the N-bit parallel data in the N flip-flops of the register. This will require that the flip-flops used have *direct set and reset* inputs (see Exercise 6.9.) Then the stored bits are shifted through, and the data become available as a series of pulses at the output of the last flip-flop in the shift register.

The Binary Number System

Before we discuss the use of flip-flop arrays to perform counting operations we shall briefly consider the binary number representation. In the decimal number representation ten symbols (0 to 9) are employed, and the position of each symbol in the decimal number determines its contribution to the value of the number. Thus the 4-digit decimal number 3,471 has the value

$$3 \times 10^3 + 4 \times 10^2 + 7 \times 10^1 + 1 \times 10^0$$

(recall that $10^0 = 1$) where 10 is called the *base* or *radix* of the number system.

In the binary number representation two symbols (0 and 1) are used; here, also, the position of each digit or bit in the binary number determines its contribution to the value of the number. Thus the 4-digit binary number 1011 (note that by convention long binary numbers are written without commas) has the value (in decimal numbers)

$$1 \times 2^3 + 0 \times 2^2 + 1 \times 2^1 + 1 \times 2^0 = 9$$

Thus binary 1011 is equivalent to decimal 9. In the binary system the radix or base is 2. Figure 6.44 lists some binary numbers and their decimal equivalents. Note that with N binary digits one can represent the numbers 0 to $2^N - 1$, a total of 2^N different numbers.

d_3	d_2	d_1	d_0	Decimal Equivalent
0	0	0	0	0
0	0	0	1	1
0	0	1	0	2
0	0	1	1	3
0	1	0	0	4
⋮	⋮	⋮	⋮	⋮
1	1	1	0	14
1	1	1	1	15

Fig. 6.44 Some 4-bit binary numbers and their decimal equivalents. Note that d_0 is the least significant bit (LSB) and d_3 is the most significant bit (MSB).

The Ripple Counter

Since a flip-flop can exist in one of two possible states, an array of N flip-flops can be in one of 2^N possible states. Such an array can therefore be made to sequence through all or some of the 2^N states under control of an input pulse signal. For instance, four flip-flops—FF_0, FF_1, FF_2, and FF_3—connected in an appropriate manner can be made to sequence through the states represented by the binary numbers in Fig. 6.44.

Let us denote the contents of these flip-flops Q_0, Q_1, Q_2, and Q_3. These four binary signals can be made to correspond to the 4 digits d_0, d_1, d_2, and d_3 of a 4-digit binary number. If initially all four flip-flops are storing zeros, then their state, represented by $Q_3 Q_2 Q_1 Q_0$, is 0000, corresponding to the binary number whose decimal equivalent is 0. If we arrange that after the first input pulse FF_0 stores a 1 and thus the state becomes 0001, then this corresponds to the binary number whose decimal equivalent is 1. Further, if we arrange that after the second input pulse FF_0 changes from 1 to 0 and FF_1 changes from 0 to 1, then the state becomes 0010, which corresponds to the binary number whose decimal equivalent is 2. Continuing in this manner, we can arrange that

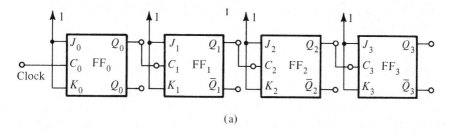

(a)

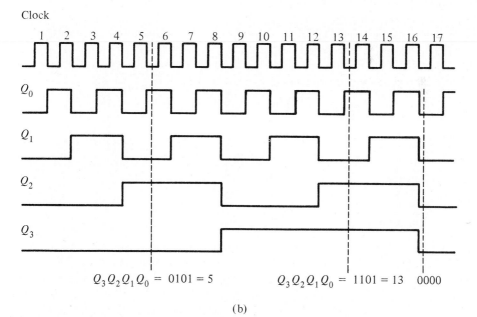

$$Q_3 Q_2 Q_1 Q_0 = 0101 = 5 \qquad Q_3 Q_2 Q_1 Q_0 = 1101 = 13 \quad 0000$$

(b)

Fig. 6.45 *A mod-16 counter (a) and its timing diagram (b).*

after the fifteenth clock pulse the state becomes 1111 and after the sixteenth clock pulse the array returns to the initial 0000 state.

An array of flip-flops connected to operate in the manner described above is called a counter. It counts the number of input pulses and displays the count in binary form. The specific counter described counts from 0 to 15 and is hence called a mod-16 counter. One such counter is depicted in Fig. 6.45a, where the pulses being counted are those of the clock. Note that the connection of the four flip-flops in Fig. 6.45a is based on the observation (from Fig. 6.44) that each digit in the binary number changes value when the preceding digit changes from 1 to 0. The reader is urged to verify the timing diagram in Fig. 6.45b, which has been prepared assuming that each flip-flop changes state as its triggering clock pulse goes back to the 0 level.

The counter in Fig. 6.45a is called a ripple counter. The name arises from the operation as the counter changes state from 1111 to 0000. First FF_0 changes from 1 to 0. This change causes FF_1 to change from 1 to 0. This, in turn, causes FF_2 to go from 1 to 0, which, finally, causes FF_3 to change from 1 to 0, thus completing the state transition. In this process the 0 is said to *ripple* through the counter.

N flip-flops can be connected to periodically sequence through any desired number of states equal to or smaller than 2^N. For instance, we can design a mod-13 or a mod-10 counter. However, we shall not explore these possibilities here.

EXERCISES

6.12   We wish to design a combinational circuit whose inputs are the outputs Q_0, Q_1, Q_2, and Q_3 of the ripple-counter flip-flops and their complements and whose outputs are the 16 lines denoted N_0 to N_{15}. The circuit, called a *decoder*, is required to raise the level of one of these 16 lines in correspondence with the count in the counter. Express N_7 and N_{13} as logic functions of Q_0, Q_1, Q_2, and Q_3.
   Ans. $N_7 = \overline{Q}_3 Q_2 Q_1 Q_0$; $N_{13} = Q_3 Q_2 \overline{Q}_1 Q_0$

6.13   Show that the counter in Fig. E6.13 is a mod-3 counter. Construct a timing diagram for this counter.

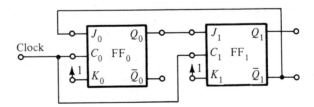

Fig. E6.13

6.7   MULTIVIBRATORS

As mentioned before, the flip-flop has two stable states and is called a bistable multivibrator. There are two other types of multivibrators, monostable and astable multivibrators. The monostable multivibrator has one stable state in which it can remain indefinitely. It has another *quasi-stable* state to which it can be triggered. The monostable can remain in the quasi-stable state for a predetermined interval T, after which it automatically reverts to the stable state. In this way the monostable generates an output

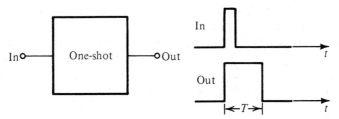

Fig. 6.46 *The monostable multivibrator (one-shot) as a black box.*

pulse of duration T. This pulse duration is in no way related to the details of the triggering pulse, as is indicated schematically in Fig. 6.46. The monostable can therefore be used as a *pulse stretcher* or, more appropriately, a *pulse standardizer*. A monostable is also referred to as a *one-shot*.

The astable multivibrator has no stable states. Rather, it has two quasi-stable states, and it remains in each for predetermined intervals T_1 and T_2. Thus after T_1 seconds in one of the quasi-stable states the astable switches to the other quasi-stable state and remains there for T_2 seconds, after which it reverts back to the original state, and so on. The astable thus oscillates with a period $T = T_1 + T_2$ or a frequency $f = 1/T$, and it can be used to generate periodic pulses such as those required for clocking.

In Section 5.8 we studied an astable multivibrator circuit that used an op amp. Op amps can also be used to implement monostables. In the following we shall discuss monostable and astable circuits using logic gates.

A Monostable Circuit

Figure 6.47 shows a simple and popular circuit for a monostable multivibrator. It is composed of two two-input NOR gates, G_1 and G_2, a capacitor of capacitance C, and a resistor of resistance R. The input source v_I supplies the triggering pulses for the monostable.

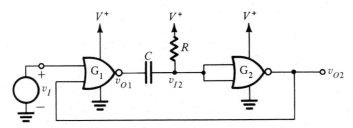

Fig. 6.47 *Monostable circuit using CMOS NOR gates.*

To make our discussion concrete we shall assume that the gates used are of the CMOS family introduced in Chapter 8 (discussed in detail in Chapter 15). Furthermore, we shall assume the CMOS gates to have the idealized transfer characteristic displayed in Fig. 6.48.

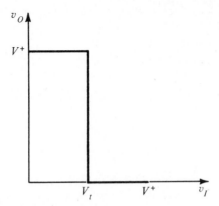

Fig. 6.48   Transfer characteristic for an ideal CMOS inverter.

CMOS gates have a special arrangement of diodes connected at their input terminals, as indicated in Fig. 6.49a. The purpose of these diodes is to prevent the input voltage signal from rising above the supply voltage V^+ (by more than one diode drop) and from falling below ground voltage (by more than one diode drop). These clamping diodes have an important effect on the operation of the monostable circuit. Specifically, we shall be interested in the effect of these diodes on the operation of the inverter-connected gate G_2. In this case each pair of corresponding diodes appears in parallel, giving rise to the equivalent circuit in Fig. 6.49b. While the diodes provide a low resistance path to the power supply for voltages exceeding the power supply limits, the input current for intermediate voltages is essentially zero.

One final property of CMOS NOR gates of interest to us here is the equivalent output circuit of the gate, illustrated in Fig. 6.50. Figure 6.50a indicates that when the gate output is low, its output resistance is R_{on}, which is normally a few hundred ohms. In this state, current can flow from the external circuit into the output terminal of the gate; the gate is said to be *sinking* current. Similarly, the equivalent output circuit in Fig. 6.50b applies when the gate output is high. In this state, current can flow from V^+

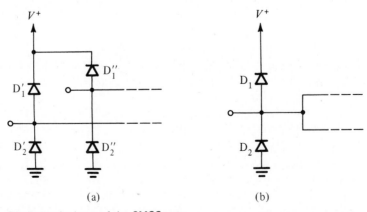

(a)   (b)

Fig. 6.49   Diodes at the input of the CMOS gate.

through the output terminal of the gate into the external circuit; the gate is said to be *sourcing* current.

To see how the monostable circuit of Fig. 6.47 operates, consider the timing diagram given in Fig. 6.51. Here a short triggering pulse of duration τ is shown in Fig. 6.51a. In the following we shall neglect the propagation delays through G_1 and G_2. These delays, however, set a lower limit on the pulse width τ, $\tau > (t_{PD1} + t_{PD2})$.

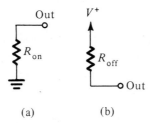

Fig. 6.50 *Output equivalent circuit of CMOS gate when (a) the output is low and (b) the output is high.*

Consider first the stable state of the monostable, that is the state of the circuit before the trigger pulse is applied. The output of G_1 is high at V^+, the capacitor is discharged, and the input voltage to G_2 is high at V^+. Thus the output of G_2 is low at ground voltage. This low voltage is fed back to G_1; since v_I is low, the output of G_1 is high, as initially assumed.

Next consider what happens as the trigger pulse is applied. The output voltage of G_1 will go low. However, because G_1 will be sinking some current and because of its finite output resistance R_{on} its output will not go all the way to 0 V. Rather, the output of G_1 drops by a value ΔV_1, which we shall shortly evaluate.

The drop ΔV_1 is coupled through C (which acts as a short circuit during the transient) to the input of G_2. Thus the input voltage of G_2 drops by an identical amount ΔV_1. Here we note that during the transient there will be an instantaneous current that flows from V^+ through R and C and into the output terminal of G_1 to ground. We thus have a voltage divider formed by R and R_{on} (note that the instantaneous voltage across C is zero) from which we can determine ΔV_1 as

$$\Delta V_1 = V^+ \frac{R}{R + R_{on}}$$

Returning to G_2, we see that the drop of voltage at its input causes its output to go high (to V^+). This signal keeps the output of G_1 low even after the triggering pulse has disappeared. The circuit is now in the quasi-stable state.

We next consider operation in the quasi-stable state. The current through R, C, and R_{on} causes C to charge, and the voltage v_{I2} rises exponentially toward V^+ with a time constant $C(R + R_{on})$, as indicated in Fig. 6.51c. The voltage v_{I2} will continue to rise until it reaches the value of the threshold voltage V_t of inverter G_2. At this time G_2 will switch to the on state; thus its output v_{O2} will go to 0 V and this will cause G_1 to turn off. The output of G_1 will attempt to rise to V^+, but, as will become obvious shortly, its instantaneous rise will be limited to an amount ΔV_2. This rise in v_{O1} is coupled faith-

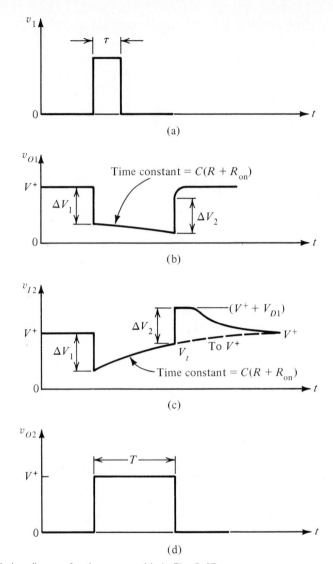

Fig. 6.51   Timing diagram for the monostable in Fig. 6.47.

fully through C to the input of G_2. Thus the input of G_2 will rise by an equal amount ΔV_2. Note here that because of diode D_1 between the input of G_1 and V^+, the voltage v_{I2} can rise only to $V^+ + V_{D1}$, where V_{D1} (approximately 0.7 V) is the drop across D_1. Thus from Fig. 6.51c we see that

$$\Delta V_2 = V^+ + V_{D1} - V_t$$

Thus it is diode D_1 that limits the size of the increment ΔV_2.

Because now v_{I2} is higher that V^+ (by V_{D1}) current will flow from the output of G_1 through C and then through the parallel combination of R and D_1. This current dis-

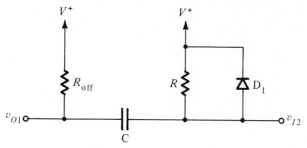

Fig. 6.52 *Circuit that applies during the discharge of C (at the end of the monostable pulse interval T).*

charges C until v_{I2} drops to V^+ and v_{O1} rises to V^+. The charging circuit is depicted in Fig. 6.52, from which we note that the existence of the diode causes the discharging to be a nonlinear process. Although the details of the transient at the end of the pulse are not of immense interest, it is important to note that the monostable cannot be retriggered until the capacitor has been discharged, since otherwise the precise output obtained will not be the standard pulse for which the one-shot is intended.

EXERCISE

6.14 Derive an expression for the pulse interval T of the monostable circuit in Fig. 6.47. Hint: use the information given in the timing diagram of Fig. 6.51.
Ans.

$$T = C(R + R_{on}) \ln \left(\frac{R}{R + R_{on}} \frac{V^+}{V^+ - V_t} \right)$$

An Astable Circuit

Figure 6.53 shows a popular astable circuit composed of two inverter-connected NOR gates, a resistor, and a capacitor. We shall consider its operation, assuming that the NOR gates are of the CMOS family. However, to simplify matters we shall make some further approximations: The finite output resistance of the CMOS gate in the on state

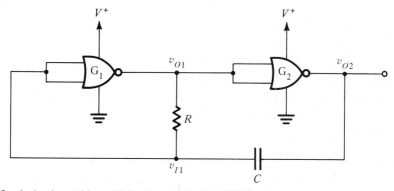

Fig. 6.53 *A simple astable multivibrator circuit using CMOS gates.*

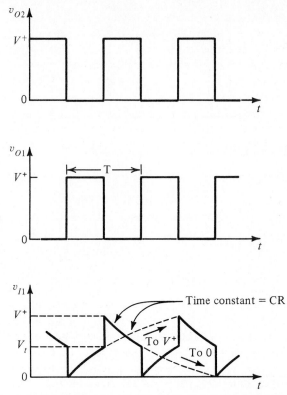

Fig. 6.54 *Waveforms for the astable circuit in Fig. 6.53. The diodes at the gate input are assumed ideal.*

will be neglected. Also, the clamping diodes will be assumed ideal (thus have zero voltage drop when conducting).

With these simplifying assumptions, the waveforms of Fig. 6.54 are obtained. The reader is urged to consider the operation of this circuit in a step-by-step manner and verify that the waveforms shown indeed apply.[3]

EXERCISE

6.15 Using the waveforms in Fig. 6.54, derive an expression for the period T of the astable multivibrator of Fig. 6.53.

Ans.

$$T = CR \ln \left(\frac{V^+}{V^+ - V_t} + \frac{V^+}{V_t} \right)$$

6.8 ANALOG-TO-DIGITAL CONVERSION

Digital Processing of Signals

Most physical signals, such as those obtained at transducer outputs, exist in analog form. Some of the processing required on such signals is most conveniently performed

---

[3]Practical circuits often use a large resistance in series with the input to G_1. This limits the effect of diode conduction and allows v_{I1} to rise to a voltage greater than V^+ and below zero.

in an analog fashion. For instance, in instrumentation systems it is quite common to use a high-input-impedance, high-gain, high-CMRR (common-mode rejection ratio) differential amplifier right at the output of the transducer. This is usually followed by a filter whose object is to eliminate interference. However, further signal processing is usually required, which can range from simply obtaining a measurement of signal strength to performing some algebraic manipulations on this and related signals to obtain the value of a particular system parameter of interest, as is usually the case in systems intended to provide a complex control function. Another example of signal processing can be found in the common need for transmission of signals to a remote receiver.

All such forms of signal processing can be performed by analog means. In previous chapters we encountered circuits for implementing a number of such tasks. However, an attractive alternative exists: It is to convert, following some initial analog processing, the signal from analog to digital form and then use economical, accurate, and convenient digital ICs to perform *digital signal processing*. Such processing can in its simplest form provide us with a measure of the signal strength as an easy-to-read number (for example, the digital voltmeter). In more involved cases the digital signal processor can perform a variety of arithmetic and logic operations that implement a *filtering algorithm*. The resulting *digital filter* does many of the same tasks that an analog filter performs, namely, eliminating interference and noise. Yet another example of digital signal processing is found in digital communications systems, where signals are transmitted as a sequence of binary pulses, with the obvious advantage that corruption of the amplitudes of these pulses by noise is to a large extent of no consequence.

Once digital signal processing has been performed, we might be content to display the result in digital form, such as a printed list of numbers. Alternatively, we might require an analog output. Such is the case in a telecommunications system, where the usual output may be speech. If an analog output is desired, then obviously we need to convert the digital signal back to an analog form.

It is not our purpose here to study the techniques of digital signal processing. Rather, we shall examine the interface circuits between the analog and digital domains. Specifically, we shall study the basic techniques and circuits employed to convert an analog signal to digital form (analog-to-digital or simply A/D conversion) and those used to convert a digital signal to analog form (digital-to-analog or simply D/A conversion).

Sampling of Analog Signals

The principle underlying digital signal processing is that of *sampling* the analog signal. Figure 6.55 illustrates in a conceptual form the process of obtaining samples of an analog signal. The switch shown closes periodically under the control of a periodic pulse signal (clock). The closure time of the switch, τ, is relatively short, and the samples obtained are stored (held) on the capacitor. The circuit of Fig. 6.55 is known as a *sample-and-hold* (S/H) circuit. In Chapter 7 we shall present a circuit implementation of the S/H building block.

Between the sampling intervals—that is, during the *hold* intervals—the voltage levels on the capacitor represent the signal samples we are after. Each of these voltage

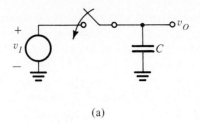

(a)

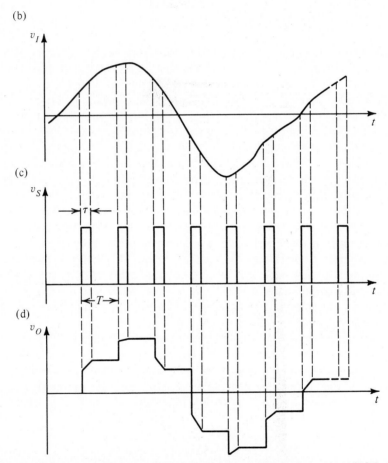

Fig. 6.55   Process of periodically sampling an analog signal. (a) Sample-and-hold (S/H) circuit. The switch closes τ seconds every period. (b) Input signal waveform. (c) Sampling signal (control signal for the switch). (d) Output signal (to be fed to A/D converter).

levels is then fed to the input of an A/D converter which provides an N-bit binary number proportional to the value of signal sample.

The fact that we can do our processing on a limited number of samples of an analog signal, while ignoring the analog signal details between samples, is based on the sampling theorem (see Ref. 6.5).

Signal Quantization

Consider an analog signal whose values range from 0 to $+10$ V. Let us assume that we wish to convert this signal to digital form and that the required output is a 4-bit[4] signal. We already know that a 4-bit binary number can represent 16 different values, 0 to 15. It follows that the *resolution* of our conversion will be 10 V/15 $= \frac{2}{3}$ V. Thus an analog signal of 0 V will be represented by 0000, $\frac{2}{3}$ V will be represented by 0001, 6 V will be represented by 1001, and 10 V will be represented by 1111.

All of the above sample numbers were multiples of the basic increment ($\frac{2}{3}$ V). A question now arises regarding the conversion of numbers that fall between these successive incremental levels. For instance, consider the case of 6.2-V analog level. This falls between 18/3 and 20/3. However, since it is closer to 18/3 we treat it as if it were 6 V and *code* it as 1001. This process is called *quantization*. Obviously errors are inherent in this process; such errors are called quantization errors. Using more bits to represent (code) an analog signal reduces quantization errors.

The A/D and D/A Converters as Black Boxes

Figure 6.56 depicts the black-box or functional representations of A/D and D/A converters. As indicated, the A/D converter accepts an analog sample v_A and produces an *N*-bit *digital word*. Conversely, the D/A converter accepts an *N*-bit digital word and

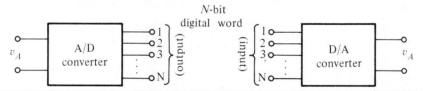

Fig. 6.56 *The A/D and D/A converters as black boxes.*

produces an analog sample. The output samples of the D/A converter are often fed to a sample-and-hold circuit. At the output of the S/H circuit a staircase waveform, such as that in Fig. 6.57, is obtained. The staircase waveform can then be smoothed by a low-pass filter, giving rise to the smooth curve shown as a broken line in Fig. 6.57. In this way an analog output signal is reconstructed.

D/A Converter Circuits

Since the D/A converter is easier to implement than the A/D converter, we shall begin by considering circuit implementations of the D/A converter first. Figure 6.58 shows a simple circuit for an *N*-bit D/A converter (or DAC, as it is sometimes called). The circuit consists of a reference voltage V_{ref}, *N* binary-weighted resistors $R, 2R, 4R, 8R,$ $\ldots, 2^{N-1}R$, *N* single-pole double-throw switches $S_1, S_2, \ldots, S_N$, and an op amp together with its feedback resistance $R_f = R/2$.

---

[4]Bit stands for *binary digit*.

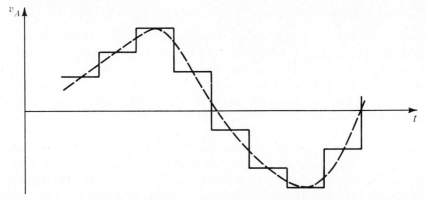

Fig. 6.57   *The analog samples at the output of a D/A converter are usually fed to a sample-and-hold circuit to obtain the staircase waveform shown. This waveform can then be filtered to obtain the smooth waveform, shown as a broken line with filter delay removed.*

The switches are controlled by an *N*-bit digital input word *D*,

$$D = \frac{b_1}{2^1} + \frac{b_2}{2^2} + \cdots + \frac{b_N}{2^N}$$

where b_1, b_2, and so on are bit coefficients that are either 1 or 0. Note that the bit b_N is the *least significant bit* (LSB) and b_1 is the *most significant bit* (MSB). In the circuit in Fig. 6.58 b_1 controls switch S_1, b_2 controls S_2, and so on. When b_i is 0 switch S_i is in position 1, and when b_i is 1 switch S_i is in position 2.

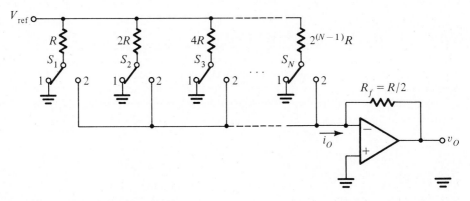

Fig. 6.58   *An N-bit D/A converter.*

Since position 1 of all switches is ground and position 2 is virtual ground, the current through each resistor remains constant. Each switch simply controls where its corresponding current goes: to ground (when the corresponding bit is 0) or to virtual ground (when the corresponding bit is 1). The currents flowing into the virtual ground add up and the sum flows through the feedback resistance R_f. The total current i_o is therefore given by

$$i_o = \frac{V_{\text{ref}}}{R} b_1 + \frac{V_{\text{ref}}}{2R} b_2 + \cdots + \frac{V_{\text{ref}}}{2^{N-1}R} b_N$$

$$= \frac{2V_{\text{ref}}}{R} \left(\frac{b_1}{2^1} + \frac{b_2}{2^2} + \cdots + \frac{b_N}{2^N} \right)$$

Thus

$$i_o = \frac{2V_{\text{ref}}}{R} D$$

and the output voltage v_O is given by

$$v_O = -i_o R_f = -V_{\text{ref}} D$$

which is directly proportional to the digital word D, as desired.

It should be noted that the accuracy of the DAC depends critically on (1) the accuracy of V_{ref}, (2) the precision of the binary-weighted resistors, and (3) the perfection of the switches. Regarding the third point, we should emphasize that these switches handle analog signals; thus their perfection is of considerable interest. While the offset voltage and the finite on resistance are not of critical significance in a digital switch, these parameters are of immense importance in *analog switches*. Circuits for analog switches will be presented in Chapters 7 and 8.

A disadvantage of the binary-weighted resistor network is that for a large number of bits ($N > 4$) the spread between the smallest and largest resistances becomes quite large. This implies difficulties in maintaining accuracy in resistor values. Other schemes exist that use a resistive network called the R-2R ladder (see Exercise 6.16).

EXERCISE

6.16 Figure E6.16 shows a DAC using a resistive network called the R-2R ladder. Because of the small spread in resistance values, this network is usually preferred to the binary-

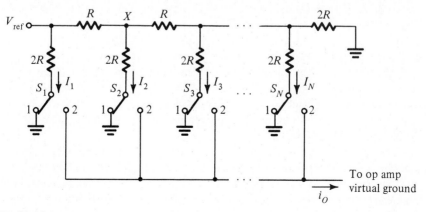

Fig. E6.16

weighted scheme, especially for $N > 4$. Show that to the right of any node, such as the node labeled X, the resistance is $2R$. Hence show that

$$I_1 = 2I_2 = 4I_3 = \cdots = 2^{N-1}I_N$$

and

$$i_O = \frac{V_{\text{ref}}}{R} D$$

where D is the input digital word.

A/D Converters

There exist a number of A/D conversion techniques varying in complexity and speed of conversion. In the following we shall discuss two simple but slow schemes and one complex (in terms of the amount of circuitry required) but extremely fast method.

Figure 6.59 shows a simple A/D converter that employs a comparator, an up-down counter, and a D/A converter. In Section 5.9 we discussed the comparator operator. An up-down counter is simply a counter such as those of Section 6.6 but one that can

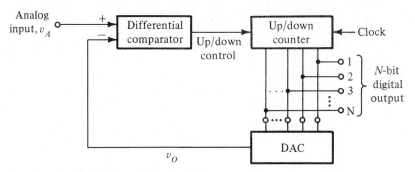

Fig. 6.59   A simple feedback-type A/D converter.

count either up or down depending on the level applied at its up-down control terminal. Because the A/D converter of Fig. 6.59 employs a DAC in its feedback loop it is usually called a feedback-type A/D converter. It operates as follows: With a 0 count in the counter, the D/A converter output will be zero and the output of the comparator will be high, instructing the counter to count the clock pulses in the up direction. As the count increases, the output of the DAC rises. The process continues until the DAC output reaches the value of the analog input signal, at which point the comparator switches and stops the counter. The counter output will simply be the digital equivalent of the input analog voltage.

Operation of the converter of Fig. 6.59 is slow if it starts from zero. This converter, however, tracks incremental changes in the input signal quite rapidly.

Another slow A/D converter, based on a different principle, is shown in Fig. 6.60. This converter operates in the following manner: Initially the counter is reset and the switch is opened. The current source I charges the capacitor C, and the voltage v_B rises

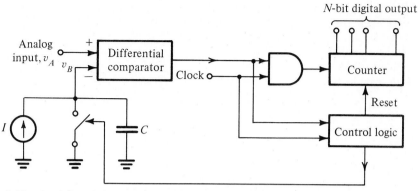

Fig. 6.60   *An A/D conversion scheme based on comparing a ramp waveform with the analog signal.*

linearly with time. As long as v_B is smaller than the input signal v_A the output of the comparator will be high and the counter is enabled to count the clock pulses. However, this situation changes as v_B reaches v_A. At this time the comparator switches and the counter is stopped. The content of the counter at this moment will be a digital word proportional to the time of counting, which in turn is proportional to v_A (because $v_A = v_B$ and v_B is linear with time). The logic block then resets the counter, closes the switch to discharge the capacitor, and then opens the switch to begin a new conversion.

The last A/D conversion scheme we shall discuss is an extremely fast one allowing an entire conversion to occur within one clock period. The method, known as *parallel* or *simultaneous* or *flash* conversion, is illustrated by the block diagram of Fig. 6.61. Conceptually it is very simple, using $2^N - 1$ comparators to compare the input signal level with each of the $2^N - 1$ possible quantization levels. The outputs of the comparators are processed by an encoding logic block to provide the N bits of the output digital word. Although fast, the method is obviously uneconomical and impractical for large N. Nevertheless, variations on this technique have been successfully employed in practice.

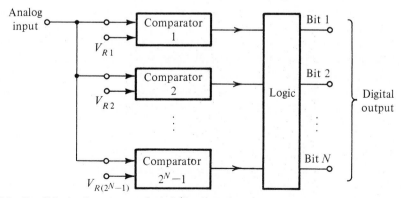

Fig. 6.61   *Parallel, simultaneous, or flash A/D conversion.*

6.17 How many clock pulses are required by the converter of Fig. 6.60 to complete the conversion of an input equal to full scale?

Ans. $2^N - 1$

6.18 Express the maximum quantization error of an N-bit A/D converter in terms of its least significant bit (LSB) and in terms of its full-scale analog input V_{FS}.

Ans. $\pm \frac{1}{2}$ LSB; $V_{FS}/2(2^N - 1)$

Finally, it should be mentioned that at present a variety of A/D and D/A converters are available in monolithic (that is, on one single chip) integrated-circuit form.

6.9 MICROPROCESSORS

We shall conclude this chapter with a brief introduction to the most important IC package in production today, the microprocessor (or μP). The microprocessor or microcomputer[5] is simply a computer on one or a few IC chip(s). To fully understand microprocessors, one needs to first learn about computers. Nevertheless, because μPs are so inexpensive one tends to treat them as electronic-circuit components. In this book we shall be interested in the electronic-circuit aspects of μPs. A brief introduction to the subject is provided here, for a detailed study the reader may consult references 6.9—6.11. It should also be mentioned that memory forms a major part of the microcomputer; in fact, the μP can be thought of as "memory with logic." Memory circuits are studied in Chapter 16.

The microprocessor is an excellent example of VLSI circuits. Its applications are limited only by the imagination of the user. A most basic application is to use the μP to replace a collection of logic circuits, such as that required for the control of an A/D converter.

The ALU and the ROM

A microprocessor consists basically of an *arithmetic-logic unit* (the ALU) under control of a *program* consisting of an ordered set of coded *instructions* to which the components of the ALU respond. The program resides in a part of the memory of the microprocessor, called read-only memory (ROM), where it is secure from change once established and operating. Each instruction is retrieved or "fetched" from ROM in turn, following completion of the activity produced in the ALU by its predecessor instruction. In addition to the arithmetic possibilities of simply adding or subtracting two numbers, it is possible also for the ALU to make logic decisions on the sign or zero value of the result. Such decisions provide the basis for a *branch* in a program; a change in the sequence of executing instructions that depends on the data provided.

Memory

In order for data to be stored during processing, memory must be available. Such memory may be general-purpose *register memory*, such as the shift registers discussed in

[5]There is a distinction between the two, but it is too subtle for our purpose at this stage.

Section 6.6, or *random-access memory* (RAM), which consists of a matrix of flip-flops organized in small connected groups called *words.* A common word length for microprocessors is currently 1 byte, or 8 binary digits (or bits).

Register memory is generally more easily accessed than RAM and is provided in smaller amounts. Registers often have additional features (for example, the ability to act as a counter of events, instructions, and so on). Thus, a register is usually used to perform the role of *instruction counter,* which keeps track of the instruction currently "being obeyed." On the other hand, RAM is organized to provide a large number of words with convenient yet economic access. Each word in the RAM has a unique *address* and is accessed via a special register, called the *address register.* The content of the address register originates either in a stored instruction or as a result of *address computation.* The latter possibility provides one important basis for the flexibility of a microprocessor. Memory circuits will be studied in Chapter 16.

Input/Output (I/O)

An overview of microprocessors cannot be complete without a glimpse at mechanisms for input and output of data (so-called *input/output,* or I/O). Suffice it to say that while many approaches to I/O are available, all in effect implement the equivalent of memory, either register or RAM, writing to which provides a means of output and reading from which implements input. (Note that one conventionally speaks of writing *to*— rather than on or in—a memory, and of reading *from* it.)

In a microprocessor having words whose length is 8 bits, input and output *ports,* said to be *parallel ports,* are also 8 bits long. Often to save pins on the IC, the input and output ports are combined or *multiplexed,* the eight leads being used for input or output depending on the value of a ninth control signal. Where pins are extremely limited, I/O may be constrained to only two pins with which information must be transferred in *serial* rather than *parallel* form.

One additional input, called *interrupt,* is often available on a microprocessor. When this signal is energized the normal program of the processor is interrupted and a new one, a so-called *interrupt-handling routine,* is begun. The initial instruction of this routine is to inquire of the cause for interruption using data provided via the normal input channel.

Microprocessor Block Diagram

The above ideas are summarized briefly in the block diagram of a microprocessor provided in Fig. 6.62. Typical capacities are indicated, with the numbers in parentheses indicating the number of wires represented by each line.

In Fig. 6.62 the arithmetic-logic unit (ALU) is shown coupled through its memory port to the *memory bus* to which ROM, RAM, and perhaps I/O are connected. Registers, in a *register file,* are shown connected to another port of the ALU. Finally, an explicit I/O system is shown connected to the ALU. Here, in addition to an interrupt line, both serial and parallel options are indicated. A single microprocessor would only rarely have all of the I/O indicated by virtue of a practical limitation on the number of pins (perhaps 40) available on the IC package.

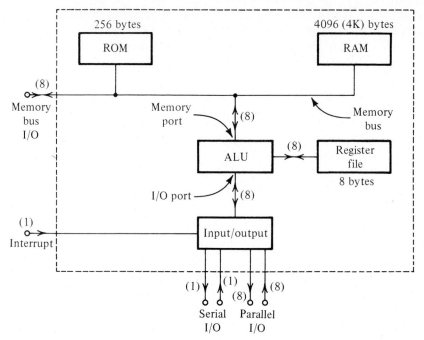

Fig. 6.62   Block diagram for microprocessor. Note that byte refers to an 8-bit word.

Finally, we should mention that while single-chip processors include all of these components, multichip systems such as the Z80 and M6800 offer increased capacity and flexibility.

6.10   CONCLUDING REMARKS

Using a black-box approach to logic and memory circuits we have studied fundamental theory and techniques of digital systems. This treatment is in a sense parallel to that of op amps in Chapters 3 and 5. Op amps are, of course, the fundamental building blocks of analog systems.

The material presented here should enable the reader to read the data sheets of digital ICs and to experiment with these circuits. However, the reader will be in a much better position to apply digital ICs after finding out "what is on the chip." In Chapters 7, 8, and 9 we shall include discussion of basic circuits suited to the implementation of a number of the building blocks considered here. A detailed and comparative study of logic circuit families will be presented in Chapter 15. Memory circuits, which are important in their own right as well as serving as examples of LSI and VLSI technology, will be studied in Chapter 16.

JUNCTION FIELD-EFFECT TRANSISTORS (JFETs)

7

In this chapter we begin our study of *three-terminal* semiconductor devices. At this stage it should be appreciated that three-terminal devices are far more useful than two-terminal ones—they can be used in a multitude of applications ranging from signal amplification to the realization of logic and memory functions. The basic principle involved is the use of the voltage between two terminals to control the current flowing in the third terminal. It is thus easy to see how a three-terminal device can be used to realize a controlled source which is the basis for amplifier design. Also, in the extreme the control signal can be used to cause the current in the third terminal to change from zero to a large value, thus allowing the device to act as a switch.

The device studied in this chapter is a special type of field-effect transistor (FET) called the *junction field-effect transistor* (JFET). The other type of FET—namely the metal-oxide-semiconductor FET, or MOSFET—will be studied in the next chapter. The name field-effect transistor arises from the fact that current flow between two of the device terminals is controlled by an electric field, which in turn is established by a voltage applied to the third terminal, as will be explained below. FETs are also called *unipolar transistors* because current is conducted by charge carriers (electrons or holes) flowing through one type of semiconductor (*n*-type in *n*-channel FETs and *p*-type in *p*-channel FETs). This is in contrast to *bipolar transistors* (which we will study in Chapter 9), where current passes through both *n*-type and *p*-type semiconductor materials in series.

Both types of FETs, and especially the MOSFET, are easier to manufacture than bipolar transistors. MOSFETs play a dominant role in digital integrated-circuit design. Microprocessors and logic and memory circuits fabricated using very-large-scale-integration (VLSI) techniques mostly employ MOS transistors.

JFETs are useful in the design of special amplifier circuits, especially those with very high input impedances. For instance, op amps with very high input impedances usually have a first stage made up of JFETs. JFETs can be also combined with bipolar transistors to provide high-performance linear circuits (called BIFET circuits). The JFET structure using a metal-semiconductor (Schottky) junction is used with gallium arsenide to form the MESFET, a device suitable for use in amplifiers and logic circuits in the gigahertz range. The JFET is also used as an analog switch and in a variety of other analog circuit applications.

7.1 PHYSICAL OPERATION

There are two types of JFET: the *n*-channel device and the *p*-channel device. In the following we will explain in detail the operation of the *n*-channel FET. The *p*-channel FET works in a similar manner except for a reversal of polarities of all currents and voltages, as will be briefly explained.

Figure 7.1 shows the basic structure[1] of the *n*-channel JFET. It consists of a slab of *n*-type semiconductor with *p*-type material diffused on both sides. The *n* region is called the *channel*, while the *p*-type regions are electrically connected together and form the *gate*. Metal contacts are made to both ends of the channel, with the terminals called the *source* (S) and the *drain* (D). Similarly, a metal contact is made to the *p*-type region to provide the *gate* terminal (G). We will always assume that the gate regions are electrically connected together, but we will not explicitly show such connections in order to keep the diagrams simple.

Figure 7.2 shows the circuit symbol for the *n*-channel JFET. Note that the gate line has an arrowhead whose direction indicates the type of the device (that is, *n*-channel or *p*-channel JFET). For the *n*-channel device we are considering, the arrow points toward the *n* channel, that is, in the forward direction of the gate-to-channel junction. Although JFETs are usually symmetric (that is, the drain and source are interchangeable), it is convenient in analyzing and designing FET circuits to indicate which terminal is the source. For this reason we will distinguish the source by drawing the gate line closer to it than to the drain.

It can be seen from Fig. 7.1 that the JFET has one *pn* junction, the gate-to-channel junction. In almost all applications this junction will be reverse-biased and hence only a very small leakage current will flow in the gate terminal. This also means that the input impedance looking into the gate will be very high.

---

[1]For actual geometries and fabrication methods the reader is referred to Appendix A. The structure discussed here is a simplified and idealized one intended only to illustrate the basic principles of operation.

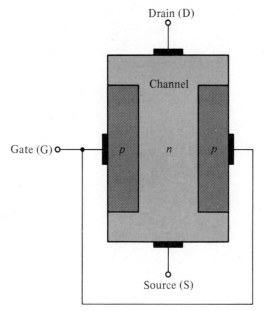

Fig. 7.1 Basic structure of n-channel JFET.

Although only one *pn* junction exists, it will be seen that the operation of the JFET is dependent on establishing different values of reverse bias at the two ends of this junction. For this reason one is sometimes tempted to speak of the gate-to-drain junction and the gate-to-source junction.

Operation When v_{DS} Is Small

Consider first the operation of the device when a small positive voltage (a fraction of a volt) v_{DS} is applied between the drain and source, as shown in Fig. 7.3. If $v_{GS} = 0$, there will be a narrow depletion region and a current i_D will flow in the channel. The value of i_D will be determined by the value of v_{DS} and the channel resistance r_{DS}. As v_{GS} is made negative, the depletion region widens and the channel narrows. Note that since v_{DS} is small, the reverse-bias voltage will be approximately the same at both ends of the channel, and the channel width will be uniform. The narrowing of the channel causes

Fig. 7.2 Circuit symbol for n-channel JFET.

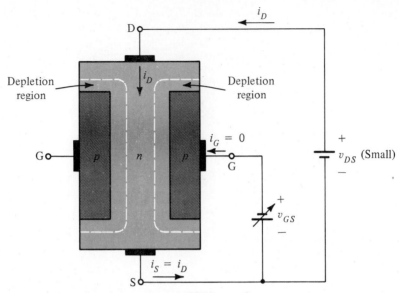

Fig. 7.3   *Physical operation of the n-channel JFET for small v_{DS}.*

its resistance to increase, and the i_D-v_{DS} characteristic remains a straight line but with a smaller slope, as shown in Fig. 7.4.

   If we keep increasing the magnitude of v_{GS} in the negative direction, a point will be reached at which the depletion region occupies the entire width of the channel. In other words, the channel will be completely depleted of charge carriers (electrons for

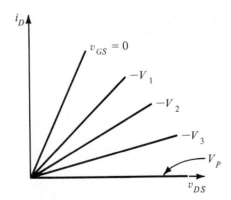

Fig. 7.4   *The i_D-v_{DS} characteristics (for various values of v_{GS}) with v_{DS} small (a fraction of a volt).*

the *n*-channel device), and hence no current will flow. This condition is called *pinch-off* and is illustrated in Fig. 7.5. The voltage v_{GS} at which pinch-off occurs is called the pinch-off voltage and is denoted by V_P,

$$V_P = v_{GS}|_{i_D=0,\ v_{DS}=\text{small}} \tag{7.1}$$

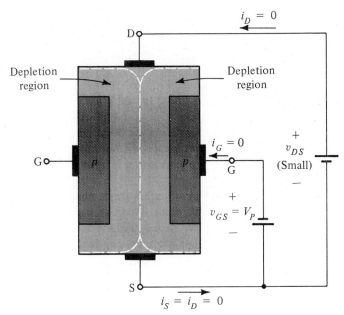

Fig. 7.5 *Pinch-off when v_{DS} is small.*

Thus for an *n*-channel device V_P is a negative number and is a parameter of the partic-ular FET. On the characteristics of Fig. 7.4 pinch-off is represented by a horizontal line at zero current level.

The JFET characteristics in Fig. 7.4 suggest that for small v_{DS} the device acts as a linear resistance r_{DS} whose value is controlled by the voltage v_{GS}. In fact, the JFET is used as a *voltage-controlled resistance* (VCR) in some applications, such as in *auto-matic gain control* (AGC) circuits which are employed in communications receivers. As will be shown, this region of operation of the JFET is not useful for linear-amplifier applications.

Operation as v_{DS} Is Increased

Consider now the characteristics as the value of v_{DS} is increased. Let v_{GS} be kept con-stant—say, at 0 V to begin with. The gate-to-channel junction will have zero voltage across it at the source end and a progressively increasing reverse bias as we move toward the drain. At the drain end the reverse bias voltage v_{DG} will be equal in magnitude to v_{DS}. It follows that the depletion region will have the tapered shape shown in Fig. 7.6, with the result that the channel will be narrowest at the drain end. Thus as v_{DS} is increased, the channel resistance will increase, causing the i_D-v_{DS} characteristic to become nonlinear, as shown in Fig. 7.7 by the curve corresponding to $v_{GS} = 0$. If we keep increasing v_{DS}, a value will be reached at which the channel is pinched off at the drain end. This should happen at the value that results in the reverse-bias voltage at the drain end being equal to the pinch-off voltage,

$$v_{DG} = -V_P \qquad (7.2)$$

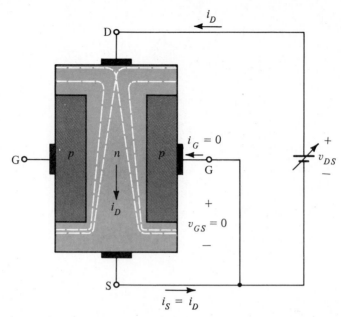

Fig. 7.6 *Effect on channel shape of increasing v_{DS} while keeping v_{GS} constant at zero volts.*

(recall that V_P is a negative number). Since we are now considering the case $v_{GS} = 0$, Eq. (7.2) implies that pinch-off at the drain end will happen when v_{DS} is given by

$$v_{DS} = -V_P$$

Any further increase in v_{DS} will not alter the channel shape, and hence the current i_D will remain constant at the value reached for $v_{DS} = -V_P$. This value of saturated drain-to-source current I_{DSS} is specified on the data sheets of the JFET. It is defined as follows:

$$I_{DSS} = i_D|_{v_{GS}=0, \, v_{DS}=-V_P} \tag{7.3}$$

Here it is important to note the difference between the case when the channel is completely pinched off and the case when pinch-off occurs only at the drain end. In the former case the channel is entirely depleted of charge carriers (electrons for the n-channel device); hence no current will flow ($i_D = 0$). In the latter case current will continue to flow through the channel; the electrons that flow through the channel will simply drift through the pinched-off region at the drain end of the channel and reach the drain terminal. The voltage across the channel and the current through it remain constant. The difference between the applied voltage v_{DS} and the value $-V_P$ appears across the depletion region at the drain end of the channel.

Consider next the case $v_{GS} = -V$, where V is a positive number smaller than $|V_P|$. For small v_{DS} the channel will be of uniform width, since the reverse-bias voltage at the drain end is almost equal to that at the source end. While as v_{DS} is increased the reverse-bias voltage at the source end remains constant, it increases at all other points along the channel. The largest magnitude of reverse bias will be at the drain end. Thus the channel will again have a tapered shape, and its resistance will increase as v_{DS} is

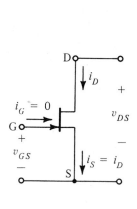

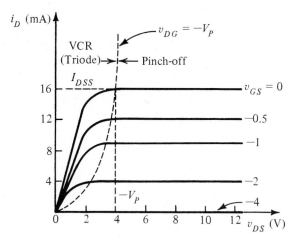

Fig. 7.7 *Complete family of i_D-v_{DS} characteristics for an n-channel JFET with $V_P = -4$ V and $I_{DSS} = 16$ mA.*

increased. Eventually, the channel will be pinched off at the drain end. This will happen at the value of the drain voltage corresponding to

$$v_{DG} = -V_P$$

Since in this case $v_{GS} = -V$, it follows that pinch-off will occur at

$$v_{DS} = -V_P - V$$

Since the channel in this case is narrower than it was for the $v_{GS} = 0$ case, the saturated value of drain current will be smaller than I_{DSS}.

Continuing in this manner for other values of v_{GS} down to $v_{GS} = V_P$, we obtain the complete family of i_D-v_{DS} characteristics shown in Fig. 7.7.

7.2 STATIC CHARACTERISTICS

The static characteristics of an *n*-channel JFET are shown in Fig. 7.7. The reason for the word static is that these characteristics are measured with dc or low-frequency signals; thus effects of internal capacitances are not observed.

As shown in Fig. 7.7, there are two distinct regions of operation: the *triode* or voltage-controlled-resistance (VCR) region, and the *pinch-off* region. The two regions are separated by a parabolic boundary, represented by the broken-line curve in Fig. 7.7.

The Triode Region

In the triode region the FET acts as a resistance (r_{DS}) whose value is controlled by the gate-to-source voltage v_{GS}. This resistance is linear for small values of v_{DS}. The i_D-v_{DS} relationship in the triode region is parabolic and is described by

$$i_D = I_{DSS}\left[2\left(1 - \frac{v_{GS}}{V_P}\right)\frac{v_{DS}}{-V_P} - \left(\frac{v_{DS}}{V_P}\right)^2\right] \tag{7.4}$$

where V_P and I_{DSS} are parameters of the JFET whose values are usually given on the device data sheets. For *n*-channel FETs V_P is a negative number. Since v_{GS} is always negative, the gate-to-channel junction remains reverse-biased at all times. This results in the gate current being almost zero. In fact, the gate current will consist mostly of the leakage current of the reverse-biased *pn* junction. For silicon devices this current is of the order of a few nanoamperes, but it increases with temperature. Temperature effects will be discussed later.

For small values of v_{DS}, Eq. (7.4) can be approximated

$$i_D \simeq \frac{2I_{DSS}}{-V_P}\left(1 - \frac{v_{GS}}{V_P}\right)v_{DS} \tag{7.5}$$

This linear relationship represents the i_D-v_{DS} characteristics near the origin. The linear resistance r_{DS} is therefore given by

$$r_{DS} = \left.\frac{v_{DS}}{i_D}\right|_{v_{DS}=\text{small}}$$

Thus

$$r_{DS} = \left[\frac{2I_{DSS}}{-V_P}\left(1 - \frac{v_{GS}}{V_P}\right)\right]^{-1} \tag{7.6}$$

The Boundary Between the Triode and Pinch-Off Regions

Pinch-off is reached when the reverse-bias voltage at the *drain end* is equal to the pinch-off voltage—that is,

$$v_{DG} = -V_P \tag{7.7}$$

This equation represents the boundary between the triode region and the pinch-off region. It can be rewritten

$$v_{DS} = v_{GS} - V_P \tag{7.8}$$

Substituting in Eq. (7.4) yields

$$i_D = I_{DSS}\left(\frac{v_{DS}}{V_P}\right)^2 \tag{7.9}$$

which is the equation of the broken-line parabola in Fig. 7.7.

Thus to operate in the triode region, for which Eq. (7.4) applies, the drain-to-gate voltage should be less than $-V_P$,

$$v_{DG} < -V_P, \qquad \text{Triode region} \tag{7.10}$$

On the other hand, operation in the pinch-off region is obtained for

$$v_{DG} > -V_P, \qquad \text{Pinch-off region}$$

which implies that *the drain voltage should be higher than the gate voltage by at least* $|V_P|$.

The Pinch-Off Region

In the pinch-off region the i_D-v_{DS} characteristics are (ideally) horizontal straight lines whose heights are determined by the value of v_{GS}. It follows that in pinch-off the JFET operates as a constant-current source with the value of the current controlled by v_{GS}. Furthermore, this constant-current source has ideally an infinite resistance (that is, looking back into the drain terminal one sees an infinite resistance). Also, the input

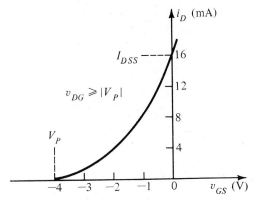

Fig. 7.8   The i_D-v_{GS} characteristic in pinch-off. For this JFET, $V_P = -4$ V and $I_{DSS} = 16$ mA.

impedance of this controlled source (looking between the control terminals G and S) is ideally infinite. The control relationship is given approximately by the square law,

$$i_D = I_{DSS} \left(1 - \frac{v_{GS}}{V_P} \right)^2 \tag{7.11}$$

This characteristic, which applies only in pinch-off, is sketched in Fig. 7.8.

The pinch-off region is useful for applications that involve using the JFET as an amplifier. For this reason the pinch-off region is also called the *active region*. JFET amplifiers will be studied in a later section.

Finite Output Resistance in Pinch-Off

It should be noted that the above model of the JFET is an approximate, idealized one. More realistic characteristic curves are shown in Fig. 7.9, from which an important second-order effect is evident that has to be taken into account in some applications. As shown, the characteristic lines in the pinch-off region have finite, nonzero slope, indicating that the output resistance of the drain current source is not infinite. Rather, the value of this resistance is finite and decreases as the current level in the device is increased.

Breakdown

Another JFET characteristic illustrated by the i_D-v_{DS} characteristics in Fig. 7.9 is that of voltage breakdown. As indicated, the JFET breakdown resembles that of a diode—

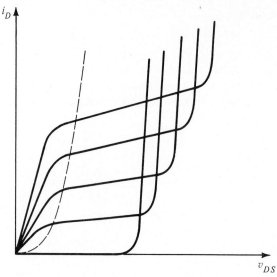

Fig. 7.9 The i_D-v_{DS} characteristics with the finite output resistance in pinch-off included. Also, the breakdown region is shown.

that is, the current in the device increases sharply with a negligible corresponding increase in voltage. JFET breakdown is usually of the avalanche type (see Section 4.13) and occurs at the drain end of the gate–channel junction. This, of course, is expected because the reverse bias is greatest at the drain end of the junction. Thus breakdown occurs when v_{DG} exceeds the rated breakdown voltage of the *pn* junction. This explains the appearance of the curves in Fig. 7.9, which show the FET breaking down at smaller v_{DS} as v_{GS} is made more negative.

Temperature Effects

Circuit designers are usually concerned with their circuits meeting specifications over a specified range of operating temperatures. Therefore the change in device parameters with temperature is of interest. For the JFET we have already mentioned one consequence of changing temperature, namely that the leakage gate current increases with temperature, roughly doubling for every $10°C$ rise in temperature. In addition, both the conductivity of the channel and the barrier voltage V_0 of the gate–channel junction are functions of temperature. The result of this can be expressed in terms of the change in gate-to-source voltage v_{GS} when temperature is changed while the drain current is held constant. It can be shown though (see reference 7.2), that there exists a particular value of drain current at which the temperature coefficient of v_{GS} is zero. Thus with proper circuit design one can arrange that the changes with temperature are very small. We shall not pursue this point any further here.

EXERCISES

In the following problems let the *n*-channel JFET have $V_P = -4$ V and $I_{DSS} = 10$ mA, and assume that in pinch-off the output resistance is infinite.

7.1   For $v_{GS} = -2$ V, find the minimum v_{DS} for the device to operate in pinch-off. Calculate i_D for $v_{GS} = -2$ V and $v_{DS} = 3$ V.
Ans.   2 V; 2.5 mA

7.2   For $v_{DS} = 3$ V, find the change in i_D corresponding to a change in v_{GS} from -2 to -1.6 V.
Ans.   1.1 mA

7.3   For small v_{DS} calculate the value of r_{DS} at $v_{GS} = 0$ V and at $v_{GS} = -3$ V.
Ans.   200 Ω; 800 Ω

7.3   THE p-CHANNEL JFET

The circuit symbol and the static characteristics of a *p*-channel JFET are shown in Fig. 7.10. The circuit symbol differs from that of the *n*-channel device only in the direction

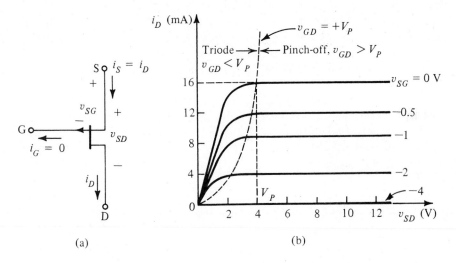

(a)   (b)

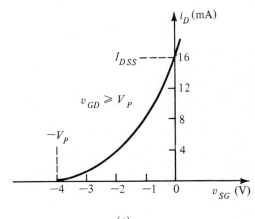

(c)

Fig. 7.10   *Circuit symbol and static characteristics for a p-channel JFET whose $V_P = +4$ V and $I_{DSS} = 16$ mA.*

of the arrowhead on the gate line. Indeed, it is the direction of the arrowhead that indicates the polarity of the device (that is, *n*-channel or *p*-channel JFET). For the *p*-channel device the arrow is pointing "outward," in the forward direction of the channel (*p*-type)-to-gate (*n*-type) junction.

Another difference in the circuit symbol is that it is drawn "upside down." This is done so that currents always flow from the top of the page down. This drawing convention, which is adopted throughout this book, enhances the ability of the reader to rapidly recognize the functional parts of a complex circuit. Needless to say, the value of this convention will be appreciated only at later stages.

We have indicated on the circuit symbol of the *p*-channel FET in Fig. 7.10 the reference directions for currents and voltages where the correlation between these reference directions and the symbols used should be noted. For instance, v_{SD} means, "how high v_S is with respect to v_D." In normal applications v_{SD} will be a positive number, which means that the source is more positive than the drain. On the other hand, v_{SG} will always be a negative number because the gate-to-channel junction has to be reverse-biased. Finally, i_D is shown flowing from source to drain, which is the actual direction of drain-current flow in a *p*-channel device. As in the *n*-channel device, the gate current is very small and ideally is zero; thus the source and drain currents are equal.

The i_D-v_{SD} characteristics look identical to those of the *n*-channel device. For operation in the pinch-off region, which is also called the active region, the drain has to be lower in potential than the gate by at least V_P volts. Note that for *p*-channel JFETs, V_P is a positive number, by convention. Thus the JFET will be in pinch-off when the following condition is satisfied:

$$v_{GD} > V_P, \qquad \text{Pinch-off region}$$

and will be in the triode region under the condition

$$v_{GD} < V_P, \qquad \text{Triode region}$$

The boundary between the two regions is characterized by

$$v_{GD} = V_P \qquad (7.12)$$

Stated in words, for the *p*-channel JFET to operate in pinch-off the drain should be lower than the gate by at least V_P volts.

The relationships governing i_D, v_{GS}, and v_{DS} are identical to those for the *n*-channel device except that here V_P is a positive number, v_{GS} is normally a positive number (because v_{SG} is negative), and v_{DS} is normally a negative number. As in the case of *n*-channel FETs, the i_D-v_{SD} characteristics in pinch-off display a finite, nonzero slope. Thus the output resistance in pinch-off is not infinite; rather it is finite with a value decreasing with the current level in the device.

EXERCISE

7.4 Consider a *p*-channel JFET with $V_P = 5$ V and $I_{DSS} = 10$ mA. If $v_{SG} = -3$ V, find i_D for $v_{SD} = 1$ V and for $v_{SD} = 2$ V.

Ans.   1.2 mA; 1.6 mA

7.4   JFET CIRCUITS AT DC

In this section we consider the analysis of some simple JFET circuits fed by dc sources. Our goal is to familiarize the reader with these devices.

Example 7.1

We want to analyze the circuit shown in Fig. 7.11 to determine the value of I_D and V_D. The FET has $V_P = -4$ V and $I_{DSS} = 10$ mA.

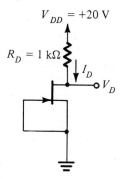

Fig. 7.11   Circuit for Example 7.1.

Solution

Since the gate is connected to the source, it follows that $V_{GS} = 0$. We do not know, however, whether the device is in the pinch-off region or in the triode region. To find out, we assume that it is in pinch-off, carry out the analysis, and then check whether our original assumption is correct or not. If the device is in pinch-off, then $I_D = I_{DSS}$. Thus

$$I_D = 10 \text{ mA}$$

The drain voltage can then be calculated from

$$V_D = V_{DD} - R_D I_D$$
$$= 20 - 1 \times 10$$

Thus

$$V_D = +10 \text{ V}$$

Since the gate is at zero volts, the drain is higher in potential than the gate by 10 V. This is greater than $|V_P|$, and hence the device is indeed in pinch-off, as originally assumed.

· · ·

The above example illustrates an interesting application of the JFET as a two-terminal constant-current device. Specifically, a JFET whose gate is connected to its source conducts a constant current equal to I_{DSS} provided that the external circuit

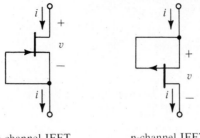

n-channel JFET   p-channel JFET

Fig. 7.12   *The application of the JFET as a constant-current device. Ideally, i is constant and equal to I_{DSS} provided that $v > |V_P|$.*

ensures that the voltage across the two-terminal device (see Fig. 7.12) is always main-tained greater than $|V_P|$. This constant-current two-terminal device is the dual of the zener diode discussed in Chapter 4. The latter provides a constant voltage as long as a certain minimum current is forced through it by the external circuit. Note, of course, that the current in the JFET will not be exactly constant, since the i_D-v_{DS} characteristic curve in pinch-off exhibits some finite slope. In other words, the current of the two-terminal device of Fig. 7.12 will be slightly dependent on the voltage that the external circuit forces across it. This again is parallel to the case of the zener diode.

Example 7.2

We wish to analyze the circuit of Fig. 7.13 to determine I_D and V_D. The JFET is spec-ified to have $V_P = -4$ V and $I_{DSS} = 10$ mA.

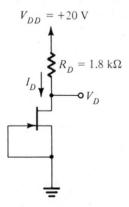

$V_{DD} = +20$ V

$R_D = 1.8$ kΩ

I_D

V_D

Fig. 7.13   *Circuit for Example 7.2.*

Solution

This circuit has a topology identical to that of Example 7.1. The difference is that here $R_D = 1.8$ kΩ. Assuming operation in pinch-off, we obtain

$$I_D = I_{DSS} = 10 \text{ mA}$$

which results in

$$V_D = 20 - 1.8 \times 10 = 2 \text{ V}$$

Thus the drain is higher than the gate by only 2 V. Since we require a minimum of 4 V for pinch-off operation, we conclude that the device is in the triode region and thus our original assumption is incorrect. We now have to repeat the analysis assuming operation in the triode region. The I_D-V_{DS} relationship is that given in Eq. (7.4),

$$I_D = I_{DSS} \left[2 \left(1 - \frac{V_{GS}}{V_P} \right) \frac{V_{DS}}{-V_P} - \left(\frac{V_{DS}}{V_P} \right)^2 \right]$$

Substituting $V_{GS} = 0$, $I_{DSS} = 10$ mA, and $V_P = -4$ V gives

$$I_D = 5V_{DS}(1 - \tfrac{1}{8}V_{DS}) \tag{7.13}$$

The other equation that governs circuit operation is

$$V_{DS} = V_{DD} - R_D I_D$$

in which we substitute $V_{DD} = 20$ V and $R_D = 1.8$ kΩ to obtain

$$I_D = \frac{20}{1.8} - \frac{V_{DS}}{1.8} \tag{7.14}$$

Equations (7.13) and (7.14) can now be solved to obtain V_{DS} and I_D. The results are

$$V_{DS} = 3 \text{ V} \qquad \text{or} \qquad 5.8 \text{ V}$$

The second answer is obviously inappropriate since it implies operation in the pinch-off region, which we have already ruled out. Thus

$$V_{DS} = 3 \text{ V}$$

and the corresponding current I_D is

$$I_D = 9.4 \text{ mA}$$

$$\cdot \quad \cdot \quad \cdot$$

Example 7.3
We desire to analyze the circuit in Fig. 7.14 to find the voltages and current. The FET has $V_P = -2$ V and $I_{DSS} = 4$ mA.

Solution
Assuming that the device is conducting, we see that the current I_D creates a voltage drop $I_D R_S$ across the source resistance R_S. Since the gate is at ground voltage, it follows that

$$V_{GS} = -I_D R_S \tag{7.15}$$

which is negative. It therefore is possible that the FET is operating either in pinch-off or in the triode region. Which of the two it will be will depend on the voltage at the drain, which we do not know. Thus to begin the analysis we have to make an assumption. We shall assume that the FET is in the pinch-off region, carry out the analysis, and then check the validity of our original assumption.

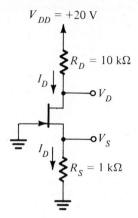

Fig. 7.14 Circuit for Example 7.3.

Assuming operation in pinch-off, the I_D-V_{GS} relationship is given by

$$I_D = I_{DSS} \left(1 - \frac{V_{GS}}{V_P} \right)^2 \tag{7.16}$$

Equations (7.15) and (7.16) can be used to determine I_D and V_{GS}. Substituting $R_S = 1\ \text{k}\Omega$, $I_{DSS} = 4\ \text{mA}$, and $V_P = -2\ \text{V}$ and combining the two equations results in the quadratic equation

$$I_D^2 - 5I_D + 4 = 0 \tag{7.17}$$

This equation has two roots, $I_D = 4\ \text{mA}$ and $I_D = 1\ \text{mA}$. The first solution is obviously inappropriate, since it equals I_{DSS}, which implies that $V_{GS} = 0$. This latter value is inconsistent with the value obtained by substituting in Eq. (7.15). It follows that

$$I_D = 1\ \text{mA}$$

and the corresponding V_{GS} is

$$V_{GS} = -1\ \text{V}$$

Thus

$$V_S = +1\ \text{V}$$
$$V_D = 20 - 1 \times 10 = +10\ \text{V}$$

Since the drain is higher than the gate by more than $|V_P|$, the device is indeed in pinch-off, as originally assumed.

$\cdot$ $\cdot$ $\cdot$

EXERCISE

7.5 In the circuit of Fig. E7.5 let the FET have $V_P = 3\ \text{V}$ and $I_{DSS} = 9\ \text{mA}$. Find the voltages V_G, V_S, and V_D.
Ans. $+10\ \text{V}$; $+8\ \text{V}$; $+7\ \text{V}$

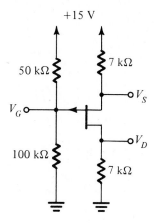

+15 V

50 kΩ

7 kΩ

V_G

V_S

100 kΩ

V_D

7 kΩ

Fig. E7.5

7.5   GRAPHICAL ANALYSIS

Since the JFET is well described by equations, it is rarely necessary to apply graphical techniques in the analysis of JFET circuits. Nevertheless, it is illustrative and instructive to consider the graphical analysis of the circuit in Fig. 7.15a. Here we assume that v_{GS} takes on negative values with a maximum of zero. In Fig. 7.15b we show the static characteristics of the FET. The operating point will be located on the curve corresponding to the specific value of v_{GS} applied. Where it will be on that curve will be determined by V_{DD} and R_D from

$$v_{DS} = V_{DD} - i_D R_D$$

which may be rewritten

$$i_D = \frac{V_{DD}}{R_D} - \frac{1}{R_D} v_{DS} \tag{7.18}$$

This is a linear equation in the variables i_D and v_{DS} and can be represented by a straight line on the i_D-v_{DS} plane. Such a line intercepts the v_{DS} axis at V_{DD} and has a slope equal to $-1/R_D$. Since when this circuit is used as an amplifier, R_D represents the *load resistance*, the straight line representing Eq. (7.18) is called the *load line*. The operating point of the JFET will lie at the intersection of the load line and the characteristic curve corresponding to the applied value of v_{GS}.

Bias Point for Amplifier Operation

In Fig. 7.15b we show two load lines, L_1 and L_2, representing two different values of R_D. Consider the steeper line (L_1), which corresponds to the smaller value of R_D. If the voltage v_{GS} is equal to $-V_1$, where V_1 is a positive number, then the operating point will be that labeled Q. Thus the voltage v_{DS} will be equal to V_Q and the current i_D will be equal to I_Q.

In a later section we shall discuss the operation of the JFET as an amplifier. It will be shown that the first step in designing an amplifier is to establish a dc operating point,

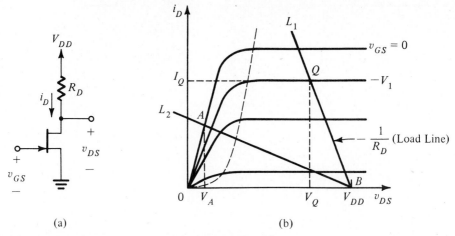

Fig. 7.15   *The graphical analysis in (b) of the circuit in (a).*

a process called *biasing*. It involves arranging the dc circuit so that a certain stable dc current, such as I_Q, flows through the FET in the absence of the signal to be amplified. Once the bias point is fixed, input signals can be applied. In the circuit of Fig. 7.15a the input signal may take the form of a time-varying voltage v_{gs} superimposed on the dc bias voltage V_{GS}. At any time the instantaneous operating point will be at the intersection of the load line and the characteristic curve corresponding to the total value of gate-to-source voltage at that instant of time, that is on the value of $v_{GS} = V_{GS} + v_{gs}$. It follows that the operating point moves along the load line. It is important to keep the operating point in the pinch-off region at all times, for if the operating point leaves the pinch-off region severe nonlinear distortion may arise. This will be illustrated later. For the time being we wish to make the point that if the device is to be used as an amplifier, then a suitable bias point (also called the *quiescent point*) is one that lies in the middle of the active region, such as point Q.

Operation as a Switch

The other load line (L_2) is useful if the FET is to be used as a switch. Specifically, if the gate-to-source voltage is a pulse having the two levels of zero and V_P volts (see Fig. 7.16a), then the operating point will be at A when $v_{GS} = 0$ and at B when $v_{GS} = V_P$. Note that at A the JFET is in the triode region, the voltage v_{DS} is small, and the resistance between drain and source (r_{DS}) is small. This corresponds to the closed position of the switch (Fig. 7.16b). In the other state, corresponding to point B, the current in the device is zero and the switch is open (Fig. 7.16c). It follows that the waveform at the drain will be a pulse whose polarity is the inverse of the input pulse, as shown in Fig. 7.16a. For this reason the circuit under consideration can be thought of as a logic inverter (see Chapter 6). Operation of the JFET as a switch will be considered later. For the time being, note that switching applications involve operating in the cutoff region and in the triode region.

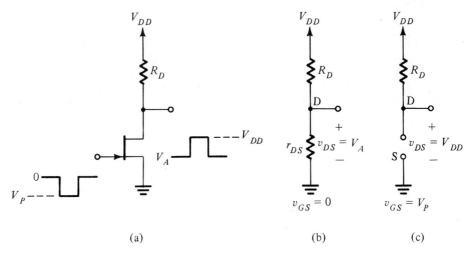

Fig. 7.16 Use of the JFET as a switch.

EXERCISES

7.6 Consider the circuit shown in Fig. 7.15a and let $V_{DD} = 15$ V, $R_D = 3$ kΩ, $I_{DSS} = 12$ mA, $V_P = -4$ V, and $V_{GS} = -2$ V. Calculate the value of I_D and V_D.
Ans. 3 mA; +6 V

7.7 For the circuit in Fig. 7.15a find the value of R_D so that when $v_{GS} = 0$, $v_{DS} = 0.1$ V. Let $V_{DD} = 15$ V, $I_{DSS} = 12$ mA, and $V_P = -4$ V.
Ans. 25 kΩ

7.6 BIASING

The first step in the design of a JFET amplifier is to establish a stable and predictable dc operating point. This operating point should be in the active region (pinch-off region) and should allow for sufficient signal swing without the device entering the triode region or the cutoff region. Later we shall discuss the signal-swing problem in some detail. The present section is devoted to techniques for establishing a stable dc operating point.

Stable Operating Point

A stable operating point is one that is almost independent of variations in the device parameters V_P and I_{DSS}. These parameters vary with temperature; more importantly, they vary considerably between different units belonging to the same device type. It is usual to see on the data sheets of a JFET a wide range specified for V_P and I_{DSS} (say, V_P from -2 to -8 V and I_{DSS} from 4 to 16 mA). A good bias design is one that ensures that I_D and V_{DS} will always be within a certain range of their nominal value (say, $\pm 20\%$) independent of the value of V_P and I_{DSS}. In other words, we wish to find a biasing arrangement by which the value of I_D (and hence V_{DS}) will not change substantially if we replace the particular FET by another of the same type.

Extreme Devices

Although the problem of biasing JFETs is difficult because of the wide variations in device parameters, there is a simplifying factor. Devices whose $|V_P|$ is large tend to also have high values of I_{DSS}, while those with low $|V_P|$ also have low I_{DSS}. Thus if a device is specified to have V_P in the range -2 to -8 V and I_{DSS} in the range 4 to 16 mA, we may assume that the two extreme devices are as follows: the "low" device, with

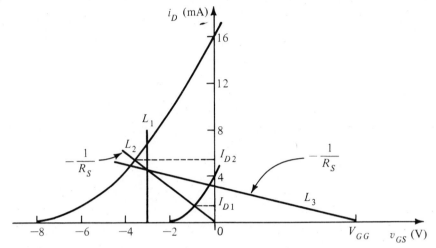

Fig. 7.17 *The i_D-v_{GS} characteristics of two devices belonging to the same JFET type. Also shown are graphical representations for fixed bias (line L_1), self-bias (line L_2), and combination of fixed bias and self-bias (line L_3).*

$V_P = -2$ V and $I_{DSS} = 4$ mA, and the "high" device, with $V_P = -8$ V and $I_{DSS} = 16$ mA. The i_D-v_{GS} characteristics of the two extreme devices are sketched in Fig. 7.17.

Fixed Bias

The simplest approach to biasing the JFET is to apply a constant dc voltage between gate and source; however, this results in very unsatisfactory performance. For instance, from the characteristics shown in Fig. 7.17 we find that a fixed voltage V_{GS} of, say, -3 V (represented graphically by the vertical line L_1) will result in the "high" device conducting a current I_D of 6.25 mA while the "low" device will be cut off! For $V_{GS} = -1$ V the "high" device will have $I_D = 12.25$ mA, while the "low" device will have $I_D = 1$ mA, a very large difference.

Self-Bias

Better results are obtained with the self-bias circuit shown in Fig. 7.18. Here we have included a resistance R_S in the source lead. The voltage $I_D R_S$ developed across this resistance will constitute the reverse bias V_{GS},

$$V_{GS} = -I_D R_S \qquad (7.19)$$

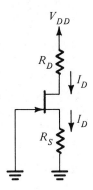

Fig. 7.18 JFET with self-bias.

The term *self-bias* arises because no external dc source is used in the gate–source circuit. Self-bias is possible only in *depletion-type* devices such as JFETs, some MOS-
FETs, and vacuum tubes. As will be seen in Chapter 9, bipolar transistors are *enhancement-type* devices, and hence self-bias is not possible.

Let us continue with the analysis of the circuit in Fig. 7.18. We may use Eq. (7.19)
together with the i_D-v_{GS} square-law relationship to write

$$I_D = I_{DSS} \left(1 + \frac{I_D R_S}{V_P} \right)^2 \tag{7.20}$$

Graphically, Eq. (7.19) can be represented by the straight line L_2 in Fig. 7.17. As can
be seen, the value of I_D will vary from I_{D1} for the "low" device to I_{D2} for the "high"
device, a less dramatic spread than in the case of fixed bias. Obviously, to minimize the
variance in I_D we should select as high a value for R_S as possible (since a high value for
R_S results in a straight line with a small slope). However, using a high value for R_S
could result in a very low current in the "low" device. In Section 7.7 it will be shown
that the gain is proportional to $\sqrt{I_D}$; thus one should attempt to use as high a value for
I_D as possible.

As an example, let $R_S = 1$ kΩ and consider the device whose two extreme characteristics are depicted in Fig. 7.17. For the "low" device we have

$$I_D = 4 \left(1 - \frac{I_D}{2} \right)^2 \tag{7.21}$$

which results in

$$I_D = 1 \text{ mA} \qquad \text{and} \qquad V_{GS} = -1 \text{ V}$$

For the high device we have

$$I_D = 16 \left(1 - \frac{I_D}{8} \right)^2 \tag{7.22}$$

which results in

$$I_D = 4 \text{ mA} \qquad \text{and} \qquad V_{GS} = -4 \text{ V}$$

Although the spread is smaller than in the case of fixed bias, it is still unacceptably large.

Combination of Fixed Bias and Self-Bias

From Fig. 7.17 it may be observed that to minimize the variance in I_D we should intersect the i_D-v_{GS} curves at a very small slope. This suggests the use of a bias circuit that results in a straight line such as L_3. This line intercepts the v_{GS} axis at a positive voltage V_{GG} and has the equation

$$v_{GS} = V_{GG} - i_D R_S \qquad (7.23)$$

Straightforward reasoning suggests the circuit shown in Fig. 7.19a. In a practical situation the voltage V_{GG} would be obtained as a fraction of V_{DD} through the use of a

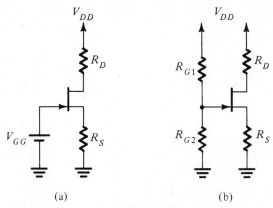

(a)   (b)

Fig. 7.19   JFET with combination of self-bias and fixed bias.

voltage divider, as shown in Fig. 7.19b. Finally, we should remind the reader that the value of R_D should be such that at all times the drain voltage is higher than the gate voltage by at least $|V_P|$, this being the condition for active-mode operation.

Example 7.4
We desire to design the bias circuit of Fig. 7.19b for the JFET whose characteristics are depicted in Fig. 7.17. Assume that the available power supply $V_{DD} = 20$ V.

Solution
To obtain a small variance in I_D and still keep the current relatively high, we should use as high a value for V_{GG} and for R_S as possible. However, the higher the value we use for V_{GG}, the smaller the voltage drop that will be available across R_D (because we have to keep D higher than G by at least $|V_P|$). For a given I_D, a smaller voltage drop across R_D implies a lower value for R_D and hence (as will be shown in Section 7.7) a lower gain. Let us choose $V_{GG} = 8$ V. For the "low" device this value implies that the minimum drain voltage should be 10 V, leaving a maximum of 10 V for the drop across R_D. For the "high" device the minimum drain voltage is 16 V, which leaves a maximum of 4 V for the drop across R_D.

We next have to select a value for R_S. In order to maximize the value of I_D we select a value for R_S that results in the maximum possible I_D for the "low" device. To avoid forward-biasing the gate-to-channel junction, we use $I_D = I_{DSS} = 4$ mA. This corresponds to $V_{GS} = 0$ for the "low" device. It is important here to note that the applied signal will cause the gate-to-source voltage to become positive. This is allowed as long as the resulting forward-bias voltage is less than about 0.5 V.

For the bias line to pass through the point $V_{GS} = 0$, $I_D = 4$ mA, it follows that

$$R_S = \frac{V_{GG}}{4 \text{ mA}} = \frac{8}{4} = 2 \text{ k}\Omega$$

We can now calculate the current in the "high" device from

$$I_D = 16 \left(1 - \frac{8 - 2I_D}{-8} \right)^2$$

which results in

$$I_D = 5.85 \text{ mA}, \qquad V_{GS} = -3.7 \text{ V}$$

Thus the spread is from 4 mA to 5.85 mA, which is much less than what we have been able to achieve with self-bias alone.

• • •

EXERCISE

7.8 For the circuit considered in Example 7.4 find the maximum value of R_D that will ensure that both the "low" and "high" devices are in pinch-off. Also calculate V_D for these extreme devices.
Ans. 683 Ω; 17.3 V, 16 V

7.7 THE JFET AS AN AMPLIFIER

Basic Amplifier Arrangement

To understand the basis for the operation of the JFET as an amplifier, consider the circuit shown in Fig. 7.20. Here we show the gate-to-source reverse bias established by a separate battery, which is not a good biasing arrangement, but our intention is to focus on the concepts involved. Superimposed on the bias voltage V_{GS} we have a signal v_{gs}; thus the total instantaneous gate-to-source voltage is given by

$$v_{GS} = V_{GS} + v_{gs} \tag{7.24}$$

Assuming that the FET will remain in pinch-off at all times, which is achieved by keeping v_D higher than v_G by at least $|V_P|$, that is,

$$v_D \geq v_{GS} + |V_P| \tag{7.25}$$

it follows that

$$i_D = I_{DSS} \left(1 - \frac{v_{GS}}{V_P} \right)^2 \tag{7.26}$$

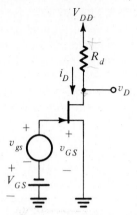

Fig. 7.20   Circuit for establishing the basis for the operation of the JFET as an amplifier.

Substituting for v_{GS} from Eq. (7.24) into Eq. (7.26) yields

$$i_D = I_{DSS}\left(1 - \frac{V_{GS}}{V_P} - \frac{v_{gs}}{V_P}\right)^2$$

$$= I_{DSS}\left(1 - \frac{V_{GS}}{V_P}\right)^2 - \frac{2I_{DSS}}{V_P}\left(1 - \frac{V_{GS}}{V_P}\right)v_{gs} + I_{DSS}\left(\frac{v_{gs}}{V_P}\right)^2 \tag{7.27}$$

Setting $v_{gs} = 0$ in Eq. (7.27) results in $i_D = I_D$, which is the dc bias current,

$$I_D = I_{DSS}\left(1 - \frac{V_{GS}}{V_P}\right)^2 \tag{7.28}$$

Substituting in Eq. (7.27) results in

$$i_D = I_D + \frac{2I_{DSS}}{-V_P}\left(1 - \frac{V_{GS}}{V_P}\right)v_{gs} + I_{DSS}\left(\frac{v_{gs}}{V_P}\right)^2 \tag{7.29}$$

If we restrict ourselves to small signals satisfying the constraint

$$\frac{v_{gs}}{V_P} \ll 1 \tag{7.30}$$

we can neglect the last term in Eq. (7.29) and obtain

$$i_D \simeq I_D + \frac{2I_{DSS}}{-V_P}\left(1 - \frac{V_{GS}}{V_P}\right)v_{gs} \tag{7.31}$$

Thus the total drain current is composed of two components: the dc bias I_D and a signal component i_d given by

$$i_d = \frac{2I_{DSS}}{-V_P}\left(1 - \frac{V_{GS}}{V_P}\right)v_{gs} \tag{7.32}$$

Thus the signal current is linearly related to the signal voltage v_{gs}, which is a requirement in a linear amplifier. This linear relationship is based on the premise that the signal v_{gs} is much smaller than $|V_P|$, which is referred to as the *small-signal approxi-mation*. Note that if the small signal assumption is not valid, the term involving v_{gs}^2 in Eq. (7.29) has to be taken into account. This obviously results in the current i_d having components harmonically related to the input signal. Such nonlinear distortion is undesirable and should be minimized in the design of linear amplifiers.

Transconductance

Assuming that the small-signal assumption is adhered to, the signal current i_d is proportional to v_{gs} with the proportionality constant called *transconductance* and denoted by g_m,

$$g_m \equiv \frac{i_d}{v_{gs}} = \frac{2I_{DSS}}{-V_P}\left(1 - \frac{V_{GS}}{V_P}\right) \tag{7.33}$$

Recalling that for *n*-channel FETs V_P is a negative number and that V_{GS} is also negative, we see that g_m is positive. A relationship for g_m that applies for both *n*- and *p*-channel devices can be written

$$g_m = \frac{2I_{DSS}}{|V_P|}\left(1 - \left|\frac{V_{GS}}{V_P}\right|\right) \tag{7.34}$$

Note that g_m is determined by the FET parameters I_{DSS} and V_P as well as by the dc operating point. We may use the square-law relationship of the JFET together with Eq. (7.34) to express g_m in the alternative form

$$g_m = \frac{2I_{DSS}}{|V_P|}\sqrt{\frac{I_D}{I_{DSS}}} \tag{7.35}$$

from which it is evident that g_m is proportional to the square root of the bias current I_D. It follows that g_m will be highest if the FET is biased at $V_{GS} = 0$ (or, equivalently, $I_D = I_{DSS}$). This maximum value of g_m is denoted g_{m0} and is given by

$$g_{m0} = \frac{2I_{DSS}}{|V_P|} \tag{7.36}$$

A graphical interpretation of the above results is shown in Fig. 7.21 which displays the i_D-v_{GS} characteristic in the pinch-off region. A triangular waveform is used for the input signal v_{gs}. Note that g_m is equal to the slope of the parabolic characteristic at the bias point Q,

$$g_m = \left.\frac{\partial i_D}{\partial v_{GS}}\right|_{i_D = I_D} \tag{7.37}$$

Also note that the small-signal approximation implies that we are operating on a segment of the graph (around Q) that is sufficiently short to approximate a straight line of slope g_m. Finally, note that the reason that one does not bias the device at $V_{GS} = 0$ is to allow for sufficient input signal swing without forward-biasing the gate-to-channel junction.

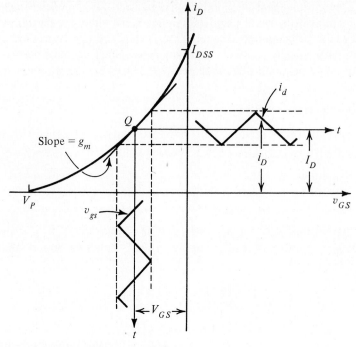

Fig. 7.21   *Operation of the JFET as an amplifier.*

Voltage Gain

Returning to the circuit in Fig. 7.20, we can now find the total voltage at the drain, v_D, as

$$\begin{aligned}
v_D &= V_{DD} - i_D R_d \\
&= V_{DD} - I_D R_d - i_d R_d \\
&= V_D - i_d R_d
\end{aligned} \tag{7.38}$$

Thus superimposed on the dc drain voltage V_D we have a signal component v_d given by

$$v_d = -i_d R_d$$

which can be written

$$v_d = -g_m v_{gs} R_d$$

The voltage gain of the FET amplifier is therefore given by

$$\frac{v_d}{v_{gs}} = -g_m R_d \tag{7.39}$$

The minus sign in Eq. (7.39) indicates that the output signal v_d is 180° out of phase with respect to the input signal v_{gs}. This is illustrated in Fig. 7.22, which shows

v_{GS} and v_D. The input signal is assumed to have a triangular waveform with an amplitude much smaller than $|V_P|$ in order to ensure linear operation. For operation in the pinch-off region the minimum value of v_D should be greater than the corresponding value of v_G by at least $|V_P|$. Also, the maximum value of v_D should be smaller than V_{DD}; otherwise the FET will enter the cutoff region and the peaks of the output signal waveform will be clipped off.

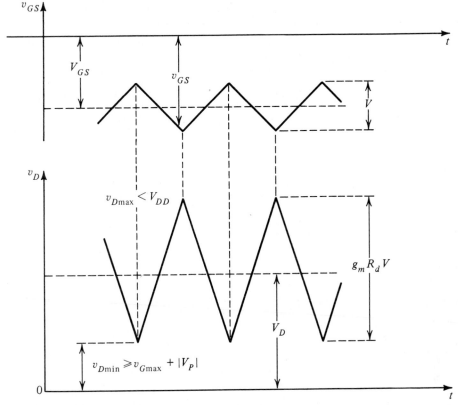

Fig. 7.22 *Total instantaneous voltages v_{GS} and v_D for the circuit in Fig. 7.20 with a triangular-wave input signal.*

Separating the DC Analysis and the Signal Analysis

From the above analysis we see that under the small-signal approximation, signal quantities are superimposed on dc quantities. For instance, the total drain current i_D equals the dc current I_D plus the signal current i_d, the total drain voltage $v_D = V_D + v_d$, and so on. It follows that the analysis and design can be greatly simplified by separating dc or bias calculations from small-signal calculations. That is, once a stable dc operating point has been established and all dc quantities calculated, we may then perform signal analysis ignoring dc quantities.

The JFET as a Voltage-Controlled Current Source

From a signal point of view the JFET behaves as a voltage-controlled current source. It accepts a signal v_{gs} between gate and source and provides a current $g_m v_{gs}$ at the drain terminal. The input resistance of this controlled source is very high and can be assumed infinite in first-order calculations. Also, the output resistance—that is, the resistance looking into the drain—is very high, and up until now we have been assuming it to be infinite. Note, of course, that the transconductance g_m is a function of the FET as well as of the dc bias point. When we perform small-signal calculations, nodes in the circuit that are connected to dc sources can be assumed grounded because an ideal dc source will not change its voltage in the presence of signals. In other words, the signal voltage at a node connected to a dc source will always be zero.

EXERCISE

7.9 In the circuit of Fig. 7.20 let the FET have $V_P = -3$ V and $I_{DSS} = 9$ mA. For $V_{GS} = -2$ V find the value of g_m. Calculate the voltage gain obtained when $R_d = 10$ kΩ. If $V_{DD} = 15$ V and the input signal is a triangular waveform of 0.2 V peak-to-peak, find the minimum and maximum voltages at the drain.
Ans.   2 mA/V; −20 V/V; 3 V; 7 V

7.8   SMALL-SIGNAL EQUIVALENT CIRCUIT MODELS

In many applications it is easy and convenient to perform signal calculations directly on the circuit diagrams. Nevertheless, it is instructive—and in some applications essential—to consider performing signal calculations on a circuit equivalent to the JFET from a small-signal point of view. Such an equivalent circuit models the operation of the JFET as an amplifier at a specific bias point.

First-Order Model

Figure 7.23 shows the simplest equivalent circuit model of the JFET. In this model the JFET is considered as an ideal voltage-controlled current source with a transconduc-

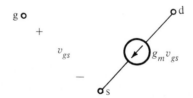

Fig. 7.23   *First-order small-signal model for the JFET.*

tance g_m. To use this model, we simply replace the JFET by this three-terminal circuit and eliminate (short-circuit) all dc supply voltages in the circuit. The resulting circuit can then be analyzed to determine signal parameters such as voltage gain.

An Alternative First-Order Model

An equally simple equivalent circuit model is shown in Fig. 7.24. This model is based on the fact that the source current is almost equal to the drain current because the gate current is very small. Thus the source current i_s is given by

$$i_s = g_m v_{gs}$$

which can be rewritten

$$i_s = \frac{v_{gs}}{1/g_m}$$

This equation says that the source current is related to the gate-to-source voltage by a resistance $1/g_m$. It follows that from a signal point of view the gate-to-source junction

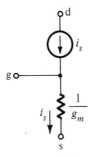

Fig. 7.24   An alternative first-order model for the JFET.

is equivalent to a resistance $1/g_m$. Note that this is the resistance between gate and source "looking into the source." The resistance between gate and source "looking into the gate" is infinite because the gate current is zero.

The model of Fig. 7.24 consists of a resistance $1/g_m$ between gate and source and a current source between drain and gate. The value of the current source is controlled by the current in the resistance $1/g_m$, with the control factor equal to unity. The reader is encouraged to prove that this circuit model is exactly equivalent to that in Fig. 7.23.

A Second-Order Model

A more complete equivalent circuit model of the JFET is shown in Fig. 7.25. This model is based on that of Fig. 7.23, but it includes three more components. The resistance r_o models the finite output resistance of the drain current source. The value of r_o can be determined directly from the static characteristics of Fig. 7.9. As previously mentioned, r_o is inversely proportional to the current level in the device, with typical values in the range of 10 to 100 kΩ.

The refined model of Fig. 7.25 includes two capacitances: the gate-to-source capacitance C_{gs} and the gate-to-drain capacitance C_{gd}. This model can be used to predict the high-frequency response of FET amplifiers, as will be demonstrated in Chapter 11. Both

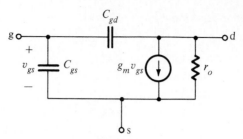

Fig. 7.25   *A second-order small-signal model for the JFET.*

C_{gs} and C_{gd} represent capacitances of the reverse-biased gate-to-channel junction and are of the order of 1 to 3 pF.

The use of equivalent circuit models in the analysis of JFET amplifiers will be demonstrated in the following sections.

7.9   *THE JFET COMMON-SOURCE AMPLIFIER*

Biasing

Figure 7.26 shows one of the most common configurations of JFET amplifiers. Here the JFET is biased using a combination of fixed bias and self-bias, as explained in Section 7.6. Resistors R_{G1} and R_{G2} form a voltage divider across the power supply V_{DD} and thus establish a voltage V_{GG} at the gate,

$$V_{GG} = V_{DD} \frac{R_{G2}}{R_{G1} + R_{G2}}$$

In order to keep the input resistance of the FET amplifier high, one should choose R_{G1} and R_{G2} as high as possible. Typically, R_{G1} and R_{G2} can be in the range 100 kΩ to a few

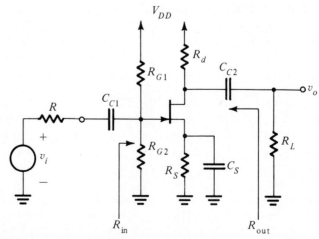

Fig. 7.26   *JFET common-source amplifier.*

megohms with no difficulty. Note, though, that the small gate current (a few nanoamps) flows through the parallel equivalent of R_{G1} and R_{G2}. Since this current increases with temperature, its value imposes an upper limit on the values of R_{G1} and R_{G2}. Otherwise, the bias point of the FET becomes temperature dependent.

Self-bias is obtained by connecting a resistance R_S in the source lead. As discussed in Section 7.6, the values of the biasing elements, including R_d, should be chosen to obtain a stable and predictable dc operating point. In the following we shall assume that this has already been accomplished, and we will concern ourselves with signal performance only.

Coupling the Source and the Load

The amplifier circuit of Fig. 7.26 is fed by a signal source with voltage v_i and resistance R. The signal is coupled to the gate of the FET through a capacitor C_{C1}. The purpose of this capacitor is to block dc so as not to disturb the bias conditions already established. The value of C_{C1} should be chosen sufficiently large so that at signal frequencies of interest it acts as a short circuit. Obviously, C_{C1} will cause the amplifier gain to drop at low frequencies and will cause the gain to be zero at dc. As mentioned in Section 2.5, amplifiers that use coupling capacitors are called ac amplifiers in order to distinguish them from direct-coupled (DC) amplifiers, which are essential in some amplications. Operational amplifiers are examples of DC amplifiers.

The output signal voltage at the drain is coupled to a load resistance R_L through another coupling capacitor C_{C2}. Again C_{C2} should be chosen large in order to act as a perfect coupler at all frequencies of interest. The load resistance R_L represents either an actual load or the input resistance of another part of the system.

Bypassing R_S

As shown in Fig. 7.26, a capacitance C_S is connected across the bias resistance R_S. The purpose of C_S is to cause the total impedance between source and ground to be very small (ideally zero) at all frequencies of interest. In other words, C_S is chosen sufficiently large so that it acts as a short circuit for signals. For this reason C_S is called a *bypass capacitor*—signal currents pass through it, thus "bypassing" the resistance R_S. As will be shown below, inclusion of bypass capacitor C_S causes the gain of the amplifier to increase considerably. Alternatively, one can say that without C_S the gain would be quite small. In Chapter 11 we shall study the effects of the capacitance C_S on the frequency response of the amplifier. It can be easily seen that at low frequencies C_S will no longer act as a perfect short circuit.

Assume for now that C_S is acting as a perfect short circuit. It follows that the source of the FET will be at signal ground. Thus as a two-port network the amplifier circuit of Fig. 7.26 has a common ground between input and output, which is the source terminal. For this reason the circuit is called a *common-source amplifier*.

Small-Signal Analysis

In the remainder of this section we shall concern ourselves with analysis of the common-source amplifier, assuming that the frequencies of interest are sufficiently high so that coupling and bypass capacitors behave as perfect short circuits. We shall also assume that the frequencies of interest are sufficiently low so that the FET internal capacitances C_{gs} and C_{gd} can be considered as open circuits. We speak of the frequency band at which both the above conditions are satisfied as *the midband*.

The first step in the analysis consists of evaluating the input resistance R_{in}, indicated in Fig. 7.26. This is the resistance seen between the gate terminal and ground. Since the input resistance of the JFET is infinite (the gate current is zero), R_{in} consists of R_{G2} in parallel with R_{G1}. (The terminal of R_{G1} connected to V_{DD} is at signal ground.) Thus

$$R_{in} = (R_{G1} \| R_{G2}) \tag{7.40}$$

Having evaluated R_{in}, we now can find the fraction of the input signal voltage that appears between gate and ground. This can be done by recognizing that R_{in} and R form a voltage divider; thus

$$v_g = v_i \frac{R_{in}}{R_{in} + R} \tag{7.41}$$

Since the source is bypassed to ground through C_S, it follows that

$$v_{gs} = v_g \tag{7.42}$$

The drain current signal i_d is given by

$$i_d = g_m v_{gs} \tag{7.43}$$

To obtain the output voltage signal v_o, which is equal to v_d, we multiply i_d by the total resistance between drain and ground. This resistance consists of R_d in parallel with R_L and in parallel with the FET output resistance r_o,

$$v_o = -i_d(R_d \| R_L \| r_o) \tag{7.44}$$

In most cases r_o will have only a small effect on gain calculations. Combining the above equations yields the voltage gain v_o/v_i of the common source amplifier as

$$\frac{v_o}{v_i} = -\frac{R_{in}}{R_{in} + R} g_m(R_d \| R_L \| r_o) \tag{7.45}$$

The output resistance R_{out} of this amplifier, exclusive of the load (see Fig. 7.26), is given by

$$R_{out} = (R_d \| r_o) \tag{7.46}$$

Example 7.5
We want to evaluate the voltage gain and the maximum allowable input signal swing of the common-source amplifier of Fig. 7.27. Ignore all capacitive effects; that is, perform the analysis at midband frequencies. The FET is specified to have $I_{DSS} = 12$ mA and $V_P = -4$ V. At $I_D = 12$ mA the output resistance $r_o = 25$ kΩ.

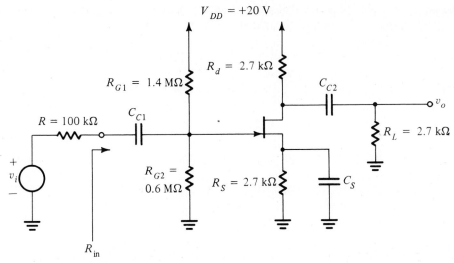

Fig. 7.27 Circuit for Example 7.5.

Solution

First we determine the dc operating point as follows:

$$V_{GG} = V_{DD} \frac{R_{G2}}{R_{G1} + R_{G2}}$$

$$= 20 \frac{0.6}{2} = 6 \text{ V}$$

$$V_{GS} = V_{GG} - I_D R_S$$
$$= 6 - 2.7 I_D$$

$$I_D = I_{DSS} \left(1 - \frac{V_{GS}}{V_P} \right)^2$$

$$= 12 \left(1 - \frac{6 - 2.7 I_D}{-4} \right)^2$$

This leads to the quadratic equation

$$I_D^2 - 7.59 I_D + 13.7 = 0$$

The appropriate solution of this equation is

$$I_D = 2.96 \text{ mA}$$

The corresponding voltage V_{GS} is

$$V_{GS} = 6 - 2.7 \times 2.96 \simeq -2 \text{ V}$$

The drain voltage will be

$$V_D = V_{DD} - I_D R_d = 20 - 2.96 \times 2.7 = 12 \text{ V}$$

Since $V_D > V_G + |V_P|$, the device is indeed in pinch-off.

The value of g_m can be evaluated as

$$g_m = \frac{2I_{DSS}}{-V_P} \sqrt{\frac{I_D}{I_{DSS}}}$$

$$= \frac{2 \times 12}{4} \sqrt{\frac{2.96}{12}}$$

Thus $g_m = 2.98$ mA/V. Since r_o is inversely proportional to I_D it follows that at $I_D = 2.96$ mA, $r_o \simeq 100$ kΩ.

The input resistance of the amplifier is given by

$$R_{in} = (R_{G1} \| R_{G2}) = \frac{1.4 \times 0.6}{1.4 + 0.6}$$

Thus $R_{in} = 0.42$ MΩ $= 420$ kΩ. The voltage gain can now be evaluated as

$$\frac{v_o}{v_i} = \frac{R_{in}}{R_{in} + R}(-g_m)(R_d \| R_L \| r_o)$$

$$= -\frac{420}{420 + 100} \times 2.96(2.7 \| 2.7 \| 100)$$

Thus $v_o/v_i = -3.2$.

To investigate the question of signal swing, we first consider the maximum amplitude of input signal for which the FET remains in pinch-off. Denote the peak value of the gate signal by $\hat{V}$. The condition for remaining in pinch-off can be expressed as

$$V_D - A\hat{V} \geq V_G + \hat{V} + |V_P|$$

where A denotes the magnitude of voltage gain between gate and drain,

$$A = g_m(R_L \| R_d \| r_o) = 3.94$$

The maximum allowable value of $\hat{V}$ can be obtained as

$$\hat{V} = \frac{V_D - V_G - |V_P|}{A + 1} = \frac{12 - 6 - 4}{4.94} = 0.4 \text{ V}$$

Since this signal is much smaller than $|V_P|$, it can be considered as a "small signal." In general, however, one can use the square-law relationship of the JFET to determine the second-harmonic distortion corresponding to this signal amplitude. Letting

$$v_{gs} = \hat{V} \sin \omega t$$

we have

$$i_D = I_{DSS} \left(1 - \frac{v_{GS}}{V_P}\right)^2$$

$$i_D = 12\left(1 - \frac{-2 + \hat{V}\sin\omega t}{-4}\right)^2$$

$$= 3 + \frac{3\hat{V}^2}{8} + 3\hat{V}\sin\omega t - \frac{3\hat{V}^2}{8}\cos 2\omega t$$

The second component on the right-hand side represents a very slight shift in the dc drain current. The last term represents the second-harmonic component of drain current. The percentage harmonic distortion is given by

$$\text{Percentage harmonic distortion} = \frac{3\hat{V}^2/8}{3\hat{V}} \times 100 \simeq 12.5\hat{V}$$

For $\hat{V} = 0.4$ V the harmonic distortion is 5%. This is quite small. If in a certain application the value of harmonic distortion is found to be unacceptably high, then the value of $\hat{V}$ should be reduced. In other words, the maximum allowable input signal swing is often determined by linearity considerations rather than by the condition that the device remain in the active mode.

Corresponding to v_{gs} having a peak value of 0.4 V, the input signal v_i will have a peak value of

$$\hat{V}_i = \hat{V}\frac{R_{in} + R}{R_{in}} \simeq 0.5 \text{ V}$$

$$\cdot \quad \cdot \quad \cdot$$

Analysis Using Equivalent Circuit Models

In all of the discussion above, small-signal analysis was performed directly on the circuit diagram. Of course, the equivalent circuit model of the FET was used, but only implicitly. What we did was possible because the model at midband frequencies is quite simple. If more exact—and hence more complicated—models are used, a different approach to signal analysis should be adopted. Specifically, the FET should be explicitly replaced by the equivalent circuit model, and all analysis should be performed on the resulting circuit.

For the common-source amplifier of Fig. 7.26, which is redrawn in Fig. 7.28a, the equivalent circuit at midband frequencies is shown in Fig. 7.28b. Analysis of this latter circuit yields results identical to those obtained in the above.

The Common-Source Amplifier with Unbypassed Source Resistance

Figure 7.29 shows an amplifier circuit similar to that of Fig. 7.26 except that part of the source resistance R_S has been left unbypassed. An unbypassed resistance in the source lead results in a reduction in gain, but it has other positive attributes, such as increasing the amplifier bandwidth (the frequency range over which the amplifier provides gain close in value to the midband gain) and increasing the maximum allowable input signal swing. The effect on bandwidth will be discussed in Chapter 11.

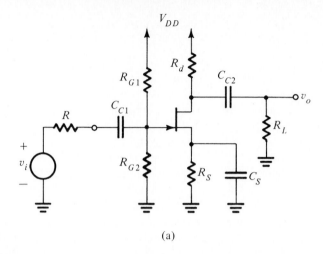

(a)

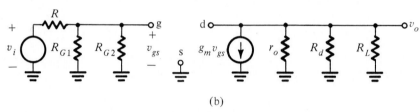

(b)

Fig. 7.28 *(a) Common-source amplifier. (b) Equivalent circuit of the common-source amplifier at midband frequencies. Note that the model used is that of Fig. 7.23.*

To evaluate the voltage gain of the circuit in Fig. 7.29 we first observe that the input resistance R_{in} remains equal to $(R_{G1} \| R_{G2})$ and thus the fraction of v_i that appears between gate and ground (v_g) remains unchanged. However, unlike the situation in the common-source amplifier, this signal does not appear between gate and source. To find the latter signal, v_{gs}, it is easiest to use the equivalent circuit of Fig. 7.24. According to this equivalent circuit the gate-to-source junction, looking into the source, is equivalent to a resistance $1/g_m$. This resistance appears in series with R_{S1}. Thus we can use the voltage-divider rule to obtain v_{gs} as

$$v_{gs} = v_g \frac{1/g_m}{R_{S1} + 1/g_m} \tag{7.47}$$

Once v_{gs} is determined the remainder of the analysis is straight-forward; the final result is

$$\frac{v_o}{v_i} = -\frac{R_{in}}{R_{in} + R} \frac{(R_d \| R_L \| r_o)}{1/g_m + R_{S1}} \tag{7.48}$$

This expression[2] is quite interesting. It consists of two factors: the first represents the gain between the signal generator and the gate terminal, and the second represents the

---

[2]In writing this expression it has been assumed that r_o appears in parallel with R_d. This is not exactly correct, since r_o appears between drain and source and the source is no longer grounded. Nevertheless, the approximation involved is good enough for our purposes at this stage.

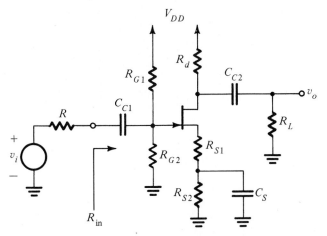

Fig. 7.29   JFET amplifier with an unbypassed source resistance R_{S1}.

gain between the gate and drain. This second factor is simply *the ratio between the total resistance in the drain lead and the total resistance in the source lead*. This observation enables one to write the gain expression directly without having to go through a step-by-step procedure.

The gain expression of Eq. (7.48) can be written in the alternative form

$$\frac{v_o}{v_i} = -\frac{R_{in}}{R_{in} + R} g_m \frac{(R_d \| R_L \| r_o)}{1 + g_m R_{S1}} \tag{7.49}$$

Comparison of this expression with that obtained for the grounded-source amplifier in Eq. (7.45) reveals that including an unbypassed resistance R_{S1} reduces the gain by the factor $1 + g_m R_{S1}$.

Since only a fraction of v_g appears between gate and source [Eq. (7.47)], it follows that one can allow v_g and hence v_i to be larger (as compared to the grounded-source case) without violating the small-signal approximation.

Power Gain

Before leaving this section we should point out that although the amplifiers considered in the examples do not exhibit large voltage gains, their power gains are quite impressive. To illustrate, let us calculate the power gain for the amplifier of Example 7.5. The input power is given by

$$P_i = \frac{v_i^2}{R + R_{in}}$$

while the output power is given by

$$P_o = \frac{v_o^2}{R_L}$$

Thus the power gain is

$$\frac{P_o}{P_i} = \frac{R + R_{in}}{R_L}\left(\frac{v_o}{v_i}\right)^2$$

Substitution of the values from the results of Example 7.5 gives

$$\frac{P_o}{P_i} = \frac{520}{2.7}(3.2)^2 = 1,972$$

In decibels the power gain is

$$10 \log \frac{P_o}{P_i} = 32.9 \text{ dB}$$

EXERCISE

7.10 Consider the amplifier circuit of Example 7.5, but let 300 Ω of the source resistance R_S be left unbypassed. Calculate the voltage gain of the modified circuit. Also calculate the maximum allowable input signal swing that corresponds to a maximum v_{gs} of 0.4 V.
Ans. -1.7 V/V; 0.94 V

7.10   THE SOURCE FOLLOWER

Because of its very high input resistance the JFET is an excellent choice for the design of buffer or isolation amplifiers. These are amplifiers whose purpose is to allow the coupling of a high-impedance source to a low-impedance load with little loss in signal level. Normally a buffer amplifier is not required to provide large voltage gain; the main objective is to provide large current and power gains.

Figure 7.30 shows a simple buffer circuit using the JFET. Here the generator signal is capacitively coupled to the gate, while the output signal is taken from the source terminal and is capacitively coupled to the load. The drain terminal is connected to the dc supply and hence is at signal ground. For this reason this circuit is referred to as *the grounded-drain or common-drain configuration*. Finally, note that the FET is self-biased using a source resistance R_S. The resistance R_G provides dc continuity for the gate ter-

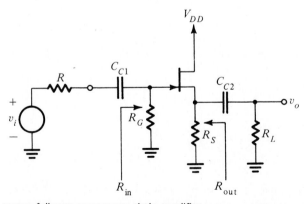

Fig. 7.30   The source follower or common-drain amplifier.

minal. Since the gate current is very small (a few nanoamps), we can safely assume that the dc voltage at the gate is zero. This is true even though one normally uses a large resistance R_G.

Input Resistance

The input resistance R_{in} of the circuit in Fig. 7.30 is simply equal to the bias resistance R_G. Since a buffer amplifier is required to have a high input resistance, one should choose as large a value for R_G as practicable. However, too high a value for R_G will cause an appreciable dc voltage drop across it. This voltage drop is due to the input leakage current of the FET and hence will increase with temperature, causing bias instability.

Gain at Midband

The midband gain of the buffer circuit of Fig. 7.30 can be evaluated in a step-by-step manner starting from the source and proceeding toward the load. The fraction of the input signal v_i that appears at the gate is given by

$$v_g = v_i \frac{R_{in}}{R_{in} + R} \tag{7.50}$$

If R_{in} is much larger than the generator resistance R, then

$$v_g \simeq v_i \tag{7.51}$$

The signal v_g appears across the series combination of the gate-to-source resistance $1/g_m$ and the resistance $(R_S \| R_L \| r_o)$. Thus to obtain the output voltage v_o we simply use the voltage-divider rule:

$$v_o = \frac{(R_S \| R_L \| r_o)}{(R_S \| R_L \| r_o) + (1/g_m)} v_g \tag{7.52}$$

If $1/g_m \ll (R_S \| R_L \| r_o)$, then

$$v_o \simeq v_g \tag{7.53}$$

Combining Eqs. (7.50) and (7.52) yields the voltage gain of the buffer as

$$\frac{v_o}{v_i} = \frac{R_{in}}{R_{in} + R} \frac{(R_S \| R_L \| r_o)}{(R_S \| R_L \| r_o) + 1/g_m} \tag{7.54}$$

and, provided that $R_{in} \gg R$ and $1/g_m \ll (R_S \| R_L \| r_o)$

$$\frac{v_o}{v_i} \simeq 1 \tag{7.55}$$

Thus the voltage gain of the common-drain amplifier is always less than unity and at the limit approaches unity.

Because the signal at the source terminal is almost equal to that at the gate, this circuit is also called a *source follower*.

Output Resistance

To find the output resistance R_{out} (see Fig. 7.30) we look back into the circuit between the source terminal and ground while the generator is shorted. Since the gate current is zero, the gate will be at signal ground. Thus it can be seen that R_{out} is made up of R_S in parallel with the gate-to-source resistance $1/g_m$ in parallel with r_o,

$$R_{out} = [R_S \| (1/g_m) \| r_o] \tag{7.56}$$

It follows that the source follower exhibits a low output resistance.

EXERCISE

7.11   Consider the source-follower circuit shown in Fig. E7.11. Let the FET have $V_P = -4$ V and $I_{DSS} = 12$ mA. Find the input resistance, voltage gain, and output resistance.
   Ans.   1 MΩ; 0.78; 307 Ω

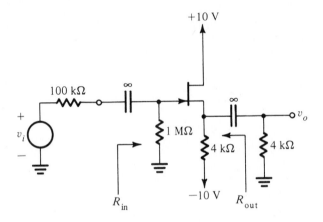

Fig. E7.11

7.11   *DIRECT-COUPLED AND MULTISTAGE AMPLIFIERS*

The JFET amplifier circuits considered thus far use coupling and bypass capacitors, which limit performance at low frequencies, as has already been mentioned. There exist applications in which gain is required down to zero frequency (dc). For such applications one requires a direct-coupled (DC) amplifier. Also, one has to seek an alternative to the biasing scheme that employs a bypass capacitor.

Another motivation to seek direct-coupled amplifier circuits is that integrated-circuit (IC) technology precludes the fabrication of large coupling and bypass capacitors. A different philosophy is thus required for the design of IC FET amplifiers. In Chapter 10 we shall study the most common configuration of DC JFET amplifier, namely, the differential pair.

Many amplifier applications require a number of stages connected in cascade. For instance, to obtain a low output resistance for the overall amplifier one might use a source follower as the last stage in the cascade. Multistage amplifiers will be encountered throughout this book. To illustrate the analysis of such amplifiers we consider a simple three-stage example.

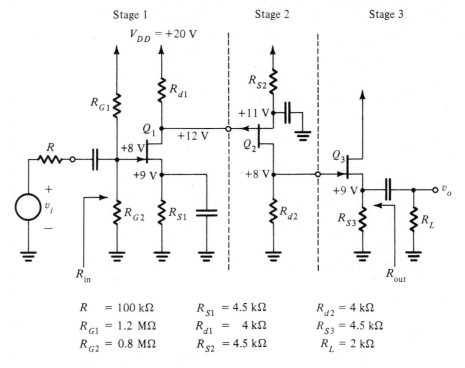

$$R = 100 \text{ k}\Omega \qquad R_{S1} = 4.5 \text{ k}\Omega \qquad R_{d2} = 4 \text{ k}\Omega$$
$$R_{G1} = 1.2 \text{ M}\Omega \qquad R_{d1} = 4 \text{ k}\Omega \qquad R_{S3} = 4.5 \text{ k}\Omega$$
$$R_{G2} = 0.8 \text{ M}\Omega \qquad R_{S2} = 4.5 \text{ k}\Omega \qquad R_L = 2 \text{ k}\Omega$$

All capacitances very large

Fig. 7.31   Three-stage amplifier for Example 7.6.

Example 7.6

Consider the three-stage amplifier shown in Fig. 7.31. The first stage uses a JFET Q_1 in the common-source configuration. The output of the first stage is directly coupled to the gate of the second-stage transistor Q_2, which is a p-channel device. Transistor Q_2 also is connected as a common-source amplifier. To obtain a low output resistance a third stage consisting of transistor Q_3 connected in the source-follower configuration is used. Although direct coupling is used between stages, this amplifier is classified as an ac amplifier (why?). In the following we shall carry out small-signal analysis at midband. Assume all FETs to have $|V_P| = 2$ V and $I_{DSS} = 8$ mA.

Bias Calculations

$$V_{G1} = V_{DD} \frac{R_{G2}}{R_{G1} + R_{G2}} = 20 \frac{0.8}{1.2 + 0.8} = +8 \text{ V}$$

$$V_{GS1} = 8 - I_{D1}R_{S1} = 8 - 4.5I_{D1}$$

Using this relationship together with the FET square-law equation gives $I_{D1} = 2$ mA, $V_{GS1} = -1$ V. Thus $V_{S1} = +9$ V and $V_{D1} = 20 - 4 \times 2 = +12$ V.

For transistor Q_2 we have

$$V_{SG2} = V_{DD} - I_{D2}R_{S2} - V_{D1} = 20 - I_{D2} \times 4.5 - 12$$

Thus $V_{SG2} = 8 - 4.5I_{D2}$. Substituting in the square-law relationship for Q_2,

$$I_D = I_{DSS} \left(1 - \frac{V_{GS2}}{V_P} \right)^2$$

where $I_{DSS} = 8$ mA and $V_P = 2$ V, and solving gives $I_{D2} = 2$ mA, $V_{SG2} = -1$ V. Thus $V_{S2} = +11$ V, $V_{D2} = I_{D2}R_{d2} = +8$ V.

For Q_3 we have

$$V_{GS3} = V_{D2} - I_{D3}R_{S3} = 8 - 4.5I_{D3}$$

Using this together with the square-law relationship gives $I_{D3} = 2$ mA, $V_{GS3} = -1$ V. Thus $V_{S3} = +9$ V.

The various dc voltages evaluated are indicated on the circuit diagram. Note that all devices are well into the pinch-off region.

Input Resistance

The amplifier input resistance is determined by the parallel equivalent of R_{G1} and R_{G2},

$$R_{in} = \frac{1.2 \times 0.8}{1.2 + 0.8} = 480 \text{ k}\Omega$$

Voltage Gain

The transmission through the input coupling circuit is given by

$$\frac{v_{g1}}{v_i} = \frac{R_{in}}{R_{in} + R} = \frac{480}{480 + 100} = 0.83 \tag{7.57}$$

To evaluate the gain of the first stage we note that its load is the input impedance of the second stage, which is very high. Thus

$$\frac{v_{d1}}{v_{g1}} = -g_{m1}R_{d1}$$

where

$$g_{m1} = \frac{2 \times 8}{2} \sqrt{\frac{2}{8}} = 4 \text{ mA/V}$$

and where we have assumed $r_o \gg R_{d1}$. Thus the gain of the first stage is given by

$$\frac{v_{d1}}{v_{g1}} = -4 \times 4 = -16 \tag{7.58}$$

The second stage is a common-source amplifier with a load resistance equal to the input resistance of Q_3, which is very high. Thus the voltage gain of the second stage will be given by

$$\frac{v_{d2}}{v_{g2}} = \frac{v_{d2}}{v_{d1}} = -g_{m2}R_{d2}$$

where

$$g_{m2} = \frac{2 \times 8}{2} \sqrt{\frac{2}{8}} = 4 \text{ mA/V}$$

Thus

$$\frac{v_{d2}}{v_{d1}} = -4 \times 4 = -16 \qquad (7.59)$$

The gain of the third stage can be evaluated from

$$\frac{v_o}{v_{d2}} = \frac{(R_{S3} \| R_L)}{(R_{S3} \| R_L) + 1/g_{m3}}$$

where $g_{m3} = 4$ mA/V and where r_{o3} has been neglected. Thus

$$\frac{v_o}{v_{d2}} = \frac{1.38}{1.38 + 0.25} = 0.84 \qquad (7.60)$$

The overall voltage gain of the amplifier can now be obtain by combining Eqs. (7.57) through (7.60):

$$\frac{v_o}{v_i} = 0.83 \times -16 \times -16 \times 0.84 = 178.5 \qquad (7.61)$$

or 45 dB.

Output Resistance
The amplifier output resistance, excluding the load resistance R_L, is equal to the output resistance of the source-follower stage,

$$R_{\text{out}} = [R_{S3} \| (1/g_{m3})]$$

Thus

$$R_{\text{out}} = (4.5 \text{ k}\Omega \| 0.25 \text{ k}\Omega) = 237 \ \Omega$$

Signal Swing
To determine the maximum allowable output signal swing we have to investigate each stage to find out the limitations it imposes. In this particular design it can be shown that the second stage is the one that limits the amplifier signal swing. Specifically, for Q_2 to remain in pinch-off the maximum drain voltage is limited to $+12 - |V_P| = +10$ V. Thus at the drain of Q_2 we can have a maximum signal swing of 2 V (that is, 4 V peak-to-peak). This means that at the output the maximum signal swing is

$$2 \times 0.84 = 1.68 \text{ V}$$

that is, 3.36 V peak-to-peak.

$$\cdot \quad \cdot \quad \cdot$$

7.12   THE JFET AS A SWITCH

Basis of Operation

For small signals the i_D-v_{DS} characteristics of the JFET are straight lines that pass through the origin, as illustrated in Fig. 7.32. Note that for small v_{SD} the characteristics extend symmetrically into the third quadrant of the i_D-v_{SD} plane and that at $i_D = 0$ the

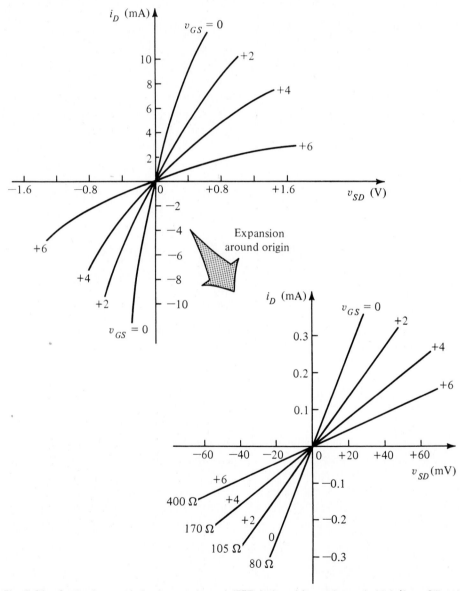

Fig. 7.32   *Static characteristics for a p-channel JFET designed for analog switching (from Siliconix data sheets, 2N 3386).*

voltage $v_{SD} = 0$. These properties make the JFET quite suitable for switching applications.

As an illustration of the operation of the JFET as a switch we consider the circuit in Fig. 7.33a. The switching or control signal is applied to the gate of the FET. When this signal is equal to 0 V the FET is on and the switch is closed. The value of the signal v and the resistance R_d should be such that the device is in the triode region. Figure 7.33b shows the i_D-v_{DS} curve for $v_{GS} = 0$ and a number of load lines corresponding to different values of v, positive and negative. Note that one can arrange that the operating point Q remain on an almost linear segment of the i_D-v_{DS} curve. In this case the FET can be considered as a resistance r_{DS} whose value is equal to the inverse of the slope of the i_D-v_{DS} line (see Fig. 7.33c). It should be mentioned that the on resistance can be as small as a few tens of ohms. Although bipolar transistors (Chapter 9) can be made to have smaller on resistances, they exhibit an undesirable offset voltage (that is, at zero current the voltage across the device is not zero as is the case with the JFET).

To turn the JFET off—that is, to open the switch—the control signal at the gate

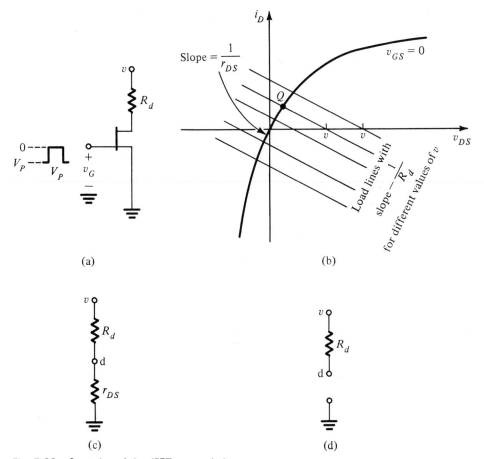

Fig. 7.33   Operation of the JFET as a switch.

should be made negative with a value higher than $|V_P|$. The current through the FET will be very close to zero, and the drain can be assumed to be disconnected from the source, as shown in Fig. 7.33d. The off resistance of the JFET switch is very high.

Before considering specific applications, we wish to once more distinguish between analog and digital switches. In a digital switch, such as that used to implement a logic inverter, one can generally tolerate an offset voltage and inaccuracies in the value of the on resistance. In an analog switch, such as those used in the design of D/A converters (Section 6.8), such nonideal properties are detrimental.

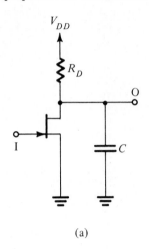

(a)

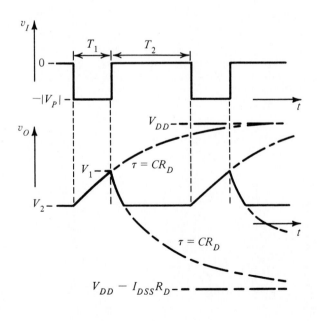

(b)

Fig. 7.34   (a) Circuit for a sawtooth-waveform generator. (b) Input and output waveforms.

A Sawtooth-Waveform Generator

As an application of the JFET switch consider the circuit of Fig. 7.34a, whose purpose is to generate a sawtooth waveform. The input terminal I is fed with a square wave whose two levels are 0 and $-|V_P|$ volts. When the input goes negative to $-|V_P|$, the FET turns off and the capacitor C charges exponentially through R_D toward the V_{DD} level (see Fig. 7.34b). If the input remains negative for a period T_1 much smaller than the time constant $\tau = CR_D$, the curve of output voltage versus time will be almost linear.

At the end of interval T_1 the input goes to 0 V and the FET turns on. Since the voltage at the drain, which is the voltage across the capacitor, cannot change instantaneously, the FET will be in the pinch-off region operating at point B in Fig. 7.35. The drain terminal will carry a constant current equal to I_{DSS} that rapidly discharges the capacitor. The equivalent circuit is shown in Fig. 7.36a, which can be simplified using Thévenin's theorem to that in Fig. 7.36b. From the latter circuit we see that the voltage

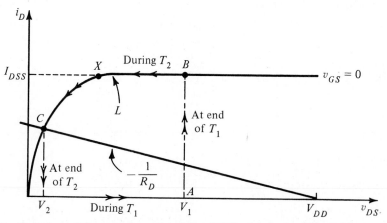

Fig. 7.35   Trajectory of the operating point of the JFET in the circuit of Fig. 7.34a.

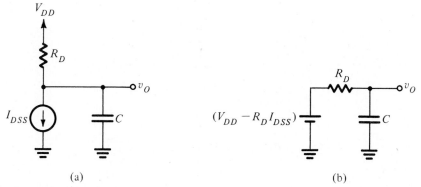

(a)   (b)

Fig. 7.36   Equivalent circuit that applies during the discharge time of C in the sawtooth-waveform generator of Fig. 7.34a.

across the capacitor will decrease exponentially toward a value of $V_{DD} - I_{DSS}R_D$, which is usually a large negative number. The discharge time constant is still CR_D. As the capacitor discharges, the operating point will move along line L in Fig. 7.35. The discharge current will slightly decrease until point X, where the device leaves the pinch-off region. In the triode region the discharge current will be smaller than I_{DSS}. Finally, the device reaches operating point C and stays there for the remainder of interval T_2. It should be noted that the equivalent circuit in Fig. 7.36 assumes that the discharge current remains constant at the value I_{DSS}, which is only approximately true.

Although the discharge time constant is still CR_D, the time taken by the output voltage to reach the low value V_2 is quite short because the exponential discharge curve is heading toward a large negative value (see Fig. 7.34b). As a result we end up with an output signal that rises almost linearly from a low value V_2 to a higher value V_1 and then quickly falls to the low value V_2. Thus the output voltage waveform is a good approximation to a sawtooth waveform.

The JFET Chopper

As another application of the JFET as a switch consider the circuit shown in Fig. 7.37a. Here v_I is a time-varying signal much smaller in magnitude than $|V_P|$. The control

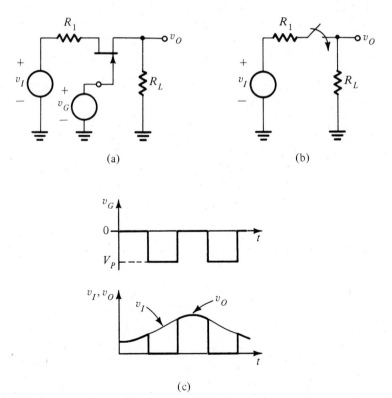

Fig. 7.37 (a) Operation of the JFET chopper, with (b) the ideal equivalent circuit and (c) the various waveforms.

signal is a square wave v_G applied to the gate terminal (that is, between gate and ground). The square wave has the two levels 0 and $-|V_P|$ and has a frequency normally much higher than that of the input signal v_I.

In this circuit the FET acts as a switch turned on and off by the control signal v_G. The ideal equivalent circuit is shown in Fig. 7.37b. When the FET is off the output voltage will be zero. On the other hand, when the FET is on, the output voltage will be

$$v_O = v_I \frac{R_L}{R_L + r_{DS} + R_1}$$

Now if R_L is much greater than the on resistance r_{DS} and the generator resistance R_1, v_O will be approximately equal to v_I. Note that because v_I is small, v_O will also be small, and the gate-to-source voltage will be approximately equal to the square-wave signal v_G.

Figure 7.37c shows the output waveform v_O, which is almost equal to the input v_I during the on intervals and is zero during the off intervals. We speak of the input signal as being "chopped" in passing through this circuit and correspondingly refer to the circuit as a *chopper*. Alternatively, we may consider this circuit as a modulator where the carrier is the square wave and the modulating signal is v_I. Choppers and modulators find extensive use in electronic systems (see Chapter 1).

A Sample-and-Hold Circuit

As a final example consider the precision sample-and-hold circuit shown in Fig. 7.38. The concept of sample-and-hold was introduced in Chapter 6. The circuit of Fig. 7.38 uses an op amp for two purposes: to buffer the input signal v_I and to enhance precision. The latter objective is achieved by placing both the JFET switch Q_1 and the JFET source follower Q_2 in the negative-feedback loop of the op amp. When the control signal v_S is at the $+15$-V level, diode D is off and thus resistance R_1 causes the gate-to-source voltage of Q_1 to be zero. Thus Q_1 turns on and presents a low series resistance between the op-amp output and capacitor C. The feedback loop around the op amp will thus be

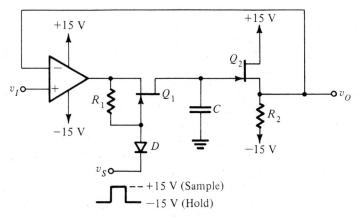

Fig. 7.38   A sample-and-hold circuit.

closed, and capacitor C charges to the value that results in the output voltage v_O being exactly (except for the op-amp offset-voltage error) equal to the sampled signal v_I (because of the virtual short circuit between the two input terminals of the op amp).

In the hold mode v_S is at -15 V, which turns diode D on. Thus the gate of JFET Q_1 will be at about -14.3 V, which turns Q_1 off. Capacitor C will thus retain its charge and voltage. Here we note the need for the source follower Q_2, which buffers C and reduces considerably the droop in output voltage. The output voltage v_O will therefore remain at the value reached just before v_S was switched. When it is properly designed this circuit can achieve output droop of less than 1 mV/s.

7.13   CONCLUDING REMARKS

The purpose of this chapter has been to introduce our first three-terminal semiconductor device, the JFET. Relying on the fundamental semiconductor concepts of Chapter 4 we provided a qualitative explanation of the physical operation of the JFET. This is generally sufficient for the understanding of device characteristics for the purpose of using the JFET in most applications.

The emphasis in the chapter has been on the application of this device both as a linear amplifier and as a switch. Since this is the first three-terminal device studied, the circuits were kept fairly simple yet practical.

A more significant device, at least from a volume-of-production point of view, is the metal-oxide-semiconductor (MOS) FET, studied in Chapter 8.

METAL-OXIDE- 8 SEMICONDUCTOR FIELD-EFFECT TRANSISTORS (MOSFETs)

Introduction

In this chapter we study a very popular class of field-effect transistor, the metal-oxide-semiconductor FET, or, simply, MOSFET. As will be seen, digital logic and memory functions can be implemented with circuits that exclusively use MOSFETs (that is, no resistors or diodes are needed). For this reason, because MOS devices can be made quite small (that is, occupying a small silicon area on the IC chip), and because their manufacturing process is relatively simple (as compared to that of bipolar transistors; see Appendix A), most very-large-scale-integrated (VLSI) circuits are currently made using MOS technology. Examples include microprocessor and memory packages. More recently (1977), MOS technology has been shown to be applicable in the design of analog integrated circuits. For instance, precision filter networks (Chapter 14) have been manufactured in integrated-circuit form using MOS op amps, MOS capacitors, and MOS analog switches. Nevertheless the bipolar-junction transistor (Chapter 9) remains the most suitable device for the design of precise linear circuits (such as op amps).

MOS transistors are either of the *p*-channel or *n*-channel type. *p*-channel devices (PMOS) were initially quite popular in digital applications, but more recently the NMOS (*n*-channel) technology has been perfected. The latter technology results in faster devices which occupy smaller silicon area. Therefore NMOS technology has become dominant in VLSI applications.

Another popular MOS technology is called

complementary-symmetry MOS (COS MOS or CMOS). As the name implies, CMOS circuits use both *n*-channel and *p*-channel devices on the same IC chip. CMOS logic circuits are currently quite popular and are sometimes referred to as the "ideal" logic family.

MOS transistors have very high input impedance and consume little static power. This makes them quite useful in the design of micropower circuits, both digital and linear. Needless to say, the MOS transistor is also very useful in the design of amplifiers with extremely high input impedance. MOS transistors can also be used as analog switches. This together with the ability to manufacture capacitors whose ratios are very accurate using MOS technology has motivated its application in the design of analog signal-processing circuits (see Chapter 14).

Before proceeding the reader is encouraged to read Appendix A in order to gain some familiarity with MOS technology.

8.1   THE DEPLETION-TYPE MOSFET

Physical Structure

Unlike JFETs, which are all of the depletion type, MOSFETs can be either depletion or enhancement type (the exact meaning of these words will become clearer shortly). The operation of the depletion-type MOSFET is very similar to that of the JFET. Let us start by considering the depletion-type MOSFET shown in Fig. 8.1. As shown, the

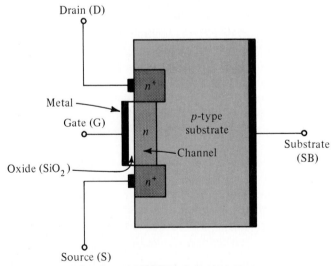

Fig. 8.1   *Physical structure of n-channel MOSFET of the depletion type.*

n-channel device is formed on a p-type silicon substrate. The two heavily doped n^+ wells form low-resistance connections between the ends of the n channel and the metal contacts of the source (S) and the drain (D). A thin oxide layer is grown on the surface of the channel, and metal (aluminum) is deposited on it to form the gate (G). It should be clear that the name MOS is derived from the structure of the device.

While the JFET is controlled by the gate–channel pn junction, no such junction exists in the MOSFET. Here the oxide layer acts as an insulator that causes the gate current to be negligibly small (10^{-12} to 10^{-15} A). This gives the MOS transistor its extremely high input resistance under all conditions.

Circuit Symbol

Figure 8.2 shows the circuit symbol of the n-channel depletion-type MOSFET. Note that the space between the line representing the gate electrode and the line representing

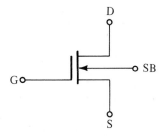

Fig. 8.2   Circuit symbol for the n-channel depletion-type MOSFET.

the channel represents the insulating oxide layer. As in the JFET symbol the gate terminal is drawn closer to the source than to the drain. The arrow on the substrate line points in the forward direction of the substrate-to-channel pn junction and hence indicates the type (or polarity) of the device (n- or p-channel device).

In most applications the substrate is electrically connected to the source. Since the drain voltage will be positive with respect to the source, the substrate-to-channel junction will always be reverse-biased and hence the substrate current will be almost zero. In the following we will normally ignore the role of the substrate in our description of MOSFET operation.

A simplified symbol for the n-channel MOSFET is shown in Fig. 8.3. Note that the polarity of the device is indicated by the direction of the arrowhead on the source line; the arrowhead points in the normal direction of current flow in the source lead.

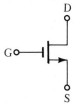

Fig. 8.3   Simplified circuit symbol for n-channel MOSFET.

Physical Operation

As mentioned before, the depletion-type MOSFET operates much like the JFET. The fundamental difference is that in the JFET the channel is depleted by reverse-biasing the gate-to-channel junction. In the MOSFET control of the channel width is effected by an electric field created by the voltage at the gate.

Consider first the triode—or voltage-controlled resistance—region of operation. Let v_{DS} be small while v_{GS} is either negative or positive (note the difference between this case and the JFET case, where v_{GS} has to be negative or at most 0.5 V positive). If v_{GS} is negative, some of the electrons in the n-channel area will be repelled from the channel, and thus a depletion region will be created below the oxide layer. The result will be a narrower channel, as shown in Fig. 8.4. From Fig. 8.4 we also note that the

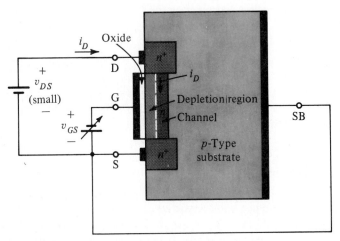

Fig. 8.4   Physical operation of the depletion-type MOSFET for small v_{DS}.

channel has a uniform width because $v_{GS} \simeq v_{GD}$. The narrower channel will have a higher resistance r_{DS}, giving rise to the set of straight-line characteristics shown in Fig. 8.5.

As v_{GS} is made more negative, a point will be reached at which the channel is completely depleted of charge carriers. The value of v_{GS} at which this happens is called the pinch-off voltage V_P,

$$V_P = v_{GS}\big|_{i_D=0,\ v_{DS}=\text{small}}$$

Obviously for an n-channel device V_P is a negative number.

Unlike the JFET case, v_{GS} can be made positive, with the result that the channel becomes wider and its resistance decreases, as illustrated in Fig. 8.5.

Next consider the operation as v_{DS} is increased. If v_{GS} is kept constant while v_{DS} is increased, the depletion region will have a constant width at the source end, but it will be progressively wider as we move toward the drain. Thus the channel will have a tapered shape, and its width will be narrowest at the drain end. It follows that as v_{DS} is increased the channel resistance increases, giving rise to a nonlinear (parabolic) i_D-

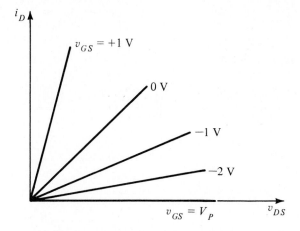

Fig. 8.5   The i_D-v_{DS} characteristics of the depletion-type MOSFET for small v_{DS}.

v_{DS} curve. Finally, if v_{DS} is increased to the value at which $v_{GD} = V_P$, the channel will be pinched off at the drain end. Any further increase in v_{DS} does not alter the channel shape, and hence the current in the channel remains constant at the value reached for $v_{GD} = V_P$ (or, correspondingly, $v_{DS} = -V_P + v_{GS}$). The result is the i_D-v_{DS} characteristic curves shown in Fig. 8.6.

Static Characteristics

The static characteristics of the *n*-channel depletion-type MOSFET take the form shown in Fig. 8.6. These characteristics are identical to those of the JFET except that in the MOSFET positive values for v_{GS} are allowed. A positive v_{GS} "enhances" the channel by attracting more electrons into it, thus increasing its width and reducing its resistance. A depletion-type MOSFET can therefore be applied in the *enhancement mode*. Figure 8.7 illustrates this point further by showing the i_D-v_{GS} characteristic curve for a device operating in the pinch-off region ($v_{DG} > |V_P|$).

The equations describing the characteristic curves of Figs. 8.6 and 8.7 are identical to those of the JFET (Chapter 7) and will not be repeated. Thus with the exceptions noted above the depletion-type MOSFET behaves much like the JFET and can be applied using circuit techniques similar to those discussed for the JFET.

The characteristic curves in Fig. 8.6 assume that in pinch-off the device acts as an ideal constant-current source. Real MOSFETs, however, have a finite output resistance, with the result that their pinch-off i_D-v_{DS} characteristics have finite nonzero slopes. Furthermore, the output resistance will decrease as the current level in the device is increased.

EXERCISE

8.1   Consider an NMOS transistor with $V_P = -2$ V and $I_{DSS} = 8$ mA. Find the minimum v_{DS} required for the device to operate in pinch-off when $v_{GS} = +1$ V. What is the corresponding value of i_D?
Ans.   3 V; 18 mA

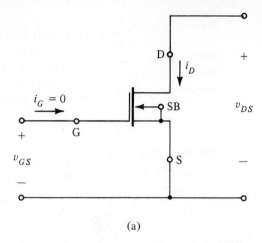

(a)

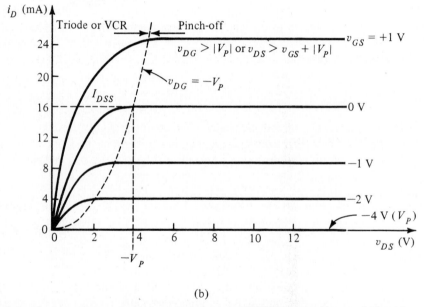

(b)

Fig. 8.6 *The i_D-v_{DS} characteristics for a depletion-type MOSFET whose V_P = -4 V and I_{DSS} = 16 mA.*

8.2   THE ENHANCEMENT-TYPE MOSFET

Structure and Physical Operation

Figure 8.8 shows an enhancement-type MOSFET. As seen, the structure looks quite similar to that of the depletion device in Fig. 8.1, with one major exception: there is no channel. It follows that if the gate is left floating, or if v_{GS} = 0, the path from drain to

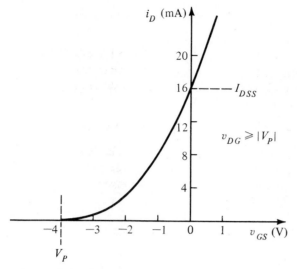

Fig. 8.7   The i_D-v_{GS} characteristic of an n-channel depletion-type MOSFET operating in pinch-off ($V_P = -4$ V, $I_{DSS} = 16$ mA).

source includes two series diodes back-to-back, which means that no current can flow. To cause current to flow from drain to source we first have to create an *n* channel. This can be done by applying a positive voltage v_{GS}. Such a positive voltage at the gate will attract electrons from the substrate and cause them to accumulate at the surface beneath the oxide layer. To attract sufficient number of electrons to form an *n* channel, the voltage v_{GS} has to become equal to or greater than a threshold value V_T. In other words, no appreciable current i_{DS} will flow until v_{GS} equals V_T.

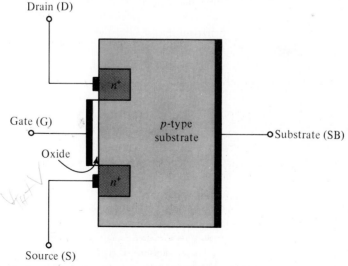

Fig. 8.8   Physical structure of n-channel enhancement-type MOSFET.

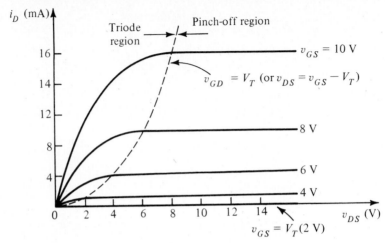

Fig. 8.9 *Ideal static i_D-v_{DS} characteristics for an n-channel enhancement-type MOSFET whose $V_T = 2$ V and $\beta = 0.5$ mA/V².*

Consider now the case where v_{DS} is small, and let v_{GS} be increased above V_T. Increasing v_{GS} will cause the created channel to widen (become *enhanced*), and thus its resistance will decrease. This process gives rise to the voltage-controlled-resistance (VCR) or triode region of operation, indicated in Fig. 8.9.

Next consider what happens if we keep v_{GS} constant—say, $v_{GS} = V_T + V$, where V is a positive voltage—and increase v_{DS}. The channel will remain at a constant width at the source end, since v_{GS} is constant. However, since increasing v_{DS} means that v_{DG} increases or, equivalently, v_{GD} decreases, the channel will become narrower at the drain end. In fact, the channel will have a tapered shape, being widest at the source end and narrowest at the drain end. It follows that increasing v_{DS} will cause the channel resistance to increase, giving rise to a nonlinear i_D-v_{DS} curve in the triode region. This process will continue until eventually the channel width becomes zero at the drain end. This pinch-off condition will occur when $v_{GD} \leq V_T$. To appreciate this fact, recall that the gate voltage had to be increased to V_T in order to create the channel. It follows that to make the channel disappear at the drain end the gate-to-drain voltage has to be reduced below V_T. In our case $v_{GS} = V_T + V$; thus pinch-off will happen when

$$v_{GD} \leq V_T \quad \text{or} \quad v_{DS} \geq V$$

Increasing the drain voltage above the value for pinch-off will not change the shape of the channel; hence the current i_D remains constant at the value reached at the onset of pinch-off.

Static Characteristics

The discussion above gives rise to the i_D-v_{DS} characteristics shown in Fig. 8.9, from which we see that there are two regions of operation: the triode region and the pinch-off region. The boundary between the two regions is determined from

$$v_{GD} = V_T \qquad \text{(8.1)}$$

(handwritten: $V_{GD} = V_{GS} - V_{DS}$)

or, equivalently,

$$v_{DS} = v_{GS} - V_T \qquad \text{(8.2)}$$

In the triode region the i_D-v_{DS} characteristics are described by the parabolic relationship

$$i_D = \beta[(v_{GS} - V_T)v_{DS} - \tfrac{1}{2}v_{DS}^2] \qquad \text{(8.3)}$$

where

$$v_{GS} \geq V_T \qquad \text{and} \qquad v_{DS} \leq v_{GS} - V_T$$

The constant β depends, among other things, on the geometry of the device.

In the pinch-off region the current i_D is constant for a given value of v_{GS}, a fact that gives this region of operation the name *saturated region*. The current i_D is given by

$$i_D = \tfrac{1}{2}\beta(v_{GS} - V_T)^2 \qquad \text{(8.4)}$$

where $v_{GS} \geq V_T$ and $v_{DS} \geq v_{GS} - V_T$. Equation (8.4) represents the graph shown in Fig. 8.10, from which the enhancement nature of the device should be evident. Specifically, at $v_{GS} = 0$ the current is almost zero, and for the device to conduct v_{GS} has to exceed the threshold voltage V_T.

Pinch-off occurs when

$$v_{DS} = v_{GS} - V_T$$

Substituting in Eq. (8.4) yields the equation of the boundary line (shown as a broken line in Fig. 8.9),

$$i_D = \tfrac{1}{2}\beta v_{DS}^2 \qquad \text{(8.5)}$$

(handwritten: equation of dashed line)

For modern MOS technology the threshold voltage V_T, which is a positive number for n-channel devices, lies in the range 1 to 3 V.

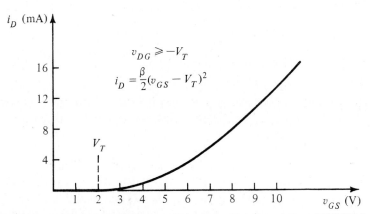

Fig. 8.10   The i_D-v_{GS} characteristic for an n-channel enhancement-type MOSFET in pinch-off ($V_T = 2$ V, $\beta = 0.5$ mA/V²).

Circuit Symbol

Figure 8.11 shows the circuit symbol of the *n*-channel enhancement-mode MOSFET. The only difference between this symbol and that of the depletion mode device is that the channel is represented by a broken line. This signifies that no channel exists in the enhancement device (until v_{GS} exceeds V_T).

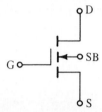

Fig. 8.11 *Circuit symbol for the n-channel enhancement-type MOSFET.*

The p-Channel Device

The operation of the *p*-channel MOSFET parallels that of the *n*-channel device described above. The characteristics are also similar except for a reversal of polarity of all currents and voltages. Also, V_T for the *p*-channel device is a negative number.

The characteristics of the *p*-channel MOSFET can be described by equations similar to Eqs. (8.1) through (8.5) except for replacing v_{GS} by v_{SG}, v_{DS} by v_{SD}, and V_T by $|V_T|$. The circuit symbol of the *p*-channel enhancement-mode MOSFET is shown in Fig. 8.12.

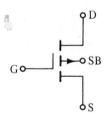

Fig. 8.12 *Circuit symbol for the p-channel enhancement-type MOSFET.*

The Role of the Substrate Bias

In most applications the substrate terminal is connected to the source terminal, which results in the *pn* junction between the substrate and the created channel being reverse-biased. In integrated circuits, however, the substrate is usually common to many MOS transistors. In order to maintain the reverse-bias condition the substrate should be connected to the most negative voltage (in the circuit) for an NMOS device and to the most positive voltage for a PMOS device. But what is the effect on device operation of the dc voltage between substrate and source? To answer this question, observe that the reverse-bias voltage between substrate and source can control the channel width in the same manner as the gate voltage in a JFET controls its operation. Thus increasing the reverse-bias voltage on the substrate causes the channel to be depleted of charge carriers. This effect can be expressed as an increase in the magnitude of threshold voltage V_T as the substrate reverse bias is increased. An approximate rule for this is

$$\Delta V_T \simeq C\sqrt{V_{SB}}$$

where V_{SB} is the source-to-substrate voltage and ΔV_T is the corresponding change in threshold voltage. The constant C depends on the doping of the substrate and generally lies in the range 0.5 to 2.0. Unless otherwise mentioned we shall assume that the substrate is connected to the source.

Temperature Effects

Both V_T and β are temperature sensitive. The magnitude of V_T decreases by about 2.5 mV for every rise of 1°C in temperature. This decrease in $|V_T|$ gives rise to a corresponding increase in drain current as temperature is increased. However, because β decreases with temperature and its effect is a dominant one, the overall observed effect of temperature increase is a *decrease* in drain current. This very interesting result is put to use in applying the MOSFET to power circuits.

Breakdown and Input Protection

In a MOSFET, breakdown occurs when the gate-to-source voltage exceeds about 100 V. However, because of the very high input impedance a small amount of static charge accumulating on the input capacitor (between gate and source) can cause this breakdown voltage to be exceeded. To make matters worse, breakdown of a MOSFET results in permanent damage to the oxide.

To prevent the accumulation of static charge on the input capacitor of a MOSFET, gate protection devices are usually included at the input of MOS integrated circuits (such as the input terminals of a CMOS logic gate). The protection mechanism invariably makes use of clamping diodes (see Fig. 6.50).

Simplified Circuit Symbols

To simplify matters, we shall adopt the simplified circuit symbols shown in Fig. 8.13. It is assumed in these symbols that the substrate is connected to the source. The arrow-

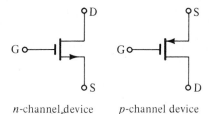

n-channel device   p-channel device

Fig. 8.13   Simplified circuit symbols for MOS transistors. The type of device (depletion or enhancement) will be made clear in each specific case.

head on the source points in the direction of normal current flow. Finally, note that the symbol for the *p*-channel device is drawn with the source up, which makes it conform with our circuit-drawing convention of currents flowing from top to bottom.

Recapitulation

Consider Fig. 8.14. For the *n*-channel enhancement device shown in the figure to conduct, v_{GS} has to be positive and greater than V_T. The device will operate in the pinch-off or active region if the drain voltage v_D is more positive than the gate voltage v_G by

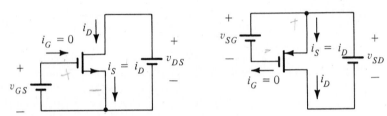

Fig. 8.14   *Normal current flow directions and voltage polarities in enhancement MOS transistors.*

at least $-V_T$. That is, the device will still be in pinch-off even if the drain voltage is lower than that of the gate by V_T volts. If the drain voltage is further reduced, the device gets out of pinch-off and goes into the triode region.

For the *p*-channel device to conduct, the source has to be made more positive than the gate by at least $|V_T|$ volts; that is, $v_{SG} \geq |V_T|$. The device will be in pinch-off as long as the drain voltage is lower than that of the gate, or even if it is higher than that of the gate by at most $|V_T|$. If v_D is increased more than $|V_T|$ volts above v_G, the device leaves pinch-off and enters the triode region.

Of course, the pinch-off region is the one suitable for amplifier application. Switching applications make use of the cutoff region and the triode region.

Finally, we should point out that real devices display a finite output resistance when operated in the pinch-off region. Figure 8.15 shows the i_D-v_{DS} characteristics of an *n*-channel MOS transistor of the enhancement type. In pinch-off the characteristic curves show finite slope that increases with the current level in the device.

EXERCISE

8.2   Consider an enhancement NMOS transistor with $V_T = 2$ V which conducts a current i_D = 1 mA when $v_{GS} = v_{DS} = 3$ V. What is the value of i_D for $v_{GS} = 4$ V and $v_{DS} = 5$ V? (Assume that in pinch-off the device acts as a current source.) Also calculate the value of the drain-to-source resistance r_{DS} for small v_{DS} and $v_{GS} = 4$ V.
Ans.   4 mA; 500 Ω

8.3   BIASING THE ENHANCEMENT MOSFET

As mentioned before, the first step in the design of a transistor amplifier involves establishing a stable and predictable dc operating point inside the active region of operation.

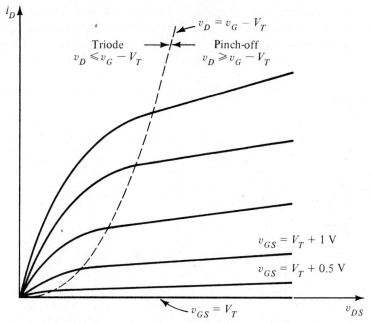

Fig. 8.15 *Actual characteristics of n-channel MOSFET, illustrating the finite nonzero slope in the pinch-off region. Note that the slope increases with the current level in the device. Thus the output resistance r_o is inversely proportional to the bias current I_D.*

In the following we shall study two popular biasing arrangements for enhancement MOSFETs.

A First Biasing Scheme

Figure 8.16 shows our first biasing arrangement; although it looks identical to that used for biasing JFETs and depletion-type MOSFETs, the principle of operation is some-

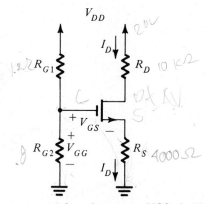

Fig. 8.16 *A popular biasing arrangement for enhancement MOS amplifiers.*

what different. Here the source resistance R_S is not a self-bias resistance. In fact, self-bias is not possible with enhancement-mode devices. Rather, as will be explained below, the sole reason for including R_S is to provide negative feedback that stabilizes the dc operating point. Of course, in depletion-type devices the self-bias resistance provides negative feedback also.

The voltage divider R_{G1}-R_{G2} supplies the gate with a constant dc voltage V_{GG},

$$V_{GG} = V_{DD} \frac{R_{G2}}{R_{G1} + R_{G2}}$$

In the absence of R_S this voltage appears directly between gate and source, and the corresponding current I_D will be highly dependent on the exact value of V_{GG} and on the device parameters β and V_T. This point is illustrated in Fig. 8.17, where we show the

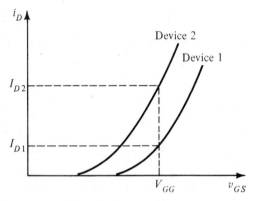

Fig. 8.17   *High dependence of the bias current on device parameters if the resistance R_S is not used. Devices 1 and 2 represent two extremes among units of the same type.*

i_D-v_{GS} characteristic curve for two extreme devices of the same type. The large difference in I_D between the two devices should be evident.

Including the resistance R_S results in the following describing equation:

$$V_{GG} = V_{GS} + I_D R_S$$

which can be rewritten

$$I_D = \frac{V_{GG}}{R_S} - \frac{1}{R_S} V_{GS}$$

This is the equation of the straight line shown in Fig. 8.18, where again we have shown the characteristics of two extreme devices. Note that the difference in the value of I_D between the two devices is much less than that observed with fixed bias alone.

To gain more insight into the stabilizing action of R_S, consider once more the circuit of Fig. 8.16. Assume that for some reason (such as a change in temperature) the drain current increases by an amount Δi_D. This incremental change in the bias current results in an incremental increase in the voltage at the source, Δv_S,

$$\Delta v_S = R_S \Delta i_D$$

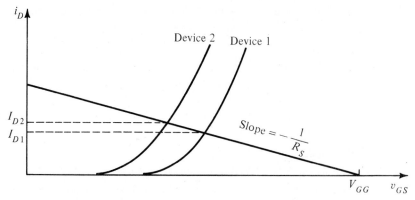

Fig. 8.18   Bias stability obtained by including resistance R_S. Devices 1 and 2 represent two extremes among units of the same type.

Since the gate is held at a constant voltage V_{GG}, an increase in the source voltage results in an equal decrease in V_{GS},

$$\Delta v_{GS} = -\Delta v_S = -R_S\Delta i_D$$

Since a decrease in V_{GS} results in a decrease in I_D, the net increase in I_D will be less than the original value Δi_D, indicating the presence of a negative-feedback mechanism. It should be mentioned again that in the JFET case also the self-bias resistance R_S provides a stabilizing negative-feedback action.

In the above it has been assumed implicitly that the value of R_d is chosen such that the device operates in the pinch-off region. This is accomplished by maintaining the drain voltage greater than $v_G - V_T$ at all times.

Example 8.1
Consider an enhancement MOSFET with $\beta = 0.5$ mA/V² and $V_T = 2$ V. We wish to bias the device at $I_D = 1$ mA using the biasing arrangement of Fig. 8.16 with $V_{DD} = 20$ V.

Solution
To determine the required value of V_{GS} we use the relationship

$$I_D = \tfrac{1}{2}\beta(V_{GS} - V_T)^2$$

For $\beta = 0.5$ mA/V², $V_T = 2$ V, and $I_D = 1$ mA, the value of V_{GS} should be $V_{GS} = 4$ V. If we choose a 4-V drop across R_S, then the gate voltage should be $V_{GG} = 8$ V. This voltage can be established by choosing $R_{G1} = 1.2$ MΩ and $R_{G2} = 0.8$ MΩ. Of course we should choose values for R_{G1} and for R_{G2} as large as practicable in order to keep the amplifier input resistance as high as possible. The value of R_S is given by

$$R_S = \frac{4}{1} = 4 \text{ k}\Omega$$

The choice of a value for R_d is governed by the required gain and signal swing. The higher the value of R_d, the higher the gain will be. However, we should ensure that the drain voltage will at no time fall below the gate voltage by more than V_T volts. For this example let us assume that a maximum signal swing of ± 4 V is required at the drain. It follows that we may choose R_d such that $V_D = +10$ V. The required value of R_d will be

$$R_d = \frac{20 - 10}{1} = 10 \text{ k}\Omega$$

$$\cdot \quad \cdot \quad \cdot$$

A Second Biasing Scheme

The second biasing arrangement we shall study is depicted in Fig. 8.19. As shown, a resistance R_G, usually quite large, is connected between drain and gate. Since the gate current is almost zero, the dc gate voltage will be equal to the dc drain voltage. This

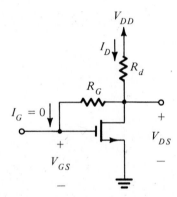

Fig. 8.19 *Another popular and simple biasing arrangement for enhancement MOS amplifiers.*

condition means that the device is still operating in the active (pinch-off) region. It should be obvious, however, that this biasing arrangment would not work with depletion-type devices.

To evaluate the dc operating point of the circuit in Fig. 8.19, consider Fig. 8.20 which shows the i_D-v_{DS} characteristics of the MOSFET. The parabolic boundary between the triode region and the active region is shown as a broken line. This boundary curve is the locus of the points at which $v_{DG} = -V_T$ or equivalently $v_{DS} = v_{GS} - V_T$. If one assumes ideal characteristics, then shifting this curve laterally by V_T volts gives the locus of the points for which $v_{DS} = v_{GS}$. Clearly the operating point Q lies on this latter curve, which is represented by the solid line in Fig. 8.20. From the circuit in Fig. 8.19 we can write

$$v_{DS} = V_{DD} - R_d i_D$$

or equivalently

$$i_D = \frac{V_{DD}}{R_d} - \frac{1}{R_d} v_{DS} \tag{8.6}$$

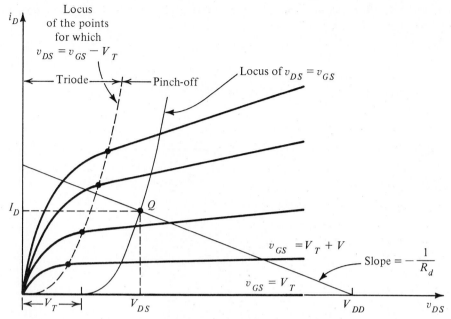

Fig. 8.20 Evaluation of the dc operating point Q of the circuit in Fig. 8.19.

which is the equation of the straight line shown in Fig. 8.20, referred to as the *load line*. The operating point Q will lie at the intersection of the load line and the solid-line parabola. Though quite illustrative, the above graphical procedure is seldom used in practice. It is far more expedient to determine i_D and v_{DS} at the operating point Q by solving Eq. (8.6) together with the equation describing the locus of the points $v_{DS} = v_{GS}$, namely,

$$i_D = \tfrac{1}{2}\beta(v_{DS} - V_T)^2 \tag{8.7}$$

It should be noted that bias stability in the circuit of Fig. 8.19 is achieved by the negative-feedback action provided by the connection of R_G. To see how this works, assume that for some reason the drain current increases by an increment Δi_D. From the circuit we see that the drain voltage will decrease by $R_d\Delta i_D$. Since the current in R_G is nearly zero, the gate voltage, and hence V_{GS}, will decrease by an equal amount, namely, $R_d\Delta i_D$. As a result of the decrease in V_{GS} the drain current will decrease. Thus the overall increase in drain current will be much smaller than the value originally assumed, Δi_D.

EXERCISES

8.3 In the circuit of Example 8.1 find the percentage change in I_D if the FET is replaced by another having the same β value but $V_T = 3$ V.
Ans. -20%

8.4 Consider the bias arrangement of Fig. 8.19. Find the value of R_d required to establish a drain current of 1 mA provided that $V_{DD} = 20$ V, $V_T = 2$ V, and $\beta = 0.5$ mA/V². What is the percentage change in I_D if the device is replaced by another having the same β value but $V_T = 3$ V?

Ans. 16 kΩ; $- 6\%$

8.4 SMALL-SIGNAL OPERATION OF THE ENHANCEMENT MOSFET AMPLIFIER

Equivalent Circuit Model

Figure 8.21 shows the small-signal equivalent circuit model of the enhancement MOS-FET. This model is identical in form to that of the JFET and the depletion-type MOS-FET. The gate-to-source and gate-to-drain capacitances C_{gs} and C_{gd} range from a frac-

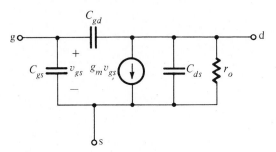

Fig. 8.21   Small-signal equivalent circuit model of the MOSFET.

tion of a picofarad to 3 pF. Capacitance C_{ds} is the drain-to-substrate (which is usually connected to the source) capacitance and is usually about 1 pF or less. From Section 2.6 we can see that the feedback capacitance C_{gd} will take part in a Miller effect, resulting in a high input capacitance in parallel with C_{gs}. The total effective input capacitance together with the resistance of the signal source feeding the amplifier constitutes a first-order low-pass filter (see Chapter 2), which reduces the gain of the amplifier at high frequencies. The high-frequency response of transistor amplifiers will be studed in detail in Chapter 11. For the time being it is sufficient to note that in calculating the gain at low frequencies, the capacitances C_{gs}, C_{gd}, and C_{ds} can be ignored (that is, assumed to act as an open circuit). The resistance r_o, which represents the finite output resistance of the MOSFET in the pinch-off region, is in the range of 10 to 100 kΩ and decreases as the current level in the device is increased.

The equivalent circuit model can be used for evaluating amplifier gain in the same manner as described for JFETs. Furthermore, for quick, approximate calculations at lower frequencies, the first-order models of Fig. 8.22 can be *implicitly* used. In the following we shall derive an expression for g_m. Then we shall present an example to illustrate the evaluation of gain in MOS amplifiers.

The Transconductance g_m

Consider the conceptual amplifier circuit in Fig. 8.23. Here the enhancement MOSFET is shown biased by a fixed dc voltage V_{GS}. A resistance R_d is connected between the

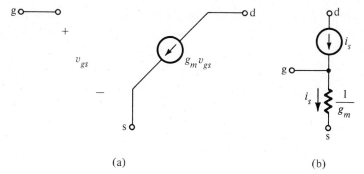

(a)   (b)

Fig. 8.22 *Simplified models that can be implicitly used for rapid approximate calculations at low and medium frequencies.*

drain and the power supply V_{DD} to establish a dc voltage V_D at the drain such that the device is in the active mode. The dc quantities I_D and V_D are given by

$$I_D = \tfrac{1}{2}\beta(V_{GS} - V_T)^2 \tag{8.8}$$
$$V_D = V_{DD} - R_d I_D \tag{8.9}$$

As shown in Fig. 8.23 a voltage signal v_{gs} is superimposed on the dc voltage V_{GS}. Thus the total instantaneous gate-to-source voltage v_{GS} is given by

$$v_{GS} = V_{GS} + v_{gs} \tag{8.10}$$

Correspondingly, the total instantaneous current i_D will be

$$
\begin{aligned}
i_D &= \tfrac{1}{2}\beta(v_{GS} - V_T)^2 \\
&= \tfrac{1}{2}\beta(V_{GS} + v_{gs} - V_T)^2 \\
&= \tfrac{1}{2}\beta(V_{GS} - V_T)^2 + \beta(V_{GS} - V_T)v_{gs} + \tfrac{1}{2}\beta v_{gs}^2
\end{aligned}
\tag{8.11}
$$

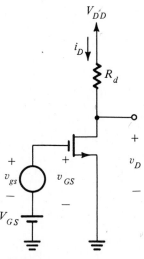

Fig. 8.23 *A conceptual MOS amplifier circuit.*

The first term on the right-hand side of Eq. (8.11) can be recognized as the dc or quiescent current I_D [Eq. (8.8)]. The second term represents a current component that is directly proportional to the input signal v_{gs}. The last term is a current component that is proportional to the square of the input signal. This last component is undesirable because it represents nonlinear distortion. To reduce the nonlinear distortion introduced by the MOSFET, the input signal should be kept small,

$$v_{gs} \ll 2(V_{GS} - V_T) \tag{8.12}$$

If this small-signal condition is satisfied, we may neglect the last term in Eq. (8.11) and express i_D as

$$i_D \simeq I_D + i_d \tag{8.13}$$

where the signal current i_d is given by

$$i_d = \beta(V_{GS} - V_T)v_{gs}$$

The constant relating i_d and v_{gs} is the transconductance g_m,

$$g_m = \beta(V_{GS} - V_T) \tag{8.14}$$

Figure 8.24 presents a graphical interpretation of the small-signal operation of the enhancement MOSFET amplifier. Note that g_m is equal to the slope of the i_D-v_{GS} characteristic at the operating point,

$$g_m = \left. \frac{\partial i_D}{\partial v_{GS}} \right|_{v_{DS} = \text{constant}} \tag{8.15}$$

The reason for maintaining v_{DS} constant is to eliminate the change in i_D due to changes in v_{DS} (which is modeled by r_o).

Voltage Gain

The total instantaneous drain voltage v_D is given by

$$v_D = V_{DD} - R_d i_D$$

Under the small-signal condition we have

$$v_D = V_{DD} - R_d(I_D + i_d)$$

which can be rewritten

$$v_D = V_D - R_d i_d$$

Thus the signal component of the drain voltage is

$$v_d = -i_d R_d$$
$$= -g_m R_d v_{gs}$$

which indicates that the voltage gain is given by

$$\frac{v_d}{v_{gs}} = -g_m R_d$$

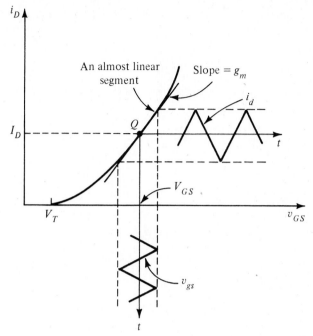

Fig. 8.24   *Small-signal operation of the enhancement MOSFET amplifier.*

It should be noted that in the above analysis we have assumed that the i_D-v_{DS} characteristics in the pinch-off region are horizontal straight lines. In other words, we have neglected the finite output resistance r_o of the voltage-controlled current source. For the amplifier circuit in Fig. 8.23 this resistance appears directly in parallel with R_d. Thus the voltage gain will be reduced to

$$\frac{v_o}{v_{gs}} = -g_m(R_d \| r_o)$$

All of the above analysis has been predicated on the assumption that the MOSFET remains in the active mode at all times, which is achieved by ensuring that the condition

$$v_D \geq v_G - V_T$$

is satisfied.

Example 8.2

Figure 8.25 shows a capacitively coupled grounded-source amplifier using an enhancement MOSFET with $V_T = 1.5$ V and $\beta = 0.25$ mA/V². We desire to evaluate the midband gain and input resistance. For this device r_o can be determined from the empirical formula

$$r_o = \frac{50}{I_D}$$

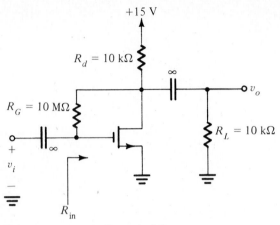

Fig. 8.25   Common-source amplifier for Example 8.2.

Solution
We first evaluate the dc operating point as follows:

$$I_D = \tfrac{1}{2} \times 0.25(V_{GS} - 1.5)^2$$

But $V_{GS} = V_D$, and thus

$$I_D = 0.125(V_D - 1.5)^2 \tag{8.16}$$

Also,

$$\begin{aligned} V_D &= 15 - R_d I_D \\ &= 15 - 10 I_D \end{aligned} \tag{8.17}$$

Solving Eqs. (8.16) and (8.17) gives

$$I_D = 1.06 \text{ mA}, \qquad V_D = 4.4 \text{ V}$$

(note that the other solution to the quadratic equation is not physically meaningful). The value of g_m is given by

$$\begin{aligned} g_m &= \beta(V_{GS} - V_T) \\ &= 0.25(4.4 - 1.5) = 0.725 \text{ mA/V} \end{aligned}$$

The output resistance r_o is given by

$$r_o = \frac{50}{I_D} = \frac{50}{1.06} = 47 \text{ k}\Omega$$

The midband voltage gain can be evaluated to a good approximation by neglecting the effect of the large resistance R_G:

$$\frac{v_o}{v_i} \simeq -g_m(R_d \| R_L \| r_o)$$

$$= -0.725(10 \| 10 \| 47) = -3.3$$

To evaluate the input resistance we apply Miller's theorem (see Section 2.6) as follows:

$$R_{in} = \frac{R_G}{1 - A}$$

where

$$A = \frac{v_d}{v_g} = \frac{v_o}{v_i} = -3.3$$

Thus

$$R_{in} = \frac{10}{1 + 3.3} = 2.33 \ \text{M}\Omega$$

· · ·

EXERCISE

8.5 Consider the circuit designed in Example 8.1. Let an input signal source be capacitively coupled to the gate and let the drain be capacitively coupled to a 10-kΩ load resistance. Also let the resistance in the source lead (R_S) be bypassed to ground by a large capacitance (as was done with the JFET amplifier in Fig. 7.26). Find the input resistance and the voltage gain (assume $r_o = 50/I_D$).
Ans. 480 kΩ; −4.54 V/V

8.5 MOS AMPLIFIER WITH ENHANCEMENT MOS LOAD

One of the major advantages of MOS technology is the ability to realize different functions with circuits that consist entirely of MOS transistors—that is, no resistors or other components are needed. Since a MOSFET requires a much smaller silicon area than a resistor, the resulting circuits are quite efficient in their use of silicon "real estate." In this section we shall introduce the idea of using the MOS transistor as a two-terminal load resistance. It will be shown that use of MOSFET loads allows for the design of amplifiers that are linear for large signals. The amplifier studied in this section will be used in Section 8.8 as a logic inverter.

The Enhancement MOSFET as a Two-Terminal Resistance

Figure 8.26 shows two common configurations for using an enhancement MOSFET as a two-terminal resistance. In Fig. 8.26a the drain is shorted to the gate and hence the

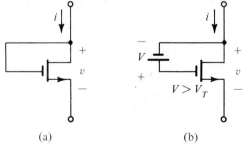

(a)   (b)

Fig. 8.26   *The two common arrangements for using the enhancement MOSFET as a two-terminal resistance. In (a) the device operates in pinch-off, while in (b) the device is in the triode region.*

device operates in the pinch-off region. In Fig. 8.26b a voltage V greater than V_T is applied between the gate and drain, causing the device to operate in the triode region.

Figure 8.27 illustrates the derivation of the i-v characteristics of the two-terminal devices of Fig. 8.26. Shown are the i_D-v_{DS} characteristics of a MOSFET with $\beta = 0.4$ mA/V^2 and $V_T = 2$ V. For simplicity it is assumed that the device has infinite resistance

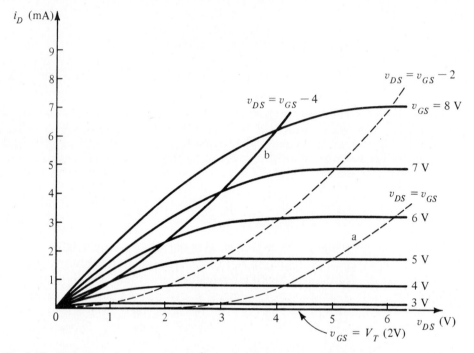

Fig. 8.27   *Derivation of the i-v characteristics of the enhancement MOS resistances of Fig. 8.26. Curve a corresponds to circuit a and curve b corresponds to circuit b.*

in the pinch-off region. To obtain the i-v characteristic of the two-terminal device of Fig. 8.26a we locate on each characteristic curve the point at which $v_{DS} = v_{GS}$. The i-v characteristic will be the locus of these points and is labeled curve a. It is easy to see that this curve is described by

$$i = \tfrac{1}{2}\beta(v - V_T)^2 \tag{8.18}$$

Thus the two-terminal circuit of Fig. 8.26a behaves as a nonlinear resistance that conducts negligible currents for $v < V_T$.

The i-v characteristic of the alternative two-terminal device of Fig. 8.26b can be determined in a similar manner. On each characteristic curve we find the point at which $v_{DS} = v_{GS} - V$ is located; the locus of these points, shown as curve b, is the i-v characteristic. We can use Eq. (8.3) together with

$$v = v_{GS} - V$$

to show that the *i-v* characteristic in question is described by

$$i = \tfrac{1}{2}\beta[v^2 + 2(V - V_T)v] \tag{8.19}$$

Thus the two-terminal circuit of Fig. 8.26b also behaves as a nonlinear resistance.

An important difference between the two circuits is that the *i-v* characteristic of circuit b passes through the origin. Also, for the same voltage *v*, circuit b conducts a much higher current. It should be noted that the "conductance" of either circuit is directly proportional to the constant β, which is determined by the geometry of the MOSFET. Specifically, the constant β is given by

$$\beta = \mu C_0 \frac{Z}{L} \tag{8.20}$$

where μ is a physical constant called the *effective electron mobility,* C_0 is the gate capacitance per unit area, and Z and L are the channel width and length, respectively. Thus by controlling the device dimensions Z and L one can design devices with different values of β on the same IC chip. Specifically, note that to obtain high resistance values Z has to be made small and L large, with the result that load devices are usually long and narrow.

The Amplifier Circuit

Figure 8.28 shows an amplifier circuit in which the enhancement MOSFET Q_1 is the amplifier transistor and Q_2 is an enhancement MOSFET used as a load. We wish to derive the transfer characteristic v_O versus v_I. This can be done graphically, as illustrated in Fig. 8.29, which shows a plot of the i_D-v_{DS} characteristics of the *driving* transistor Q_1. Note that the current i_{D1} is the same current that flows in the load device Q_2. Also note that $v_{DS1} = v_O$, that each of the characteristic curves corresponds to a constant value of v_{GS1}, and that $v_{GS1} = v_I$.

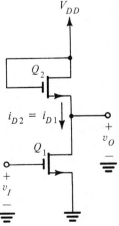

Fig. 8.28 *An enhancement MOS amplifier with an enhancement load operated in pinch-off.*

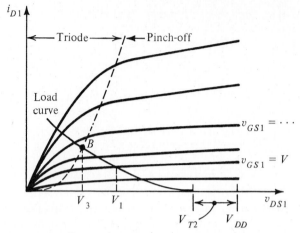

Fig. 8.29 Graphical derivation of the transfer characteristic v_O versus v_I of the MOS amplifier of Fig. 8.28.

Superimposed on the static characteristics of Q_1 is the *load curve,* which is drawn in the same manner used to draw a load line—namely, we locate the V_{DD} point on the v_{DS1} axis and draw a mirror image of the *i-v* characteristic of the load resistance. Now the transfer characteristic is determined by the intersection points of the load curve and the i_{D1}-v_{DS1} characteristic curves. For instance, for $v_I = V$ we find the intersection of the curve corresponding to $v_{GS1} = V$ and the load curve. As shown, at this point $v_{DS1} = V_1$; thus $v_O = V_1$. This process should be repeated for all possible values of v_I. The result is the transfer characteristic shown in Fig. 8.30.

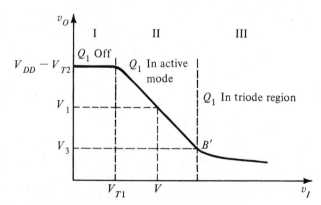

Fig. 8.30 The transfer characteristic of the amplifier of Fig. 8.28.

The Amplifier Transfer Characteristic

The characteristic curve of Fig. 8.30 displays three well defined regions. In region I the driving transistor Q_1 is off, since $v_I < V_{T1}$. Nevertheless, Q_2 is in the pinch-off region and is conducting a negligible current; thus the voltage across Q_2 is equal to V_{T2}, and hence the output voltage is $V_{DD} - V_{T2}$. (Note that in fact Q_2 is always in pinch-off.) In region II, Q_1 is conducting and is operating in pinch-off, and, as will be shown analytically, the transfer curve in region II is linear. Therefore this region is very useful for amplifier operation. Finally, in region III, Q_1 leaves the active mode and enters the triode region. The onset of this region, point B', corresponds to the intersection of the load curve and the boundary curve between the pinch-off and triode regions (point B in Fig. 8.29).

We shall now derive the equation describing the transfer characteristic under the assumption that both devices have infinite resistance (that is, horizontal characteristic lines) in pinch-off. Furthermore, the two devices will be assumed to have equal threshold voltages V_T but different values of β, β_1 and β_2, which corresponds to normal practice.

When Q_1 is in pinch-off we have

$$i_{D1} = \tfrac{1}{2}\beta_1(v_{GS1} - V_T)^2$$

Since $i_{D1} = i_{D2} = i_D$ and $v_{GS1} = v_I$, this equation can be rewritten

$$i_D = \tfrac{1}{2}\beta_1(v_I - V_T)^2 \tag{8.21}$$

The operation of Q_2 is described by

$$i_D = \tfrac{1}{2}\beta_2(v_{GS2} - V_T)^2$$

Since $v_{GS2} = V_{DD} - v_O$, this equation can be rewritten

$$i_D = \tfrac{1}{2}\beta_2(V_{DD} - v_O - V_T)^2 \tag{8.22}$$

Combining Eqs. (8.21) and (8.22) and with some simple manipulations we obtain

$$v_O = \left(V_{DD} - V_T + \sqrt{\frac{\beta_1}{\beta_2}}\, V_T \right) - \sqrt{\frac{\beta_1}{\beta_2}}\, v_I \tag{8.23}$$

which is a linear equation between v_O and v_I. This is obviously the equation of the straight-line portion of the transfer characteristic [region II] of Fig. 8.30.

From Eq. (8.23) we see that the circuit behaves as a linear amplifier for large signals. The gain of the amplifier is $-\sqrt{\beta_1/\beta_2}$, which is determined by the geometries of the two devices. Such circuits have been built with typically $\beta_1 = 10\beta_2$.

It is left to the reader as an exercise to determine the boundary of the linear region of operation.

It is interesting and instructive to derive an expression for the small-signal gain of the amplifier circuit in Fig. 8.28. Using the equivalent circuit of Fig. 8.22b we can replace the load transistor Q_2 by a resistance $1/g_{m2}$. It then follows that the small-signal gain will be $-g_{m1}/g_{m2}$, which can be shown to be equal to the large-signal gain derived above, $-\sqrt{\beta_1/\beta_2}$.

Example 8.3

Consider the capacitively coupled amplifier shown in Fig. 8.31a. Resistance R_G establishes a dc operating point on the linear segment of the transfer curve. Figure 8.31b illustrates the process of determining the bias point Q, where the 45° straight line rep-

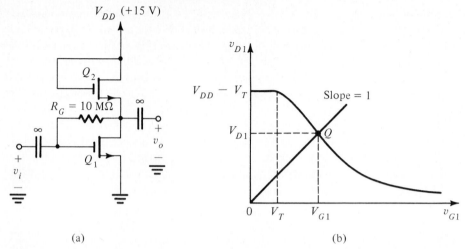

Fig. 8.31   *(a) A capacitively coupled MOS amplifier with MOS load. (b) Illustrating the determination of the dc operating point Q.*

resents the constraint that R_G imposes, $V_{D1} = V_{G1}$. Thus the operating point lies at the intersection of this straight line with the transfer curve. Analytically, the straight-line portion of the transfer curve is described by Eq. (8.23). Thus

$$V_{D1} = \left(V_{DD} - V_T + \sqrt{\frac{\beta_1}{\beta_2}} V_T \right) - \sqrt{\frac{\beta_1}{\beta_2}} V_{G1}$$

Substitution of $V_{G1} = V_{D1}$ gives

$$V_{D1} = \frac{V_{DD} - V_T + V_T \sqrt{\beta_1/\beta_2}}{1 + \sqrt{\beta_1/\beta_2}} \tag{8.24}$$

Consider the case where $V_T = 2$ V, $V_{DD} = +15$ V, $\beta_1 = 270$ μA/V^2, and $\beta_2 = 30$ μA/V^2. The value of V_{D1} will be 4.75 V, and the dc drain current I_D will be given by

$$I_D = 0.27(4.75 - 2)^2 \simeq 2 \text{ mA}$$

At this operating point the voltage gain will be

$$\frac{v_o}{v_i} = - \sqrt{\frac{\beta_1}{\beta_2}} = -3$$

The actual voltage gain will be slightly lower than this value because of the finite output resistance r_o of each of the two MOSFETs. Also, the existence of an external load resistance will reduce the voltage gain.

. . .

8.6 Figure E8.6 shows a voltage-divider circuit composed of three MOSFETs connected as resistors. If Q_1 and Q_3 are identical with $\beta = 70$ $\mu A/V^2$, while β of Q_2 is 0.87 $\mu A/V^2$, and assuming that $V_T = 1.5$ V for all devices, calculate V_1, V_2, and I.
Ans. $+11.18$ V; -11.18 V; 188 μA

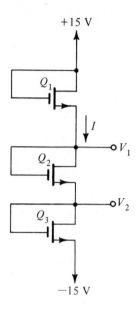

Fig. E8.6

8.6 BIASING IN ANALOG NMOS INTEGRATED CIRCUITS[1]

NMOS technology has been recently employed in the design of analog integrated circuits. Although this technology is not as suited to analog-circuit design as bipolar-transistor technology, there is considerable motivation to use it in the design of such analog building blocks as the op amp. In this way one potentially has the attractive possibility of realizing analog and digital circuits on the same IC chip. In this section, we discuss some of the techniques employed for dc biasing in an integrated-circuit NMOS op amp.

Figure 8.32 shows part of the circuit of an IC op amp designed using *n*-channel enhancement-type MOSFETs. The table accompanying the circuit diagram lists the dimensions Z and L of all transistors. Also given in the table are the β values, which have been evaluated with the use of Eq. (8.20) from the given dimensions, with $\mu C_0 = 20$ $\mu A/V^2$, a value that corresponds to practical situations.

Since the threshold voltage V_T varies considerably from one IC wafer (see Appendix A) to another, an important design objective is to minimize the dependence on the exact value of V_T. However, one can take advantage of the matching and tracking properties of MOSFETs on the same chip. This strategy has been extensively employed in

---

[1]In a first reading the reader may prefer to skip this section and study it later in conjunction with Chapter 10.

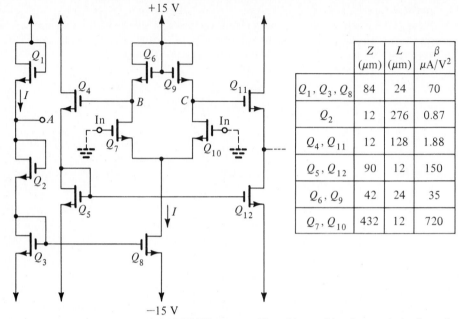

	Z (μm)	L (μm)	β μA/V^2
Q_1, Q_3, Q_8	84	24	70
Q_2	12	276	0.87
Q_4, Q_{11}	12	128	1.88
Q_5, Q_{12}	90	12	150
Q_6, Q_9	42	24	35
Q_7, Q_{10}	432	12	720

Fig. 8.32 *Part of the circuit of an IC NMOS op amp. The table provides the transistor dimensions and β values calculated using Eq. (8.20) with $\mu C_o = 20$ μA/V^2 (Ref. 8.4).*

the design of the circuit in Fig. 8.32, as will be explained. Let us first examine the circuit to identify the function of its various parts.

Transistors Q_1, Q_2, and Q_3, which are connected as two-terminal resistors, constitute a voltage divider. In this divider note (from the table in Fig. 8.32) that Q_1 and Q_3 are identical; thus $V_{GS1} = V_{GS3}$. Transistor Q_3 is connected in parallel with *an identical* device Q_8. Thus Q_8 will conduct a current I equal to that in the voltage-divider string. Transistor Q_8, in fact, acts as a constant-current source supplying the *differential amplifier pair* Q_7 and Q_{10} with a constant bias current I.

The differential pair is a very important (perhaps the most important!) circuit configuration employed in the design of integrated-circuit amplifiers, both bipolar and MOS. We shall study differential pairs in great detail in later chapters. For our present purposes it is sufficient to note that the differential pair consists of two identical and matched transistors Q_7 and Q_{10}. The input signal is applied between the two gates. For bias calculations we will assume that the two input terminals are grounded, as indicated in Fig. 8.32.

Transistors Q_6 and Q_9, which are identical and connected as resistances, are used as load devices for the differential pair Q_7 and Q_{10}. The output signal voltage, taken between the drains of Q_7 and Q_{10}, is applied to the second stage of the op amp, composed of Q_4 and Q_{11} with Q_5 and Q_{12} used for biasing. Explanation in detail of the signal operation of this second stage is beyond the scope of this text at this point. For bias calculations, however, we should note that this second stage is symmetric—that is, Q_4 is identical to Q_{11} and Q_5 is identical to Q_{12}.

Let us now look at bias details. If the current through the string Q_1, Q_2, and Q_3 is denoted by I, Q_8 will conduct an equal current (I). This latter current is supplied to Q_7 and Q_{10}, which, being identical and for zero input signal, will each conduct a current $I/2$. Thus the two identical devices Q_6 and Q_9 will each conduct a current of $I/2$. Now from the table of dimensions in Fig. 8.32 we observe that $\beta_6 = \beta_9 = \beta_1/2$; thus $V_{GS6} = V_{GS9} = V_{GS1}$ independent of the value of V_T. It follows that node B will be at the same potential as node A. Furthermore, since $\beta_2/\beta_3 = \beta_4/\beta_5$, one can show that $V_{GS4} = V_{GS2}$ and $V_{GS5} = V_{GS3}$. Finally, we note that node C will be at the same potential as node B and that $V_{DG12} = V_{DG5} = 0$ V. Again, all of this is independent of the value of V_T.

EXERCISE

8.7 Assuming that all the transistors in the circuit of Fig. 8.32 have $V_T = 1.5$ V and using the values given in the table of Fig. 8.32, calculate I_D and V_{GS} for each transsistor.
Ans.

	Q_1	Q_2	Q_3	Q_4	Q_5	Q_6	Q_7	Q_8	Q_9	Q_{10}	Q_{11}	Q_{12}
V_{GS} (V)	3.82	22.36	3.82	22.36	3.82	3.82	2.0	3.82	3.82	2.0	22.36	3.82
I_D (mA)	0.2	0.2	0.2	0.4	0.4	0.1	0.1	0.2	0.1	0.1	0.4	0.4

8.7 MOS DIGITAL CIRCUITS: AN OVERVIEW

Logic circuits implemented using n-channel MOS devices are referred to as *NMOS logic,* while those using p-channel devices are called *PMOS logic.* Because of earlier technological difficulties in manufacturing n-channel devices in IC form, PMOS logic was initially more popular. These difficulties have since been overcome, and NMOS has virtually replaced PMOS in digital-circuit applications. The NMOS technology produces faster circuits that occupy less silicon area and use lower-voltage dc supplies. The latter property is a result of our ability to manufacture n-channel devices with V_T in the range of 1 to 2 V.

Although NMOS technology is most suitable for implementing LSI digital circuits such as memory and microprocessor chips, it is not convenient for producing general-purpose logic circuits. This term refers to packages containing a number of logic gates or flip-flops of various kinds. The reason that NMOS is not applicable in this area is that its circuits do not have sufficient capability to drive capacitive loads. General-purpose logic circuits will be studied in detail in Chapter 15. For the time being it is sufficient to know that another MOS technology, complementary-symmetry MOS (COS-MOS or CMOS), is extremely suitable for general-purpose logic circuit design. The basics of CMOS circuits will be discussed in Section 8.11, with more details on this popular family of logic circuits presented in Chapter 15.

In Sections 8.8 through 8.10 we shall study the fundamentals of the NMOS family of digital circuits.[2] At this point the reader is advised to review pertinent material from Chapter 6.

---

[2]In a first reading the reader may prefer to skip the remaining sections of this chapter and study them later in conjunction with Chapter 15.

8.8 THE NMOS INVERTER WITH ENHANCEMENT LOAD

The basis for any family of digital circuits is the logic inverter. In NMOS technology there are three approaches for implementing the logic inverter. Two of these approaches are illustrated in Fig. 8.33. Both inverters use an enhancement MOSFET as a load. The difference between the two circuits is that in Fig. 8.33a the MOS load is operated in the pinch-off region, while in Fig. 8.33b the load transistor is biased to operate in the triode region. We have already considered the transfer characteristics of both inverter circuits. These characteristics are displayed in Fig. 8.33. Note that for $v_I < V_T$ the output of inverter b is V_{DD} and that of inverter a is $V_{DD} - V_T$. From this point of view circuit b is superior to circuit a. Also, circuit b is capable of supplying higher currents to a load. Circuit b, however, requires an additional power supply. In the following we shall consider only circuit a.

It should be noted that although it is not explicitly shown, the two transistors in the NMOS inverter have a common substrate that is connected to ground. This causes the two devices to have different threshold voltages. Nevertheless, to simplify matters we shall assume the two threshold voltages to be equal.

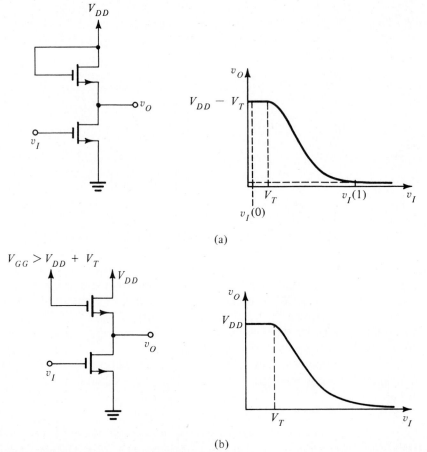

(a)

(b)

Fig. 8.33 NMOS inverters and their transfer characteristics.

Static Operation

To see how the circuit in Fig. 8.33a behaves as a logic inverter, consider a positive-logic system in which logic 0 is represented by a voltage smaller than V_T and close to zero volts and logic 1 is represented by a voltage close to the power-supply voltage V_{DD}. From Fig. 8.33a we see that if v_I is logic 0, $v_I(0) < V_T$, then the output is $V_{DD} - V_T$, which is logic 1; if v_I is logic 1, $v_I(1) \simeq V_{DD}$, then the output is a small voltage close to zero. For proper logic operation this latter voltage should be smaller than V_T.

Dynamic Operation

We next consider the dynamic operation of the inverter circuit of Fig. 8.33a. The circuit is redrawn in Fig. 8.34a with a capacitive load C that represents the input capacitance of another gate, or a number of gates, driven by the inverter under study. We wish to find the output waveform $v_O(t)$ if the input is driven by the pulse waveform $v_I(t)$ shown in Fig. 8.34b. Note that the two levels of the input pulse correspond to logic 0 and logic 1.

When the input is high the output voltage v_O is equal to V_{on}, which is quite small. Thus transistor Q_1 will be operating at point A in Fig. 8.34c. As the input drops to the logic 0 level (that is, $v_I = V_{on} \simeq 0$), Q_1 will immediately[3] turn off and the circuit becomes equivalent to that shown in Fig. 8.34d. Transistor Q_2 will supply C with current i to charge it up to the logic 1 level, $V_{DD} - V_T$. The operating point moves along the load curve from point A to point B (see Fig. 8.34c). Since the current available from Q_2 is relatively low, the charging process will be relatively slow. This is reflected in the slow rising edge of the output voltage waveform, shown in Fig. 8.34b. An expression for the rise time of the output waveform can be easily derived as follows:

Consider the circuit of Fig. 8.34d and assume that the original value V_{on} of the output voltage is approximately zero. The instantaneous value of the charging current i will be given by

$$i = \tfrac{1}{2}\beta_2(V_{DD} - v_O - V_T)^2 \tag{8.25}$$

The charging operation of the capacitor can be described by

$$i \, dt = C \, dv_O$$

which can be expressed as

$$dt = C \frac{dv_O}{i}$$

Substituting for i from Eq. (8.25) gives

$$dt = 2C \frac{dv_O}{\beta_2(V_{DD} - v_O - V_T)^2}$$

Integrating both sides from $t = t_1$, which corresponds to $v_O = 0.1(V_{DD} - V_T)$, to $t = t_2$, which corresponds to $v_O = 0.9(V_{DD} - V_T)$, gives

---

[3]We are assuming the turn-on and turn-off times of transistor Q_1 to be much shorter than the finite rise and fall times of the output waveform due to the capacitive load.

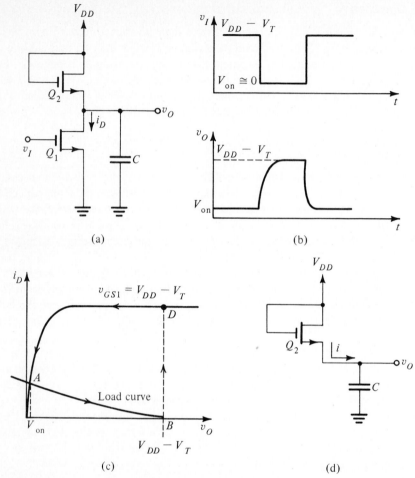

Fig. 8.34 *Dynamic operation of the NMOS inverter with enhancement load. (a) Inverter circuit with a capacitive load; (b) input and output waveforms; (c) trajectory of operating point on the i_D-v_{DS} characteristics of the driving transistor Q_1; and (d) the equivalent circuit during the time v_I is low.*

$$\int_{t_1}^{t_2} dt = \int_{0.1(V_{DD}-V_T)}^{0.9(V_{DD}-V_T)} \frac{2C}{\beta_2} \frac{dv_O}{(V_{DD} - v_O - V_T)^2}$$

Thus we obtain the rise time t_r as

$$t_r = t_2 - t_1 = \frac{2C}{\beta_2} \left[\frac{1}{V_{DD} - v_O - V_T} \right]_{v_O = 0.1(V_{DD}-V_T)}^{v_O = 0.9(V_{DD}-V_T)}$$

The result is

$$t_r = \frac{17.8C}{\beta_2(V_{DD} - V_T)} \tag{8.26}$$

As an example, let us calculate the rise time for $\beta_2 = 3 \ \mu A/V^2$, $C = 1$ pF, $V_T = 1$ V, and $V_{DD} = 10$ V. The result is

$$t_r = 659 \text{ ns}$$

which is rather high.

Unlike the turn-off or rise time, the turn-on or fall time is relatively short. To see that this is the case, consider the inverter circuit in Fig. 8.34a as the input v_I goes from V_{on} to $V_{DD} - V_T$. Transistor Q_1 will turn on and the operating point will instantaneously move from B to D (Fig. 8.34c). It is seen that Q_1 will be in the pinch-off region and thus will conduct an appreciable current I_D,

$$I_D = \tfrac{1}{2}\beta_1(V_{DD} - V_T - V_T)^2 \tag{8.27}$$

This large current will cause C to discharge rapidly. As C discharges the operating point moves along the characteristic curve corresponding to $v_{GS1} = V_{DD} - V_T$ (see Fig. 8.34c). Eventually Q_1 will get out of the pinch-off region, and its current will decrease. The process stops when the operating point reaches point A.

An approximate expression can be derived for the fall time as follows: Assume that $V_{on} \simeq 0$ and that Q_1 remains in the active mode during the entire discharge time. (For a more accurate expression see Problem 8.34.) Thus C will discharge at a constant current I_D given by Eq. (8.27). Therefore the time t_f for v_O to fall from $0.9(V_{DD} - V_T)$ to $0.1(V_{DD} - V_T)$ will be

$$t_f = \frac{C \times 0.8(V_{DD} - V_T)}{I_D}$$

which can be approximated by

$$t_f \simeq \frac{1.6C}{\beta_1(V_{DD} - 2V_T)} \tag{8.28}$$

As an example, let us calculate the fall time for $\beta_1 = 30 \ \mu A/V^2$, $V_T = 1$ V, $V_{DD} = 10$ V, and $C = 1$ pF. The result is $t_f = 6.7$ ns, which is quite small.

We conclude that the enhancement inverter with an enhancement load has a long turn-off time and a short turn-on time.

EXERCISE

8.8   For the inverter in Fig. 8.34a let $V_{T1} = V_{T2} = 1$ V, $\beta_1 = 30 \ \mu A/V^2$, $\beta_2 = 3 \ \mu A/V^2$, and $V_{DD} = 10$ V. Calculate the logic 0 level (V_{on}) and the logic 1 level.
Ans.   0.47 V; 9 V

8.9   THE ENHANCEMENT NMOS INVERTER WITH DEPLETION LOAD

A much sharper transfer curve and a faster turn-off time are obtained by using a depletion-type MOSFET as a load. The resulting inverter circuit is shown in Fig. 8.35a, from which we note that the depletion device Q_2 has its gate connected to its source. This results in the load transistor having the *i-v* characteristic shown in Fig. 8.35b. To obtain the transfer characteristic v_O versus v_I of the inverter we superimpose the load curve on

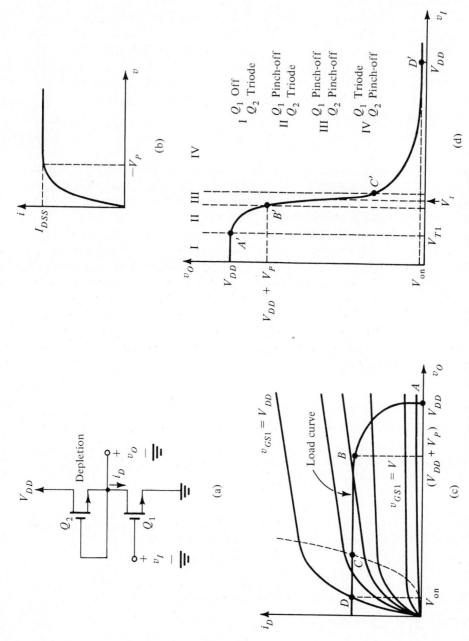

Fig. 8.35 The NMOS inverter with depletion load: (a) circuit, (b) i-v characteristic of the depletion load, (c) graphical construction to determine the inverter static characteristic, and (d) static characteristic.

the i_D-v_{DS} characteristics of the driving enhancement transistor, as shown in Fig. 8.35c. The transfer curve can thus be determined by the points of intersection between the load curve and the Q_1 characteristic curves. It should be noted that each of the Q_1 characteristic curves corresponds to a constant value of v_I. The result is the transfer curve shown in Fig. 8.35d, which is much sharper than that obtained for either of the two inverters that use an enhancement load. This is due to the fact that the depletion load acts almost as a constant-current source and hence has a high resistance. The reader will recall from Chapter 6 that from a noise-immunity point of view it is desirable to have a transfer characteristic with a sharp transition.

Static Characteristics

As indicated in Fig. 8.35d, the transfer characteristic has four distinct regions labeled I, II, III, and IV. In each region the mode of operation of each of the two transistors is indicated. The boundaries of these regions are the points A', B', and C' which correspond to points A, B, and C in Fig. 8.35c.

If the pinch-off i-v characteristics of both transistors are assumed ideal (that is, horizontal straight lines), the segment $B'C'$ of the transfer characteristic becomes a vertical straight line, thus reducing the width of region III to zero. In this ideal case we can easily derive an expression for the *inverter switching threshold* V_t as follows: When the input voltage v_I is at this value V_t the current i_{D1} will be

$$i_{D1} = \tfrac{1}{2}\beta_1(V_t - V_{T1})^2 \qquad (8.29)$$

where β_1 and V_{T1} are the parameters of the enhancement device Q_1. This same current will flow through the depletion load device which will be operating in pinch-off. Thus we have

$$i_{D2} = I_{DSS}\left(1 - \frac{v_{GS2}}{V_P}\right)^2$$

where V_P (negative number) is the pinch-off voltage of the depletion device Q_2. At this point we should mention that the i_D-v_{GS} relationship of the depletion MOSFET can be expressed in the form

$$i_{D2} = \tfrac{1}{2}\beta_2(v_{GS2} - V_{T2})^2$$

where

$$V_{T2} = V_P \qquad \text{and} \qquad I_{DSS} = \tfrac{1}{2}\beta_2 V_{T2}^2$$

This form is similar to that for enhancement devices. It should be noted, however, that here the voltage V_T is a *negative number* for the *n*-channel device.

Let us return to the inverter of Fig. 8.35a. Since $v_{GS2} = 0$, then

$$i_{D2} = \tfrac{1}{2}\beta_2 V_{T2}^2$$

Equating i_{D2} to i_{D1} in Eq. (8.29) gives

$$V_t = V_{T1} + \frac{|V_{T2}|}{\sqrt{\beta_1/\beta_2}} \tag{8.30}$$

Finally, note that the logic 0 level at the output V_{on} can be calculated by finding the v_{DS} voltage of Q_1 (in the triode region) at which the drain current, corresponding to $v_{GS1} = V_{DD}$, is equal to I_{DSS} of Q_2 (that is, $\frac{1}{2}\beta_2 V_{T2}^2$) (see Fig. 8.35c).

Dynamic Operation

To study the dynamic operation of the inverter with depletion load, consider Fig. 8.36a which shows the circuit with a load capacitance C. In Fig. 8.36b we show an input pulse waveform v_I whose two levels are $V_{on} \simeq 0$ (logic 0) and V_{DD} (logic 1). Also shown is

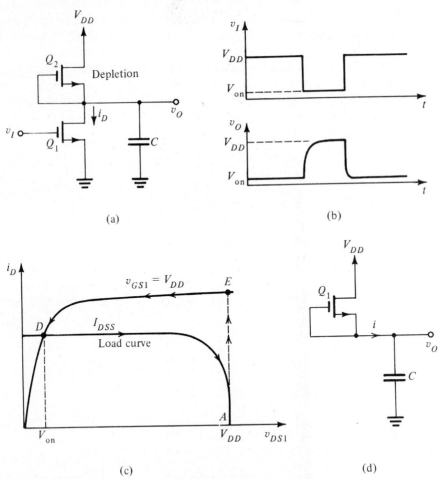

(a)   (b)

(c)   (d)

Fig. 8.36   Dynamic operation of the inverter with depletion load: (a) circuit with a load capacitance C, (b) input and output waveforms, (c) trajectory of the operating point on the static characteristic, and (d) equivalent circuit during the interval in which v_I is low.

the corresponding output waveform, which, like that of the enhancement-load inverter, has a short fall time but a relatively long rise time. The reasons for this response shape are explained in the following.

When $v_I = V_{DD}$ the driving transistor Q_1 is on and operating at point D (Fig. 8.36c). As v_I goes down to the low level V_{on}, Q_1 turns off and the equivalent circuit becomes that shown in Fig. 8.36d. Transistor Q_1 supplies a current i that charges C up to the high level V_{DD}. During this charging interval the operating point moves along the load curve toward point A (Fig. 8.36c). Note that during most of the charging interval the current i is almost constant and equal to I_{DSS} of the depletion device. However, because $I_{DSS} = \frac{1}{2}\beta_2 |V_{T2}|^2$ and because β_2 of the load device is usually small the charging time is usually relatively long. Specifically, if we make the gross approximation that during the entire charging interval the current supplied by Q_2 remains constant and equal to I_{DSS}, we obtain for the 10% to 90% rise time t_r the formula

$$t_r \simeq \frac{1.6 C V_{DD}}{\beta_2 |V_{T2}|^2}$$

For $\beta_2 = 3 \ \mu A/V^2$, $V_{T2} = -3$ V, $V_{DD} = 10$ V, and $C = 1$ pF, this formula gives $t_r = 592$ ns, which is slightly less than the value obtained for the enhancement-load inverter. We should note, however, that since the gain here is higher than in the enhancement-load inverter, β_2 can be made larger, resulting in a proportional reduction in t_r.

Consider next the turn-on process. When v_I increases to the high level V_{DD}, Q_1 turns on immediately. The operating point moves from A to E (Fig. 8.36c), and Q_1 operates in the pinch-off region. Thus Q_1 "pulls" a relatively large current, which causes C to rapidly discharge. As C discharges and v_O decreases the operating point moves along the Q_1 characteristic curve corresponding to $v_{GS1} = V_{DD}$ until it reaches point D. Note that during most of the discharge interval Q_1 is sinking a relatively high current. Thus the turn-on time and correspondingly the fall time of the output waveform are quite short and are close in value to the figures obtained for the enhancement-load inverter.

In conclusion, as compared to the enhancement-load inverter the depletion-load device offers a much sharper transition curve and hence improved noise immunity. The dynamic response is also improved; but fabrication is slightly more difficult.

EXERCISES

In the following consider the inverter circuit of Fig. 8.35a with $V_{T1} = 1$ V, $\beta_1 = 30 \ \mu A/V^2$, $V_{T2} = -3$ V, $\beta_2 = 3 \ \mu A/V^2$, and $V_{DD} = 10$ V.

8.9 Calculate an approximate value for the inverter switching voltage V_t. Also calculate the logic 0 level, V_{on}.
Ans.   1.95 V; 0.05 V

8.10 Owing to the finite output resistance r_o of the depletion load Q_2, its i-v curve in Fig. 8.35b will show a finite nonzero slope in pinch-off. Also, Q_1 has a finite output resistance. If for both devices $r_o \simeq 50/I_D$, find the small-signal gain of this inverter at $v_I = 2$ V. [*Hint:* gain $= -g_{m1}(r_{o1} \| r_{o2})$.]
Ans.   −55.5

8.10   REPRESENTATIVE NMOS LOGIC CIRCUITS

An NMOS NOR Gate

As an example of NMOS logic we show in Fig. 8.37 a three-input NOR gate realized using the enhancement-load technology. For positive logic with level 0 close to zero volts and level 1 close to V_{DD}, the logic gate operates as follows: If any of the inputs goes

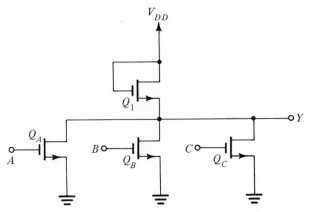

Fig. 8.37 *An NMOS three-input NOR gate.*

high, its corresponding transistor turns on and the output goes low ($V_Y = V_{on} \simeq 0$ V). For the output Y to be high the three inputs should be simultaneously low. This last statement can be described by the Boolean expression (see Chapter 6)

$$Y = \overline{A}\,\overline{B}\,\overline{C}$$

which can be rewritten

$$Y = \overline{A + B + C}$$

Hence the gate performs the NOR function.

An SR Flip-Flop Using NMOS

In Chapter 6 we learned that the basic bistable or memory element can be formed by cross-coupling two inverters. Further, to provide the means for setting and resetting the flip-flop (that is, storing 1 and 0) we may use a pair of two-input NOR or NAND gates with the two extra input terminals used as set and reset inputs. An SR flip-flop constructed from two NMOS NOR gates is shown in Fig. 8.38. The heart of the bistable is the two cross-coupled inverters formed by Q_1, Q_5 and Q_2, Q_6. We have arbitrarily labeled one of the outputs Y and the other $\overline{Y}$. The flip-flop is said to store a 1 if Y is high (that is, Y is a logic 1 and $V_Y = V_{DD} - V_T$) and is said to store a 0 if Y is low (that is, Y is a logic 0 and $V_Y = V_{on} \simeq 0$ V). In the rest state of the flip-flop the set (S) and reset (R) terminals should be at logic 0 levels. To set the flip-flop, terminal S should be raised momentarily to a logic 1 level ($\simeq V_{DD}$). Similarly, to reset the flip-flop

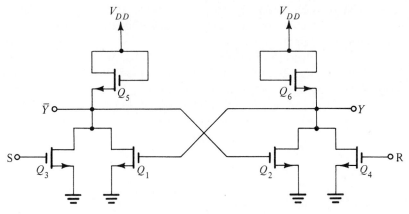

Fig. 8.38   An SR flip-flop formed by cross-coupling two NMOS NOR gates.

terminal R should be raised momentarily to a logic 1 level ($\simeq V_{DD}$). As an example assume that the flip-flop is storing a 0. In this case Q_2 will be on and Q_1 will be off. By raising S to a voltage close to V_{DD}, Q_3 turns on and $\overline{Y}$ goes to a low voltage. This causes Q_2 to turn off, and Y rises to $V_{DD} - V_T$. When Y rises in potential, Q_1 turns on, thus keeping $\overline{Y}$ at the logic 0 level V_{on}. The triggering signal at S, which started the *positive-feedback* action, can obviously be removed, and the flip-flop remains in the 1 state with Q_1 on and Q_2 off.

8.11   COMPLEMENTARY-SYMMETRY MOS LOGIC

The CMOS family of logic circuits is currently (1981) one of the two most popular ones in the design of general-purpose digital systems. In this section we shall study the fundamentals that underlie the operation of CMOS circuits. Further details will be given in Chapter 15.

The CMOS Inverter

Figure 8.39 shows the basic CMOS inverter. It utilizes two enhancement-type MOS-FETs, one with an *n* channel and the other with a *p* channel. To understand the basis

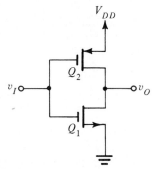

Fig. 8.39   The basic CMOS inverter.

of operation of the CMOS inverter we shall first consider the two extreme cases, when v_I is at logic 0 level, which is approximately 0 V, and when v_I is at logic 1 level, which is approximately V_{DD} volts. In both cases we shall consider the n-channel device Q_1 to be the driving transistor and the p-channel device Q_2 to be the load. However, since the circuit is completely symmetric, this assumption is obviously arbitrary, and the reverse would lead to identical results.

Figure 8.40 illustrates the case when $v_I = V_{DD}$, showing the i_D-v_{DS} characteristic curve for Q_1 with $v_{GS1} = V_{DD}$. (Note that $i_D = i$ and $v_{DS1} = v_O$.) Superimposed on the

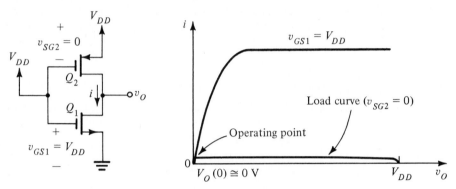

Fig. 8.40   Operation of the CMOS inverter when v_I is high.

Q_1 characteristic curve is the load curve, which is the i_D-v_{SD} curve of Q_2 for the case $v_{SG2} = 0$ V. Since $v_{SG2} < V_T$, the load curve will be a horizontal straight line at almost zero current level. The operating point will be at the intersection of the two curves, where we note that the output voltage is nearly zero (actually less than 10 mV) and the current through the two devices is also nearly zero. This means that the power dissipation in the circuit is very small (typically a fraction of a microwatt).

The other extreme case, when $v_I = 0$ V, is illustrated in Fig. 8.41. In this case Q_1 is operating at $v_{GS1} = 0$; hence its i_D-v_{DS} characteristic is almost a horizontal straight

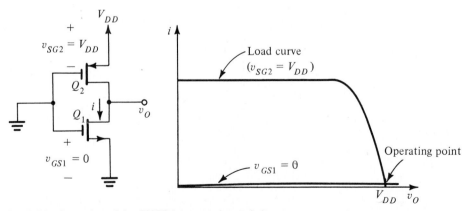

Fig. 8.41   Operation of the CMOS inverter when v_I is low.

line at zero current level. The load curve is the i_D-v_{SD} characteristic of the p-channel device with $v_{SG2} = V_{DD}$. As shown, at the operating point the output voltage is almost equal to V_{DD} (typically less than 10 mV below V_{DD}), and the current in the two devices is still nearly zero. Thus the power dissipation in the circuit is very small in both extreme states.

From the above we conclude that the basic CMOS logic inverter behaves as an "ideal" logic element: the output voltage is almost equal to zero volts or V_{DD} volts and the power dissipation is almost zero.

To obtain the transfer curve for the CMOS inverter we simply repeat the procedure used in the two extreme cases for all intermediate values of v_I. The result will be a curve such as that shown in Fig. 8.42. This characteristic curve is achieved if the two

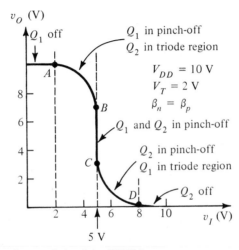

Fig. 8.42 *The transfer characteristic of the CMOS inverter.*

devices are matched, in which case the inverter switches state at half the power-supply voltage V_{DD}. This can be analytically shown using the i-v relationships of the two devices. From the discussion in Chapter 6 we see that this transfer characteristic is a very desirable one as the noise immunity is maximized.

From the transfer curve in Fig. 8.42 we note that the transition between the two states is "very sharp." In other words, the gain of the inverter is very high. This should come as no surprise, since in the transition region each device sees the drain of the other device as a load. Since in the active region the drain acts almost as a constant-current source, each of the two devices sees a very high load resistance. We will have more to say about CMOS inverters in Chapter 15.

Dynamic Performance of the CMOS Inverter

Unlike NMOS inverters, the CMOS inverter features equally fast turn-on and turn-off times in the presence of capacitive loads. To illustrate, consider the capacitively loaded pulse-driven CMOS inverter shown in Fig. 8.43a with input and output waveforms as shown in Fig. 8.43b. Since the circuit is symmetric, the rise and fall times of the output waveform should be equal. Therefore it is sufficient to consider either the turn-on or the turn-off case.

In Fig. 8.43c we show the trajectory of the operating point as the input pulse goes from zero to V_{DD} volts. At time $t = 0-$ (that is, just prior to the transition in the input waveform) the output equals V_{DD} and the capacitor C is charged to this voltage. At $t = 0$, v_I rises to V_{DD}, causing Q_2 to turn off immediately. From there on the circuit is equivalent to that shown in Fig. 8.43d with the initial value of v_O equal to V_{DD}. Thus the operating point at $t = 0+$ is point E in Fig. 8.43c, at which it is seen that Q_1 will be in the active mode and conducting a large current. Furthermore, the current remains relatively high during most of the discharge interval as the operating point moves from

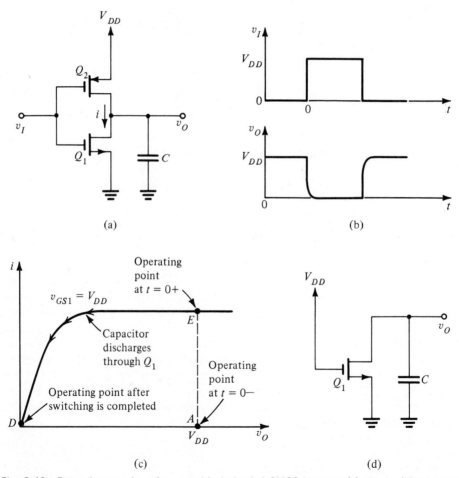

Fig. 8.43 *Dynamic operation of a capacitively loaded CMOS inverter: (a) circuit, (b) input and output waveforms, (c) trajectory of operating point as the input goes from low to high at t = 0, (d) equivalent circuit at t = 0+.*

E to D. Thus C will be discharged quite rapidly and the fall time of v_O will be reasonably short.

At this juncture we wish to draw the reader's attention to an important point that frequently goes unappreciated. The power dissipation in a CMOS inverter, or gate, is quite low in both extreme states, but this is not the case during the transitions between states. As can be seen from the above discussion the inverter supplies relatively large currents during transitions for the purpose of charging and discharging the inevitable capacitive loads. Therefore while the *static* power dissipation is very low, the *dynamic* power dissipation can be quite significant. The latter is, of course, proportional to the frequency at which the gate is being switched.

A CMOS NOR Gate

As a representative example of CMOS logic circuits we show in Fig. 8.44 a two-input NOR gate. As for the inverter, the logic levels are zero volts for logic 0 and V_{DD} for

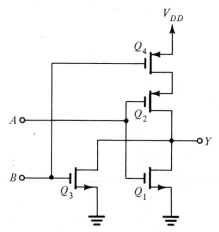

Fig. 8.44 A two-input CMOS NOR gate.

logic 1. The operation of the gate is quite straightforward. If A or B is high, its associated transistor (Q_1 or Q_3) will be on and Y will be low. The output will be high only when A and B are simultaneously low. Thus

$$Y = \overline{A}\,\overline{B} \qquad \text{or} \qquad Y = \overline{A + B}$$

The basic structure of the gate circuit should be noted. It is based on the inverter formed by Q_1 and Q_2, with Q_3 added in parallel with Q_1 and with Q_4 added in series with Q_2. The extension to more inputs should be obvious.

EXERCISES

8.11 Using the i-v relationship of both the p-channel and the n-channel devices of the CMOS inverter of Fig. 8.39 and assuming that the two devices have equal β and V_T, show that the transfer curve of Fig. 8.42 is obtained when $V_{DD} = 10$ V and $V_T = 2$ V.

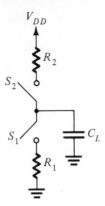

Fig. E8.12

8.12 Figure E8.12 shows a representation of the CMOS inverter as two switches S_1 and S_2, of which one is closed and the other is open depending on the input signal to the gate. Resistances R_1 and R_2 represent the conducting paths through the two MOSFETs, and C_L is a load capacitance. Show that if the gate is switched at a frequency f, then the dynamic power dissipation in watts is

$$P = C_L V_{DD}^2 f$$

8.12 MOS ANALOG SWITCHES

Like the JFET, the MOSFET is well suited for analog switching applications because of the almost linear i_D-v_{DS} characteristics in the region around the origin. Furthermore, these characteristic lines pass through the origin—that is, there is no voltage offset as in the case of bipolar transistors (Chapter 9)—and extend symmetrically in the third quadrant. As explained in Chapter 6, and in conjunction with the JFET in Chapter 7, analog switches are useful in many applications such as sample-and-hold circuits, chopper circuits, and digital-to-analog converters. A specific application of MOS switches is to be found in the design of "switched-capacitor filters," discussed in Chapter 14.

An NMOS Switch

When the MOSFET is used for switching or gating analog signals, a specific configuration using CMOS circuitry is the most convenient. To appreciate this fact, consider first the analog switch formed by a single enhancement NMOS transistor, as shown in Fig. 8.45. Here v_A is the analog input signal and is assumed to be in the range -5 V to $+5$ V. The load to which this analog signal is to be connected is assumed to be a resistance R_L in parallel with a capacitance C_L. In order to keep the substrate-to-source and substrate-to-drain pn juctions reverse-biased at all times, the substrate terminal is connected to -5 V.

The purpose of the control signal v_C is to turn the switch on and off. Assume that the device has a threshold voltage $V_T = 2$ V. It follows that in order to turn the transistor on for all possible input signal levels the high value of v_C should be at least $+7$ V. Similarly, to turn the transistor off for all possible input signal levels the low value

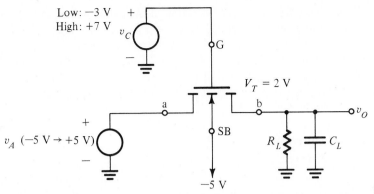

Fig. 8.45   *Application of the enhancement MOSFET as an analog switch. The parallel combination of R_L and C_L represents the load.*

of v_C should be a maximun of -3 V. Note, however, that these levels are not sufficient in practice, since the transistor will be barely on and barely off at the limits. In any case we see that the range of the control voltage is at least equal to the range of the analog input signal being switched. Unfortunately, though, the on resistance of the switch will depend to a large extent on the value of the analog input signal. Thus the transient response of the switch will depend on the value of input signal. Another obvious disadvantage is the rather inconvenient levels required for v_C.

The reader will observe that we have not indicated in Fig. 8.45 which terminal is the source and which is the drain. We have avoided labeling them first of all because the MOSFET is a symmetric device, with the source and the drain interchangeable. More importantly, the operation of the device as a switch is based on this interchangeability of roles. Specifically, if the analog input signal is positive, say, $+4$ V, then it is most convenient to think of terminal a as the drain and of terminal b as the source. For this case the circuit (when v_C is high) takes the familiar form shown in Fig. 8.46a. It

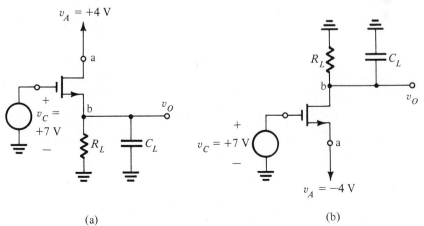

Fig. 8.46   *Method for visualizing the operation of the circuit in Fig. 8.45 when v_c is high: (a) v_A is positive; (b) v_A is negative.*

is easy to see that the device will be operating in the triode region and that v_O will be very close to the input analog signal level of $+4$ V. On the other hand, if the input signal is negative, say, -4 V, then it is most convenient to think of terminal a as the source and of terminal b as the drain. In this case the circuit (in the case of v_C high) can be redrawn in the familiar form of Fig. 8.46b. Here again it should be clear that the device operates in the triode region, and v_O will be only slightly higher than the input analog signal level of -4 V.

The CMOS Transmission Gate

Let us consider next the more elaborate CMOS analog switch shown in Fig. 8.47. This circuit is commonly known as a *transmission gate*. As in the previous case, we are assuming that the analog signal being switched lies in the range -5 V to $+5$ V. In order to prevent the substrate junctions from becoming forward-biased at any time, the substrate of the p-channel device is connected to the most positive voltage level ($+5$ V) and that of the n-channel device is connected to the most negative voltage level (-5 V). The transistor gates are controlled by two complementary signals denoted v_C and $\overline{v_C}$. Unlike the single NMOS switch, here the levels of v_C can be the same as the extremes of the analog signal, $+5$ V and -5 V. When v_C is at the low level, the gate of the n-channel device will be at -5 V, thus preventing the n-channel device from conducting for any value of v_A (in the range -5 V to $+5$ V). Simultaneously, the gate of the p-channel device will be at $+5$ V, which prevents that device from conducting for any value of v_A (in the range -5 V to $+5$ V). Thus with v_C low the switch is open.

In order for us to close the switch, we have to raise the control signal v_C to the high level of $+5$ V. Correspondingly, the n-channel device will have its gate at $+5$ V and

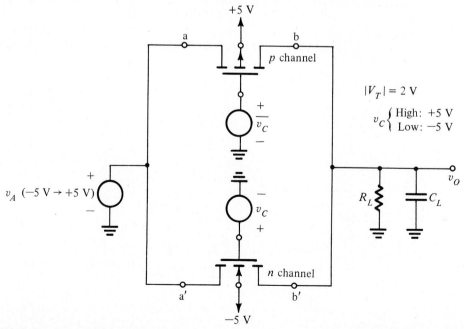

Fig. 8.47   The CMOS transmission gate.

will thus conduct for any value of v_A in the range of -5 V to $+3$ V. Simultaneously, the p-channel device will have its gate at -5 V and will thus conduct for any value of v_A in the range -3 V to $+5$ V. We thus see that for v_A less than -3 V only the n-channel device will be conducting, while for v_A greater than $+3$ V only the p-channel device will be conducting. For the range $v_A = -3$ V to $+3$ V both devices will be conducting. Furthermore, we can see that as one device conducts more heavily, conduction in the other device is reduced. Thus as the resistance r_{DS} of one device decreases, the resistance of the other device increases, with the parallel equivalent, which is the on resistance of the switch, remaining approximately constant. This is clearly an advantage of the CMOS transmission gate over the single NMOS switch previously considered. This advantage operates in addition to the obviously more convenient levels required for the control signal in the CMOS switch.

The operation of the transmission gate (in the closed position) can be better understood from the two equivalent circuits shown in Fig. 8.48. Here the circuit in Fig. 8.48a

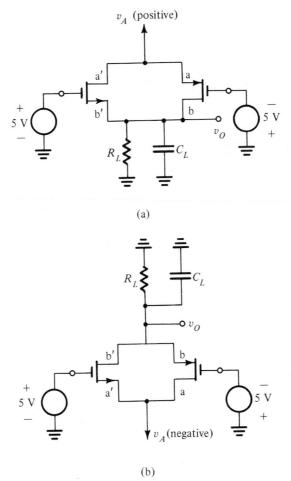

(a)

(b)

Fig. 8.48 Equivalent circuits for visualizing the operation of the transmission gate in the closed (on) position: (a) v_A is positive; (b) v_A is negative.

applies when v_A is positive, while that in Fig. 8.48b applies when v_A is negative. Note the interchangeability of the roles played by the source and drain of each of the two devices. Also note that in thinking about the operation of the CMOS gate one should consider that the drain of one device is connected to the source of the other, and vice versa.

EXERCISE

8.13 Consider the CMOS transmission gate and its equivalent circuit shown in Fig. 8.48b. Let the two devices have $|V_T| = 2$ V and $\beta = 0.1$ mA/V^2, and let $R_L = 50$ kΩ. For (a) $v_A = -5$ V, (b) $v_A = -2$ V, and (c) $v_A = 0$ V, calculate v_O and the total resistance of the switch. [*Hint:* Since both devices will be operating in the triode region and $|v_{DS}|$ will be small, use $i_D \simeq \beta(v_{GS} - V_T)v_{DS}$ for the *n*-channel device and $i_D \simeq \beta(v_{SG} - |V_T|)v_{SD}$ for the *p*-channel device.]
Ans. (a) -4.84 V, 1.624 kΩ; (b) -1.963 V, 1.649 kΩ; (c) 0 V, 1.650 kΩ

8.13 CONCLUDING REMARKS

The purpose of this chapter has been to introduce the MOSFET and to explain its physical operation and the resulting terminal characteristics. Because the depletion-mode MOSFET is quite like the JFET studied in Chapter 7 we have devoted little space to this MOSFET type. Much of the chapter has been concerned with the enhancement device and its applications. In addition to studying the application of the MOSFET as an amplifier, we considered in detail its digital applications. Specifically, both the NMOS and CMOS inverters were studied in detail. We have also shown an NMOS NOR gate, an NMOS flip-flop, and a CMOS NOR gate. We shall return to the subject of MOS digital circuits and their applications in Chapters 15 and 16.

As mentioned in the introduction to this chapter, MOS technology is currently making inroads into the analog signal-processing area. Such applications require MOS op amps, such as the circuit briefly considered in Section 8.6, and MOS analog switches, which we studied in detail in Section 8.12. We will have more to say about MOS analog signal-processing circuits in Chapter 14.

BIPOLAR JUNCTION TRANSISTORS (BJTs)

9

Introduction

In this chapter we shall study the characteristics and basic applications of another three-terminal solid-state device, the bipolar junction transistor (BJT). The BJT, which is usually referred to simply as "the transistor," is by far the most widely used single solid-state device. It is used in discrete circuits as well as in integrated circuits (ICs), both analog and digital. The device characteristics are so well understood that one is able easily to design transistor circuits whose performance is remarkably predictable and quite insensitive to variations in device parameters.

We shall start by presenting a simple qualitative description of the operation of the transistor. Although simple, this physical description provides considerable insight into the performance of the transistor as a circuit element. We will quickly move from describing current flow in terms of holes and electrons to a study of transistor terminal characteristics. First-order models for transistor operation in different modes will be developed and utilized in the analysis of transistor circuits. One of the main objectives of this chapter is to develop in the reader a high degree of familiarity with the transistor. Thus by the end of the chapter the reader should be able to perform rapid first-order analysis of transistor circuits.

9.1 PHYSICAL STRUCTURE AND MODES OF OPERATION

Figure 9.1 shows a fictitious structure for a BJT. Although real transistors do not look like this, this structure will serve to illustrate the essence of transistor operation. Practical transistor structures are described in Appendix A, which deals with fabrication technology.

As shown in Fig. 9.1, the BJT consists of three semiconductor regions: the emitter region (*n* type), the base region (*p* type), and the collector region (*n* type). Such a

351

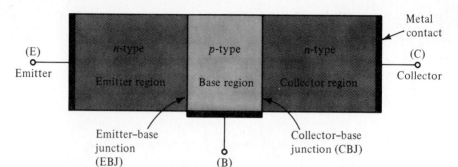

Fig. 9.1 *A fictitious (simplified) structure of the npn transistor.*

transistor is called an *npn* transistor. Another transistor, a dual of the *npn* as shown in Fig. 9.2, has a *p*-type emitter, an *n*-type base, and a *p*-type collector and is appropriately called a *pnp* transistor.

Each of the three semiconductor regions of a transistor has a terminal connecting it to the outside world. The terminals are labeled *emitter* (E), *base* (B), and *collector* (C).

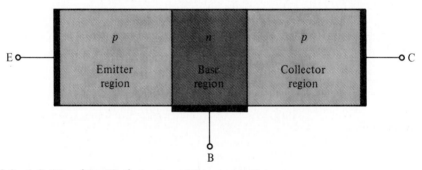

Fig. 9.2 *A fictitious (simplified) structure of the pnp transistor.*

The transistor consists of two *pn* junctions, the emitter–base junction (EBJ) and the collector–base junction (CBJ). Depending on the bias condition (forward or reverse) of each of these junctions, different modes of operation of the BJT are obtained, as shown in Table 9.1.

The active mode is the one used if the transistor is to operate as an amplifier.

TABLE 9.1 *BJT Modes of Operation*

Mode	EBJ	CBJ
Cutoff	Reverse	Reverse
Active	Forward	Reverse
Saturation	Forward	Forward

Switching applications (for example, logic circuits) utilize both the cutoff and the saturation modes.

9.2 OPERATION OF THE npn TRANSISTOR IN THE ACTIVE MODE

Let us start by considering the physical operation of the transistor in the active mode. This situation is illustrated in Fig. 9.3 for the *npn* transistor. Two external voltage sources (shown as batteries) are used to establish the required bias conditions for active

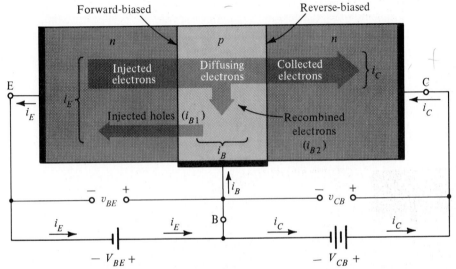

Fig. 9.3 *Current flow in an npn transistor biased to operate in the active mode. (Reverse-current components due to drift of thermally generated minority carriers are not shown.)*

mode operation. The voltage V_{BE} causes the *p*-type base to be higher in potential than the *n*-type emitter, thus forward-biasing the emitter–base junction. The collector–base voltage V_{CB} causes the *n*-type collector to be higher in potential than the *p*-type base, thus reverse-biasing the collector–base junction.

Current Flow

In the following description of current flow only diffusion-current components are considered. Drift currents due to thermally generated minority carriers are usually very small and can be neglected. We will have more to say about these reverse-current components at a later stage.

The forward bias on the emitter–base junction will cause current to flow across this junction. Current will consist of two components: electrons injected from the emitter into the base, and holes injected from the base into the emitter. As will become apparent shortly, it is highly desirable to have the first component (electrons from emitter to base) at a much higher level than the second component (holes from base to emitter). This can be accomplished by fabricating the device with a heavily doped emitter and a lightly doped base; that is, the device is designed to have a high density of electrons in the emitter and a low density of holes in the base.

The current that flows across the emitter–base junction will constitute the emitter current i_E, as indicated in Fig. 9.3. The direction of i_E is "out of" the emitter lead, which is in the direction of the hole current and opposite to the direction of the electron current, with the emitter current i_E being equal to the sum of these two components. However, since the electron component is much larger than the hole component, the emitter current will be dominated by the electron component.

Let us now consider the electrons injected from the emitter into the base. These electrons will be minority carriers in the p-type base region. Because the base is usually very thin, in the steady state the excess minority carrier (electron) concentration in the base will have an almost straight-line profile as indicated by the solid straight line in Fig. 9.4. The concentration will be highest [denoted by n_b (0)] at the emitter side and

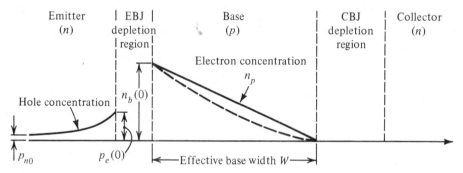

Fig. 9.4 *Profiles of minority carrier concentrations in the base and emitter of an npn transistor operating in the active mode; $v_{BE} > 0$ and $v_{CB} \geq 0$.*

lowest (zero) at the collector side. As in the case of any forward-biased pn junction (Chapter 4), the concentration n_b (0) will be proportional to e^{v_{BE}/V_T}, where v_{BE} is the forward-bias voltage and V_T is the thermal voltage, which is equal to approximately 25 mV at room temperature. The reason for the zero concentration at the collector side of the base is that the positive collector voltage v_{CB} causes the electrons at that end to be swept across the CBJ depletion region.

The tapered minority carrier concentration profile (Fig. 9.4) causes the electrons injected into the base to diffuse through the base region toward the collector. This diffusion current is directly proportional to the slope of the straight-line concentration profile. Thus the diffusion current will be proportional to the concentration n_b (0) and inversely porportional to the base width W.

Some of the electrons that are diffusing through the base region will combine with holes, which are the majority carriers in the base. However, since the base is usually very thin, the percentage of electrons "lost" through this recombination process will be quite small. Nevertheless, the recombination in the base region causes the excess minority carrier concentration profile to deviate from a straight line and take the slightly concave shape indicated by the broken line in Fig. 9.4. The slope of the concentration profile at the EBJ is slightly higher than that at the CBJ, with the difference accounting for the small number of electrons lost in the base region through recombination.

From the above we see that most of the diffusing electrons will reach the boundary of the collector–base depletion region. Because the collector is more positive than the base (by v_{CB} volts) these successful electrons will be swept across the CBJ depletion region into the collector. They will thus get "collected" to constitute the collector current i_C. By convention the direction of i_C will be opposite to that of electron flow; thus i_C will flow *into* the collector terminal.

Another important observation to make here is that the magnitude of i_C is independent of v_{CB}. That is, as long as the collector is positive with respect to the base, the electrons that reach the collector side of the base region will be swept into the collector and register as collector current.

The Collector Current

From the above discussion we can see that the collector current i_C may be expressed as

$$i_C = I_S e^{v_{BE}/V_T} \tag{9.1}$$

where I_S is a constant called the saturation current and V_T is the thermal voltage. The reason for the exponential dependence is that the electron diffusion current is proportional to the minority carrier concentration $n_b(0)$, which in turn is proportional to e^{v_{BE}/V_T}. The saturation current I_S is inversely porportional to the base width W and is directly proportional to the area of the EBJ. Typically I_S is in the range of 10^{-12} to 10^{-15} A (depending on the size of the device) and is a function of temperature.

Because I_S is directly proportional to the junction area (that is, the device size) it will be referred to as the *current scale factor*. Two transistors that are identical except that one has an EBJ area, say, twice that of the other will have saturation currents with that same ratio (2). Thus for the same value of v_{BE} the larger device will have a collector current twice that in the smaller device. This concept is frequently employed in integrated-circuit design.

The Base Current

The base current i_B is composed of two components. The first and dominant component i_{B1} is due to the holes injected from the base region into the emitter region. This current component is proportional to e^{v_{BE}/V_T}. The second component of base current, i_{B2}, is due to holes that have to be supplied by the external circuit in order to replace the holes lost from the base through the recombination process. The number of electrons (and hence the number of holes) taking part in the recombination process is proportional to the concentration $n_b(0)$ and to the base width W. Thus i_{B2} will be proportional to e^{v_{BE}/V_T} and to W.

From the above discussion we conclude that the total base current i_B ($= i_{B1} + i_{B2}$) will be proportional to $e^{(v_{BE}/V_T)}$. We may therefore express i_B as a fraction of i_C,

$$i_B = \frac{i_C}{\beta} \tag{9.2}$$

Thus

$$i_B = \frac{I_S}{\beta} e^{v_{BE}/V_T}$$ (9.3)

where β is a constant for the particular transistor. For modern *npn* transistors β is in the range 100 to 200, but it can be as high as 1,000 for special devices. For reasons that will become clear later, the constant β is called the *common-emitter current gain*.

As may be seen from the above, the value of β is highly influenced by two factors: the width of the base region and the relative dopings of the emitter region and the base region. To obtain a high β (which is, of course, desirable) the base should be thin and lightly doped and the emitter heavily doped. The discussion thus far assumes an idealized situation, where β is independent of the current level in the device.

The Emitter Current

Since the current that enters a transistor should leave it, it can be seen from Fig. 9.3 that the emitter current i_E is equal to the sum of the collector current i_C and the base current i_B,

$$i_E = i_C + i_B$$ (9.4)

Use of Eqs. (9.2) and (9.4) gives

$$i_E = \frac{\beta + 1}{\beta} i_C$$ (9.5)

that is,

$$i_E = \frac{\beta + 1}{\beta} I_S e^{v_{BE}/V_T}$$ (9.6)

Alternatively, we can express Eq. (9.5) in the form

$$i_C = \alpha i_E$$ (9.7)

where the constant α is related to β by

$$\alpha = \frac{\beta}{\beta + 1}$$ (9.8)

or

$$\beta = \frac{\alpha}{1 - \alpha}$$ (9.9)

It can be seen from Eq. (9.8) that α is a constant (for the particular transistor) less than but very close to unity. For instance, if $\beta = 100$, then $\alpha \simeq 0.99$. Equation (9.9) reveals an important fact: small changes in α correspond to very large changes in β. This mathematical observation manifests itself physically, with the result that transistors of the same type may have widely different values of β. For reasons that will become apparent later, α is called the *common-base current gain*.

Recapitulation

We have presented a first-order model for the operation of the *npn* transistor in the active mode. Basically, the forward-bias voltage v_{BE} causes an exponentially related current i_C to flow in the collector terminal. The collector current i_C is independent of the value of the collector voltage as long as the collector–base junction remains reverse-biased, that is, $v_{CB} \geq 0$. Thus in the active mode the collector terminal behaves as an ideal constant-current source where the value of the current is determined by v_{BE}. The base current i_B is a factor $1/\beta$ of the collector current, and the emitter current is equal to the sum of the collector and base currents. Since i_B is much smaller than i_C (that is, $\beta \gg 1$), $i_E \simeq i_C$. More precisely, the collector current is a fraction α of the emitter current, with α smaller than but close to unity.

Equivalent Circuit Models

The first-order model of transistor operation described above can be represented by either of the two equivalent circuits shown in Fig. 9.5. The reader is urged to verify that this is indeed true. The model of Fig. 9.5a suggests that the transistor can be used as a

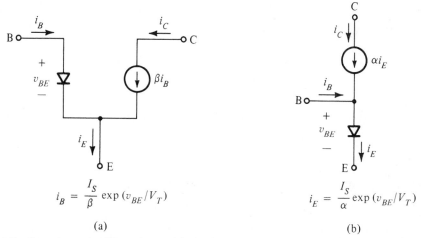

$$i_B = \frac{I_S}{\beta} \exp\left(v_{BE}/V_T\right)$$

$$i_E = \frac{I_S}{\alpha} \exp\left(v_{BE}/V_T\right)$$

(a)   (b)

Fig. 9.5   *Two alternative first-order models for the operation of the npn BJT in the active mode.*

two-port network with port 1 between B and E and port 2 between C and E. Because the emitter terminal is common to both ports this configuration is called the *common-emitter configuration*. The input current, at port 1, is i_B, the output current, at port 2, is i_C. Thus the current gain of this two-port network is equal to β, which gives β the name *common-emitter current gain*.

The circuit model of Fig. 9.5b suggests that the transistor may be used as a two-port network with port 1 between emitter and base and port 2 between collector and base. Since the base terminal is common to both ports, this configuration is called the *common-base configuration*. Here the input current is i_E and the output current is i_C. Thus the current gain is equal to α, which gives α the name *common-base current gain*.

The Constant n

In the diode equation (Chapter 4) we used a constant n in the exponential and mentioned that its value is between 1 and 2. For modern bipolar junction transistors the constant n is close to unity except in special cases: (1) at high currents (that is, high relative to the normal current range of the particular transistor) the i_C-v_{BE} relationship exhibits a value for n that is close to 2, and (2) at low currents, the i_B-v_{BE} relationship shows a value for n of approximately 2. Note that for our purposes we shall assume always that $n = 1$.

The Collector–Base Reverse Current (I_{CBO})

In our discussion of current flow in transistors we ignored the small reverse currents carried by thermally generated minority carriers. Although such currents can be safely neglected in modern transistors, the reverse current across the collector–base junction deserves some mention. This current, denoted I_{CBO}, is the reverse current flowing from collector to base with the emitter open-circuited (hence the subscript O). This current is usually in the nanoampere range, a value which is many times higher than its theoretically predicted value. As with the diode reverse current, I_{CBO} stems mainly from leakage effects, and its value is dependent on v_{CB}. I_{CBO} depends strongly on temperature, approximately doubling for every $10°C$ rise.

EXERCISES

9.1 Consider an *npn* transistor with $v_{BE} = 0.7$ V at $i_C = 1$ mA. Find v_{BE} at $i_C = 0.1$ mA and 10 mA.

   Ans. 0.64 V; 0.76 V

9.2 Transistors of a certain type are specified to have β values in the range 50 to 150. Find the range of their α values.

   Ans. 0.980 to 0.993

9.3 THE pnp TRANSISTOR

The *pnp* transistor operates in a manner similar to that of the *npn* device described in Section 9.2. Figure 9.6 shows a *pnp* transistor biased to operate in the active mode. Here the voltage V_{EB} causes the *p*-type emitter to be higher in potential than the *n*-type base, thus forward-biasing the base–emitter junction. The collector–base junction is reverse-biased by the voltage V_{BC}, which keeps the *n*-type base higher in potential than the *p*-type collector.

Unlike the *npn* transistor, current in the *pnp* device is mainly conducted by holes injected from the emitter into the base by the forward-bias voltage V_{EB}. Since the component of emitter current contributed by electrons injected from base to emitter is kept small by using a lightly doped base, most of the emitter current will be due to holes. The electrons injected from base to emitter give rise to the dominant component of base current, i_{B1}. Also, a number of the holes injected into the base will recombine with the majority carriers in the base (electrons) and will thus be lost. The disappearing base electrons will have to be replaced from the external circuit, giving rise to the second

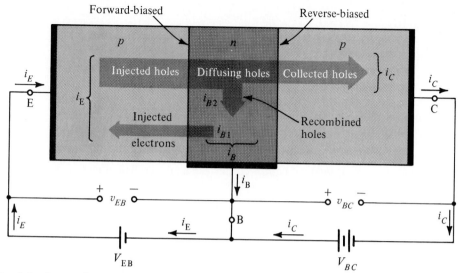

Fig. 9.6   *Current flow in a pnp transistor biased to operate in the active mode.*

component of base current, i_{B2}. The holes that succeed in reaching the boundary of the depletion region of the collector–base junction will be attracted by the negative voltage on the collector. Thus these holes will be swept across the depletion region into the collector and appear as collector current.

It can easily be seen from the above description that the current–voltage relationships of the *pnp* transistor will be identical to those of the *npn* transistor except that v_{BE} has to be replaced by v_{EB}. Figure 9.7 shows two alternative first-order models for the *pnp* transistor in the active mode.

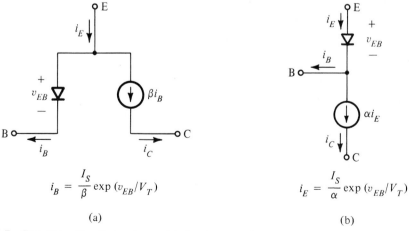

$$i_B = \frac{I_S}{\beta} \exp (v_{EB}/V_T)$$

(a)

$$i_E = \frac{I_S}{\alpha} \exp (v_{EB}/V_T)$$

(b)

Fig. 9.7   *Two alternative first-order models for the operation of the pnp BJT in the active mode.*

EXERCISE
9.3   For a *pnp* transistor with $I_S = 10^{-14}$ A and $\beta = 100$, calculate i_C, i_B, and i_E for $v_{EB} = 0.7$ V, assuming that $v_{CB} \le 0$.
Ans.   1.446 mA; 14.46 µA; 1.460 mA

9.4   CIRCUIT SYMBOLS AND CONVENTIONS

The physical structure used thus far to explain transistor operation is rather cumbersome to employ in drawing the schematic of a multitransistor circuit. Fortunately, a very descriptive and convenient circuit symbol exists for the BJT. Figure 9.8a shows the

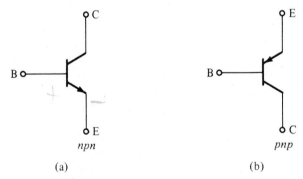

Fig. 9.8   *Circuit symbols for BJTs.*

symbol for the *npn* transistor while the *pnp* symbol is given in Fig. 9.8b. In both symbols the emitter is distinguished by an arrowhead. This distinction is important because practical BJTs are not symmetric devices. That is, interchanging the emitter and collector will result in a different, and much lower, value for α, a value called the *inverse or reverse* α.[1]

The polarity of the device—*npn* or *pnp*—is indicated by the direction of the arrowhead on the emitter. This arrowhead points in the direction of normal current flow in the emitter. Since we have adopted a drawing convention by which currents flow from top to bottom, we will always draw *pnp* transistors in the manner shown in Fig. 9.8 (that is, with their emitters on top).

Figure 9.9 shows *npn* and *pnp* transistors biased to operate in the active mode. It should be mentioned in passing that the biasing arrangement shown, utilizing two dc sources, is not a usual one and is used merely to illustrate operation. Practical biasing schemes will be presented in Chapter 10. Figure 9.9 also indicates the reference and actual directions of current flow throughout the transistor. Our convention will be to

[1]The inverse mode of operation of the BJT will be studied in Chapter 15.

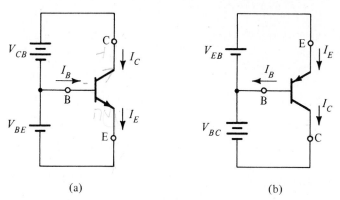

(a)   (b)

Fig. 9.9   Voltage polarities and current flow in transistors biased in the active mode.

take the reference direction to coincide with the actual direction of current flow. Hence we should never encounter a negative value for i_E, i_B, or i_C.

The convenience of the circuit drawing convention that we have adopted should be obvious from Fig. 9.9. Note that currents flow from top to bottom and that voltages are higher at the top and lower at the bottom. The arrowhead on the emitter also implies the polarity of the emitter–base voltage that should be applied in order to forward-bias the emitter–base junction. Just a glance at the circuit symbol of the *pnp* transistor, for example, indicates that we should make the emitter higher in voltage than the base (by v_{EB}) in order to cause current to flow into the emitter (downward). Note that the symbol v_{EB} means the voltage by which the emitter (E) is higher than the base (B). Thus for a *pnp* transistor operating in the active mode v_{EB} is a positive number, while in an *npn* transistor v_{BE} is positive.

From the discussion of Section 9.3 it follows that an *npn* transistor whose EBJ is forward-biased will operate in the active mode *as long as the collector is higher in potential than the base.* Active-mode operation will be maintained even if the collector voltage falls to equal that of the base, since a silicon *pn* junction is essentially nonconducting when the voltage across it is zero. The collector voltage should not be allowed, however, to fall below that of the base if active-mode operation is required. If it does fall below the base voltage, the collector–base junction could become forward-biased, with the transistor entering a new mode of operation, saturation. We will discuss the saturation mode later.

In a parallel manner the *pnp* transistor will operate in the active mode *if the collector is lower (or equal) in potential than the base.* The collector voltage should not be allowed to rise above that of the base if active-mode operation is to be maintained.

EXERCISE

9.4   In the circuit shown in Fig. E9.4 the voltage at the emitter was measured and found to be
-0.7 V. If $\beta = 50$, find I_E, I_B, I_C, and V_C.
Ans.   0.93 mA; 18.2 μA; 0.91 mA; $+5.44$ V

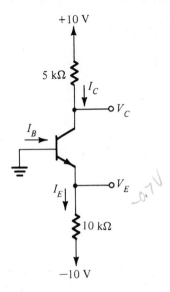

Fig. E9.4

9.5   GRAPHICAL REPRESENTATION OF TRANSISTOR CHARACTERISTICS

It is sometimes useful to describe the transistor i-v characteristics graphically. Figure 9.10 shows the i_C-v_{BE} characteristic, which is the exponential relationship

$$i_C = I_S e^{v_{BE}/V_1}$$

which is identical (except for the value of constant n) to the diode i-v relationship. The i_E-v_{BE} and i_B-v_{BE} characteristics appear quite similar to that of the i_C-v_{BE} curve of Fig. 9.10. Since the constant of the exponential characteristic, $1/V_T$, is quite high ($\simeq 40$), the curve rises very sharply. For v_{BE} smaller than about 0.5 V the current is negligibly small. Also, over most of the normal current range v_{BE} lies in the range 0.6 to 0.8 V. In

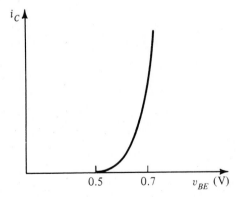

Fig. 9.10   The i_C-v_{BE} characteristic for an npn transistor.

performing rapid first-order dc calculations we normally will assume that $V_{BE} \simeq 0.7$ V, which is similar to the approach used in the analysis of diode circuits (Chapter 4). For a *pnp* transistor the i_C-v_{EB} characteristic will look identical to that of Fig. 9.10.

As in silicon diodes, the voltage across the emitter–base junction decreases by about 2 mV for each rise of 1 °C in temperature, provided that the junction is operating at a constant current. Figure 9.11 illustrates this temperature dependence by depicting i_C-v_{BE} curves at three different temperatures for an *npn* transistor.

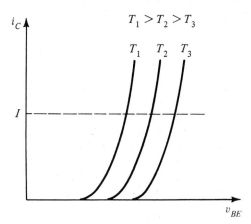

Fig. 9.11 *Effect of temperature on the i_c-v_{BE} characteristic. At a constant emitter current (dotted line) v_{BE} changes by* −*2 mV/° C.*

Figure 9.12b shows the i_C versus v_{CB} characteristics of an *npn* transistor for various values of the emitter current i_E. These characteristics can be measured using the circuit shown in Fig. 9.12a. Only active-mode operation is shown, since only the portion of the characteristics for $v_{CB} \geq 0$ is drawn. As can be seen, the curves are horizontal straight lines, corroborating the fact that the collector behaves as a constant-current source. In this case the value of the collector current is controlled by that of the emitter current

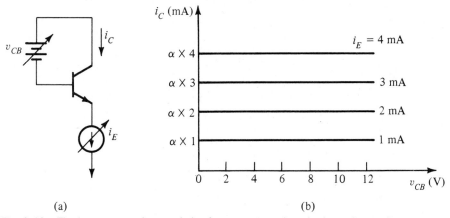

(a)   (b)

Fig. 9.12 *The i_C versus v_{CB} characteristics for an npn transistor in the active mode.*

($i_C = \alpha i_E$), and the transistor may be thought of as a current-controlled current source (see the model of Fig. 9.5b).

9.5 Consider a *pnp* transistor with $v_{EB} = 0.7$ V at $i_E = 1$ mA. Let the base be grounded, the emitter be fed by a 2-mA constant-current source and the collector be connected to a -5-V supply through a 1-kΩ resistance. If the temperature increases by $30°$C, find the changes in emitter and collector voltages.

Ans. -60 mV; 0 V

9.6 DC ANALYSIS OF TRANSISTOR CIRCUITS

We are now ready to consider the analysis of some simple transistor circuits to which only dc voltages are applied. In the following examples we will use the simple constant-V_{BE} model, which is similar to that discussed in Section 4.8 for the junction diode. Specifically we will assume that $V_{BE} = 0.7$ V irrespective of the exact value of current. If it is desired, this approximation can be refined using techniques similar to those employed in the diode case. The emphasis here, however, is on the essence of transistor circuit analysis.

Example 9.1

Consider the circuit shown in Fig. 9.13a, which is redrawn in Fig. 9.13b to remind the reader of the convention employed throughout this book for indicating connections to dc sources. We wish to analyze this circuit to determine all node voltages and branch currents. We will assume that β is specified to be 100.

Solution

We do not know apriori whether the transistor is in the active mode or not. A simple approach would be to assume that the device is in the active mode, proceed with the solution, and finally check whether or not the transistor is in fact in the active mode. If we find that the conditions for active-mode operation are met, then our work is completed. Otherwise, the device is in another mode of operation and we have to solve the problem again. Obviously, at this stage we have learned only about the active mode of operation and thus will not be able to deal with circuits that we find are not in the active mode.

Glancing at the circuit in Fig. 9.13a, we note that the base is connected to $+4$ V and the emitter is connected to ground through a resistance R_E. It therefore is safe to conclude that the base–emitter junction will be forward-biased. Assuming that this is the case and assuming that V_{BE} is approximately 0.7 V, it follows that the emitter voltage will be

$$V_E = 4 - V_{BE} \simeq 4 - 0.7 = 3.3 \text{ V}$$

We are now in an opportune position: we know the voltages at the two ends of R_E and thus can determine the current through it, which is I_E,

$$I_E = \frac{V_E - 0}{R_E} = \frac{3.3}{3.3} = 1 \text{ mA}$$

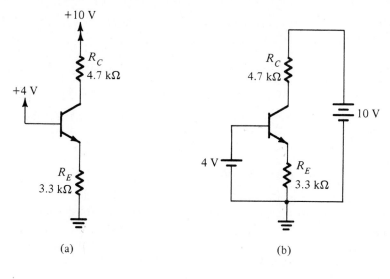

(a)   (b)

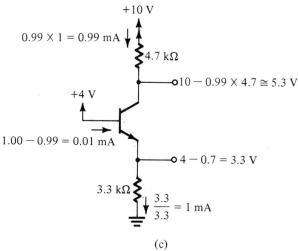

(c)

Fig. 9.13   Circuit for Example 9.1. Analysis of the circuit is shown in (c).

Since the collector is connected through R_C to the $+10$-V power supply, it appears possible that the collector voltage will be higher than the base voltage, which is essential for active-mode operation. Assuming that this is the case, we can evaluate the collector current from

$$I_C = \alpha I_E$$

The value of α is obtained from

$$\alpha = \frac{\beta}{\beta + 1} = \frac{100}{101} \simeq 0.99$$

Thus I_C will be given by

$$I_C = 0.99 \times 1 = 0.99 \text{ mA}$$

We are now in a position to use Ohm's law to determine the collector voltage V_C,

$$V_C = 10 - I_C R_C = 10 - 0.99 \times 4.7 \simeq +5.3 \text{ V}$$

Since the base is at $+4$ V, the collector–base junction is reverse-biased by 1.3 V, and the transistor is indeed in the active mode as assumed.

It remains only to determine the base current I_B,

$$I_B = \frac{I_E}{\beta + 1} = \frac{1}{101} \simeq 0.01 \text{ mA}$$

Before leaving this example we wish to strongly emphasize the value of carrying out the analysis directly on the circuit diagram. Only in this way will one be able to analyze complex circuits in a finite length of time. Figure 9.13c illustrates the above analysis on the circuit diagram.

$$\cdot \quad \cdot \quad \cdot$$

Example 9.2
We wish to analyze the circuit of Fig. 9.14a to determine the voltages at all nodes and the currents through all branches. Note that this circuit is identical to that of Fig. 9.13 except that the voltage at the base is now $+6$ V.

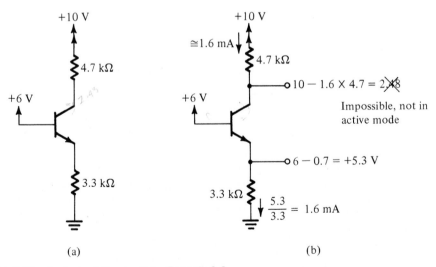

(a)   (b)

Fig. 9.14   Analysis of the circuit for Example 9.2.

Solution
Assuming active-mode operation, we have

$$V_E = +6 - V_{BE} \simeq +6 - 0.7 = 5.3 \text{ V}$$

$$I_E = \frac{5.3}{3.3} = 1.6 \text{ mA}$$

$$V_C = +10 - 4.7 \times I_C \simeq 10 - 7.52 = 2.48 \text{ V}$$

Since the collector voltage calculated appears to be less than the base voltage by 3.52 V, it follows that our original assumption of active-mode operation is incorrect. In fact, the transistor has to be in the *saturation* mode. Since we have not yet studied the saturation mode of operation, we shall defer the analysis of this circuit to a later section.

The details of the analysis performed above are illustrated in Fig. 9.14b.

• • •

Example 9.3

We wish to analyze the circuit in Fig. 9.15a to determine the voltages at all nodes and the currents through all branches. Note that this circuit is identical to that considered in Examples 9.1 and 9.2 except that now the base voltage is zero.

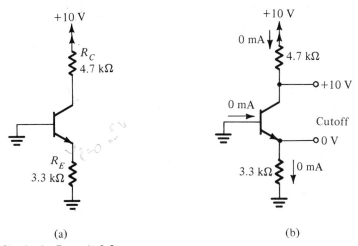

(a)   (b)

Fig. 9.15   Circuits for Example 9.3.

Solution

Since the base is at zero volts, the emitter–base junction cannot conduct and the emitter current is zero. Also, the collector–base junction cannot conduct, since the n-type collector is connected through R_C to the positive power supply while the p-type base is at ground. It follows that the collector current will be zero. The base current will also have to be zero, and the transistor is in the *cutoff* mode of operation.

The emitter voltage will obviously be zero, while the collector voltage will be equal to $+10$ V, since the voltage drop across R_C is zero. Figure 9.15b shows the analysis details.

• • •

Example 9.4

We desire to analyze the circuit of Fig. 9.16a to determine the voltages at all nodes and the currents through all branches.

Solution

The base of this *pnp* transistor is grounded, while the emitter is connected to a positive supply ($V^+ = +10$ V) through R_E. It follows that the emitter–base junction will be forward-biased with

$$V_E = V_{EB} \simeq 0.7 \text{ V}$$

Thus the emitter current will be given by

$$I_E = \frac{V^+ - V_E}{R_E} = \frac{10 - 0.7}{2} = 4.65 \text{ mA}$$

Since the collector is connected to a negative supply (more negative than the base voltage) through R_C, is it *possible* that this transistor is operating in the active mode. Assuming this to be the case, we obtain

$$I_C = \alpha I_E$$

Since no value for β has been given, we shall assume $\beta = 100$, which results in $\alpha = 0.99$. Since large variations in β result in small differences in α, this assumption will not be critical as far as determining the value of I_C is concerned. Thus

$$I_C = 0.99 \times 4.65 = 4.6 \text{ mA}$$

The collector voltage will be

$$\begin{aligned} V_C &= V^- + I_C R_C \\ &= -10 + 4.6 \times 1 = -5.4 \text{ V} \end{aligned}$$

Thus the collector–base junction is reverse-biased by 5.4 V and the transistor is indeed in the active mode, which supports our original assumption.

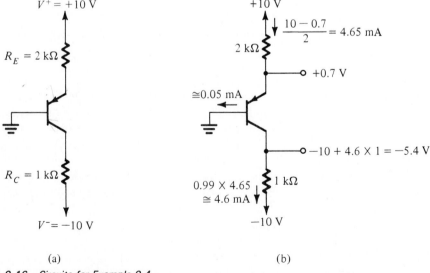

(a)   (b)

Fig. 9.16   Circuits for Example 9.4.

It remains only to calculate the base current,

$$I_B = \frac{I_E}{\beta + 1} = \frac{4.65}{101} \simeq 0.0465 \text{ mA}$$

Obviously, the value of β critically affects the base current. Note, however, that in this circuit the value of β will have no effect on the mode of operation of the transistor. Since β is generally an ill-specified parameter, this circuit represents a good design. As a rule one should strive to *design the circuit such that its performance is as insensitive to the value of β as possible*. The analysis details are illustrated in Fig. 9.16b.

. . .

Example 9.5
We want to analyze the circuit in Fig. 9.17a to determine the voltages at all nodes and the currents in all branches. Assume $\beta = 100$.

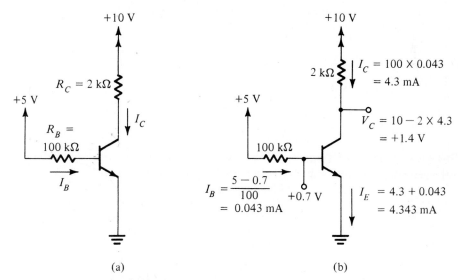

(a)   (b)

Fig. 9.17   Circuits for Example 9.5.

Solution
The base–emitter junction is clearly forward-biased. Thus

$$I_B = \frac{+5 - V_{BE}}{R_B} \simeq \frac{5 - 0.7}{100} = 0.043 \text{ mA}$$

Assume that the transistor is operating in the active mode. We now can write

$$I_C = \beta I_B = 100 \times 0.043 = 4.3 \text{ mA}$$

The collector voltage can now be determined as

$$V_C = +10 - I_C R_C = 10 - 4.3 \times 2 = +1.4 \text{ V}$$

Since the base voltage V_B is

$$V_B = V_{BE} \simeq +0.7 \text{ V}$$

it follows that the collector–base junction is reverse-biased by 0.7 V and the transistor is indeed in the active mode. The emitter current will be given by

$$I_E = (\beta + 1)I_B = 101 \times 0.043 \simeq 4.3 \text{ mA}$$

We note from this example that the collector and emitter currents depend critically on the value of β. In fact, if β were 10% higher, the transistor would leave the active mode and enter saturation. Therefore this clearly is a bad design. The analysis details are illustrated in Fig. 9.17b.

· · ·

Example 9.6
We want to analyze the circuit of Fig. 9.18a to determine the voltages at all nodes and the currents through all branches. Assume $\beta = 100$.

Solution
The first step in the analysis consists of simplifying the base circuit using Thévenin's theorem. The result is shown in Fig. 9.18b, where

$$V_{BB} = +15 \frac{R_{B2}}{R_{B1} + R_{B2}} = 15 \frac{50}{100 + 50} = +5 \text{ V}$$
$$R_{BB} = (R_{B1} \parallel R_{B2}) = (100 \parallel 50) = 33.3 \text{ k}\Omega$$

To evaluate the base or the emitter current we have to write a loop equation around the loop marked L in Fig. 9.18b. Note, though, that the current through R_{BB} is different from the current through R_E. The loop equation will be

$$V_{BB} = I_B R_{BB} + V_{BE} + I_E R_E$$

Substituting for I_B by

$$I_B = \frac{I_E}{\beta + 1}$$

and rearranging the equation gives

$$I_E = \frac{V_{BB} - V_{BE}}{R_E + [R_{BB}/(\beta + 1)]}$$

For the numerical values given we have

$$I_E = \frac{5 - 0.7}{3 + (33.3/100)} = 1.29 \text{ mA}$$

The base current will be

$$I_B = \frac{1.29}{101} = 0.0128 \text{ mA}$$

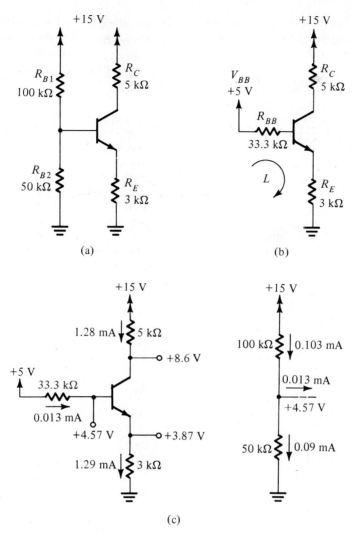

Fig. 9.18   Circuits for Example 9.6.

The base voltage is given by

$$V_B = V_{BE} + I_E R_E$$
$$= 0.7 + 1.29 \times 3 = 4.57 \text{ V}$$

Assume active-mode operation. We can evaluate the collector current as

$$I_C = \alpha I_E = 0.99 \times 1.29 = 1.28 \text{ mA}$$

The collector voltage can now be evaluated as

$$V_C = +15 - I_C R_C = 15 - 1.28 \times 5 = 8.6 \text{ V}$$

It follows that the collector is higher in potential than the base by 4.03 V, which means that the transistor is in the active mode, as had been assumed. The results of the analysis are given in Fig. 9.18c.

· · ·

Example 9.7
We wish to analyze the circuit in Fig. 9.19a to determine the voltages at all nodes and the currents through all branches.

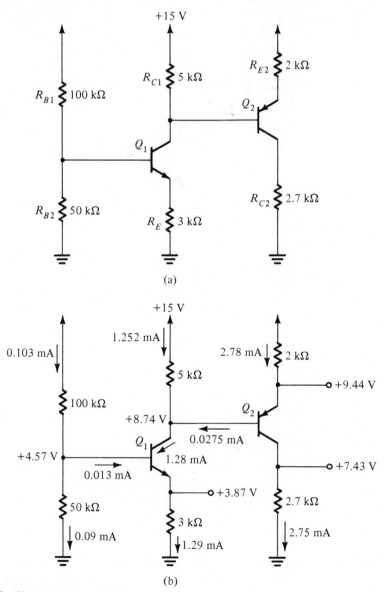

(a)

(b)

Fig. 9.19 *Circuits for Example 9.7.*

Solution

We first recognize that part of this circuit is identical to the circuit we analyzed in Example 9.6, namely, the circuit of Fig. 9.18a. The difference, of course, is that in the new circuit we have an additional transistor Q_2 together with its associated resistors R_{E2} and R_{C2}. Assume that Q_1 is still in the active mode. The following values will be identical to those obtained in the previous example:

$$V_{B1} = +4.57 \text{ V}$$
$$I_{B1} = 0.0128 \text{ mA}$$
$$I_{E1} = 1.29 \text{ mA}$$
$$I_{C1} = 1.28 \text{ mA}$$

However, the collector voltage will be different than previously calculated, since part of the collector current I_{C1} will flow in the base lead of Q_2 (I_{B2}). As a first approximation we may assume that I_{B2} is much smaller than I_{C1}; that is, we may assume that the current through R_{C1} is almost equal to I_{C1}. This will enable us to calculate V_{C1}:

$$V_{C1} \simeq +15 - I_{C1}R_{C1}$$
$$= 15 - 1.28 \times 5 = +8.6 \text{ V}$$

Thus Q_1 is in the active mode, as had been assumed.

As far as Q_2 is concerned, we note that its emitter is connected to $+15$ V through R_{E2}. It is therefore safe to assume that the emitter–base junction of Q_2 will be forward-biased. Thus the emitter of Q_2 will be at a voltage V_{E2} given by

$$V_{E2} = V_{C1} + V_{EB}|_{Q_2} \simeq 8.6 + 0.7 = +9.3 \text{ V}$$

The emitter current of Q_2 may now be calculated as

$$I_{E2} = \frac{+15 - V_{E2}}{R_{E2}} = \frac{15 - 9.3}{2} = 2.85 \text{ mA}$$

Since the collector of Q_2 is returned to ground via R_{C2}, it is possible that Q_2 is operating in the active mode. Assume this to be the case. We now find I_{C2} as

$$I_{C2} = \alpha_2 I_{E2}$$
$$= 0.99 \times 2.85 = 2.82 \text{ mA} \qquad \text{(assuming } \beta_2 = 100\text{)}$$

The collector voltage of Q_2 will be

$$V_{C2} = I_{C2}R_{C2} = 2.82 \times 2.7 = 7.62 \text{ V}$$

which is lower than V_{B2} by 0.98 V. Thus Q_2 is in the active mode, as assumed.

It is important at this stage to find the magnitude of the error incurred in our calculation by the assumption that I_{B2} is negligible. The value of I_{B2} is given by

$$I_{B2} = \frac{I_{E2}}{\beta_2 + 1} = \frac{2.85}{101} = 0.028 \text{ mA}$$

which is indeed much smaller than I_{C1} (1.28 mA). If desired, we can obtain more accurate results by iterating one more time, assuming I_{B2} to be 0.028 mA. The new values will be

$$\text{Current in } R_{C1} = I_{C1} - I_{B2} = 1.28 - 0.028 = 1.252 \text{ mA}$$

$$V_{C1} = 15 - 5 \times 1.252 = 8.74 \text{ V}$$

$$V_{E2} = 8.74 + 0.7 = 9.44 \text{ V}$$

$$I_{E2} = \frac{15 - 9.44}{2} = 2.78 \text{ mA}$$

$$I_{C2} = 0.99 \times 2.78 = 2.75 \text{ mA}$$

$$V_{C2} = 2.75 \times 2.7 = 7.43 \text{ V}$$

$$I_{B2} = \frac{2.78}{101} = 0.0275 \text{ mA}$$

Note that the new value of I_{B2} is very close to the value used in our iteration, and no further iterations are warranted. The final results are indicated in Fig. 9.19b.

The reader justifiably might be wondering about the necessity of using an iterative scheme in solving a linear (or linearized) problem. Indeed, we can obtain the exact solution (if we can call anything we are doing with a first-order model exact!) by writing appropriate equations. The reader is encouraged to find this solution and compare the results with those obtained above. It is important to emphasize, however, that in most such problems it is quite sufficient to obtain an approximate solution, provided that we can obtain it quickly and, of course, correctly.

$$\bullet \quad \bullet \quad \bullet$$

Important Note

In Examples 9.1 through 9.7 we frequently used the value of α to calculate the collector current. Since $\alpha \simeq 1$, the error in such calculations will be very small if one assumes $\alpha = 1$ and $i_C = i_E$. Therefore except in calculations that depend critically on the value of α (such as the calculation of base current) one usually assumes $\alpha \simeq 1$.

EXERCISE

9.6 For the circuit shown in Fig. E9.6 assume β to be very high and find I_E, V_E, and V_C.

  Ans. 1 mA; +6.8 V; +10.7 V

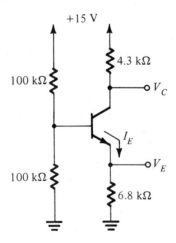

Fig. E9.6

9.7 THE TRANSISTOR AS AN AMPLIFIER

To operate as an amplifier a transistor must be biased in the active region. The biasing problem is that of establishing a constant dc current in the emitter (or the collector). This current should be predictable and insensitive to variations in temperature, value of β, and so on. While deferring the study of different biasing techniques to Chapter 10, we will demonstrate in the following the need to bias the transistor at a constant collector current. This requirement stems from the fact that the operation of the transistor as an amplifier is highly influenced by the value of the quiescent (or bias) current, as shown below.

DC Conditions

To understand how the transistor operates as an amplifier, consider the conceptual circuit shown in Fig. 9.20. Here the bias current I_C is assumed to be established by a dc voltage V_{BE} (battery), clearly an impractical arrangement. The reverse bias of the CBJ

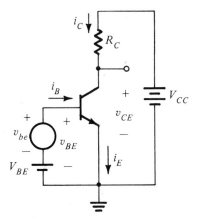

Fig. 9.20 Operation of the transistor as an amplifier.

is established by connecting the collector to another power-supply voltage V_{CC} through a resistor R_C. Consider first the dc conditions; that is, assume that the signal v_{be} is zero. We have the following dc quantities:

$$I_C = I_S e^{V_{BE}/V_T} \tag{9.10}$$

$$I_E = \frac{I_C}{\alpha}$$

$$I_B = \frac{I_C}{\beta}$$

$$V_C = V_{CE} = V_{CC} - I_C R_C \tag{9.11}$$

Obviously, for active-mode operation V_C should be greater than V_B by an amount that allows for a reasonable signal swing at the collector yet maintains the transistor in the active region at all times. We shall return to this point later.

Transconductance

If a signal v_{be} is applied as shown in Fig. 9.20, the total instantaneous base–emitter voltage v_{BE} becomes

$$v_{BE} = V_{BE} + v_{be}$$

Correspondingly, the collector current becomes

$$
\begin{aligned}
i_C &= I_S e^{v_{BE}/V_T} \\
&= I_S e^{(V_{BE}+v_{be})/V_T} \\
&= I_S e^{(V_{BE}/V_T)} e^{(v_{be}/V_T)}
\end{aligned}
\tag{9.12}
$$

Use of Eq. (9.10) yields

$$i_C = I_C e^{v_{be}/V_T} \tag{9.13}$$

Now, if $v_{be} \ll V_T$, we may approximate Eq. (9.13) as

$$i_C \simeq I_C \left(1 + \frac{v_{be}}{V_T} \right) \tag{9.14}$$

Here we have expanded the exponential in Eq. (9.13) in a series and retained only the first two terms. This approximation, which is valid only for v_{be} less than about 10 mV, is referred to as the *small-signal approximation*. Under this approximation the total collector current is given by Eq. (9.14) and can be rewritten

$$i_C = I_C + \frac{I_C}{V_T} v_{be} \tag{9.15}$$

Thus the collector current is composed of the dc bias value I_C and a signal component i_c,

$$i_c = \frac{I_C}{V_T} v_{be} \tag{9.16}$$

The quantity I_C/V_T relates the signal current in the collector to the base–emitter signal voltage and hence is called the *transconductance* denoted by g_m,

$$g_m = \frac{I_C}{V_T} \tag{9.17}$$

Equation (9.16) can now be rewritten

$$i_c = g_m v_{be} \tag{9.18}$$

Note that the transconductance is directly proportional to the collector bias current I_C. BJTs have relatively high transconductances (as compared to FETs): for instance, at $I_C = 1$ mA, $g_m \simeq 40$ mA/V. It should be obvious from Eq. (9.17) that to obtain a constant, predictable value for g_m we need a constant, predictable I_C.

A graphical interpretation for g_m is given in Fig. 9.21, where it is shown that g_m is equal to the slope of the i_C-v_{BE} characteristic curve at $i_C = I_C$. Thus

$$g_m = \left. \frac{\partial i_C}{\partial v_{BE}} \right|_{i_C - I_C} \tag{9.19}$$

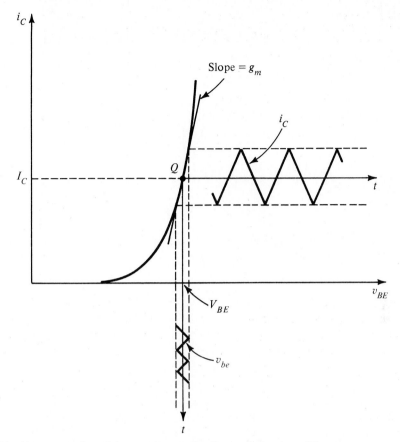

Fig. 9.21   *Linear operation of the transistor under the small-signal condition.*

The small-signal approximation implies keeping the signal amplitude sufficiently small so that operation is restricted to an almost linear segment of the i_C-v_{BE} exponential curve. Increasing the signal amplitude will result in the collector current having components nonlinearly related to v_{be}. Similar types of approximation were employed for diodes (Section 4.9), JFETs (Section 7.7) and MOSFETs (Section 8.4).

The Input Resistance at the Base

The analysis above suggests that for small signals ($v_{be} \ll V_T$) the transistor behaves as a voltage-controlled current source. The input port of this controlled source is between base and emitter, and the output port is between collector and emitter. The transconductance of the controlled source is g_m, and the output resistance is infinite. This latter ideal property is a result of our first-order model of transistor operation in which the collector voltage has no effect on the collector current in the active mode. It now remains to determine the input resistance of this controlled source.

To determine the resistance seen by v_{be}, we first evaluate the total base current i_B using Eq. (9.15),

$$i_B = \frac{i_C}{\beta} = \frac{I_C}{\beta} + \frac{1}{\beta}\frac{I_C}{V_T}v_{be}$$

Thus

$$i_B = I_B + i_b \tag{9.20}$$

where I_B is equal to I_C/β and the signal component i_b is given by

$$i_b = \frac{1}{\beta}\frac{I_C}{V_T}v_{be} \tag{9.21}$$

Substituting for I_C/V_T by g_m gives

$$i_b = \frac{g_m}{\beta}v_{be} \tag{9.22}$$

The small-signal input resistance between base and emitter, *looking into the base*, is denoted by r_π and is defined

$$r_\pi \equiv \frac{v_{be}}{i_b} \tag{9.23}$$

Using Eq. (9.22) gives

$$r_\pi = \frac{\beta}{g_m} \tag{9.24}$$

Thus r_π is directly dependent on β and is inversely proportional to the bias current I_C.

The Hybrid-π Model

For small signals the BJT can be represented by the equivalent circuit model shown in Fig. 9.22a. Note that this equivalent circuit applies at a particular bias point, since the two parameters g_m and r_π depend on the value of I_C. In order to be consistent with the literature we use v_π to denote the base–emitter signal voltage v_{be}.

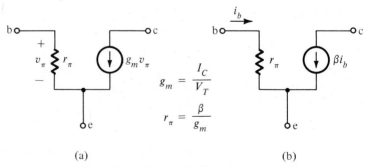

(a)   (b)

Fig. 9.22 *A first-order small-signal model for the BJT in two equivalent forms. This model is called the hybrid-π model and is applicable to both npn and pnp transistors.*

Since the output current $g_m v_\pi$ can be written

$$g_m v_\pi = g_m r_\pi i_b = \beta i_b$$

the equivalent circuit model can be converted to the current-controlled current-source form depicted in Fig. 9.22b. Note that this equivalent circuit is the incremental, or small-signal, version of the model of Fig. 9.5a, which can easily be verified by showing that r_π is the incremental resistance of the base–emitter diode at the quiescent current I_B (recall that $n = 1$ for BJTs).

For reasons that will become apparent in Chapter 11, the equivalent circuit model of Fig. 9.22 is called the *simplified hybrid-π model*. It should be noted that this model applies to both *npn* and *pnp* transistors with *no change* in polarities required.

The Emitter Resistance r_e

The total emitter current i_E can be determined using Eq. (9.15) and

$$i_E = \frac{i_C}{\alpha}$$

Thus

$$i_E = I_E + i_e$$

where I_E is equal to I_C/α, and the signal current i_e is given by

$$i_e = \frac{I_C}{\alpha V_T} v_{be} = \frac{I_E}{V_T} v_{be}$$

This relationship can be expressed in the form

$$i_e = \frac{v_{be}}{r_e} \tag{9.25}$$

where r_e, called the *emitter resistance*, is defined as

$$r_e \equiv \frac{V_T}{I_E} \tag{9.26}$$

Comparison with Eq. (9.17) reveals that

$$r_e = \frac{\alpha}{g_m} \simeq \frac{1}{g_m} \tag{9.27}$$

An extremely useful interpretation for r_e is that it is the resistance between base and emitter *looking into the emitter*. This is to distinguish it from the resistance between base and emitter *looking into the base*, which is r_π. It can be shown that r_e and r_π are related by

$$r_\pi = (\beta + 1) r_e$$

which is intuitively obvious because $\beta + 1$ is the ratio of i_e (which is the current flowing through r_e) and i_b (which is the current flowing through r_π).

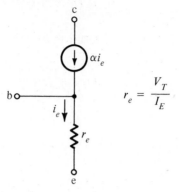

Fig. 9.23   An alternative small-signal model for the BJT.

In the equivalent circuit models of Fig. 9.22, r_e is not explicitly shown. Nevertheless it is implicitly included, as can easily be demonstrated. An alternative small-signal model that explicitly employs r_e is given in Fig. 9.23. It can be shown that this model is the incremental version of the model given in Fig. 9.5b.

Voltage Gain

Up until now we have established only that the transistor senses the base–emitter signal v_{be} or v_π and causes a proportional current $g_m v_\pi$ to flow in the collector lead at a high (ideally infinite) impedance level. In this way the transistor is acting as a voltage-controlled current source. To obtain an output voltage signal we may force this current to flow through a resistor, as is done in Fig. 9.20. Then the total collector voltage v_C will be

$$
\begin{aligned}
v_C &= V_{CC} - i_C R_C \\
&= V_{CC} - (I_C + i_c) R_C \\
&= (V_{CC} - I_C R_C) - i_c R_C \\
&= V_C - i_c R_C
\end{aligned}
\tag{9.28}
$$

Here the quantity V_C is the dc bias voltage at the collector, and the signal voltage is given by

$$
\begin{aligned}
v_c &= -i_c R_C \\
&= -g_m v_{be} R_C \\
&= (-g_m R_C) v_{be}
\end{aligned}
\tag{9.29}
$$

Thus the voltage gain of this amplifier is

$$
\text{Voltage gain} \equiv \frac{v_c}{v_{be}} = -g_m R_C
\tag{9.30}
$$

Here again we note that because g_m is directly proportional to the collector (or emitter) bias current the gain will be as stable as the collector bias current is made.

Use of Small-Signal Equivalent Circuits

From the above analysis we conclude that under the small-signal approximation each of the currents and voltages in the circuit will be composed of a dc bias component and a signal component. We may therefore simplify matters and consider the two components separately. That is, we first establish a suitable dc operating point and subsequently ignore all dc quantities while we carry out a signal analysis. From a signal point of view the transistor may be replaced by one of the models in Figs. 9.22 and 9.23. All these models are, of course, equivalent.

Although equivalent circuit models are indispensible in the analysis of complicated circuits, it is far quicker and certainly more insightful to analyze simple circuits with the model *implicitly* assumed. This is especially true if one is interested only in obtaining approximate values using first-order models such as those developed above. At a later stage we will add to the hybrid-π model components that represent second-order effects.

Example 9.8

We wish to analyze the transistor amplifier shown in Fig. 9.24a. Assume $\beta = 100$.

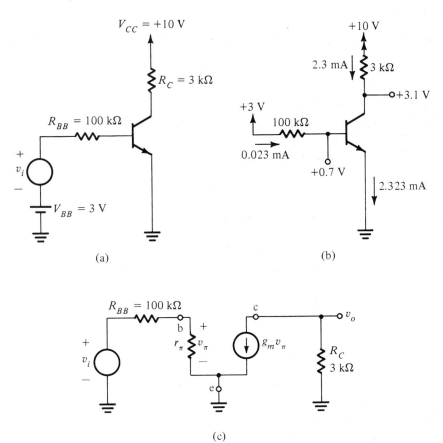

Fig. 9.24   Example 9.8: (a) circuit; (b) dc analysis; (c) small-signal model.

Solution

The first step in the analysis consists of determining the quiescent operating point. For this purpose we assume that $v_i = 0$. The dc base current will be

$$I_B = \frac{V_{BB} - V_{BE}}{R_{BB}}$$

$$\simeq \frac{3 - 0.7}{100} = 0.023 \text{ mA}$$

The dc collector current will be

$$I_C = \beta I_B = 100 \times 0.023 = 2.3 \text{ mA}$$

The dc voltage at the collector will be

$$V_C = V_{CC} - I_C R_C$$
$$= +10 - 2.3 \times 3 = +3.1 \text{ V}$$

Since $V_B = +0.7$ V, it follows that in the quiescent condition the transistor will be operating in the active mode.

Having determined the operating point, we may now proceed to determine the small-signal model parameters:

$$r_e = \frac{V_T}{I_E} = \frac{25 \text{ mV}}{(2.3/0.99) \text{ mA}} = 10.8 \ \Omega$$

$$g_m = \frac{I_C}{V_T} = \frac{2.3 \text{ mA}}{25 \text{ mV}} = 92 \text{ mA/V}$$

$$r_\pi = \frac{\beta}{g_m} = \frac{100}{92} = 1.09 \text{ k}\Omega$$

To carry out a small-signal analysis, it is convenient to employ the first-order hybrid-π equivalent circuit model of Fig. 9.22a. The resulting small-signal equivalent circuit model of the amplifier circuit is given in Fig. 9.24c. Note that no dc quantities are included in this equivalent circuit. It is most important to note that the dc supply voltage V_{CC} has been replaced by a *short circuit* in the signal equivalent circuit because the circuit terminal connected to V_{CC} will always have a constant voltage. That is, the signal voltage at this terminal will be zero. In other words, *a circuit terminal connected to a constant dc source can always be considered as a signal ground.*

Analysis of the equivalent circuit in Fig. 9.24c proceeds as follows:

$$v_\pi = v_i \frac{r_\pi}{r_\pi + R_{BB}}$$

$$= v_i \frac{1.09}{101.09} = 0.011 v_i \tag{9.31}$$

The output voltage v_o is given by

$$v_o = -g_m v_\pi R_C$$
$$= -92 \times 0.011 v_i \times 3 = -3.04 v_i$$

Thus the voltage gain will be

$$\frac{v_o}{v_i} = -3.04 \qquad (9.32)$$

where the minus sign indicates a phase reversal.

To gain more insight into the behavior of this circuit, let us consider the waveforms at various points. For this purpose we shall assume v_i to have a triangular waveform. The first question facing us will be, How high can the amplitude of v_i become? One constraint on signal amplitude is the small-signal approximation on which all of the above analysis is based. This constraint stipulates that v_{be} (that is, v_π) not exceed about 10 mV. If we take the triangular waveform v_π to be 20 mV peak-to-peak and work backward, Eq. (9.31) can be used to determine the maximum possible peak of v_i,

$$\hat{V}_i = \frac{\hat{V}_\pi}{0.011} = 0.91 \text{ V}$$

To check whether or not the transistor remains in the active mode with v_i having a peak value $\hat{V}_i = 0.91$ V, we have to evaluate the collector voltage. The voltage at the collector will consist of a triangular wave v_c superimposed on the dc value $V_C = 3.1$ V. The peak voltage of the triangular waveform will be

$$\hat{V}_c = \hat{V}_i \times \text{Gain} = 0.91 \times 3.04 = 2.77 \text{ V}$$

It follows that when the output swings negative the collector voltage reaches a minimum of $3.1 - 2.77 = 0.33$ V, which is lower than the base voltage $\simeq 0.7$ V. Thus the transistor will not remain in the active mode with v_i having a peak value of 0.91 V. We can easily determine, though, the maximum value of the peak of the input signal such that the transistor remains active at all times. This can be done by finding the value of $\hat{V}_i$ that corresponds to the minimum value of the collector voltage being equal to the base voltage, which is approximately 0.7 V. Thus

$$\hat{V}_i = \frac{3.1 - 0.7}{3.04} = 0.79 \text{ V}$$

Let us choose $\hat{V}_i$ to be approximately 0.8 V and complete the analysis of this problem. The signal current in the base will be triangular, with a peak value $\hat{I}_b$ of

$$\hat{I}_b = \frac{\hat{V}_i}{R_{BB} + r_\pi} = \frac{0.8}{100 + 1.09} = 0.008 \text{ mA}$$

This triangular-wave current will be superimposed on the quiescent base current I_B, as shown in Fig. 9.25b. The base–emitter voltage will consist of a triangular-wave component superimposed on the dc V_{BE} that is approximately 0.7 V. The peak value of the triangular waveform will be

$$\hat{V}_\pi = \hat{V}_i \frac{r_\pi}{r_\pi + R_{BB}} = 0.8 \frac{1.09}{100 + 1.09} = 8.6 \text{ mV}$$

The total v_{BE} is sketched in Fig. 9.25c.

The signal current in the collector will be triangular in waveform, with a peak value $\hat{I}_c$ given by

$$\hat{I}_c = \beta \hat{I}_b = 100 \times 0.008 = 0.8 \text{ mA}$$

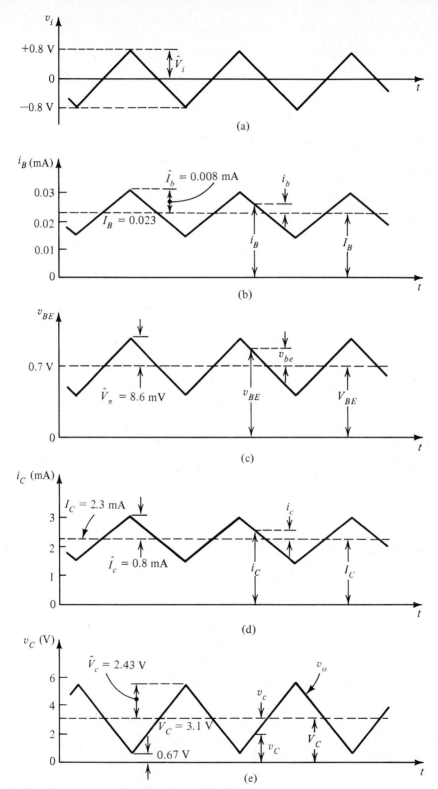

Fig. 9.25 Signal waveforms in the circuit of Fig. 9.24.

This current will be superimposed on the quiescent collector current I_C ($= 2.3$ mA), as shown in Fig. 9.25d.

Finally, the signal voltage at the collector can be obtained by multiplying v_i by the voltage gain,

$$\hat{V}_c = 3.04 \times 0.8 = 2.43 \text{ V}$$

Figure 9.25e shows a sketch of the total collector voltage v_C versus time. Note the phase reversal between the input signal v_i and the output signal v_c. Also note that although the minimum collector voltage is slightly lower than the base voltage, the transistor will remain in the active mode. In fact, BJTs remain in the active mode even with their collector-base junctions forward-biased by as much as 0.1 or 0.2 V.

Before we leave this example, we wish to make an important observation. Although we have drawn the equivalent circuit in Fig. 9.24b, and used it in the analysis, this step could have been avoided. In fact, for first-order calculations the equivalent circuit model is usually used implicitly, as will be illustrated in Example 9.9.

· · ·

Example 9.9
We need to analyze the circuit of Fig. 9.26a to determine the voltage gain and the signal waveforms at various points. The capacitor C is a coupling capacitor whose purpose is to couple the signal v_i to the emitter while blocking dc. In this way the dc bias established by V^+ and V^- together with R_E and R_C will not be disturbed when the signal v_i is connected. For the purpose of this example, C will be assumed to be infinite—that is, acting as a perfect short circuit at signal frequencies of interest.

Solution
We shall start by determining the dc operating point as follows:

$$I_E = \frac{+10 - V_E}{R_E} \simeq \frac{+10 - 0.7}{10} = 0.93 \text{ mA}$$

Assuming $\beta = 100$, then $\alpha = 0.99$ and

$$I_C = 0.99 I_E = 0.92 \text{ mA}$$
$$V_C = -10 + I_C R_C$$
$$= -10 + 0.92 \times 5 = -5.4 \text{ V}$$

Thus the transistor is in the active mode. Furthermore, the collector signal can swing from -5.4 V to zero (which is the base voltage) without the transistor going into saturation. However, a 5.4-V swing in the collector voltage will (theoretically) cause the minimum collector voltage to be -10.8 V, which is more negative than the power-supply voltage. It follows that if we attempt to apply an input that results in such an output signal, the transistor will cut off and the negative peaks of the output signal will be clipped off, as illustrated in Fig. 9.27. The waveform in Fig. 9.27, however, is shown to be linear (except for the clipped peak)—that is, the effect of the nonlinear i_C-v_{BE} characteristic is not taken into account. This is not correct, since if we are driving the transistor into cutoff at the negative signal peaks, then for sure we are exceeding the small-signal limit, as will be shown later.

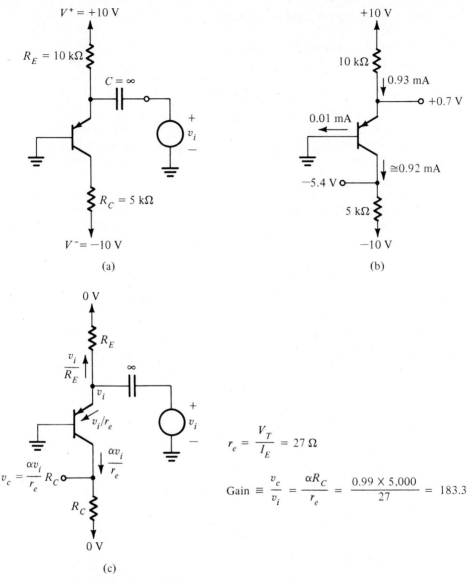

Fig. 9.26   Example 9.9: (a) circuit; (b) dc analysis; (c) signal analysis.

Let us now proceed to determine the small-signal voltage gain. Toward that end, we evaluate the emitter resistance r_e,

$$r_e = \frac{V_T}{I_E} = \frac{25}{0.93} \simeq 27\ \Omega$$

We also note that

$$v_{eb} = v_i$$

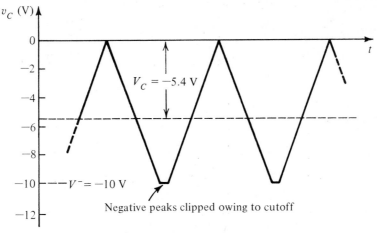

Fig. 9.27   *Distortion in output signal due to transistor cutoff. Note that it is assumed that no distortion due to transistor nonlinear characteristics is occurring.*

Thus for small-signal operation the peak value of v_i should be limited to less than 10 mV; otherwise nonlinear distortion will occur.

The signal current in the emitter can be easily obtained as

$$i_e = \frac{v_{eb}}{r_e} = \frac{v_i}{r_e}$$

The signal current in the collector will be

$$i_c = \alpha i_e = \frac{\alpha v_i}{r_e}$$

The signal component of the collector voltage can now be evaluated assuming the negative power supply (V^-) to be a signal ground. Thus

$$v_c = i_c R_C = \frac{\alpha v_i}{r_e} R_C$$

which results in the voltage gain

$$\frac{v_c}{v_i} = \frac{\alpha R_C}{r_e} = g_m R_C$$

$$= \frac{0.99 \times 5,000}{27} = 183.3$$

Note that the voltage gain is positive, indicating that the output is in phase with the input signal. This property is due to the fact that the input signal is applied to the emitter rather than the base, as was done in Example 9.8. We should emphasize that the positive gain has nothing to do with the fact that the transistor used in this example is of the *pnp* type.

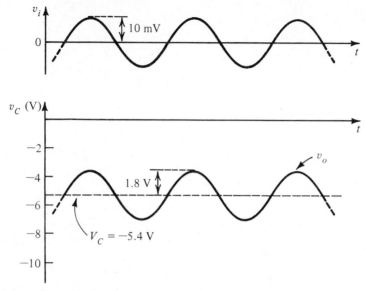

Fig. 9.28   *Input and output waveforms for the circuit of Fig. 9.26.*

Assuming an input sine wave signal with a peak of 10 mV, as shown in Fig. 9.28, it follows that the signal component of the collector voltage will have a peak value $\hat{V}_c$,

$$\hat{V}_c = 183.3 \times 0.01 = 1.833 \text{ V}$$

Thus as expected the output signal voltage is limited by the maximum signal that we can apply at the input without running the risk of nonlinear distortion.

· · ·

Transistor Capacitances

In our study of the *pn* junction in Chapter 4 we learned that two capacitive effects exist. The first such effect is the change in charge stored in the depletion region as the voltage across the junction changes, which is modeled by the space-charge or depletion-layer capacitance C_j. The second capacitive effect relates to the change in the excess minority carrier charge stored in the *p* and *n* materials as the voltage across the junction is changed, which is modeled by the diffusion capacitance C_d. The diffusion capacitance C_d is proportional to the current flowing across the junction and is zero for a reverse-biased junction.

Applying this to the BJT, we see that for a device operating in the active mode the forward-biased EBJ displays two parallel capacitances: a depletion layer capacitance C_{je} and a diffusion capacitance C_{de}. On the other hand, because the CBJ is reverse-biased it displays only a depletion capacitance C_{jc}. Figure 9.29 shows the small-signal equivalent circuit model of a BJT operating in the active mode with the junction capacitances included. The EBJ capacitances have been added together to obtain the total capacitance C_π. Also, in order to conform with the literature, the CBJ depletion capac-

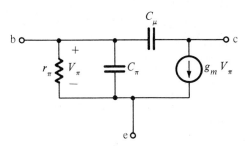

Fig. 9.29 The BJT equivalent circuit model, including the two junction capacitances C_π and C_μ.

itance has been denoted by C_μ. The values of C_π and C_μ depend on the transistor type, junction areas, and, of course, the dc bias point. We will have more to say about these capacitances in Chapter 11, when we deal with amplifier frequency response. For the time being it should be mentioned that while C_π can range from few picofarads to a few tens of picofarads, C_μ is usually about 1 or 2 picofarads. Nevertheless, C_μ plays an important role in determining the effective input capacitance of a transistor amplifier through the Miller multiplication effect (see Chapter 2). It will be shown in Chapter 11 that C_π and C_μ determine the high-frequency response of transistor amplifiers. For gain calculations at low and medium frequencies these capacitances can be ignored.

EXERCISES

9.7 For a transistor with $\beta = 100$ operating at $I_C = 0.5$ mA, find g_m, r_e, and r_π.
Ans. 20 mA/V; 50 Ω; 5 kΩ

9.8 In Exercise 9.6 the dc bias point of the circuit in Fig. E9.8 was found. We now wish to find r_e. Assume $\alpha \simeq 1$ and use the equivalent circuit of Fig. 9.23 to calculate the voltage gains v_{o1}/v_i and v_{o2}/v_i.
Ans. 25 Ω; 0.996 V/V; −0.63 V/V

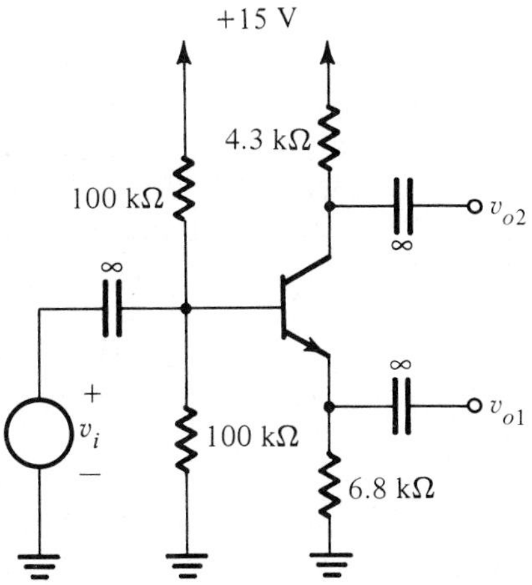

Fig. E9.8

9.8 THE TRANSISTOR AS A SWITCH: CUTOFF AND SATURATION

Having studied the active mode of operation in detail, we are now ready to complete the picture by considering what happens when the transistor leaves the active region. At one end the transistor will enter the cutoff region, while at the other extreme the transistor enters the saturation region. These two extreme modes of operation are very useful if the transistor is to be used as a switch, such as in digital logic circuits. In the present section we shall study the cutoff and saturation modes of operation. Examples of the application of the BJT in switching circuits are presented in Section 9.9, and a detailed study of the saturation mode of operation and of BJT digital circuits is included in Chapter 15. Before proceeding further, the reader is advised to review pertinent material from Chapter 6.

Cutoff Region

To help introduce cutoff and saturation, we consider the simple circuit shown in Fig. 9.30, which is fed with a voltage source v_I. We wish to analyze this circuit for different values of v_I.

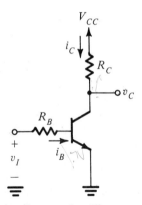

Fig. 9.30 *A simple circuit used to illustrate the different modes of operation of the BJT.*

If v_I is smaller than about 0.5 V, the EBJ will conduct negligible current. In fact, the EBJ could be considered "reverse" biased, and the device will be in the cutoff mode. It follows that

$$i_B = 0, \qquad i_E = 0, \qquad i_C = 0, \qquad v_C = V_{CC}$$

Note that the CBJ is, of course, reverse-biased.

Active Region

To turn the transistor on, we have to increase v_I above 0.5 V. In fact, for appreciable currents to flow v_{BE} should be about 0.7 V and v_I should be higher. For $v_I > 0.7$ V we have

$$i_B = \frac{v_I - V_{BE}}{R_B} \tag{9.33}$$

which may be approximated by

$$i_B \simeq \frac{v_I - 0.7}{R_B} \tag{9.34}$$

provided that $v_I \gg 0.7$ V (for example, ≥ 2 V) and that the resulting collector current is in the normal range for this particular transistor. The collector current is given by

$$i_C = \beta i_B \tag{9.35}$$

which applies only if the device is in the active mode. How do we know that the device is in the active mode? We do not know; therefore we assume that it is in the active mode, calculate i_C using Eq. (9.35) and v_C from

$$v_C = V_{CC} - R_c i_C \tag{9.36}$$

and then check whether $v_{CB} \geq 0$ or not. In our case we merely check whether $v_C \geq 0.7$ V or not. If $v_C \geq 0.7$ V, then our original assumption is correct and we have completed the analysis for the particular value of v_I. On the other hand, if v_C is found to be less than 0.7 V, then the device has left the active region and entered the saturation region.

Obviously as v_I is increased i_B will increase [Eq. (9.34)], i_C will correspondingly increase [Eq. (9.35)], and v_C will decrease [Eq. (9.36)]. Eventually, v_C will become less than v_B (0.7 V) and the device will enter the saturation region.

Saturation Region

Saturation occurs when we attempt to force a current in the collector higher than the collector circuit can support while maintaining active-mode operation. For the circuit in Fig. 9.30 the maximum current that the collector "can take" without the transistor leaving the active mode can be evaluated by setting $v_{CB} = 0$, which results in

$$\hat{I}_C = \frac{V_{CC} - V_B}{R_C} \simeq \frac{V_{CC} - 0.7}{R_C} \tag{9.37}$$

This collector current is obtained by forcing a base current $\hat{I}_B$, given by

$$\hat{I}_B = \frac{\hat{I}_C}{\beta}$$

and the corresponding required value of v_I can be obtained from Eq. (9.34). Now, if we increase i_B above $\hat{I}_B$, the collector current will increase and the collector voltage will fall below that of the base. This will continue until the CBJ becomes forward-biased with a forward-bias voltage of about 0.4 to 0.5 V. This situation is referred to as saturation, since any further increase in the base current will result in a very small increase in the collector current and a corresponding small decrease in the collector voltage. This means that in saturation the *incremental* β ($\Delta i_C / \Delta i_B$) is negligibly small. Any "extra" current that we force into the base terminal mostly will flow through the emitter terminal. Thus the ratio of the collector current to the base current of a saturated transistor is *not* equal to β and can be set to any desired value—smaller than β—simply by pushing more current into the base.

Let us now return to the circuit of Fig. 9.30, which we have redrawn in Fig. 9.31 with the assumption that the transistor is in saturation. The value of V_{BE} of a saturated transistor is usually slightly higher than that of the device operating in the active mode.

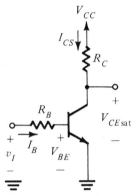

Fig. 9.31 *A saturated transistor.*

The increase in V_{BE} is due to the increased base current producing a sizable ohmic (IR) voltage drop across the bulk resistance of the base region. In other words, part of V_{BE} will appear across the base semiconductor material as an IR drop, while the remainder will appear across the EBJ. For simplicity, we shall assume V_{BE} to remain around 0.7 V even if the device is in saturation.

Since in saturation the base voltage is higher than the collector voltage by about 0.4 or 0.5 V, it follows that the collector voltage will be higher than the emitter voltage by 0.3 or 0.2 V. This latter voltage is referred to as V_{CEsat}, and we will normally assume that $V_{CEsat} \simeq 0.3$ V. Note, however, that if we push more current into the base, we drive the transistor "deeper" into saturation and the CBJ forward-bias increases, which means that V_{CEsat} will decrease.

The value of the collector current in saturation will be almost constant. We denote this value by I_{CS}. It follows that for the circuit in Fig. 9.31 we have

$$I_{CS} = \frac{V_{CC} - V_{CEsat}}{R_C} \tag{9.38}$$

In order to ensure that the transistor is driven into saturation, we have to force a base current of at least

$$I_{BS} = \frac{I_{CS}}{\beta} \tag{9.39}$$

Normally one designs the circuit such that I_B is higher than I_{BS} by a factor of 2 to 10 (called the *overdrive factor*). The ratio of I_{CS} to I_B is called the *forced β* (β_{forced}), since its value can be set at will,

$$\beta_{forced} = \frac{I_{CS}}{I_B}$$

Transistor Inverter

Figure 9.32 shows the transfer characteristic of the circuit of Fig. 9.30. The shape of this curve should be obvious to the reader who has followed the above discussion. The three regions of operation—cutoff, active, and saturation—are indicated in Fig. 9.32. For the transistor to be operated as an amplifier, it should be biased somewhere in the active region, such as at the point marked X. The voltage gain of the amplifier is equal to the slope of the transfer characteristic at this point.

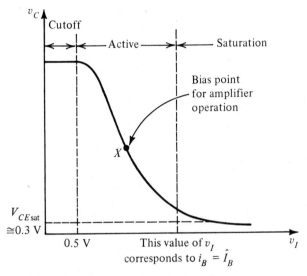

Fig. 9.32 *Transfer characteristic for the circuit in Fig. 9.30.*

   For switching applications the transistor is usually operated in cutoff and saturation.[2] That is, one state of the switch will correspond to the transistor being cutoff, and the other state corresponds to the transistor in saturation. There are a number of reasons for choosing these two extreme modes of operation. One reason is that in both cutoff and saturation the currents and voltages in the transistor are well defined and do not depend on such ill-specified parameters as β. Another reason is that in both cutoff and saturation the power dissipated in the transistor is minimal. Although this is obvious for the cutoff mode, it is also the case in the saturation mode, since the voltage V_{CEsat} is small.

   The circuit that we have been using as an example is in fact the basic transistor logic inverter. To see how it works as a logic inverter, consider a positive-logic system where voltages around zero represent logic 0 and voltages around the power-supply voltage represent logic 1. If the input is a logic 1, then v_I will be in the vicinity of V_{CC}, and—assuming that R_B and R_C are chosen such that the transistor is saturated—the output voltage will be equal to V_{CEsat}, which is close to zero volts and represents logic 0. For a logic 0 input v_I will be around zero, but could be as high as V_{CEsat}. This small

---

[2] An exception to this is found in emitter-coupled logic (ECL) which is studied in detail in Chapter 15.

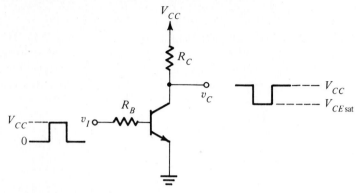

Fig. 9.33 The basic transistor logic inverter.

input voltage will be insufficient to turn the transistor on, and the device will be cut off with $v_C = V_{CC}$, a logic 1. Figure 9.33 illustrates the above discussion.

Model for Saturated BJT

From the discussion above we obtain a simple model for transistor operation in the saturation mode, as shown in Fig. 9.34. Normally, we use such a model implicitly in the analysis of a given circuit.

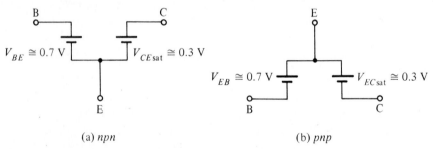

(a) npn   (b) pnp

Fig. 9.34 Model for the saturated BJT.

For quick approximate calculations one may consider V_{BE} and V_{CEsat} to be zero and use the three-terminal short circuit shown in Fig. 9.35 to model a saturated transistor.

Finally it should be mentioned that a more elaborate large-signal model for the BJT will be studied in Chapter 15.

Fig. 9.35 An approximate model for the saturated BJT.

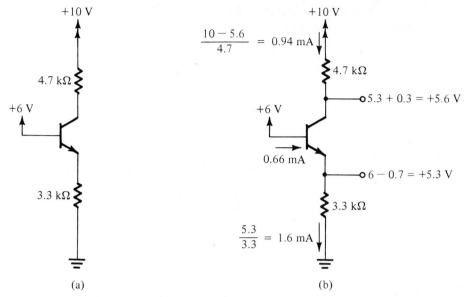

Fig. 9.36   Circuit for Example 9.10.

Example 9.10
We wish to analyze the circuit in Fig. 9.36a to determine the voltages at all nodes and the currents in all branches. Assume the transistor β is specified to be *at least* 50.

Solution
We have already considered this circuit in Example 9.2 and discovered that the transistor has to be in saturation. Assuming this to be the case, we have

$$V_E = +6 - 0.7 = +5.3 \text{ V}$$
$$V_C = V_E + V_{CEsat} \simeq +5.3 + 0.3 = +5.6 \text{ V}$$
$$I_E = \frac{V_E}{3.3} = \frac{5.3}{3.3} = 1.6 \text{ mA}$$
$$I_C = \frac{+10 - 5.6}{4.7} = 0.94 \text{ mA}$$
$$I_B = I_E - I_C = 1.6 - 0.94 = 0.66 \text{ mA}$$

Thus the transistor is operating at a forced β of

$$\beta_{\text{forced}} = \frac{I_C}{I_B} = \frac{0.94}{0.66} = 1.4$$

Since β_{forced} is less than the *minimum* specified value of β, the transistor is indeed saturated. We should emphasize here that in testing for saturation the minimum value of β should be used. By the same token, if we are designing a circuit in which a transistor is to be saturated, the design should be based on the minimum specified β. Obviously, if a transistor with this minimum β is saturated, then transistors with higher values of β will also be saturated. The details of the analysis are shown in Fig. 9.36b.

· · ·

Example 9.11

The transistor in Fig. 9.37 is specified to have β in the range 50 to 150. Find the value of R_B that results in saturation with an overdrive factor of at least 10.

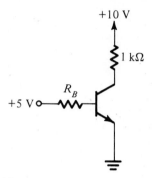

Fig. 9.37   *Circuit for Example 9.11.*

Solution

When the transistor is saturated the collector voltage will be

$$V_C = V_{CEsat} \simeq 0.3 \text{ V}$$

Thus the collector current is given by

$$I_{CS} = \frac{+10 - 0.3}{1} = 9.7 \text{ mA}$$

To saturate the transistor with the lowest β we need to provide a base current of at least

$$I_{BS} = \frac{I_{CS}}{\beta_{min}} = \frac{9.7}{50} = 0.194 \text{ mA}$$

For an overdrive factor of 10, base current should be

$$I_B = 10 \times 0.194 = 1.94 \text{ mA}$$

Thus we require a value of R_B such that

$$\frac{+5 - 0.7}{R_B} = 1.94$$

$$R_B = \frac{4.3}{1.94} = 2.2 \text{ k}\Omega$$

· · ·

Example 9.12

We want to analyze the circuit of Fig. 9.38 to determine the voltages at all nodes and the currents through all branches. The minimum value of β is specified to be 30.

Solution

A quick glance at this circuit reveals that the transistor will be either active or saturated. Assuming active-mode operation and neglecting the base current, we see that the

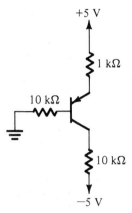

Fig. 9.38   Circuit for Example 9.12.

base voltage will be approximately zero volts, the emitter voltage will be approximately $+0.7$ V, and the emitter current will be approximately 4.3 mA. Since the maximum current that the collector can support while the transistor remains in the active mode is approximately 0.5 mA, it follows that the transistor is saturated for sure.

Assuming that the transistor is saturated and denoting the voltage at the base by V_B, it follows that

$$V_E = V_B + V_{EB} \simeq V_B + 0.7$$

$$V_C = V_E - V_{ECsat} \simeq V_B + 0.7 - 0.3 = V_B + 0.4$$

$$I_E = \frac{+5 - V_E}{1} = \frac{5 - V_B - 0.7}{1} = 4.3 - V_B \quad \text{mA}$$

$$I_B = \frac{V_B}{10} = 0.1 V_B \quad \text{mA}$$

$$I_C = \frac{V_C - (-5)}{10} = \frac{V_B + 0.4 + 5}{10} = 0.1 V_B + 0.54 \quad \text{mA}$$

Using the relationship

$$I_E = I_B + I_C$$

we obtain

$$4.3 - V_B = 0.1 V_B + 0.1 V_B + 0.54$$

which results in

$$V_B = \frac{3.76}{1.2} \simeq 3.13 \text{ V}$$

Substituting in the equations above, we obtain

$$V_E = 3.83 \text{ V}$$
$$V_C = 3.53 \text{ V}$$
$$I_E = 1.17 \text{ mA}$$
$$I_C = 0.853 \text{ mA}$$
$$I_B = 0.313 \text{ mA}$$

(note that I_E does not exactly equal $I_B + I_C$ because the value of V_B is approximate). It is clear that the transistor is saturated, since the value of forced β is

$$\beta_{\text{forced}} = \frac{0.853}{0.313} \simeq 2.7$$

which is much smaller than the specified minimum β.

$\cdot$ $\cdot$ $\cdot$

Example 9.13
We desire to evaluate the voltages at all nodes and the currents through all branches in the circuit of Fig. 9.39a. Assume $\beta = 100$.

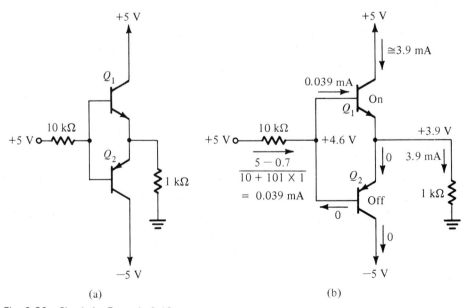

(a)

(b)

Fig. 9.39 *Circuit for Example 9.13.*

Solution

By examining the circuit we conclude that the two transistors Q_1 and Q_2 cannot be simultaneously conducting. Thus if Q_1 is on, Q_2 will be off, and vice versa. Assume that Q_2 is on. It follows that current will flow from ground through the 1-kΩ load resistor into the emitter of Q_2. Thus the base of Q_2 will be at a negative voltage, and base current will be flowing out of the base through the 10-kΩ resistor and into the +5-V supply. This is impossible, since if the base is negative, current in the 10-kΩ resistor will have to flow into the base. Thus we conclude that our original assumption—that Q_2 is on—is incorrect. It follows that Q_2 will be off and Q_1 will be on.

The question now is whether Q_1 is active or saturated. The answer in this case is obvious. Since the base is fed with a +5-V supply and since base current flows into the base of Q_1, it follows that the base of Q_1 will be at a voltage lower than +5 V. Thus the collector–base junction of Q_1 is reverse-biased and Q_1 is in the active mode. It remains only to determine the currents and voltages using techniques already described in detail. The results are given in Fig. 9.39b.

$$\bullet \quad \bullet \quad \bullet$$

Transistor Switching Times

Because of their internal capacitive effects, transistors do not switch in zero time. Figure 9.40 illustrates this point by displaying the waveform of the collector current i_C of a simple transistor inverter together with the waveforms for the input voltage v_I and the base current i_B. As indicated, when the input voltage v_I rises from the negative (or zero) level V_1 to the positive level V_2 the collector current does not respond immediately. Rather, a finite delay time t_d elapses before any appreciable collector current begins to flow. This delay time is required mainly for the EBJ depletion capacitance[3] to charge up to the forward-bias voltage V_{BE} (approximately 0.7 V). After this charging process is completed the collector current begins an exponential rise toward a final value of βI_{B2}, where I_{B2} is the current pushed into the base,

$$I_{B2} = \frac{V_2 - V_{BE}}{R_B}$$

The time constant of the exponential rise is determined by the junction capacitances. In fact, it is during the interval of the rising edge of i_C that the excess minority carrier charge is being stored in the base region.

Although the exponential rise of i_C is heading toward βI_{B2}, this value will never be reached, since the transistor will saturate and the collector current will be limited to I_{CS}. A measure of the BJT switching speed is the *rise time t_r*, indicated in Fig. 9.40c. Another, more popular measure is the *turn-on time,* also indicated in Fig. 9.40c.

Figure 9.41a shows the profile of excess minority carrier charge stored in the base of a saturated transistor. Unlike the active mode case, the excess minority concentration is not zero at the edge of the CBJ. This is because the CBJ is now forward-biased. Of special interest here is the extra charge stored in the base and represented by the

---

[3] Since during t_d the current is zero, the diffusion capacitance will be zero.

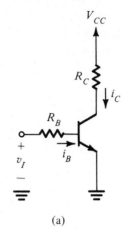

(a)

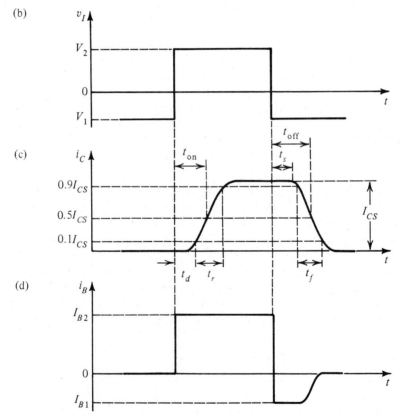

Fig. 9.40   Switching times of the BJT in the simple inverter circuit of (a) when the input v_I has the pulse waveform in (b).

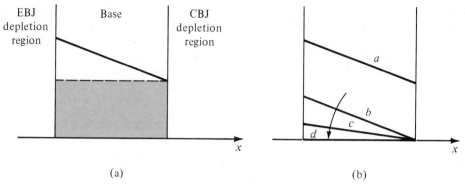

Fig. 9.41 *(a) Profile of excess minority carriers in the base of a saturated transistor. The shaded area represents the extra (or saturating) charge. (b) As the transistor is turned off the extra stored charge has to be removed first. During this interval the profile changes from line a to line b. Then the profile decreases toward zero (line d), and the collector current falls exponentially to zero.*

shaded area in Fig. 9.41a. Because this charge does not contribute to the slope of the concentration profile it does not result in a corresponding collector current component. Rather, this extra stored charge arises from the pushing of more current into the base than is required to saturate the transistor. The higher the overdrive factor used, the greater the amount of extra charge stored in the base. In fact the extra charge Q_x, called *saturating charge*, is proportional to the excess base drive $I_{B2} - I_{CS}/\beta$, that is

$$Q_x = \tau_s \left(I_{B2} - I_{CS}/\beta \right)$$

where τ_s is a transistor parameter known as the *storage time constant.*

Let us now consider the turn-off process. When the input voltage v_I returns to the low level V_1 the collector current does not respond and remains almost constant for a time t_s (Fig. 9.40c). This is the time required to remove the saturating charge from the base. During the time t_s, called the *storage time,* the profile of stored minority carriers will change from that of line *a* to that of line *b* in Fig. 9.41b. As indicated in Fig. 9.40d, the base current reverses direction because v_{BE} remains approximately 0.7 V while v_I is at the negative (or zero) level V_1. The reverse current I_{B1} helps to "discharge the base" and remove the extra stored charge; in the absence of the reverse base current I_{B1}, the saturating charge has to be removed entirely by recombination. It can be shown (see Ref. 9.5) that the storage time t_s is given by

$$t_s = \tau_s \frac{I_{B2} - I_{CS}/\beta}{I_{B1} + I_{CS}/\beta} \tag{9.40}$$

Once the extra stored charge has been removed, the collector current begins to fall exponentially with a time constant determined by the junction capacitances. During the fall time the slope of excess charge profile decreases toward zero, as indicated in Fig. 9.41b. Finally, note that the reversed base current eventually decreases to zero as the EBJ capacitance charges up to the reverse-bias voltage V_1.

Typically t_d, t_r, and t_f are of the order of few nanoseconds to a few tens of nanoseconds. The storage time t_s, however, is much larger and usually constitutes the limiting factor on the switching speed of the transistor. As mentioned before, t_s increases with the overdrive factor (that is, with how deep the transistor is driven into saturation). It follows that if one desires high-speed digital circuits, then operation in the saturation region should be avoided. This is the idea behind a nonsaturating form of digital logic circuits called emitter-coupled logic, to be discussed later.

EXERCISES

9.9 Solve the problem in Example 9.13 with the voltage feeding the bases changed to $+10$ V. Assume that $\beta_{\min} = 30$, and find V_E, V_B, I_{C1}, and I_{C2}.
Ans. $+4.7$ V; $+5.4$ V; 4.24 mA; 0

9.10 We wish to use the transistor equivalent circuit of Fig. 9.29 to find an expression for the delay time t_d of a simple transistor inverter fed by a step voltage source with a resistance R_B. Since during the delay time the transistor is not conducting, the resistance r_x is infinite and C_x will consist entirely of the depletion capacitance C_{je}. As an approximation C_{je} can be assumed to remain constant during t_d. Also, since the collector voltage does not change during t_d, the collector can be considered grounded. Assume that the two levels of v_I are V_1 and V_2 and that the end of t_d can be taken as the time at which $v_x = 0.7$ V.
Ans. $t_d = R_B (C_{je} + C_\mu) \ln [(V_2 - V_1)/(V_2 - 0.7)]$

9.9   THE BJT IN DIGITAL CIRCUITS

Having studied the operation of the BJT as a switch, we shall now illustrate its application in saturated logic and memory circuits. In this section examples of such circuits will be presented, with the detailed study of BJT digital circuits deferred to Chapter 15.

The Basic Logic Inverter

Figure 9.42a shows the basic logic inverter, which we studied in detail in Section 9.8. The transfer characteristic of the inverter is shown in Fig. 9.42b, with the three regions

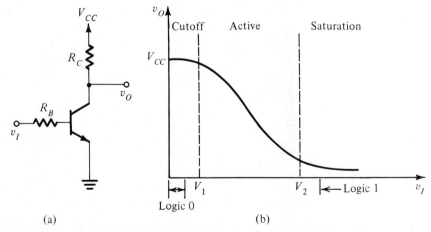

(a)   (b)

Fig. 9.42   *The basic transistor logic inverter and its transfer characteristic.*

of operation of the BJT indicated. If the circuit is to be used as a saturated logic inverter, the input levels, for a positive-logic system, should be as follows:

$$\text{Logic 0:} \qquad v_I < V_1$$
$$\text{Logic 1:} \qquad v_I > V_2$$

Note that for a logic 0 input the transistor will be off and the output will be equal to V_{CC}, a logic 1. On the other hand, a logic 1 at the input saturates the transistor and results in an output voltage V_{CEsat} of approximately 0.2 to 0.3 V, a logic 0.

In order to provide the inverter with some noise immunity, the allowable logic levels are usually limited to narrower regions, as indicated in Fig. 9.42b, thus leaving a margin between the maximum allowed value of the logic 0 and the switching threshold V_1. Similarly, a margin is left between the minimum allowed value of the logic 1 and the switching threshold V_2.

Resistor-Transistor Logic (RTL)

Figure 9.43 shows a two-input NOR gate of the RTL family. The circuit works as follows: If one of the inputs—say, X_1—is high (logic 1), then the corresponding transistor (Q_1) will be on and saturated. This will result in $v_Y \simeq 0.3$ V, which is low (logic

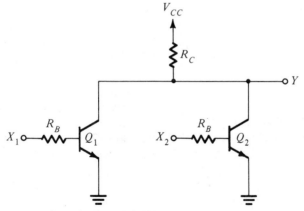

Fig. 9.43 A two-input NOR gate of the RTL family of logic circuits.

0). If the other input (X_2) is also high (logic 1), then the corresponding transistor (Q_2) will be on and saturated, thus keeping the output low (logic 0). It can be seen that for the output to be high ($v_Y = V_{CC}$) both Q_1 and Q_2 have to be off simultaneously. Clearly, this is obtained if both X_1 and X_2 are simultaneously low. That is, a logic 1 will appear at the output only in one case: X_1 and X_2 are low. We may therefore write the Boolean expression

$$Y = \overline{X}_1 \overline{X}_2$$

which can be also written

$$Y = \overline{X_1 + X_2}$$

which is a NOR function.

Diode-Transistor Logic (DTL)

The basic circuit for a NAND gate of the DTL family is shown in Fig. 9.44. The circuit operates as follows: Let input X_2 be left open. If a logic 0 signal ($\simeq$ 0 V) is applied to X_1, diode D_1 will conduct and the voltage at node A will be one diode drop (0.7 V) above the logic 0 value. The two diodes D_3 and D_4 will be conducting, thus causing the base of transistor Q to be two diode drops below the voltage at node A. Thus the base will be at a small negative voltage, and hence Q will be off and $V_Y = v_{CC}$ (logic 1).

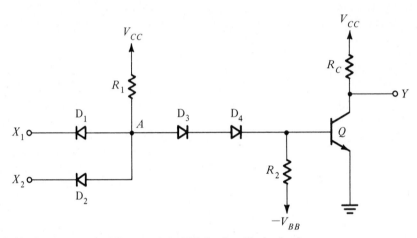

Fig. 9.44   *A two-input NAND gate of the DTL family of logic circuits.*

Consider now increasing the voltage v_{X1}. It can be seen that diode D_1 will remain conducting and node A will keep rising in potential. Diodes D_3 and D_4 will remain conducting, and hence the base will also rise in potential. This situation will continue until the voltage at the base reaches about 0.5 V, at which point the transistor will start to conduct. This will occur when the voltage at X_1 is

$$v_{X1} \simeq 0.5 + V_{D4} + V_{D3} - V_{D1} \simeq 1.2 \text{ V}$$

Small increases in v_{X1} above this threshold value will appear as increases in v_{BE} and hence in i_C. In this range the transistor will be in the active region. Eventually the voltage at the base will reach 0.7 V and the transistor will be fully conducting. At this point the voltage at A will be clamped to two diode drops above V_{BE}, and any further increases in v_{X1} will reverse-bias D_1. It can be seen that the current in D_1 will begin to decrease when v_{X1} reaches about 1.4 V. As D_1 stops conducting, all the current through R_1 will be diverted through D_3 and D_4 into the base of the transistor. The circuit is normally designed such that the current into the base will be sufficient to drive the transistor into saturation. Thus when X_1 is at logic 1 level the transistor will be saturated and v_Y will be equal to V_{CEsat} ($\simeq$ 0.3 V), which is a logic 0 output.

Extrapolating from the above discussion, it can be seen that if either or both of the inputs is low the corresponding diode (D_1, D_2, or both) will be conducting, the transistor will be off, and the output Y will be high. The output will be low if the transistor is on, which will happen for only one input combination, when all the inputs are simultaneously high. We may therefore write the Boolean expression

$$\overline{Y} = X_1 X_2$$

which can be rewritten

$$Y = \overline{X_1 X_2}$$

which is a NAND function. This should come as no surprise, since the DTL circuit consists of a diode AND gate formed by D_1, D_2, and R_1 (see Chapter 6) followed by a transistor inverter.

An SR Flip-Flop

The circuits considered thus far are all of the nonregenerative type—that is, they do not have memory (see Chapter 6). As an example of regenerative circuits we shall consider a set/reset flip-flop formed of two RTL gates. The concept of constructing a bistable circuit by cross-coupling two logic inverters was introduced in Chapter 6. Also, we learned that by cross-coupling two NOR or NAND gates the two "extra" inputs can be used for setting (storing a 1) and resetting (storing a 0).

Figure 9.45 shows an SR flip-flop formed by cross-coupling two NOR gates of the type shown in Fig. 9.43. We have arbitrarily labeled as Y the output variable represented by the voltage at the collector of Q_3, with the output on the other side labeled $\overline{Y}$. Thus the flip-flop is said to be storing a 1 [$Y = 1$] if it is in the state where Q_1 is on and saturated and Q_3 is off. The other state, that in which Q_1 is off and Q_3 is on and saturated, corresponds to $Y = 0$ and $\overline{Y} = 1$, which implies a 0 being stored.

In the rest or inactive state the reset (R) and set (S) input terminals are left open or grounded. To see how the flip-flop can be triggered to change state, consider the situation when the circuit is storing a 0 and we wish to set it (store a 1). This can be done by applying a positive pulse to terminal S. This pulse will turn Q_2 on, causing

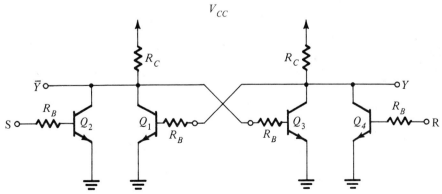

Fig. 9.45 An SR flip-flop formed by cross-coupling two NOR gates of the type shown in Fig. 9.43.

terminal $\overline{Y}$ to fall in voltage. This in turn will cause the current in Q_3 to decrease, and hence its collector (terminal Y) will rise in potential. The increase in voltage at Y will cause Q_1 to turn on, and hence its collector $\overline{Y}$ will fall in voltage. This positive feedback will continue until the circuit switches to the other state, in which Q_1 is on and saturated and Q_3 is off. Note that the triggering pulse at S serves only to start the positive-feed-back action. Thus the triggering pulse need be present only for a very short time.

Resetting the flip-flop is accomplished by applying a short positive pulse at R.

9.10 SECOND-ORDER EFFECTS AND COMPLETE STATIC CHARACTERISTICS

We conclude this chapter with a discussion of some of the second-order effects that occur in the BJT and their influence on the device static characteristics. Each of these second-order effects limits the performance of transistor circuits in some way. Examples of such limitations will appear in subsequent chapters. This section also presents the graphical analysis of transistor circuits.

Common-Base Characteristics

Figure 9.46 shows the complete set of i_C-v_{CB} characteristic curves for an *npn* transistor. As mentioned in Section 9.5, the i_C-v_{CB} characteristics are measured at constant values of emitter current i_E (see Fig. 9.12a). Since in such an arrangement the base is con-nected to a constant voltage, the i_C-v_{CB} curves are called the *common base characteristics*.

The curves in Fig. 9.46 differ from those presented in Fig. 9.12 in three aspects. First, the avalanche breakdown of the CBJ at large voltages is indicated and will be

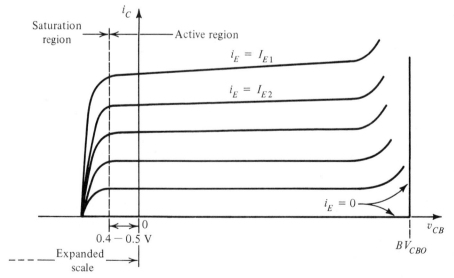

Fig. 9.46   The i_C-v_{CB} or common-base characteristics of an npn transistor. Note that in the active region there is a slight dependence of i_C on the value of v_{CB}. The result is a finite output resistance that decreases as the current level in the device is increased.

briefly explained at a later stage. Second, the saturation-region characteristics are included. As indicated, as v_{CB} goes negative the CBJ becomes forward-biased and the collector current decreases. Since for each curve i_E is held constant, the decrease in i_C results in an equal increase in i_B. The large effect that v_{CB} has on the collector current in saturation is evident from Fig. 9.46 and is consistent with our earlier description of the saturation mode of operation.

The third difference between the characteristic curves of Fig. 9.46 and the curves presented earlier is that in the active region the characteristic curves are shown to have a very small slope. That is, in a real transistor the value of the collector current in the active mode depends slightly on the value of collector-to-base voltage. This is a second-order effect that we have neglected in our first-order description of transistor operation. The reason for the dependence of i_C on v_{CB} is that as v_{CB} increases the reverse bias of the CBJ increases. This causes an increase in the width of the CBJ depletion region and a corresponding decrease in the *effective base width W*. As the base width decreases the number of electrons that disappear in the base by recombination decreases, causing a decrease in the base current and a corresponding, equal increase in the collector current (because the emitter current is held constant). This effect is known as *base-width modulation* or as the *Early effect* (in reference to its discoverer). We will not provide a detailed explanation of the Early effect in this book; however, it is important to point out that the i_C-v_{CB} characteristic curves in the active region can be approximated by straight lines whose slope increases in proportion to the value of the emitter current I_E.

Common-Emitter Characteristics

An alternative way of graphically displaying the transistor characteristics is shown in Fig. 9.47, where i_C is plotted versus v_{CE} for various values of the base current i_B. These

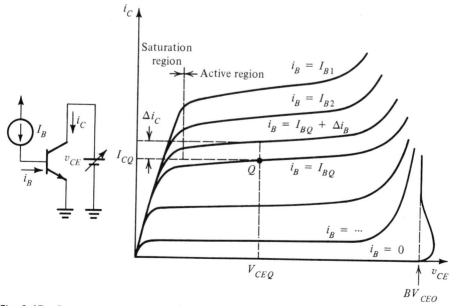

Fig. 9.47   *Common-emitter characteristics.*

characteristics can be obtained using the conceptual circuit given in Fig. 9.47, in which we note that the emitter is connected to a constant voltage. Thus these curves are called *common-emitter characteristics*. The different regions of operation of the BJT (including the breakdown region, which will be explained later) are indicated on the characteristics of Fig. 9.47.

The effect of the collector voltage on the value of the collector current in the active region is much more evident in the common-emitter characteristics than in the common-base characteristics. That is, the common-emitter curves display higher slopes, with the slope increasing with the current level in the device.

The slope of the characteristic curves in the active region is still much smaller than that exhibited in the saturation region. Thus while the transistor is in the active region the collector acts as a current source with a high but finite output resistance, but in saturation the transistor behaves as a "closed switch" with a small "closure resistance" R_{CEsat}. Since the characteristic curves are all "bunched together" in saturation, we show an expanded view of the saturation portion of the characteristics in Fig. 9.48. Note that

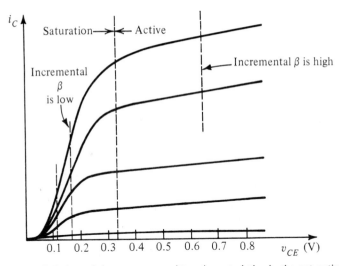

Fig. 9.48 *An expanded view of the common-emitter characteristics in the saturation region.*

the curves do not pass through the origin. In fact, for a given value of i_B the i_C-v_{CE} characteristic in saturation can be approximated by a straight line intersecting the v_{CE} axis at a point V_{CEoff}, as illustrated in Fig. 9.49. The voltage V_{CEoff} is called the *offset voltage* of the transistor switch. Field-effect transistors (Chapters 7 and 8) do not exhibit such offset voltages and hence make superior switches. FETs, however, display higher values of closure resistance.

Modeling the Effects of v_C on i_B and i_C

Figure 9.50 shows the hybrid-π model of Fig. 9.22a with two additional resistances r_μ and r_o. Resistance r_μ models the effect of a change in the collector voltage on the base

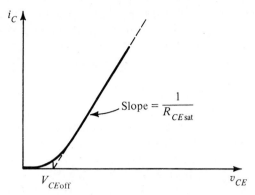

Fig. 9.49 *The i_c-v_{CE} characteristic in the saturation region. Note that the characteristic can be modeled by an offset voltage V_{CEO} and a small resistance R_{CEsat}.*

current. The effect of a change in the collector voltage on the collector current is modeled by r_o. Typically $r_\mu \gg r_o$, and both are (approximately) inversely proportional to the current level in the device (that is, I_C or I_E). We will present a more complete discussion of the hybrid-π model in Chapter 11.

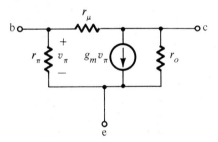

Fig. 9.50 *The hybrid-π model, including the two resistances r_o and r_μ which model the effects of v_c on i_c and i_b.*

The Transistor β

Earlier we defined β as the ratio of the total current in the collector to the total current in the base when the transistor is operating in the active mode. Let us be more specific. Assume that the transistor is operating at a base current I_{BQ}, a collector current I_{CQ}, and a collector-to-emitter voltage V_{CEQ}. These quantities define the operating or bias point Q in Fig. 9.47. The ratio of I_{CQ} to I_{BQ} is called the dc β or h_{FE} (the reason for the latter name will be explained in Chapter 11),

$$h_{FE} = \beta_{dc} = \frac{I_{CQ}}{I_{BQ}}$$

When the transistor is used as an amplifier it is first biased at a point such as Q. Applied signals then cause incremental changes in i_B, i_C, and v_{CE} around the bias point. We may therefore define an incremental or ac β as follows: Let the collector-to-emitter voltage be maintained constant at V_{CEQ} (in order to eliminate the Early effect), and change the base current by an increment Δi_B. If the collector current changes by an increment Δi_C (see Fig. 9.47), then β_{ac} (or h_{fe}, as it is usually called) at the operating point Q is defined as

$$h_{fe} = \beta_{ac} = \left.\frac{\Delta i_C}{\Delta i_B}\right|_{V_{CE} = \text{constant}}$$

The fact that V_{CE} is held constant implies that the incremental voltage v_{ce} is zero; therefore h_{fe} is called the *short-circuit current gain*.

When we perform small-signal analysis, the β should be the ac β (h_{fe}). On the other hand, if we are analyzing or designing a switching circuit, β_{dc} (h_{FE}) is the appropriate β. The difference in value between β_{dc} and β_{ac} is usually small, and we will not

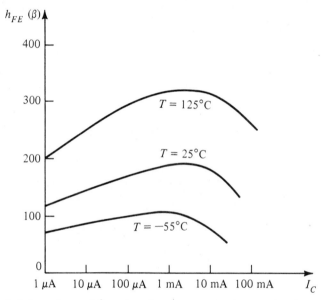

Fig. 9.51 *Typical dependence of β on i_C and on temperature in a modern integrated-circuit npn silicon transistor.*

normally distinguish between the two. A point worth mentioning, however, is that the value of β depends on the current level in the device, and the relationship takes the form shown in Fig. 9.51.

Transistor Breakdown

The maximum voltages that can be applied to a BJT are limited by the EBJ and CBJ breakdown effects that follow the avalanche multiplication mechanism described in Section 4.13. Consider first the common-base configuration. The i_C-v_{CB} characteristics in Fig. 9.46 indicate that for $i_E = 0$ (that is, with the emitter open-circuited) the collector–

base junction breaks down at a voltage denoted by BV_{CBO}. For $i_E > 0$ breakdown occurs at voltages smaller than BV_{CBO}. Typically BV_{CBO} is greater than 50 V.

Next consider the common-emitter characteristics of Fig. 9.47, which shows break-down occurring at a voltage BV_{CEO}. Here although breakdown is still of the avalanche type, the effects on the characteristics are more complex than that in the common-base configuration. We will not explain these details; it is sufficient to point out that typically BV_{CEO} is about half BV_{CBO}. On the transistor data sheets BV_{CEO} is sometimes referred to as the *sustaining voltage* LV_{CEO}.

Breakdown of the CBJ in either the common-base or common-emitter configura-tion is not destructive as long as the power dissipation in the device is kept within safe limits. This, however, is not the case with the breakdown of the emitter–base junction. The EBJ breaks down in an avalanche manner at a voltage BV_{EBO} much smaller than BV_{CBO}. Typically BV_{EBO} is in the range 6 to 8 V, and the breakdown is destructive in the sense that the β of the transistor is reduced. This does not prevent use of the EBJ as a zener diode to generate reference voltages in IC design. In such applications, how-ever, one is not concerned with the β-degradation effect. A circuit arrangement to pre-vent EBJ breakdown in IC amplifiers will be discussed in Chapter 13.

Graphical Analysis of Transistor Circuits

Although it is of little practical value in the analysis and design of most transistor cir-cuits, it is illustrative to portray with graphical techniques the operation of a simple transistor circuit. Consider the simple circuit shown in Fig. 9.52, which we have already analyzed in a number of examples. A graphical analysis of the circuit may be performed as follows: First, we have to determine the base current I_{BQ} using the technique illus-trated in Fig. 9.53 (we have employed this technique in the analysis of diode circuits in Chapter 4). We next move to the i_C-v_{CE} characteristics, shown in Fig. 9.54. We know that the operating point will lie on the curve corresponding to the value of base current

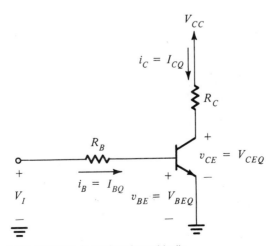

Fig. 9.52   Circuit whose operation is analyzed graphically.

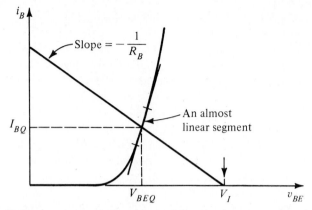

Fig. 9.53 *Graphical construction for the determination of the base current in the circuit of Fig. 9.52.*

(I_{BQ}) which we have just determined. Where it lies on the curve will be determined by the collector circuit. Specifically, the collector circuit imposes the constraint

$$v_{CE} = V_{CC} - i_C R_C$$

which can be rewritten

$$i_C = \frac{V_{CC}}{R_C} - \frac{1}{R_C} v_{CE}$$

and which represents a linear relationship between v_{CE} and i_C. This relationship can be represented by a straight line, as shown in Fig. 9.54. The operating point Q will be at the intersection of this straight line, called the load line, and the characteristic curve corresponding to the base current I_{BQ}.

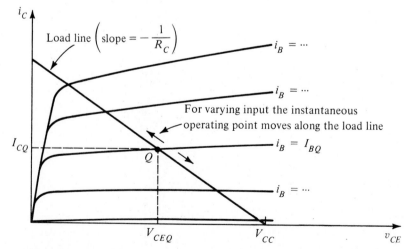

Fig. 9.54 *Graphical construction for determining the collector current and the collector-to-emitter voltage in the circuit of Fig. 9.52.*

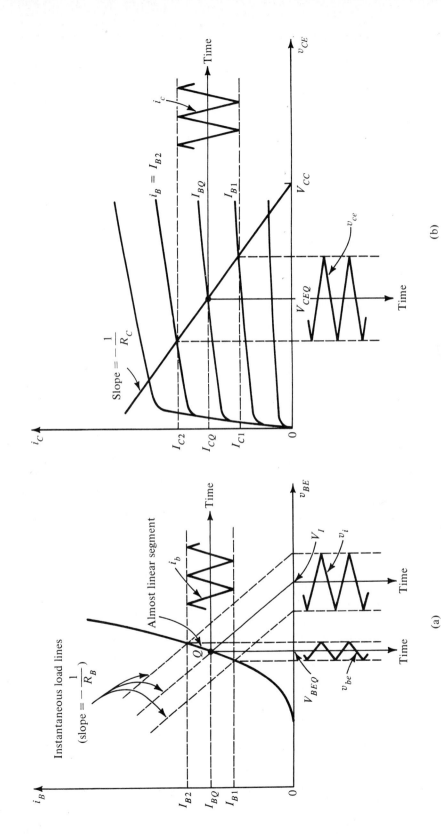

Fig. 9.55 Graphical determination of the signal components v_{be}, i_b, i_c, and v_{ce} when a signal component v_i is superimposed on the dc voltage V_I (see Fig. 9.52).

If the load line intersects the characteristic curve corresponding to I_{BQ} at a point in the active region, the transistor will be operating in the active mode. This is the case illustrated in Fig. 9.54. If an input signal v_i is superimposed on V_I, there will be a corresponding base current signal i_b and a base–emitter voltage signal v_{be}. As shown in Fig. 9.55a, if v_i is "small enough," the instantaneous operating point will move along a linear segment of the i_B-v_{BE} exponential curve. On the i_C-v_{CE} characteristics the instantaneous operating point will move along the load line, as illustrated in Fig. 9.55b.

Finally, Fig. 9.56 shows a load line that intersects the i_C-v_{CE} characteristics at a point in the saturation region. Note that in this case changes in the base current result in very small changes in i_C and v_{CE} and that in saturation the incremental β (β_{ac}) is negligibly small.

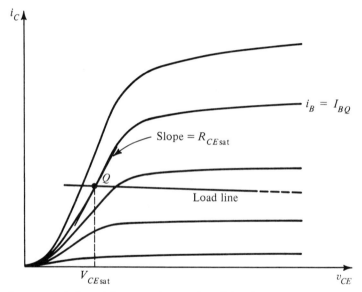

Fig. 9.56 *An expanded view of the saturation portion of the characteristics together with a load line that results in operation at a point Q in the saturation region.*

9.11 CONCLUDING REMARKS

The objective of this chapter has been to introduce the BJT, explain its terminal i-v characteristics, and illustrate its application as an amplifier and as a switch in digital circuit design. It is most important that at this stage the reader be thoroughly familiar with the BJT. This intimate familiarity can be attained through analysis of many transistor circuits of the types included in the examples, the exercises, and the problems. Once more we should reiterate the importance of being able to perform rapid approximate circuit analysis.

At this point we have completed our study of the basic electronic devices (diodes, FETs, and BJTs) and are ready to consider more elaborate circuit applications. Such applications will begin in the next chapter with linear amplifier circuits.

TRANSISTOR AMPLIFIERS 10

Introduction

Having studied the three basic semiconductor amplifying devices—the JFET, the MOSFET, and the BJT—we are now ready to consider their application in the design of amplifier circuits. This chapter is the first of four devoted to this topic.

We shall begin by studying the classical techniques employed in biasing the BJT in discrete (as opposed to integrated) circuit design. Following this, the analysis and design of classical single-stage BJT amplifiers are presented. This material builds on the knowledge acquired in Chapter 9 and parallels the development of single-stage FET amplifiers in Chapters 7 and 8. Next, a very useful circuit configuration, the emitter follower, is presented and discussed in detail. The emitter follower is similar to the source follower studied in Chapter 7 and is widely employed in both discrete and integrated circuit design.

The techniques employed for biasing the BJT in IC design are discussed in Section 10.4. This leads naturally to the study of the most important analog circuit configuration, *the differential pair*. Sections 10.5 and 10.6 are concerned with the BJT differential pair. The JFET differential pair, studied in Section 10.7, is employed in the design of the input stage of high-input-impedance op amps. The chapter concludes with a brief introduction to the design of practical multistage amplifiers. The analysis of such circuits is illustrated by a detailed example.

Because the BJT is the dominant device in analog integrated circuit design, most of the circuits studied in this chapter use BJTs. Finally, it should be pointed out that all the analysis of this chapter is performed at midband frequencies (as well as the low-frequency band in the case of dc amplifiers). Frequency-response analysis is the subject of Chapter 11.

10.1 BIASING THE BJT FOR DISCRETE CIRCUIT DESIGN

The Biasing Problem

Since—as demonstrated in Chapter 9—the performance of the BJT amplifier depends directly on the value of the dc current in its emitter, we shall begin by considering the problem of biasing BJTs. The biasing problem is that of establishing a constant dc current in the emitter of the BJT. This current has to be calculable, predictable, and insensitive to variations in temperature and to the large variations in the value of β encountered among transistors of the same type. In this section we shall deal with the classical approach to solving the bias problem in transistor circuits designed from discrete devices. Bias methods for integrated-circuit design are presented in Section 10.4.

Bias Arrangement

Figure 10.1a shows the arrangement most commonly used for biasing a transistor amplifier if only a single power supply is available. The technique, which is identical to that used with enhancement MOSFETs, consists of supplying the base of the transistor

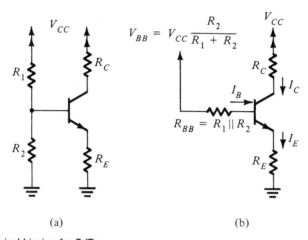

(a)   (b)

Fig. 10.1   Classical biasing for BJTs.

with a fraction of the supply voltage V_{CC} through the voltage divider R_1, R_2. In addition, a resistor R_E is connected to the emitter.

Figure 10.1b shows the same circuit with the voltage-divider network replaced by its Thévenin equivalent,

$$V_{BB} = \frac{R_2}{R_1 + R_2} V_{CC} \tag{10.1}$$

$$R_B = \frac{R_1 R_2}{R_1 + R_2} \tag{10.2}$$

The current I_E can be determined by writing a Kirchhoff loop equation for the base-emitter-ground loop and substituting $I_B = I_E/(\beta + 1)$:

$$I_E = \frac{V_{BB} - V_{BE}}{R_E + R_B/(\beta + 1)} \tag{10.3}$$

To make I_E insensitive to temperature and β variations, we design the circuit to satisfy the following two constraints:

$$V_{BB} \gg V_{BE} \tag{10.4}$$

$$R_E \gg \frac{R_B}{\beta + 1} \tag{10.5}$$

Condition (10.4) ensures that small variations in V_{BE} (around 0.7 V) will be swamped by the much larger V_{BB}. Typically one designs for V_{BB} from 3 to 5 V; a rule of thumb states that V_{BB} should be chosen approximately equal to $V_{CC}/3$.

Condition (10.5) makes I_E insensitive to variations in β and could be satisfied by selecting R_B small. This in turn is achieved by using low values for R_1 and R_2. Lower values for R_1 and R_2, however, will mean a higher current drain from the power supply and normally will result in a lowering of the input resistance of the amplifier, which is the trade-off involved in this design problem. It should be noted that Condition (10.5) means that we want to make the base voltage independent of the value of β and determined solely by the voltage divider. This will obviously be satisfied if the current in the divider is made much larger than the base current. Typically one selects R_1 and R_2 such that their current is in the range of I_E to $0.1I_E$.

The Common-Emitter Amplifier

Figure 10.2 shows the complete circuit of a classical single-stage transistor amplifier employing the bias arrangement just described. This circuit is quite similar to the classical JFET and MOSFET amplifiers studied in Chapters 7 and 8. As indicated in Fig. 10.2, the signal source v_s has a resistance R_s and is coupled to the base of the transistor through capacitor C_{C1}. The value of C_{C1} should be chosen sufficiently large so that it acts almost as a short circuit over the frequency band of interest. The output signal at the collector is coupled to a load resistance R_L through another coupling capacitor C_{C2} that as C_{C1} should be chosen sufficiently large. Finally, the effect of the emitter bias resistance R_E on the signal performance of the amplifier is eliminated by connecting a large capacitor C_E across R_E. This capacitance acts as a short circuit at signal frequen-

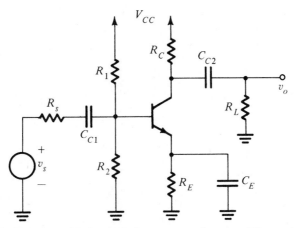

Fig. 10.2   Circuit for a capacitively coupled common-emitter amplifier employing the biasing arrangement of Fig. 10.1.

cies of interest such that R_E is effectively eliminated as far as signals are concerned. An alternative way of visualizing the action of C_E is to note that it presents a low-impedance path between the emitter and ground. Thus while the dc emitter current will continue to flow through R_E, the signal current i_e will flow through C_E, *bypassing* R_E. For this reason C_E is called an *emitter bypass capacitor* and the circuit is called a *grounded-emitter* or *common-emitter amplifier*. (Recall that we have used bypass capacitors in the design of FET amplifiers.) The effect of leaving part of R_E unbypassed will be discussed in Section 10.2.

At this point it is important to note that the amplifier circuit in Fig. 10.2 is an ac amplifier whose gain deteriorates at low frequencies. This is due, of course, to the finite values of C_{C1}, C_{C2}, and C_E. The analysis of amplifier frequency response will be considered in Chapter 11; for the time being we shall assume that these capacitances are of infinite value. Finally, note that the design of ac amplifiers is easier than that of dc amplifiers because in the former case we may separate the problem of designing the dc bias circuit from that of small-signal design and analysis.

Example 10.1

We wish to design the bias network of the amplifier in Fig. 10.2 to establish a current $I_E = 1$ mA using a power supply $V_{CC} = +12$ V.

Solution

As a rule of thumb one should allocate one-third of the supply voltage to the voltage drop across R_2 and another third to the voltage drop across R_C, leaving one-third for possible signal swing at the collector. Following this rule we obtain

$$V_B = +4 \text{ V}$$
$$V_E = 4 - V_{BE} \simeq 3.3 \text{ V}$$

Thus R_E should be selected to be

$$R_E = \frac{V_E}{I_E} = 3.3 \text{ k}\Omega$$

From the discussion above we select a voltage-divider current of $0.1I_E$. Neglecting the base current, we find

$$R_1 + R_2 = \frac{12}{0.1I_E} = 120 \text{ k}\Omega$$

$$\frac{R_2}{R_1 + R_2} V_{CC} = 4 \text{ V}$$

Thus $R_2 = 40$ kΩ and $R_1 = 80$ kΩ.

At this point it is desirable to find a more accurate estimate for I_E, taking into account the finite base current. Using Eq. (10.3) (assuming that β is specified to be 100) we obtain

$$I_E = \frac{3.3}{3.3 + 0.267} = 0.93 \text{ mA}$$

We could, of course, have obtained a value much closer to the desired 1 mA by designing with exact equations. However, since our work is based on first-order models, it does not make sense to strive for accuracy better than 5% or 10%.

It should be noted that if we are willing to draw a higher current from the power supply and if we are prepared to accept a lower input resistance for the amplifier, then we may use a voltage-divider current equal, say, to I_E, resulting in $R_1 = 8$ kΩ and $R_2 = 4$ kΩ. The effect of this on the amplifier input resistance will be analyzed in Section 10.2. We shall refer to the circuit using these latter values as design 2, for which the actual value of I_E will be

$$I_E = \frac{3.3}{3.3 + 0.026} \simeq 1 \text{ mA}$$

The value of R_C can be determined from

$$R_C = \frac{12 - V_C}{I_C}$$

Thus for design 1 we have

$$R_C = \frac{12 - 8}{0.99 \times 0.93} = 4.34 \text{ k}\Omega$$

whereas for design 2 we have

$$R_C = \frac{12 - 8}{0.99 \times 1} = 4.04 \text{ k}\Omega$$

For simplicity we shall select $R_C = 4$ kΩ for both designs.

$\bullet \quad \bullet \quad \bullet$

EXERCISE

10.1   For design 1 in Example 10.1 calculate the expected range of I_E if the transistor used has β in the range 50 to 150. Repeat for design 2.
  Ans.   For design 1, 0.86 to 0.95 mA; for design 2, 0.98 to 0.995 mA

10.2   CLASSICAL SINGLE-STAGE AMPLIFIERS

In this section the amplifier circuit of Fig. 10.3 will be analyzed and the various trade-offs involved in its design will be pointed out. This circuit is a more general version of the common-emitter amplifier of Fig. 10.2 with part of the emitter bias resistance R_E being left unbypassed.

Input Resistance

In order to find the fraction of the input signal that appears at the transistor base, v_b, we first need to evaluate the input resistance R_{in} (see Fig. 10.3). Specifically, the transmission from the source to the transistor base is given by

$$\frac{v_b}{v_s} = \frac{R_{\text{in}}}{R_s + R_{\text{in}}} \tag{10.6}$$

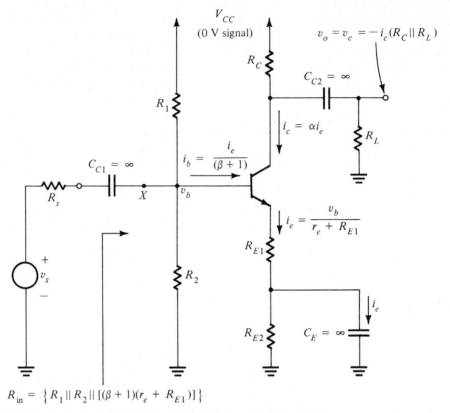

$$R_{in} = \left\{ R_1 \| R_2 \| [(\beta + 1)(r_e + R_{E1})] \right\}$$

Fig. 10.3   Small-signal analysis of a classical single-stage BJT amplifier.

To evaluate R_{in} imagine yourself standing at point X (Fig. 10.3) and looking to the right. Between X and ground you will see R_2 in parallel with R_1 (note: a power supply is considered a signal ground) in parallel with the input resistance looking into the base of the transistor. The latter resistance is defined as

$$R_b \equiv \frac{v_b}{i_b}$$

But we have

$$v_b = i_e r_e + i_e R_{E1}$$

and

$$i_b = \frac{i_e}{\beta + 1}$$

leading to

$$R_b = (\beta + 1)(r_e + R_{E1})$$

This is an important result: it says that *the input resistance looking into the base of a transistor is equal to the total resistance in its emitter multiplied by the factor $\beta + 1$.* This factor obviously arises because the base current is $\beta + 1$ times smaller than the emitter current. [Are you wondering where r_π has gone? It is still there since $(\beta + 1)r_e = r_\pi$.]

In summary, the input resistance is given by

$$R_{\text{in}} = \{R_1 \parallel R_2 \parallel [(\beta + 1)(r_e + R_{E1})]\} \tag{10.7}$$

From this formula it is obvious that selection of smaller values for the bias resistances R_1 and R_2, as was done in design 2 in Example 10.1, results in a lowering of the input resistance R_{in} and hence a smaller transmission ratio from source to base. Thus to avoid excessive loss of gain in coupling a signal to the amplifier we wish to keep $R_{\text{in}} \gg R_s$ and therefore choose higher values for R_1 and R_2.

Another important observation we can make from Eq. (10.7) relates to the effect of the unbypassed emitter resistance R_{E1} in increasing the input resistance. This attribute of emitter resistance is exploited in the design of circuits with high input resistance, as will be illustrated later on.

EXERCISES

10.2 Calculate the input resistance of the transistor amplifier in Fig. 10.3 for both design 1 and design 2 (see Example 10.1) for (a) $R_{E1} = 0$ and (b) $R_{E1} = 425\ \Omega$. Assume $\beta = 100$.

    Ans.   Design 1: (a) 2.5 kΩ; (b) 16.8 kΩ; Design 2: (a) 1.3 kΩ; (b) 2.5 kΩ

10.3 If $R_s = 4$ kΩ calculate the magnitude of transmission v_b/v_s for both design 1 and design 2 in the two cases (a) and (b) considered in Exercise 10.2.

    Ans.   Design 1: (a) 0.38; (b) 0.81; Design 2: (a) 0.25; (b) 0.38

Voltage Gain

Figure 10.3 illustrates some of the details of calculating the voltage gain v_o/v_s. Starting from the signal at the transistor base v_b we find the signal current in the emitter i_e as

$$i_e = \frac{v_b}{r_e + R_{E1}}$$

Thus the collector signal current i_c will be

$$i_c = \alpha i_e$$
$$= \frac{\alpha v_b}{r_e + R_{E1}}$$

To obtain the output voltage v_o we multiply i_c by the total resistance between collector and ground, which is seen to be $(R_C \parallel R_L)$; thus

$$v_o = v_c = -\alpha i_e (R_C \parallel R_L)$$
$$= \frac{-\alpha v_b}{r_e + R_{E1}} (R_C \parallel R_L)$$

Since $\alpha \simeq 1$, the voltage gain between base and collector is given by

$$\frac{v_o}{v_b} \simeq -\frac{(R_C \parallel R_L)}{r_e + R_{E1}}$$

which is simply *the ratio of the total resistance in the collector lead to the total resistance in the emitter lead*. This is a handy, easy-to-remember rule for calculating gain. The overall voltage gain v_o/v_s can be obtained by multiplying v_o/v_b by the transmission from the source to the base, v_b/v_s,

$$\text{Gain} \equiv \frac{v_o}{v_s} = \frac{v_b}{v_s} \frac{v_o}{v_b}$$

EXERCISE
10.4 If $R_L = 4$ kΩ, calculate the overall voltage gain for design 1 and design 2 for the two cases (a) and (b) of Exercise 10.2. Use the results of Exercise 10.3.
Ans. Design 1: (a) -28.0 V/V; (b) -3.6 V/V; Design 2: (a) -19.8 V/V; (b) -1.7 V/V

Comments

From the results of Exercise 10.4 we observe that the gains obtained with design 2 are considerably smaller than these obtained with design 1. Thus the price paid for the more stable operating point of design 2 is a reduction in gain (in addition to the increased current drain on the supply). The reduction in gain obtained as a result of including an unbypassed emitter resistance is also observed. Not apparent from these results, however, is the fact that including the emitter resistance R_{E1} enables us to apply a larger input signal to the amplifier without running the risk of nonlinear distortion because the signal that appears between base and emitter (v_{be} or v_π), which has to be limited to 10 mV or so, is only a fraction of the signal at the base,

$$v_{be} = v_b \frac{r_e}{r_e + R_{E1}}$$

The use of R_{E1} to control the amplifier input resistance has been already mentioned. This control is significant because there are many applications where it is important to avoid loss of gain in coupling the signal to the amplifier.

EXERCISES
10.5 If the amplitude of v_{be} is to be limited to 10 mV, find the maximum allowed amplitude of v_s for design 1 in both cases (a) and (b) considered in Exercise 10.4.
Ans. (a) 26.3 mV; (b) 208 mV
10.6 Find the output signal amplitude corresponding to the two cases considered in Exercise 10.5.
Ans. (a) 0.74 V; (b) 0.74 V

A Possible Design Optimization

The numerical results obtained in Exercise 10.6 indicate that our design is far from being optimum. Having allowed 4 V for the collector signal swing, we are forced by considerations of nonlinear distortion to operate with only a fraction of this (0.74 V). Obviously we can increase R_C and thus obtain higher gain and larger output signal swing. The maximum value that we may use for R_C is that which causes the transistor to be at the verge of leaving the active mode of operation.

The Common-Base Configuration

Figure 10.4 shows a *pnp* transistor used in the *common-base* amplifier configuration. The biasing arrangement is identical to that employed in the common-emitter configuration, and the base is shorted to ground through a large capacitance (shown as infinite

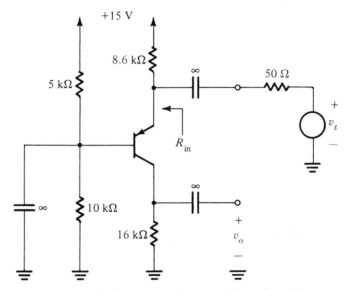

Fig. 10.4 A pnp transistor used in the common-base amplifier configuration.

in value). There is no advantage at low and medium frequencies of the common-base configuration over the common-emitter configuration; in fact, there is a disadvantage— the input resistance is lower. The major advantage of the common-base configuration lies in its superior high-frequency performance, which is a result of the absence of the Miller effect which severely limits the performance of common-emitter amplifiers. This point will be considered in detail in Chapter 11.

EXERCISE

10.7 Consider the common-base amplifier of Fig. 10.4. Calculate the dc bias current, the input resistance, the voltage gain, and the maximum allowable amplitude of v_s. Assume $\beta \gg 1$.
Ans. 0.5 mA; 50 Ω; 160 V/V; 12.5 mV

10.3   THE EMITTER FOLLOWER

In this section we shall study an extremely useful single-transistor amplifier circuit: the *emitter follower,* or the *common-collector configuration.* Like the source follower, studied in Section 7.10, the emitter follower is characterized by a high input resistance and a low output resistance. It therefore is useful as an isolation or buffer amplifier to connect a high-resistance source to a low-resistance load.

Figure 10.5 shows a capacitively coupled emitter follower utilizing a biasing arrangement similar to that discussed in Section 10.1. The emitter follower can be used as a direct-coupled amplifier stage, as will be illustrated later. The results derived in this section apply also to the direct-coupled emitter follower.

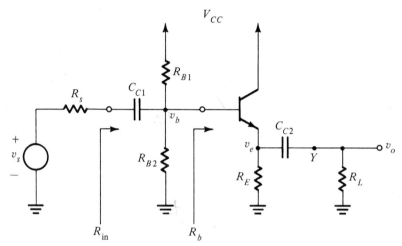

Fig. 10.5   *A capacitively coupled emitter follower.*

As shown in Fig. 10.5, the input signal is applied to the base and the output signal is taken from the emitter. Since the collector is connected to the dc supply, it will have zero signal voltage and thus act as a signal ground, giving rise to the name *grounded-collector* or *common-collector configuration.* The name *emitter follower* arises because the voltage at the emitter follows that at the input, as will become apparent shortly.

DC Analysis

Analysis of the emitter-follower circuit to determine dc bias quantities is straightforward and follows the techniques previously explained. It should be pointed out, however, that in designing an emitter follower one normally chooses large values for the bias resistances R_{B1} and R_{B2} in order to keep the input resistance high and in spite of the resulting increased dependence of I_E on the value of β. It will be shown that this dependence is not critical in emitter-follower design because the voltage gain is not highly dependent on the value of I_E.

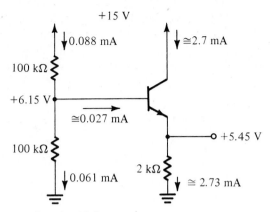

Fig. E10.8   Answer to Exercise 10.8.

10.8   Find all dc voltages and currents in the circuit of Fig. 10.5 for $V_{CC} = +15$ V, $R_{B1} = R_{B2} = 100$ kΩ, $R_E = 2$ kΩ, and $\beta = 100$.
Ans.   See Fig. E10.8

Voltage Gain

We now wish to find the voltage gain of the emitter follower driven by a source with resistance R_s and connected to a load R_L, as shown in Fig. 10.5. To obtain the input resistance R_b between base and ground, we use the resistance reflection rule formulated in Section 10.1—that is, the total resistance in the emitter lead is multiplied by $\beta + 1$ to obtain R_b:

$$R_b = (\beta + 1)[r_e + (R_E \parallel R_L)]$$

Using this value for R_b we can obtain R_{in} from

$$R_{in} = (R_{B1} \parallel R_{B2} \parallel R_b) \qquad (10.8)$$

Having determined R_{in}, we can now find the transmission between the source and the base as

$$\frac{v_b}{v_s} = \frac{R_{in}}{R_{in} + R_s} \qquad (10.9)$$

The output voltage v_o is equal to the emitter voltage v_e and can be determined using the voltage-divider rule relating v_e and v_b:

$$\frac{v_e}{v_b} = \frac{(R_E \parallel R_L)}{(R_E \parallel R_L) + r_e} \qquad (10.10)$$

Combining Eqs. (10.9) and (10.10) and replacing v_e by v_o gives the gain expression

$$\frac{v_o}{v_s} = \frac{R_{in}}{R_{in} + R_s} \frac{(R_E \parallel R_L)}{(R_E \parallel R_L) + r_e} \qquad (10.11)$$

where R_{in} is given by Eq. (10.8). We observe that the voltage gain is less than unity. However, since R_{in} is usually large (because of the β multiplication effect) and since r_e is usually quite small, the gain is usually close to unity. The advantage of the emitter follower, of course, lies in its high input resistance, which enables us to couple a high-resistance source to a low-resistance load without loss of signal strength.

EXERCISE

10.9 Consider the emitter follower whose dc analysis was carried out in Exercise 10.8. If $R_s = 5$ kΩ and $R_L = 1$ kΩ, find the input resistance R_{in} and the voltage gain.
Ans. 28.9 kΩ; 0.84 V/V

Use of the Equivalent Circuit Model

Thus far in this chapter we have been performing signal analysis directly on the circuit—that is, with the equivalent circuit model *implicitly* assumed. It is worthwhile at this point to consider the explicit use of equivalent circuit models. Figure 10.6 shows

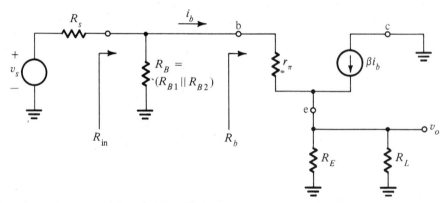

Fig. 10.6 The emitter follower of Fig. 10.5 with the transistor replaced by its simplified hybrid-π model.

the equivalent circuit model of the emitter follower where the transistor has been replaced by its simplified hybrid-π model. The equivalent circuit of Fig. 10.6 can be used to find R_{in} and the voltage gain by performing straightforward circuit analysis. The reader is urged to carry out such an analysis and verify that the results are identical to those obtained above. Specifically, we urge the reader to derive the resistance reflection rule.

Output Resistance

An alternative way to describe the performance of the emitter follower is to specify its open-circuit voltage gain and its output resistance—that is, to specify the Thévenin equivalent circuit at the output of the follower. The open-circuit voltage gain A_v can be obtained from the formula derived earlier with R_L set equal to ∞:

$$A_v = \frac{R_{\text{in}}}{R_{\text{in}} + R_s} \frac{R_E}{R_E + r_e}$$

where

$$R_{\text{in}} = [R_{B1} \| R_{B2} \| (\beta + 1)(r_e + R_E)]$$

To obtain the output resistance R_o we will make use of the equivalent circuit of Fig. 10.6. In this way we shall develop a simple rule for reflecting resistances from the base side to the emitter side. As will be seen, this rule is the inverse of the one that we have been using to reflect resistances from the emitter side to the base side.

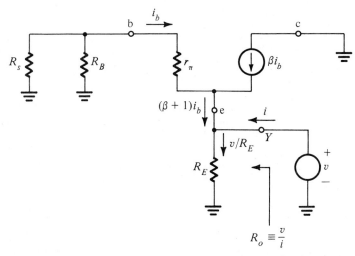

Fig. 10.7   Illustrating the method for evaluating the output resistance R_o of the emitter follower.

Figure 10.7 shows the equivalent circuit of the emitter follower with the voltage source v_s reduced to zero and the source resistance R_s left in the circuit. To obtain the output resistance R_o we have applied a voltage source v. Our task now is to find the current i drawn from this source, since by definition

$$R_o \equiv \frac{v}{i}$$

From the circuit in Fig. 10.7 we find

$$i = \frac{v}{R_E} - (\beta + 1)i_b \qquad (10.12)$$

We now need a relationship between i_b and v. This is easily obtained. Note that v is the voltage across r_π in series with $(R_B \| R_s)$,

$$i_b = -\frac{v}{r_\pi + (R_B \| R_s)} \qquad (10.13)$$

where the minus sign comes about because of the reference directions of v and i_b. Substituting for i_b from Eq. (10.13) into Eq. (10.12) gives

$$i = \frac{v}{R_E} + \frac{(\beta + 1)v}{r_\pi + (R_B \parallel R_s)} \tag{10.14}$$

from which we obtain

$$\frac{i}{v} = \frac{1}{R_E} + \frac{1}{r_\pi/(\beta + 1) + (R_B \parallel R_s)/(\beta + 1)}$$

Thus R_o is given by

$$R_o = R_E \parallel [r_\pi/(\beta + 1) + (R_B \parallel R_s)/(\beta + 1)]$$

Since $r_e = r_\pi/(\beta + 1)$, we may write R_o as

$$R_o = R_E \parallel [r_e + (R_B \parallel R_s)/(\beta + 1)] \tag{10.15}$$

which is the final result.

Clearly the above analysis is lengthy, and we have to find a simple rule that enables us to write the final result directly. The rule used is based on the fact that the emitter current is $\beta + 1$ times the base current. Therefore *all resistances on the base side may be reflected to the emitter side after dividing their values by $\beta + 1$*. To apply this rule to the evaluation of the output resistance of the emitter follower, "grab hold" of the point marked Y in Fig. 10.5 and look to the left. Between Y and ground we see the resistance R_E in parallel with a path through the emitter of the transistor. This latter path consists of r_e in series with the resistance between the base and ground reflected to the emitter side. Between the base and ground we see three parallel resistances: R_{B1}, R_{B2}, and R_s. Thus the second path between Y and ground has a resistance $r_e + (R_B \parallel R_s)/(\beta + 1)$, and the output resistance of the emitter follower will be given by Eq. (10.15).

Examination of Eq. (10.15) reveals that the output resistance of the emitter follower is usually quite small. The reason for the low output resistance is that the source resistance appears divided by $\beta + 1$.

EXERCISE

10.10 Calculate the open-circuit voltage gain and the output resistance of the emitter follower considered in Exercises 10.8 and 10.9.
Ans. 0.885 V/V; 52.7 Ω

The open-circuit voltage gain A_v and output resistance R_o can be used to evaluate the gain for any load resistance R_L using

$$\text{Voltage gain} = A_v \frac{R_L}{R_L + R_o}$$

The result obtained this way should be identical to that found from direct analysis [Eq. (10.11)].

Signal Swing

Finally, we consider the problem of the maximum allowed input signal swing in the emitter-follower circuit. Since only a small fraction of the input signal appears across the base–emitter junction, the emitter follower exhibits linear performance for a large range of input signal amplitude. In fact, the upper limit on the value of the input signal amplitude is usually imposed by transistor cutoff. To see how this comes about, consider as an example an input sine-wave signal (refer to Fig. 10.5). As the input goes negative, the output v_o will also go negative, and the current in R_L will be flowing from ground into the emitter terminal. The total voltage at the emitter will remain positive but will decrease, since $v_E = V_E + v_e$ and since we are considering the time at which v_e is negative. Thus the current in R_E will still be flowing from emitter to ground, but its value will be reduced.

Now, by writing a node equation at the emitter we can find the value of v_e at which the emitter current will be reduced to zero (that is, the transistor cuts off). Denote the peak value of v_e by $\hat{V}_e$. We may write

$$\frac{V_E - \hat{V}_e}{R_E} = \frac{\hat{V}_e}{R_L} \tag{10.16}$$

from which $\hat{V}_e$ can be obtained as

$$\hat{V}_e = \frac{V_E}{1 + R_E/R_L} \tag{10.17}$$

The corresponding value of the amplitude of v_s can be obtained from

$$\hat{V}_s = \frac{\hat{V}_e}{\text{Gain}} \tag{10.18}$$

Increasing the amplitude of v_s above this value results in the transistor becoming cut off, and the negative peaks of the output signal waveform will be clipped off.

> **EXERCISE**
>
> **10.11**   For the emitter follower considered in Exercises 10.8, 10.9, and 10.10 find the input signal amplitude that causes the transistor to cut off.
> *Ans.*   2.1 V

10.4   INTEGRATED-CIRCUIT BIASING TECHNIQUES

The biasing techniques discussed thus far are not suitable for the design of IC amplifiers. This shortcoming stems from the need for a large number of resistors (three per amplifier stage) as well as large coupling and bypass capacitors. With present IC technology it is almost impossible to create large capacitances, and it is uneconomical to manufacture large resistances. On the other hand, IC technology provides the designer with the possibility of using many transistors, which can be produced cheaply. Furthermore, it is easy to make transistors with matched characteristics that track with changes in environmental conditions. We will have more to say in later chapters about the IC technology constraints and the response of circuit designers to these constraints. In this section we shall study some of the techniques employed in biasing IC transistor amplifier stages.

Basically, biasing in integrated-circuit design is based on the use of constant-current sources. On an IC chip with a number of amplifier stages a constant dc current is generated at one location and is then reproduced at various other locations for biasing the various amplifier stages. This approach has the advantage that the bias currents of the various stages track each other in case of changes in power-supply voltage or in temperature.

We will study circuits for current sources, current sinks, and current steering. The circuits presented, of course, can be used in discrete circuit design as well, but with reduced performance since device matching and tracking is more difficult.

Current Sinks

Figure 10.8 shows a transistor current source or, more appropriately, a *current sink*, since the output current in the collector lead, I_O, is being pulled from a load connected to a positive voltage. For this circuit to provide the load with a constant current the transistor should be maintained in the active mode at all times. This will occur if the collector is kept at a voltage equal to or higher than that of the base.

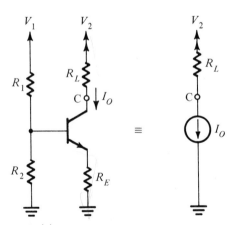

Fig. 10.8 A constant-current sink.

Assume active-mode operation. We see that this circuit is identical to the classical biasing circuit discussed previously. It follows that the output current of the current sink will be given by

$$I_O = \alpha \frac{V_1[R_2/(R_1 + R_2)] - V_{BE}}{R_E + (R_1 \parallel R_2)/(\beta + 1)} \qquad (10.19)$$

In order for I_O to be stable and predictable, its dependence on β and V_{BE} should be minimized by use of the design guidelines given in Section 10.1, namely,

$$\frac{(R_1 \parallel R_2)}{\beta + 1} \ll R_E \qquad (10.20a)$$

$$V_1 \frac{R_2}{R_1 + R_2} \gg V_{BE} \qquad (10.20b)$$

A Temperature-Compensated Current Sink

It is possible to design a current sink whose output current is, to first order, independent of V_{BE} and hence independent of temperature. Figure 10.9 shows such a circuit, where

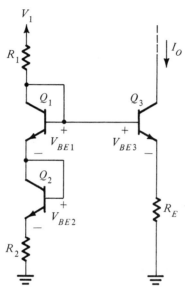

Fig. 10.9 A temperature-compensated current sink. For $R_1 = R_2 \simeq R_E$, $I_o \simeq \alpha V_1/2R_E$.

Q_1 and Q_2 are two transistors connected as diodes by shorting the collector to the base. This connection results in a two-terminal device (see Fig. 10.10) whose i-v characteristic is identical to the i_E-v_{BE} characteristic of an identical transistor. The reason for the identical characteristics is that the diode-connected transistor still behaves internally as a transistor operating in the active mode (recall that $v_{CB} = 0$ means operation in the active mode). Thus the collector and base currents of a diode-connected transistor have a ratio equal to β, as illustrated in Fig. 10.10.

In the circuit of Fig. 10.9 the three transistors Q_1, Q_2, and Q_3 are usually identical devices fabricated on the same IC silicon chip. Thus the three devices can be assumed to have matched characteristics. That is, for equal emitter currents the three devices will have equal base–emitter voltages. Furthermore, the circuit will be designed such that the current in Q_1 and Q_2 is equal to the current in Q_3, which is the output current. Assume $\beta \gg 1$; thus the base current of Q_3 can be neglected. It now can be shown that

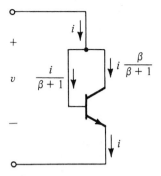

Fig. 10.10   A diode-connected transistor.

the condition for equal currents in all transistors and for the output current I_O to be independent of V_{BE} is

$$R_1 = R_2 = R_E \left(1 - \frac{2V_{BE}}{V_1} \right)$$

which is approximately

$$R_1 = R_2 \simeq R_E$$

Under this condition the output current I_O will be given by

$$I_O \simeq \alpha \frac{V_1}{2R_E}, \qquad \beta \gg 1$$

Since α is a relatively constant parameter, this current sink will draw a stable and predictable current.

Current Sources

The circuits discussed thus far are more appropriately described as current sinks, since their output terminal pulls a constant current from a load connected to a more positive supply voltage. Specifically, the output terminal should always be kept at a potential higher than that of the base. In many applications we require the dual circuit that pushes a constant current into a load connected to a more negative voltage. Such a circuit, appropriately called a *current source*, can be easily realized using a *pnp* transistor, as shown in Fig. 10.11. Here the output current I_O will be given by

$$I_O = \alpha \frac{V_1[R_2/(R_1 + R_2)] - V_{EB}}{R_E + (R_1 \parallel R_2)/(\beta + 1)} \tag{10.21}$$

This current source will operate as long as the transistor remains in the active mode (that is, as long as the collector voltage remains lower or equal to that of the base).

As with the current-sink circuit discussed above, the current source can be made insensitive to the value of β by making $(R_1 \parallel R_2)/(\beta + 1)$ much smaller than R_E. Also, the technique of using two additional diode-connected transistors to eliminate the V_{EB} term in Eq. (10.21) can be employed in the design of the current source.

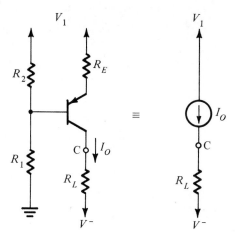

Fig. 10.11 *A current source.*

The Current Mirror

In an integrated circuit containing a number of amplifier stages (such as the circuit of an op amp) a constant dc current is usually generated at one location and reproduced at many other locations for biasing the different transistors in the circuit. A popular circuit building block for accomplishing current reproduction is the *current mirror,* shown in Fig. 10.12. It consists of two matched transistors Q_1 and Q_2 with their bases and emitters connected together. In addition, Q_1 is connected as a diode by shorting its collector to its base. The current mirror is shown fed by a current source i_I, and the output current is taken from the collector of Q_2.

Of course, the circuit fed by the collector of Q_2 should ensure active-mode operation for Q_2 (by keeping its collector voltage higher than that of the base) at all times.

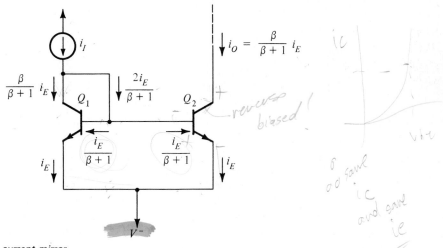

Fig. 10.12 *The current mirror.*

As will become apparent shortly, the performance of the current mirror is independent of the value of the voltage V^- as long as Q_2 is maintained in the active mode.

Analysis of the current mirror proceeds as follows: Since Q_1 and Q_2 are matched and since they have equal v_{BE} (because their EBJs are in parallel), their emitter currents will be equal. This is the key point. The rest of the analysis is straightforward and is illustrated in Fig. 10.12. It follows that

$$i_O = \frac{\beta}{\beta + 1} i_E$$

$$i_I = \frac{\beta + 2}{\beta + 1} i_E$$

Thus the current gain of the mirror is given by

$$\frac{i_O}{i_I} = \frac{\beta}{\beta + 2} = \frac{1}{1 + 2/\beta} \tag{10.22}$$

$$\frac{i_O}{i_I} \simeq 1 \qquad \text{for } \beta \gg 1 \tag{10.23}$$

As an application of the current-mirroring technique, consider the circuit of Fig. 10.13. It can be seen that

$$I = \frac{V_{CC} - V_{BE}}{R} \simeq \frac{V_{CC} - 0.7}{R}$$
$$I_{C1} = I_{C2} = I_{C3} \simeq I, \qquad \text{for } \beta \gg 1$$
$$V_{C1} = V_{CC} - IR_{C1}$$
$$V_{C2} = V_{CC} - 2IR_{C2}$$

It should be mentioned that for current mirrors with multiple outputs, such as that of Fig. 10.13, the errors due to the base currents accumulate and the current transfer ratio of the mirror will not be given by Eq. (10.23).

Before leaving this section we wish to mention that although we have here distinguished between current sources and current sinks, in the future we will refer to both as current sources. Also, in many cases it will be found easier to draw a diode-connected transistor simply as a diode.

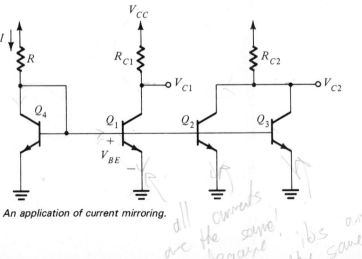

Fig. 10.13   An application of current mirroring.

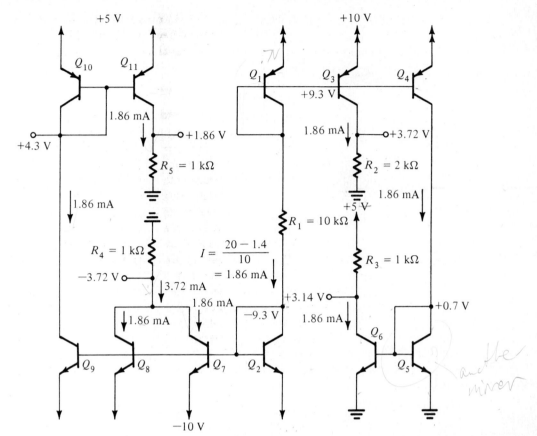

Fig. 10.14   *Circuit for Example 10.2.*

Example 10.2

Evaluate the voltages at all nodes and the currents through all branches in the circuit of Fig. 10.14. Assume $\beta \gg 1$ and that appropriate transistors are matched.

Solution

To start the solution, we have to locate a branch in which the current can be evaluated directly. The only such branch is formed by Q_1, R_1, and Q_2; both Q_1 and Q_2 are connected as diodes in series with R_1. The current through this branch can be easily evaluated as

$$I = \frac{10 - (-10) - 1.4}{10} = 1.86 \text{ mA}$$

Next we recognize that Q_1, Q_3, and Q_4 form a two-output current mirror. Thus

$$I_3 \simeq I = 1.86 \text{ mA} \qquad I_4 \simeq I = 1.86 \text{ mA}$$

Now, I_3 passes through R_2, causing V_{C3} to be

$$V_{C3} = I_3 R_2 = 1.86 \times 2 = 3.72 \text{ V}$$

Since the base of Q_3 is at $+9.3$ V, it follows that Q_3 is in the active mode, as had been assumed.

The collector current of Q_4 feeds the diode-connected transistor Q_5, which together with Q_6 forms another mirror. Thus

$$I_6 = I_5 = 1.86 \text{ mA}$$
$$V_{C6} = +5 - 1.86 \times 1 = +3.14 \text{ V}$$

Since the collector of Q_5, and hence the collector of Q_4, is at $+0.7$ V, it follows that Q_4 is active, as assumed. Also, the collector of Q_6 (at $+3.14$ V) is at higher voltage than its base (at $+0.7$ V), which keeps Q_6 active, as assumed.

Proceeding further, we recognize the three-output mirror formed by Q_2, Q_7, Q_8, and Q_9. Thus

$$I_7 = I_8 = I_9 = I = 1.86 \text{ mA}$$

The collectors of Q_7 and Q_8 are tied together, causing the sum of their currents to flow in R_4. Thus

$$V_{C8} = 0 - 2 \times 1.86 \times 1 = -3.72 \text{ V}$$

This voltage is higher than the voltage at the base (-9.3 V), and thus both Q_7 and Q_8 are in the active mode, as assumed. At this point we should note that Q_7 and Q_8 are connected in parallel, with the object being to produce a current in R_4 equal to twice the value of I. In an IC a single transistor with double the area would be used in place of the combination Q_7, Q_8.

The current in Q_9 is pulled through the diode-connected transistor Q_{10}, which together with Q_{11} forms another mirror. Thus

$$I_{10} = I_{11} = I_9 = 1.86 \text{ mA}$$
$$V_{C11} = 1.86 \times 1 = +1.86 \text{ V}$$

Since the collector of Q_9 is at $+4.3$ V, as determined by the base of Q_{10}, it follows that Q_9 is active, as assumed. Finally, the collector of Q_{11} (at $+1.86$ V) is lower in potential than the base (at $+4.3$ V), keeping Q_{11} in the active mode, as assumed.

This completes our solution. Note the simplicity with which we have been able to handle this relatively complex circuit with 11 transistors! Granted, our analysis is approximate (since β was assumed very large and V_{BE} was assumed to be 0.7 V throughout); nevertheless we obtained a quick solution that will be quite satisfactory in many applications. The reader should at this point appreciate the value of the circuit-drawing convention adopted in this text.

· · ·

EXERCISE

10.12   Figure E10.12 shows a current-mirror circuit (known as the Wilson circuit) in which the dependence on β is much smaller than that in the circuit of Fig. 10.12. Assume matched transistors and find the current gain i_o/i_I.
Ans. $[1 + 1/(\beta^2 + 2\beta)]^{-1}$

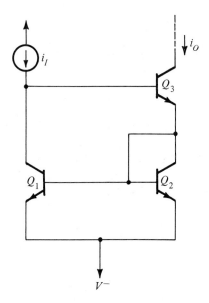

Fig. E10.12

10.5   THE DIFFERENTIAL PAIR

The differential pair, which can be implemented using BJTs, JFETs, or MOSFETs, is one of the most important circuit configurations employed in the design of integrated circuits. For instance, the first stage of every operational amplifier is essentially a differential-pair configuration. The differential pair is also used in digital applications and forms the basis for a high-speed logic family called *emitter-coupled logic* (ECL), which will be studied in Chapter 15. Therefore a thorough understanding of the operation of the differential pair is essential to the understanding and intelligent application of modern IC packages.

In this section we shall provide an introduction to the BJT differential pair. The small-signal operation of the BJT differential amplifier is given in Section 10.6, and the JFET differential pair is studied in Section 10.7.

Qualitative Description of Operation

Figure 10.15 shows the basic BJT differential-pair configuration. It consists of two matched transistors, Q_1 and Q_2, whose emitters are joined together and biased by a constant-current source I. The latter is usually implemented by a transistor current source such as that described in Section 10.4. Although each collector is connected to the positive supply voltage V_{CC} through a resistance R_C, this connection is not essential to the operation of the differential pair—that is, in some applications the two collectors may be connected to other transistors rather than to resistive loads. It is essential, though, that the collector circuits be such that Q_1 and Q_2 never enter the saturation mode.

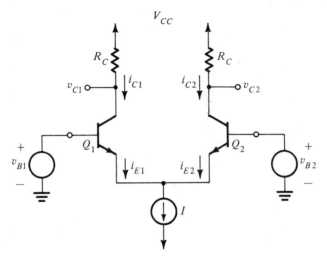

Fig. 10.15 *The basic differential-pair configuration.*

To see how the differential pair works, consider first the case where the two bases are joined together and connected to a voltage v_{CM}, called the *common-mode voltage*. That is, as shown in Fig. 10.16a, $v_{B1} = v_{B2} = v_{CM}$. Since Q_1 and Q_2 are matched, it follows from symmetry that the current I will divide equally between the two devices. Thus $i_{E1} = i_{E2} = I/2$, and the voltage at the emitters will be $v_{CM} - V_{BE}$ where V_{BE} is the voltage corresponding to an emitter current of $I/2$. The voltage at each collector will be $V_{CC} - \frac{1}{2}\alpha I R_C$, and the difference in voltage between the two collectors will be zero.

Now let us vary the value of the common-mode input signal v_{CM}. Obviously, as long as Q_1 and Q_2 remain in the active region the current I will still divide equally between Q_1 and Q_2, and the voltages at the collectors will not change. Thus the differential pair does not respond to *(rejects)* common-mode input signals.

As another experiment, let the voltage v_{B2} be set to a constant value, say, zero (by grounding $B2$), and let $v_{B1} = +1$ V (see Fig. 10.16b). With a bit of reasoning it can be seen that Q_1 will be on and conducting all of the current I and that Q_2 will be off. For Q_1 to be on the emitter has to be at $+0.3$ V, which keeps the EBJ of Q_2 reverse-biased. The collector voltages will be $v_{C1} = V_{CC} - \alpha I R_C$ and $v_{C2} = V_{CC}$.

Let us now change v_{B1} to -1 V (Fig. 10.16c). Again with some reasoning it can be seen that Q_1 will turn off and Q_2 will carry all the current I. The common emitter will be at -0.7 V, which means that the EBJ of Q_1 will be reverse-biased by 0.3 V. The collector voltages wil be $v_{C1} = V_{CC}$ and $v_{C2} = V_{CC} - \alpha I R_C$.

From the above we see that the differential pair certainly responds to *difference-mode or differential signals*. In fact, with relatively small difference voltages we are able to steer the entire bias current from one side of the pair to the other. This current-steering property of the differential pair allows it to be used in logic circuits, as will be demonstrated in Chapter 15.

To use the differential pair as a linear amplifier, we apply a very small differential signal (a few millivolts), which will result in one of the transistors conducting a current

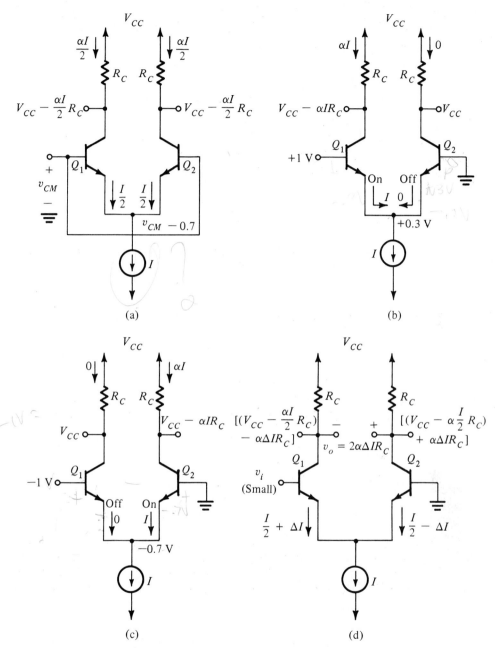

Fig. 10.16   Different modes of operation of the differential pair. (a) The differential pair with a common-mode input signal v_{CM}. (b) The differential pair with a "large" differential input signal. (c) The differential pair with a large differential input signal of polarity opposite to that in (b). (d) The differential pair with a small differential input signal v_i.

of $I/2 + \Delta I$; the current in the other transistor will be $I/2 - \Delta I$, with ΔI being proportional to the difference input voltage (see Fig. 10.16d). The output voltage taken between the two collectors will be $2\alpha\Delta IR_C$, which is proportional to the differential input signal v_i. The small-signal operation of the differential pair will be studied in Section 10.6.

EXERCISE
10.13 Find v_E, v_{C1}, and v_{C2} in the circuit of Fig. E10.13.
   Ans. $+0.7$ V; -5 V; -0.7 V

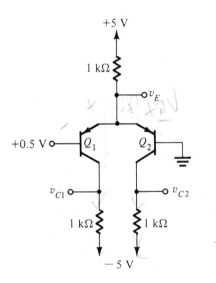

Fig. E10.13

Example 10.3
Figure 10.17 shows the circuit of a simple operational amplifier. Terminals 1 and 2, shown connected to ground, are the op-amp's input terminals, and terminal 3 is the output terminal.

   (a) Perform an approximate dc analysis (assuming $\beta \gg 1$) to calculate the dc currents and voltages everywhere in the circuit.

   (b) Calculate the quiescent power dissipation in this circuit.

   (c) If transistors Q_1 and Q_2 have $\beta = 100$, calculate the input bias current of the op amp.

   (d) What is the common-mode range of this op amp? *Note:* the *common-mode range* is the range of input common-mode voltage over which the input differential stage operates linearly, that is, remains in the active mode.

Solution
   (a) The values of all dc currents and voltages are indicated on the circuit diagram. These values were calculated ignoring the base current of every transistor; that is, assuming β to be very high. The analysis starts at the base of Q_3 where the voltage can

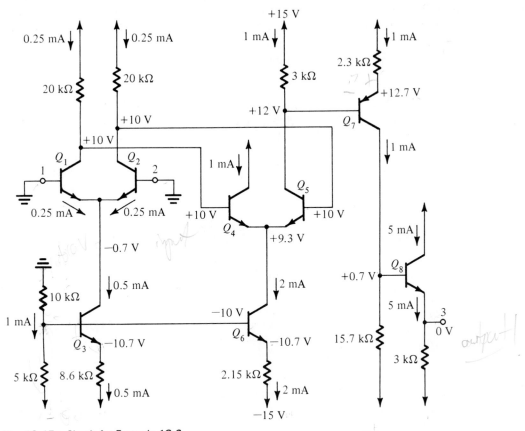

Fig. 10.17 Circuit for Example 10.3.

be calculated [applying the voltage divider rule to the (10 K, 5 K) resistors] to be −10 V. Thus the emitter of Q_3 will be at −10.7 V and the emitter current can be calculated as $[-10.7 - (-15)]/8.6 = 0.5$ mA. The current-source transistor Q_3 feeds the differential pair (Q_1, Q_2) with 0.5 mA. Thus each of Q_1 and Q_2 will be biased at 0.25 mA. The collectors of Q_1 and Q_2 will be at $[+15 - 0.25 \times 20] = +10$ V.

Proceeding to the second differential stage formed by Q_4 and Q_5, we find the voltage at their emitters to be $[+10 - 0.7] = 9.3$ V. This differential pair is biased by the current-source transistor Q_6 and we already know the voltage at the base of Q_6 (−10 V); thus the emitter voltage will be −10.7 V and the emitter current will be $[-10.7 - (-15)]/2.15 = 2$ mA. It should be noted that in a practical circuit transistor Q_6 should be constructed to have four times the area of Q_3. This ensures that Q_3 and Q_6 will have equal V_{BE} drops when Q_6 is carrying four times the current of Q_3. Furthermore, in an integrated circuit the resistors in the emitters of Q_3 and Q_6 would be eliminated and the current mirror configuration would be employed.

The current source transistor Q_6 supplies the differential pair Q_4, Q_5 with 2 mA; thus Q_4 and Q_5 will each be biased at 1 mA. We can now calculate the voltage at the collector of Q_5 as $+15 - 1 \times 3 = +12$ V. This will cause the voltage at the emitter of the *pnp* transistor Q_7 to be $+12.7$ V, and the emitter current of Q_7 will be $(+15 - 12.7)/2.3 = 1$ mA.

The collector current of Q_7, 1 mA, causes the voltage at the collector to be $-15 + 1 \times 15.7 = +0.7$ V. The emitter of Q_8 will be 0.7 V below the base; thus output terminal 3 will be at 0 V. Finally, the emitter current of Q_8 can be calculated to be $[0 - (-15)]/3 = 5$ mA.

(b) To calculate the power dissipated in the circuit in the quiescent state (that is, with zero input signal) we simply evaluate the dc current that the circuit draws from each of the two power supplies. From the $+15$-V supply the dc current is $I^+ = 0.25 + 0.25 + 1 + 1 + 1 + 5 = 8.5$ mA. Thus the power supplied by the positive power supply is $P^+ = 15 \times 8.5 = 127.5$ mW. The -15-V supply provides a current I^- given by $I^- = 1 + 0.5 + 2 + 1 + 5 = 9.5$ mA. Thus the power provided by the negative supply is $P^- = 15 \times 9.5 = 142.5$ mW. Adding P^+ and P^- provides the total power dissipated in the circuit P_D: $P_D = P^+ + P^- = 270$ mW.

(c) The input bias current of the op amp is the average of the dc currents that flow in the two input terminals (that is, in the bases of Q_1 and Q_2). These two currents are equal (because we have assumed matched devices); thus the bias current is given by

$$I_B = \frac{I_{E1}}{\beta + 1} \simeq 2.5 \ \mu A$$

(d) The upper limit on the input common-mode voltage is determined by the voltage at which Q_1 and Q_2 leave the active mode and enter saturation. This will happen if the input voltage equals or exceeds the collector voltage, which is $+10$ V. Thus the upper limit of the common-mode range is $+10$ V.

The lower limit of the input common-mode range is determined by the voltage at which Q_3 leaves the active mode and thus ceases to act as a constant-current source. This will happen if the collector voltage of Q_3 goes below the voltage at its base, which is -10 V. It follows that the input common-mode voltage should not go lower than $-10 + 0.7 = -9.3$ V. Thus the common-mode range is -9.3 to $+10$ V.

· · ·

Large-Signal Operation of the Differential Pair

We now present a general analysis of the BJT differential pair of Fig. 10.15. If we denote the voltage at the common emitter by v_E, the exponential relationship applied to each of the two transistors may be written

$$i_{E1} = \frac{I_S}{\alpha} e^{(v_{B1} - v_E)/V_T} \tag{10.24}$$

$$i_{E2} = \frac{I_S}{\alpha} e^{(v_{B2} - v_E)/V_T} \tag{10.25}$$

These two equations can be combined to obtain

$$\frac{i_{E1}}{i_{E2}} = e^{(v_{B1} - v_{B2})/V_T} \tag{10.26}$$

which can be manipulated to yield

$$\frac{i_{E1}}{i_{E1} + i_{E2}} = \frac{1}{1 + e^{(v_{B2} - v_{B1})/V_T}} \tag{10.27}$$

$$\frac{i_{E2}}{i_{E1} + i_{E2}} = \frac{1}{1 + e^{(v_{B1} - v_{B2})/V_T}} \tag{10.28}$$

The circuit imposes the additional constraint

$$i_{E1} + i_{E2} = I \tag{10.29}$$

Using Eq. (10.29) together with Eqs. (10.27) and (10.28) gives

$$i_{E1} = \frac{I}{1 + e^{(v_{B2} - v_{B1})/V_T}} \tag{10.30}$$

$$i_{E2} = \frac{I}{1 + e^{(v_{B1} - v_{B2})/V_T}} \tag{10.31}$$

Large Signal

The collector currents i_{C1} and i_{C2} can be obtained simply by multiplying the emitter currents in Eqs. (10.30) and (10.31) by α, which is normally very close to unity.

The fundamental operation of the differential amplifier is illustrated by Eqs. (10.30) and (10.31). First, note that the amplifier responds only to the difference voltage $v_{B1} - v_{B2}$. That is, if $v_{B1} = v_{B2} = v_{CM}$, the current I divides equally between the two transistors irrespective of the value of the common-mode voltage v_{CM}. This is the essence of the differential-amplifier operation, which also gives rise to its name.

Another important observation is that a relatively small difference voltage $v_{B1} - v_{B2}$ will cause the current I to flow almost entirely in one of the two transistors. Figure 10.18 shows a plot of the two collector currents (assuming $\alpha \simeq 1$) as a function of the difference signal. This is a normalized plot that can be used universally. Note that a difference voltage of about $4V_T$ ($\simeq 100$ mV) is sufficient to switch the current almost entirely to one side of the pair.

The nonlinear transfer characteristics of the differential pair, shown in Fig. 10.18, will not be utilized any further in this chapter. In the following we shall be interested specifically in the application of the differential pair as a small-signal amplifier. For this purpose the difference input signal is limited to less than about $V_T/2$ in order that we may operate on a linear segment of the characteristics around the midpoint x.

EXERCISE

10.14 Find the value of input differential signal sufficient to cause $i_{E1} = 0.99I$.
Ans. 115 mV

10.6 SMALL-SIGNAL OPERATION OF THE DIFFERENTIAL AMPLIFIER

In this section we shall study the application of the differential pair in small-signal amplification. Figure 10.19 shows the differential pair with a difference voltage signal v_d applied between the two bases. Implied is that the dc level at the input—that is, the

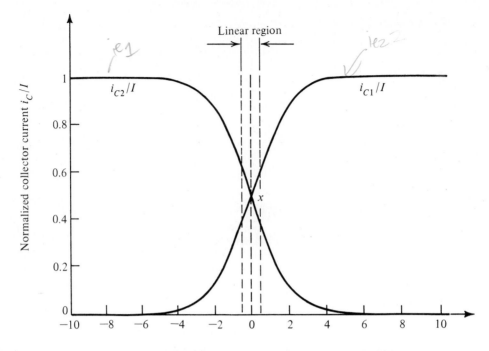

Normalized differential input voltage $(v_{B1} - v_{B2})/V_T$

Fig. 10.18 *Transfer characteristics of the BJT differential pair of Fig. 10.15.*

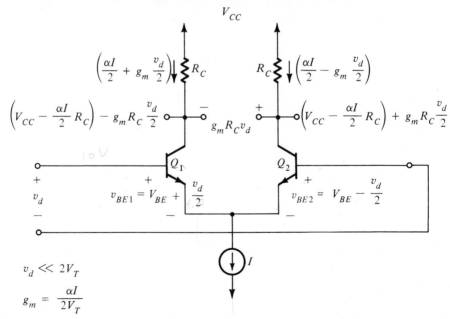

Fig. 10.19 *The currents and voltages in the differential amplifier when a small difference signal v_d is applied.*

common-mode input signal—has been somehow established. For instance, one of the two input terminals can be grounded and v_d applied to the other input terminal. Alternatively, the differential amplifier may be fed from the output of another differential amplifier. In this case the voltage at one of the input terminals will be $v_{CM} + v_d/2$ while that at other input terminals will be $v_{CM} - v_d/2$. We will consider common-mode operation at a later stage.

The Collector Currents When v_d Is Applied

Return now to the circuit of Fig. 10.19. We may use Eqs. (10.30) and (10.31) to find the total currents i_{C1} and i_{C2} as functions of the differential signal v_d by substituting $v_{B1} - v_{B2} = v_d$.

$$i_{C1} = \frac{\alpha I}{1 + e^{-v_d/V_T}} \tag{10.32}$$

$$i_{C2} = \frac{\alpha I}{1 + e^{v_d/V_T}} \tag{10.33}$$

Multiplying the numerator and the denominator of the right-hand side of Eq. (10.32) by $e^{(v_d/2V_T)}$ gives

$$i_{C1} = \frac{\alpha I e^{(v_d/2V_T)}}{e^{(v_d/2V_T)} + e^{(-v_d/2V_T)}} \tag{10.34}$$

Assume that $v_d \ll 2V_T$. We may thus expand the exponential $e^{(\pm v_d/2V_T)}$ in a series and retain only the first two terms:

$$i_{C1} \simeq \frac{\alpha I(1 + v_d/2V_T)}{1 + v_d/2V_T + 1 - v_d/2V_T}$$

Thus

$$i_{C1} = \frac{\alpha I}{2} + \frac{\alpha I}{2V_T} \frac{v_d}{2} \tag{10.35}$$

Similar manipulations can be applied to Eq. (10.33) to obtain

$$i_{C2} = \frac{\alpha I}{2} - \frac{\alpha I}{2V_T} \frac{v_d}{2} \tag{10.36}$$

Equations (10.35) and (10.36) tell us that when $v_d = 0$ the bias current I divides equally between the two transistors of the pair. Thus each transistor is biased at an emitter current of $I/2$. When a "small-signal" v_d is applied differentially (that is, between the two bases) the collector current of Q_1 increases by an increment i_c and that of Q_2 decreases by an equal amount. This ensures that the sum of the total currents in Q_1 and Q_2 remains constant, as constrained by the current-source bias. The incremental or signal current component i_c is given by

$$i_c = \frac{\alpha I}{2V_T} \frac{v_d}{2} \tag{10.37}$$

Equation (10.37) has an easy interpretation. First, note from the symmetry of the circuit (Fig. 10.19) that the differential signal v_d should divide equally between the base–emitter junctions of the two transistors. Thus the total base–emitter voltages will be

$$v_{BE}|_{Q_1} = V_{BE} + \frac{v_d}{2}$$

$$v_{BE}|_{Q_2} = V_{BE} - \frac{v_d}{2}$$

where V_{BE} is the dc EB voltage corresponding to an emitter current of $I/2$. Therefore the collector current of Q_1 will increase by $g_m v_d/2$ and the collector current of Q_2 will decrease by $g_m v_d/2$. Here g_m denotes the transconductance of Q_1 and of Q_2, which are equal and given by

$$g_m = \frac{I_C}{V_T} = \frac{\alpha I/2}{V_T}$$

Thus Eq. (10.37) simply states that $i_c = g_m v_d/2$.

An Alternative Viewpoint

There is an extremely useful alternative interpretation of the above results. Assume the current source I to be ideal. Its incremental resistance then will be infinite. Thus the voltage v_d appears across a total resistance of $2r_e$, where

$$r_e = \frac{V_T}{I_E} = \frac{V_T}{I/2}$$

Correspondingly there will be a signal current i_e, as illustrated in Fig. 10.20, given by

$$i_e = \frac{v_d}{2r_e}$$

Thus the collector of Q_1 will exhibit a current increment i_c and the collector of Q_2 will exhibit a current decrement i_c:

$$i_c = \alpha i_e = \frac{\alpha v_d}{2r_e} = g_m \frac{v_d}{2}$$

Note that in Fig. 10.20 we have shown signal quantities only. It is implied, of course, that each transistor is biased at an emitter current of $I/2$.

This method of analysis is particularly useful when resistances are included in the emitters, as shown in Fig. 10.21. For this latter circuit we have

$$i_e = \frac{v_d}{2r_e + 2R_E}$$

Input Differential Resistance

The input differential resistance is the resistance seen between the two bases; that is, it is the resistance seen by the differential input signal v_d. For the differential amplifier in

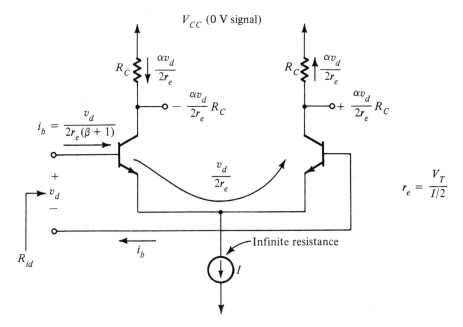

Fig. 10.20 Simple technique for determining the signal currents in a differential amplifier excited by a differential voltage signal v_d; dc quantities are not shown.

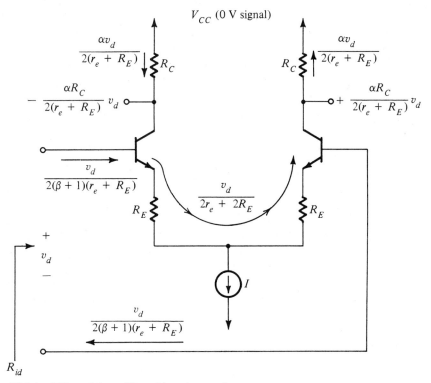

Fig. 10.21 Differential amplifier with emitter resistances.

Figs. 10.19 and 10.20 it can be seen that the base current of Q_1 shows an increment i_b and the base current of Q_2 shows an equal decrement,

$$i_b = \frac{i_e}{\beta + 1} = \frac{v_d/2r_e}{\beta + 1}$$

Thus the differential input resistance R_{id} is given by

$$R_{id} \equiv \frac{v_d}{i_b} = (\beta + 1)2r_e = 2r_\pi$$

This result is just a restatement of the familiar resistance reflection rule; namely, *the resistance seen between the two bases is equal to the total resistance in the emitter circuit multiplied by $\beta + 1$.* We can employ this rule to find the input differential resistance for the circuit in Fig. 10.21 as

$$R_{id} = (\beta + 1)(2r_e + 2R_E)$$

Differential Voltage Gain

We have established that for small difference input voltages ($v_d \ll 2V_T$; that is, v_d smaller that about 20 mV) the collector currents are given by

$$i_{C1} = I_C + g_m \frac{v_d}{2}$$

$$i_{C2} = I_C - g_m \frac{v_d}{2}$$

where

$$I_C = \frac{\alpha I}{2}$$

Thus the total voltages at the collectors will be

$$v_{C1} = (V_{CC} - I_C R_C) - g_m R_C \frac{v_d}{2}$$

$$v_{C2} = (V_{CC} - I_C R_C) + g_m R_C \frac{v_d}{2}$$

The quantities in parentheses are simply the dc voltages at each of the two collectors.

The output voltage signal of a differential amplifier can be taken either *differentially* (that is, between the two collectors) or *single-ended* (that is, between one collector and ground). If the output is taken differentially, then the differential gain (as opposed to the common-mode gain) of the differential amplifier will be

$$A_d = \frac{v_{c1} - v_{c2}}{v_d} = -g_m R_C \qquad (10.38)$$

On the other hand, if we take the output single ended (say, between the collector of Q_1 and ground), then the differential gain will be given by

$$A_d = \frac{v_{c1}}{v_d} = -\tfrac{1}{2}g_m R_C \qquad (10.39)$$

For the differential amplifier with resistances in the emitter leads (Fig. 10.21) the differential gain when the output is taken differentially is given by

$$A_d = -\frac{\alpha(2R_C)}{2r_e + 2R_E} \simeq -\frac{R_C}{r_e + R_E} \tag{10.40}$$

This equation is a familiar one: it states that *the voltage gain is equal to the ratio of the total resistance in the collector circuit* $(2R_C)$ *to the total resistance in the emitter circuit* $(2r_e + 2R_E)$.

Equivalence of the Differential Amplifier to a Common-Emitter Amplifier

The above analysis and results are quite similar to those obtained in the case of a common-emitter amplifier stage. That the differential amplifier is in fact equivalent to a common-emitter amplifier is illustrated in Fig. 10.22. Figure 10.22a shows a differential

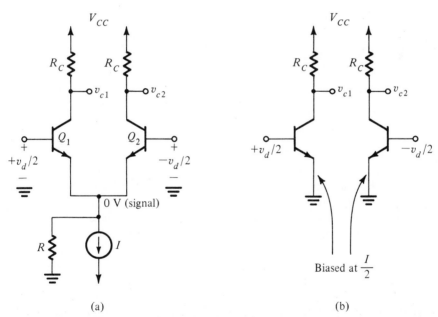

(a)   (b)

Fig. 10.22 *Equivalence of the differential amplifier in (a) to the two common-emitter amplifiers in (b). This equivalence applies only for differential input signals. Either of the two common-emitter amplifiers in (b) can be used to evaluate the differential gain, input differential resistance, frequency response, etc., of the differential amplifier.*

amplifier fed by a differential signal v_d with the differential signal applied in a *complementary (push-pull)* manner. That is, while the base of Q_1 is raised by $v_d/2$ the base of Q_2 is lowered by $v_d/2$. We have also included the output resistance R of the bias current source. From symmetry it follows that the signal voltage at the common emitter will be zero. Thus the circuit is equivalent to the two common-emitter amplifiers shown in Fig. 10.22b, where each of the two transistors is biased at an emitter current of $I/2$. Note that the finite output resistance R of the current source will have no effect on the operation. The equivalent circuit in Fig. 10.22b is valid for differential operation only. This equivalence is useful in frequency-response evaluation, as will be illustrated in Chapter 11.

In many applications the differential amplifier is not fed in a complementary fashion; rather, the input signal may be applied to one of the input terminals while the other terminal is grounded, as shown in Fig. 10.23. In this case the signal voltage at the

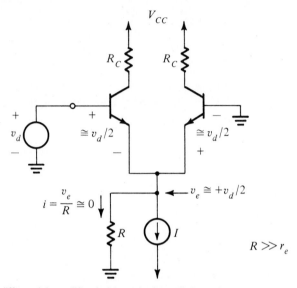

Fig. 10.23   *The differential amplifier fed in a single-ended manner.*

emitters will not be zero, and thus the resistance R will have an effect on the operation. Nevertheless, if R is large ($R \gg r_e$), as is usually the case, then v_d will still divide equally (approximately) between the two junctions, as shown in Fig. 10.23. Thus the operation of the differential amplifier in this case will be almost identical to that in the case of symmetric feed, and the common-emitter equivalence can still be employed.

Common-Mode Gain

Figure 10.24a shows a differential amplifier fed by a common-mode voltage signal v_{CM}. The resistance R is the incremental output resistance of the bias current-source. From symmetry it can be seen that the circuit is equivalent to that shown in Fig. 10.24b, where each of the two transistors Q_1 and Q_2 is biased at an emitter current $I/2$ and has a resistance $2R$ in its emitter lead. Thus the common-mode output voltage v_{c1} will be

$$v_{c1} = -v_{CM}\frac{\alpha R_C}{2R + r_e} \simeq -v_{CM}\frac{\alpha R_C}{2R}$$

At the other collector we have an equal common-mode signal v_{c2},

$$v_{c2} \simeq -v_{CM}\frac{\alpha R_c}{2R}$$

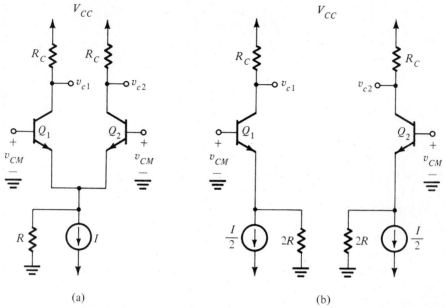

(a)   (b)

Fig. 10.24 (a) The differential amplifier fed by a common-mode voltage signal. (b) Equivalent "half-circuits" for common-mode calculations.

Now, if the output is taken differentially, then the output common-mode voltage $v_{c1} - v_{c2}$ will be zero and the common-mode gain also will be zero. On the other hand, if the output is taken single-ended, the common mode gain A_{cm} will be finite and given by

$$A_{cm} = -\frac{\alpha R_C}{2R} \tag{10.41}$$

Since in this case the differential gain is

$$A_d = \tfrac{1}{2} g_m R_C$$

the common-mode rejection ratio (CMRR) will be

$$\text{CMRR} = \left| \frac{A_d}{A_{cm}} \right| \simeq g_m R, \quad \alpha \simeq 1 \tag{10.42}$$

Normally the CMRR is expressed in dB:

$$\text{CMRR} = 20 \log \left| \frac{A_d}{A_{cm}} \right|$$

The above analysis assumes that the circuit is perfectly symmetric. Actual circuits are not perfectly symmetric, with the result that the common-mode gain will not be zero even if the output is taken differentially. To illustrate, consider the case of perfect symmetry except for a mismatch ΔR_C in the collector resistances. That is, let the collector of Q_1 have a load resistance R_C and that of Q_2 have a load resistance $R_C + \Delta R_C$. It follows that

$$v_{c1} = -v_{CM} \frac{\alpha R_C}{2R + r_e}$$

$$v_{c2} = -v_{CM} \frac{\alpha(R_C + \Delta R_C)}{2R + r_e}$$

Thus the common-mode signal at the output will be

$$v_o = v_{c1} - v_{c2} = v_{CM} \frac{\alpha \Delta R_C}{2R + r_e}$$

and the common-mode gain will be

$$A_{cm} = \frac{\alpha \Delta R_C}{2R + r_e} \simeq \frac{\Delta R_C}{2R}$$

This expression can be rewritten

$$A_{cm} = \frac{R_C}{2R} \frac{\Delta R_C}{R_C} \tag{10.43}$$

Compare the common-mode gain in Eq. (10.43) with that for the case of single-ended output in Eq. (10.41). We see that the common-mode gain is much smaller in the case of differential output. Therefore the input differential stage of an op amp is usually a balanced one, with the output taken differentially. This ensures that the op amp will have a low common-mode gain or, equivalently, a high CMRR.

Input Common-Mode Resistance

The definition of the *common-mode input resistance* R_{icm} is illustrated in Fig. 10.25a. Figure 10.25b shows the equivalent half-circuit (from a common-mode point of view).

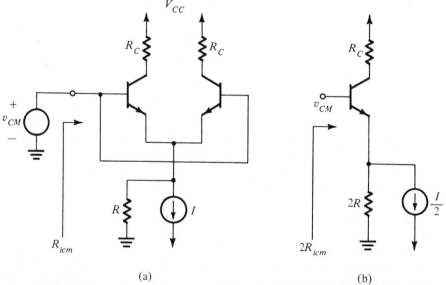

(a)   (b)

Fig. 10.25   (a) Definition of the input common-mode resistance R_{icm}. (b) The equivalent common-mode half-circuit.

The input common-mode resistance R_{icm} is the parallel equivalent of the input resistances of two identical circuits each of which is like the circuit shown in Fig. 10.25b. Thus

$$2R_{icm} = (\beta + 1)(2R + r_e) \simeq 2(\beta + 1)R$$

which gives

$$R_{icm} \simeq (\beta + 1)R$$

Since $(\beta + 1)R$ is usually very high, the resistance r_μ between base and collector (Section 9.10) could have a significant effect on the value of R_{icm}. Since one end of r_μ (the collector end) is at an almost constant voltage, r_μ appears effectively in parallel with the input in the circuit of Fig. 10.25b. Thus

$$R_{icm} \simeq [(\beta + 1)R \parallel r_\mu/2]$$

We will have more to say about r_μ in Chapter 11.

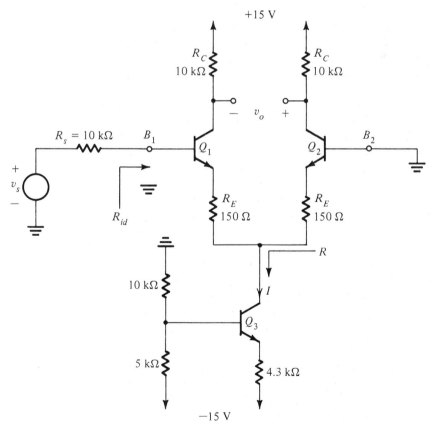

Fig. E10.15

EXERCISE

10.15 The differential amplifier in Fig. E10.15 uses transistors with $\beta = 100$. Evaluate the following:

(a)   the dc current *I*;
(b)   the input bias current;
(c)   the input common-mode range;
(d)   the input differential resistance R_{id};
(e)   the differential gain $A_d \equiv v_o/v_s$;
(f)   the worst-case common-mode gain if the two collector resistances are accurate to within ±1% and if the output resistance of Q_3 is 5.9 MΩ;
(g)   the CMRR, in dB;
(h)   the input common-mode resistance (assuming r_μ = 50 MΩ).

Ans.   (a) 1 mA; (b) 5 μA; (c) −9.3 to +10 V; (d) 40 kΩ; (e) 40 V/V; (f) 1.7 × 10⁻⁵ V/V; (g) 127 dB; (h) 24 MΩ

10.7   THE JFET DIFFERENTIAL PAIR

Because of their high input resistance FETs are becoming popular in the design of the differential input stage of op amps. FET-input amplifiers are currently available commercially and exhibit very small input bias currents (in the picoamp range). In this section we shall study the operation of the JFET differential amplifier.

Figure 10.26 shows the basic JFET differential amplifier fed by two voltage sources v_{G1} and v_{G2} and biased by a constant-current source *I*. We wish to derive expressions

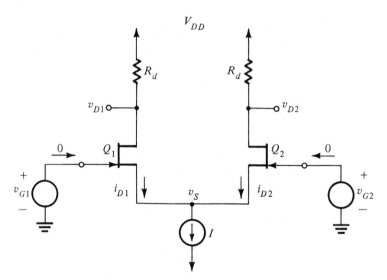

Fig. 10.26   The JFET differential pair.

for the currents i_{D1} and i_{D2} in terms of v_{G1} and v_{G2}. To do this we will assume that the JFETs are operating in the pinch-off region. Therefore we may use the square-law i_D-v_{GS} relationship

$$i_D = I_{DSS}\left(1 - \frac{v_{GS}}{V_P}\right)^2$$

where for n-channel devices V_P is a negative number. Furthermore, we will assume that the two FETs are perfectly matched. For transistor Q_1 we have

$$i_{D1} = I_{DSS}\left(1 - \frac{v_{G1} - v_S}{V_P}\right)^2$$

which can be rewritten

$$\sqrt{i_{D1}} = \sqrt{I_{DSS}}\left(1 - \frac{v_{G1}}{V_P} + \frac{v_S}{V_P}\right) \tag{10.44}$$

Similarly for Q_2 we have

$$i_{D2} = I_{DSS}\left(1 - \frac{v_{G2} - v_S}{V_P}\right)^2$$

which can be rewritten

$$\sqrt{i_{D2}} = \sqrt{I_{DSS}}\left(1 - \frac{v_{G2}}{V_P} + \frac{v_S}{V_P}\right) \tag{10.45}$$

Subtracting Eq. (10.45) from Eq. (10.44) gives

$$\sqrt{i_{D1}} - \sqrt{i_{D2}} = \sqrt{I_{DSS}}\frac{v_{G1} - v_{G2}}{-V_P}$$

Substituting $v_{G1} - v_{G2} = v_{id}$, where v_{id} is the differential input voltage, gives

$$\sqrt{i_{D2}} = \sqrt{i_{D1}} - \sqrt{I_{DSS}}\frac{v_{id}}{-V_P} \tag{10.46}$$

The current-source bias imposes the following constraint:

$$i_{D1} + i_{D2} = I \tag{10.47}$$

Substituting for i_{D2} from Eq. (10.46) into Eq. (10.47) yields a quadratic equation that can be solved to yield ultimately

$$i_{D1} = \frac{I}{2} + v_{id}\frac{I}{-2V_P}\sqrt{2\frac{I_{DSS}}{I} - \left(\frac{v_{id}}{V_P}\right)^2\left(\frac{I_{DSS}}{I}\right)^2} \tag{10.48}$$

$$i_{D2} = \frac{I}{2} - v_{id}\frac{I}{-2V_P}\sqrt{2\frac{I_{DSS}}{I} - \left(\frac{v_{id}}{V_P}\right)^2\left(\frac{I_{DSS}}{I}\right)^2} \tag{10.49}$$

Equations (10.48) and (10.49) indicate that the differential pair responds to a difference signal v_{id} by changing the proportion in which the bias current I divides between the two transistors. For $v_{id} = 0$, $i_{D1} = i_{D2} = I/2$, as should be expected. If v_{id} is positive, the current i_{D1} increases by the amount given by the second term on the right-hand side of Eq. (10.48), and i_{D2} decreases by an equal amount. The maximum value of the increment in i_{D1} (the decrement in i_{D2}) is limited to $I/2$ and is obtained when v_{id} is of such a value that

$$v_{id} \frac{I}{-2V_P} \sqrt{2 \frac{I_{DSS}}{I} - \left(\frac{v_{id}}{V_P}\right)^2 \left(\frac{I_{DSS}}{I}\right)^2} = \frac{I}{2}$$

which results in

$$\left|\frac{v_{id}}{V_P}\right| = \sqrt{\frac{I}{I_{DSS}}} \tag{10.50}$$

That is, the value of v_{id} given by Eq. (10.50) results in the current I being entirely carried by one of the two transistors (Q_1 for positive v_{id} and Q_2 for negative v_{id}). At this point we should note that the bias current I should be smaller than I_{DSS}; otherwise one of the two FETs could carry a current greater than I_{DSS}, which would result in its gate–channel junction becoming forward biased.

As in the case of the BJT differential pair the input common-mode range is determined at the low end by the current-source device (which can be a FET or a BJT) leaving the active region and at the high end by Q_1 and Q_2 leaving the active (pinch-off) region. This upper limit on v_{CM} will therefore be $|V_P|$ volts below the quiescent voltage at the drain, $V_{DD} - (I/2)R_d$.

Small-Signal Operation

Small-signal analysis of the JFET differential amplifier can be carried out in a manner identical to that given in detail for the BJT circuit. For instance, to evaluate the small-signal differential gain we use Eqs. (10.48) and (10.49) and assume that the terms involving v_{id}^2 are negligibly small; that is

$$\left|\frac{v_{id}}{V_P}\right| \ll \sqrt{\frac{2I}{I_{DSS}}} \tag{10.51}$$

Note, however, that unlike the BJT case, where v_{id} is limited to about 20 mV, here the small-signal condition limits v_{id} to a volt or so. Under this condition Eqs. (10.48) and (10.49) can be approximated to

$$i_{D1} \simeq \frac{I}{2} + i_d \tag{10.52}$$

$$i_{D2} \simeq \frac{I}{2} - i_d \tag{10.53}$$

where the current signal i_d is given by

$$i_d = \frac{v_{id}}{2}\left(\frac{2I_{DSS}}{-V_P}\sqrt{\frac{I/2}{I_{DSS}}}\right) \tag{10.54}$$

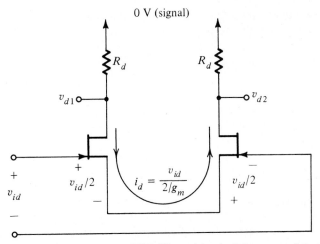

Fig. 10.27 *Small-signal analysis of the JFET differential pair. It is assumed that each transistor is biased at a drain current of I/2.*

We immediately recognize the quantity in parentheses as g_m of Q_1 and of Q_2. Thus Eq. (10.54) simply reaffirms our expectations that the input difference voltage v_{id} divides equally between the two transistors,

$$v_{gs1} = v_{sg2} = \frac{v_{id}}{2} \qquad (10.55)$$

and thus the signal current in Q_1 is $g_m v_{id}/2$ and that in Q_2 is $-g_m v_{id}/2$. Alternatively, we can think of the voltage v_{id} as appearing across a total source resistance of $2/g_m$; thus a current

$$i_d = \frac{v_{id}}{2/g_m}$$

flows, as illustrated in Fig. 10.27.

At the drain of Q_1 the voltage signal v_{d1} will be

$$v_{d1} = -i_d R_d = -g_m \frac{v_{id}}{2} R_d \qquad (10.56)$$

and at the drain of Q_2 we have

$$v_{d2} = +i_d R_d = +g_m \frac{v_{id}}{2} R_d \qquad (10.57)$$

If the output is taken differentially (that is, between the two drains), then

$$v_o = v_{d1} - v_{d2} = -g_m R_d v_{id} \qquad (10.58)$$

and the differential gain will be

$$\frac{v_o}{v_{id}} = -g_m R_d \qquad (10.59)$$

Finally, it should be pointed out that the current source supplying the bias current I will inevitably have a finite output resistance, which causes the differential amplifier to have a finite nonzero common-mode gain. The common-mode gain can be evaluated in a manner identical to that used for the BJT case.

EXERCISE

10.16 Consider the circuit in Fig. 10.26 with $V_{DD} = +15$ V, $I = 1$ mA, $R_d = 10$ kΩ, $I_{DSS} = 2$ mA, and $V_P = -2$ V. Find the value of v_{id} required to switch the current entirely to Q_1. Also calculate the small-signal voltage gain if the output is taken differentially.
Ans. 1.4 V; 10 V/V

10.8 MULTISTAGE AMPLIFIERS

Practical transistor amplifiers usually consist of a number of stages connected in cascade. In addition to providing gain, the first or input stage is usually required to provide a high input resistance in order to avoid loss of signal level when the amplifier is fed with a high-resistance source. In a differential amplifier the input stage must also provide large common-mode rejection. The function of the middle stages of an amplifier cascade is to provide the bulk of the voltage gain. In addition, the middle stages provide such other functions as the conversion of the signal from differential mode to single-ended mode and the shifting of the dc level of the signal. These two functions and others will be illustrated later in this section and in greater detail in Chapter 13.

Finally, the main function of the last or output stage of an amplifier is to provide a low output resistance in order to avoid loss of gain when a low-valued load resistance is connected to the amplifier. Also, the output stage should be able to supply the current required by the load in an efficient manner—that is, without dissipating an unduly large amount of power in the output transistors. We have already studied one type of amplifier configuration suitable for implementing output stages, namely, the source follower and the emitter follower. It will be shown in Chapter 13 that the source and emitter followers are not optimum from the point of view of power efficiency and that other, more appropriate circuit configurations exist for output stages.

To illustrate the structure and method of analysis of multistage amplifiers, we will conclude this chapter with a detailed example. The amplifier circuit to be analyzed is shown in Fig. 10.28. The dc analysis of this simple op-amp circuit was presented in Example 10.3, which we urge the reader to review before studying the following material.

The op-amp circuit in Fig. 10.28 consists of four stages. The input stage is *differential in, differential out* and consists of transistors Q_1 and Q_2, which are biased by current source Q_3. The second stage is also a differential-input amplifier, but its output is taken single ended at the collector of Q_5. This stage is formed by Q_4 and Q_5, which are biased by the current source Q_6. Note that the conversion from differential to single ended as performed by the second stage results in a loss of gain of a factor of 2. A more elaborate method for accomplishing this conversion will be given in Chapter 13.

In addition to providing some voltage gain, the third stage, consisting of the *pnp*

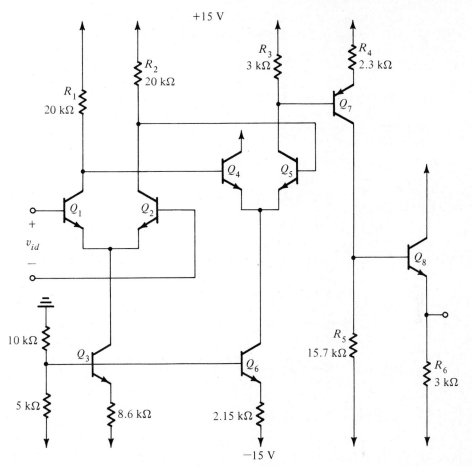

Fig. 10.28 A multistage amplifier circuit (Example 10.4).

transistor Q_7, provides the essential function of *shifting the dc level* of the signal. Thus while the signal at the collector of Q_5 is not allowed to swing below the voltage at the base of Q_5 ($+10$ V), the signal at the collector of Q_7 can swing negative (and positive, of course). From our study of op amps in Chapter 3 we know that the output terminal of the op amp should be capable of positive and negative voltage swings. Therefore every op-amp circuit includes a *level-shifting* stage. Although the use of the complementary *pnp* transistor provides a simple solution to the level-shifting problem, other forms of level shifters exist, one of which will be discussed in Chapter 13.

Finally, we note that the output stage consists of emitter follower Q_8 and that ideally the dc level at the output is zero volts (as was calculated in Example 10.3).

Example 10.4

Use the dc bias quantities evaluated in Example 10.3 and analyze the circuit in Fig. 10.28 to determine the input resistance, the voltage gain, and the output resistance.

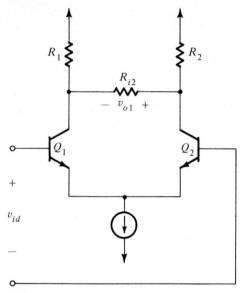

Fig. 10.29 *Equivalent circuit for calculating the gain of the input stage of the amplifier in Fig. 10.28.*

Solution

The input differential resistance R_{id} is given by

$$R_{id} = r_{\pi 1} + r_{\pi 2}$$

Since Q_1 and Q_2 are each operating at an emitter current of 0.25 mA, it follows that

$$r_{e1} = r_{e2} = \frac{25}{0.25} = 100 \ \Omega$$

Assume $\beta = 100$; then

$$r_{\pi 1} = r_{\pi 2} = 101 \times 100 = 10.1 \ k\Omega$$

Thus $R_{id} = 20.2 \ k\Omega$.

To evaluate the gain of the first stage we first find the input resistance of the second stage, R_{i2},

$$R_{i2} = r_{\pi 4} + r_{\pi 5}$$

Q_4 and Q_5 are each operating at an emitter current of 1 mA; thus

$$r_{e4} = r_{e5} = 25 \ \Omega$$
$$r_{\pi 4} = r_{\pi 5} = 101 \times 25 = 2.525 \ k\Omega$$

Thus $R_{i2} = 5.05 \ k\Omega$. This resistance appears between the collectors of Q_1 and Q_2, as shown in Fig. 10.29. Thus the gain of the first stage will be

$$A_1 \equiv \frac{v_{o1}}{v_{id}} = \frac{\text{Total resistance in collector circuit}}{\text{Total resistance in emitter circuit}}$$

$$= \frac{[R_{i2} \| (R_1 + R_2)]}{r_{e1} + r_{e2}}$$

$$= \frac{(5.05 \text{ k}\Omega \| 40 \text{ k}\Omega)}{200 \ \Omega} = 22.4 \text{ V/V}$$

Figure 10.30 shows an equivalent circuit for calculating the gain of the second stage. As indicated, the input voltage to the second stage is the output voltage of the first stage, v_{o1}. Also shown is the resistance R_{i3}, which is the input resistance of the third stage formed by Q_7. The value of R_{i3} can be found by multiplying the total resistance in the emitter of Q_7 by $\beta + 1$:

$$R_{i3} = (\beta + 1)(R_4 + r_{e7})$$

Since Q_7 is operating at an emitter current of 1 mA,

$$r_{e7} = \frac{25}{1} = 25 \ \Omega$$

$$R_{i3} = 101 \times 2.325 = 234.8 \text{ k}\Omega$$

We can now find the gain A_2 of the second stage as the ratio of the total resistance in the collector circuit to the total resistance in the emitter circuit:

$$A_2 \equiv \frac{v_{o2}}{v_{o1}} = -\frac{(R_3 \| R_{i3})}{r_{e4} + r_{e5}}$$

$$= -\frac{(3 \text{ k}\Omega \| 234.8 \text{ k}\Omega)}{50 \ \Omega} = -59.25 \text{ V/V}$$

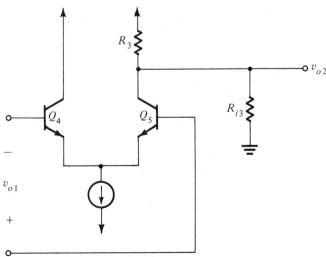

Fig. 10.30 *Equivalent circuit for calculating the gain of the second stage of the amplifier in Fig. 10.28.*

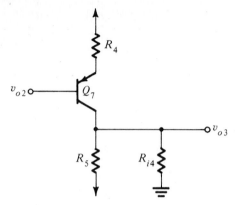

Fig. 10.31 Equivalent circuit for evaluating the gain of the third stage in the amplifier circuit of Fig. 10.28.

To obtain the gain of the third stage we refer to the equivalent circuit shown in Fig. 10.31, where R_{i4} is the input resistance of the output stage formed by Q_8. Using the resistance reflection rule we calculate the value of R_{i4} as

$$R_{i4} = (\beta + 1)(r_{e8} + R_6)$$

where

$$r_{e8} = \frac{25}{5} = 5\ \Omega$$
$$R_{i4} = 101(5 + 3,000) = 303.5\ \text{k}\Omega$$

The gain of the third stage is given by

$$A_3 \equiv \frac{v_{o3}}{v_{o2}} = -\frac{(R_5\ \|\ R_{i4})}{r_{e7} + R_4}$$
$$= -\frac{(15.7\ \text{k}\Omega\ \|\ 303.5\ \text{k}\Omega)}{2.325\ \text{k}\Omega} = -6.42\ \text{V/V}$$

Finally, to obtain the gain A_4 of the output stage we refer to the equivalent circuit in Fig. 10.32 and write

$$A_4 \equiv \frac{v_o}{v_{o3}} = \frac{R_6}{R_6 + r_{e8}}$$
$$= \frac{3,000}{3,000 + 5} = 0.998 \simeq 1$$

The overall voltage gain of the amplifier can then be obtained as follows:

$$\frac{v_o}{v_{id}} = A_1 A_2 A_3 A_4 = 8,520.6\ \text{V/V}$$

or 78.6 dB.

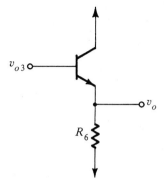

Fig. 10.32   The output stage of the amplifier circuit of Fig. 10.28.

To obtain the output resistance R_o we "grab hold" of the output terminal in Fig. 10.28 and look into the circuit. By inspection we find

$$R_o = \{R_6 \parallel [r_{e8} + R_5/(\beta + 1)]\}$$

which gives $R_o = 152 \ \Omega$.

$\cdot \quad \cdot \quad \cdot$

EXERCISE

10.17 Use the results of Example 10.4 to calculate the voltage gain of the amplifier in Fig. 10.28 when it is connected to a source having a resistance of 10 kΩ and a load of 1 kΩ. *Ans.* 4,947.2 V/V

10.9 CONCLUDING REMARKS

The purpose of this chapter has been to study the basic configurations of transistor amplifiers, both discrete and integrated. At this point the reader should be fluent in the analysis of relatively complex amplifier circuits for the purpose of determining small-signal gain. Also, the material studied should enable the reader to design his or her own amplifier circuits.

In order to focus attention on the fundamental methods and concepts we used first-order transistor models throughout this chapter. In most cases first-order models are sufficient for an initial design. To more accurately evaluate the design as well as to refine it one then resorts to more elaborate models, such as the hybrid-π model introduced briefly in Chapter 9. With the more elaborate models, however, analysis by hand becomes tedious and one normally resorts to computer-aided analysis.

The complete hybrid-π model will be described and applied in Chapter 11. Chapter 11 also will provide a detailed study of the frequency limitations of transistor amplifiers.

FREQUENCY RESPONSE 11

Introduction

In Chapter 2 we introduced the topic of amplifier frequency response and briefly mentioned the various frequency-response shapes encountered. In addition, the frequency, step, and pulse responses of single-time-constant (STC) networks were studied in detail. This material was found to be directly applicable to the evaluation of the frequency response of op-amp circuits (Chapter 3). Apart from this and the occasional mention of limitations imposed on amplifier frequency response, the detailed study of this important topic has been deferred to the present chapter. This was done for a number of reasons, the most important of which is the need for circuit theoretic concepts and methods that the reader might not have had at the beginning of the book. We refer specifically to the complex frequency variable s and associated concepts such as poles and zeros. These topics are normally found in introductory texts on circuit analysis (see, for example, references 11.1 and 11.2). In the following it will be assumed that the reader is familiar with this material.

After a brief review of s-domain analysis (Section 11.1) and a study of amplifier transfer functions (Section 11.2), the frequency response of the classical common-source amplifier is presented (Section 11.3). The BJT hybrid-π model is studied in Section 11.4 and used to find the frequency response of the common-emitter amplifier (Section 11.5), the common-base amplifier (Section 11.6), and the common-collector amplifier (Section 11.7). A number of two-stage amplifier configurations having the advantage of extended bandwidth are also studied in this chapter. Also, the differential amplifier is considered in detail.

The single-stage and two-stage amplifiers analyzed in this chapter can be used either singly or as building blocks of more complex multistage amplifiers. Frequency-response analysis of multistage amplifiers uses the methods studied in this chapter and will be illustrated further in Chapter 13 in the context of analyzing an op-amp circuit.

Although the emphasis in this chapter is on analysis, the material is readily applicable to design. This is achieved by keeping the analysis relatively simple and thus focusing attention on the mechanisms that limit frequency response and on methods for extending amplifier bandwidth.

11.1   s-DOMAIN ANALYSIS

Most of our work in this chapter will be concerned with finding amplifier gain (voltage gain or current gain) as a transfer function of frequency, $T(\omega)$. We introduced the concepts of transfer function and frequency response in Chapter 2. Here, however, we will carry out our circuit analysis in terms of the *complex frequency variable s*. Thus a capacitance C is replaced by an admittance sC, or equivalently an impedance $1/sC$, and an inductance L is replaced by an impedance sL. Then, using usual circuit-analysis techniques, one derives the network transfer function of interest, $T(s)$.

EXERCISE

11.1   Find the voltage transfer function $T(s) \equiv V_o(s)/V_i(s)$ for the STC network shown in Fig. E11.1.

Ans.
$$T(s) = \frac{1/CR_1}{s + 1/C(R_1 \| R_2)}$$

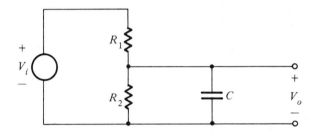

Fig. E11.1

Once the transfer function $T(s)$ is obtained, it can be evaluated for *physical frequencies* by replacing s by $j\omega$. The resulting transfer function $T(j\omega)$ is in general a complex quantity whose magnitude gives the magnitude response (or transmission) and whose angle gives the phase response of the amplifier. [Note that in Chapter 2 we denoted transfer functions by $T(\omega)$; in the following we shall denote the transfer function by $T(j\omega)$ in order to emphasize the relationship to $T(s)$.]

In many cases it will not be necessary to substitute $s = j\omega$ and evaluate $T(j\omega)$; rather, the form of $T(s)$ will reveal many useful facts about the circuit performance. In general, for all the circuits dealt with in this chapter, $T(s)$ can be expressed in the form

$$T(s) = \frac{a_m s^m + a_{m-1} s^{m-1} + \cdots + a_0}{s^n + b_{n-1} s^{n-1} + \cdots + b_0} \tag{11.1}$$

where the coefficients a and b are real numbers and the order m of the numerator is smaller than or equal to the order n of the denominator; the latter is the *order of the network*. Furthermore, for description of a *stable* circuit—that is, one that does not generate signals on its own—the denominator coefficients should be such that *the roots of the denominator polynomial all have negative real parts.* We shall study the problem of amplifier stability in Chapter 12.

Poles and Zeros

An alternative form for expressing $T(s)$ is

$$T(s) = a_m \frac{(s - Z_1)(s - Z_2) \cdots (s - Z_m)}{(s - P_1)(s - P_2) \cdots (s - P_n)} \tag{11.2}$$

where a_m is a multiplicative constant (the coefficient of s^m in the numerator), $Z_1, Z_2, \ldots, Z_m$ are the roots of the numerator polynomial, and $P_1, P_2, \ldots, P_n$ are the roots of the denominator polynomial. $Z_1, Z_2, \ldots, Z_m$ are called the *transfer function zeros* or *transmission zeros,* and $P_1, P_2, \ldots, P_n$ are *the transfer function poles* or *the natural modes* of the network. A transfer function is completely specified in terms of its poles and zeros together with the value of the multiplicative constant.

The poles and zeros can be either real or complex numbers. However, since the a and b coefficients are real numbers, the complex poles (or zeros) must occur in *conjugate pairs.* That is, if $5 + j3$ is a zero, then $5 - j3$ also must be a zero. A zero that is pure imaginary $(\pm j\omega_Z)$ causes the transfer function $T(j\omega)$ to be exactly zero at $\omega = \omega_Z$. Thus the "trap" one places at the input of a television set is a circuit that has a transmission zero at the particular interfering frequency. Real zeros, on the other hand, do not produce transmission nulls (why not?). Finally, note that the transfer function in Eq. (11.1) has $n - m$ zeros at $s = \infty$.

First-Order Functions

All the transfer functions encountered in this chapter have real poles and zeros and can therefore be written as the product of first-order transfer functions of the form

$$T(s) = \frac{a_1 s + a_0}{s + \omega_0} \tag{11.3}$$

where $-\omega_0$ is the location of the real pole. The quantity ω_0 is called the *pole frequency* and is equal to the inverse of the time constant of this single-time-constant (STC) network (see Chapter 2). The constants a_0 and a_1 determine the type of STC network. Specifically, we studied in Chapter 2 two types of STC networks, low pass and high pass. For the low-pass first-order network we have

$$T(s) = \frac{a_0}{s + \omega_0} \tag{11.4}$$

In this case the dc gain is a_0/ω_0, and ω_0 is the corner or 3-dB frequency. Note that this transfer function has one zero at $s = \infty$. On the other hand, the first-order high-pass transfer function has a zero at dc and can be written

$$T(s) = \frac{a_1 s}{s + \omega_0} \tag{11.5}$$

At this point the reader is strongly urged to review the material on STC networks and their frequency and pulse responses in Chapter 2. Of specific interest are the plots of the magnitude and phase responses of the two special kinds of STC networks. Such plots can be employed to generate the magnitude and phase plots of a high-order transfer function, as explained briefly below.

Bode Plots

A simple technique exists for obtaining an approximate plot of the magnitude and phase of a transfer function given its poles and zeros. The technique is particularly useful in the case of real poles and zeros. The method was developed by H. Bode, and the resulting diagrams are called *Bode plots*.

A transfer function of the form depicted in Eq. (11.2) consists of a product of factors of the form $s + a$, where such a factor appears on top if it corresponds to a zero and on the bottom if it corresponds to a pole. It follows that the magnitude response in decibels of the network can be obtained by summing together terms of the form $20 \log_{10} \sqrt{a^2 + \omega^2}$, and the phase response can be obtained by summing terms of the form $\tan^{-1}(\omega/a)$. In both cases the terms corresponding to poles are summed with negative signs. For convenience we can extract the constant a and write the typical magnitude term in the form $20 \log \sqrt{1 + (\omega/a)^2}$. On a plot of decibels versus log frequency this term gives rise to the curve and straight-line asymptotes shown in Fig. 11.1. Here the low-frequency asymptote is a horizontal straight line at 0 dB level and the high-frequency asymptote is a straight line with a slope of 6 dB/octave or, equivalently, 20 dB/decade. The two asymptotes meet at the frequency $\omega = |a|$, which is called *the corner frequency*. As indicated, the actual magnitude plot differs slightly from the value

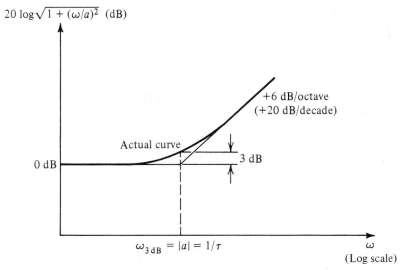

Fig. 11.1   Bode plot for the typical magnitude term. The curve shown applies for the case of a zero. For a pole the high-frequency asymptote should be drawn with a −6 dB/octave slope.

given by the asymptotes; the maximum difference is 3 dB and occurs at the corner frequency.

For $a = 0$—that is, a pole or a zero at $s = 0$—the plot is simply a straight line of 6 dB/octave slope intersecting the 0-dB line at $\omega = 1$.

In summary, to obtain the Bode plot for the magnitude of a transfer function, the asymptotic plot for each pole and zero is first drawn. The slope of the high-frequency asymptote of the curve corresponding to a zero is $+20$ dB/decade, while that for a pole is -20 dB/decade. The various plots are then added together, and the overall curve is shifted vertically by an amount determined by the multiplicative constant of the transfer function.

Example 11.1
An amplifier has the voltage transfer function

$$T(s) = \frac{10s}{(1 + s/10^2)(1 + s/10^5)}$$

Find the poles and zeros and sketch the magnitude of the gain versus frequency. Find approximate values for the gain at $\omega = 10$, 10^3, and 10^6 rad/s.

Solution
The zeros are as follows: one at $s = 0$ and one at $s = \infty$. The poles are as follows: one at $s = -10^2$ and one at $s = -10^5$.

Figure 11.2 shows the asymptotic Bode plots of the different factors of the transfer function. Curve 1, which is a straight line with $+20$ dB/decade slope, corresponds to the s term (that is, the zero at $s = 0$) in the numerator. The pole at $s = -10^2$ results in curve 2, which consists of two asymptotes intersecting at $\omega = 10^2$. Similarly, the pole

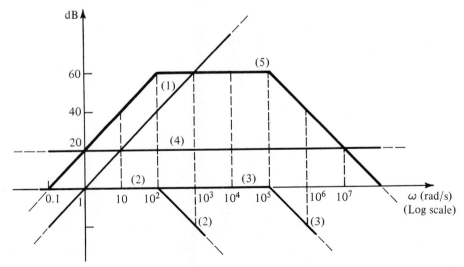

Fig. 11.2   *Bode plots for Example 11.1.*

at $s = -10^5$ is represented by curve 3, where the intersection of the asymptotes is at $\omega = 10^5$. Finally, curve 4 represents the multiplicative constant of value 10.

Adding the four curves results in the asymptotic Bode diagram of the amplifier gain (curve 5). Note that since the two poles are widely separated, the gain will be very close to 10^3 (60 dB) over the frequency range 10^2 to 10^5 rad/s. At the two corner frequencies (10^2 and 10^5 rad/s) the gain will be approximately 3 dB below the maximum of 60 dB. At the three specific frequencies the values of the gain as obtained from the Bode plot and from exact evaluation are as follows:

ω	**Approximate Gain**	**Exact Gain**
10	40 dB	39.96 dB
10^3	60 dB	59.96 dB
10^6	40 dB	39.96 dB

• • •

We next consider the phase plot. Figure 11.3 shows a plot of the typical phase term $\tan^{-1}(\omega/a)$, assuming that a is negative. Also shown is an asymptotic straight-line approximation of the arctan function. The asymptotic plot consists of three straight

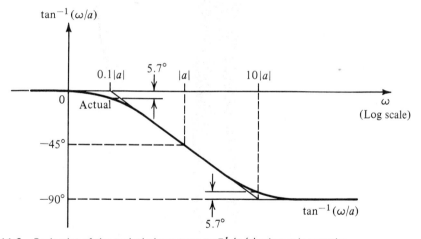

Fig. 11.3 *Bode plot of the typical phase term $\tan^{-1}(\omega/a)$ when a is negative.*

lines. The first is horizontal at $\phi = 0$ and extends up to $\omega = 0.1|a|$. The second line has a slope of $-45°$/decade and extends from $\omega = 0.1|a|$ to $\omega = 10|a|$. The third line has a zero slope and a level of $\phi = -90°$. The complete phase response can be obtained by summing the asymptotic Bode plots of the phase of all poles and zeros.

Example 11.2
Find the Bode plot for the phase of the transfer function of the amplifier considered in Example 11.1.

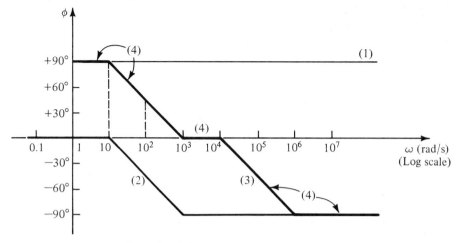

Fig. 11.4   Phase plots for Example 11.2.

Solution
The zero at $s = 0$ gives rise to a constant $+90°$ phase function represented by curve 1 in Fig. 11.4. The pole at $s = -10^2$ gives rise to the phase function

$$\phi_1 = -\tan^{-1} \frac{\omega}{10^2}$$

(the leading minus sign is due to the fact this singularity is a pole). The asymptotic plot for this function is given by curve 2 in Fig. 11.4. Similarly, the pole at $s = -10^5$ gives rise to the phase function

$$\phi_2 = -\tan^{-1} \frac{\omega}{10^5}$$

whose asymptotic plot is given by curve 3. The overall phase response (curve 4) is obtained by direct summation of the three plots.

· · ·

11.2   THE AMPLIFIER TRANSFER FUNCTION

The amplifiers considered in this chapter have voltage-gain functions of either of the two forms shown in Fig. 11.5. Figure 11.5a applies for direct-coupled or dc amplifiers and Fig. 11.5b for capacitively coupled or ac amplifiers. The only difference between the two types is that the gain of the ac amplifier falls off at low frequencies. In the following we shall study the more general response shown in Fig. 11.5b. The response of the dc amplifier follows as a special case.

The Three Frequency Bands

As can be seen from Fig. 11.5b the amplifier gain is almost constant over a wide frequency range called the midband. In this frequency range all capacitances (coupling, bypass, and transistor internal capacitances) have negligible effects and can be ignored

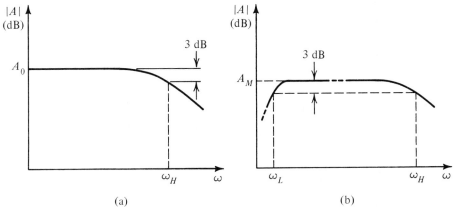

Fig. 11.5   Frequency response for (a) a dc amplifier and (b) a capacitively coupled amplifier.

in gain calculations. At the high-frequency end of the spectrum the gain drops owing to the effect of the internal capacitances of the device. On the other hand, at the low-frequency end of the spectrum the coupling and bypass capacitances no longer act as perfect short circuits and thus cause the gain to drop. The extent of the midband is usually defined by the two frequencies ω_L and ω_H. These are the frequencies at which the gain drops by 3 dB below the value at midband. The amplifier bandwidth is usually defined as

$$\text{BW} = \omega_H - \omega_L$$

and, since $\omega_L \ll \omega_H$,

$$\text{BW} \simeq \omega_H$$

A figure-of-merit for the amplifier is its *gain–bandwidth product,* defined as

$$GB \equiv A_M \omega_H$$

where A_M is the magnitude of midband gain. As will be shown in later sections it is generally possible to trade off gain for bandwidth.

The Gain Function A(s)

The amplifier gain as a function of the complex frequency s can be expressed in the general form

$$A(s) = A_M F_L(s) F_H(s) \tag{11.6}$$

where $F_L(s)$ and $F_H(s)$ are functions that account for the dependence of gain on frequency in the low-frequency band and in the high-frequency band, respectively. For frequencies ω much greater than ω_L the function $F_L(s)$ approaches unity. Similarly, for frequencies ω much smaller than ω_H the function $F_H(s)$ approaches unity. Thus for $\omega_L \ll \omega \ll \omega_H$,

$$A(s) \simeq A_M$$

as should have been expected. It also follows that the gain of the amplifier in the low-frequency band, $A_L(s)$, can be expressed as

$$A_L(s) \simeq A_M F_L(s) \tag{11.7}$$

and the gain in the high-frequency band can be expressed as

$$A_H(s) \simeq A_M F_H(s) \tag{11.8}$$

The function $F_L(s)$ takes the general form

$$F_L(s) = \frac{(s + \omega_{Z1})(s + \omega_{Z2}) \cdots (s + \omega_{Zn_L})}{(s + \omega_{P1})(s + \omega_{P2}) \cdots (s + \omega_{Pn_L})} \tag{11.9}$$

where $\omega_{P1}, \omega_{P2}, \ldots, \omega_{Pn_L}$ are positive numbers representing the frequencies of the n_L low-frequency poles and $\omega_{Z1}, \omega_{Z2}, \ldots, \omega_{Zn_L}$ are positive, negative, or zero numbers representing the n_L zeros. It should be noted from Eq. (11.9) that as s approaches infinity (in fact, as $s = j\omega$ approaches midband frequencies), $F_L(s)$ approaches unity.

In many cases the zeros are at such low frequencies (much smaller than ω_L) as to be of little importance in determining the lower 3-dB frequency ω_L. Also, usually one of the poles—say, ω_{P1}—has a much higher frequency than all other poles. It follows that for frequencies ω close to the midband, $F_L(s)$ can be approximated by

$$F_L(s) \simeq \frac{s}{s + \omega_{P1}} \tag{11.10}$$

which is the transfer function of a first-order high-pass network. In this case the low-frequency response of the amplifier is *dominated* by the pole at $s = -\omega_{P1}$ and the lower 3-dB frequency is approximately equal to ω_{P1},

$$\omega_L \simeq \omega_{P1}$$

If this *dominant-pole approximation* holds, it becomes a simple matter to determine ω_L. Otherwise one has to find the complete Bode plot for $|F_L(j\omega)|$ and thus determine ω_L. As a rule of thumb the dominant-pole approximation can be made if the highest-frequency pole is separated from the nearest pole or zero by at least two octaves.

Consider next the high-frequency end. The function $F_H(s)$ can be expressed in the general form

$$F_H(s) = \frac{(1 + s/\omega_{Z1})(1 + s/\omega_{Z2}) \cdots (1 + s/\omega_{Zn_H})}{(1 + s/\omega_{P1})(1 + s/\omega_{P2}) \cdots (1 + s/\omega_{Pn_H})} \tag{11.11}$$

where $\omega_{P1}, \omega_{P2}, \ldots, \omega_{Pn_H}$ are positive numbers representing the frequencies of the n_H real poles and $\omega_{Z1}, \omega_{Z2}, \ldots, \omega_{Zn_H}$ are positive, negative, or infinite numbers representing the frequencies of the n_H zeros. Note from Eq. (11.11) that as s approaches 0 (in fact as $s = j\omega$ approaches midband frequencies), $F_H(s)$ approaches unity.

In many cases the zeros are either at infinity or at such high frequencies as to be of little significance in determining the upper 3-dB frequency ω_H. If in addition one of the high-frequency poles—say, ω_{P1}—is of much lower frequency than any of the other poles, then the high-frequency response of the amplifier will be *dominated* by this pole and the function $F_H(s)$ can be approximated by

$$F_H(s) \simeq \frac{1}{1 + s/\omega_{P1}} \tag{11.12}$$

which is the transfer function of a first-order low-pass network. It follows that if a dominant high-frequency pole exists, then the determination of ω_H is greatly simplified:

$$\omega_H \simeq \omega_{P1} \tag{11.13}$$

Using Superposition for the Approximate Determination of ω_L and ω_H

If the various poles and zeros characterizing the amplifier transfer function can be determined easily, then we can determine ω_L and ω_H either from a complete Bode diagram or by using the dominant-pole approximation (provided that a dominant pole exists). In many cases, however, it is not a simple matter to determine the poles and zeros. In such cases an approximate superposition method yields quick and reasonable results.

Consider first the high-frequency response. The function $F_H(s)$ of Eq. (11.11) can be expressed in the alternate form

$$F_H(s) = \frac{1 + a_1 s + a_2 s^2 + \cdots + a_n s^n}{1 + b_1 s + b_2 s^2 + \cdots + b_n s^n} \tag{11.14}$$

where the a and b coefficients are related to the zero and pole frequencies, respectively. Specifically, the coefficient b_1 is given by

$$b_1 = \frac{1}{\omega_{P1}} + \frac{1}{\omega_{P2}} + \cdots + \frac{1}{\omega_{Pn_H}} \tag{11.15}$$

It can be shown that the value of b_1 can be obtained by considering the various capacitances in the circuit one at a time. That is, to obtain the contribution of capacitance C_i we reduce all other capacitances to zero and determine the resistance R_i seen by C_i. The process is then repeated for all other capacitances in the circuit. The value of b_1 is computed by summing the individual time constants:

$$b_1 = \sum_{i=1}^{n_H} C_i R_i \tag{11.16}$$

where we have assumed that there are n_H capacitances in the circuit.

This method of determining b_1 is exact; the approximation comes about in using the value of b_1 to determine ω_H. Specifically, if the zeros are not dominant and if one of the poles—say, ω_{P1}—is dominant, then from Eq. (11.15)

$$b_1 \simeq \frac{1}{\omega_{P1}}$$

and the upper 3-dB frequency will be approximately equal to ω_{P1}, which in turn is approximately equal to $1/b_1$:

$$\omega_H \simeq 1/b_1 \qquad (11.17)$$

Here it should be pointed out that in complex circuits we usually do *not* know whether or not a dominant pole exists. Nevertheless, using the computed value of b_1 to determine ω_H via Eq. (11.17) normally yields remarkably good results even if a dominant pole does not exist. This point will be illustrated by an example.

Example 11.3

Consider a dc amplifier with low-frequency gain A_0 (note that in the case of a dc amplifier $A_M = A_0$) and a high-frequency response characterized by two poles,

$$A(s) = \frac{A_0}{(1 + s/\omega_{P1})(1 + s/\omega_{P2})}$$

For each of the following four cases we wish to determine the error incurred in computing ω_H using the superposition method described above:

  (a) $\omega_{P2} = 10\omega_{P1}$
  (b) $\omega_{P2} = 4\omega_{P1}$
  (c) $\omega_{P2} = 2\omega_{P1}$
  (d) $\omega_{P2} = \omega_{P1}$

Solution

By substituting $s = j\omega$ and taking magnitudes, the exact value of ω_H can be determined from

$$|A(j\omega_H)| = \frac{A_0}{\sqrt{2}} = \frac{A_0}{\sqrt{[1 + (\omega_H/\omega_{P1})^2][1 + (\omega_H/\omega_{P2})^2]}} \qquad (11.18)$$

On the other hand, the superposition method yields

$$b_1 = \frac{1}{\omega_{P1}} + \frac{1}{\omega_{P2}}$$

Use of this value to determine ω_H results in

$$\omega_H \simeq \left(\frac{1}{\omega_{P1}} + \frac{1}{\omega_{P2}}\right)^{-1} \qquad (11.19)$$

For the four cases given we use Eqs. (11.18) and (11.19) to obtain the exact and approximate values, respectively, of ω_H (as a fraction of ω_{P1}):

Case	Exact ω_H	Approximate ω_H	Error
(a) $\omega_{P2} = 10\omega_{P1}$	$0.99\omega_{P1}$	$0.91\omega_{P1}$	-8%
(b) $\omega_{P2} = 4\omega_{P1}$	$0.95\omega_{P1}$	$0.8\omega_{P1}$	-16%
(c) $\omega_{P2} = 2\omega_{P1}$	$0.84\omega_{P1}$	$0.67\omega_{P1}$	-20%
(d) $\omega_{P2} = \omega_{P1}$	$0.64\omega_{P1}$	$0.5\omega_{P1}$	-22%

Note that even in the extreme case of identical poles the error incurred using the simple superposition method is only 22%. Furthermore, the error introduced results in a more conservative estimate of bandwidth.

. . .

The superposition method has an added advantage: it tells the circuit designer which of the various capacitances is significant in determining the amplifier frequency response. Specifically, the relative contribution of the various capacitances to the effective time constant b_1 is immediately obvious.

Next we outline the use of the superposition method for determining an approximate value for ω_L. The function $F_L(s)$ of Eq. (11.9) can be expressed in the alternate form

$$F_L(s) = \frac{s^{n_L} + d_1 s^{n_L-1} + \cdots}{s^{n_L} + e_1 s^{n_L-1} + \cdots} \tag{11.20}$$

where the coefficients d and e are related to the zero and pole frequencies, respectively. Specifically, the coefficient e_1 is given by

$$e_1 = \omega_{P1} + \omega_{P2} + \cdots + \omega_{Pn_{L-1}} \tag{11.21}$$

It can be shown that the exact value of e_1 can be obtained by analyzing the circuit, taking one capacitance at a time. To be more specific, if capacitance C_i is under consideration, then all other capacitances (that contribute to the low-frequency transfer function) are set *equal to infinity* (in order to reduce the contribution of their reactances to zero) and the resistance R_i seen by C_i is determined. This process is then repeated for all other capacitances, and the value of e_1 is computed from

$$e_1 = \sum_{i=1}^{n_L} \frac{1}{C_i R_i} \tag{11.22}$$

Provided that none of the zeros is dominant and that a dominant pole exists, the exact value of e_1 can be used to compute an approximate value for ω_L. Thus if ω_{P1} is a dominant pole, then its frequency is much higher than ω_{P2}, ω_{P3}, and so on. From Eq. (11.21) we have

$$e_1 \simeq \omega_{P1}$$

We also know that in this case

$$\omega_L \simeq \omega_{P1}$$

which leads to

$$\omega_L \simeq e_1 \tag{11.23}$$

EXERCISE

11.2 Consider an amplifier whose low-frequency response is characterized by two real poles with frequencies of 0.5 rad/s and 1 rad/s, a zero at $s = 0$, and a zero at $s = -0.25$ rad/s. Find the exact value of ω_L. Also find the approximate value of ω_L obtained using the superposition method.
Ans. 1.4 rad/s; 1.5 rad/s

11.3   FREQUENCY RESPONSE OF THE COMMON-SOURCE AMPLIFIER

In this section the frequency response of the classical capacitor-coupled common-source amplifier stage, shown in Fig. 11.6, will be analyzed. The analysis given applies equally well to MOSFETs. Furthermore, the limitations on the high-frequency response of direct-coupled common-source amplifiers are identical to those of the capacitively coupled circuit.

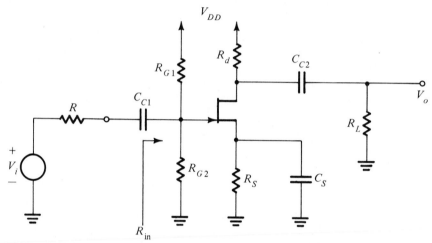

Fig. 11.6   *The classical capacitively coupled common-source amplifier.*

Figure 11.7 shows the equivalent circuit model that applies to FETs of all types. We studied this model in Chapters 7 and 8. Typically, capacitances C_{gs}, C_{gd}, and C_{ds} are in the range of 1 to 3 pF. Although quite small, C_{gd} gets multiplied by a large number (Miller effect) and effectively appears in parallel with C_{gs}. The resulting large input capacitance together with the resistance of the input signal source form a low-pass filter that limits the high-frequency response. Capacitance C_{ds} appears in parallel with the load resistance, thus forming another low-pass filter. It follows that the three internal capacitances C_{gs}, C_{gd}, and C_{ds} cause the amplifier gain to drop at the high-

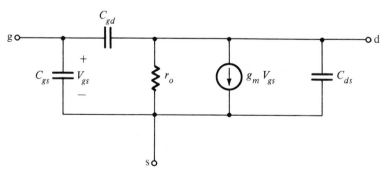

Fig. 11.7   *Equivalent circuit for JFET and MOSFET.*

frequency end of the spectrum. At midband, and certainly in the low-frequency band, these capacitances have very high reactances and hence can be neglected.

The dependence of gain on frequency in the low-frequency band is caused by the coupling and bypass capacitances exhibiting relatively large reactances. Thus at low frequencies these capacitances can no longer be assumed to act as perfect short circuits. This assumption, however, is safe to make in the midband and certainly in the high-frequency band.

Analysis in the Low-Frequency Band

To find the low-frequency gain of the circuit in Fig. 11.6 we shall start at the signal source and proceed toward the load in a step-by-step manner. Using the voltage-divider rule at the input side we can find the voltage V_g (between gate and ground) as

$$V_g(s) = V_i(s) \frac{R_{in}}{R_{in} + R + 1/sC_{C1}}$$

where $R_{in} = (R_{G1} \parallel R_{G2})$. Thus the transfer function from the input to the gate is given by

$$\frac{V_g(s)}{V_i(s)} = \frac{R_{in}}{R_{in} + R} \frac{s}{s + 1/C_{C1}(R_{in} + R)} \qquad (11.24)$$

which is a high-pass function indicating that C_{C1} introduces a zero at zero frequency (dc) and a real pole with a frequency ω_{P1},

$$\omega_{P1} = \frac{1}{C_{C1}(R_{in} + R)} \qquad (11.25)$$

Note that we could have arrived at this result by inspection of the input circuit of the amplifier using the techniques of STC network analysis of Chapter 2. Specifically, the input circuit is a high-pass STC network with a time constant equal to C_{C1} multiplied by the total resistance seen by C_{C1}; ω_{P1} is simply the inverse of this time constant.

The next step in the analysis is to find the drain current $I_d(s)$:

$$I_d(s) = I_s(s) = \frac{V_g(s)}{1/g_m + Z_S} \qquad (11.26)$$

where we have made use of the fact that the resistance between gate and source is equal to $1/g_m$; thus the total impedance between gate and ground, on the source side, is $1/g_m$ in series with Z_S, which denotes the parallel equivalent of R_S and C_S. Equation (11.26) can be rewritten

$$I_d(s) = g_m V_g(s) \frac{Y_S}{g_m + Y_S}$$

where

$$Y_S = \frac{1}{Z_S} = \frac{1}{R_S} + sC_S$$

Thus,

$$I_d(s) = g_m V_g(s) \frac{1/R_S + sC_S}{g_m + 1/R_S + sC_S}$$

that is,

$$I_d(s) = g_m V_g(s) \frac{s + 1/C_S R_S}{s + (g_m + 1/R_S)/C_S} \tag{11.27}$$

which indicates that the bypass capacitor C_S introduces a real zero and a real pole. The real zero has a frequency ω_Z,

$$\omega_Z = \frac{1}{C_S R_S} \tag{11.28}$$

while the frequency of the real pole is given by

$$\omega_{P2} = \frac{g_m + 1/R_S}{C_S} = \frac{1}{C_S(R_S \parallel 1/g_m)} \tag{11.29}$$

It thus can be seen that ω_Z will always be lower in value than ω_{P2}.

It is instructive to interpret physically the above results regarding the effects of C_S. C_S introduces a zero at the value of s that makes Z_S infinite, which makes physical sense because an infinite Z_S will cause I_d and hence V_o to be zero. The pole frequency is the inverse of the time constant formed by multiplying C_S by the resistance seen by the capacitor. To evaluate the latter resistance we ground the signal source (note that the network poles, or natural modes, are not a function of the excitation) and grab hold of the terminals of C_S. The resistance seen by C_S will be R_S in parallel with the resistance between source and gate, which is $1/g_m$.

Having determined $I_d(s)$, we can now obtain the output voltage using the output equivalent circuit in Fig. 11.8a. There is a slight approximation in this equivalent circuit: the resistance r_o is shown connected between drain and ground rather than between drain and source in spite of the fact that the source terminal is no longer at

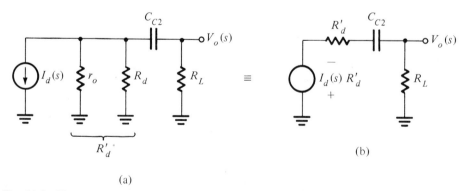

(a)

(b)

Fig. 11.8   The output equivalent circuit (at low frequency) for the amplifier in Fig. 11.6.

ground potential because C_S is not acting as a perfect bypass. However, since the effect of r_o is small anyway, this approximation is valid.

Figure 11.8b shows the equivalent output circuit after application of Thévenin's theorem. From this figure we obtain

$$V_o(s) = -I_d(s)(R_d\|r_o\|R_L)\frac{s}{s + 1/C_{C2}[R_L + (R_d\|r_o)]} \tag{11.30}$$

Thus C_{C2} introduces a zero at zero frequency (dc) and a real pole with a frequency ω_{P3},

$$\omega_{P3} = \frac{1}{C_{C2}[R_L + (R_d\|r_o)]} \tag{11.31}$$

Again the frequency of this pole could have been found by inspection: it equals the inverse of the time constant found by multiplying C_{C2} by the total resistance seen by that capacitor.

The low-frequency amplifier gain $A_L(s)$ can be found by combining Eqs. (11.24), (11.27), and (11.30):

$$A_L(s) = \frac{V_o(s)}{V_i(s)} = A_M\frac{s}{s + \omega_{P1}}\frac{s + \omega_Z}{s + \omega_{P2}}\frac{s}{s + \omega_{P3}} \tag{11.32}$$

where the midband gain A_M is given by

$$A_M = -\frac{R_{in}}{R_{in} + R}g_m(R_d\|r_o\|R_L) \tag{11.33}$$

and where ω_{P1}, ω_Z, ω_{P2}, and ω_{P3} are given by Eqs. (11.25), (11.28), (11.29), and (11.31), respectively. Note from Eq. (11.32) that as the frequency $s = j\omega$ becomes much larger in magnitude than ω_{P1}, ω_{P2}, ω_{P3}, and ω_Z, the gain approaches the midband value A_M.

Having determined the low-frequency poles and zeros we can employ the techniques of Section 11.2 to find the lower 3-dB frequency ω_L.

Analysis in the High-Frequency Band

The high-frequency equivalent circuit of the common-source amplifier is shown in Fig. 11.9a, where R_{in} denotes $(R_{G1}\|R_{G2})$. A simplified version of the equivalent circuit is shown in Fig. 11.9b, which is obtained by applying Thévenin's theorem at the input side and by combining the three resistances r_o, R_d, and R_L.

Since C_{gd} is usually small, one can assume that the current through it is much smaller than the current of the controlled source, g_mV_{gs}. The same approximation can be done with the current through C_{ds}, with the result that the current through R_L' is approximately equal to g_mV_{gs}, thus

$$V_o \simeq -g_mR_L'V_{gs} \tag{11.34}$$

Thus we know the ratio of the voltages at the two ends of C_{gd}, which enables us to apply Miller's theorem (see Section 2.6) to obtain the equivalent circuit shown in Fig. 11.9c. In the output part of this circuit $C_{gd} + C_{ds}$ can be normally neglected because the pole

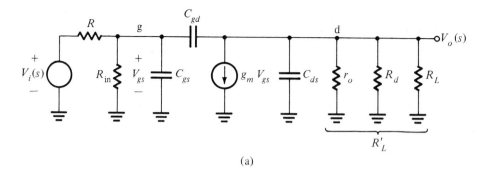

(a)

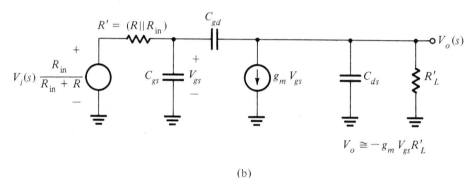

(b)

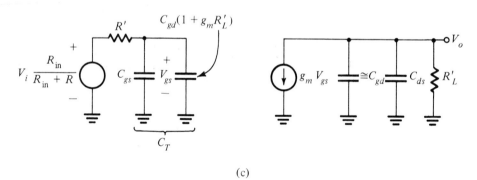

(c)

Fig. 11.9 Equivalent circuits for evaluating the high-frequency response of the amplifier of Fig. 11.6.

caused by $C_{gd} + C_{ds}$ in parallel with R_L usually occurs at a very high frequency. If we neglect $C_{gd} + C_{ds}$, then the output voltage will be given by Eq. (11.34), which is consistent with our previous assumptions. At the input side we recognize a first-order low-pass filter section formed by the total input capacitance

$$C_T = C_{gs} + C_{gd}(1 + g_m R_L')$$

and the resistance R',

$$R' = (R \| R_{in})$$

Thus the high-frequency response of the amplifier will be characterized by a dominant pole caused by the input low-pass network. The upper 3-dB frequency ω_H will be approximately equal to the frequency of the dominant pole,

$$\omega_H \simeq \frac{1}{C_T R'} \tag{11.35}$$

We may therefore express the high-frequency gain as

$$A_H(s) = A_M \frac{1}{1 + s/\omega_H} \tag{11.36}$$

where as before A_M denotes the midband gain given in Eq. (11.33). Note from Eq. (11.36) that as the frequency $s = j\omega$ becomes much smaller in magnitude than ω_H the gain approaches the midband value A_M.

Before leaving this section we wish to draw the reader's attention to the important role played by the small feedback capacitance C_{gd} in determining the high-frequency response of the common-source amplifier. Because the voltages at the two ends of C_{gd} are in the ratio of $-g_m R'_L$, which is a large number approximately equal to the midband gain, C_{gd} gives rise to a large capacitance, $C_{gd}(1 + g_m R'_L)$, across the input terminals of the amplifier. This is the Miller effect, discussed in Section 2.6. It follows that to increase the upper 3-dB or cutoff frequency of the amplifier one has to eliminate or reduce the Miller effect. This is achieved in the grounded-gate—or *common-gate*—configuration and in the *common-base* configuration. These and other special configurations for *wideband amplifiers* will be studied in later sections.

Example 11.4
We wish to determine the complete frequency response of the amplifier in Fig. 11.6 for $V_{DD} = 20$ V, $R = 100$ kΩ, $R_{G1} = 1.4$ MΩ, $R_{G2} = 0.6$ MΩ, $R_S = 3.5$ kΩ, $R_d = 5$ kΩ, $R_L = 10$ kΩ, $C_{C1} = 0.1$ μF, $C_S = 10$ μF, and $C_{C2} = 1$ μF. The JFET has $V_P = -2$ V, $I_{DSS} = 8$ mA, $C_{gs} = 3$ pF, $C_{gd} = 1$ pF, and $C_{ds} = 1$ pF.

Solution
With the methods of Chapter 7 the following dc operating point is determined:

$$I_D = 2 \text{ mA}; \qquad V_{GS} = -1 \text{ V}; \qquad V_D = +10 \text{ V}$$

At this operating point the transconductance is

$$g_m = \frac{2 I_{DSS}}{-V_P} \sqrt{\frac{I_D}{I_{DSS}}}$$

Thus

$$g_m = \frac{2 \times 8}{2} \sqrt{\frac{2}{8}} = 4 \text{ mA/V}$$

The midband gain can be determined as follows: The input resistance R_{in} is given by

$$R_{in} = \frac{R_{G1} R_{G2}}{R_{G1} + R_{G2}} = \frac{1.4 \times 0.6}{2} = 420 \text{ k}\Omega$$

Since no value has been given for r_o, we shall assume it to be large enough to have a negligible effect on the value of A_M. Thus we can write

$$A_M = \frac{R_{in}}{R_{in} + R} \times -g_m(R_d \| R_L)$$

$$A_M = -\frac{420}{520} \times 4 \times \frac{5 \times 10}{5 + 10} = -10.8$$

Thus the amplifier has a midband gain of 20.6 dB.

We next determine the low-frequency poles and zeros as follows: Coupling capacitor C_{C1} causes a zero at dc and a real pole at a frequency f_{P1},

$$f_{P1} = \frac{1}{2\pi C_{C1}(R + R_{in})}$$

$$= \frac{1}{2\pi \times 0.1 \times 10^{-6} \times (100 + 420) \times 10^3}$$

Thus $f_{P1} \simeq 3$ Hz.

The bypass capacitor C_S introduces a real zero with a frequency f_Z,

$$f_Z = \frac{1}{2\pi C_S R_S} = \frac{1}{2\pi \times 10 \times 10^{-6} \times 3.5 \times 10^3} \simeq 4.5 \text{ Hz},$$

and a real pole with a frequency f_{P2},

$$f_{P2} = \frac{1}{2\pi C_S(R_S \| 1/g_m)} = \frac{1}{2\pi \times 10 \times 10^{-6}(3.5\text{K} \| 0.25\text{K})} \simeq 68.2 \text{ Hz}$$

Coupling capacitor C_{C2} introduces a zero at dc and a real pole at f_{P3},

$$f_{P3} = \frac{1}{2\pi C_{C2}(R_d + R_L)} = \frac{1}{2\pi \times 10^{-6} \times (5 + 10) \times 10^3} \simeq 10.6 \text{ Hz}$$

Examination of the values of the low-frequency singularities computed above reveals that the amplifier has a dominant low-frequency pole at f_{P2}. Thus the gain drops by 3 dB at a frequency f_L almost equal to f_{P2}, $f_L \simeq 68.2$ Hz.

To determine the dominant pole at high frequency, we first evaluate the total input capacitance C_T,

$$C_T = C_{gs} + C_{gd}(1 + g_m R_L')$$

where $R_L' = (R_L \| R_d)$. Thus

$$C_T = 3 + 1\left(1 + 4 \times \frac{5 \times 10}{15}\right) \simeq 15.3 \text{ pF}$$

This capacitance together with the equivalent source resistance R',

$$R' = \frac{RR_{\text{in}}}{R + R_{\text{in}}} = \frac{100 \times 420}{100 + 420} = 80.8 \text{ k}\Omega$$

forms a low-pass filter at the input. The dominant pole at high frequency is due to this low-pass filter. Thus the upper 3-dB frequency is approximately

$$f_H = \frac{1}{2\pi C_T R'} = \frac{1}{2\pi \times 15.3 \times 10^{-12} \times 80.8 \times 10^3} = 128.7 \text{ kHz}$$

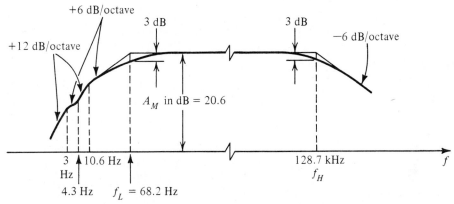

Fig. 11.10   *Bode diagram for the magnitude of the gain of the amplifier in Example 11.4.*

Figure 11.10 shows a sketch of the Bode plot of the amplifier considered. Since the amplifier displays a dominant low-frequency pole and a dominant high-frequency pole, we may neglect the other singularities and express the transfer function as

$$A(s) = A_M \frac{s}{s + \omega_L} \frac{1}{1 + s/\omega_H}$$

Note that at frequencies $\omega_L \ll \omega \ll \omega_H$, $A(s) \simeq A_M$.

· · ·

Example 11.5
If the amplifier analyzed in Example 11.4 is excited by a pulse of 0.2 V height and 0.2 ms width, find the height, the rise and fall times, and the sag of the output pulse.

Solution
The output pulse will have a height V,

$$V = A_M \times 0.2 = 10.8 \times 0.2 = 2.16 \text{ V}$$

The rise and fall times of the output pulse will be determined by the dominant high-frequency pole (see Chapter 2),

$$t_r = t_f \simeq 2.2\tau_H$$

where

$$\tau_H = \frac{1}{\omega_H} = \frac{1}{2\pi f_H} = \frac{1}{2\pi \times 128.7 \times 10^3} \simeq 1.2 \; \mu s$$

Thus

$$t_r = t_f = 2.64 \; \mu s$$

On the other hand, the sag or decay in the amplitude of the output pulse will be determined by the dominant low-frequency pole (see Chapter 2). Specifically, if the loss in pulse height is denoted by ΔV, then

$$\text{sag} = \frac{\Delta V}{V} = \frac{t_p}{\tau_L} = \omega_L t_p = 2\pi f_L t_p$$

Thus

$$\text{sag} = 2\pi \times 68.2 \times 0.2 \times 10^{-3} = 0.086 = 8.6\%$$

$\bullet \quad \bullet \quad \bullet$

EXERCISES

11.3 Use the equivalent circuit in Fig. 11.9b to find expressions for the resistance R_{gs} seen by C_{gs} (with $C_{gd} = C_{ds} = 0$), the resistance R_{gd} seen by C_{gd} (with $C_{gs} = C_{ds} = 0$), and the resistance R_{ds} seen by C_{ds} (with $C_{gs} = C_{gd} = 0$). (*Hint:* Figure E11.3 shows the circuit for determining R_{gd}.)

Ans. $R_{gs} = R'$; $R_{gd} = R' + (1 + g_m R')R_L'$; $R_{ds} = R_L'$

11.4 Use the results of Exercise 11.3 to find f_H of the amplifier in Example 11.4. Use the superposition method. What is the percentage contribution of C_{gd} to the effective time constant?

Ans. 113 kHz; 82.5%

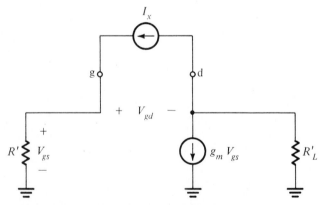

Fig. E11.3   Circuit for determining $R_{gd} \equiv v_{gd}/I_x$, where I_x is a test current source.

11.4   THE HYBRID-π EQUIVALENT CIRCUIT MODEL

Before considering the frequency response analysis of BJT amplifiers we shall take a closer look at the hybrid-π model and give methods for the determination of the model parameters from terminal measurements or from data sheet specifications.

The Low-Frequency Model

Figure 11.11 shows the complete low-frequency hybrid-π equivalent circuit model. In addition to the intrinsic model parameters r_π and g_m this model includes three resistances r_o, r_μ, and r_x.

Resistance r_o models the slight effect of the collector voltage on the collector current in the active region of operation. Typically r_o is in the range of tens to hundreds of kΩ and its value is inversely proportional to the dc bias current. Since g_m is directly

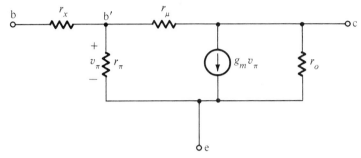

Fig. 11.11   Complete low-frequency hybrid-π model.

proportional to the bias current, the product $g_m r_o$, denoted μ, is a constant for a given transistor, with a value of a few thousand. In most of our previous analysis we ignored r_o. Including r_o is essential in some applications, as will be demonstrated shortly.

The resistance r_μ models the effect of the collector voltage on the base current, and its value is usually larger than r_o. It can be shown from physical considerations of device operation that r_μ is at least equal to $\beta_0 r_o$ (β_0 denotes the value of β at low frequencies). For modern IC transistors r_μ is closer to $10\beta_0 r_o$. Because of its extremely large value and because including r_μ in the equivalent circuit model destroys its unilateral character and thus complicates the analysis, one usually ignores r_μ. There are very special situations, however, where one has to include r_μ. We have already encountered one such case (in evaluating the common-mode input resistance of a differential amplifier).

The resistance r_x models the resistance of the silicon material of the base region between the base terminal b and a fictitious internal, or intrinsic, base terminal b'. The latter node represents the base side of the emitter–base junction. Typically r_x is a few tens of ohms, and its value depends on the current level in a rather complicated manner. Since r_x is much smaller than r_π, the effect of r_x is negligible at low frequencies. Its presence is felt, however, at high frequencies, where the input impedance of the transistor becomes highly capacitive. This point will become apparent later in this section. In conclusion, r_x can usually be neglected in low-frequency applications.

EXERCISE

11.5   Each of the common-emitter i_C-v_{CE} curves is measured with the value of base current held constant. This is equivalent to assuming that the signal current $i_b = 0$; that is, from a signal point of view the base is open-circuited. Use the equivalent circuit in Fig. E11.5 to determine the slope of the i_C-v_{CE} curves in the active region.

Ans. $i_c/v_{ce} \simeq 1/r_o + \beta_0/r_\mu$

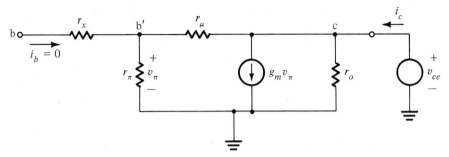

Fig. E11.5

Example 11.6
Derive an expression for the output resistance of the current source shown in Fig. 11.12a. Evaluate the output resistance assuming that the following values are specified for the BJT: $\beta_0 = 100$, $\mu = 4,000$, and $r_\mu = 5\beta_0 r_o$.

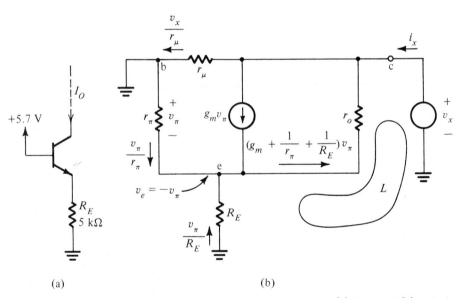

(a)   (b)

Fig. 11.12 *The evaluation of the output resistance of a current source; (a) circuit and (b) equivalent circuit with a test voltage v_x applied; $R_o \equiv v_x/i_x$.*

Solution
Figure 11.12b shows the equivalent circuit of the current source for the purpose of finding the output resistance. Note that the base is shown grounded, since it is fed with a constant dc source. The voltage source v_x applies a test signal, with the output resistance obtained as

$$R_o \equiv \frac{v_x}{i_x}$$

Some of the analysis is shown in Fig. 11.12b. A loop equation for loop L results in

$$v_x = -r_o \left(g_m + \frac{1}{r_\pi} + \frac{1}{R_E} \right) v_\pi - v_\pi$$

which can be arranged to yield

$$v_\pi = \frac{-v_x}{1 + r_o (g_m + 1/r_\pi + 1/R_E)} \tag{11.37}$$

A node equation at the collector provides

$$i_x = -\left(g_m + \frac{1}{r_\pi} + \frac{1}{R_E} \right) v_\pi + g_m v_\pi + \frac{v_x}{r_\mu} \tag{11.38}$$

Substituting for v_π from Eq. (11.37) and with some manipulations, we obtain

$$i_x = \frac{v_x}{r_o} \frac{1 + R_E/r_\pi}{1 + R_E/r_e + R_E/r_o} + \frac{v_x}{r_\mu}$$

Thus the output resistance is given by

$$R_o = \left(r_o \frac{1 + R_E/r_e + R_E/r_o}{1 + R_E/r_\pi} \bigg\| r_\mu \right)$$

Since $r_o \gg r_e$,

$$R_o \simeq \left(r_o \frac{1 + R_E/r_e}{1 + R_E/r_\pi} \bigg\| r_\mu \right) \tag{11.39}$$

Note that since $r_\pi \gg r_e$, the term

$$\frac{1 + R_E/r_e}{1 + R_E/r_\pi}$$

will be greater than unity. That is, including a resistance R_E in the emitter results in an increase in the collector output resistance. This is further illustrated by substituting the numerical values given. The dc voltage at the emitter is approximately 5 V resulting in $I_E \simeq 1$ mA. Thus $r_e \simeq 25\ \Omega$, $g_m = 40$ mA/V, $r_\pi = \beta_0/g_m = 2.5$ kΩ, $r_o = \mu/g_m = 100$ kΩ, and $r_\mu = 5\beta_0 r_o = 50$ MΩ. Substituting these values in Eq. (11.39) gives $R_o = 5.9$ MΩ, which is much larger than r_o.

· · ·

The use of an emitter resistance R_E to increase the output resistance is a powerful technique that is widely employed. In Chapter 12 we will see that this increase happens because of the negative-feedback action of R_E.

Determination of the Low-Frequency Model Parameters

The transistor is a three-terminal device that can be converted into a two-port network by grounding one of its terminals. It therefore can be characterized by one of the various two-port parameter sets studied in Section 2.7. For the BJT at low frequencies, the h

parameters have been found to be the most convenient. In the following we briefly discuss methods for measuring the *h* parameters for a transistor biased to operate in the active region. We also derive formulas relating the hybrid-π model parameters to the measured *h* parameters. These formulas allow us to determine the hybrid-π parameters. It should be emphasized that because the hybrid-π model is closely related to the physical operation of the transistor, its use provides the circuit designer with considerable insight into circuit operation. Thus our interest in *h* parameters is solely for the purpose of determining the hybrid-π component values.

If the emitter of a transistor biased in the active mode is grounded, port 1 is defined to be between base and emitter, and port 2 is defined to be between collector and emitter, then for small signals around the given bias point we can write

$$v_b = h_{ie}i_b + h_{re}v_c \qquad (11.40)$$
$$i_c = h_{fe}i_b + h_{oe}v_c \qquad (11.41)$$

These are the defining equations of the common-emitter *h* parameters where rather than using the notation h_{11}, h_{12}, and so on, we have assigned more descriptive subscripts to the *h* parameters: *i* means input, *r* means reverse, *f* means forward, *o* means output, and the added *e* denotes a common emitter.

For measurement of h_{ie} and h_{fe} the circuit shown in Fig. 11.13 can be used. Here R_B is a large resistance that together with V_{BB} determines I_B. The resistance R_C is used

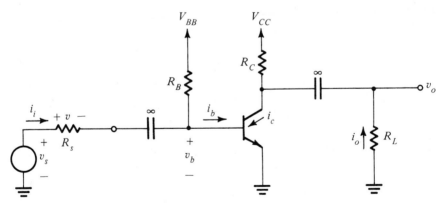

Fig. 11.13   Circuit for measuring h_{ie} and h_{fe}.

to establish the desired dc voltage at the collector, and a very small resistance R_L is used to enable measuring the signal current in the collector. Since R_L is small, the collector is effectively short-circuited to ground and

$$i_c \simeq i_o = -\frac{v_o}{R_L}$$

The input signal current i_i is determined by measuring the voltage v across a known resistance R_s. If R_B is large, then

$$i_b \simeq i_i = \frac{v}{R_s}$$

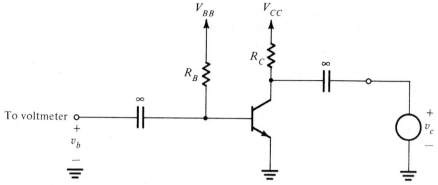

Fig. 11.14 Circuit for measuring h_{re}.

The input signal voltage v_b can be measured directly at the base. Using these measured values one can compute h_{ie} and h_{fe}:

$$h_{ie} = \frac{v_b}{i_b} \qquad \text{and} \qquad h_{fe} = \frac{i_c}{i_b}$$

To measure h_{re} we use the circuit shown in Fig. 11.14. Here again R_B should be large (much larger than r_π), and the voltmeter used to measure v_b should have a high input resistance to ensure that the base is effectively open-circuited. The value of h_{re} can then be determined:

$$h_{re} = \frac{v_b}{v_c}$$

Finally, from Eq. (11.41) we note that h_{oe} is the output conductance with the base open-circuited, which is the definition of the slope of the i_C-v_{CE} characteristic curves. Thus h_{oe} can be most conveniently determined from the static common-emitter characteristics.

Expressions for h_{ie} and h_{fe} in terms of the hybrid-π model parameters can be derived by analyzing the equivalent circuit in Fig. 11.15. This analysis yields

$$h_{ie} = r_x + (r_\pi \| r_\mu)$$

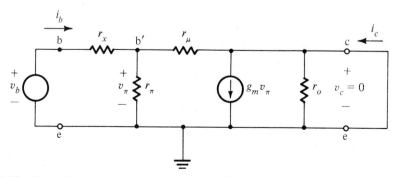

Fig. 11.15 Circuit for deriving expressions for h_{ie} and h_{fe}.

which can be approximated as

$$h_{ie} \simeq r_x + r_\pi \tag{11.42}$$

$$h_{fe} = g_m r_\pi \tag{11.43}$$

Here we should note that by definition h_{fe} is identical to the ac β or β_{ac} and that the value given by Eq. (11.43) indeed corresponds to the formula we used earlier for the low-frequency β, β_0.

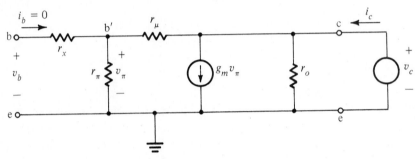

Fig. 11.16   Circuit for deriving expressions for h_{re} and h_{oe}.

To derive expressions for h_{re} and h_{oe} we use the equivalent circuit model in Fig. 11.16 and obtain

$$h_{re} = \frac{r_\pi}{r_\pi + r_\mu}$$

which can be approximated as

$$h_{re} \simeq \frac{r_\pi}{r_\mu} \tag{11.44}$$

$$h_{oe} \simeq \frac{1}{r_o} + \frac{\beta_0}{r_\mu} \tag{11.45}$$

This last expression is identical to that given for the slope of the i_C-v_{CE} characteristics (Exercise 11.5).

The above expressions can be used to determine the values of the hybrid-π model parameters from the measured h parameters as follows:

$$g_m = \frac{I_C}{V_T} \tag{11.46}$$

$$r_\pi = \frac{h_{fe}}{g_m} \tag{11.47}$$

$$r_x = h_{ie} - r_\pi \tag{11.48}$$

$$r_\mu = \frac{r_\pi}{h_{re}} \tag{11.49}$$

$$r_o = \left(h_{oe} - \frac{h_{fe}}{r_\mu} \right)^{-1} \tag{11.50}$$

It should be noted, however, that since normally $r_x \ll r_\pi$, Eq. (11.48) does not provide an accurate determination of r_x. In fact, there is no accurate way to determine r_x at low frequencies, which should come as no surprise, since r_x plays a minor role at low frequencies.

EXERCISE

11.6   The following parameters were measured on a transistor biased at $I_C = 1$ mA: $h_{ie} = 2.6$ kΩ, $h_{fe} = 100$, $h_{re} = 0.5 \times 10^{-4}$, $h_{oe} = 1.2 \times 10^{-5}$ A/V. Determine the values of g_m, r_π, r_x, r_μ, and r_o.
      Ans.   40 mA/V; 2.5 kΩ; 100 Ω; 50 MΩ; 100 kΩ

The High-Frequency Model

With the exception of the extremely large resistance r_μ, the high-frequency hybrid-π model shown in Fig. 11.17 includes all the resistances of the low-frequency model as well as two capacitances: the emitter–base capacitance C_π and the collector–base capacitance C_μ. The resistance r_μ is omitted because even at moderate frequencies the

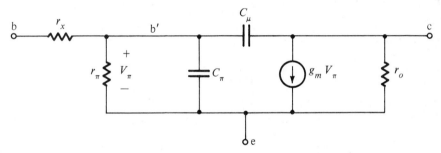

Fig. 11.17   The high-frequency hybrid-π model.

reactance of C_μ is much smaller than r_μ. As mentioned in Section 9.7, the emitter–base capacitance C_π is composed of two parts: a diffusion capacitance, which is proportional to the dc bias current, and a depletion-layer capacitance, which depends on the value of V_{BE}. The collector–base capacitance C_μ is entirely a depletion capacitance, and its value depends on V_{CB}. Typically C_π is in the range of a few picofarads to a few tens of picofarads, and C_μ is in the range of a fraction of a picofarad to a few picofarads.

The Cutoff Frequency

The transistor data sheets do not usually specify the value of C_π. Rather, the behavior of h_{fe} versus frequency is normally given. In order to determine C_π and C_μ we shall derive an expression for h_{fe} as a function of frequency in terms of the hybrid-π components. For this purpose consider the circuit shown in Fig. 11.18, in which the collector is shorted to the emitter. The short-circuit collector current I_c is

$$I_c = (g_m - sC_\mu)V_\pi \qquad (11.51)$$

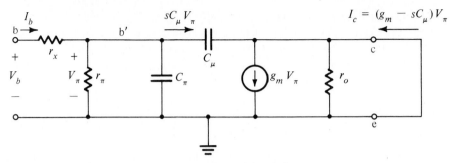

Fig. 11.18   Circuit for deriving an expression for $h_{fe}(s) \equiv I_c/I_b$.

A relationship between V_π and I_b can be established by multiplying I_b by the impedance seen between b′ and e:

$$V_\pi = I_b(r_\pi \| C_\pi \| C_\mu) \tag{11.52}$$

Thus h_{fe} can be obtained by combining Eqs. (11.51) and (11.52):

$$h_{fe} \equiv \frac{I_c}{I_b} = \frac{g_m - sC_\mu}{1/r_\pi + s(C_\pi + C_\mu)}$$

At the frequencies for which this model is valid, $g_m \gg \omega C_\mu$, resulting in

$$h_{fe} \simeq \frac{g_m r_\pi}{1 + s(C_\pi + C_\mu)r_\pi}$$

Thus

$$h_{fe} = \frac{\beta_0}{1 + s(C_\pi + C_\mu)r_\pi} \tag{11.53}$$

Thus h_{fe} has a single-pole response with a 3-dB frequency at $\omega = \omega_\beta$,

$$\omega_\beta = \frac{1}{(C_\pi + C_\mu)r_\pi} \tag{11.54}$$

Figure 11.19 shows a Bode plot for $|h_{fe}|$. From the −6 dB/octave slope it follows that the frequency at which $|h_{fe}|$ drops to unity, which is called the *unity-gain bandwidth* ω_T, is given by

$$\omega_T = \beta_0 \omega_\beta \tag{11.55}$$

Thus

$$\omega_T = \frac{g_m}{C_\pi + C_\mu} \tag{11.56}$$

The unity-gain bandwidth f_T is usually specified on the data sheets of the transistor. In some cases f_T is given as a function of I_C and V_{CE}. To see how f_T changes with I_C, recall that g_m is directly proportional to I_C but only part of C_π (the diffusion capacitance) is directly proportional to I_C. It follows that f_T decreases at low currents, as

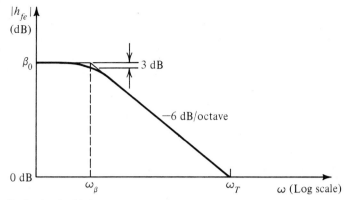

Fig. 11.19   Bode plot for $|h_{fe}|$.

shown in Fig. 11.20. However, the decrease in f_T at high currents, also shown in Fig. 11.20, cannot be explained by this argument; it rather is due to the same phenomenon that causes β_0 to decrease at high currents. In the region where f_T is almost constant, C_π is dominated by the diffusion part.

Typically, f_T is in the range of 100 to 1,000 MHz, with 400 MHz being a common figure for IC *npn* transistors operating at normal current levels. The value of f_T can be used in Eq. (11.56) to determine $C_\pi + C_\mu$. The capacitance C_μ is usually determined separately by measuring the capacitance between base and collector at the desired reverse-bias voltage V_{CB}.

Before leaving this section we should mention that the hybrid-π model of Fig. 11.17 characterizes the transistor operation fairly accurately up to a frequency of about $0.2\omega_T$. At higher frequencies one has to add other parasitic elements to the model as well as refine the model to account for the fact that the transistor is in fact a distributed-parameter network that we are trying to model with a lumped-component circuit. One such refinement consists of splitting r_x into a number of parts and replacing C_μ by a number

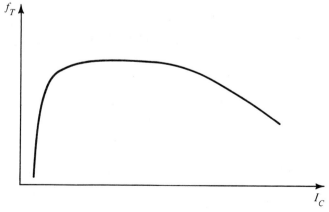

Fig. 11.20   Variation of f_T with I_C.

of capacitors each connected between the collector and one of the taps of r_x. This topic is beyond the scope of this book.

An important observation to make from the high-frequency model of Fig. 11.17 is that at frequencies above $5\omega_\beta$ or $10\omega_\beta$ one may ignore the resistance r_π. It can be seen then that r_x becomes the only resistive part of the input impedance at high frequencies. Thus r_x plays an important role in determining the frequency response of transistors at high frequencies. It follows that an accurate determination of r_x should be made from a high-frequency measurement.

11.7 For the same transistor considered in Exercise 11.6 and at the same bias point, determine C_π if $C_\mu = 2$ pF and $|h_{fe}| = 10$ at 50 MHz.
Ans.   10.7 pF

11.5   FREQUENCY RESPONSE OF THE COMMON-EMITTER AMPLIFIER

Analysis of the frequency response of the classical common-emitter amplifier stage in Fig. 11.21 follows a procedure identical to that used for the FET common-source amplifier in Section 11.3. Thus the overall gain can be written in the form

$$A(s) = A_M \frac{1}{1 + s/\omega_H} \frac{s^2(s + \omega_Z)}{(s + \omega_{P1})(s + \omega_{P2})(s + \omega_{P3})} \qquad (11.57)$$

where the midband gain A_M is evaluated by ignoring all capacitive effects, and where ω_H is the frequency of the dominant high-frequency pole obtained in a manner identical to that used in the FET case, and ω_Z, ω_{P1}, ω_{P2}, and ω_{P3} are the zero and three poles introduced in the low-frequency band by the coupling and the bypass capacitors. Because of the finite input resistance of the BJT, determination of the low-frequency

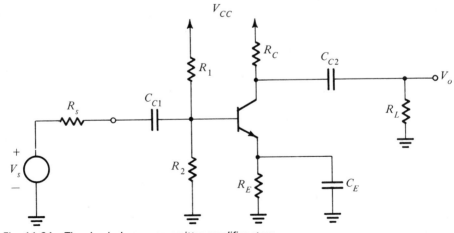

Fig. 11.21   The classical common-emitter amplifier stage.

singularities is more complicated than it is for the FET case. This can be seen from the low-frequency equivalent circuit shown in Fig. 11.22. Although we can certainly analyze this circuit and determine its transfer function and hence the poles and zeros, the expressions derived will be too complicated to yield useful insights. Rather, we will make use of the superposition method, described in Section 11.2, to obtain an approximate estimate of the lower 3-dB frequency ω_L.

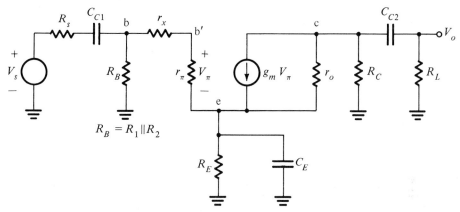

Fig. 11.22   *Equivalent circuit for the amplifier of Fig. 11.21 in the low-frequency band.*

The determination of ω_L proceeds as follows: First, we set C_E and C_{C2} to infinity and find the resistance R_{C1} seen by C_{C1}. From the equivalent circuit in Fig. 11.22, with C_E set to ∞, we find

$$R_{C1} = R_s + [R_B \| (r_x + r_\pi)]$$

Next, we set C_{C1} and C_{C2} to ∞ and determine the resistance R'_E seen by C_E. Again from the equivalent circuit in Fig. 11.22, or simply using the rule for reflecting resistances from the base to the emitter circuit, we obtain

$$R'_E = R_E \left\| \frac{r_\pi + r_x + (R_B \| R_s)}{\beta_0 + 1} \right.$$

Finally, we set both C_{C1} and C_E to infinity and obtain the resistance seen by C_{C2}:

$$R_{C2} = R_L + (R_C \| r_o)$$

An approximate value for the lower 3-dB frequency can now be determined from

$$\omega_L \simeq \frac{1}{C_{C1} R_{C1}} + \frac{1}{C_E R'_E} + \frac{1}{C_{C2} R_{C2}} \tag{11.58}$$

At this point we should note that the zero introduced by C_E is at the value of s that makes $Z_E = 1/R_E + sC_E$ infinite,

$$s_Z = -\frac{1}{C_E R_E}$$

The frequency of the zero is usually much lower than ω_L, justifying the approximation involved in using the superposition method.

If we assume that the common-emitter amplifier is indeed properly characterized by a dominant low-frequency pole, Eq. (11.57) can be approximated by

$$A(s) \simeq A_M \frac{1}{1 + s/\omega_H} \frac{s}{s + \omega_L}$$

11.8 Consider the amplifier circuit in Fig. 11.21 with $R_s = 4$ kΩ, $R_1 = 8$ kΩ, $R_2 = 4$ kΩ, $R_E = 3.3$ kΩ, $R_C = 6$ kΩ, $R_L = 4$ kΩ, $C_{C1} = 1$ μF, $C_{C2} = 1$ μF, $C_E = 10$ μF, and $V_{CC} = 12$ V. The dc emitter current can be shown to be $I_E \simeq 1$ mA. At this current the transistor has $\beta_0 = 100$, $C_\pi = 13.9$ pF, $C_\mu = 2$ pF, $r_o = 100$ kΩ, and $r_x = 50$ Ω. Find A_M, f_H, f_L, and the frequency of the transmission zero due to C_E.
Ans. -22.3 V/V; 796 kHz; 446 Hz; 5 Hz

11.9 The amplifier of Exercise 11.8 is excited by a pulse of 44.8 mV height and 10 μs width. Find the rise time, the height, and the magnitude of sag of the output pulse waveform.
Ans. 0.44 μs; 1 V; 0.028 V

11.6 THE COMMON-BASE AND CASCODE CONFIGURATIONS

In the previous sections it was shown that the high-frequency response of the common-source amplifier and the common-emitter amplifier is limited by the Miller effect introduced by the feedback capacitance (C_{gd} in the FET and C_μ in the BJT). It follows that to extend the upper frequency limit of a transistor amplifier stage one has to reduce or eliminate the Miller capacitance multiplication. In the following we shall show that this can be achieved in the common-base amplifier configuration. An almost identical analysis can be applied to the common-gate configuration.

We shall also study a two-transistor amplifier configuration, called the cascode configuration, and show that it combines the advantages of the common-emitter and the common-base circuits (the common-source and the common-gate circuits in the FET case). To be general and to show the parallels with the common-emitter amplifier, we shall present the circuits in their capacitively coupled form. The high-frequency analysis, however, applies directly to direct-coupled circuits.

Analysis of Common-Base Amplifier

Figure 11.23 shows a common-base amplifier stage of the capacitively coupled type. Recall that such a configuration was analyzed at midband frequencies (that is, with all capacitive effects neglected) in Chapter 9. In the following we shall be interested specifically in the high-frequency analysis; the low-frequency response can be determined using techniques similar to those of the previous sections.

The equivalent circuit, which applies at medium and high frequencies, of the common-base circuit is shown in Fig. 11.24a. To simplify matters and focus attention on the special features of the common-base circuit, r_o and r_x have been omitted.

In the circuit of Fig. 11.24a we observe that the voltage at the emitter terminal V_e

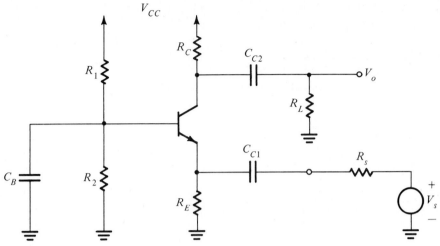

Fig. 11.23 *Common-base amplifier stage.*

is equal to $-V_\pi$. We can write a node equation at the emitter terminal that enables us to express the emitter current I_e as

$$I_e = -V_\pi\left(\frac{1}{r_\pi} + sC_\pi\right) - g_m V_\pi = V_e\left(\frac{1}{r_\pi} + g_m + sC_\pi\right)$$

Thus the input admittance looking into the emitter is

$$\frac{I_e}{V_e} = \frac{1}{r_\pi} + g_m + sC_\pi = \frac{1}{r_e} + sC_\pi$$

Therefore at the input of the circuit we may replace the transistor by this input admittance, as shown in Fig. 11.24b.

At the output side (Fig. 11.24a) we see that V_o is determined by the current source $g_m V_\pi$ feeding $(R_C \| R_L \| C_\mu)$. This observation is used in drawing the output part in the simplified equivalent circuit in Fig. 11.24b.

The simplified equivalent circuit of Fig. 11.24b clearly shows the most important feature of the common-base configuration: the absence of an internal feedback capacitance. Unlike the common-emitter circuit, here C_μ has one terminal grounded, and no Miller effect is present. We therefore expect that the upper cutoff frequency will be much higher than that of the common-emitter configuration.

The high-frequency poles can be directly determined from the equivalent circuit of Fig. 11.24b. At the input side we have a pole whose frequency ω_{P1} can be written by inspection as

$$\omega_{P1} = \frac{1}{C_\pi(r_e \| R_E \| R_s)}$$

Since r_e is usually very small, the frequency ω_{P1} will be quite high. At the output side there is a pole with frequency ω_{P2} given by

$$\omega_{P2} = \frac{1}{C_{\mu}(R_C \| R_L)}$$

Since C_{μ} is quite small, ω_{P2} also will be quite high.

The question now arises as to the accuracy of the above analysis. Since we are dealing with poles at much higher frequencies, we should take into account effects normally thought to be negligible. For instance, the parasitic capacitance usually present

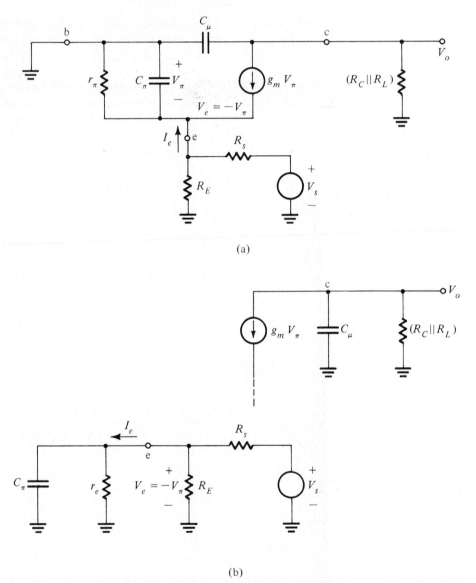

(a)

(b)

Fig. 11.24   (a) Equivalent circuit of the common-base amplifier in Fig. 11.23; (b) simplified version of the circuit in (a).

between collector and substrate (ground) in an IC transistor will obviously have a considerable effect on the value of ω_{P2}. Also, it is not clear that r_x could be neglected. It follows that for the accurate determination of the high-frequency response of a common-base amplifier a more elaborate transistor model should be used, and one normally employs a computer circuit-analysis program. Nevertheless, the point that we wish to emphasize here is that the common-base amplifier has a much higher upper cutoff frequency than that of the common-emitter circuit.

The principal disadvantage of the common-base circuit is its low input resistance. As can be seen from the equivalent circuit in Fig. 11.24b, the input resistance at midband is approximately equal to r_e. Thus unless R_s is small there will be a significant loss of gain in coupling the signal source to the emitter. The midband gain can be obtained by inspection of the circuit in Fig. 11.24b as

$$A_M \simeq g_m(R_C \| R_L) \frac{r_e}{r_e + R_s}$$

The Cascode Configuration

The cascode configuration combines the advantages of the common-emitter and the common-base circuits. Figure 11.25 shows a capacitively coupled cascode amplifier designed using bipolar transistors. The following analysis applies equally well to FET cascode circuits and to mixed circuits (that is, a cascode in which Q_1 is an FET and Q_2 is a BJT).

In the cascode circuit, Q_1 is connected in the common-emitter configuration and therefore presents a relatively high input resistance to the signal source. The collector

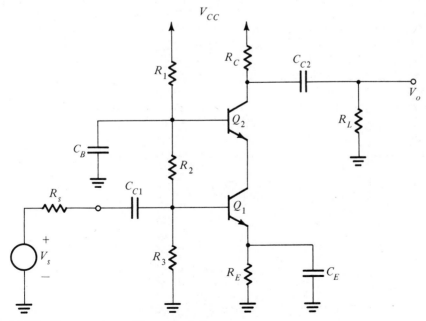

Fig. 11.25 *The cascode amplifier configuration.*

signal current of Q_1 is fed to the emitter of Q_2, which is connected in the common-base configuration. Thus the load resistance seen by Q_1 is simply the input resistance r_e of Q_2. This low load resistance of Q_1 considerably reduces the Miller multiplier effect of $C_{\mu 1}$ and thus extends the upper cutoff frequency. This is achieved without reducing the midband gain, since the collector of Q_2 carries a current almost equal to the collector current of Q_1. Furthermore, since it is in the common-base configuration, Q_2 does not suffer from the Miller effect and hence does not limit the high-frequency response. Transistor Q_2 acts essentially as a current buffer or an impedance transformer, faithfully passing on the signal current to the load while presenting a low load resistance to the amplifying device Q_1.

Detailed analysis of the cascode amplifier of Fig. 11.25 will now be presented: Figure 11.26a shows an equivalent circuit that applies at the middle- and high-frequency bands. To simplify matters r_{x2} and r_{o2} have been omitted. Although the two transistors are operating at equal bias currents and therefore their corresponding parameters are equal, we have for clarity kept the identity of the two sets of parameters separate.

Application of Thévenin's theorem enables us to reduce the circuit to the left of line xx' (Fig. 11.26a) to a source V_s' and a resistance R_s', as shown in Fig. 11.26b, where

$$V_s' = V_s \frac{(R_2 \| R_3)}{R_s + (R_2 \| R_3)} \frac{r_{\pi 1}}{r_{\pi 1} + r_{x1} + (R_2 \| R_3 \| R_s)} \tag{11.59}$$

$$R_s' = \{ r_{\pi 1} \| [r_{x1} + (R_3 \| R_2 \| R_s)] \} \tag{11.60}$$

Another important simplification included in the circuit of Fig. 11.26b is the replacement of the current source $g_{m2}V_{\pi 2}$ by a resistance $1/g_{m2}$ (see the source absorption theorem in Section 2.6). This resistance is then combined with the parallel resistance $r_{\pi 2}$ to obtain r_{e2}. Since $r_{e2} \ll r_{o1}$, we see that between the collector of Q_1 and ground the total resistance is approximately r_{e2}. Capacitance $C_{\pi 2}$ together with resistance r_{e2} produces a transfer-function pole with a frequency

$$\omega_2 = \frac{1}{C_{\pi 2} r_{e2}} \simeq \omega_T \tag{11.61}$$

which is much higher than the frequency of the pole that arises due to the interaction of R_s' and the input capacitance of Q_1. It follows that at the frequency range of interest $C_{\pi 2}$ can be ignored in calculating the voltage at the collector of Q_1; that is,

$$V_{c1} \simeq -g_{m1}V_{\pi 1}r_{e2} \simeq -V_{\pi 1}$$

Thus the gain between b_1' and c_1 (Fig. 11.26b) is approximately -1, and we can employ Miller's theorem to replace the bridging capacitance $C_{\mu 1}$ by a capacitance $2C_{\mu 1}$ between b_1' and ground and a capacitance $2C_{\mu 1}$ between c_1 and ground. The resulting equivalent circuit is shown in Fig. 11.26c, from which we can now evaluate the frequency of the pole due to the RC low-pass circuit at the input as

$$\omega_1 = \frac{1}{R_s'(C_{\pi 1} + 2C_{\mu 1})} \tag{11.62}$$

This frequency will normally be much lower than both ω_2 [Eq. (11.61)] and the frequency of the pole produced by the output part of the circuit,

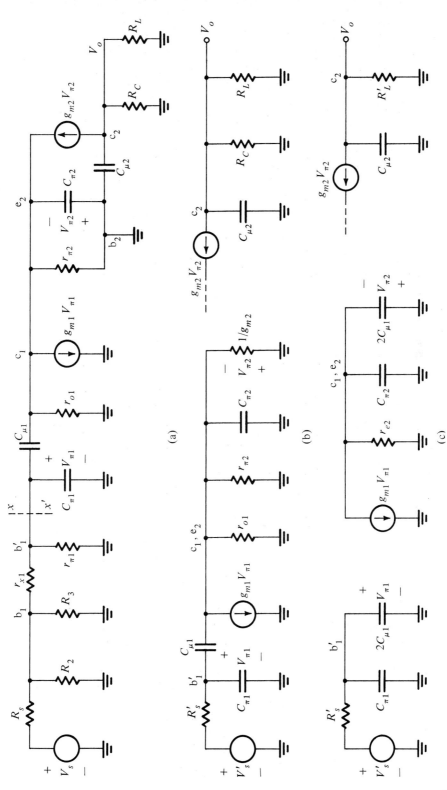

Fig. 11.26 High-frequency analysis of the cascode amplifier in Fig. 11.25.

$$\omega_3 = \frac{1}{C_{\mu 2} R'_L} \tag{11.63}$$

That is, the input circuit produces a dominant high-frequency pole, and the upper 3-dB frequency ω_H is given by

$$\omega_H \simeq \omega_1$$

It should be pointed out that the superposition method could have been applied directly to the circuit in Fig. 11.26b.

The midband gain can be easily evaluated by ignoring the capacitances in the equivalent circuit of Fig. 11.26c together with substituting for V'_s and R'_s from Eqs. (11.59) and (11.60):

$$A_M = \frac{V_o}{V_i} = -g_m(R_L \| R_C) \frac{(R_2 \| R_3)}{(R_2 \| R_3) + R_s} \frac{r_\pi}{r_\pi + (R_2 \| R_3 \| R_s)} \tag{11.64}$$

This expression is identical in form to the expression for the gain of a common-emitter circuit.

EXERCISE

11.10 Consider the cascode circuit in Fig. 11.25 with the following component values: $R_s = 4$ kΩ, $R_1 = 18$ kΩ, $R_2 = 4$ kΩ, $R_3 = 8$ kΩ, $R_E = 3.3$ kΩ, $R_C = 6$ kΩ, $R_L = 4$ kΩ, $C_{C1} = 1$ μF, $C_{C2} = 1$ μF, $C_B = 10$ μF, $C_E = 10$ μF, and $V_{CC} = +15$ V. Show that each transistor is operating at $I_E \simeq 1$ mA. Note that this design is identical to that of the common-emitter amplifier in Exercise 11.8. If we thus assume that the transistors are of the same type as that used in Exercise 11.8, we are able to compare results and draw conclusions. Calculate A_M, f_1, f_2, f_3, and f_H.
Ans. -22.3 V/V; 8.8 MHz; 400 MHz; 33 MHz; 8.8 MHz

11.7 FREQUENCY RESPONSE OF THE EMITTER FOLLOWER

The emitter-follower or common-collector configuration was studied in Section 10.3. In the following we consider the high-frequency response of this important circuit configuration. The results apply, with some modifications, to the FET source follower.

Consider the direct-coupled emitter-follower circuit shown in Fig. 11.27a, where R_s represents the source resistance and R_E represents the combination of emitter-biasing resistance and load resistance. The high-frequency equivalent circuit is shown in Fig. 11.27b and is redrawn in a slightly different form in Fig. 11.27c. Analysis of this circuit results in the emitter-follower transfer function $V_o(s)/V_s(s)$, which can be shown to have two poles and one real zero:

$$\frac{V_o(s)}{V_s(s)} = A_M \frac{1 + s/\omega_Z}{(1 + s/\omega_{P1})(1 + s/\omega_{P2})} \tag{11.65}$$

where A_M denotes the value of the gain at low and medium frequencies. Unfortunately, though, symbolic analysis of the circuit will not reveal whether or not one of the poles is dominant. To gain more insight we shall take an alternative route.

Writing a node equation at the emitter (Fig. 11.27b) results in

$$V_o = (g_m + y_\pi) R_E V_\pi \tag{11.66}$$

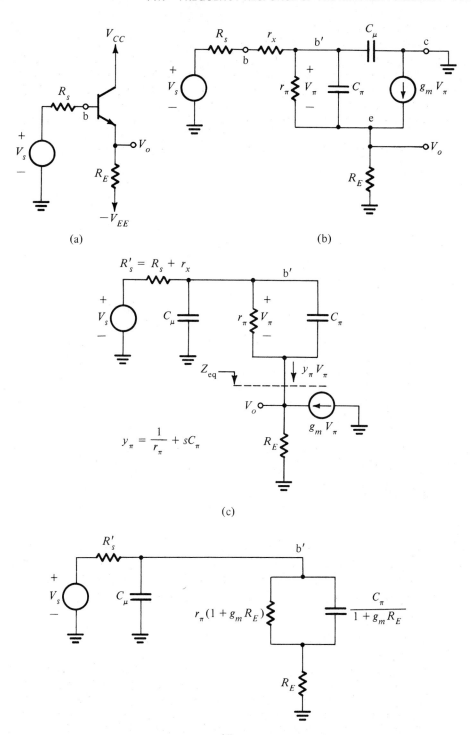

Fig. 11.27 *High-frequency analysis of the emitter follower.*

where

$$y_\pi = \frac{1}{r_\pi} + sC_\pi$$

Thus V_o will be zero at the value of s that makes $V_\pi = 0$ and at the value of s that makes $g_m + y_\pi = 0$. In turn, V_π will be zero at the value of s that makes $z_\pi = 0$ or equivalently $y_\pi = \infty$, namely, $s = \infty$. The fact that a transmission zero exists at $s = \infty$ correlates with Eq. (11.65). The other transmission zero is obtained from

$$g_m + y_\pi = 0$$

that is,

$$g_m + \frac{1}{r_\pi} + s_Z C_\pi = 0$$

which yields

$$s_Z = -\frac{g_m + 1/r_\pi}{C_\pi} = -\frac{1}{C_\pi r_e} \simeq -\omega_T \tag{11.67}$$

Since the frequency of this zero is quite high, it will normally play a minor role in determining the high-frequency response of the emitter follower.

Next we consider the poles. Whether or not one of the two poles is dominant will depend on the particular application—specifically, on the values of R_s and R_E. In most applications R_s is large, and it together with the input capacitance provides a dominant pole. To see this more clearly, consider the equivalent circuit of Fig. 11.27c. By invoking the source-absorption theorem (Section 2.6), we can replace the circuit below the dotted line by its equivalent impedance Z_{eq},

$$Z_{eq} \equiv \frac{V_o}{y_\pi V_\pi}$$

Thus

$$Z_{eq} = \frac{(g_m + y_\pi) R_E}{y_\pi} \tag{11.68}$$

Note that Z_{eq} is simply R_E reflected to the base side with use of a generalized form of the reflection rule: R_E is multiplied by $(h_{fe} + 1)$. The total impedance between b' and ground is

$$Z_{b'} = \frac{1}{y_\pi} + Z_{eq} = \frac{1 + g_m R_E}{y_\pi} + R_E$$

As shown in Fig. 11.27d, this impedance can be represented by a resistance R_E in series with an RC network consisting of a resistance $(1 + g_m R_E) r_\pi$ in parallel with a capacitance $C_\pi/(1 + g_m R_E)$. Since the impedance of the parallel RC circuit is usually much larger than R_E, we may neglect the latter impedance and obtain a simple STC low-pass network. From this STC circuit it follows that a pole exists at

$$\omega_P = \left[\left(C_\mu + \frac{C_\pi}{1 + g_m R_E} \right) [R_s' \| (1 + g_m R_E) r_\pi] \right]^{-1}$$

Even though this pole is usually dominant, its frequency is normally quite high, giving the emitter follower a wide bandwidth.

An alternative approach for finding an approximate value of the 3-dB frequency ω_H is to use the superposition method on the equivalent circuit in Fig. 11.27d.

EXERCISE

11.11 For an emitter follower biased at $I_C = 1$ mA and having $R_s = R_E = 1$ kΩ and using a transistor specified to have $f_T = 400$ MHz, $C_\mu = 2$ pF, $r_x = 100$ Ω, and $\beta_0 = 100$, evaluate the midband gain A_M and the frequency of the dominant high-frequency pole.
Ans. 0.97 V/V; 62.5 MHz

11.8 *THE COMMON-COLLECTOR COMMON-EMITTER CASCADE*

The excellent high-frequency response of the emitter follower is due to the absence of the Miller capacitance multiplication effect. The problem with it, though, is that it does not provide voltage gain. It appears possible that we can obtain gain and wide bandwidth by using a cascade of a common-collector and a common-emitter stages, as shown in Fig. 11.28. Here the emitter-follower transistor Q_1 is shown biased by a current

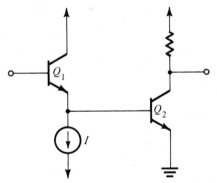

Fig. 11.28 *The common-collector common-emitter cascade amplifier.*

source I, as is usually the case in integrated-circuit design. Because the collector of Q_1 is at a signal ground, $C_{\mu 1}$ does not get multiplied by the stage gain, as is the case in the common-emitter amplifier. Thus the pole caused by the interaction of the source resistance and the input capacitance will be at a high frequency.

The voltage gain is provided by the common-emitter transistor Q_2. This transistor suffers from the Miller effect; that is, the total effective capacitance between its base and ground will be large. Nevertheless, this will not be detrimental; the resistance seen by that capacitance will be small because of the low output resistance of emitter follower Q_1.

Before considering a numerical example, we wish to draw the reader's attention to the similarities between the circuit of Fig. 11.28 and the cascode amplifier studied in Section 11.6. Both circuits employ a common-emitter amplifier to obtain voltage gain. Both circuits achieve wider bandwidth (than that obtained in a common-emitter amplifier) through minimizing the effect of the Miller multiplier. In the cascode circuit this is achieved by isolating the load resistance from the collector of the common-emitter stage by a low-input-resistance common-base stage. In the present circuit, while Miller multiplication occurs, the resulting large capacitance is isolated from the source resistance by an emitter follower.

Example 11.7

Figure 11.29 shows a capacitively coupled amplifier designed as cascade of a common-collector stage and a common-emitter stage. Assume that the transistors used have $f_T = 400$ MHz and $C_\mu = 2$ pF and neglect r_x and r_o. We wish to evaluate the midband

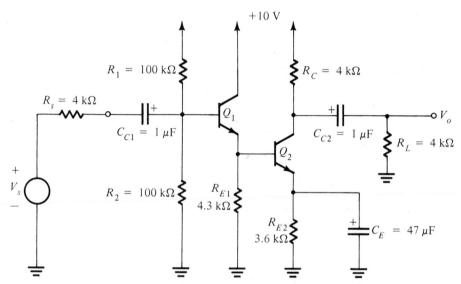

Fig. 11.29   A capacitively coupled amplifier using the common-collecter common-emitter cascade configuration.

gain and the high-frequency response of this circuit. Note that the load and source resistances and the transistor parameters are identical to those used in the common-emitter case (Exercise 11.8) and in the cascode case (Exercise 11.10); hence comparisons can be made. ($\beta = 100$)

Solution

We first determine the dc bias currents as follows:

$$V_{B1} \simeq 5 \text{ V}, \qquad V_{E1} \simeq 4.3 \text{ V}; \qquad I_{E1} = \frac{4.3 \text{ V}}{4.3 \text{ k}\Omega} = 1 \text{ mA}$$

$$V_{E2} \simeq 3.6 \text{ V}; \qquad I_{E2} = \frac{3.6 \text{ V}}{3.6 \text{ k}\Omega} = 1 \text{ mA}$$

Thus both transistors are operating at emitter currents of approximately 1 mA, and both are in the active mode. At this operating point the equivalent circuit components are

$$g_m \simeq 40 \text{ mA/V}, \qquad r_e \simeq 25 \ \Omega, \qquad r_\pi \simeq 2.5 \text{ k}\Omega$$

$$C_\pi + C_\mu = \frac{g_m}{\omega_T} = 15.9 \text{ pF}, \qquad C_\mu = 2 \text{ pF}, \qquad C_\pi = 15.9 - 2 = 13.9 \text{ pF}$$

To evaluate the midband gain we shall first determine the value of the input resistance R_{in}. Toward that end note that the input resistance between the base of Q_2 and ground is equal to $r_{\pi 2}$. Thus in the emitter circuit of Q_1 we have R_{E1} in parallel with $r_{\pi 2}$. The input resistance R_{in} will therefore be given by

$$R_{\text{in}} = (R_1 \| R_2 \| \{ (\beta_1 + 1) [r_{e1} + (R_{E1} \| r_{\pi 2})] \})$$

which leads to $R_{\text{in}} \simeq 38 \text{ k}\Omega$. The transmission from the input to the base of Q_1 is

$$\frac{V_{b1}}{V_s} = \frac{R_{\text{in}}}{R_{\text{in}} + R_s} = 0.9 \tag{11.69}$$

Next the gain of the emitter follower Q_1 can be obtained as

$$\frac{V_{e1}}{V_{b1}} = \frac{(R_{E1} \| r_{\pi 2})}{(R_{E1} \| r_{\pi 2}) + r_{e1}} = 0.98 \tag{11.70}$$

Finally, the gain of the common-emitter amplifier Q_2 can be evaluated as

$$\frac{V_o}{V_{e1}} = -g_{m2}(R_C \| R_L) = -80 \tag{11.71}$$

The overall voltage gain can be obtained by combining Eqs. (11.69) through (11.71):

$$\frac{V_o}{V_s} = -0.9 \times 0.98 \times 80 = -70.6 \text{ V/V}$$

The high-frequency response can be determined from the equivalent circuit shown in Fig. 11.30a. We apply Miller's theorem to the second stage and perform a number of other simplifications, and we obtain the circuit in Fig. 11.30b, where

$$R_s' = (R_s \| R_1 \| R_2)$$
$$V_s' = V_s \frac{(R_1 \| R_2)}{(R_1 \| R_2) + R_s}$$
$$C_T = C_{\pi 2} + C_{\mu 2}(1 + g_{m2} R_L')$$
$$R_L' = (R_L \| R_C)$$

This circuit is still quite complex, and an exact pencil-and-paper analysis would be quite tedious. Alternatively, we employ the superposition technique discussed in Section 11.2 to determine the dominant pole as follows: Capacitor $C_{\mu 1}$ sees a resistance $R_{\mu 1}$ given by

$$R_{\mu 1} = (R_s' \| R_{\text{in}}) = (3.7 \| 38) = 3.4 \text{ k}\Omega$$

It can be shown that $C_{\pi 1}$ sees a resistance $R_{\pi 1}$ given by

$$R_{\pi 1} = \left(r_{\pi 1} \middle\| \frac{R_s' + R_{E1}'}{1 + g_m R_{E1}'} \right)$$

where $R_{E1}' = (R_{E1} \| r_{\pi 2})$. Thus $R_{\pi 1} = 80 \ \Omega$.

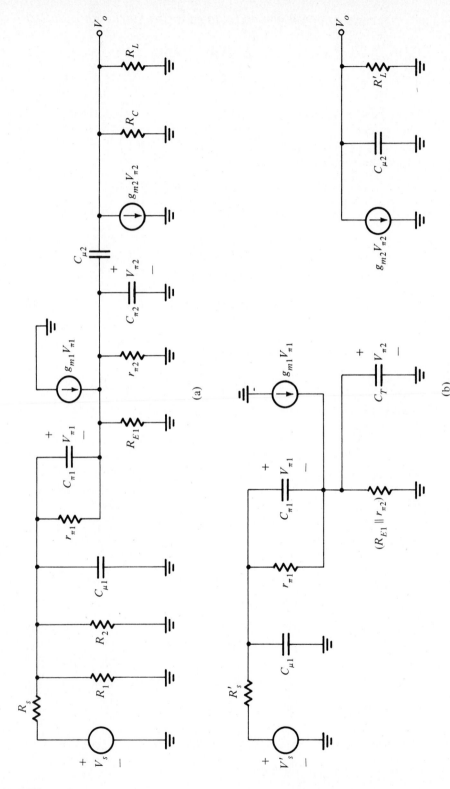

Fig. 11.30 Equivalent circuits for the determination of the high-frequency response of the amplifier in Fig. 11.29.

Capacitance C_T, which is equal to 271.7 pF, sees a resistance R_T given by

$$R_T = \left(R'_{E1} \left\| \frac{r_{\pi 1} + R'_s}{\beta_1 + 1} \right. \right) = 59 \ \Omega$$

Capacitance $C_{\mu 2}$ sees a resistance R'_L given by

$$R'_L = (R_L \| R_C) = 2 \ k\Omega$$

Thus the effective time constant is given by

$$\tau = C_{\mu 1}R_{\mu 1} + C_{\pi 1}R_{\pi 1} + C_T R_T + C_{\mu 2}R'_L = 27.94 \ \text{ns}$$

which corresponds to an upper 3-dB frequency of

$$f_H \simeq \frac{1}{2\pi\tau} = 5.7 \ \text{MHz}$$

Although the upper 3-dB frequency is not as high as the value obtained for the cascode amplifier (8.8 MHz), the midband gain here, 70.6, is higher than that found for the cascode (22.3). A figure of merit for an amplifier is its gain–bandwidth product. For the common-collector common-emitter circuit this value is quite close to the f_T of the transistors used.

· · ·

EXERCISES

11.12 We wish to use the superposition method to determine an approximate value of the lower 3-dB frequency f_L of the amplifier circuit in Fig. 11.29. Also find the frequency of the zero introduced by C_E.
Ans. 158.7 Hz; 0.94 Hz

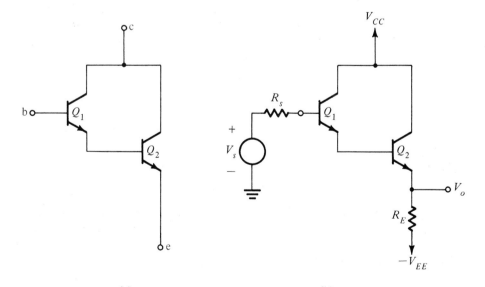

(a)   (b)

Fig. E11.13 *(a) The Darlington configuration; (b) voltage follower using the Darlington configuration.*

11.13   The common-collector common-emitter configuration studied in this section is a modified version of the composite device obtained by connecting two transistors in the form shown in Fig. E11.13a. This configuration, known as the *Darlington configuration,* is equivalent to a single transistor with $\beta = \beta_1\beta_2$. It can therefore be used as a high-performance follower, as illustrated in Fig. E11.13b. For the latter circuit assume that Q_2 is biased at $I_E = 5$ mA and let $R_s = 100$ kΩ, $R_E = 1$ kΩ, and $\beta_1 = \beta_2 = 100$. Find R_{in}, V_o/V_s, and R_{out}.

Ans.   10.3 MΩ; 0.98 V/V; 20 Ω

11.9   FREQUENCY RESPONSE OF THE DIFFERENTIAL AMPLIFIER

The differential pair studied in Chapter 10 is the most important building block in analog integrated circuits. In this section we analyze its frequency response. Although only the BJT differential pair is considered, the method applies equally well to the FET pair.

Frequency Response in the Case of Symmetric Excitation

Consider the differential amplifier shown in Fig. 11.31a. The input signal V_s is applied in a complementary (push-pull) fashion, and the source resistance R_s is equally distrib-

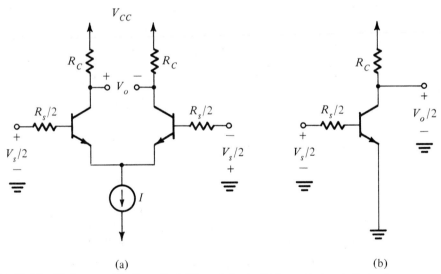

(a)   (b)

Fig. 11.31   *(a) A symmetrically excited differential pair; (b) its equivalent half-circuit.*

uted between the two sides of the pair. This situation arises, for instance, if the differential amplifier is fed from the output of another differential stage.

Since the circuit is symmetric and is fed in a complementary fashion, its frequency response will be identical to that of the equivalent common-emitter circuit shown in Fig. 11.31b. We have analyzed the common-emitter circuit in detail in Section 11.5. Since the differential pair is a direct-coupled amplifier, its gain will extend down to zero frequency with a low-frequency value of

$$\frac{V_o}{V_s} = -\frac{r_\pi}{r_\pi + R_s/2} g_m R_C \qquad (11.72)$$

The high-frequency response will be dominated by a real pole with a frequency ω_P,

$$\omega_P = \frac{1}{[(R_s/2)\|r_\pi][C_\pi + C_\mu (1 + g_m R_C)]} \qquad (11.73)$$

Denote the low-frequency gain given in Eq. (11.72) by A_0. The transfer function of the differential pair will be given by

$$\frac{V_o}{V_s} = \frac{A_0}{1 + s/\omega_P} \qquad (11.74)$$

Thus the plot of gain versus frequency will have the standard single-pole shape shown in Fig. 11.32. The 3-dB frequency ω_H is equal to the pole frequency,

$$\omega_H = \omega_P$$

In the above analysis r_x was neglected. It can be easily included simply by replacing $R_s/2$ by $R_s/2 + r_x$ in Eqs. (11.72) and (11.73).

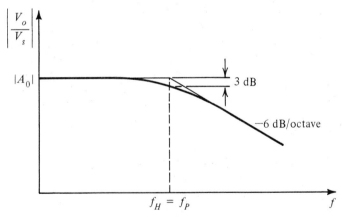

Fig. 11.32   Frequency response of the differential amplifier.

Frequency Response in the Case of Single-Ended Excitation

Figure 11.33a shows a differential amplifier driven in a single-ended fashion. Although it is not obvious, the frequency response in this case is almost identical to that of the symmetrically driven amplifier considered above. A proof of this assertion is illustrated by a series of equivalent circuits given in Fig. 11.33. Figure 11.33b shows the complete equivalent circuit. We can write a node equation at node X and obtain

$$V_{\pi 1}\left(\frac{1}{r_\pi} + sC_\pi + g_m\right) + V_{\pi 2}\left(\frac{1}{r_\pi} + sC_\pi + g_m\right) = 0$$

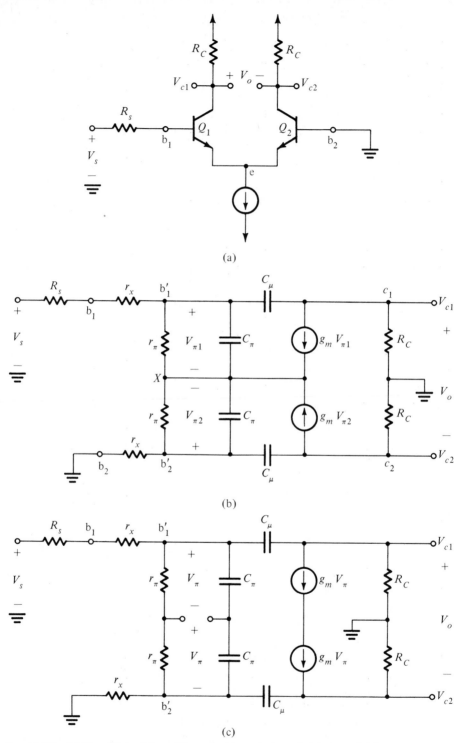

Fig. 11.33 (a) The differential amplifier exited in a single-ended fashion. (b) Equivalent circuit of the amplifier in (a). (c) Simplified equivalent circuit using the fact that $V_{\pi 1} = -V_{\pi 2}$.

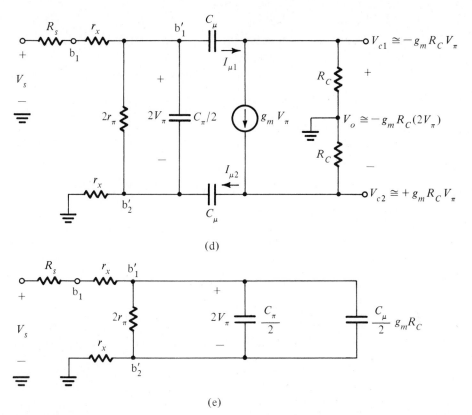

(d)

(e)

Fig. 11.33   (d) A further simplification of the equivalent circuit. (e) The input equivalent circuit after use of Miller's theorem.

Thus $V_{\pi 1} = -V_{\pi 2}$. This leads to the equivalent circuit in Fig. 11.33c, which is further simplified in Fig. 11.33d.

Consider Fig. 11.33d. If we neglect $I_{\mu 1}$ and $I_{\mu 2}$ in comparison with $g_m V_\pi$, it follows that

$$V_{c1} \simeq -g_m R_C V_\pi \qquad V_{c2} \simeq +g_m R_C V_\pi$$

Thus $V_o \simeq -g_m R_C(2V_\pi)$. If we further assume that r_x is small and that $g_m R_C \gg 2$, then Miller's theorem now can be applied to obtain the simplified input equivalent circuit shown in Fig. 11.33e, from which we see that the high-frequency response is dominated by a pole at $s = -\omega_P$, where ω_P is given by

$$\omega_P = \frac{1}{[2r_\pi \| (R_s + 2r_x)][C_\pi/2 + (C_\mu/2)(g_m R_C)]} \qquad (11.75)$$

The low-frequency gain A_0 is given by

$$A_0 = \frac{V_o}{V_s} = -g_m R_C \frac{2r_\pi}{2r_\pi + R_s + 2r_x} \qquad (11.76)$$

These results are almost identical to those obtained in the case of symmetric excitation.

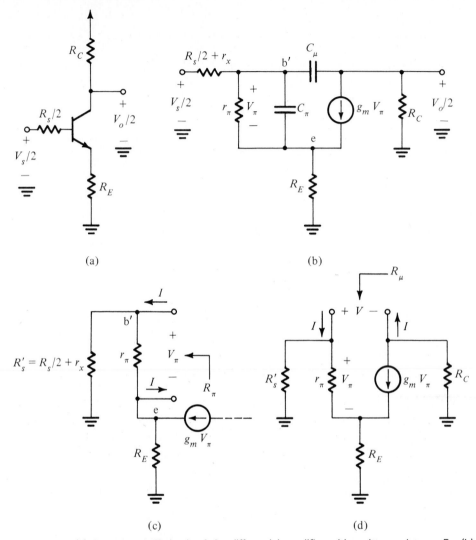

(a)

(b)

(c)

(d)

Fig. 11.34 *(a) Equivalent half-circuit of the differential amplifier with emitter resistance R_E. (b) Equivalent circuit of the half-circuit in (a). (c) Circuit for determining the resistance R_π seen by C_π. (d) Circuit for determining the resistance R_μ seen by C_μ.*

Effect of Emitter Resistance on the Frequency Response

The bandwidth of the differential amplifier can be widened (that is, ω_H can be increased) by including two equal resistances R_E in the emitters. This is achieved at the expense of a reduction in the low-frequency gain. To evaluate the effect of the emitter resistances on frequency response, consider the equivalent half-circuit shown in Fig. 11.34a. The low-frequency gain is given by

$$A_0 \equiv \frac{V_o}{V_s} = \frac{-(\beta + 1)(r_e + R_E)}{R_s/2 + (\beta + 1)(r_e + R_E)} \frac{\alpha R_C}{R_E + r_e} \tag{11.77}$$

The high-frequency equivalent circuit is shown in Fig. 11.34b. Since it is no longer convenient to apply Miller's theorem, we shall use the superposition technique explained in Section 11.2. The method proceeds as follows: We first eliminate C_μ and determine the resistance seen by C_π, which we shall call R_π. Figure 11.34c shows the circuit for finding R_π,

$$R_\pi = \left(r_\pi \left\| \frac{R_s' + R_E}{1 + g_m R_E} \right. \right) \tag{11.78}$$

Next, the resistance R_μ seen by C_μ can be determined from the circuit in Fig. 11.34d,

$$R_\mu = R_C + \frac{1 + R_E/r_e + g_m R_C}{1/r_\pi + (1/R_s')(1 + R_E/r_e)} \tag{11.79}$$

The overall effective time constant will be given by

$$\tau = C_\pi R_\pi + C_\mu R_\mu \tag{11.80}$$

and the 3-dB frequency ω_H will be

$$\omega_H \simeq \frac{1}{\tau} \tag{11.81}$$

EXERCISES

11.14 Consider a differential amplifier biased with a current source $I = 1$ mA and having $R_C = 10$ kΩ. Let the amplifier be fed with a source having $R_s = 10$ kΩ. Also let the transistors be specified to have $\beta_0 = 100$, $C_\pi = 6$ pF, $C_\mu = 2$ pF, and $r_x = 50$ Ω. Find the dc differential gain A_0, the 3-dB frequency f_H, and the gain–bandwidth product. *Ans.*   100 V/V (40 dB); 156 kHz; 15.6 MHz

11.15 Consider the differential amplifier of Exercise 11.14 but with a 150-Ω resistance included in each emitter lead. Find A_0, R_π, R_μ, f_H, and the gain–bandwidth product. *Ans.*   40 V/V (32 dB); 1 kΩ; 214 kΩ; 367 kHz; 14.7 MHz

(*Note:* The difference in the gain–bandwidth product is due mainly to the different approximations made.)

Variation of the CMRR With Frequency

The common-mode rejection ratio (CMRR) of a differential amplifier falls off at high frequencies because of a number of factors, the most important of which is the increase of the common-mode gain with frequency. To see how this comes about, consider the common-mode equivalent half-circuit shown in Fig. 11.35. Here the resistance R is the output resistance and the capacitance C is the output capacitance of the bias current source. From our study of the frequency response of the common-emitter amplifier in Section 11.5 we know that the components $2R$ and $C/2$ will introduce a zero in the common-mode gain function. This zero will be at a frequency f_Z,

$$f_Z = \frac{1}{2\pi(2R)(C/2)} = \frac{1}{2\pi RC}$$

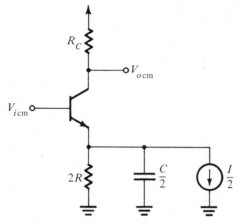

Fig. 11.35   The equivalent common-mode half circuit.

Since R is usually very large, even a very small output capacitance C will result in f_Z having a relatively low value. The result is that the common-mode gain will start increasing with a slope of $+6$ dB/octave at a relatively low frequency, as shown in Fig. 11.36a. The common-mode gain falls off at higher frequencies because of the internal capacitances C_π and C_μ.

The behavior of the common-mode gain shown in Fig. 11.36a together with the high-frequency rolloff of the differential gain (Fig. 11.36b) results in the CMRR having the frequency response shown in Fig. 11.36c.

11.10   THE DIFFERENTIAL PAIR AS A WIDEBAND AMPLIFIER: THE COMMON-COLLECTOR COMMON-BASE CONFIGURATION

A slight modification of the differential-amplifier circuit results in a configuration with a much higher bandwidth. The resulting circuit, shown in Fig. 11.37, is obtained by simply eliminating the collector resistance of Q_1. It can be easily seen that this eliminates the Miller capacitance multiplication of $C_{\mu 1}$. Furthermore, since $C_{\mu 2}$ has one of its terminals grounded, there will be no Miller effect in Q_2 either. We should therefore expect the circuit of Fig. 11.37 to have an extended frequency response, and we may add it to our repertoire of wideband amplifiers.

Low-Frequency Gain

An alternative way of looking at the circuit of Fig. 11.37 is to consider it as a common-collector stage (Q_1) followed by a common-base stage (Q_2). The common-collector transistor Q_1 has in its emitter circuit the emitter resistance r_e of Q_2. Thus the signal current in the emitter of Q_1 is given by

$$i_e = \frac{v_i}{r_{e1} + r_{e2}} = \frac{v_i}{2r_e}$$

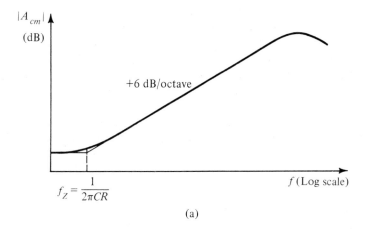

(a)

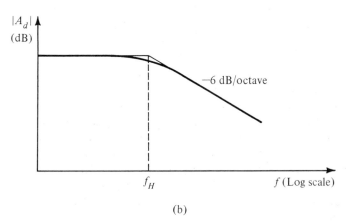

(b)

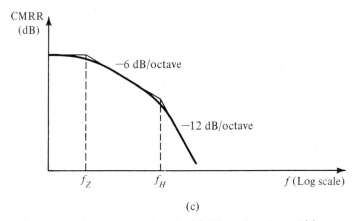

(c)

Fig. 11.36   *Variation of (a) common-mode gain, (b) differential gain, and (c) common-mode rejection ratio with frequency.*

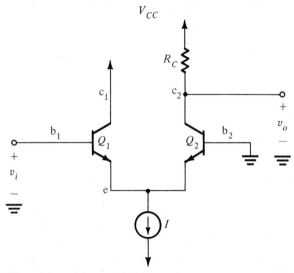

Fig. 11.37   *The differential amplifier modified for wideband amplification. Eliminating the collector resistance of Q_1 eliminates the Miller capacitance multiplication.*

because

$$r_{e1} = r_{e2} = r_e = \frac{V_T}{I/2}$$

This signal current flows in the emitter of Q_2 and appears in its collector multiplied by α,

$$i_{c2} = \alpha i_e = \frac{\alpha v_i}{2r_e}$$

Thus the output signal voltage at the collector of Q_2 will be

$$v_o = i_{c2} R_C = \frac{\alpha v_i}{2r_e} R_C$$

and the voltage gain will be

$$\frac{v_o}{v_i} = \frac{\alpha R_C}{2r_e}$$

Frequency Response

To simplify matters we shall neglect the effect of r_x and thus obtain the equivalent circuit shown in Fig. 11.38a. It is assumed that the circuit is fed with a signal source having a voltage V_s and a resistance R_s. A node equation at node X reveals that

$$V_{\pi1} = -V_{\pi2}$$

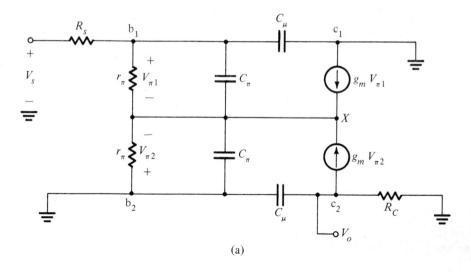

(a)

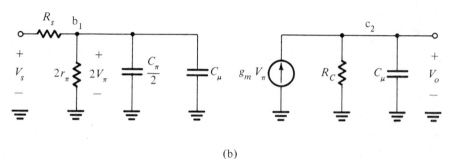

(b)

Fig. 11.38   (a) Equivalent circuit for the modified differential amplifier in Fig. 11.37. (b) Simplified equivalent circuit.

This allows us to simplify the circuit to that shown in Fig. 11.38b, where $V_\pi = V_{\pi1} = -V_{\pi2}$. It can be seen that there are two real poles, one at the input with a frequency f_{P1},

$$f_{P1} = \frac{1}{2\pi(R_s\|2r_\pi)(C_\pi/2 + C_\mu)}$$

and one at the output with a frequency f_{P2},

$$f_{P2} = \frac{1}{2\pi R_C C_\mu}$$

Whether one of these poles is dominant or not will depend on the particular application. If no dominant pole exists, then one has to use the overall transfer function to evaluate the 3-dB frequency f_H. This is demonstrated in Example 11.8.

Example 11.8

Consider the modified differential amplifier discussed above and let $I = 1$ mA, $R_C = 10$ kΩ, $R_s = 10$ kΩ, $f_T = 400$ MHz, and $C_\mu = 2$ pF. Evaluate the low-frequency gain and the 3-dB frequency f_H. ($\beta = 100$)

Solution

Each transistor is biased at an emitter current of 0.5 mA. Thus

$$r_e \simeq 50 \ \Omega, \qquad g_m \simeq 20 \text{ mA/V}, \qquad r_\pi \simeq 5 \text{ k}\Omega$$

$$C_\pi + C_\mu = \frac{g_m}{\omega_T} = \frac{20 \times 10^{-3}}{2\pi \times 400 \times 10^6} \simeq 8 \text{ pF}$$

$$C_\mu = 2 \text{ pF}, \qquad C_\pi = 6 \text{ pF}$$

The low-frequency gain is given by

$$A_0 = \frac{2r_\pi}{R_s + 2r_\pi} \frac{\alpha R_C}{2r_e} \simeq \frac{10}{10 + 10} \frac{10}{0.1}$$

Thus $A_0 = 50$, or 34 dB. The pole at the input has a frequency ω_{P1},

$$\omega_{P1} = \frac{1}{(R_s \| 2r_\pi)(C_\pi/2 + C_\mu)}$$

$$= \frac{1}{5 \times 10^3 (3 + 2) \times 10^{-12}} = 40 \text{ Mrad/s}$$

The pole at the output has a frequency ω_{P2},

$$\omega_{P2} = \frac{1}{C_\mu R_C} = \frac{1}{2 \times 10^{-12} \times 10 \times 10^3} = 50 \text{ Mrad/s}$$

Since the two poles are very close to each other, we have to use the overall transfer function to evaluate the 3-dB frequency f_H. The amplifier transfer function is given by

$$A(s) \equiv \frac{V_o}{V_s} = \frac{A_0}{(1 + s/\omega_{P1})(1 + s/\omega_{P2})}$$

Thus

$$|A(j\omega)| = \frac{A_0}{\sqrt{(1 + \omega^2/\omega_{P1}^2)(1 + \omega^2/\omega_{P2}^2)}}$$

At $\omega = \omega_H$, $|A(j\omega_H)| = A_0/\sqrt{2}$. Thus

$$2 = \left(1 + \frac{\omega_H^2}{\omega_{P1}^2}\right)\left(1 + \frac{\omega_H^2}{\omega_{P2}^2}\right)$$

which leads to $\omega_H = 28.53$ Mrad/s. Thus the upper 3-dB frequency f_H is $f_H = \omega_H/2\pi = 4.54$ MHz.

$\cdot \quad \cdot \quad \cdot$

Before leaving this section we should point out a disadvantage of the modified differential-amplifier circuit. Since the output is taken single-ended, the CMRR will be much lower than that of the balanced differential amplifier. For this reason the first stage of an operational amplifier is usually a balanced one. The modified circuit can be used in subsequent stages or in wideband amplifiers, where the requirement of a high CMRR is not important.

11.11 CONCLUDING REMARKS

In this chapter we studied methods for determining the frequency response of transistor amplifiers. Both capacitively coupled amplifier stages, whose frequency response falls off at low frequencies, and direct-coupled stages, useful in integrated circuit design, were treated. The emphasis has been on approximate techniques, since these techniques are suitable for pencil-and-paper analysis and yield insights into the performance of the circuit. Through such insights the designer can identify the reasons for the limited frequency response of his or her amplifier and hence can devise methods for extending the amplifier bandwidth. For example, it is such insight that led to the invention of the cascode configuration and the other broadband techniques described.

Once an initial design is obtained, computer programs can be employed for an accurate determination of frequency response. The results obtained this way can be used to further "fine tune" the design.

The circuits analyzed in this chapter are used on their own and as building blocks of more complex multistage amplifiers. Frequency-response analysis of multistage amplifiers follows the procedures studied in this chapter. As an example, we shall consider the frequency response of an op amp circuit in Chapter 13.

FEEDBACK 12

Most physical systems embody some form of feedback. It is interesting to note, though, that the theory of negative feedback has been developed by electronics engineers. In his search for methods for the design of amplifiers with stable gain for use in telephone repeaters, Harold Black, an electronics engineer with the Western Electric Company, invented the feedback amplifier in 1928. Since then the technique has been so widely used that it is almost impossible to think of electronic circuits without some form of feedback, either implicit or explicit. Furthermore, the concept of feedback and its associated theory is currently used in areas other than engineering, such as in the modeling of biological systems.

Feedback can be either *negative* (degenerative) or *positive* (regenerative). In amplifier design negative feedback is applied to effect one or more of the following properties:

1. *Desensitize the gain;* that is, make the value of the gain less sensitive to variations in the value of circuit components, such as variations that might be caused by changes in temperature.
2. *Reduce nonlinear distortion;* that is, make the output proportional to the input (in other words, make the gain constant independent of signal level).
3. *Reduce the effect of noise* (unwanted electrical signals generated by the circuit components) and extraneous interference.
4. *Control the input and output impedances;* as will be shown, by selecting an appropriate feedback topology one can cause the input and output impedances to increase or decrease as desired.
5. *Extend the bandwidth* of the amplifier.

All of the above desirable properties are obtained at the expense of a reduction in gain. It will be shown that the gain-reduction factor, called the *amount of feedback,* is the factor by which the circuit is desensitized, by which the input impedance of a voltage amplifier is increased, by which the bandwidth is extended, and so on. In short, the basic idea of negative feedback is to trade off gain for other desirable properties. This chapter is devoted to the study of negative-feedback amplifiers: their analysis, design, and characteristics.

Under certain conditions the negative feedback in an amplifier can become positive and of such a magnitude as to cause oscillations. The reader will recall the use of positive feedback in design of oscillators and bistable circuits (Chapters 5 and 6). Furthermore, in Chapter 14 we shall employ positive feedback in the design of sinusoidal oscillators. In this chapter, however, we are interested in the design of stable amplifiers. We shall therefore study the stability problems of negative-feedback amplifiers.

It should not be implied, however, that positive feedback always leads to instability. Positive feedback is useful in a number of applications, such as the design of active filters, which are studied in Chapter 14.

Before we begin our study of negative feedback we wish to remind the reader that we have already encountered negative feedback in a number of applications. Almost all op-amp circuits employ negative feedback. Another popular application of negative feedback is the use of the emitter resistance R_E to stabilize the bias point of bipolar transistors. In addition, the emitter follower and the source follower employ a large amount of negative feedback. The question then arises as to the need for a formal study of negative feedback. As will be appreciated by the end of this chapter, the formal study of feedback provides an invaluable tool for the analysis and design of electronic circuits. Also, the insight gained by thinking in terms of feedback is extremely profitable.

12.1 THE GENERAL FEEDBACK STRUCTURE

Figure 12.1 shows the basic structure of a feedback amplifier. Rather than showing voltages and currents, Fig. 12.1 is a *signal-flow* diagram, where each x can represent either a voltage or a current signal. The *open-loop* amplifier has a gain A; thus its output x_o is related to the input x_i by

$$x_o = A x_i \tag{12.1}$$

The output x_o is fed to the load as well as to a feedback network, which produces a sample of the output. This sample x_f is related to x_o by the feedback factor β,

$$x_f = \beta x_o \tag{12.2}$$

The feedback signal x_f is *subtracted* from the source signal x_s, which is the input to the complete feedback amplifier, to produce the signal x_i, which is the input to the basic amplifier,

$$x_i = x_s - x_f \tag{12.3}$$

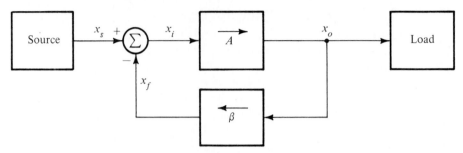

Fig. 12.1   *General structure of the feedback amplifier. This is a signal flow diagram, and the quantities x represent either voltage or current signals.*

Here we note that it is this subtraction that makes the feedback negative. In essence, negative feedback reduces the signal that appears at the input of the basic amplifier.

Implicit in the above description is that the source, the load, and the feedback network *do not* load the basic amplifier. That is, the gain *A* does not depend on any of these three networks. In practice this will not be the case, and we shall have to find a method for casting a real circuit into the ideal structure depicted in Fig. 12.1. Also implicit in Fig. 12.1 is that the forward transmission occurs entirely through the basic amplifier and the reverse transmission occurs entirely through the feedback network.

The gain of the feedback amplifier can be obtained by combining Eqs. (12.1) through (12.3):

$$A_f \equiv \frac{x_o}{x_s} = \frac{A}{1 + A\beta} \tag{12.4}$$

The quantity $A\beta$ is called the *loop gain,* a name that follows from Fig. 12.1. For the feedback to be negative, the loop gain $A\beta$ should be positive; that is, the feedback signal x_f should have the same sign as x_s, thus resulting in a smaller difference signal x_i. Equation (12.4) indicates that for positive $A\beta$ the gain with feedback will be smaller than the open-loop gain A by the quantity $1 + A\beta$, which is called *the amount of feedback.*

If, as is the case in many circuits, the loop gain $A\beta$ is large,

$$A\beta \gg 1$$

then from Eq. (12.4) it follows that

$$A_f \simeq \frac{1}{\beta}$$

which is a very interesting result: the gain of the feedback amplifier is almost entirely determined by the feedback network. Since the feedback network usually consists of passive components, which can be chosen to be as accurate as one wishes, the advantage of negative feedback in obtaining accurate, predictable, and stable gain should be apparent. In other words the overall gain will have very little dependence on the gain of the basic amplifier, *A*, a desirable property because the gain *A* is usually a function of many parameters, some of which might have wide tolerances. We have seen a dramatic illustration of all of these results in op-amp circuits, where the *closed-loop gain* (which is another name for the gain with feedback) is almost entirely determined by the feedback elements.

Equations (12.1) through (12.3) can be combined to obtain the following expression for the feedback signal x_f:

$$x_f = \frac{A\beta}{1 + A\beta} x_s$$

Thus for $A\beta \gg 1$ we see that

$$x_f \simeq x_s$$

which implies that the signal x_i at the input of the basic amplifier is reduced to almost zero. Thus if a large amount of negative feedback is employed, the feedback signal x_f becomes an almost identical replica of the input signal x_s. An outcome of this property is the tracking of the two input terminals of an op amp. The difference between x_s and x_f, which is x_i, is sometimes referred to as the "error signal". Accordingly the input differencing circuit is often also called a *comparator*. (It is also known as a *mixer*.)

EXERCISE

12.1 The noninverting op-amp configuration shown in Fig. E12.1 provides a direct implementation of the feedback loop of Fig. 12.1
  a. Assume that the op amp has infinite input resistance and zero output resistance. Find an expression for the feedback factor β.
  b. If $A = 10^4$, find R_2/R_1 to obtain a closed-loop gain A_f of 10.
  c. What is the amount of feedback in decibels?
  d. If $V_s = 1$V, find V_o, V_f, and V_i.
  e. If A decreases by 20%, what is the corresponding decrease in A_f?
  Ans. (a) $\beta = R_1/(R_1 + R_2)$; (b) 9.01; (c) 60 dB; (d) 10 V, 0.999 V, 0.001 V; (e) 0.02%

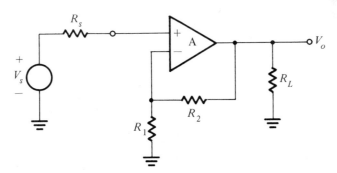

Fig. E12.1

12.2   SOME PROPERTIES OF NEGATIVE FEEDBACK

The properties of negative feedback were mentioned in the Introduction. In the following we shall consider some of these properties in more detail.

Gain Desensitivity

The effect of negative feedback on desensitizing the closed-loop gain was demonstrated in Exercise 12.1, where we saw that a 20% reduction in the gain of the basic amplifier gave rise to only 0.02% reduction in the gain of the closed-loop amplifier. This sensitivity reduction property can be analytically established as follows:

Assume that β is constant. Taking differentials of both sides of Eq. (12.4) results in

$$dA_f = \frac{dA}{(1 + A\beta)^2} \tag{12.5}$$

Dividing Eq. (12.5) by Eq. (12.4) yields

$$\frac{dA_f}{A_f} = \frac{1}{(1 + A\beta)} \frac{dA}{A} \tag{12.6}$$

which says that the percentage change in A_f (due to variations in some circuit parameter) is smaller than the change in A by the amount of feedback. For this reason the amount of feedback, $1 + A\beta$, is also known as the *desensitivity factor*.

Bandwidth Extension

Consider an amplifier whose high-frequency response is characterized by a single pole. Its gain can be expressed as

$$A(s) = \frac{A_M}{1 + s/\omega_H} \tag{12.7}$$

where A_M denotes the midband gain and ω_H is the upper 3-dB frequency. Application of negative feedback, with a frequency-independent factor β, around this amplifier results in a closed-loop gain $A_f(s)$ given by

$$A_f(s) = \frac{A(s)}{1 + \beta A(s)}$$

Substituting for $A(s)$ from Eq. (12.7) results in

$$A_f(s) = \frac{A_M/(1 + A_M\beta)}{1 + s/\omega_H(1 + A_M\beta)}$$

Thus the feedback amplifier will have a midband gain of $A_M/(1 + A_M\beta)$ and an upper 3-dB frequency ω_{Hf} given by

$$\omega_{Hf} = \omega_H(1 + A_M\beta) \tag{12.8}$$

It follows that the upper 3-dB frequency is increased by a factor equal to the amount of feedback.

Similarly, it can be shown that if the open-loop gain is characterized by a dominant low-frequency pole giving rise to a lower 3-dB frequency ω_L, then the feedback amplifier will have a lower 3-dB frequency ω_{Lf},

$$\omega_{Lf} = \frac{\omega_L}{1 + A_M\beta} \tag{12.9}$$

12.2 Consider the noninverting op-amp circuit of Exercise 12.1. Let the open-loop gain A have a low frequency value of 10^4 and a uniform -6 dB/octave rolloff at high frequencies with a 3-dB frequency of 100 Hz. Find the low-frequency gain and the upper 3-dB frequency of a closed-loop amplifier with $R_1 = 1$ kΩ and $R_2 = 9$ kΩ.

Ans. 9.99 V/V; 100.1 kHz

Noise Reduction

Negative feedback can be employed to reduce the noise or interference in an amplifier or, more precisely, to increase the ratio of signal to noise. However, as we shall now explain, this noise-reduction process is possible only under certain conditions. Consider the situation illustrated in Fig. 12.2. Figure 12.2a shows an amplifier with gain A_1, an input signal V_s, and noise or interference V_n. It is assumed that for some reason this amplifier suffers from noise and that the noise can be assumed to be introduced at the input of the amplifier. The *signal-to-noise ratio* for this amplifier is

$$S/N = \frac{V_s}{V_n}$$

Consider next the circuit in Fig. 12.2b. Here we assume that it is possible to build another amplifier stage with gain A_2 that does not suffer from the noise problem. If this

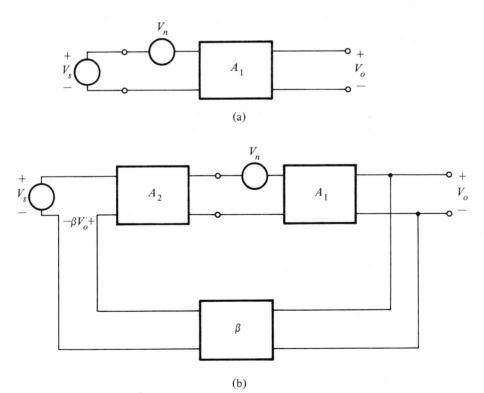

(a)

(b)

Fig. 12.2 Illustrating the application of negative feedback to improve the signal-to-noise ratio in amplifiers.

is the case, then we may precede our original amplifier A_1 by the *clean* amplifier A_2 and apply negative feedback around the overall cascade of such an amount as to keep the overall gain constant. The output voltage of the circuit in Fig. 12.2b can be found by superposition:

$$V_o = V_s \frac{A_1 A_2}{1 + A_1 A_2 \beta} + V_n \frac{A_1}{1 + A_1 A_2 \beta}$$

Thus the signal-to-noise ratio at the output becomes

$$S/N = \frac{V_s}{V_n} A_2$$

which is A_2 times higher than in the original case.

We should emphasize once more that the improvement in signal-to-noise ratio by the application of feedback is possible only if one can precede the noisy stage by a (relatively) noise-free stage. This situation, however, is not uncommon in real life. The best example is found in the output power-amplifier stage of an audio amplifier. Such a stage usually suffers from a problem known as *power-supply hum*. The problem arises because of the large currents that this stage draws from the power supply and the difficulty in providing adequate power-supply filtering inexpensively.

The power output stage is required to provide large power gain but little or no voltage gain. We may therefore precede the power output stage by a small-signal amplifier that provides large voltage gain and apply a large amount of negative feedback, thus restoring the voltage gain to its original value. Since the small-signal amplifier can be fed from another, less hefty (and hence better regulated) power supply, it will not suffer from the hum problem. The hum at the output will then be reduced by the amount of the voltage gain of this added *preamplifier*.

EXERCISE

12.3 Consider a power output stage with voltage gain $A_1 = 1$, an input signal $V_s = 1$ V, and a hum V_n of 1 V. Assume that this power stage is preceded by a small-signal stage with gain $A_2 = 100$ V/V and that overall feedback with $\beta = 1$ is applied. If V_s and V_n remain unchanged, find the signal and noise voltages at the output and hence the improvement in S/N.
Ans. $\simeq 1$ V; $\simeq 0.01$ V; 100 (40 dB)

Reduction in Nonlinear Distortion

Curve a in Fig. 12.3 shows the transfer characteristic of an amplifier. As indicated, the characteristic is piecewise linear, with the gain changing from 1,000 to 100 and then to 0. This nonlinear transfer characteristic will result in this amplifier generating a large amount of nonlinear distortion.

The amplifier transfer characteristic can be considerably *linearized* (that is, made less nonlinear) through the application of negative feedback. That this is possible should not be too surprising, since we have already seen that negative feedback reduces the dependence of the overall closed-loop amplifier gain on open-loop gain of the basic amplifier. Thus large changes in open-loop gain (1,000 to 100 in this case) give rise to much smaller corresponding changes in the closed-loop gain.

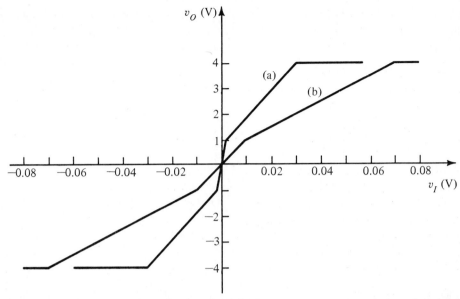

Fig. 12.3 *Illustrating the application of negative feedback to reduce the nonlinear distortion in amplifiers. Curve (a) shows the amplifier transfer characteristic without feedback. Curve (b) shows the characteristic with negative feedback ($\beta = 0.01$) applied.*

To illustrate, let us apply negative feedback with $\beta = 0.01$ to the amplifier whose open-loop transfer characteristic is depicted in Fig. 12.3. The resulting transfer characteristic of the closed-loop amplifier is shown in Fig. 12.3 as curve b. Here the slope of the steepest segment is given by

$$A_{f1} = \frac{1{,}000}{1 + 1{,}000 \times 0.01} = 90.9$$

and the slope of the next segment is given by

$$A_{f2} = \frac{100}{1 + 100 \times 0.01} = 50$$

Thus the order-of-magnitude change in slope has been considerably reduced. The price paid, of course, is a reduction in gain. Thus if the overall gain has to be restored, then a preamplifier should be added. This preamplifier should not present a severe nonlinear distortion problem, since it will be dealing with smaller signals.

Finally, it should be noted that negative feedback does nothing about amplifier saturation, since in saturation the gain is very small (almost zero) and hence the amount of feedback is also very small (almost zero).

12.3 THE FOUR BASIC FEEDBACK TOPOLOGIES

Based on the quantity to be amplified (voltage or current) and on the desired form of output (voltage or current), amplifiers can be classified into four categories. These categories were discussed in Chapter 2. In the following we shall review this amplifier classification and point out the feedback topology appropriate in each case.

Voltage Amplifiers

Voltage amplifiers are intended to amplify an input voltage signal and provide an output voltage signal. The voltage amplifier is essentially a voltage-controlled voltage source. The input impedance is required to be high, and the output impedance is required to be low. Since the signal source is essentially a voltage source, it is convenient to represent it in terms of a Thévenin's equivalent circuit. In a voltage amplifier the output quantity of interest is the output voltage. It follows that the feedback network should *sample* the output *voltage*. Also, because of the Thévenin's representation of the source, the feedback signal x_f should be a voltage that can be *mixed* with the source voltage in *series*.

A suitable feedback topology for the voltage amplifier is the *voltage-sampling series-mixing* one shown in Fig. 12.4a. As will be shown, this topology not only stabilizes the voltage gain but also results in a higher input resistance and a lower output resistance, which are desirable properties for a voltage amplifier. The noninverting op-amp configuration of Fig. E12.1 is an example of this feedback topology. Finally, it should be mentioned that this feedback topology is also known as the *series-shunt feedback*, where series refers to the connection at input and shunt refers to the connection at output.

Current Amplifiers

Here the input signal is essentially a current, and thus the signal source is most conveniently represented by its Norton's equivalent. The output quantity of interest is current; hence the feedback network should *sample* the output *current*. The feedback signal should be in current form so that it may be *mixed* in *shunt* with the source current. Thus the feedback topology suitable for a current amplifier is the *current-sampling shunt-mixing* topology, illustrated in Fig. 12.4b. As will be shown, this topology not only stabilizes the current gain but also results in a lower input resistance and a higher output resistance, desirable properties for a current amplifier.

An example of the current-sampling shunt-mixing feedback topology is given in Fig. 12.5. Note that the bias details are not shown. Also note that the current being sampled is not the output current but the almost equal emitter current of Q_2. This is done for circuit design convenience and is quite usual in circuits involving current sampling.

The reference direction indicated in Fig. 12.5 for the feedback current I_f is such that it subtracts from I_s. This reference notation will be followed in all circuits in this chapter, since it is consistent with the notation used in the general feedback structure of Fig. 12.1. Therefore, in all circuits for the feedback to be negative the loop gain $A\beta$ should be positive. The reader is urged to verify that in the circuit of Fig. 12.5 A is negative and β is negative.

It is of utmost importance to be able to qualitatively ascertain the feedback polarity (positive or negative). This can be done by "following the signal around the loop." For

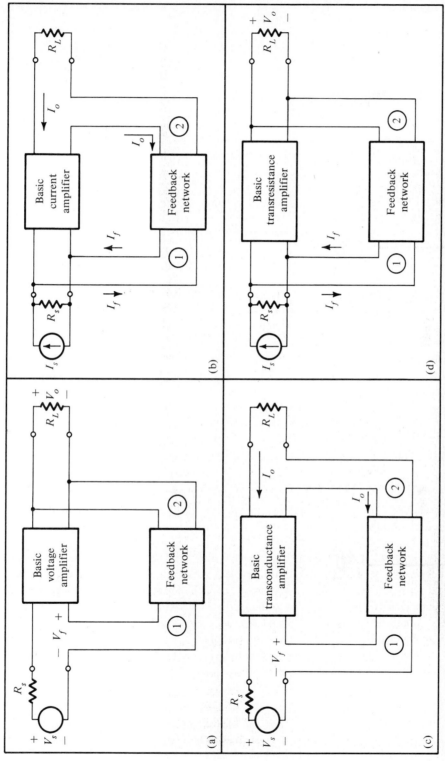

Fig. 12.4 The four basic feedback topologies: (a) voltage-sampling series-mixing (series-shunt) topology; (b) current-sampling shunt-mixing (shunt-series) topology; (c) current-sampling series-mixing (series-series) topology; (d) voltage-sampling shunt-mixing (shunt-shunt) topology.

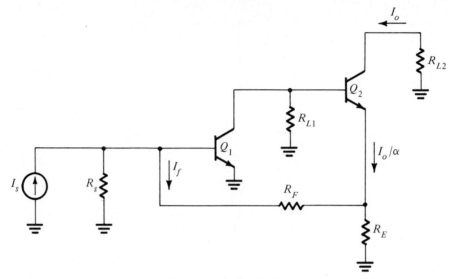

Fig. 12.5 *A transistor amplifier with shunt-series feedback.*

instance, let the current I_s in Fig. 12.5 increase. We see that the base current of Q_1 will increase, and thus its collector current will also increase. This will cause the collector voltage of Q_1 to decrease, and thus the collector current of Q_2, I_o, will decrease. From the feedback network we see that if I_o/α decreases, then I_f (in the direction shown) will increase. The increase in I_f will subtract from I_s, causing a smaller increment to be seen by the amplifier. Hence the feedback is negative.

Finally, we should mention that this feedback topology is also known as *shunt-series* feedback.

Transconductance Amplifiers

Here the input signal is voltage and the output signal is current. It follows that the appropriate feedback topology is the *current-sampling series-mixing* topology, illustrated in Fig. 12.4c.

An example of this feedback topology is given in Fig. 12.6. Here note that as in the circuit of Fig. 12.5 the current sampled is not the output current but the almost equal emitter current of Q_3. In addition, the mixing loop is not a conventional one; it is not a simple series connection, since the feedback signal developed across R_{E1} is in the emitter circuit of Q_1, while the source is in the base circuit of Q_1. These two approximations are done for convenience of circuit design.

Finally, it should be mentioned that the current-sampling series-mixing feedback topology is also known as the *series-series* feedback configuration.

Transresistance Amplifiers

Here the input signal is current and the output signal is voltage. It follows that the appropriate feedback topology is of the *voltage-sampling current-mixing* type, shown in Fig. 12.4d.

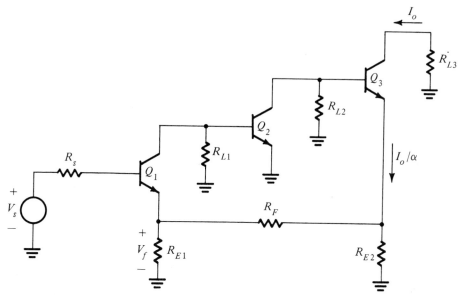

Fig. 12.6   An example of the series-series feedback topology.

An example of this feedback topology is found in the inverting op-amp configuration of Fig. 12.7a. The circuit is redrawn in Fig. 12.7b with the source converted to Norton's form.

This feedback topology is also known as *shunt-shunt* feedback.

12.4   ANALYSIS OF THE SERIES-SHUNT FEEDBACK AMPLIFIER

The Ideal Situation

The ideal structure of the series-shunt feedback amplifier is shown in Fig. 12.8a. It consists of a *unilateral* open-loop amplifier (the *A* circuit) and an ideal voltage-sampling

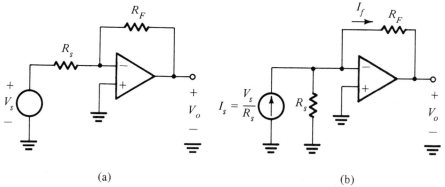

(a)   (b)

Fig. 12.7   The inverting op-amp configuration as an example of shunt-shunt feedback.

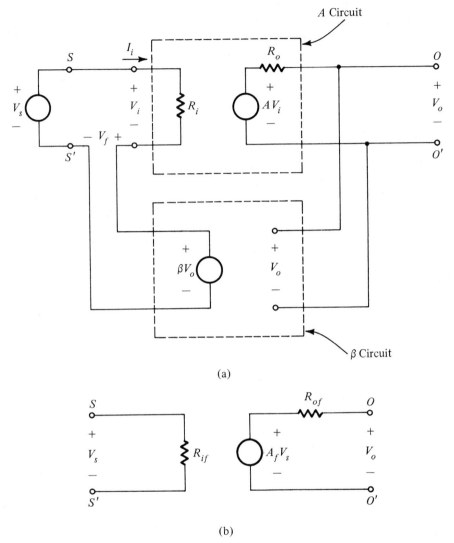

Fig. 12.8   *The series-shunt feedback amplifier: (a) ideal structure; (b) equivalent circuit.*

series-mixing feedback network (the β circuit). The A circuit has an input resistance R_i, a voltage gain A, and an output resistance R_o. It is assumed that the source and load resistances have been included inside the A circuit. Furthermore, note that the β circuit does *not* load the A circuit; that is, connecting the β circuit does not change the value of A (defined $A \equiv V_o/V_i$).

The circuit of Fig. 12.8a exactly follows the ideal feedback model of Fig. 12.1. Therefore the closed-loop gain A_f is given by

$$A_f \equiv \frac{V_o}{V_s} = \frac{A}{1 + A\beta}$$

The equivalent circuit model of the series-shunt feedback amplifier is shown in Fig. 12.8b. Here R_{if} and R_{of} denote the input and output resistances with feedback. The relationship between R_{if} and R_i can be established by considering the circuit in Fig. 12.8a:

$$R_{if} \equiv \frac{V_s}{I_i} = \frac{V_s}{V_i/R_i}$$

$$= R_i \frac{V_s}{V_i}$$

$$= R_i \frac{V_i + \beta A V_i}{V_i}$$

Thus

$$R_{if} = R_i(1 + A\beta) \tag{12.10}$$

That is, the negative feedback in this case increases the input resistance by a factor equal to the amount of feedback. Since the above derivation does not depend on the method of sampling (shunt or series), it follows that the relationship between R_{if} and R_i is a function only of the method of mixing. We shall discuss this point further in later sections.

Note, however, that this result is not surprising and is physically intuitive: Since the feedback voltage V_f subtracts from V_s, the voltage that appears across R_i—that is, V_i—becomes quite small. Thus the input current I_i becomes correspondingly small and the resistance seen by V_s becomes large. Finally, it should be pointed out that Eq. (12.10) can be generalized to the form

$$Z_{if}(s) = Z_i(s)[1 + A(s)\beta(s)] \tag{12.11}$$

To find the output resistance, R_{of}, of the feedback amplifier in Fig. 12.8a we reduce V_s to zero and apply a test voltage V_t at the output, as shown in Fig. 12.9,

$$R_{of} \equiv \frac{V_t}{I}$$

From Fig. 12.9 we can write

$$I = \frac{V_t - A V_i}{R_o}$$

and since $V_s = 0$ it follows from Fig. 12.8a that

$$V_i = -V_f = -\beta V_o = -\beta V_t$$

Thus

$$I = \frac{V_t + A\beta V_t}{R_o}$$

leading to

$$R_{of} = \frac{R_o}{1 + A\beta} \tag{12.12}$$

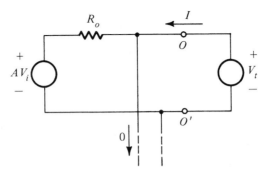

Fig. 12.9 Measuring the output resistance of the feedback amplifier of Fig. 12.8a.

That is, the negative feedback in this case reduces the output resistance by a factor equal to the amount of feedback. With a little thought one can see that the derivation of Eq. (12.12) does not depend on the method of mixing. Thus the relationship between R_{of} and R_o depends only on the method of sampling. Again this result is not surprising and is physically intuitive: With reference to Fig. 12.9 we see that since the feedback samples V_t and is negative, the controlled voltage source will be $-A\beta V_t$, independent of the method of mixing. This large negative voltage causes the current I to be large, indicating that the effective output resistance is small. Finally, we note that Eq. (12.12) can be generalized to

$$Z_{of}(s) = \frac{Z_o(s)}{1 + A(s)\beta(s)} \tag{12.13}$$

The Practical Situation

In a practical series-shunt feedback amplifier the feedback network will not be an ideal voltage-controlled voltage source. Rather, the feedback network is usually passive and hence will load the basic amplifier and thus affect the values of A, R_i, and R_o. In addition, the source and load resistances will affect these three parameters. Thus the problem we have is as follows: given a series-shunt feedback amplifier represented by the block diagram of Fig. 12.10a, find the A circuit and the β circuit.

Our problem essentially involves representing the amplifier of Fig. 12.10a by the ideal structure of Fig. 12.8a. As a first step toward that end we observe that the source and load resistances should be lumped with the basic amplifier. This, together with representing the two-port feedback network in terms of its h parameters (see Section 2.7), is illustrated in Fig. 12.10b. The choice of h parameters is based on the fact that this is the only parameter set that represents the feedback network by a series network at port 1 and a parallel network at port 2. Such a representation is obviously convenient in view of the series connection at the input and the parallel connection at the output.

Examination of the circuit in Fig. 12.10b reveals that the current source $h_{21}I_1$ represents the forward transmission of the feedback network. Since the feedback network is usually passive, its forward transmission can be neglected by comparison to the much larger forward transmission of the basic amplifier. We will therefore assume that h_{21} of the feedback network is approximately zero.

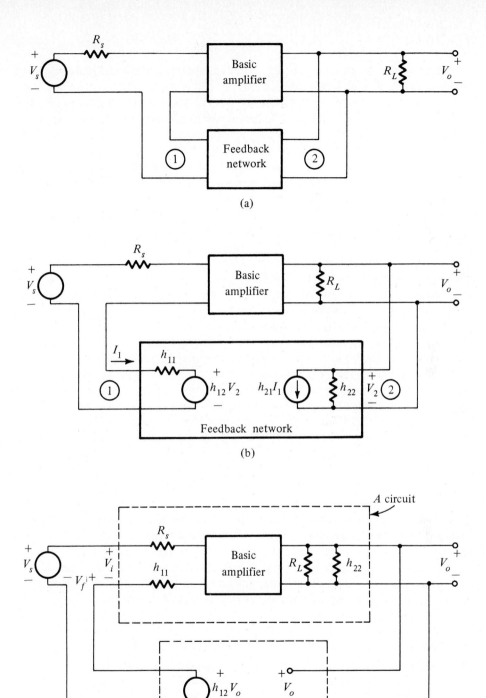

Fig. 12.10 *Derivation of the A circuit and β circuit for the series-shunt feedback amplifier. (a) Block diagram of a practical series-shunt feedback amplifier. (b) The circuit in (a) with the feedback network represented by its h parameters. (c) The circuit in (b) after neglecting h_{21}.*

Compare the circuit of Fig. 12.10b (after eliminating the current source $h_{21}I_1$) to the ideal circuit of Fig. 12.8a. We see that by including h_{11} and h_{22} with the basic amplifier we obtain the circuit shown in Fig. 12.10c, which is very similar to the ideal circuit. Now, if the basic amplifier is unilateral (or almost unilateral), then the circuit of Fig. 12.10c is equivalent (or approximately equivalent) to the ideal circuit. It follows then that the A circuit is obtained by augmenting the basic amplifier at the input with the source impedance R_s and the impedance h_{11} of the feedback network, and augmenting it at the output with the load impedance R_L and the admittance h_{22} of the feedback network.

We conclude that the loading effect of the feedback network on the basic amplifier is represented by the components h_{11} and h_{22}. From the definitions of the h parameters in Chapter 2 we see that h_{11} is the impedance looking into port 1 of the feedback network with port 2 short-circuited. Since port 2 of the feedback network is connected in *shunt* with the output port of the amplifier, short-circuiting port 2 destroys the feedback. Similarly, h_{22} is the admittance looking into port 2 of the feedback network with port 1 open-circuited. Since port 1 of the feedback network is connected in *series* with the amplifier input, open-circuiting port 1 destroys the feedback.

These observations suggest a simple rule for finding the loading effects of the feedback network on the basic amplifier. The loading effect is found by looking into the appropriate port of the feedback network while the other port is open-circuited or short-circuited so as to destroy the feedback. If the connection is a shunt one, we short-circuit the port; if it is a series one, we open-circuit it. In Sections 12.5 and 12.6 it will be seen that this simple rule applies also to the other three feedback topologies.

We next consider the determination of β. From Fig. 12.10c we see that β is equal to h_{12} of the feedback network,

$$\beta = h_{12} \equiv \left. \frac{V_1}{V_2} \right|_{I_1=0}$$

Thus to measure β one applies a voltage to port 2 and measures the voltage that appears at port 1 while the latter port is open-circuited. This result is intuitively appealing because the object of the feedback network is to sample the output voltage ($V_2 = V_o$) and provide a voltage signal ($V_1 = V_f$) that is mixed in series with the input source. The series connection at the input suggests that (as in the case of finding the loading effects of the feedback network) β should be found with port 1 open-circuited.

Summary

A summary of the rules for finding the A circuit and β for a given series-shunt feedback amplifier of the form in Fig. 12.10a is given in Fig. 12.11.

Example 12.1

Figure 12.12a shows an op amp connected in the noninverting configuration. The op amp has an open-loop gain μ, a differential input resistance R_{id}, a common-mode input resistance R_{icm}, and an output resistance r_o. Find expressions for A, β, the closed-loop gain V_o/V_s, the input resistance R'_{if} (see Fig. 12.12a) and the output resistance R'_{of}.

(a) The A circuit is

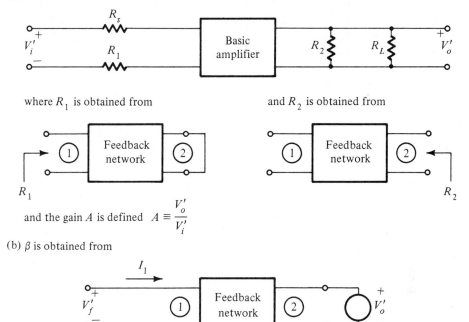

where R_1 is obtained from

and R_2 is obtained from

R_1

R_2

and the gain A is defined $A \equiv \dfrac{V_o'}{V_i'}$

(b) β is obtained from

$$\beta \equiv \left. \dfrac{V_f'}{V_o'} \right|_{I_1 = 0}$$

Fig. 12.11 Summary of the rules for finding the A circuit and β for the voltage-sampling series-mixing case of Fig. 12.10a.

Also find numerical values, given $\mu = 10^4$, $R_{id} = 100$ kΩ, $R_{icm} = 10$ MΩ, $r_o = 1$ kΩ, $R_L = 2$ kΩ, $R_1 = 1$ kΩ, $R_2 = 1$ MΩ, and $R_s = 10$ kΩ.

Solution
We first observe that the existence of the two resistances labeled $2R_{icm}$ between the op-amp inputs and ground will complicate the analysis.[1] This problem, however, can be easily overcome, as shown in Fig. 12.12b. We now observe that the feedback network consists of R_2 and R_1'. This network samples the output voltage V_o and provides a voltage signal (across R_1') that is mixed in series with the input source V_s'.

The A circuit is easily obtained with the rules of Fig. 12.11 and is shown in Fig. 12.12c. For this circuit we can write by inspection

$$A \equiv \dfrac{V_o'}{V_i'} = \mu \dfrac{[R_L \| (R_1' + R_2)]}{[R_L \| (R_1' + R_2)] + r_o} \dfrac{R_{id}}{R_{id} + R_s' + (R_1' \| R_2)}$$

[1]Refer to Fig. 12.10c and let the lower terminal of V_s be grounded. We can see that none of the input terminals of the A circuit can be grounded. Thus the resistance R_i is simply the resistance between the two input terminals of the A circuit. If, however, there exist inside the A circuit resistances with grounded terminals, the determination of R_i becomes quite lengthy.

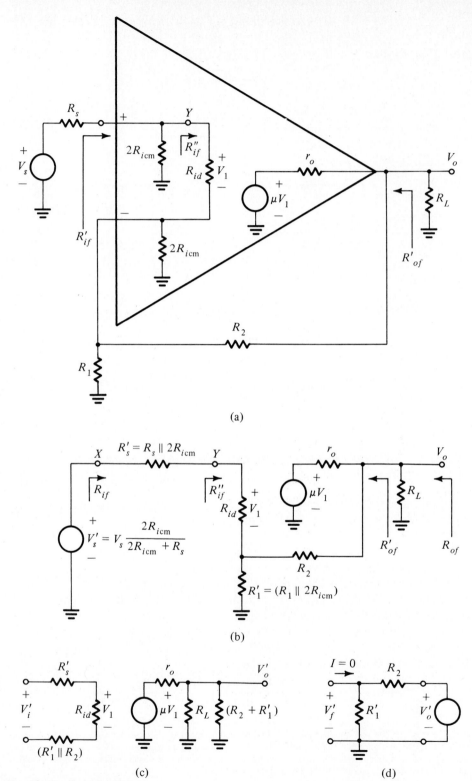

(a)

(b)

(c)   (d)

Fig. 12.12   Example 12.1.

For the values given we find that $R_1' \simeq R_1 = 1$ kΩ and $R_s' \simeq R_s = 10$ kΩ, leading to $A \simeq 6{,}000$.

The circuit for obtaining β is shown in Fig. 12.12d, from which we obtain

$$\beta \equiv \frac{V_f'}{V_o'} = \frac{R_1'}{R_1' + R_2} \simeq 10^{-3}$$

The gain with feedback is now obtained as

$$A_f \equiv \frac{V_o}{V_s'} = \frac{A}{1 + A\beta} = \frac{6{,}000}{7} = 857 \text{ V/V}$$

This can be used to obtain V_o/V_s:

$$\frac{V_o}{V_s} = \frac{V_o}{V_s'} \frac{2R_{icm} + R_s}{2R_{icm}} \simeq \frac{V_o}{V_s'} = 857 \text{ V/V}$$

The input resistance R_{if} given by the feedback equations is the resistance seen by the external source (in this case V_s'); that is, it is the resistance between node X and ground in Fig. 12.12b. This resistance is given by

$$R_{if} = R_i(1 + A\beta)$$

where R_i is the input resistance of the A circuit in Fig. 12.12c:

$$R_i = R_s' + R_{id} + (R_1' \| R_2)$$

For the values given, $R_i \simeq 111$ kΩ, resulting in

$$R_{if} = 111 \times 7 = 777 \text{ k}\Omega$$

This, however, is not the resistance asked for. What is required is R_{if}', indicated in Fig. 12.12a. To obtain R_{if}' we first subtract R_s' from R_{if} and find R_{if}'', indicated in Fig. 12.12b and in Fig. 12.12a. From the latter figure we see that R_{if}' can be obtained by including $2R_{icm}$ in parallel with R_{if}'':

$$R_{if}' = [2R_{icm} \| (R_{if} - R_s')]$$

For the values given, $R_{if}' = 739$ kΩ. The resistance R_{of} given by the feedback equations is the output resistance of the feedback amplifier, including the load resistance R_L, as indicated in Fig. 12.12b. R_{of} is given by

$$R_{of} = \frac{R_o}{1 + A\beta}$$

where R_o is the output resistance of the A circuit. R_o can be obtained by inspection of Fig. 12.12c as

$$R_o = [r_o \| R_L \| (R_2 + R_1')]$$

For the values given $R_o \simeq 667$ Ω and

$$R_{of} = \frac{667}{7} = 95.3 \ \Omega$$

The resistance asked for, R'_{of}, is the output resistance of the feedback amplifier excluding R_L. From Fig. 12.12b we see that

$$R_{of} = (R'_{of} \| R_L)$$

Thus

$$R'_{of} \simeq 100 \ \Omega$$

$$\cdot \quad \cdot \quad \cdot$$

EXERCISES
12.4 If the op amp of Example 12.1 has a uniform -6 dB/octave high-frequency rolloff with $f_{3dB} = 1$ kHz, find the 3-dB frequency of the closed-loop gain V_o/V_s.
  Ans.   7 kHz

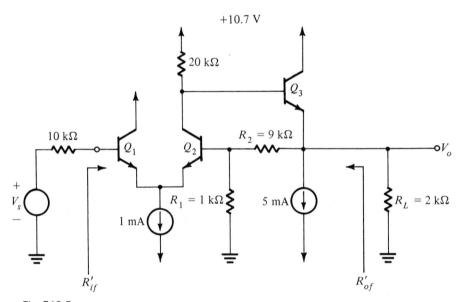

Fig. E12.5

12.5 The circuit shown in Fig. E12.5 consists of a differential stage followed by an emitter follower and with series-shunt feedback supplied by the resistors R_1 and R_2. Assuming that the dc component of V_s is zero, find the dc operating current of each of the three transistors and show that the dc voltage at the output is approximately zero. Then find the values of A, β, $A_f \equiv V_o/V_s$, R'_{if}, and R'_{of}. Assume that the transistors have $\beta = 100$.
  Ans.   85.7 V/V; 0.1 V/V; 8.96 V/V; 191 kΩ; 19.1 Ω

12.5   ANALYSIS OF THE SERIES-SERIES FEEDBACK AMPLIFIER

The Ideal Case

As mentioned in Section 12.3, the series-series feedback topology stabilizes I_o/V_s and is therefore best suited for transconductance amplifiers. Figure 12.13a shows the ideal structure for the series-series feedback amplifier. It consists of a unilateral open-loop

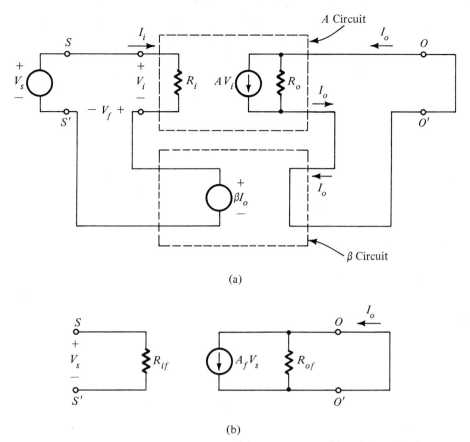

(a)

(b)

Fig. 12.13   *The series-series feedback amplifier: (a) ideal structure; (b) equivalent circuit.*

amplifier (the A circuit) and an ideal feedback network. Note that in this case A is a transconductance,

$$A \equiv \frac{I_o}{V_i}$$

while β is a transresistance. Thus the loop gain $A\beta$ remains a dimensionless quantity, as it should always be.

In the ideal structure of Fig. 12.13a the load and source resistances have been absorbed inside the A circuit, and the β circuit does not load the A circuit. Thus the circuit follows the ideal feedback model of Fig. 12.1, and we can write

$$A_f \equiv \frac{I_o}{V_s} = \frac{A}{1 + A\beta}$$

This transconductance with feedback is included in the equivalent circuit model of the feedback amplifier shown in Fig. 12.13b. In this model R_{if} is the input resistance with feedback. Using an analysis similar to that in Section 12.5 we can show that

$$R_{if} = R_i(1 + A\beta)$$

This relationship is identical to that obtained in the case of series-shunt feedback. This confirms our earlier observation that the relationship between R_{if} and R_i is a function only of the method of mixing. Series mixing therefore always increases the input resistance.

To find the output resistance R_{of} of the series-series feedback amplifier of Fig. 12.13a we reduce V_s to zero and break the output circuit to apply a test current I_t, as shown in Fig. 12.14:

$$R_{of} \equiv \frac{V}{I_t}$$

In this case: $V_i = -V_f = -\beta I_o = -\beta I_t$. Thus for the circuit in Fig. 12.14 we obtain

$$V = (I_t - AV_i)R_o = (I_t + A\beta I_t)R_o$$

Hence

$$R_{of} = (1 + A\beta)R_o$$

That is, in this case the negative feedback increases the output resistance. This should have been expected, since the negative feedback tries to make I_o constant in spite of

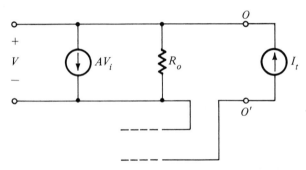

Fig. 12.14   *Measuring the output resistance R_{of} of the series-series feedback amplifier.*

changes in the output voltage, which means increased output resistance. This result also confirms our earlier observation; the relationship between R_{of} and R_o is a function only of the method of sampling. While voltage (shunt) sampling reduces the output resistance, current (series) sampling increases it.

The Practical Case

Figure 12.15a shows in block diagram form a practical series-series feedback amplifier. To be able to apply the feedback equations to this amplifier we have to represent it by the ideal structure of Fig. 12.13a. Our objective therefore is to devise a simple method for finding A and β.

The series-series amplifier of Fig. 12.15a is redrawn in Fig. 12.15b with R_s and R_L shown closer to the basic amplifier and the two-port feedback network represented by its z parameters. This parameter set has been chosen because it provides a representation of the feedback network with a series circuit at the input and a series circuit as the

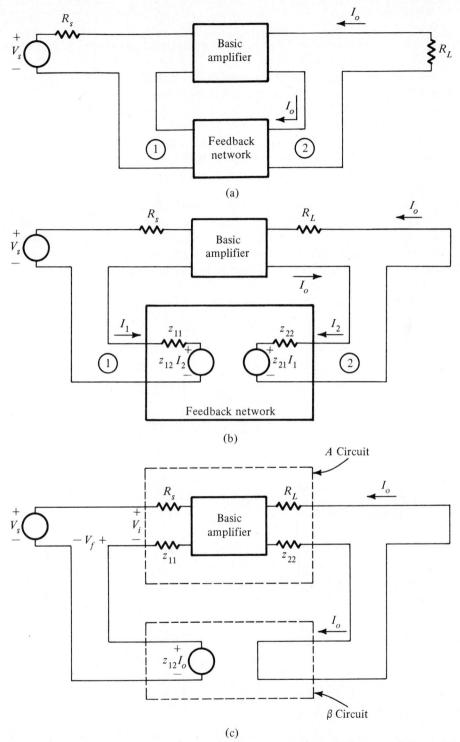

Fig. 12.15 Derivation of the A circuit and the β circuit for series-series feedback amplifiers. (a) A series-series feedback amplifier. (b) The circuit of (a) with the feedback network represented by its z parameters. (c) A redrawing of the circuit in (b) after neglecting z_{21}.

output. This is obviously convenient in view of the series connections at input and output.

As we have done in the case of the series-shunt amplifier, we shall assume that the forward transmission through the feedback network is negligible as compared with that through the basic amplifier. This enables us to assume that z_{21} of the feedback network is almost zero, and thus we may dispense with the voltage source $z_{21}I_1$ in Fig. 12.15b. Doing this and redrawing the circuit to include z_{11} and z_{22} with the basic amplifier results in the circuit in Fig. 12.15c. Now if the basic amplifier is unilateral (or almost unilateral), we see that this circuit is equivalent (or almost equivalent) to the ideal circuit of Fig. 12.13a.

It follows that the A circuit is composed of the basic amplifier augmented at the input with R_s and z_{11} and augmented at the output with R_L and z_{22}. Since z_{11} and z_{22} are the impedances looking into ports 1 and 2, respectively, of the feedback network with the other port open-circuited, we see that finding the loading effects of the feedback network on the basic amplifier follows the rule formulated in Section 12.4. That is, we look into one port of the feedback network while the other port is open-circuited or short-circuited so as to destroy the feedback (open if series and short if shunt).

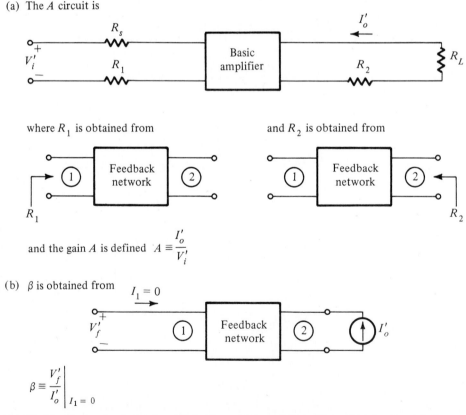

(a) The A circuit is

where R_1 is obtained from   and R_2 is obtained from

and the gain A is defined $A \equiv \dfrac{I_o'}{V_i'}$

(b) β is obtained from $I_1 = 0$

$$\beta \equiv \dfrac{V_f'}{I_o'}\bigg|_{I_1 = 0}$$

Fig. 12.16   Finding the A circuit and β for the current-sampling series-mixing (series-series) case.

From Fig. 12.15c we see that β is equal to z_{12} of the feedback network,

$$\beta = z_{12} \equiv \left.\frac{V_1}{I_2}\right|_{I_1=0}$$

This result is intuitively appealing. Recall that in this case the feedback network samples the output current $[I_2 = I_o]$ and provides a voltage $[V_f = V_1]$ that is mixed in series with the input source. Again, the series connection at the input suggests that β is measured with port 1 open.

Summary

For future reference we present in Fig. 12.16 a summary of the rules for finding A and β for a given series-series feedback amplifier of the type shown in Fig. 12.15a.

Example 12.2

Figure 12.17a shows a voltage-controlled current-source circuit. It embodies feedback of the current-sampling series-mixing type. However, for practical reasons the current

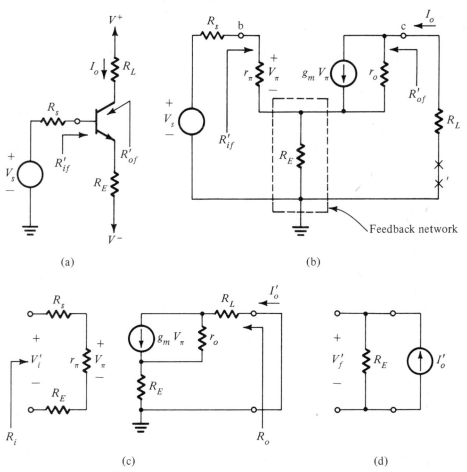

(a)   (b)

(c)   (d)

Fig. 12.17   Example 12.2.

being sampled is *not* the output current in the collector but rather the almost equal emitter current. We wish to carry out small-signal analysis to determine I_o/V_s, R'_{if}, and R'_{of} (see Fig. 12.17a).

Solution
The small-signal equivalent circuit is shown in Fig. 12.17b, from which we see that the series-series feedback is provided by resistance R_E. The rules of Fig. 12.16 can be directly applied to obtain the A circuit shown in Fig. 12.17c. For this circuit we can write by inspection

$$A \equiv \frac{I'_o}{V_i} = \frac{g_m V_\pi r_o}{r_o + R_L + R_E} \frac{1}{V'_i}$$

$$= \frac{g_m r_o}{r_o + R_L + R_E} \frac{r_\pi}{r_\pi + R_s + R_E}$$

The circuit for determining β is shown in Fig. 12.17d, from which we obtain

$$\beta \equiv \frac{V'_f}{I'_o} = R_E$$

The overall transconductance A_f is then obtained using

$$A_f = \frac{A}{1 + A\beta}$$

Before substituting for A and β we shall make the approximation

$$r_o \gg R_L + R_E$$

which reduces A to

$$A \simeq \frac{g_m r_\pi}{r_\pi + R_s + R_E}$$

Correspondingly we have for A_f

$$A_f \equiv \frac{I_o}{V_s} \simeq \frac{g_m r_\pi}{R_E(1 + g_m r_\pi) + r_\pi + R_s}$$

Substituting $g_m r_\pi = h_{fe}$ and

$$\frac{r_\pi}{h_{fe} + 1} = r_e$$

we obtain

$$\frac{I_o}{V_s} = \frac{h_{fe}}{R_E(1 + h_{fe}) + r_\pi + R_s}$$

a result that we could have written by inspection from our knowledge of transistor circuit analysis. Thus in simple circuits it is usually expedient to do the analysis directly.

Nevertheless, the feedback method provides a quicker way for finding the output resistance, as will be demonstrated shortly.

The input resistance of the A circuit is obtained from Fig. 12.17c as

$$R_i = R_s + r_\pi + R_E$$

The input resistance with feedback R_{if} can be obtained as

$$R_{if} = (1 + A\beta)R_i = R_s + r_\pi + R_E + (g_m r_\pi)R_E$$

This is the input resistance of the feedback amplifier including R_s. To obtain the resistance required, R'_{if}, we simply subtract R_s from R_{if}:

$$R'_{if} = r_\pi + R_E(1 + g_m r_\pi) = r_\pi + R_E(1 + h_{fe})$$

an expression that we could have written by inspecting the circuit of Fig. 12.17a.

Finally, the output resistance of the A circuit can be obtained from Fig. 12.17c (after reducing the input V'_i to zero) as

$$R_o = R_L + r_o + R_E$$

The output resistance of the feedback circuit, R_{of}, is given by

$$R_{of} = (1 + A\beta)R_o$$

Use of the exact expression for A and substituting $\beta = R_E$ gives

$$R_{of} = R_L + R_E + r_o + \frac{g_m r_\pi R_E}{r_\pi + R_s + R_E} r_o$$

This resistance includes R_L. To obtain the output resistance R'_{of} excluding R_L we simply subtract R_L from R_{of}:

$$R'_{of} = R_E + r_o + \frac{g_m r_\pi R_E}{r_\pi + R_s + R_E} r_o$$

$$\simeq r_o\left(1 + \frac{h_{fe} R_E}{r_\pi + R_s + R_E}\right)$$

$$\cdot \ \cdot \ \cdot$$

EXERCISE

12.6 Because negative feedback extends the amplifier bandwidth, it is commonly used in the design of broadband amplifiers. One such amplifier is the MC1553. Part of the circuit of the MC1553 is shown in Fig. E12.6. The circuit shown (called a *feedback triple*) is composed of three gain stages with series-series feedback provided by the network composed of R_{E1}, R_F, and R_{E2}. Assume that the bias circuit, which is not shown, causes $I_{C1} = 0.6$ mA, $I_{C2} = 1$ mA, and $I_{C3} = 4$ mA. Using these values and assuming that $h_{fe} = 100$ and $r_o = \infty$, find the open-loop gain A, the feedback factor β, the closed-loop gain $A_f \equiv I_o/V_s$, the voltage gain V_o/V_s, and the input resistance R_{if}. (*Hint:* In the A circuit you will find resistances in the emitters of Q_1 and Q_3. Do not use

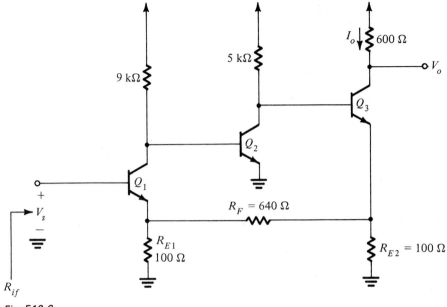

Fig. E12.6

the feedback method to deal with these resistances; rather, it is simpler to use direct analysis.)

Ans. $A = 20.8$ A/V; $\beta = 12\ \Omega$; $A_f = 0.083$ A/V; $V_o/V_s = -49.8$; $R_{if} = 3.26\ \mathrm{M}\Omega$

12.6   ANALYSIS OF THE SHUNT-SHUNT AND THE SHUNT-SERIES FEEDBACK AMPLIFIERS

In this section we shall extend—without proof—the method of Sections 12.4 and 12.5 to the two remaining feedback topologies.

The Shunt-Shunt Configuration

Figure 12.18 shows the ideal structure for a shunt-shunt feedback amplifier. Here the A circuit has an input resistance R_i, a transresistance A, and an output resistance R_o. The β circuit is a voltage-controlled current source, and β is a transconductance. The closed-loop gain A_f is defined

$$A_f \equiv \frac{V_o}{I_s}$$

and is given by

$$A_f = \frac{A}{1 + A\beta}$$

The input resistance with feedback is given by

$$R_{if} = \frac{R_i}{1 + A\beta}$$

where we note that the shunt connection at the input results in a reduced input resistance. Also note that the resistance R_{if} is the resistance seen by the source I_s, and it includes any source resistance.

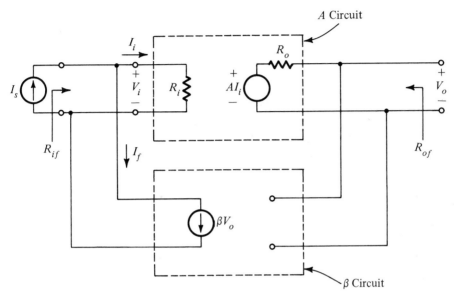

Fig. 12.18 *Ideal structure for the shunt-shunt feedback amplifier.*

The output resistance with feedback is given by

$$R_{of} = \frac{R_o}{1 + A\beta}$$

where we note that the shunt connection at the output results in a reduced output resistance. This resistance includes any load resistance.

Given a practical shunt-shunt feedback amplifier having the block diagram of Fig. 12.19, we use the method given in Fig. 12.20 to obtain the A circuit and the circuit for

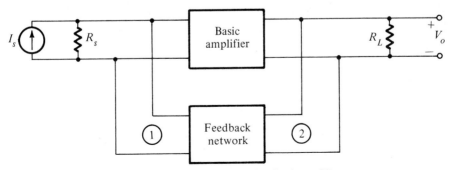

Fig. 12.19 *Block diagram for a practical shunt-shunt feedback amplifier.*

(a) The A circuit is

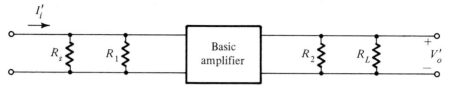

where R_1 is obtained from

and R_2 is obtained from

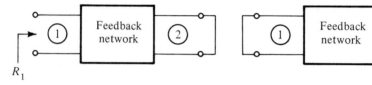

and the gain A is defined $A \equiv \dfrac{V'_o}{I'_i}$

(b) β is obtained from

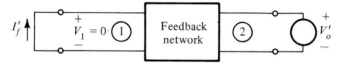

$$\beta \equiv \left. \frac{I'_f}{V'_o} \right|_{V_1=0}$$

Fig. 12.20 *Finding the A circuit and β for the voltage-sampling current-mixing (shunt-shunt) case.*

determining β. As in Sections 12.4 and 12.5 the method of Fig. 12.20 assumes that the basic amplifier is almost unilateral and that the forward transmission through the feedback network is negligibly small.

Example 12.3
We want to analyze the circuit of Fig. 12.21a to determine the small-signal voltage gain V_o/V_s, the input resistance R'_{if}, and the output resistance R_{of}. The transistor has $\beta =$ 100.

Solution
First we determine the transistor dc operating point. The dc analysis is illustrated in Fig. 12.21b, from which we can write

$$V_C = 0.7 + (I_B + 0.07)\,47 = 3.99 + 47I_B$$

and

$$\frac{12 - V_C}{4.7} = (\beta + 1)I_B + 0.07$$

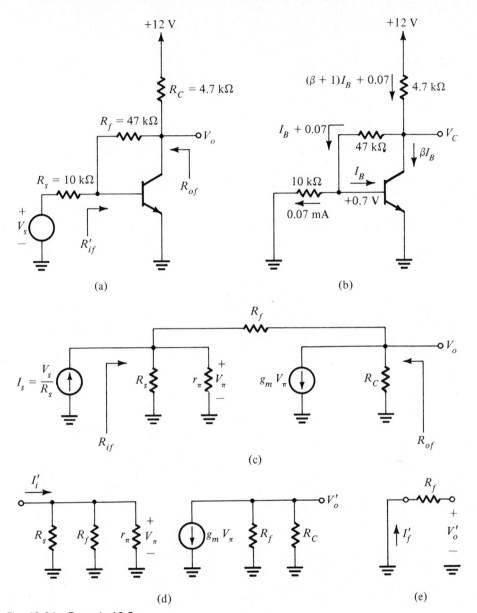

Fig. 12.21 Example 12.3.

These two equations can be solved to obtain $I_B \simeq 0.015$ mA, $I_C \simeq 1.5$ mA, and $V_C = 4.7$ V.

To carry out small-signal analysis we first recognize that the feedback is provided by R_f, which samples the output voltage V_o and feeds back a current that is mixed with the source current. Thus it is convenient to use the Norton's source representation, as is shown in Fig. 12.21c. The A circuit can be easily obtained using the rules of Fig. 12.20, and it is shown in Fig. 12.22d. For the A circuit we can write by inspection

$$V_\pi = I_i'(R_s \| R_f \| r_\pi)$$
$$V_o' = -g_m V_\pi(R_f \| R_C)$$

Thus

$$A = \frac{V_o'}{I_i'} = -g_m(R_f \| R_C)(R_s \| R_f \| r_\pi)$$
$$= -358.7 \text{ k}\Omega$$

The input and output resistances of the A circuit can be obtained from Fig. 12.21d as

$$R_i = (R_s \| R_f \| r_\pi) = 1.4 \text{ k}\Omega$$
$$R_o = (R_C \| R_f) = 4.27 \text{ k}\Omega$$

The circuit for determining β is shown in Fig. 12.21e, from which we obtain

$$\beta \equiv \frac{I_f'}{V_o'} = -\frac{1}{R_f} = -\frac{1}{47 \text{ k}\Omega}$$

Note that as usual the reference direction for I_f has been selected so that I_f subtracts from I_s. The resulting negative sign of β should cause no concern, since A is also negative, keeping the loop gain $A\beta$ positive, as it should be for the feedback to be negative.

We can now obtain A_f (for the circuit in Fig. 12.21c) as

$$A_f \equiv \frac{V_o}{I_s} = \frac{A}{1 + A\beta}$$
$$\frac{V_o}{I_s} = \frac{-358.7}{1 + 358.7/47} = \frac{-358.7}{8.63} = -41.56 \text{ k}\Omega$$

To find the voltage gain V_o/V_s we note that

$$V_s = I_s R_s$$

Thus

$$\frac{V_o}{V_s} = \frac{V_o}{I_s R_s} = \frac{-41.56}{10} \simeq -4.16 \text{ V/V}$$

The input resistance with feedback is given by

$$R_{if} = \frac{R_i}{1 + A\beta}$$

Thus

$$R_{if} = \frac{1.4}{8.63} = 162.2 \ \Omega$$

This is the resistance seen by the current source I_s in Fig. 12.21c. To obtain the input resistance of the feedback amplifier excluding R_s (that is, the required resistance R_{if}') we subtract $1/R_s$ from $1/R_{if}$ and invert the result; thus $R_{if}' = 165 \ \Omega$. Finally, the amplifier output resistance R_{of} is evaluated using

$$R_{of} = \frac{R_o}{1 + A\beta} = \frac{4.27}{8.63} = 495 \ \Omega$$

. . .

The Shunt-Series Configuration

Figure 12.22 shows the ideal structure of the shunt-series feedback amplifier. It is a current amplifier whose gain with feedback is defined as

$$A_f \equiv \frac{I_o}{I_s} = \frac{A}{1 + A\beta}$$

The input resistance with feedback is the resistance seen by the current source I_s and is given by

$$R_{if} = \frac{R_i}{1 + A\beta}$$

Again we note that the shunt connection at the input reduces the input resistance. The output resistance with feedback is the resistance seen by breaking the output circuit,

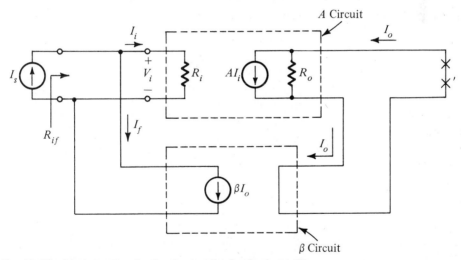

Fig. 12.22   Ideal structure for the shunt-series feedback amplifier.

such as between X and X', and looking between the two terminals thus generated (that is, between X and X'). This resistance R_{of} is given by

$$R_{of} = R_o(1 + A\beta)$$

where we note that the increase in output resistance is due to the current (series) sampling.

Given a practical shunt-series feedback amplifier, such as that represented by the block diagram of Fig. 12.23, we follow the method given in Fig. 12.24 in order to obtain A and β.

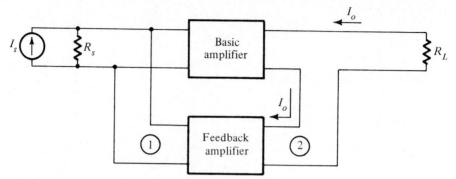

Fig. 12.23   Block diagram for a practical shunt-series feedback amplifier.

Example 12.4

Figure 12.25 shows a feedback circuit of the shunt-series type. Find I_{out}/I_{in}, R_{in}, and R_{out}. Assume the transistors to have $\beta = 100$ and $r_o = 100$ kΩ.

Solution

We begin by determining the dc operating points. In this regard we note that the feedback signal is capacitively coupled; thus the feedback has no effect on dc. The dc analysis proceeds as follows:

$$V_{B1} = 12\,\frac{15}{100 + 15} = 1.57 \text{ V}$$

$$V_{E1} \simeq 1.57 - 0.7 = 0.87 \text{ V}$$

$$I_{E1} = \frac{0.87}{0.87} = 1 \text{ mA}$$

$$V_{C1} \simeq 12 - 10 \times 1 = 2 \text{ V}$$

$$V_{E2} \simeq 2 - 0.7 = 1.3 \text{ V}$$

$$I_{E2} \simeq \frac{1.3}{1.3} \simeq 1 \text{ mA}$$

$$V_{C2} \simeq 12 - 1 \times 8 = 4 \text{ V}$$

The amplifier equivalent circuit is shown in Fig. 12.25b, from which we note that the feedback network is composed of R_{E2} and R_f. The feedback network samples the emitter current of Q_2, which is approximately equal to the collector current I_o. Also note that the required current gain, I_{out}/I_{in}, will be slightly different than the closed-loop current gain $A_f \equiv I_o/I_s$.

The A circuit is shown in Fig. 12.25c, where we have obtained the loading effects of the feedback network using the rules of Fig. 12.24. For the A circuit we can write

$$V_{\pi 1} = I_i'[R_s \| (R_{E2} + R_f) \| R_B \| r_{\pi 1}]$$

$$V_{b2} = -g_{m1}V_{\pi 1}\{r_{o1} \| R_{C1} \| [r_{\pi 2} + (\beta + 1)(R_{E2} \| R_f)]\}$$

$$V_{\pi 2} = V_{b2}\,\frac{r_{\pi 2}}{r_{\pi 2} + (\beta + 1)(R_{E2} \| R_f)}$$

$$I_o' \simeq g_{m2}V_{\pi 2}$$

(a) The A circuit is

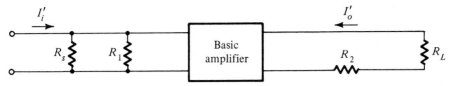

where R_1 is obtained from   and R_2 is obtained from

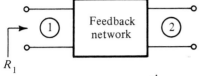   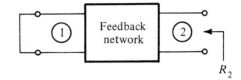

The gain A is defined as $A \equiv \dfrac{I'_o}{I'_i}$

(b) β is found from

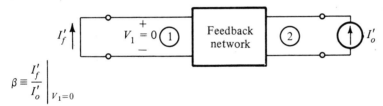

$$\beta \equiv \dfrac{I'_f}{I'_o}\bigg|_{V_1=0}$$

Fig. 12.24 *Finding the A circuit and β for the current-sampling shunt-mixing (shunt-series) case.*

These equations can be combined to obtain the open-loop current gain A,

$$A \equiv \frac{I'_o}{I'_i} \simeq -427.7$$

The input resistance R_i is given by

$$R_i = [R_s \| (R_{E2} + R_f) \| R_B \| r_{\pi 1}] = 1.5 \ \text{k}\Omega$$

The output resistance R_o is that found by looking into the output loop of the A circuit with the input excitation I'_i set to zero. It can be shown that

$$R_o = R'_L + r_{o2} + (1 + g_{m2}r_{o2})(R_{E2} \| R_f \| r_{\pi 2} \| R_{C1} \| r_{o1})$$
$$= 3 \ \text{M}\Omega$$

The circuit for determining β is shown in Fig. 12.27d, from which we find

$$\beta \equiv \frac{I'_f}{I'_o} = -\frac{R_{E2}}{R_{E2} + R_f} = -\frac{1.3}{11.3} = -0.115$$

Thus

$$1 + A\beta = 50.2$$

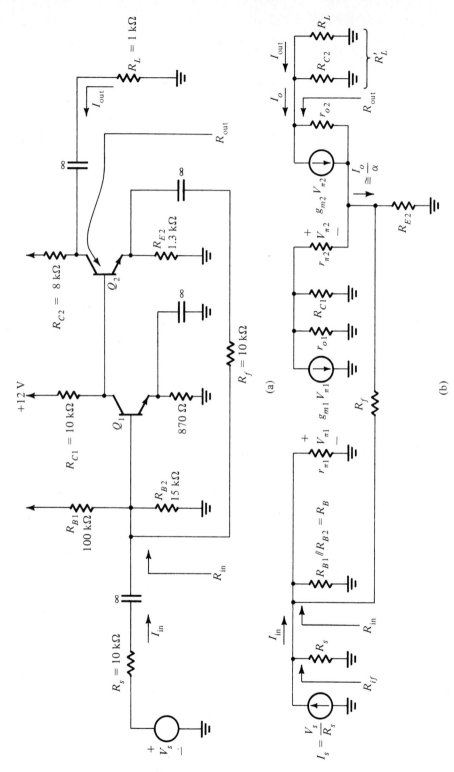

(a)

(b)

Fig. 12.25 Example 12.4

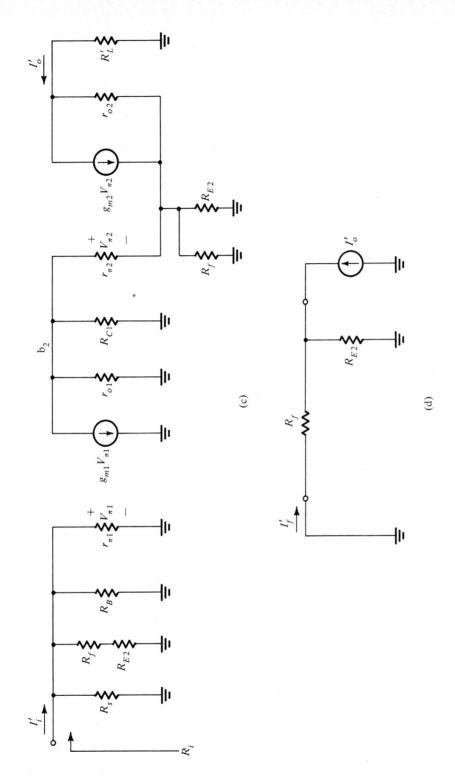

Fig. 12.25 Example 12.4 (continued).

The input resistance R_{if} is given by

$$R_{if} = \frac{R_i}{1 + A\beta} = 29.8 \ \Omega$$

The required input resistance R_{in} is given by

$$R_{in} = \frac{1}{1/R_{if} - 1/R_s} = 29.9 \ \Omega$$

Since $R_{in} \simeq R_{if}$, it follows from Fig. 12.25b that $I_{in} \simeq I_s$. The current gain A_f is given by

$$A_f \equiv \frac{I_o}{I_s} = \frac{A}{1 + A\beta} = -8.5$$

Note that because $A\beta \gg 1$ the closed-loop gain is approximately equal to $1/\beta$.
    Now the required current gain is given by

$$\frac{I_{out}}{I_{in}} \simeq \frac{I_{out}}{I_s} = \frac{R_{C2}}{R_L + R_{C2}} \frac{I_o}{I_s}$$

Thus $I_{out}/I_{in} = -7.55$.
    Finally, the output resistance R_{of} is given by

$$R_{of} = R_o(1 + A\beta) \simeq 150 \ \text{M}\Omega$$

The required output resistance R_{out} can be obtained by subtracting R'_L from R_{of}:

$$R_{out} \simeq R_{of} = 150 \ \text{M}\Omega$$

· · ·

EXERCISE

12.7   Use the feedback method to find the voltage gain V_o/V_s, the input resistance R'_{if}, and the output resistance R'_{of} of the inverting op-amp configuration of Fig. E12.7. Let the op amp have open-loop gain $\mu = 10^4$, $R_{id} = 100 \ \text{k}\Omega$, $R_{icm} = 10 \ \text{M}\Omega$, and $r_o = 1 \ \text{k}\Omega$. (*Hint*: The feedback is of the shunt-shunt type.)
    Ans. $-870 \ \text{V/V}$; $150 \ \Omega$; $91 \ \Omega$

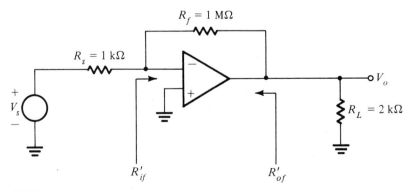

Fig. E12.7

12.7 DETERMINING THE LOOP GAIN

We have already seen that the loop gain $A\beta$ is a very important quantity that characterizes a feedback loop. Furthermore, in the following sections it will be shown that $A\beta$ determines whether the feedback amplifier is stable (as opposed to oscillatory). In this section we shall describe an alternative approach to the determination of loop gain.

The method is simple and consists of first reducing the network excitation to zero: external voltage sources are short-circuited and external current sources are open-circuited. Then the loop is broken at a convenient point, such as the terminals XX' in Fig. 12.26. The left-hand side of the loop should be terminated in an impedance Z_t equal to

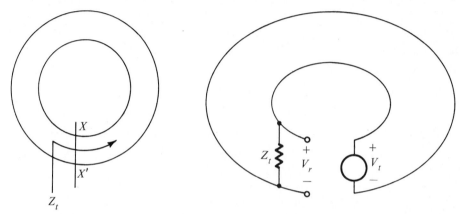

Fig. 12.26 A conceptual feedback loop is broken at XX' and a test voltage V_t is applied. The impedance Z_t is that previously seen looking to the right of XX'; $L \equiv A\beta = -V_r/V_t$.

that seen looking to the right of XX' before opening the loop. A test signal V_t is then applied to the terminals of the right-hand side of the loop and the *returned voltage V_r* (see Fig. 12.26) is measured (or calculated). The loop gain L is obtained from

$$L = A\beta = -\frac{V_r}{V_t}$$

As an illustration of this procedure we consider the feedback loop shown in Fig. 12.27a. This feedback loop represents both the inverting and the noninverting op-amp configurations. Using a simple equivalent circuit model for the op amp we obtain the circuit of Fig. 12.27b. Examination of this latter circuit reveals that a convenient place to break the loop is at the input terminals of the op amp. The loop broken in this manner is shown in Fig. 12.27c with a test signal V_t applied to the right-hand-side terminals and a resistance R_{id} terminating the left-hand-side terminals. The returned voltage V_r is found by inspection as

$$V_r = -\mu V_1 \frac{\{R_L\|[R_2 + R_1\|(R_{id} + R)]\}}{\{R_L\|[R_2 + R_1\|(R_{id} + R)]\} + r_o} \cdot \frac{[R_1\|(R_{id} + R)]}{[R_1\|(R_{id} + R)] + R_2} \cdot \frac{R_{id}}{R_{id} + R}$$

This equation can be used directly to find the loop gain $L = A\beta = -V_r/V_t = -V_r/V_1$.

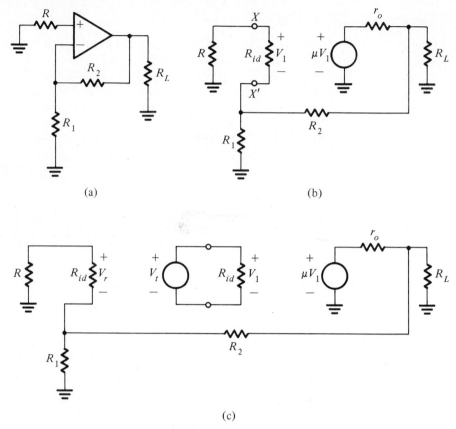

Fig. 12.27 Determination of the loop gain of the feedback loop in (a).

Since the loop gain L is generally a function of frequency, it is usual to call it *loop transmission* and denote it by $L(s)$ or $L(j\omega)$.

Equivalence of Circuits from a Feedback-Loop Point of View

A given feedback loop may be used to generate a number of circuits having the same poles but different transmission zeros. Such circuits are said to be equivalent from a feedback-loop point of view. The poles are the same because these are the roots of the characteristic equation, which is dependent only on the feedback loop. On the other hand, the form of gain function and the transmission zeros depend on how the input signal is injected into the loop.

As an example consider the feedback loop of Fig. 12.27a. This loop can be used to generate the noninverting op-amp circuit by feeding the input voltage signal to the terminal of R, which is connected to ground; that is, we lift this terminal off ground and connect it to V_s. The same feedback loop can be used to generate the inverting op-amp circuit by feeding the input voltage signal to the terminal of R_1, which is connected to ground.

Recognition of the fact that two or more circuits are equivalent from a feedback-loop point of view is very useful because (as will be shown in Section 12.8) stability is a function of the loop. Thus one needs to perform the stability analysis only once for a given loop.

In Chapter 14 we shall employ the concept of loop equivalence in the synthesis of active filters.

12.8 Assume ideal op amps and find the transfer functions of the two circuits in Fig. E12.8 and thus show that they have the same poles. Then verify with the techniques discussed in this section that the two circuits indeed have the same feedback loop.

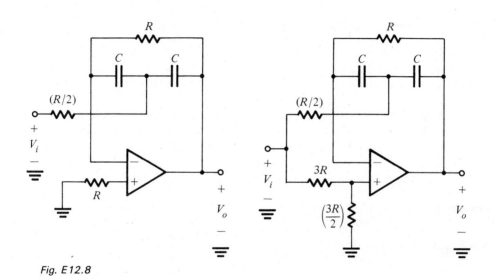

Fig. E12.8

12.8 THE STABILITY PROBLEM

In a feedback amplifier such as that represented by the general structure of Fig. 12.1, the open-loop gain A is generally a function of frequency, and it should therefore be more accurately called the *open-loop transfer function*, $A(s)$. Also, we have been assuming for the most part that the feedback network is resistive and hence that the feedback factor β is constant, but this need not be always the case. In fact, we have already encountered two circuits (Exercise 12.8) in which the feedback factor is a general function of frequency. We shall therefore assume that in the general case the *feedback transfer function* is $\beta(s)$. It follows that the *closed-loop transfer function* $A_f(s)$ is given by

$$A_f(s) = \frac{A(s)}{1 + A(s)\beta(s)} \tag{12.14}$$

To focus attention on the points central to our discussion in this part, we shall assume that the amplifier is direct-coupled with constant dc gain A_0 and with poles and zeros occurring in the high-frequency band. Also, for the time being let us assume that at low frequencies $\beta(s)$ reduces to a constant value. Thus at low frequencies the loop gain $A(s)\beta(s)$ becomes a constant, which should be a positive number—otherwise the feedback would not be negative. The question then arises as to what happens at higher frequencies.

For physical frequencies $s = j\omega$, Eq. (12.14) becomes

$$A_f(j\omega) = \frac{A(j\omega)}{1 + A(j\omega)\beta(j\omega)} \tag{12.15}$$

Thus the loop gain $A(j\omega)\beta(j\omega)$ is a complex number that can be represented by its magnitude and phase,

$$L(j\omega) \equiv A(j\omega)\beta(j\omega) \tag{12.16}$$
$$= |A(j\omega)\beta(j\omega)|\, e^{j\phi(\omega)}$$

It is the manner in which the loop-gain varies with frequency that determines the stability or instability of the feedback amplifier. To appreciate this fact, consider the frequency at which the phase angle $\phi(\omega)$ becomes $180°$. At this frequency, ω_{180}, the loop gain $A(j\omega)\beta(j\omega)$ will be a real number with a negative sign. Thus at this frequency the feedback will become positive. If at $\omega = \omega_{180}$ the magnitude of the loop-gain is less than unity, then from Eq. (12.15) we see that the closed-loop gain $A_f(j\omega)$ will be greater than the open-loop gain $A(j\omega)$, since the denominator of Eq. (12.15) will be smaller than unity. Nevertheless, the feedback amplifier will be stable.

On the other hand, if at the frequency ω_{180} the magnitude of the loop-gain is equal to unity, it follows from Eq. (12.15) that $A_f(j\omega)$ will be infinite. This means that the amplifier will have an output for zero input, which is by definition an *oscillator*. To visualize how this feedback loop may oscillate, consider the general loop of Fig. 12.1 with the external input x_s set to zero. Any disturbance in the circuit, such as the closure of the power supply switch, will generate a signal $x_i(t)$ at the input to the amplifier. Such a noise signal usually contains a wide range of frequencies, and we shall now concentrate on the component with frequency $\omega = \omega_{180}$, that is, the signal $X_i \sin(\omega_{180}t)$. This input signal will result in a feedback signal given by

$$X_f = A(j\omega_{180})\beta(j\omega_{180})X_i = -X_i$$

Since X_f is further multiplied by -1 in the summer block at the input, we see that the feedback causes the signal X_i at the amplifier input to be *sustained*. That is, from this point on there will be sinusoidal signals at the amplifier input and output of frequency ω_{180}. Thus the amplifier is said to oscillate at the frequency ω_{180}.

The question now is, What happens if at ω_{180} the magnitude of the loop gain is greater than unity? The answer, which is not obvious from Eq. (12.15), is that the circuit will oscillate, and the oscillations will grow in amplitude until some nonlinearity (which is always present in some form) reduces the magnitude of the loop gain to exactly unity, at which point sustained oscillations will be obtained. This mechanism for starting oscillations by using positive feedback with a loop gain greater than unity and then using a nonlinearity to reduce the loop gain to unity at the desired amplitude will be exploited in the design of sinusoidal oscillators in Chapter 14. Our objective here is just the opposite: now that we know how oscillations could occur in a negative-feedback amplifier, we wish to find methods to prevent their occurrence.

The Nyquist Plot

The Nyquist plot is a formalized approach for testing for stability based on the above discussion. It is simply a polar plot of loop gain with frequency used as a parameter. Figure 12.28 shows such a plot. Note that the radial distance is $|A\beta|$ and the angle is the phase angle ϕ. The solid-line plot is for positive frequencies. Since the loop gain—and for that matter any gain function of a physical network—has a magnitude that is an even function of frequency and a phase that is an odd function of frequency, the $A\beta$ plot for negative frequencies can be drawn as a mirror image through the Re axis and is shown in Fig. 12.28 as a broken line.

The Nyquist plot intersects the negative real axis at the frequency ω_{180}. Thus if this intersection occurs to the left of the point $(-1,0)$, we know that the magnitude of loop gain at this frequency is greater than unity and the amplifier will be unstable. On the other hand, if the intersection occurs to the right of the point $(-1,0)$ the amplifier will be stable. It follows that if the Nyquist plot encircles the point $(-1,0)$ then the amplifier will be unstable. It should be mentioned, however, that this statement is a simplified version of the *Nyquist criterion;* nevertheless, it applies to all the circuits in which we are interested. For the full theory behind the Nyquist method and for details on its application, consult reference 12.2.

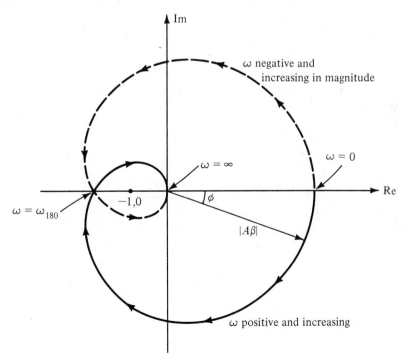

Fig. 12.28   The Nyquist plot.

EXERCISE

12.9 Consider a feedback amplifier for which the open-loop transfer function $A(s)$ is given by

$$A(s) = \left(\frac{10}{1 + s/10^4} \right)^3$$

Let the feedback factor β be a constant independent of frequency. Find the frequency ω_{180} at which the phase shift is 180°. Then, show that the feedback amplifier will be stable if the feedback factor β is less than a critical value β_{cr} and unstable if $\beta \geq \beta_{cr}$, and find the value of β_{cr}.

Ans. $\omega_{180} = \sqrt{3} \times 10^4$ rad/s; $\beta_{cr} = 0.008$

12.9   EFFECT OF FEEDBACK ON THE AMPLIFIER POLES

The amplifier frequency response and stability are determined directly by its poles. We shall therefore investigate the effect of applying feedback on the poles of the amplifier.

Stability and Pole Location

We shall begin by considering the relation between stability and pole location. For an amplifier or any other system to be stable, its poles should lie in the left half of the s plane. A pair of complex conjugate poles on the $j\omega$ axis gives rise to sustained sinusoidal oscillations. Poles in the right half of the s plane give rise to growing oscillations.

To verify the above statement, consider an amplifier with a pole pair at

$$s = \sigma_0 \pm j\omega_n$$

If this amplifier is subjected to a disturbance, such as that caused by closure of the power-supply switch, its transient response will contain terms of the form

$$v(t) = e^{\sigma_0 t}[e^{+j\omega_n t} + e^{-j\omega_n t}] = 2e^{\sigma_0 t} \cos(\omega_n t)$$

This is a sinusoidal signal with an envelope $e^{\sigma_0 t}$. Now if the poles are in the left half of the s plane, then σ_0 will be negative and the oscillations will decay exponentially toward zero, as shown in Fig. 12.29a, indicating that the system is stable. If on the other hand the poles are in the right half-plane, then σ_0 will be positive and the oscillations will grow exponentially (until some nonlinearity limits their growth), as shown in Fig. 12.29b. Finally, if the poles are on the $j\omega$ axis, then σ_0 will be zero and the oscillations will be sustained, as shown in Fig. 12.29c.

Although the above discussion is in terms of complex conjugate poles, it can be shown that the existence of any right-half-plane poles results in instability. Here it is interesting to note that instability does not necessarily lead to sinusoidal oscillations. For instance, the bistable circuits studied in Chapters 5 and 6 have an unstable state in which they have right-half-plane poles. Nevertheless, they do not show explicit oscillations. They can, however, be thought of as oscillating at zero frequency.

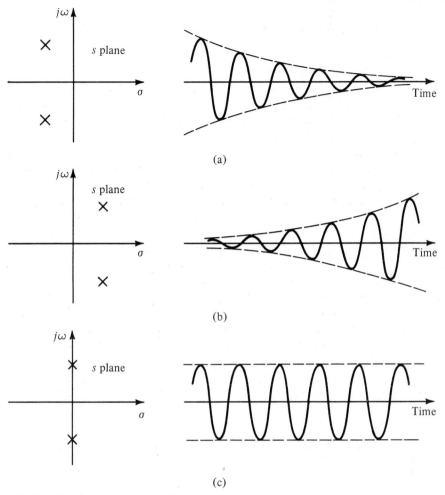

Fig. 12.29   *Relationship between pole location and transient response.*

Poles of the Feedback Amplifier

From the closed-loop transfer function in Eq. (12.14) we see that the poles of the feedback amplifier are the zeros of $1 + A(s)\beta(s)$. That is, the feedback amplifier poles are obtained by solving the equation

$$1 + A(s)\beta(s) = 0$$

which is called the *characteristic equation* of the feedback loop. It should therefore be apparent that applying feedback to an amplifier changes its poles.

In the following we shall consider how feedback affects the amplifier poles. For this purpose we shall assume that the open-loop amplifier has real poles and no zeros (that is, all the zeros are at $s = \infty$). This will simplify the analysis and enable us to focus our attention on the fundamental concepts involved. We shall also assume that the feedback factor β is independent of frequency.

Amplifier With Single-Pole Response

Consider first the case of an amplifier whose open-loop transfer function is characterized by a single pole.

$$A(s) = \frac{A_0}{1 + s/\omega_P}$$

The closed-loop transfer function is given by

$$A_f(s) = \frac{A_0/(1 + A_0\beta)}{1 + s/\omega_P(1 + A_0\beta)} \tag{12.17}$$

Thus the feedback moves the pole along the negative real axis to a frequency ω_{Pf},

$$\omega_{Pf} = \omega_P(1 + A_0\beta) \tag{12.18}$$

This process is illustrated in Fig. 12.30a. Figure 12.30b shows Bode plots for $|A|$ and $|A_f|$. Note that while at low frequencies the difference between the two plots is

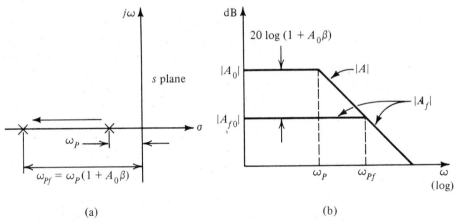

(a) (b)

Fig. 12.30 Effect of feedback on (a) the pole location, and (b) the frequency response of an amplifier having a single-pole open-loop response.

$20 \log (1 + A_0\beta)$, the two curves coincide at high frequencies. One can show that this indeed is the case by approximating Eq. (12.17) for frequencies $\omega \gg \omega_P(1 + A_0\beta)$:

$$A_f(s) \simeq \frac{A_0\omega_P}{s} \simeq A(s) \tag{12.19}$$

Physically speaking, at such high frequencies the loop gain is much smaller than unity and the feedback is ineffective.

Since the pole of the closed-loop amplifier never enters the right half of the s plane, the single-pole amplifier is stable for any value of β. Thus this amplifier is said to be *unconditionally stable*. This result, however, is hardly surprising, since the phase lag associated with a single-pole response can never be greater than $90°$. Thus the loop gain never achieves the $180°$ phase shift required for the feedback to become positive.

Amplifier With Two-Pole Response

Consider next an amplifier whose open-loop transfer function is characterized by two real-axis poles:

$$A(s) = \frac{A_0}{(1 + s/\omega_{P1})(1 + s/\omega_{P2})} \tag{12.20}$$

In this case the closed-loop poles are obtained from

$$1 + A(s)\beta = 0$$

which leads to

$$s^2 + s(\omega_{P1} + \omega_{P2}) + (1 + A_0\beta)\omega_{P1}\omega_{P2} = 0 \tag{12.21}$$

Thus the closed-loop poles are given by

$$s = -\tfrac{1}{2}(\omega_{P1} + \omega_{P2}) \pm \tfrac{1}{2} \sqrt{(\omega_{P1} + \omega_{P2})^2 - 4(1 + A_0\beta)\omega_{P1}\omega_{P2}} \tag{12.22}$$

From this equation we see that as the loop gain $A_0\beta$ is increased from zero the poles are brought closer together. Then a value of loop gain is reached at which the poles become coincident. If the loop gain is further increased the poles become complex conjugate and move along a vertical line. Figure 12.31 shows the locus of the poles for increasing loop gain. This plot is called *root locus diagram*, where root refers to the fact that the poles are the roots of the characteristic equation.

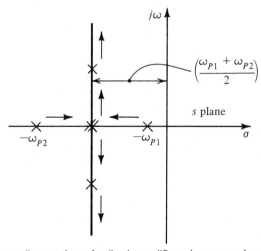

Fig. 12.31 Root locus diagram for a feedback amplifier whose open-loop transfer function has two real poles.

From the root locus diagram of Fig. 12.31 we see that this feedback amplifier also is unconditionally stable. Again this result should come as no surprise; the maximum phase shift of $A(s)$ in this case is $180°$ ($90°$ per pole), but this value is reached at $\omega = \infty$. Thus there is no finite frequency at which the phase shift reaches $180°$.

Another observation to make on the root locus diagram of Fig. 12.31 is that the open-loop amplifier might have a dominant pole but this is not necessarily the case for the closed-loop amplifier. The response of the closed-loop amplifier can, of course, always be plotted once the poles are found from Eq. (12.22). As is the case with second-order responses, the closed-loop response can show a peak (see Chapter 14). To be more specific, the characteristic equation of a second-order network can be written in the standard form

$$s^2 + s\frac{\omega_0}{Q} + \omega_0^2 = 0 \tag{12.23}$$

where ω_0 is called the *pole frequency* and Q is called the *pole Q factor*. The poles are complex if Q is greater than 0.5. A geometric interpretation for ω_0 and Q of a pair of complex conjugate poles is given in Fig. 12.32, from which we note that ω_0 is the radial

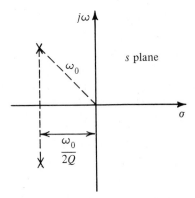

Fig. 12.32   *Definition of ω_0 and Q of a pair of complex conjugate poles.*

distance of the poles and that Q indicates the distances of the poles from the $j\omega$ axis. Poles on the $j\omega$ axis have $Q = \infty$.

By comparing Eqs. (12.21) and (12.23) we obtain the Q factor for the poles of the feedback amplifier as

$$Q = \frac{\sqrt{(1 + A_0\beta)\omega_{P1}\omega_{P2}}}{(\omega_{P1} + \omega_{P2})} \tag{12.24}$$

From the study of second-order network responses in Chapter 14 it will be seen that the response of the feedback amplifier under consideration shows no peak for $Q \leq 0.707$. The boundary case corresponding to $Q = 0.707$ results in the *maximally flat* response. Figure 12.33 shows a number of possible responses obtained for various values of Q (or, correspondingly, various values of $A_0\beta$).

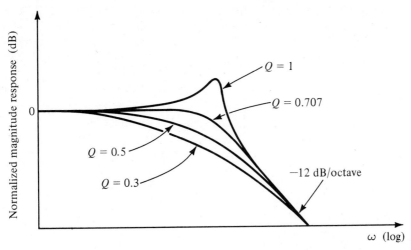

Fig. 12.33 *Normalized magnitude response of a two-pole feedback amplifier for various values of Q.*

Example 12.5

As an illustration of some of the ideas discussed above we consider the feedback circuit shown in Fig. 12.34a. Find the loop transmission $L(s)$ and the characteristic equation. Sketch a root locus diagram for varying K and find the value of K that results in the maximally flat response and the value of K that makes the circuit oscillate. Assume that the amplifier has infinite input impedance and zero output impedance.

Solution

To obtain the loop transmission we short-circuit the signal source and break the loop at the amplifier input. We then apply a test voltage V_t and find the returned voltage V_r, as indicated in Fig. 12.34b. The loop transmission $L(s) \equiv A(s)\beta(s)$ is given by

$$L(s) = -\frac{V_r}{V_t} = -KT(s)$$

where $T(s)$ is the transfer function of the two-port RC network shown inside the dotted box in Fig. 12.34b:

$$T(s) \equiv \frac{V_r}{V_1} = \frac{s(1/CR)}{s^2 + s(3/CR) + (1/CR)^2}$$

Thus

$$L(s) = \frac{-s(K/CR)}{s^2 + s(3/CR) + (1/CR)^2}$$

The characteristic equation is

$$1 + L(s) = 0$$

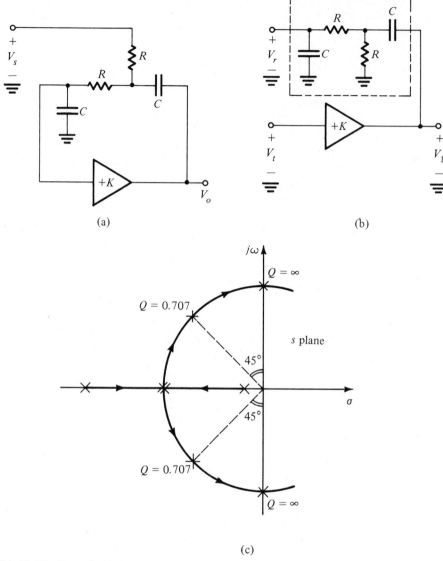

(a)

(b)

(c)

Fig. 12.34   Example 12.5.

that is,

$$s^2 + s\frac{3}{CR} + \left(\frac{1}{CR}\right)^2 - s\frac{K}{CR} = 0$$

$$s^2 + s\frac{3 - K}{CR} + \left(\frac{1}{CR}\right)^2 = 0$$

By comparing this equation to the standard form of second-order characteristic equation [Eq. (12.23)] we see that the pole frequency ω_0 is given by

$$\omega_0 = \frac{1}{CR}$$

and the Q factor is

$$Q = \frac{1}{3 - K}$$

Thus for $K = 0$ the poles have $Q = \frac{1}{3}$ and are therefore located on the negative real axis. As K is increased the poles are brought closer together and eventually coincide ($Q = 0.5$, $K = 1$). Increasing K further results in the poles becoming complex and conjugate. The root locus is then a circle because the radial distance ω_0 remains constant independent of the value of K.

The maximally flat response is obtained when $Q = 0.707$, which results when $K = 1.586$. In this case the poles are at $45°$ angles, as indicated in Fig. 12.34c. The poles cross the $j\omega$ axis into the right half of the s plane at the value of K that results in $Q = \infty$, that is, $K = 3$. Thus for $K \geq 3$ this circuit becomes unstable. This might appear to contradict our earlier conclusion that the feedback amplifier with a second-order response is unconditionally stable. Note, however, that the circuit in this example is quite different from the negative-feedback amplifier that we have been studying. Here we have an amplifier with a positive gain K and a feedback network whose transfer function $T(s)$ is frequency dependent. This feedback is in fact *positive,* and the circuit will oscillate at the frequency for which the phase of $T(j\omega)$ is zero.

· · ·

Example 12.5 illustrates the use of feedback (positive feedback in this case) to move the poles of an RC network from their negative real-axis locations to complex conjugate locations. One can accomplish the same task using negative feedback, as the root locus diagram of Fig. 12.31 demonstrates. The process of pole control is the essence of *active filter design,* as will be discussed in Chapter 14.

Amplifiers With Three or More Poles

Figure 12.35 shows the root locus diagram for a feedback amplifier whose open-loop response is characterized by three poles. As indicated, increasing the loop gain from zero moves the highest-frequency pole outwardly while the two other poles are brought closer together. As $A_0\beta$ is increased further, the two poles become coincident and then become complex and conjugate. A value of $A_0\beta$ exists at which this pair of complex conjugate poles enters the right half of the s plane, thus causing the amplifier to become unstable.

This result is not entirely unexpected, since an amplifier with three poles has a phase shift that reaches $-270°$ as ω approaches ∞. Thus there exists a finite frequency, ω_{180}, at which the loop gain has $180°$ phase shift.

From the root locus diagram of Fig. 12.35 we observe that one can always maintain

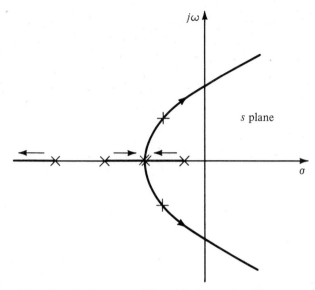

Fig. 12.35   Root locus diagram for an amplifier with three poles. The arrows indicate the pole movement as $A_0\beta$ is increased.

amplifier stability by keeping the loop gain $A_0\beta$ smaller than the value corresponding to the poles entering the right half-plane. In terms of the Nyquist diagram, the critical value of $A_0\beta$ is that for which the diagram passes through the $(-1,0)$ point. Reducing $A_0\beta$ below this value causes the Nyquist plot to shrink and thus intersect the negative real axis to the right of the $(-1,0)$ point, indicating stable amplifier performance. On the other hand, increasing $A_0\beta$ above the critical value causes the Nyquist plot to expand, thus encircling the $(-1,0)$ point and indicating unstable performance.

For a given open-loop gain the above conclusions can be stated in terms of the feedback factor β. That is, there exists a *maximum value* for β above which the feedback amplifier becomes unstable. Alternatively, we can state that there exists a *minimum value* for the closed-loop gain A_{f0} below which the amplifier becomes unstable. To obtain lower values of closed-loop gain one needs therefore to alter the loop transfer function $L(s)$. This is the process known as *frequency compensation*. We shall study the theory and techniques of frequency compensation in Section 12.11.

Before leaving this section we should point out that construction of the root locus diagram for amplifiers having three or more poles as well as finite zeros is an involved process for which a systematic procedure exists. However, such a procedure will not be presented here, and the interested reader can consult reference 12.2. Although the root locus diagram provides the amplifier designer with considerable insight, other, simpler techniques based on Bode plots can be effectively employed, as will be explained in Section 12.10.

EXERCISE

12.10   Consider a feedback amplifier for which the open-loop transfer function $A(s)$ is given by

$$A(s) = \left(\frac{10}{1 + s/10^4} \right)^3$$

Let the feedback factor β be frequency independent. Find the closed-loop poles as functions of β and hence show that the root locus is that of Fig. E12.10. Also find the value of β at which the amplifier becomes unstable. (*Note:* This is the same amplifier that was considered in Exercise 12.9.)

Ans.   See Fig. E12.10; $\beta_{critical} = 0.008$

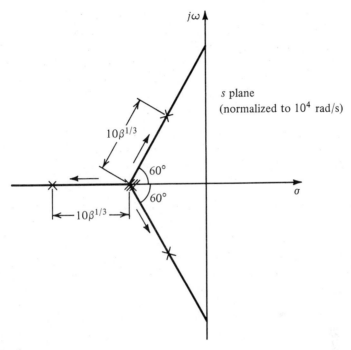

Fig. E12.10

12.10   STABILITY STUDY USING BODE PLOTS

Gain and Phase Margins

From Sections 12.8 and 12.9 we know that one can determine whether the feedback amplifier is stable or not by examining the loop gain $A\beta$ as a function of frequency. One of the simplest and most effective means for doing this is through the use of a Bode plot for $A\beta$, such as the one shown in Fig. 12.36. (Note that because the phase approaches $-360°$ the network examined is a fourth-order one.) The feedback amplifier whose loop gain is plotted in Fig. 12.36 will be stable, since at the frequency of $180°$ phase shift, ω_{180}, the magnitude of loop gain is less than unity (negative dB). The difference between the value of $|A\beta|$ at ω_{180} and unity is called the *gain margin* and is usually expressed in dB. The gain margin represents the amount by which the loop gain can be increased while stability is maintained. Feedback amplifiers are usually designed

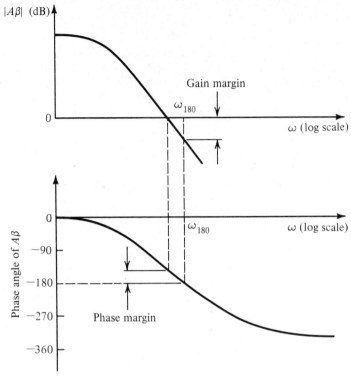

Fig. 12.36 Bode plot for the loop-gain $A\beta$ illustrating the definitions of the gain and phase margins.

to have sufficient gain margin to allow for the inevitable changes in loop gain with temperature, time, and so on.

Another way to investigate the stability and to express its degree is to examine the Bode plot at the frequency for which $|A\beta| = 1$, which is the point at which the magnitude plot crosses the 0-dB line. If at this frequency the phase angle is less (in magnitude) than 180°, then the amplifier is stable. This is the situation illustrated in Fig. 12.36. The difference between the phase angle at this frequency and 180° is termed the *phase margin*. On the other hand, if at the frequency of unity loop gain magnitude the phase lag is in excess of 180°, the amplifier will be unstable.

EXERCISE

12.11 Consider an op amp having a single-pole open-loop response with $A_0 = 10^5$ and $f_P = 10$ Hz. Let the op amp be ideal otherwise (infinite input impedance, zero output impedance, and so on). If this amplifier is connected in the noninverting configuration with a nominal low-frequency closed-loop gain of 100, find the frequency at which $|A\beta| = 1$. Also, find the phase margin.
Ans. 10^4 Hz; 90°

Effect of Phase Margin on Closed-Loop Response

Feedback amplifiers are normally designed with a phase margin of at least $45°$. The amount of phase margin has a profound effect on the shape of the closed-loop magnitude response. To see this relationship, consider a feedback amplifier with a large low-frequency loop gain, $A_0\beta \gg 1$. It follows that the closed-loop gain at low frequencies is approximately $1/\beta$. Denoting the frequency at which the magnitude of loop gain is unity by ω_1 we have

$$A(j\omega_1)\beta = 1 \times e^{-j\theta} \qquad (12.25)$$

where

$$\theta = 180° - \text{phase margin}$$

At ω_1 the closed-loop gain is

$$A_f(j\omega_1) = \frac{A(j\omega_1)}{1 + A(j\omega_1)\beta}$$

Substituting from Eq. (12.25) gives

$$A_f(j\omega_1) = \frac{(1/\beta)e^{-j\theta}}{1 + e^{-j\theta}}$$

Thus the magnitude of gain at ω_1 is

$$|A_f(j\omega_1)| = \frac{1/\beta}{|1 + e^{-j\theta}|}$$

For a phase margin of $45°$ we obtain

$$|A_f(j\omega_1)| = 1.3 \frac{1}{\beta}$$

That is, the gain peaks by a factor of 1.3 above the low-frequency value of $1/\beta$. This peaking increases as the phase margin is reduced, eventually reaching ∞ when the phase margin is zero. Zero phase margin, of course, implies that the amplifier can sustain oscillations [poles on the $j\omega$ axis; Nyquist plot passes through $(-1,0)$].

EXERCISE

12.12 Find the closed-loop gain at ω_1 relative to the low-frequency gain when the phase margin is $30°$, $60°$, and $90°$.
Ans. 1.93; 1; 0.707

An Alternative Approach

Investigating stability by constructing Bode plots for the loop gain $A\beta$ can be a tedious and time-consuming process, especially if we have to investigate the stability of a given amplifier for a variety of feedback networks. An alternative approach, which is much simpler, is to construct a Bode plot for the open-loop gain $A(j\omega)$ only. Assuming for the

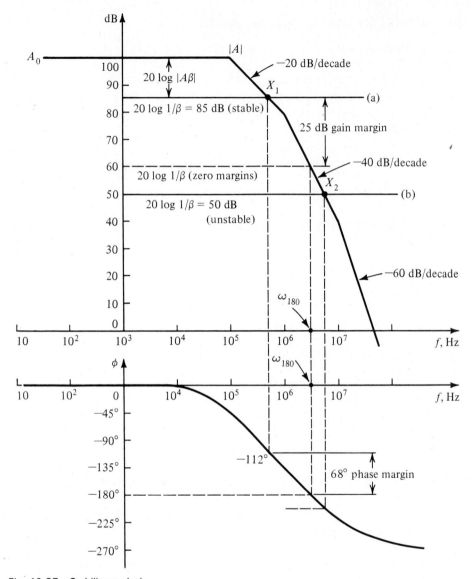

Fig. 12.37   Stability analysis.

time being that β is independent of frequency, we can plot $20 \log(1/\beta)$ as a horizontal straight line on the same plane used for $20 \log|A|$. The difference between the two curves will be

$$20 \log|A(j\omega)| - 20 \log \frac{1}{\beta} = 20 \log|A\beta|$$

which is the loop gain expressed in dB. We may therefore study stability by examining the difference between the two plots. If we wish to evaluate stability for a different feedback factor we simply draw another horizontal straight line at the level $20 \log (1/\beta)$.

To illustrate, consider an amplifier whose open-loop transfer function is characterized by three poles. For simplicity let the three poles be widely separated—say, at 0.1, 1, and 10 MHz, as shown in Fig. 12.37. Note that because the poles are widely separated the phase is approximately $-45°$ at the first pole frequency, $-135°$ at the second, and $-225°$ at the third. The frequency at which the phase of $A(j\omega)$ is $-180°$ lies on the -40 dB/decade segment, as indicated in Fig. 12.37.

Consider next the horizontal straight line labeled a. This line represents a feedback factor for which $20 \log(1/\beta) = 85$ dB, which corresponds to a closed-loop gain of approximately 85 dB. The loop gain is the difference between the $|A|$ curve and the $1/\beta$ line. The point of intersection X_1 corresponds to the frequency at which $|A\beta| = 1$. At this frequency we see from the phase plot that the phase is $-112°$. Thus the closed-loop amplifier whose $20 \log(1/\beta) = 85$ dB will be stable with a phase margin of $68°$. The gain margin can be easily obtained from Fig. 12.37; it is 25 dB.

Next, suppose that we wish to use this amplifier to obtain a closed-loop gain of 50 dB nominal value. Since $A_0 = 100$ dB, we see that $A_0\beta \gg 1$ and $20 \log(A_0\beta) \simeq 50$ dB, resulting in $20 \log(1/\beta) \simeq 50$ dB. To see whether this closed-loop amplifier is stable or not, we draw line b in Fig. 12.37 with a height of 50 dB. This line intersects the open-loop gain curve at point X_2, where the corresponding phase is greater than $180°$. Thus the closed-loop amplifier with 50 dB gain will be unstable.

In fact, it can easily be seen from Fig. 12.37 that the *minimum* value of $20 \log(1/\beta)$ that can be used with the resulting amplifier being stable is 60 dB. In other words, the minimum value of stable closed-loop gain obtained with this amplifier is approximately 60 dB. At this value of gain, however, the amplifier may still oscillate, since no margin is left to allow for possible changes in gain.

Since the $180°$ phase point always occurs on the -40 dB/decade segment of the Bode plot for $|A|$, a rule of thumb to guarantee stability is as follows: The closed-loop amplifier will be stable if the $20 \log(1/\beta)$ line intersects the $20 \log|A|$ curve at a point on the -20 dB/decade segment. Following this rule ensures that a phase margin of about $45°$ is obtained. For the example of Fig. 12.37, the rule implies that the maximum value of β is 10^{-4}, which corresponds to a closed-loop gain of approximately 80 dB.

The above rule of thumb can be generalized for the case in which β is a function of frequency. The general rule states that at the intersection of $20 \log[1/|\beta(j\omega)|]$ and $20 \log|A(j\omega)|$ the difference of slopes (called the *rate of closure*) should not exceed 20 dB/decade.

EXERCISE

12.13 Consider an op amp whose open-loop gain is identical to that of Fig. 12.37. Assume that the op amp is ideal otherwise. Let the op amp be connected as a differentiator. Use the above rule of thumb to show that for stable performance the differentiator time constant should be greater than 1 s.

12.11 FREQUENCY COMPENSATION

In this section we shall discuss methods for modifying the open-loop transfer function $A(s)$ of an amplifier having three or more poles so that the closed-loop amplifier is stable for any desired value of closed-loop gain.

Theory

The simplest method of frequency compensation consists of introducing a new pole in the function $A(s)$ at a sufficiently low frequency, f_D, such that the modified open-loop gain, $A'(s)$, intersects the $20 \log(1/|\beta|)$ curve with a slope difference of 20 dB/decade. As an example, let it be required to compensate the amplifier whose $A(s)$ is shown in Fig. 12.38 such that closed-loop amplifiers with β as high as 10^{-2} (that is, closed-loop gains as low as approximately 40 dB) will be stable. First, we draw a horizontal straight line at the 40-dB level to represent $20 \log(1/\beta)$, as shown in Fig. 12.38. We then locate point Y on this line at the frequency of the first pole, f_{P1}. From Y we draw a line with

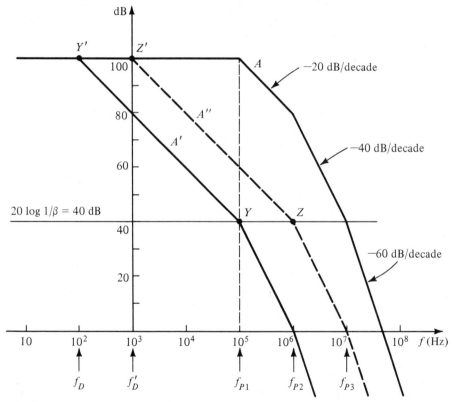

Fig. 12.38 Frequency compensation for $\beta = 10^{-2}$. The response labeled A' is obtained by introducing an additional pole at f_D. The A" response is obtained by moving the low-frequency pole to f'_D.

-20 dB/decade slope and determine the point at which this line intersects the dc gain line, point Y'. This latter point gives the frequency f_D of the new pole that has to be introduced in the open-loop transfer function.

The compensated open-loop response $A'(s)$ is indicated in Fig. 12.38. It has four poles: at f_D, f_{P1}, f_{P2}, and f_{P3}. Thus $|A'|$ begins to roll off with a slope of -20 dB/decade at f_D. At f_{P1} the slope changes to -40 dB/decade, at f_{P2} it changes to -60 dB/decade, and so on. Since the 20 log$(1/\beta)$ line intersects the 20 log$|A'|$ curve at point Y on the -20 dB/decade segment, the closed-loop amplifier with this β value (or lower values) will be stable.

A serious disadvantage of this compensation method is that at most frequencies the open-loop gain has been drastically reduced. This means that at most frequencies the amount of feedback available will be small. Since all the advantages of negative feedback are directly proportional to the amount of feedback, the performance of the compensated amplifier has been impaired.

Careful examination of Fig. 12.38 shows that the reason the gain $A'(s)$ is low is the pole at f_{P1}. If we can somehow eliminate this pole, then—rather than locating point Y, drawing YY', and so on—we can now start from point Z (at the frequency of the second pole) and draw the line ZZ'. This would result in the open-loop curve $A''(s)$, which shows considerably higher gain than $A'(s)$.

Although it is not possible to eliminate the pole at f_{P1}, it is usually possible to shift that pole from $f = f_{P1}$ to $f = f_D'$. This will be explained next.

Implementation

We shall now address the question of implementing the frequency-compensation scheme discussed above. The amplifier circuit normally consists of a number of cascaded gain stages, with each stage responsible for one or more of the transfer-function poles. Through manual and/or computer analysis of the circuit, one identifies which stage introduces each of the important poles f_{P1}, f_{P2}, and so on. For the sake of our discussion, assume that the first pole f_{P1} is introduced at the interface between the two cascaded differential stages shown in Fig. 12.39a. In Fig. 12.39b we show a simple small-signal model of the circuit at this interface. Current source I_x represents the output current of the $Q_1 - Q_2$ stage. Resistance R_x and capacitance C_x represent the total resistance and capacitance between the two nodes B and B'. It follows that the pole f_{P1} is given by

$$f_{P1} = \frac{1}{2\pi C_x R_x}$$

Let us now connect the compensating capacitor C_C between nodes B and B'. This will result in the modified equivalent circuit shown in Fig. 12.39c, from which we see that the pole introduced will no longer be at f_{P1}; rather the pole can be at any desired lower frequency f_D'

$$f_D' = \frac{1}{2\pi(C_x + C_C)R_x}$$

We thus conclude that one can select an appropriate value for C_C so as to shift the pole frequency from f_{P1} to the value f_D' determined by point Z' in Fig. 12.38.

At this juncture it should be pointed out that adding the capacitor C_C will usually result in changes in the location of the other poles (those at f_{P2} and f_{P3}). One might therefore need to calculate the new location of f_{P2} and perform a few iterations to arrive at the required value for C_C.

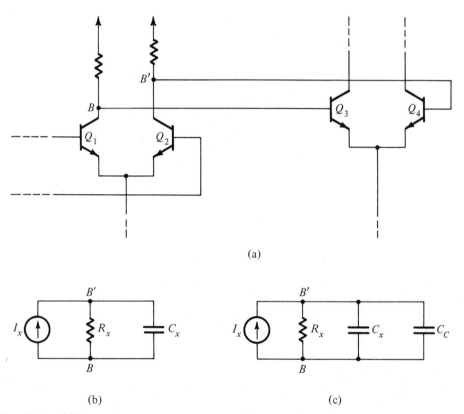

(a)

(b)   (c)

Fig. 12.39   (a) Two cascaded gain stages of a multistage amplifier. (b) Equivalent circuit for the interface between the two stages in (a). (c) Same as in (b) but with a compensating capacitor C_c added.

A disadvantage of this implementation method is that the required value of C_C is usually quite large. Thus if the amplifier to be compensated is an IC op amp, it will be difficult and probably impossible to include this compensating capacitor on the IC chip. (As pointed out in Chapter 13 and in Appendix A, the maximum practical size of a monolithic capacitor is about 100 pF.) An elegant solution to this problem is to connect the compensating capacitor in the feedback path of an amplifier stage. Because of the Miller effect, the compensating capacitance will be multiplied by the stage gain, resulting in a much larger effective capacitance. Furthermore, as explained below, another unexpected benefit accrues.

Miller Compensation and Pole Splitting

Figure 12.40a shows one gain stage in a multistage amplifier. For simplicity, the stage is shown as a common-emitter amplifier, but in practice it can be a more elaborate circuit. In the feedback path of this common-emitter stage we have placed a compensating capacitor C_f.

Figure 12.40b shows a simplified equivalent circuit of the gain stage of Fig. 12.40a. Here R_1 and C_1 represent the total resistance and total capacitance between node b and

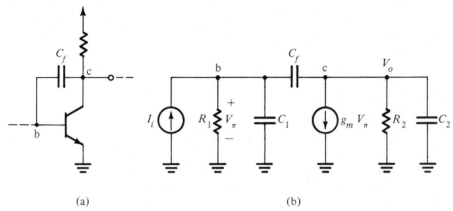

(a) (b)

Fig. 12.40 *(a) A gain stage in a multistage amplifier with a compensating capacitor connected in the feedback loop, and (b) equivalent circuit.*

ground. Similarly, R_2 and C_2 represent the total resistance and total capacitance between node c and ground. Furthermore, it is assumed that C_1 and C_2 include the Miller components due to capacitance C_μ. Finally, I_i represents the output signal current of the preceding stage.

In the absence of the compensating capacitor C_f, we can see from Fig. 12.40b that there are two poles—one at the input and one at the output. Let us assume that these two poles are f_{P1} and f_{P2} of Fig. 12.38; thus

$$f_{P1} = \frac{1}{2\pi C_1 R_1} \qquad f_{P2} = \frac{1}{2\pi C_2 R_2}$$

With C_f present, analysis of the circuit yields the transfer function

$$\frac{V_o}{I_i} = \frac{(sC_f - g_m)R_1R_2}{1 + s[C_1R_1 + C_2R_2 + C_f(g_mR_1R_2 + R_1 + R_2)] + s^2[C_1C_2 + C_f(C_1 + C_2)]R_1R_2}$$

$$(12.26)$$

The zero is usually at a much higher frequency than the dominant pole, and we shall neglect its effect. The denominator polynomial $D(s)$ can be written in the form

$$D(s) = \left(1 + \frac{s}{\omega'_{P1}}\right)\left(1 + \frac{s}{\omega'_{P2}}\right) = 1 + s\left(\frac{1}{\omega'_{P1}} + \frac{1}{\omega'_{P2}}\right) + \frac{s^2}{\omega'_{P1}\omega'_{P2}}$$

where ω'_{P1} and ω'_{P2} are the new frequencies of the two poles. Normally one of the poles will be dominant; $\omega'_{P1} \ll \omega'_{P2}$. Thus

$$D(s) \simeq 1 + \frac{s}{\omega'_{P1}} + \frac{s^2}{\omega'_{P1}\omega'_{P2}} \tag{12.27}$$

Equating the coefficients of s in the denominator of Eq. (12.26) and in Eq. (12.27) results in

$$\omega'_{P1} = \frac{1}{C_1 R_1 + C_2 R_2 + C_f(g_m R_1 R_2 + R_1 + R_2)}$$

which can be approximated by

$$\omega'_{P1} \simeq \frac{1}{g_m R_2 C_f R_1} \tag{12.28}$$

To obtain ω'_{P2} we equate the coefficients of s^2 in the denominators of Eqs. (12.26) and (12.27) and use Eq. (12.28):

$$\omega'_{P2} \simeq \frac{g_m C_f}{C_1 C_2 + C_f(C_1 + C_2)} \tag{12.29}$$

From Eqs. (12.28) and (12.29) we see that as C_f is increased, ω'_{P1} is reduced and ω'_{P2} is increased. This is referred to as *pole splitting*. Note that the increase in ω'_{P2} is highly beneficial; it allows us to move point Z (see Fig. 12.38) further to the right, thus resulting in higher compensated open-loop gain. Finally, note from Eq. (12.28) that C_f is multiplied by the Miller-effect factor $g_m R_2$, thus resulting in a much larger capacitance, $g_m R_2 C_f$. In other words, the required value of C_f will be much smaller than that of C_C in Fig. 12.39.

Example 12.6

Consider an op amp whose open-loop transfer function is identical to that shown in Fig. 12.37. We wish to compensate this op amp so that the closed-loop amplifier with resistive feedback is stable for any gain (that is, for β up to unity). Assume that the op-amp circuit includes a stage such as that of Fig. 12.40 with $C_1 = 100$ pF, $C_2 = 5$ pF, and $g_m = 40$ mA/V, that the pole at f_{P1} is caused by the input circuit of that stage, and that the pole at f_{P2} is introduced by the output circuit. Find the value of the compensating capacitor if it is connected either between the input node b and ground or in the feedback path of the transistor.

Solution

First we determine R_1 and R_2 from

$$f_{P1} = 0.1 \text{ MHz} = \frac{1}{2\pi C_1 R_1}$$

Thus

$$R_1 = \frac{10^5}{2\pi} \, \Omega$$

$$f_{P2} = 1 \text{ MHz} = \frac{1}{2\pi C_2 R_2}$$

Thus

$$R_2 = \frac{10^5}{\pi} \, \Omega$$

First, if a compensating capacitor C_C is connected across the input terminals of the transistor stage, then the frequency of the first pole changes from f_{P1} to f'_D:

$$f'_D = \frac{1}{2\pi(C_1 + C_C)R_1}$$

The second pole remains unchanged. The required value for f'_D is determined by drawing a -20 dB/decade line from the 1-MHz frequency point on the $20 \log(1/\beta) = 20 \log 1 = 0$ dB line. This line will intersect the 100-dB dc gain line at 10 Hz. Thus

$$f'_D = 10 \text{ Hz} = \frac{1}{2\pi(C_1 + C_C)R_1}$$

which results in $C_C \simeq 1 \, \mu F$, which is quite large and which certainly cannot be included on the IC chip.

Next, if a compensating capacitor C_f is connected in the feedback path of the transistor, then both poles change location to the values given by Eqs. (12.28) and (12.29):

$$f'_{P1} \simeq \frac{1}{2\pi g_m R_2 C_f R_1} \tag{12.30}$$

$$f'_{P2} \simeq \frac{g_m C_f}{2\pi[C_1 C_2 + C_f(C_1 + C_2)]}$$

To determine where we should locate the first pole we need to know the value of f'_{P2}. As an approximation let us assume that $C_f \gg C_2$, which enables us to obtain

$$f'_{P2} \simeq \frac{g_m}{2\pi(C_1 + C_2)} = 60.6 \text{ MHz}$$

Thus it appears that this pole will move to a frequency higher than f_{P3} (which is 10 MHz). Let us therefore assume that the second pole will be at f_{P3}. This requires that the first pole be located at 100 Hz:

$$f'_{P1} = 100 \text{ Hz} = \frac{1}{2\pi g_m R_2 C_f R_1}$$

which results in $C_f = 78.5$ pF. Although this value is indeed much greater than C_2, we can determine the location of the pole f'_{P2} from Eq. (12.30) which yields $f'_{P2} = 57.2$ MHz, confirming the fact that this pole has indeed been moved past f_{P3}.

We conclude that using Miller compensation not only results in a much smaller compensating capacitor but owing to pole splitting also enables us to place the dominant pole a decade higher in frequency. This results in a wider bandwidth for the compensated op amp.

· · ·

12.12   CONCLUDING REMARKS

The material studied in this chapter will be found particularly useful in Chapters 13 and 14 as well as in the more advanced topics of electronic circuits. There is little doubt that the concept and techniques of feedback are very important tools in the analysis, synthesis, and design of active circuits.

ANALOG INTEGRATED CIRCUITS

13

Analog ICs include operational amplifiers, analog multipliers, A/D and D/A converters, phase-lock loops, and a variety of other, more specialized functional blocks. Over the years designers of analog ICs have introduced many ingenious circuit techniques. In this chapter we shall provide an introduction to the art of analog IC design and analysis. Specifically, we shall study in some detail the circuit of the most popular analog IC in production today, the 741 internally compensated op amp. This op amp was introduced in 1966 and is currently produced by almost every analog semiconductor manufacturer.

Introduction

In addition to introducing the reader to some of the analog IC design techniques, study of the 741 circuit will serve to tie together many of the ideas and concepts of the previous chapters.

13.1 THE 741 OP-AMP CIRCUIT

We shall begin with a qualitative study of the 741 op-amp circuit, which is shown in Fig. 13.1. Note that in keeping with the IC design philosophy the circuit uses a large number of transistors but relatively few resistors and only one capacitor. This philosophy is dictated by the economics (silicon area, ease of fabrication, quality of realized component) of the fabrication of active and passive components in IC form (see Appendix A).

As is the case with most modern IC op amps, the 741 requires two power supplies, $+V_{CC}$ and $-V_{EE}$. Normally, $V_{CC} = V_{EE} = 15$ V, but the circuit operates satisfactorily with the power supplies reduced to much lower values (such as ± 5 V). It is important to observe that no circuit node is connected to ground, the common terminal of the two supplies.

With a relatively large circuit such as that in Fig. 13.1, the first step in the analysis is the identification of its recognizable parts and their functions. This can be done as follows.

Bias Circuit

The reference bias current of the 741 circuit, I_{REF}, is generated in the branch at the extreme left, consisting of the two diode-connected transistors Q_{11} and Q_{12} and the

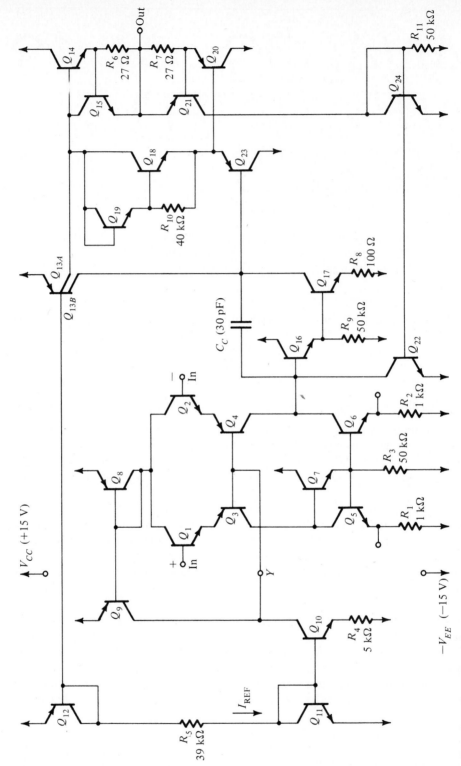

Fig. 13.1 The 741 op-amp circuit.

resistance R_5. Through a "modified current mirror" formed by Q_{11}, Q_{10}, and R_4, bias current for the first stage is generated in the collector of Q_{10}. Another current mirror formed by Q_8 and Q_9 takes part in biasing the first stage.

The reference bias current I_{REF} is used to provide two proportional currents in the collectors of Q_{13}. This double-collector *lateral*[1] *pnp* transistor can be thought of as two transistors whose base–emitter junctions are connected in parallel. Thus Q_{12} and Q_{13} form a two-output current mirror: one output, the collector of Q_{13B}, provides bias current for Q_{17}, and the other output, the collector of Q_{13A}, provides bias current for the output stage of the op amp.

Two more transistors, Q_{18} and Q_{19}, take part in the dc bias process. The purpose of Q_{18} and Q_{19} is to establish two V_{BE} drops between the bases of the output transistors Q_{14} and Q_{20}.

Short-Circuit Protection Circuitry

The 741 circuit includes a number of transistors that are normally off and that conduct only in the event that one attempts to draw a large current from the op-amp output terminal. This would happen if the output terminal is short-circuited to one of the two supplies. The short-circuit protection network consists of R_6, R_7, Q_{15}, Q_{21}, Q_{24}, and Q_{22}. In the following we shall assume that these transistors are off. Operation of the short-circuit protection network will be explained in Section 13.6.

The Input Stage

The 741 circuit consists of three stages: an input differential stage, an intermediate single-ended high-gain stage, and an output-buffering stage. The input stage consists of transistors Q_1 through Q_7, with biasing performed by Q_8, Q_9, and Q_{10}. Transistors Q_1 and Q_2 act as emitter followers, causing the input resistance to be high and delivering the differential input signal to the differential common-base amplifier formed by Q_3 and Q_4.

Transistors Q_5, Q_6, and Q_7 and resistors R_1, R_2, and R_3 form the load circuit of the input stage. This is an elaborate load circuit, which we will discuss in detail in Section 13.3. It will be shown that this load circuit not only provides a high resistance load but also converts the signal from differential to single-ended with no loss in gain or common-mode rejection. The output of the input stage is taken single-endedly at the collector of Q_6.

As mentioned in Section 10.8, every op-amp circuit includes a *level shifter* whose function is to shift the dc level of the signal so that the signal at the op-amp output can swing positive and negative. In the 741, level shifting is done in the first stage using the lateral *pnp* transistors Q_3 and Q_4. Although lateral *pnp* transistors have poor high-frequency performance, their use in the common-base configuration (which is known to have good high-frequency response) does not seriously impair the op-amp frequency response.

---

[1]See Appendix A for a description of lateral *pnp* transistors.

The use of the lateral *pnp* transistors Q_3 and Q_4 in the first stage results in an added advantage: protection of the input-stage transistors Q_1 and Q_2 against emitter–base junction breakdown. Since the emitter–base junction of an *npn* transistor breaks down at about 7 V of reverse bias (see Section 9.10), regular *npn* differential stages would suffer such a breakdown if, say, the supply voltage is accidently connected between the input terminals. Lateral *pnp* transistors, however, have high emitter–base breakdown voltages (about 50 V) and because they are connected in series with Q_1 and Q_2, they provide protection of the 741 input transistors, Q_1 and Q_2.

The Second Stage

The second or intermediate stage is composed of Q_{16}, Q_{17}, Q_{13B}, and the two resistors R_8 and R_9. Transistor Q_{16} acts as an emitter follower, thus giving the second stage a high input resistance. This minimizes the loading on the input stage and avoids loss of gain. Transistor Q_{17} acts as a common-emitter amplifier with a 100-Ω resistance in the emitter. Its load is composed of the high output resistance of the *pnp* current source Q_{13B} in parallel with the input resistance of the output stage (seen looking into the base of Q_{23}). Using a transistor current source as a load resistance is a technique called *active load*. It enables one to obtain high gain without resorting to the use of high load resistances, which would occupy large chip area.

The output of the second stage is taken at the collector of Q_{17}. Capacitor C_C is connected in the feedback path of the second stage to provide frequency compensation using the Miller compensation technique studied in Section 12.11. It will be shown in Section 13.7 that the relatively small capacitor C_C gives the 741 a dominant pole at about 4 Hz. Furthermore, pole splitting causes other poles to be shifted to much higher frequencies, giving the op amp a uniform -20 dB/decade gain rolloff with a unity-gain bandwidth of about 1 MHz. It should be pointed out that although C_C is small in value, the chip area that it occupies is about thirteen times that of a standard *npn* transistor!

The Output Stage

The purpose of the output stage is to provide the amplifier with a low output resistance. In addition, the output stage should be able to supply relatively large load currents without dissipating an unduly large amount of power in the IC.

The simplest output stage is the emitter follower, studied in detail in Chapter 10. A drawback of the emitter follower, however, is its low efficiency. Specifically, the ratio of the maximum power supplied to a load by an emitter-follower output stage to the power drawn from the supply is only about 25%. The 741 uses an alternative, more efficient output stage, which we shall study in detail in Section 13.6.

The output stage consists of the complementary pair Q_{14} and Q_{20}, where Q_{20} is a *substrate pnp* (see Appendix A). Transistors Q_{18} and Q_{19} are fed by current source Q_{13A} and bias the output transistors Q_{14} and Q_{20}. Transistor Q_{23} (which is another substrate *pnp*) acts as an emitter follower, thus minimizing the loading of the output stage on the second stage.

Device Parameters

In the following sections we shall carry out a detailed analysis of the 741 circuit. For the standard *npn* and *pnp* transistors the following parameters will be used:

$$npn: \quad I_S = 10^{-14} \text{ A}, \qquad \beta = 200, \qquad \mu = 5{,}000$$
$$pnp: \quad I_S = 10^{-14} \text{ A}, \qquad \beta = 50, \qquad \mu = 2{,}000$$

In the 741 circuit the nonstandard devices are Q_{13}, Q_{14}, and Q_{20}. Transistor Q_{13} will be assumed to be equivalent to two transistors, Q_{13A} and Q_{13B}, with parallel base–emitter junctions and with the following saturation currents:

$$I_{SA} = 0.25 \times 10^{-14} \text{ A}, \qquad I_{SB} = 0.75 \times 10^{-14} \text{ A}$$

Transistors Q_{14} and Q_{20} will be assumed to each have an area three times that of a standard device. Output transistors usually have relatively large areas in order to be able to supply large load currents and dissipate relatively large amounts of power with moderate increases in the device temperature.

EXERCISES

13.1 For the standard *npn* transistor whose parameters are given above, find approximate values for the following parameters if $I_C = 1$ mA: V_{BE}, g_m, r_e, r_π, r_o, r_μ. (*Note:* Assume $r_\mu = 10\beta r_o$.)
Ans.   633 mV; 40 mA/V; 25 Ω; 5 kΩ; 125 kΩ; 250 MΩ

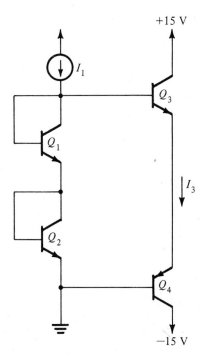

Fig. E13.2

13.2 For the circuit in Fig. E13.2 neglect base currents and use the exponential $i_C - v_{BE}$ relationship to show that

$$I_3 = I_1 \sqrt{\frac{I_{S3}I_{S4}}{I_{S1}I_{S2}}}$$

13.2   DC ANALYSIS OF THE 741

In this section we shall carry out a dc analysis of the 741 circuit to determine the bias point of each device. For dc analysis of an op-amp circuit the input terminals are grounded. Theoretically speaking, this should result in zero dc voltage at the output. However, because the op amp has very large gain, any slight approximation in the analysis will show that the output voltage is far from being zero and is close to either $+V_{CC}$ or $-V_{EE}$. In actual practice an op amp left open loop will have an output voltage saturated close to one of the two supplies. To overcome this problem in the dc analysis it will be assumed that the op amp is connected in a negative-feedback loop that stabilizes the output dc voltage to zero volts.

Reference Bias Current

The reference bias current I_{REF} is generated in the branch composed of the two diode-connected transistors Q_{11} and Q_{12} and resistor R_5. With reference to Fig. 13.1 we can write

$$I_{REF} = \frac{V_{CC} - V_{EB12} - V_{BE11} - (-V_{EE})}{R_5}$$

For $V_{CC} = V_{EE} = 15$ V and $V_{BE11} = V_{EB12} \cong 0.7$ V we have $I_{REF} = 0.73$ mA.

Input Stage Bias

Transistor Q_{11} is biased by I_{REF}, and the voltage developed across it is used to bias Q_{10}, which has a series emitter resistance R_4. This part of the circuit is redrawn in Fig. 13.2. This arrangement is known as the *Widlar current source* [Ref. 13.7], and its function is to generate a much smaller current (compared to I_{REF}) in the collector of Q_{10} without using large-value resistors. The collector current of Q_{10} can be determined as follows (see Fig. 13.2): Neglecting base currents we can write

$$V_{BE11} = V_T \ln \frac{I_{REF}}{I_{S11}}$$

$$V_{BE10} = V_T \ln \frac{I_{C10}}{I_{S10}}$$

For identical transistors $I_{S10} = I_{S11}$, and these two equations can be combined to yield

$$V_{BE11} - V_{BE10} = V_T \ln \frac{I_{REF}}{I_{C10}}$$

But from the circuit we have

$$V_{BE11} - V_{BE10} = I_{C10}R_4$$

Thus

$$V_T \ln \frac{I_{REF}}{I_{C10}} = I_{C10}R_4 \tag{13.1}$$

If we are given values for I_{REF} and R_4, this equation can be solved by trial and error to determine I_{C10}. For our case the result is $I_{C10} = 19 \ \mu A$.

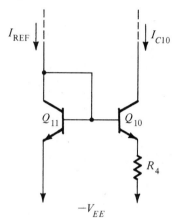

Fig. 13.2   Widlar current source.

EXERCISE

13.3   Design the Widlar current source of Fig. 13.2 to generate a current $I_{C10} = 10 \ \mu A$ given that $I_{REF} = 1$ mA. If at a collector current of 1 mA $V_{BE} = 0.7$ V, find V_{BE11} and V_{BE10}.
   Ans. $R_4 = 11.5$ kΩ; $V_{BE11} = 0.7$ V; $V_{BE10} = 0.585$ V

Having determined I_{C10}, we proceed to determine the dc current in each of the input-stage transistors. Part of the input stage is redrawn in Fig. 13.3. From symmetry we see that

$$I_{C1} = I_{C2}$$

Denote this current by I. We see that if the *npn* β is high, then

$$I_{E3} = I_{E4} \cong I$$

and the base currents of Q_3 and Q_4 are equal with a value of $I/(\beta_P + 1) \cong I/\beta_P$, where β_P denotes β of the *pnp* devices.

   The current mirror formed by Q_8 and Q_9 is fed by an input current of $2I$. Using the result in Eq. (10.22) we can express the output current of the mirror as

$$I_{C9} = \frac{2I}{1 + 2/\beta_P}$$

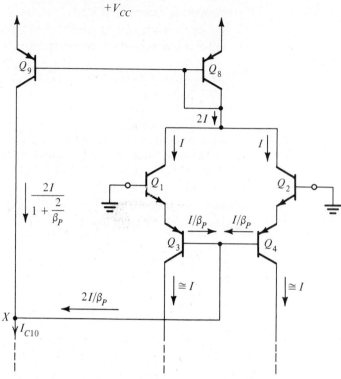

Fig. 13.3   The dc analysis of the 741 input stage.

We can now write a node equation for node X in Fig. 13.3 and thus determine the value of I. If $\beta_P \gg 1$, then this node equation gives

$$2I \cong I_{C10}$$

For the 741, $I_{C10} = 19 \ \mu A$; thus $I \cong 9.5 \ \mu A$. We have thus determined that

$$I_{C1} = I_{C2} \cong I_{C3} = I_{C4} = 9.5 \ \mu A$$

At this point we should note that transistors Q_1 through Q_4, Q_8, and Q_9 form a *negative-feedback loop,* which works to stabilize the value of I at approximately $I_{C10}/2$. To appreciate this fact, assume that for some reason the current I in Q_1 and Q_2 increases. This will cause the current pulled from Q_8 to increase, and the output current of the Q_8-Q_9 mirror will correspondingly increase. However, since I_{C10} remains constant, node X forces the combined base currents of Q_3 and Q_4 to decrease. This in turn will cause the emitter currents of Q_3 and Q_4, and hence the collector currents of Q_1 and Q_2, to decrease. This is opposite in direction to the originally assumed change. Hence the feedback is negative, and it stabilizes the value of I.

Figure 13.4 shows the remainder of the 741 input stage. If we neglect the base current of Q_{16} then

$$I_{C6} \cong I$$

Similarly, neglecting the base current of Q_7 we obtain

$$I_{C5} \cong I$$

The bias current of Q_7 can be determined from

$$I_{C7} \cong I_{E7} = \frac{2I}{\beta_N} + \frac{V_{BE6} + IR_2}{R_3} \qquad (13.2)$$

where β_N denotes β of the *npn* transistors. To determine V_{BE6} we use the transistor exponential relationship and write

$$V_{BE6} = V_T \ln \frac{I}{I_S}$$

Substituting $I_S = 10^{-14}$ A and $I = 9.5\ \mu\text{A}$ results in $V_{BE6} = 517$ mV. Then substituting in Eq. (13.2) yields $I_{C7} = 10.5\ \mu\text{A}$. Note that the base current of Q_7 is indeed negligible compared to the value of I, as has been assumed.

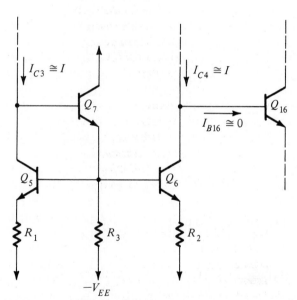

Fig. 13.4   The dc analysis of the 741 input stage continued.

Input Bias and Offset Currents

The *input bias current* of an op amp is defined as (Chapter 3)

$$I_B = \frac{I_{B1} + I_{B2}}{2}$$

For the 741 we obtain

$$I_B = \frac{I}{\beta_N}$$

Using $\beta_N = 200$ yields $I_B = 47.5$ nA. Note that this value is reasonably small and is typical of general-purpose op amps that use BJTs in the input stage. Much lower input bias currents (in the picoamp range) can be obtained using a FET input stage. Also, there exist techniques for reducing the input bias current of bipolar-input op amps.

Because of possible mismatches in the β values of Q_1 and Q_2 the input base currents will not be equal. One can calculate the *input offset current*, defined as

$$I_{\text{off}} = |I_{B1} - I_{B2}|$$

Input Offset Voltage

The input offset voltage of an op amp is determined mostly by mismatches in the load circuit of the input stage. Refer to Fig. 13.1 and let the input voltage be zero. It follows that the currents in Q_3 and Q_4 will be equal and their value will be I. If the load circuit is balanced, as was assumed in the dc analysis above, then the currents in Q_5 and Q_6 also are equal (and equal to I) and the output current of the first stage is zero. On the other hand, suppose now that a mismatch exists between R_1 and R_2. It follows that Q_5 and Q_6 will no longer have equal currents. Rather, while the current in Q_5 will remain equal to I (neglecting the base current of Q_7), that in Q_6 will be $I + \Delta I$, where ΔI is a function of the mismatch between R_1 and R_2. Now, since $I_{C4} = I$, we see that the first stage will produce an output current of ΔI. This output current will cause the output voltage of the first stage to change. To reduce the change in output voltage to zero a differential signal has to be applied at the input. By definition, this input voltage is equal to the *input offset voltage*. Since evaluation of the input offset voltage requires knowledge of the input stage gain, we shall postpone making these calculations to Section 13.3.

Input Common-Mode Range

The *input common-mode range* is the range of input common-mode voltages over which the input stage remains in the linear active mode. Refer to Fig. 13.1. We see that in the 741 circuit the input common-mode range is determined at the upper end by saturation of Q_1 and Q_2 and at the lower end by saturation of Q_3 and Q_4.

> **EXERCISE**
> **13.4** Neglect the voltage drops across R_1 and R_2 and assume that $V_{CC} = V_{EE} = 15$ V. Show that the input common-mode range of the 741 is approximately -12.6 to $+14.4$ V. (Assume that $V_{BE} \cong 0.6$ V.)

Second-Stage Bias

If we neglect the base current of Q_{23} then we see from Fig. 13.1 that the collector current of Q_{17} is approximately equal to the current supplied by current source Q_{13B}. Because Q_{13B} has a scale current 0.75 times that of Q_{12}, its collector current will be

$$I_{C13B} \cong 0.75 I_{\text{REF}}$$

where we have assumed that $\beta_P \gg 1$. Thus $I_{C13B} = 550\ \mu A$ and $I_{C17} \cong 550\ \mu A$. At this current level the base–emitter voltage of Q_{17} is

$$V_{BE17} = V_T \ln \frac{I_{C17}}{I_S} = 618\ \text{mV}$$

The collector current of Q_{16} can be determined from

$$I_{C16} \cong I_{E16} = I_{B17} + \frac{I_{E17}R_8 + V_{BE17}}{R_9}$$

This calculation yields $I_{C16} = 16.2\ \mu A$. Note that the base current of Q_{16} will indeed be negligible compared to the input stage bias I, as we have previously assumed.

Output-Stage Bias

Figure 13.5 shows the output stage of the 741 with the short-circuit protection circuitry omitted. Current source Q_{13A} delivers a current of $0.25I_{REF}$ (because I_S of Q_{13A} is 0.25 times the I_S of Q_{12}) to the network composed of Q_{18}, Q_{19}, and R_{10}. If we neglect the base

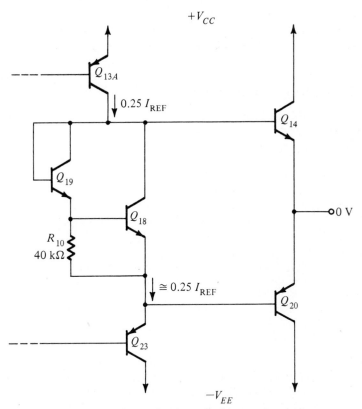

Fig. 13.5   The 741 output stage without the short-circuit protection devices.

currents of Q_{14} and Q_{20}, then the emitter current of Q_{23} will also be equal to $0.25 I_{REF}$. Thus

$$I_{C23} \cong I_{E23} \cong 0.25 I_{REF} = 180 \ \mu A$$

We thus see that the base current of Q_{23} is only $180/50 = 3.6 \ \mu A$, which is negligible compared to I_{C17}, as we have previously assumed.

If we assume that V_{BE18} is approximately 0.6 V, we can determine the current in R_{10} as 15 μA. The emitter current of Q_{18} is therefore

$$I_{E18} = 180 - 15 = 165 \ \mu A, \qquad I_{C18} \cong I_{E18} = 165 \ \mu A$$

At this value of current we find that $V_{BE18} = 588$ mV, which is quite close to the value assumed. The base current of Q_{18} is $165/200 = 0.8 \ \mu A$, which can be added to the current in R_{10} to determine the Q_{19} current as

$$I_{C19} \cong I_{E19} = 15.8 \ \mu A$$

The voltage drop across the base–emitter junction of Q_{19} can now be determined as

$$V_{BE19} = V_T \ln \frac{I_{C19}}{I_S} = 530 \text{ mV}$$

As mentioned in Section 13.1, the purpose of the Q_{18}-Q_{19} network is to establish two V_{BE} drops between the bases of the output transistors Q_{14} and Q_{20}. This voltage drop, V_{BB}, can be now calculated as

$$V_{BB} = V_{BE18} + V_{BE19} = 588 + 530 = 1.118 \text{ V}$$

Since V_{BB} appears across the series combination of the base–emitter junctions of Q_{14} and Q_{20}, we can write

$$V_{BB} = V_T \ln \frac{I_{C14}}{I_{S14}} + V_T \ln \frac{I_{C20}}{I_{S20}}$$

Using the calculated value of V_{BB} and substituting $I_{S14} = I_{S20} = 3 \times 10^{-14}$ A, we determine the collector currents as

$$I_{C14} = I_{C20} = 154 \ \mu A$$

Summary

For future reference, Table 13.1 provides a listing of the values of the collector bias currents of the 741 transistors.

EXERCISE

13.5 If in the circuit of Fig. 13.5 the Q_{18}-Q_{19} network is replaced by two diode-connected transistors, find the current in Q_{14} and Q_{20}. (*Hint:* Use the result of Exercise 13.2.) *Ans.* 540 μA

TABLE 13.1   DC Collector Currents of the 741 Circuit (μA)

Q_1	9.5	Q_8	19	Q_{13B}	550	Q_{19}	15.8
Q_2	9.5	Q_9	19	Q_{14}	154	Q_{20}	154
Q_3	9.5	Q_{10}	19	Q_{15}	0	Q_{21}	0
Q_4	9.5	Q_{11}	730	Q_{16}	16.2	Q_{22}	0
Q_5	9.5	Q_{12}	730	Q_{17}	550	Q_{23}	180
Q_6	9.5	Q_{13A}	180	Q_{18}	165	Q_{24}	0
Q_7	10.5						

(*Note:*   One of the reasons for using the Q_{18}-Q_{19} network is to obtain a much smaller bias current in the output transistors and thus minimize the power dissipation. Physically speaking, why does the Q_{18}-Q_{19} network result in a smaller bias current than that obtained with two diode-connected transistors?)

13.3 SMALL-SIGNAL ANALYSIS OF THE 741 INPUT STAGE

Figure 13.6 shows part of the 741 input stage for the purpose of performing small-signal analysis. Note that since the collectors of Q_1 and Q_2 are connected to a constant dc

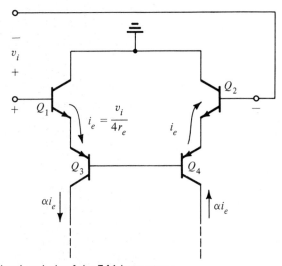

Fig. 13.6   Small-signal analysis of the 741 input stage.

voltage, they are shown grounded. Also, the constant-current biasing of the bases of Q_3 and Q_4 is equivalent to having the common base terminal open-circuited.

The differential signal v_i applied between the input terminals effectively appears across four equal emitter resistances connected in series—those of Q_1, Q_2, Q_3, and Q_4. As a result, emitter signal currents flow as indicated in Fig. 13.6 with

$$i_e = \frac{v_i}{4r_e} \tag{13.3}$$

where r_e denotes the emitter resistance of each of Q_1 through Q_4. Thus

$$r_e = \frac{V_T}{I} = \frac{25 \text{ mV}}{9.5 \text{ } \mu\text{A}} = 2.63 \text{ k}\Omega$$

Thus the four transistors Q_1 through Q_4 supply the load circuit with a pair of complementary current signals αi_e, as indicated in Fig. 13.6.

The input differential resistance of the op amp can be obtained from Fig. 13.6 as

$$R_{id} = 4r_\pi = 4(\beta_N + 1)r_e$$

For $\beta_N = 200$ we obtain $R_{id} = 2.1 \text{ M}\Omega$.

Proceeding with the input stage analysis, we show in Fig. 13.7 the load circuit fed with the complementary pair of current signals found above. Neglecting the signal current in the base of Q_7, we see that the collector signal current of Q_5 is approximately

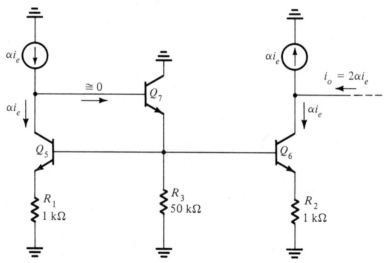

Fig. 13.7 *The load circuit of the input stage fed by the two complementary current signals generated by Q_1 through Q_4 in Fig. 13.6.*

equal to the input current αi_e. Now, since Q_5 and Q_6 are identical and their bases are tied together, and since equal resistances are connected in their emitters, it follows that their collector signal currents must be equal. Thus the signal current in the collector of Q_6 is forced to be equal to αi_e. In other words, the load circuit functions as a *current mirror*.

Now consider the output node of the input stage. The output current i_o is given by

$$i_o = 2\alpha i_e \tag{13.4}$$

The factor of two in this equation indicates that conversion from differential to single-ended is performed without losing half the signal. The trick, of course, is the use of the current mirror to invert one of the current signals and then add the result to the other current signal.

Equations (13.3) and (13.4) can be combined to obtain the transconductance of the input stage, G_{m1}:

$$G_{m1} \equiv \frac{i_o}{v_i} = \frac{\alpha}{2r_e} \ \text{'}$$

Substituting $r_e = 2.63 \text{ k}\Omega$ and $\alpha \cong 1$ yields $G_{m1} = 1/5.26 \text{ mA/V}$.

13.6   For the circuit in Fig. 13.7 find in terms of i_e:
   (a)   The signal voltage at the base of Q_6;
   (b)   The signal current in the emitter of Q_7;
   (c)   The signal current in the base of Q_7;
   (d)   The signal voltage at the base of Q_7;
   (e)   The input resistance seen by the left-hand-side signal current source αi_e.
   (*Note:*   For simplicity assume that $I_{C7} \cong I_{CS} = I_{C6}$.)
   Ans.   (a) 3.63 kΩ × i_e; (b) 0.08i_e; (c) 0.0004i_e; (d) 3.74 kΩ × i_e; (e) 3.74 kΩ

To complete modeling the 741 input stage we must find its output resistance R_{o1}. This is the resistance seen "looking back" at the collector terminal of Q_8 in Fig. 13.7. Thus R_{o1} is the parallel equivalent of the output resistance of the current source supplying the signal current αi_e and the output resistance of Q_6. The first component is the resistance looking into the collector of Q_4 in Fig. 13.6. Finding this resistance is considerably simplified if we assume that the common bases of Q_3 and Q_4 are at a *virtual ground*. This of course happens only when the input signal v_i is applied in a complementary fashion. Nevertheless, this assumption does not result in a large error.

Assuming that the base of Q_4 is at virtual ground, the resistance we are after is R_{o4}, indicated in Fig. 13.8a. This is the output resistance of a common-base transistor

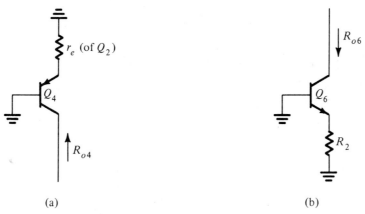

(a)   (b)

Fig. 13.8   *Simplified circuit for finding the two components of the output resistance R_{o1} of the first stage.*

that has a resistance (r_e of Q_2) in its emitter. To find R_{o4} we may use the expression developed in Example 11.6:

$$R_o \doteq r_o \left(\frac{1 + R_E/r_e}{1 + R_E/r_\pi} \right) \| r_\mu \qquad (13.5)$$

Substituting $R_E = r_e$ and $r_o = \mu/g_m$ and neglecting the very large r_μ results in $R_{o4} = 10.5$ MΩ.

The second component of the output resistance is that seen looking into the collector of Q_6 in Fig. 13.7. Although the base of Q_6 is not at signal ground, we shall assume that the signal voltage at the base is sufficiently small to make this approximation valid. The circuit then takes the form in Fig. 13.8b, and R_{o6} can be determined using Eq. (13.5) with $R_E = R_2$. Thus $R_{o6} \cong 18.2$ MΩ.

Finally, we combine R_{o4} and R_{o6} in parallel to obtain the output resistance of the input stage, R_{o1}, $R_{o1} = 6.7$ MΩ.

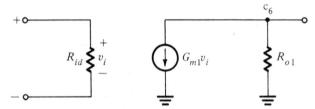

Fig. 13.9   Small-signal equivalent circuit for the input stage of the 741 op amp.

Figure 13.9 shows the equivalent circuit that we have derived for the input stage. This is a simplified version of the y-parameter model of a two-port network with y_{12} assumed negligible.

Example 13.1
We wish to find the input offset voltage resulting from a 2% mismatch between the resistances R_1 and R_2 in Fig. 13.1.

Solution
Consider first the situation when both input terminals are grounded, as shown in Fig. 13.10. We assume that $R_1 = R$ and $R_2 = R + \Delta R$, where $\Delta R/R = 0.02$. From Fig. 13.10 we see that while Q_5 still conducts a current equal to I, the current in Q_6 will be smaller by ΔI. The value of ΔI can be found from

$$V_{BE5} + IR = V_{BE6} + (I - \Delta I)(R + \Delta R)$$

Thus

$$V_{BE5} - V_{BE6} = I\Delta R - \Delta I(R + \Delta R) \qquad (13.6)$$

The quantity on the left-hand side is the change in V_{BE} due to a change in I_E of ΔI. We may therefore write

$$V_{BE5} - V_{BE6} \cong \Delta I r_e \qquad (13.7)$$

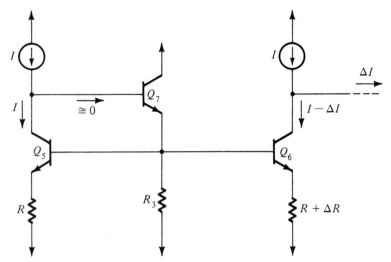

Fig. 13.10   Input stage with both inputs grounded and a mismatch ΔR between R_1 and R_2.

Equations (13.6) and (13.7) can be combined to obtain

$$\frac{\Delta I}{I} = \frac{\Delta R}{R + \Delta R + r_e}$$

Substituting $R = 1$ kΩ and $r_e = 2.63$ kΩ shows that a 2% mismatch between R_1 and R_2 gives rise to an output current $\Delta I = 5.5 \times 10^{-3}I$. To reduce this output current to zero we have to apply an input voltage V_{off} given by

$$V_{off} = \frac{\Delta I}{G_{m1}} = \frac{5.5 \times 10^{-3}I}{G_{m1}}$$

Substituting $I = 9.5$ μA and $G_{m1} = 1/5.26$ mA/V results in the offset voltage $V_{off} \cong 0.3$ mV.

It should be pointed out that the offset voltage calculated is only one component of the input offset voltage of the 741. Other components arise because of mismatches in transistor characteristics. The 741 offset voltage is specified to be typically 2 mV.

· · ·

EXERCISES

13.7   For the current mirror with emitter resistances, shown in Fig. E13.7, show that the small-signal current gain is given by

$$\frac{i_o}{i_i} \cong 1 - \frac{\Delta R}{R + r_e}$$

where r_e is the small-signal resistance of each of the diode and the transistor emitter, and $\Delta R \ll R + r_e$.

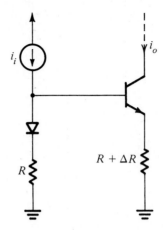

Fig. E13.7

13.8 For the purpose of finding the *common-mode transconductance*, Fig. E13.8 shows a simplified equivalent circuit of the 741 input stage. Note that the load circuit has been replaced by a current mirror, and the resistances R_1 and R_2 are assumed to have a mismatch ΔR. Resistor R_o represents the output resistance of the current source that biases

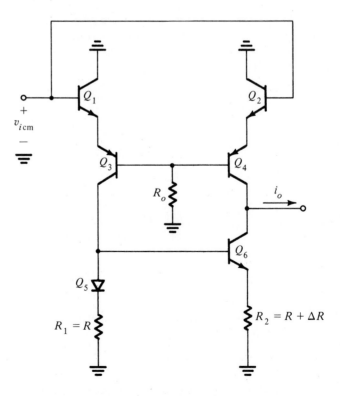

Fig. E13.8

the common bases of Q_3 and Q_4. Using the result of Exercise 13.7 show that the common-mode transconductance G_{mcm} is given by

$$G_{mcm} \equiv \frac{i_o}{v_{icm}} \cong \frac{\beta_P}{2R_o} \frac{\Delta R}{R + r_e}$$

where r_e is the resistance of the diode and the transistor in the current mirror and where it is assumed that $R_o \gg \beta_P r_{e1}$.

13.9 (a) Refer to Fig. 13.1 and assume that the bases of Q_9 and Q_{10} are at constant dc voltages. Find the resistance R_o seen by looking to the left of node Y.

(b) Use this value in the expression derived in Exercise 13.8 to determine G_{mcm} and hence the common-mode rejection ratio defined, $20 \log (G_{m1}/G_{mcm})$. Assume that $\Delta R/R = 2\%$.

Ans.   (a) 2.4 MΩ; (b) $G_{mcm} = 0.058 \ \mu A/V$, CMRR $= 70$ dB

(*Note:*   The CMRR calculated in Exercises 13.7 through 13.9 is smaller than that specified for the 741 (80–90 dB). The reason for this is that the CMRR is increased by the negative-feedback loop formed by Q_1 through Q_4, Q_8, Q_9, and Q_{10}. Just as this feedback loop stabilizes bias, it minimizes the signal current in Q_1 through Q_4 due to a common-mode input signal. This feedback is in fact referred to as *common-mode feedback*.)

13.4   SMALL-SIGNAL ANALYSIS OF THE 741 SECOND STAGE

Figure 13.11 shows the 741 second stage prepared for small-signal analysis. In this section we shall analyze the second stage to determine the values of the parameters of the equivalent circuit shown in Fig. 13.12. Again, this is a simplified y-parameter equivalent circuit with y_{12} neglected.

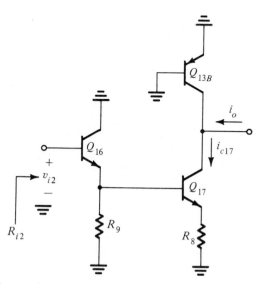

Fig. 13.11   *The 741 second stage prepared for small-signal analysis.*

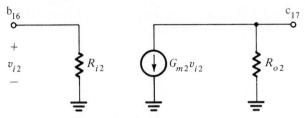

Fig. 13.12   Small-signal equivalent circuit model of the second stage.

Input Resistance

The input resistance R_{i2} can be found by inspection to be

$$R_{i2} = (\beta_{16} + 1)[r_{e16} + R_9\|(\beta_{17} + 1)(r_{e17} + R_8)]$$

Substituting the appropriate parameter values yields $R_{i2} \cong 4$ MΩ.

Transconductance

From the equivalent circuit of Fig. 13.12 we see that the transconductance G_{m2} is the ratio of the *short-circuit output current* to the input voltage. Short-circuiting the output terminal of the second stage (Fig. 13.11) to ground makes the signal current through the output resistance of Q_{13B} zero, and the output short-circuit current becomes equal to the collector signal current of Q_{17} (i_{c17}). This latter current can be easily related to v_{i2} as follows:

$$i_{c17} = \frac{\alpha v_{b17}}{r_{e17} + R_8}$$

$$v_{b17} = v_{i2} \frac{(R_9\| R_{i17})}{(R_9\| R_{i17}) + r_{e16}}$$

$$R_{i17} = (\beta_{17} + 1)(r_{e17} + R_8)$$

These equations can be combined to obtain

$$G_{m2} \equiv \frac{i_{c17}}{v_{i2}}$$

which for the 741 parameter values is found to be $G_{m2} = 6.53$ mA/V.

Output Resistance

To determine the output resistance R_{o2} of the second stage in Fig. 13.11 we ground the input terminal and find the resistance looking back into the output terminal. It follows that R_{o2} is given by

$$R_{o2} = (R_{o13B}\| R_{o17}) \tag{13.8}$$

where R_{o13B} is the resistance looking into the collector of Q_{13B} while its base and emitter are connected to ground. It can be easily shown that

$$R_{o13B} = r_{o13B}$$

For the 741 component values we obtain $R_{o13B} = 90.9$ kΩ.

The second component in Eq. (13.8), R_{o17}, is the resistance seen looking into the collector of Q_{17}, as indicated in Fig. 13.13. Since the resistance between the base of Q_{17}

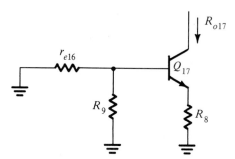

Fig. 13.13   Definition of R_{o17}.

and ground is relatively small, one can considerably simplify matters by assuming that the base is grounded. Doing this, we can use Eq. (13.5) to determine R_{o17}. For our case the result is $R_{o17} \cong 728$ kΩ. Combining R_{o13B} and R_{o17} in parallel yields $R_{o2} = 81$ kΩ.

Thévenin Equivalent Circuit

The second-stage equivalent circuit can be converted to the Thévenin form, as shown in Fig. 13.14. Note that the stage open-circuit voltage gain is $-G_{m2}R_{o2}$.

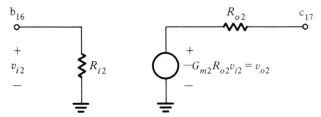

Fig. 13.14   Thévenin form of the small-signal model of the second stage.

13.5   OUTPUT STAGES

As mentioned before, the function of the output stage is twofold: to provide a low output impedance and to supply relatively large load currents in an efficient manner. In this section various types of output stage circuits are studied.

Class A Output Stage

The simplest type of output stage is the emitter follower, studied in detail in Section 10.3. Figure 13.15 shows an emitter-follower output stage where the emitter-follower transistor Q_1 is biased by a constant current I provided by current source transistor Q_2. In this circuit the bias current I is greater than the maximum load current, and the circuit is called a class A output stage.

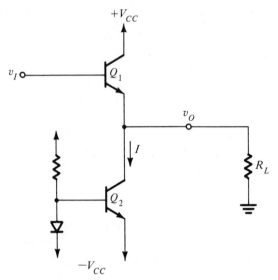

Fig. 13.15 *Emitter-follower output stage (class A).*

It will be assumed that the quiescent dc voltage at the output is zero. Thus in the quiescent state the power dissipated in Q_1 is $V_{CC}I$, and an equal amount of power is dissipated in Q_2.

Consider now the operation of the emitter follower with an input sine wave. As v_I goes positive, v_O follows and the load current flows *out* of the transistor, causing the total emitter current of Q_1 to increase. If the amplitude of v_I is increased, this mode of operation continues until v_I reaches a value that causes Q_1 to saturate. Thus the maximum possible positive value of v_O is $V_{CC} - V_{CEsat}$, which we shall assume to be approximately equal to V_{CC}.

Next consider the situation for negative v_I. In this case v_O will be negative and the load current will flow *into* the transistor, causing the total emitter current of Q_1 to decrease. If the amplitude of v_I is increased, this mode of operation continues until either Q_2 saturates or Q_1 cuts off. Transistor Q_2 saturates when v_O reaches $-V_{CC} + V_{CEsat}$ or approximately $-V_{CC}$. Transistor Q_1 cuts off when the load current equals the bias current I. It will be shown that maximum efficiency is obtained when these two constraints are reached simultaneously—that is, when the load resistance R_L is given by

$$R_L = V_{CC}/I$$

If R_L satisfies this relationship, then

$$v_{O\max} = -v_{O\min} \cong V_{CC}$$

and the waveforms of Fig. 13.16 are ideally obtained. We urge the reader to verify these waveforms.

Of special interest is the waveform of the *instantaneous power dissipation* in Q_1, which is shown in Fig. 13.16d. Note that the maximum power dissipation is $V_{CC}I$, which is equal to the quiescent power dissipation. We conclude that the emitter follower dissipates the largest amount of power when no input signal is applied.

The efficiency η of an output stage is defined as

$$\eta \equiv \frac{\text{Load power}}{\text{Supply power}}$$

Assume that the output voltage is a sinusoid with peak value $\hat{V}_o$. The load power is given by

$$P_L = \frac{1}{2}\frac{\hat{V}_o^2}{R_L} \tag{13.9}$$

Since the current in Q_2 is constant, the power drawn from the negative supply is $V_{CC}I$. The *average* current in Q_1 is also equal to I, and thus the average power from the positive supply is $V_{CC}I$. Thus the supply power is

$$P_S = 2V_{CC}I \tag{13.10}$$

Equations (13.9) and (13.10) can be combined to obtain

$$\eta = \frac{1}{4}\frac{\hat{V}_o^2}{IR_LV_{CC}}$$

Since $\hat{V}_o \leq V_{CC}$ and $\hat{V}_o \leq IR_L$, maximum efficiency is obtained when

$$\hat{V}_o = V_{CC} = IR_L$$

and the maximum efficiency obtained is 25%. This is rather low, which implies that for a given amount of output power, the class A stage draws a large amount of power from the supply. This has two disadvantages:

1. The supply current is large, which might rule out battery operation of the IC.
2. The power dissipated in the IC is large, which raises the temperature of the chip and could possibly result in burnout.

EXERCISE

13.10 Consider a class A output stage such as that shown in Fig. 13.15 with $V_{CC} = 15$ V, $I = 10$ mA, and $R_L = 1$ kΩ. If the output voltage is a 5-V-peak sinusoid find the following:
(a) The power drawn from the supplies;
(b) The maximum instantaneous power dissipation in Q_1;
(c) The efficiency.
Ans. (a) 300 mW; (b) 150 mW; (c) 4.2%

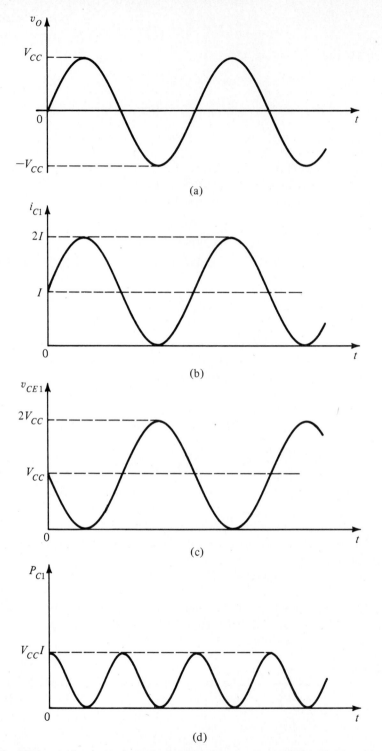

Fig. 13.16 Maximum signal waveforms in the class A output stage of Fig. 13.15 under the condition $R_L = V_{CC}/I$.

Class B Output Stage

An alternative output stage is shown in Fig. 13.17. It consists of a complementary pair of transistors connected in such a way that both cannot be on simultaneously. The circuit operates as follows: When the input voltage v_I is zero, both transistors are cut off and the output voltage v_O is zero. If v_I is positive and greater than about 0.5 V, Q_N conducts and operates as an emitter follower. In this case, v_O follows v_I and Q_N supplies

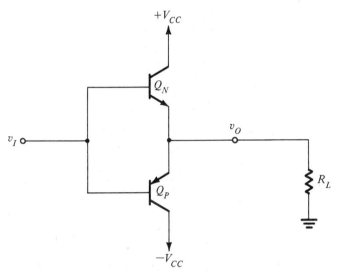

Fig. 13.17 *Class B output stage.*

the current required by the load. Meanwhile, the emitter–base junction of Q_P will be reverse-biased by the 0.7-V V_{BE} drop of Q_N. Thus Q_P will be off.

On the other hand, if the input goes negative by more than about 0.5 V, Q_P turns on and acts as an emitter follower. Again v_O follows v_I, but in this case Q_P supplies the load current and Q_N will be off.

Thus we see that unlike the class A stage of Fig. 13.15, in which the emitter-follower transistor Q_1 is biased at a current greater than the maximum load current, the transistors in the circuit of Fig. 13.17 are biased at zero current and turn on only when the input signal is present. This mode of operation is called class B, and the circuit of Fig. 13.17 is called a *class B output stage*. It is also known as a *push-pull circuit* because Q_N *pushes* current into the load (during the positive half-cycles) and Q_P *pulls* current from the load (during the negative half-cycles).

The transfer characteristic of the class B output stage is sketched in Fig. 13.18. Note that there exists a range of v_I centered around zero where both transistors remain off and the output remains zero. This *dead band* results in the *crossover distortion* illustrated in Fig. 13.19 for the case when the input is sinusoidal. Obviously, the crossover distortion will be most severe when the amplitude of the input signal is small.

Before discussing the method used in the 741 output stage to eliminate crossover

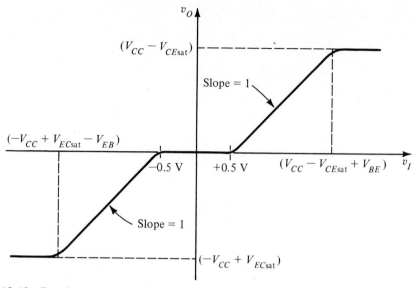

Fig. 13.18   Transfer characteristic for the class B output stage in Fig. 13.17.

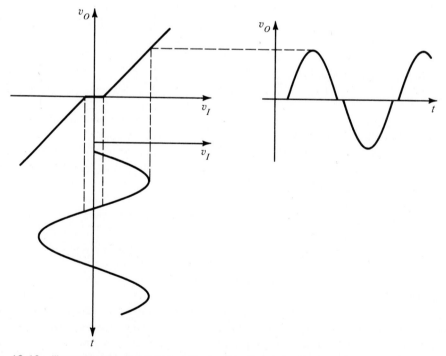

Fig. 13.19   Illustrating how the dead band in the class B transfer characteristic results in crossover distortion.

distortion, we shall derive an expression for the efficiency of the class B stage of Fig. 13.17. For this purpose let the input signal be sinusoidal and let us neglect the crossover distortion. If the output sinusoid has a peak amplitude $\hat{V}_o$, then the load power is

$$P_L = \frac{1}{2} \frac{\hat{V}_o^2}{R_L} \tag{13.11}$$

The current drawn from the positive supply consists of half-sinusoids with peak value of $\hat{V}_o/R_L$. Thus the average current is $(1/\pi)\, \hat{V}_o/R_L$, and the power from the positive supply is

$$P_{S+} = \frac{1}{\pi} \frac{\hat{V}_o}{R_L} V_{CC}$$

Similarly, the power from the negative supply is

$$P_{S-} = \frac{1}{\pi} \frac{\hat{V}_o}{R_L} V_{CC}$$

and the total supply power is given by

$$P_S = \frac{2}{\pi} \frac{\hat{V}_o}{R_L} V_{CC} \tag{13.12}$$

Equations (13.11) and (13.12) can be combined to obtain

$$\eta = \frac{\pi}{4} \frac{\hat{V}_o}{V_{CC}}$$

It follows that maximum efficiency is obtained when $\hat{V}_o$ is at its maximum. This maximum is limited by the saturation of Q_N and Q_P to $V_{CC} - V_{CEsat} \cong V_{CC}$. At this value of output, a maximum efficiency of 78.6% is obtained.

EXERCISE

13.11 Consider a class B output stage such as that of Fig. 13.17 with $V_{CC} = 15$ V and $R_L = 1$ kΩ. If the output voltage is a 5-V-peak sinusoid, neglect crossover distortion and find the following:
(a) The power drawn from the supply;
(b) The maximum instantaneous power dissipation in Q_N;
(c) The efficiency.
Ans. (a) 47.7 mW; (b) 50 mW; (c) 26.2%
(Note: These results should be compared to those of Exercise 13.10.)

Class AB Output Stage

Crossover distortion can be virtually eliminated by biasing the output transistors at a small but finite current. This is usually accomplished by using two additional diodes (or diode-connected transistors) fed from a current source, as shown in Fig. 13.20. The resulting circuit is called a *class AB output stage,* since it is like class A in the sense that the output devices are always on and like class B in the sense that the output

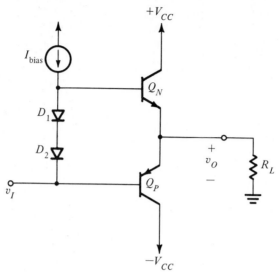

Fig. 13.20   Class AB output stage.

devices are biased at a much smaller current than the peak current delivered to the load.

The class AB stage operates in a manner quite similar to that of the class B stage. When v_I goes positive, the voltage at the base of Q_N increases and Q_N supplies the load current. Thus the base–emitter voltage of Q_N increases. This causes the emitter–base voltage of Q_P to decrease by an equal amount (since the voltage difference between the bases is constant). Thus although Q_P remains conducting, its current will be negligible compared to the current in Q_N. The reverse happens when v_I goes negative. The major difference between this circuit and the class B circuit is that here both devices are on when v_I is zero, and the dead band in the transfer characteristic is eliminated.

Finally, it should be pointed out that the *npn* and *pnp* devices in a class AB or class B stage should be matched in order for the two halves of the output waveform to be identical. However, since the devices are used as emitter followers, their operation is not highly dependent on device parameters, and the matching requirement is not a critical one. This is especially true in the 741 output stage, which is driven by another emitter follower, thus minimizing the dependence of the op-amp gain on the β values of the output devices. The 741 output stage is analyzed in Section 13.6.

13.6   ANALYSIS OF THE 741 OUTPUT STAGE

The 741 output stage is shown in Fig. 13.21 without the short-circuit protection circuit. The stage is shown driven by the second-stage transistor Q_{17} and loaded with a 2-kΩ resistance. The circuit is of the AB class, with the network composed of Q_{18}, Q_{19}, and R_{10} providing the bias of the output transistors Q_{14} and Q_{20}. The use of this network rather than two diode-connected transistors in series enables biasing the output transistors at a low current (0.15 mA) in spite of the fact that the output devices are about

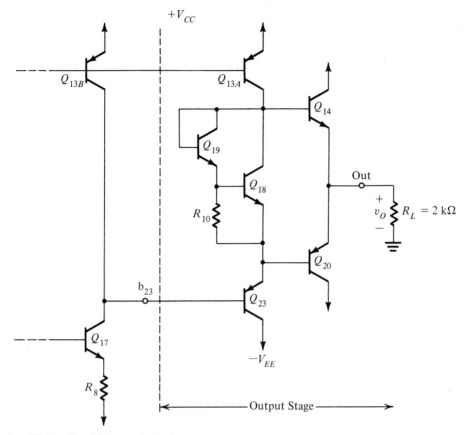

Fig. 13.21   *The 741 output stage.*

three times as large as the standard devices. This is obtained by arranging that the current in Q_{19} is very small and thus its V_{BE} is also small. We analyzed the dc bias in Section 13.2.

Another feature of the 741 output stage worth noting is that the stage is driven by an emitter follower Q_{23}. As will be shown, this emitter follower provides added buffering, which makes the op-amp gain almost independent of the parameters of the output transistors.

Output Voltage Limits

The maximum positive output voltage is limited by the saturation of current source transistor Q_{13A}. Thus

$$v_{O\text{max}} = V_{CC} - V_{CE\text{sat}} - V_{BE14}$$

which is about 1 V below V_{CC}. The minimum output voltage (that is, maximum negative amplitude) is limited by the saturation of Q_{17}. Neglecting the voltage drop across R_8 we obtain

$$v_{O\min} = -V_{CC} + V_{CE\text{sat}} + V_{EB23} + V_{EB20}$$

which is about 1.5 V above $-V_{CC}$.

Small-Signal Model

We shall now carry out a small-signal analysis of the output stage with the objective of determining the values of the parameters of the equivalent circuit model shown in Fig. 13.22. The model is shown fed by v_{o2}, which is the open-circuit output voltage of the second stage. From Fig. 13.14 v_{o2} is given by

$$v_{o2} = -G_{m2}R_{o2}v_{i2}$$

and G_{m2} and R_{o2} were previously determined as $G_{m2} = 6.53$ mA/V and $R_{o2} = 81$ kΩ. Resistance R_{i3} is the input resistance of the output stage determined with the amplifier

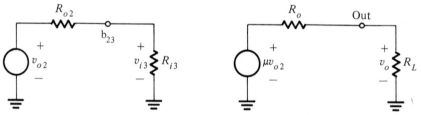

Fig. 13.22   Model for the 741 output stage.

loaded with R_L. Although the effect of loading an amplifier stage is negligible in the input and second stages, this is not the case in general in an output stage. Defining R_{i3} in this manner enables correct evaluation of the voltage gain of the second stage, A_2, as

$$A_2 \equiv \frac{v_{i3}}{v_{i2}} = -G_{m2}R_{o2}\frac{R_{i3}}{R_{i3} + R_{o2}} \tag{13.13}$$

To determine R_{i3} assume that one of the two output transistors—say, Q_{20}—is conducting a current of, say, 5 mA. It follows that the input resistance looking into the base of Q_{20} is approximately $\beta_{20}R_L$. Assuming $\beta_{20} = 50$ and for $R_L = 2$ kΩ the input resistance of Q_{20} is 100 kΩ. This resistance appears in parallel with the series combination of the output resistance of Q_{13A} ($r_{o13A} \cong 2.5$ MΩ) and the resistance of the Q_{18}-Q_{19} network. This series combination will be much greater than 100 kΩ, and we may assume that the total resistance in the emitter of Q_{23} is approximately 100 kΩ. Thus the input resistance R_{i3} is given by

$$R_{i3} \cong \beta_{23} \times 100 \text{ kΩ}$$

which for $\beta_{23} = 50$ is $R_{i3} \cong 5$ MΩ. Since $R_{o2} = 81$ kΩ, we see that $R_{i3} \gg R_{o2}$, and the value of R_{i3} will have little effect on the performance of the op amp. We can use the value obtained for R_{i3} to determine the gain of the second stage in Eq. (13.13) as $A_2 = -520.5$. The value of A_2 will be needed in Section 13.7 in connection with frequency-response analysis.

Continuing with the determination of the equivalent circuit model parameters, we note from Fig. 13.22 that μ is the *open-circuit voltage gain* of the output stage,

$$\mu = \frac{v_o}{v_{o2}}\bigg|_{R_L=\infty}$$

With $R_L = \infty$ the gain of the emitter-follower output transistor (Q_{14} or Q_{20}) will be nearly unity. Also, with $R_L = \infty$ the resistance in the emitter of Q_{23} will be very large. This means that the gain of Q_{23} will be nearly unity and the input resistance of Q_{23} will be very large. We thus conclude that $\mu \cong 1$.

Next we shall find the value of the output resistance of the op amp, R_o. For this purpose refer to the circuit shown in Fig. 13.23. In accordance with the definition of R_o,

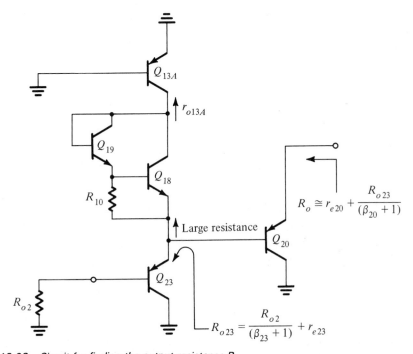

Fig. 13.23   Circuit for finding the output resistance R_o.

the input source feeding the output stage is grounded but its resistance (which is the output resistance of the second stage, R_{o2}) is included. We have assumed that the output voltage v_O is negative, and thus Q_{20} is conducting most of the current; transistor Q_{14} has therefore been eliminated. The exact value of the output resistance will of course depend on which transistor (Q_{14} or Q_{20}) is conducting and on the value of load current. Nevertheless, we wish to find an estimate of R_o.

As indicated in Fig. 13.23 the resistance seen looking into the emitter of Q_{23} is

$$R_{o23} = \frac{R_{o2}}{\beta_{23} + 1} + r_{e23}$$

Substituting $R_{o2} = 81$ kΩ, $\beta_{23} = 50$, and $r_{e23} = 25/0.18 = 139$ Ω yields $R_{o23} = 1.73$ kΩ. This resistance appears in parallel with the series combination of r_{o13A} and the resistance of the Q_{18}-Q_{19} network. Since r_{o13A} alone (2.5 MΩ) is much larger than R_{o23}, the effective resistance between the base of Q_{20} and ground is approximately equal to R_{o23}. Now we can find the output resistance R_o as

$$R_o = \frac{R_{o23}}{\beta_{20} + 1} + r_{e20}$$

For $\beta_{20} = 50$ the first component of R_o is 34 Ω. The second component depends critically on the value of output current. For an output current of 5 mA, r_{e20} is 5 Ω and R_o is 39 Ω. To this value we must add the resistance R_7 (27 Ω) (see Fig. 13.1), which is included for short-circuit protection. The output resistance of the 741 is specified to be typically 75 Ω.

EXERCISES

13.12 Using a simple (r_x, g_m) model for each of the two transistors Q_{18} and Q_{19} in Fig. E13.12, find the small signal resistance between A and A'.
(*Note:* From Table 13.1 $I_{C18} = 165$ μA and $I_{C19} \cong 16$ μA.)
Ans. 165 Ω

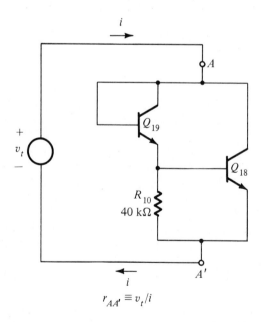

Fig. E13.12

13.13 Figure E13.13 shows the circuit for determining the op-amp output resistance when v_O is positive and Q_{14} is conducting most of the current. Using the resistance of the Q_{18}-Q_{19} network calculated in Exercise 13.12 and neglecting the output resistance of Q_{13A}, find R_o.
Ans. 14.6 Ω

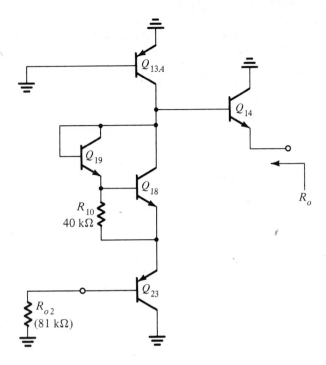

Fig. E13.13

Output Short-Circuit Protection

If the op-amp output terminal is short-circuited to one of the power supplies, one of the two output transistors could conduct a large amount of current. Such a large current can result in sufficient heating to cause burnout of the IC. To guard against this possibility, the 741 op amp is equipped with a special circuit for short-circuit protection. The function of this circuit is to limit the current in the output transistors in the event of a short circuit.

Refer to Fig. 13.1. Resistance R_6 together with transistor Q_{15} limits the current that would flow out of Q_{14} in the event of a short circuit. Specifically, if the current in the emitter of Q_{14} exceeds about 20 mA, the voltage drop across R_6 exceeds 540 mV, which turns Q_{15} on. As Q_{15} turns on, its collector robs some of the current supplied by Q_{13A}, thus reducing the base current of Q_{14}. This mechanism thus limits the maximum current that the op amp can source (that is, supply from the output terminal in the outward direction) to about 20 mA.

Limiting of the maximum current that the op amp can sink, and hence the current through Q_{20}, is done by a mechanism similar to the one discussed above. The relevant circuit is composed of R_7, Q_{21}, Q_{24}, and Q_{22}.

13.7   GAIN AND FREQUENCY RESPONSE OF THE 741

In this section we shall evaluate the overall small-signal voltage gain of the 741 op amp. We shall then consider the op amp's frequency response and its slew-rate limitation.

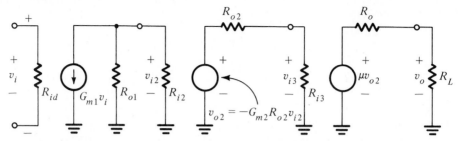

Fig. 13.24   *Cascading the small-signal equivalent circuits of the individual stages for the evaluation of the overall voltage gain.*

Small-Signal Gain

The overall small-signal gain can be easily found from the cascade of the equivalent circuits derived in the previous sections for the three op-amp stages. This cascade is shown in Fig. 13.24, loaded with $R_L = 2$ kΩ, which is the typical value used in measuring and specifying the 741 data. The overall gain can be expressed as

$$\frac{v_o}{v_i} = \frac{v_{i2}}{v_i}\frac{v_{o2}}{v_{i2}}\frac{v_o}{v_{o2}}$$

$$= -G_{m1}(R_{o1}\|R_{i2})(-G_{m2}R_{o2})\mu\,\frac{R_L}{R_L + R_o}$$

Using the values found in previous sections yields

$$\frac{v_o}{v_i} = -475.9 \times (-528.9) \times 0.97 = 243{,}545 \text{ V/V}$$

$$= 107.7 \text{ dB}$$

Frequency Response

The 741 is an internally compensated op amp. It employs the Miller compensation technique, studied in Section 12.11, to introduce a dominant low-frequency pole. Specifically, a 30-pF capacitor (C_C) is connected in the negative-feedback path of the second stage. An approximate estimate of the frequency of the dominant pole can be obtained as follows:

Using Miller's theorem (Section 2.6) the effective capacitance due to C_C between the base of Q_{16} and ground is (see Fig. 13.1)

$$C_i = C_C(1 + |A_2|)$$

where A_2 is the second-stage gain. Use of the value calculated for A_2 in Section 13.6, $A_2 = -520.5$, results in $C_i = 15{,}645$ pF. Since this capacitance is quite large, we shall neglect all other capacitances between the base of Q_{16} and signal ground. The total resistance between this node and ground is

$$R_t = (R_{o1}\|R_{i2})$$

$$= (6.7 \text{ M}\Omega\|4 \text{ M}\Omega) = 2.5 \text{ M}\Omega$$

Thus the dominant pole has a frequency f_P given by

$$f_P = \frac{1}{2\pi C_i R_t} = 4.1 \text{ Hz}$$

It should be noted that this approach is equivalent to using the approximate formula in Eq. (12.28).

As discussed in Section 12.11, Miller compensation produces an additional advantageous effect, pole splitting. As a result, the other poles of the circuit are moved to very high frequencies. This has been confirmed by computer-aided analysis (Ref. 13.1).

Assuming that all nondominant poles are at high frequencies, the calculated values give rise to the Bode plot shown in Fig. 13.25. The unity-gain bandwidth f_t can be calculated from

$$f_t = A_0 f_{3dB}$$

Thus

$$f_t = 243,545 \times 4.1 \cong 1 \text{ MHz}$$

Although this Bode plot implies that the phase shift at f_t is $-90°$ and thus that the phase margin is $90°$, in practice a phase margin of about $80°$ is obtained. The excess phase shift (about $10°$) is due to the nondominant poles. This phase margin is sufficient to provide stable operation of closed-loop amplifiers with any value of feedback factor β. This convenience of use of the internally compensated 741 is achieved at the expense of a great reduction in open-loop gain and hence in the amount of negative feedback. In other words, if one requires a closed-loop amplifier with a gain of 1,000, then the 741 is overcompensated for such an application and one would be much better off designing her or his own compensation (assuming, of course, the availability of an op amp that is not internally compensated).

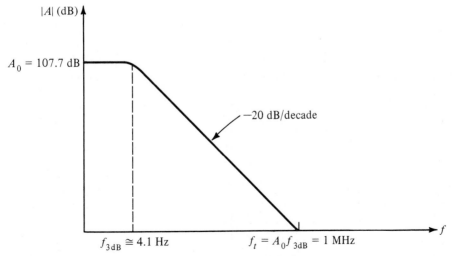

Fig. 13.25   Bode plot for the 741 gain, neglecting nondominant poles.

A Simplified Model

Figure 13.26 shows a simplified model of the 741 op amp in which the high-gain second stage with its feedback capacitance C_C is modeled by an ideal integrator. In this model the gain of the second stage is assumed sufficiently large so that a virtual ground appears at its input. For this reason the output resistance of the input stage and the

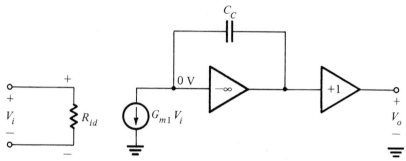

Fig. 13.26   *A simple model for the 741 based on modeling the second stage as an integrator.*

input resistance of the second stage have been omitted. Furthermore, the output stage is assumed to be an ideal unity-gain follower.

Analysis of the model in Fig. 13.26 gives

$$A(s) \equiv \frac{V_o(s)}{V_i(s)} = \frac{G_{m1}}{sC_C} \tag{13.14}$$

Thus

$$A(j\omega) = \frac{G_{m1}}{j\omega C_C} \tag{13.15}$$

and the magnitude of gain becomes unity at $\omega = \omega_t$, where

$$\omega_t = \frac{G_{m1}}{C_C} \tag{13.16}$$

Substituting $G_{m1} = 1/5.26$ mA/V and $C_C = 30$ pF yields

$$f_t = \frac{\omega_t}{2\pi} \cong 1 \text{ MHz}$$

which is equal to the value calculated before. It should be pointed out, however, that this model is valid at frequencies $f \gg f_{3dB}$. At such frequencies the gain falls off with a slope of -20 dB/decade, just like an integrator.

Slew Rate

The slew-rate limitation of op amps is discussed in Chapter 3. We shall now point out the origin of the slewing phenomenon.

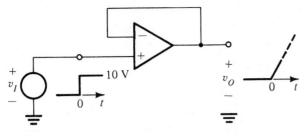

Fig. 13.27 A unity-gain follower with a large step input. Since the output voltage cannot change immediately, a large differential voltage appears between the op-amp input terminals.

Consider the unity-gain follower of Fig. 13.27 with a step of, say, 10 V applied at the input. Because of amplifier dynamics, its output will not change in zero time. Thus immediately after the input is applied almost the entire value of the step will appear as a differential signal between the two input terminals. This large input voltage causes the input stage to be *overdriven*, and its small-signal model no longer applies. Rather, half the stage cuts off and the other half conducts all the current. Specifically, reference to Fig. 13.1 shows that a large differential input voltage causes Q_1 and Q_3 to conduct all the available bias current ($2I$) while Q_2 and Q_4 will be cut off. The current mirror Q_5, Q_6, and Q_7 will still function, and Q_6 will produce a collector current of $2I$.

Using the above observations and modeling the second stage as an ideal integrator results in the model of Fig. 13.28. From this circuit we see that the output voltage is a ramp with a slope of $2I/C_C$:

$$v_O(t) = \frac{2I}{C_C} t$$

Thus the slew rate SR is given by

$$SR = \frac{2I}{C_C} \tag{13.17}$$

For the 741, $I = 9.5\ \mu A$ and $C_C = 30$ pF, resulting in SR $= 0.63$ V/μs.

It should be pointed out that this is a rather simplified model of the slewing process. More detail can be found in reference 13.1.

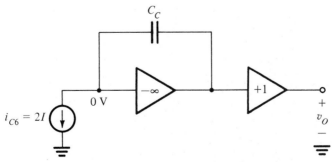

Fig. 13.28 Model for the 741 op amp when a large differential signal is applied.

EXERCISE
13.14   Use the value of the slew rate calculated above and find the full-power bandwidth f_M of the 741 op amp. Assume that the maximum output is ± 10 V.
Ans.   10 kHz

Relationship Between f_t and SR

A simple relationship exists between the unity-gain bandwidth f_t and the slew rate SR. This relationship is obtained from Eqs. (13.16) and (13.17) together with

$$G_{m1} = 2\frac{1}{4r_e}$$

where r_e is the emitter resistance of each of Q_1 through Q_4. Thus

$$r_e = \frac{V_T}{I}$$

and

$$G_{m1} = \frac{I}{2V_T} \tag{13.18}$$

Substituting in Eq. (13.16) results in

$$\omega_t = \frac{I}{2C_C V_T} \tag{13.19}$$

Substituting for I/C_C from Eq. (13.17) gives

$$\omega_t = \frac{SR}{4V_T} \tag{13.20}$$

which can be expressed in the alternative form

$$SR = 4\omega_t V_T \tag{13.21}$$

As a check, for the 741 we have

$$SR = 4 \times 2\pi \times 10^6 \times 25 \times 10^{-3} = 0.63 \text{ V/}\mu s$$

which is the result obtained previously.

EXERCISE
13.15   If a resistance R_E is included in each of the emitter leads of Q_3 and Q_4 show that $SR = 4\omega_t(V_T + IR_E/2)$. Hence find the value of R_E that would double the 741 slew rate.
Ans.   5.26 kΩ
(*Note:*   This is a viable technique for increasing slew rate. It is referred to as the G_m reduction method.)

13.8   CONCLUDING REMARKS

Near the beginning of the text (Chapter 3) we studied the terminal characteristics of the op-amp and some of its basic applications. More op-amp applications were studied in Chapter 5 and still two more important applications will be studied in Chapter 14. In our previous work, the op amp was treated as a black box. Although such an approach is quite useful, it has limitations; we believe that the material studied in this chapter should provide the reader with greater appreciation of the nonideal performance of op amps and thus enhance his or her abilities in using the op amp as a circuit element.

An equally important objective of this chapter is to introduce the reader to some of the circuit design techniques employed in analog ICs. Due to space limitations we focussed on one example only: the popular 741 op-amp. Much more about analog ICs can be found in the references.

The analog IC that we discussed uses BJTs exclusively. JFETs and MOSFETs are currently used in the input stages in a number of popular op-amp circuits. Furthermore, at the time of this writing (1981) MOS technology is making considerable inroads into the area of analog ICs. This is a very significant development because it has opened the door to the design of very-large-scale integrated (VLSI) analog circuits. Such chips actually incorporate both analog and digital circuits. We shall consider one application of analog MOS technology in Chapter 14. Digital MOS circuits were introduced in Chapter 8 and will be studied in greater detail in Chapters 15 and 16. Finally, the reader who is interested in pursuing the subject of analog MOS ICs should consult reference 13.9.

FILTERS AND OSCILLATORS

14

In this chapter we shall study two different but related functional circuit blocks: the electronic filter and the sinusoidal oscillator. Both employ feedback to produce complex conjugate poles, which in the case of the filter lie in the left half of the s plane and in the case of the oscillator lie on the $j\omega$ axis.

A filter is a two-port network whose purpose is to pass signals having a specific frequency spectrum and stop signals whose frequency spectra differ from this specific one. Filters therefore are *frequency-selective networks*. They are important building blocks of many electronic systems. Although their most extensive use is in communications systems, almost every electronic instrument contains some sort of a filter.

Different technologies exist for realizing filters. The oldest technology uses inductors and capacitors, and the resulting filters are called *passive LC filters*. The problem with LC filters is that in low-frequency applications (dc to 100 kHz) the required inductors are bulky in size, and their characteristics are quite nonideal. Furthermore, such inductors are impossible to manufacture in monolithic form and are incompatible with any of the modern techniques for assembling electronic systems. Of the various possible types of *inductorless filters,* we shall study *active-RC filters* and *switched-capacitor filters*.

We have already encountered first-order RC filters on a number of occasions. The selectivity of RC filters is limited because their poles are restricted to lie on the negative real axis in the s plane. As we shall see, the idea behind active-RC filters is to employ an active device to move the circuit poles from the negative real axis to complex locations. In this way more selective filter responses are obtained. From our study of feedback in Chapter 12 we know that this pole movement can be accomplished by connecting the RC network in the feedback path of the active element, which is usually an op amp. We shall study the use of this technique to design second-order filters of various types. Filters of higher order can be realized by cascading second-order sections.

The switched-capacitor technique enables us to design fully integrated monolithic filters. We shall provide an introduction to this new and exciting filter technology.

The second part of the chapter is concerned with the design of *linear sinusoidal oscillators.* The word "linear" is used to distinguish this type of sinusoidal generator from the ones discussed in Chapter 5, which create sine waves by nonlinear shaping of triangular waveforms. Because of limitations of space, we shall emphasize oscillator circuits that employ op amps, resistors, and capacitors.

14.1   SECOND-ORDER FILTER FUNCTIONS

The general second-order, biquadratic, or simply *biquad,* transfer function can be written in the form

$$T(s) = \frac{n_2 s^2 + n_1 s + n_0}{s^2 + s(\omega_0/Q) + \omega_0^2} \tag{14.1}$$

Here ω_0 and Q are parameters that determine the location of the poles. Figure 14.1 shows a pair of complex conjugate poles and illustrates the definition of ω_0 and Q. As

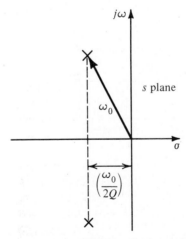

Fig. 14.1   *Definition of the parameters ω_0 and Q of a pair of complex conjugate poles.*

indicated, the *pole frequency* ω_0 is the radial distance of the poles from the origin, and the pole Q factor, or simply *pole-Q*, determines how far the poles are from the $j\omega$ axis. It can be easily shown that Q less than 0.5 means that the poles are on the negative real axis, $Q = 0.5$ means that the poles are coincident and real, $Q > 0.5$ means that the poles are complex conjugate, and $Q = \infty$ means the poles are on the $j\omega$ axis.

The numerator coefficients, n_0, n_1, and n_2, determine the transmission zeros and thus the shape of the magnitude response and the type of filter (low pass, high pass, etc.). Special cases of interest are as follows:

1. *Low-Pass (LP) Filter.* In this case $n_1 = n_2 = 0$ and the transfer function becomes

$$T(s) = \frac{n_0}{s^2 + s(\omega_0/Q) + \omega_0^2} \tag{14.2}$$

Thus the two transmission zeros are at $s = \infty$. Figure 14.2a shows the magnitude response, $20 \log |T(j\omega)|$, for a second-order low-pass filter with unity dc gain ($n_0 = \omega_0^2$). Note that the response exhibits a peak. It can be shown that this peak occurs if Q is greater than 0.707. The case $Q = 0.707$ results in the *maximally flat* magnitude response. Also note that at high frequencies the gain falls off with an asymptotic slope of -40 dB/decade.

2. *High-Pass (HP) Filter.* In this case $n_0 = n_1 = 0$ and the transfer function becomes

$$T(s) = \frac{n_2 s^2}{s^2 + s(\omega_0/Q) + \omega_0^2} \tag{14.3}$$

Thus the two transmission zeros are at $s = 0$ (dc). Figure 14.2b shows the magnitude response for a second-order high-pass filter with unity high-frequency gain ($n_2 = 1$). The effect of the value of Q on the response shape is similar to that in

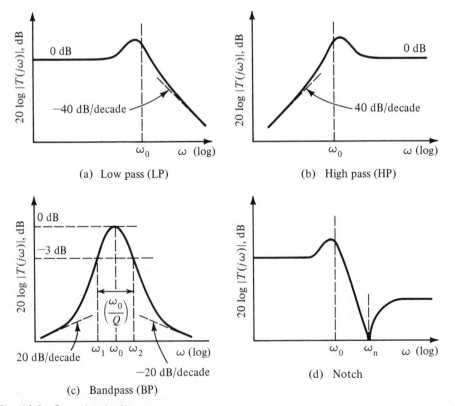

(a) Low pass (LP)

(b) High pass (HP)

(c) Bandpass (BP)

(d) Notch

Fig. 14.2 *Second-order filter responses.*

the low-pass case. Note that at low frequencies the transmission falls off with an asymptotic slope of 40 dB/decade.

3. *Bandpass (BP) Filter.* In this case $n_0 = n_2 = 0$ and the transfer function becomes

$$T(s) = \frac{n_1 s}{s^2 + s(\omega_0/Q) + \omega_0^2} \qquad (14.4)$$

Thus there is one transmission zero at $s = 0$ (dc) and one at $s = \infty$. Figure 14.2c shows the magnitude response of a second-order bandpass filter with unity center-frequency gain ($n_1 = \omega_0/Q$). Note that the response peaks at $\omega = \omega_0$; thus ω_0 is also called the *center frequency*. The pole-Q determines how narrow (selective) the bandpass filter is. Specifically, the two frequencies ω_1 and ω_2 at which the transmission drops by 3 dB below the value at ω_0 are separated by ω_0/Q. This distance is called the *3-dB bandwidth*. As Q is increased, the bandwidth decreases and the filter becomes more selective. Finally, note that the gain falls off at very low and high frequencies with a slope of 20 dB/decade.

4. *Notch Filter.* In this case $n_1 = 0$ and the transfer function can be written in the form

$$T(s) = n_2 \frac{s^2 + \omega_n^2}{s^2 + s(\omega_0/Q) + \omega_0^2} \qquad (14.5)$$

Thus the zeros are at $s = \pm j\omega_n$, and the transmission becomes zero at $\omega = \omega_n$. Figure 14.2d shows the response of a second-order notch filter for which $\omega_n > \omega_0$. This type of response is referred to as a *low-pass notch* (LPN).

5. *All-Pass Filter.* In this case the transfer function is given by

$$T(s) = n_2 \frac{s^2 - s(\omega_0/Q) + \omega_0^2}{s^2 + s(\omega_0/Q) + \omega_0^2} \qquad (14.6)$$

Thus the zeros are in the right half-plane at mirror-image locations of the poles. The magnitude response is constant independent of frequency; hence the name all-pass. Such networks are used as phase shifters or to modify the phase response of a system in a desired manner.

We show in Fig. 14.3 RLC circuit realizations of the five special second-order transfer functions discussed above. Note that in all cases when V_i is reduced to zero (that is, when the input voltage source is short-circuited) the circuit reduces to a parallel LCR tuned circuit. Hence all circuits shown have the same poles. The reader is urged as an exercise to derive the transfer functions for the networks of Fig. 14.3.

In the following sections we shall study circuits that use op amps and RC networks to realize second-order transfer functions. Such circuits are usually referred to as biquadratic circuits or simply biquads.

14.2 SINGLE-AMPLIFIER BIQUADS

The simplest realization of second-order filter functions utilizes a single op amp and is known as a *single-amplifier biquad* (SAB). In this section we shall study the synthesis of SABs following a two step approach:

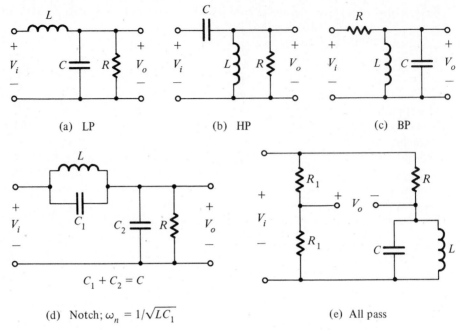

(a) LP   (b) HP   (c) BP

(d) Notch; $\omega_n = 1/\sqrt{LC_1}$ (e) All pass

Fig. 14.3   Passive realizations of special second-order filter functions. For all circuits $\omega_o = 1/\sqrt{LC}$ **and** $Q = \omega_o CR$.

1.   Synthesis of a feedback loop that realizes a pair of complex conjugate poles characterized by a frequency ω_0 and a Q factor Q.
2.   Injecting the input signal in a way that realizes the desired transmission zeros.

Synthesis of the Feedback Loop

Consider the circuit shown in Fig. 14.4a, which consists of a two-port RC network n placed in the negative-feedback path of an op amp. We shall assume that, except for having a finite gain A, the op amp is ideal. We shall denote by $t(s)$ the open-circuit voltage transfer function of the RC network n, where the definition of $t(s)$ is illustrated in Fig. 14.4b. The transfer function $t(s)$ can in general be written as the ratio of two polynomials $N(s)$ and $D(s)$:

$$t(s) = \frac{N(s)}{D(s)}$$

The roots of $N(s)$ are the transmission zeros of the RC network, and the roots of $D(s)$ are its poles. Study of network theory shows that while the poles of an RC network are restricted to lie on the negative real axis, the zeros can in general lie anywhere in the s plane.

The loop gain $L(s)$ of the feedback circuit in Fig. 14.4a can be determined using

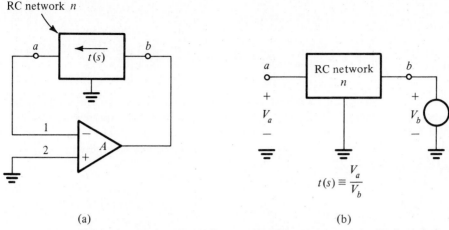

(a)   (b)

Fig. 14.4 (a) Feedback loop obtained by placing two-port RC network n in the feedback path of an op amp. (b) Definition of the open-circuit transfer function t(s) of the RC network.

the method of Section 12.7. It is simply the product of the op amp gain A and the transfer function $t(s)$,

$$L(s) = At(s) = \frac{AN(s)}{D(s)} \tag{14.7}$$

Substituting for $L(s)$ into the characteristic equation

$$1 + L(s) = 0 \tag{14.8}$$

results in the poles s_P of the closed-loop circuit obtained as solutions to the equation

$$t(s_P) = -\frac{1}{A} \tag{14.9}$$

In the ideal case $A = \infty$ and the poles are obtained from

$$N(s_P) = 0 \tag{14.10}$$

That is, the poles are identical to the zeros of the RC network.

Since our objective is to realize a pair of complex conjugate poles, we should select an RC network that has complex conjugate transmission zeros. The simplest such networks are the bridged-T networks shown in Fig. 14.5 together with their transfer function $t(s)$ from b to a with a open-circuited. As an example, consider the circuit generated by placing the bridged-T of Fig. 14.5a in the negative-feedback path of an op amp, as shown in Fig. 14.6. The pole polynomial of the active-filter circuit will be equal to the numerator polynomial of the bridged-T network; thus

$$s^2 + s\frac{\omega_0}{Q} + \omega_0^2 = s^2 + s\left(\frac{1}{C_1} + \frac{1}{C_2}\right)\frac{1}{R_3} + \frac{1}{C_1 C_2 R_3 R_4}$$

which enables us to obtain ω_0 and Q as

$$\omega_0 = \frac{1}{\sqrt{C_1 C_2 R_3 R_4}} \tag{14.11}$$

$$Q = \left[\frac{\sqrt{C_1 C_2 R_3 R_4}}{R_3} \left(\frac{1}{C_1} + \frac{1}{C_2} \right) \right]^{-1} \tag{14.12}$$

If we are designing this circuit, ω_0 and Q are given and Eqs. (14.11) and (14.12) can be used to determine C_1, C_2, R_3, and R_4. It follows that there are two degrees of freedom. Let us exhaust one of these by selecting $C_1 = C_2 = C$. Let us also denote $R_3 = R$ and $R_4 = R/m$. By substituting in Eqs. (14.11) and (14.12) and with some manipulations we obtain

$$m = 4Q^2 \tag{14.13}$$

$$CR = \frac{2Q}{\omega_0} \tag{14.14}$$

Thus if we are given the value of Q, Eq. (14.13) can be used to determine the ratio of the two resistances R_3 and R_4. Then the given values of ω_0 and Q can be substituted in Eq. (14.14) to determine the time constant CR. There remains one degree of freedom—the value of C or R can be arbitrarily chosen. In an actual design this value, which sets

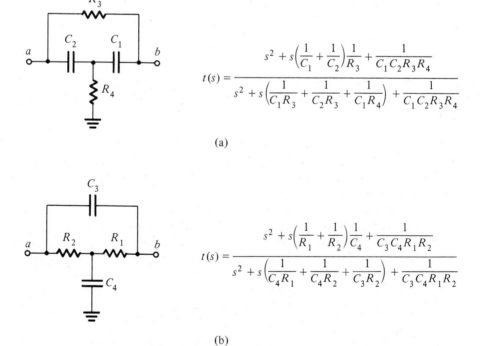

$$t(s) = \frac{s^2 + s\left(\frac{1}{C_1} + \frac{1}{C_2} \right)\frac{1}{R_3} + \frac{1}{C_1 C_2 R_3 R_4}}{s^2 + s\left(\frac{1}{C_1 R_3} + \frac{1}{C_2 R_3} + \frac{1}{C_1 R_4} \right) + \frac{1}{C_1 C_2 R_3 R_4}}$$

(a)

$$t(s) = \frac{s^2 + s\left(\frac{1}{R_1} + \frac{1}{R_2} \right)\frac{1}{C_4} + \frac{1}{C_3 C_4 R_1 R_2}}{s^2 + s\left(\frac{1}{C_4 R_1} + \frac{1}{C_4 R_2} + \frac{1}{C_3 R_2} \right) + \frac{1}{C_3 C_4 R_1 R_2}}$$

(b)

Fig. 14.5   Two RC networks (called bridged-T networks) which have complex conjugate transmission zeros. The transfer functions given are from b to a with a open-circuited.

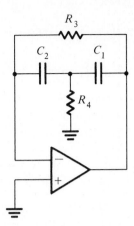

Fig. 14.6   An active filter feedback loop generated using the bridged-T network of Fig. 14.5a.

the *impedance level* of the circuit, should be chosen so that the resulting component values are practical.

EXERCISES

14.1   Design the circuit of Fig. 14.6 to realize a pair of poles with $\omega_0 = 10^4$ rad/s and $Q = 1$. Select $C_1 = C_2 = 1$ nF.
   Ans. $R_3 = 200$ kΩ; $R_4 = 50$ kΩ

14.2   For the circuit designed in Exercise 14.1 find the location of the poles of the RC network in the feedback loop.
   Ans. -0.382×10^4 and -2.618×10^4 rad/s

Injecting the Input Signal

Having synthesized a feedback loop that realizes a given pair of poles, we now consider connecting the input signal source to the circuit. We wish to do this, of course, without altering the poles.

Since for the purpose of finding the poles of a circuit an ideal voltage source is equivalent to a short circuit, it follows that any circuit node that is connected to ground can instead be connected to the input voltage source without causing the poles to change. Thus the method of injecting the input signal into the feedback loop is to simply disconnect a component (or a number of components) which is (are) connected to ground and connect it (them) to the input source. Depending on the component(s) through which the input signal is injected, different transmission zeros are obtained.

As an example, consider the feedback loop of Fig. 14.6. Here we have two grounded nodes (one terminal of R_4 and the positive input terminal of the op amp) that can serve for injecting the input signal. Figure 14.7 shows the circuit with the input signal injected through part of the resistance R_4. Note that the two resistances R_4/α and $R_4/(1 - \alpha)$ have a parallel equivalent of R_4. It can be shown that this circuit realizes the second-order bandpass function and that the value of α ($0 < \alpha \leq 1$) can be used to obtain the desired center-frequency gain.

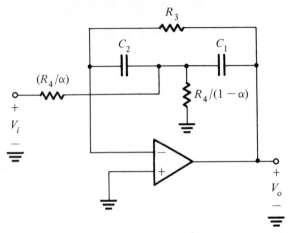

Fig. 14.7 *The loop of Fig. 14.6 with the input signal injected through part of resistance R_4. This circuit realizes the bandpass function.*

EXERCISE

14.3 Assume the op amp to be ideal and analyze the circuit in Fig. 14.7 to obtain its transfer function. Thus show that the circuit realizes the bandpass function and that its poles are identical to the zeros of $t(s)$ in Fig. 14.5a. Using the component values of Exercise 14.1 find the values of R_4/α and $R_4/(1 - \alpha)$ to obtain unity center-frequency gain.
Ans. 100 kΩ; 100 kΩ.

The versatility of a feedback loop in generating filters with different transmission zeros increases with the number of grounded nodes. In addition, the complementary transformation can be applied to a loop to generate an equivalent feedback loop having the same poles but the ability to realize different transmission zeros. This transformation will be studied next.

Generation of Equivalent Feedback Loops

The *complementary transformation* of feedback loops is based on the property of linear networks illustrated in Fig. 14.8 for the case of the two-port RC network *n*. (This property is briefly discussed and applied in Chapter 3.) Application of the complementary transformation to a feedback loop to generate an equivalent feedback loop is a two-step process:

1. Nodes of the feedback network and any of the op amp inputs that are connected to ground should be disconnected from ground and connected to the op-amp output. Conversely, those nodes that were connected to the op-amp output should be now connected to ground. That is, we simply interchange the op-amp output terminal with ground.
2. The two input terminals of the op amp should be interchanged.

The feedback loop generated by this transformation has the same characteristic equation and hence the same poles as the original loop.

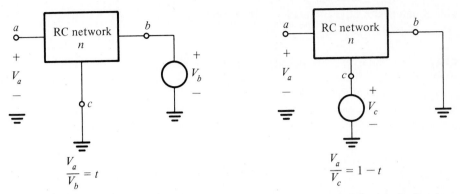

Fig. 14.8 Interchanging input and ground, resulting in the complement of the transfer function.

To illustrate, we show in Fig. 14.9a the feedback loop formed by connecting a two-port RC network in the negative-feedback path of an op amp. Application of the complementary transformation to this loop results in the feedback loop of Fig. 14.9b. Note that in the latter loop the op amp is used in the unity-gain follower configuration. We shall now show that the two loops of Fig. 14.9 are equivalent.

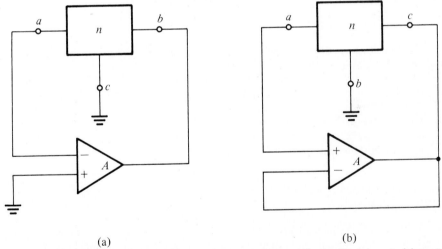

(a)   (b)

Fig. 14.9 Application of the complementary transformation to the feedback loop in (a) results in the equivalent (same poles) loop in (b).

If the op amp has an open-loop gain A, the follower in the circuit of Fig. 14.9b will have a gain of $A/(A + 1)$. This together with the fact that the transfer function of network n from c to a is $1 - t$ (see Fig. 14.8) enables us to write for the circuit in Fig. 14.9b the characteristic equation

$$1 - \frac{A}{A + 1}(1 - t) = 0$$

This equation can be manipulated to the form

$$1 + At = 0$$

which is the characteristic equation of the loop in Fig. 14.9a. As an example, consider the application of the complementary transformation to the feedback loop of Fig. 14.6: the feedback loop of Fig. 14.10a results. Injecting the input signal through C_1 results in the circuit in Fig. 14.10b, which can be shown to realize a second-order high-pass function.

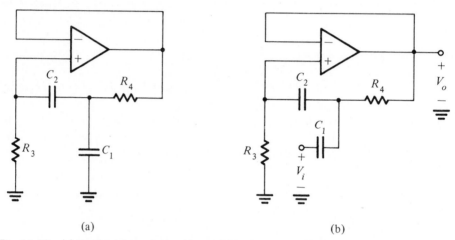

(a)   (b)

Fig. 14.10   (a) Feedback loop obtained by applying the complementary transformation to the loop in Fig. 14.6. (b) Injecting the input signal through C_1 realizes the high-pass function.

EXERCISE

14.4   Use the bridged-T network of Fig. 14.5b in the negative-feedback path of an op amp to realize a pair of complex poles. Then apply the complementary transformation to this circuit and show that the resulting circuit can be used to realize the second-order low-pass function.

14.3   SENSITIVITY

Because of the tolerances in component values and because of the finite op-amp gain the response of the actual assembled filter will deviate from the ideal response. As a means for predicting such deviations, the filter designer employs the concept of *sensitivity*. Specifically, for second-order filters one is normally interested in finding how *sensitive* their poles are relative to variations (both initial tolerances and future changes) in RC component values and amplifier gain. These sensitivities can be quantified using the *classical sensitivity function* S_x^y, defined

$$S_x^y \equiv \lim_{\Delta x \to 0} \frac{\Delta y/y}{\Delta x/x} \tag{14.15}$$

Thus

$$S_x^y = \frac{\partial y}{\partial x}\frac{x}{y} \tag{14.16}$$

Here x denotes the value of a component (a resistor, a capacitor, or amplifier gain) and y denotes an output parameter of interest (say, ω_0 or Q). For small changes

$$S_x^y \cong \frac{\Delta y/y}{\Delta x/x} \tag{14.17}$$

Thus we can use the value of S_x^y to determine the per-unit change in y due to a given per-unit change in x. For instance, if the sensitivity of Q relative to a particular resistance R_1 is 5, then a 1% increase in R_1 results in a 5% increase in the value of Q.

Example 14.1
For the feedback loop of Fig. 14.6 find the sensitivities of ω_0 and Q relative to all the passive components and the op-amp gain. Evaluate these sensitivities for the design considered in the previous section for which $C_1 = C_2$.

Solution
To find the sensitivities with respect to the passive components, called *passive sensitivities*, we assume that the op-amp gain is infinite. In this case ω_0 and Q are given by Eqs. (14.11) and (14.12). Thus for ω_0 we have

$$\omega_0 = \frac{1}{\sqrt{C_1 C_2 R_3 R_4}}$$

which can be used together with the sensitivity definition of Eq. (14.16) to obtain

$$S_{C_1}^{\omega_0} = S_{C_2}^{\omega_0} = S_{R_3}^{\omega_0} = S_{R_4}^{\omega_0} = -\tfrac{1}{2}$$

For Q we have

$$Q = \left[\sqrt{C_1 C_2 R_3 R_4} \left(\frac{1}{C_1} + \frac{1}{C_2} \right) \frac{1}{R_3} \right]^{-1}$$

on which we apply the sensitivity definition to obtain

$$S_{C_1}^Q = \frac{1}{2} \left(\sqrt{\frac{C_2}{C_1}} - \sqrt{\frac{C_1}{C_2}} \right) \left(\sqrt{\frac{C_2}{C_1}} + \sqrt{\frac{C_1}{C_2}} \right)^{-1}$$

For the design with $C_1 = C_2$ we see that $S_{C_1}^Q = 0$. Similarly, we can show that

$$S_{C_2}^Q = 0, \qquad S_{R_3}^Q = \tfrac{1}{2}, \qquad S_{R_4}^Q = -\tfrac{1}{2}$$

It is important to remember that the sensitivity expression should be derived *before* substituting values corresponding to a particular design.

Next we consider the sensitivities relative to the amplifier gain. Assuming the op amp to have a finite gain A, the characteristic equation for the loop becomes

$$1 + At(s) = 0 \tag{14.18}$$

where $t(s)$ is given in Fig. 14.5a. To simplify matters we can substitute for the passive components by their design values. This causes no errors in evaluating sensitivities, since we are now finding the sensitivity with respect to the amplifier gain. Using the design values previously obtained; namely, $C_1 = C_2 = C$, $R_3 = R$, $R_4 = R/4Q^2$, and $CR = 2Q/\omega_0$; gives

$$t(s) = \frac{s^2 + s(\omega_0/Q) + \omega_0^2}{s^2 + s(\omega_0/Q)(2Q^2 + 1) + \omega_0^2} \tag{14.19}$$

where ω_0 and Q denote the nominal or design values of the pole frequency and Q factor. The actual values are obtained by substituting for $t(s)$ in Eq. (14.18):

$$s^2 + s\frac{\omega_0}{Q}(2Q^2 + 1) + \omega_0^2 + A\left(s^2 + s\frac{\omega_0}{Q} + \omega_0^2\right) = 0$$

Assuming the gain A to be real and dividing both sides by $A + 1$ gives

$$s^2 + s\frac{\omega_0}{Q}\left(1 + \frac{2Q^2}{A + 1}\right) + \omega_0^2 = 0 \tag{14.20}$$

From this equation we see that the actual pole frequency, ω_{0a}, and pole-Q, Q_a, are

$$\omega_{0a} = \omega_0 \tag{14.21}$$

$$Q_a = \frac{Q}{1 + 2Q^2/(A + 1)} \tag{14.22}$$

Thus,

$$S_A^{\omega_{0a}} = 0$$

$$S_A^{Q_a} = -\frac{A}{A + 1}\frac{2Q^2/(A + 1)}{1 + 2Q^2/(A + 1)}$$

For $A \gg 2Q^2$ and $A \gg 1$ we obtain

$$S_A^{Q_a} \cong -\frac{2Q^2}{A}$$

It is usual to drop the subscript a in this expression and write

$$S_A^Q \cong -\frac{2Q^2}{A} \tag{14.23}$$

Note that if Q is high ($Q \geq 5$) its sensitivity relative to the amplifier gain can be quite high.[1]

· · ·

14.4   MULTIPLE-AMPLIFIER BIQUADS

The results of Example 14.1 indicate a serious disadvantage of single-amplifier biquads—the sensitivity of Q relative to the amplifier gain is quite high. Although there

---

[1]Because the open-loop gain A of op amps usually has wide tolerance, it is important to keep $S_A^{\omega_0}$ and S_A^Q very small.

exists a technique for reducing S_A^Q in SABs, this is done at the expense of increased passive sensitivities. Nevertheless, the resulting SABs are extensively used in many applications. In the following, however, we shall present two circuits that utilize two and three op amps, respectively, to realize second-order filter functions with reduced sensitivities and increased versatility.

GIC-Based Biquads

In Example 3.4 we introduced the generalized impedance converter (GIC) and considered its use in realizing an inductance. (The reader is urged to review Example 3.4 at this point.) Figure 14.11 shows the GIC circuit with the components selected according

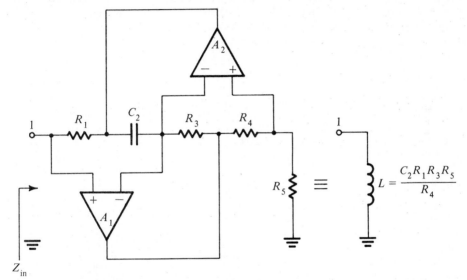

Fig. 14.11 The GIC used to realize an inductance.

to case a of Example 3.4. Assume that $A_1 = A_2 = \infty$. It can be shown that the input impedance Z_{in} between node 1 and ground is given by

$$Z_{in}(s) = sC_2 \frac{R_1 R_3 R_5}{R_4} \tag{14.24}$$

Thus between node 1 and ground we have an inductance of value L,

$$L = \frac{C_2 R_1 R_3 R_5}{R_4} \tag{14.25}$$

Resonance can be obtained by connecting a capacitor C_6 between node 1 and ground. The input signal can then be fed to this *resonator* through a resistor R_7. Figure 14.12a shows the resulting circuit, and Fig. 14.12b shows the equivalent circuit. This resonator has a pole frequency ω_0 given by

$$\omega_0 = \frac{1}{\sqrt{LC_6}}$$

Thus

$$\omega_0 = \frac{1}{\sqrt{C_2 C_6 R_1 R_3 R_5 / R_4}} \qquad (14.26)$$

The Q factor is determined by the resistance R_7 according to

$$Q = \omega_0 C_6 R_7 = R_7 \sqrt{\frac{C_6 R_4}{C_2 R_1 R_3 R_5}} \qquad (14.27)$$

If the output is taken across the tuned circuit—that is, between node 1 and ground—the function realized is a bandpass one. In fact, this circuit is equivalent to the passive realization shown in Fig. 14.3c. Node 1, however, is a high-impedance node and is not suitable as an output node. In other words, connecting a load impedance between node 1 and ground changes the function realized. Fortunately, however, a low-imped-ance node exists where the voltage is directly proportional to that at node 1. This is the output terminal of op amp A_1, where, as the reader can easily verify, the voltage is

$$V_{o1} = V_1 \frac{R_4 + R_5}{R_5}$$

Thus the filter output is usually taken as V_{o1}.

The analysis of this circuit to determine its sensitivity relative to the gains of the

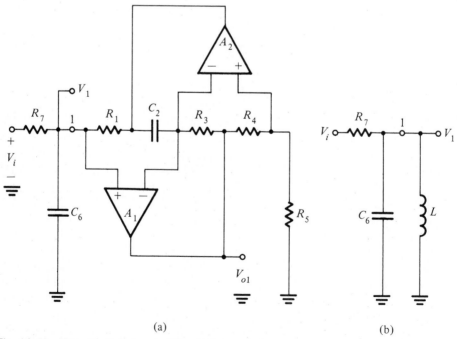

(a)   (b)

Fig. 14.12   Using the inductance of Fig. 14.11 to obtain resonance.

op amps is straightforward but quite tedious (see Ref. 14.1). The results indicate that S_{A1}^{Q} and S_{A2}^{Q} are proportional to the value of Q rather than to Q^2, as in the case of SABs.

Finally, it should be mentioned that a complete set of circuits based on the GIC has been developed for the realization of the various second-order filter functions (see Ref. 14.1).

EXERCISE

14.5 Design the biquad circuit of Fig. 14.12a to realize a bandpass function with center frequency $f_0 = 10$ kHz and $Q = 20$. Select $C_2 = C_6 = C$ and $R_1 = R_3 = R_4 = R_5 = 10$ kΩ. Find the values of C and R_7. Also find the value of the center-frequency gain.
Ans. 1.59 nF; 200 kΩ; 2

Two-Integrator-Loop Biquads

The circuits discussed next utilize two integrators in a feedback loop. To develop this two-integrator-loop realization consider the high-pass transfer function

$$\frac{V_{hp}}{V_i} = \frac{n_2 s^2}{s^2 + s\,(\omega_0/Q) + \omega_0^2} \tag{14.28}$$

Crossmultiplying and dividing both sides by s^2 gives

$$V_{hp} = -\frac{1}{Q}\frac{\omega_0}{s}\,V_{hp} - \frac{\omega_0^2}{s^2}\,V_{hp} + n_2 V_i \tag{14.29}$$

Utilizing two inverting integrators and a summer, Fig. 14.13a shows a block diagram realization of this equation. An actual circuit is shown in Fig. 14.13b. The reader is urged to verify the correspondence between the circuit and the block diagram and to use this correspondence to show that

$$CR = \frac{1}{\omega_0}, \qquad \frac{R_3}{R_2} = 2Q - 1, \qquad \text{and} \qquad n_2 = 2 - \frac{1}{Q}.$$

It should be noted that the function available at the output of the summer is high pass, that at the output of the first integrator is bandpass, and that at the output of the second integrator is low pass. The simultaneous availability of three different filtering functions makes this circuit suitable for application as a universal filter. Other filtering functions can be obtained by employing an additional summer that forms a weighted sum of the outputs of the three op amps.

An alternative two-integrator-loop realization in which all three op amps are used in a single-ended mode can be developed as follows: Rather than using the input summer to add signals with positive and negative coefficients, we can introduce an additional inverter, as shown in Fig. 14.14a. Now the summer has all positive coefficients, and we may dispense with the summer amplifier altogether and perform the summation at the virtual-ground input of the first integrator. The resulting circuit is shown in Fig. 14.14b, from which we observe that the high-pass function is no longer available. This is the price paid for obtaining a circuit that utilizes the op amps in a single-ended mode.

Two-integrator-loop circuits display low sensitivities and considerable flexibility and versatility. Although the circuits shown here suffer drastically from the effects of

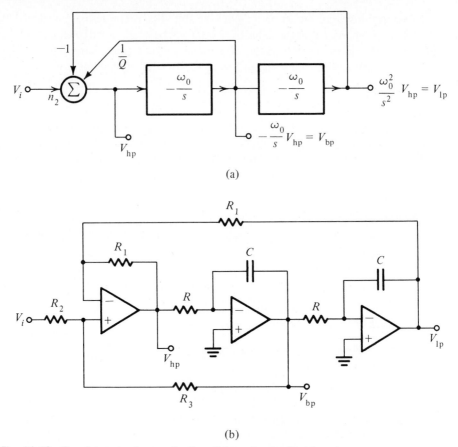

Fig. 14.13 *Two-integrator-loop realization of second-order filter functions.*

the finite op-amp bandwidth, simple modifications desensitize the circuits to these effects. This topic, however, is beyond the scope of this book.

EXERCISE

14.6 Use the circuit of Fig. 14.14b to realize a bandpass filter with $f_0 = 10$ kHz, $Q = 20$, and unity center-frequency gain. If $R = 10$ kΩ, give the values of C, R_d, and R_g.
Ans. 1.59 nF; 200 kΩ; 200 kΩ.

14.5 SWITCHED-CAPACITOR FILTERS

The active-RC filter circuits presented above have two properties that make their production in monolithic IC form difficult and practically impossible; these are the need for large-valued capacitors and the requirement of accurate RC time constants. The search therefore continued for a method of filter design that would lend itself more naturally to IC implementation. In this section we shall introduce one such method. At the time of this writing (1981) it appears to be the best contender for the task.

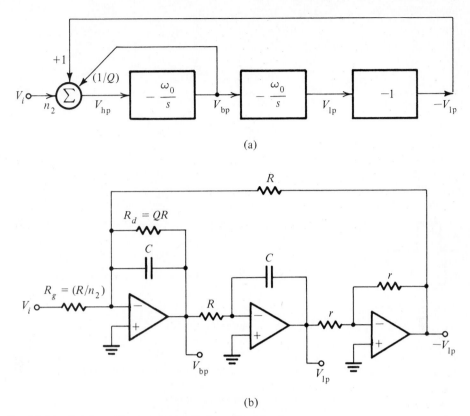

(a)

(b)

Fig. 14.14   An alternative two-integrator-loop filter circuit.

The Basic Principle

The switched-capacitor filter technique is based on the realization that a capacitor switched between two circuit nodes at a sufficiently high rate is equivalent to a resistor connecting these two nodes. To be specific, consider the active-RC integrator of Fig. 14.15a. This is the familiar Miller integrator, which we used in the two-integrator-loop biquad in Section 14.4. In Fig. 14.15b we have replaced the input resistor R_1 by a grounded capacitor C_1 together with two MOS transistors acting as switches. In some circuits more elaborate switch configurations are used, but such details are beyond our present need.

The two MOS switches in Fig. 14.15b are driven by a *nonoverlapping* two-phase clock. Figure 14.15c shows the clock waveforms. We shall assume in this introductory exposition that the clock frequency f_c ($f_c = 1/T_c$) is much higher than the frequency of the signal being filtered. Thus during clock phase ϕ_1, when C_1 is connected across the input signal source v_i, the variations in the input signal are negligibly small. It follows that during ϕ_1 capacitor C_1 charges up to the voltage v_i,

$$q_{C1} = C_1 v_i$$

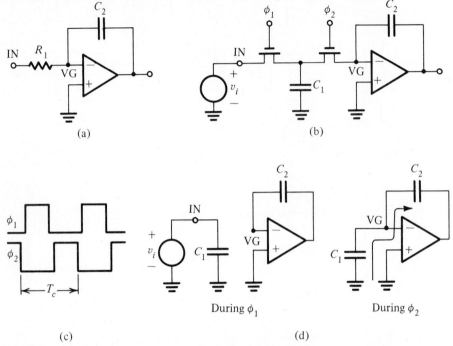

Fig. 14.15 *Basic principle of the switched-capacitor filter technique. (a) Active-RC integrator. (b) Switched-capacitor integrator. (c) Two-phase clock (nonoverlapping). (d) During ϕ_1, C_1 charges up to the current value of v_i and then, during ϕ_2, discharges into C_2.*

Then during clock phase ϕ_2 capacitor C_1 is connected to the virtual-ground input of the op amp, as indicated in Fig. 14.15d. Capacitor C_1 is thus forced to discharge, and its previous charge q_{C1} is transferred to C_2.

From the above description we see that during each clock period T_c an amount of charge $q_{C1} = C_1 v_i$ is extracted from the input source and supplied to the integrator capacitor C_2. Thus the average current flowing between the input node (IN) and the virtual ground node (VG) is

$$i_{av} = \frac{C_1 v_i}{T_c}$$

If T_c is sufficiently short, one can think of this process as almost continuous and define an equivalent resistance R_{eq} that is in effect present between nodes IN and VG:

$$R_{eq} \equiv \frac{v_i}{i_{av}}$$

Thus

$$R_{eq} = \frac{T_c}{C_1} \qquad\qquad (14.30)$$

Using R_{eq} we obtain an equivalent time constant for the integrator:

$$\text{Time constant} = C_2 R_{eq} = T_c \frac{C_2}{C_1} \qquad (14.31)$$

Thus the time constant that determines the frequency response of the filter is determined by the clock period T_c and the capacitor ratio C_2/C_1. Both of these parameters can be well controlled in an IC process. Specifically, note the dependence on capacitor ratios rather than on absolute values of capacitors. The accuracy of capacitor ratios in MOS technology can be controlled to within 0.1%.

Another point worth observing is that with reasonable clocking frequency (such as 100 kHz) and not-too-large capacitor ratios (say, 10) one can obtain reasonably large time constants (such as 10^{-4} s) suitable for audio applications. Since capacitors typically occupy relatively large areas on the IC chip, it is important to mention that the ratio accuracies quoted above are obtainable with the smaller capacitor value as low as 0.2 pF.

Actual Circuits

The switched-capacitor (SC) circuit in Fig. 14.15b realizes a negative integrator. As we saw in Section 14.4 a two-integrator-loop active filter is composed of one inverting and one noninverting integrator.[2] To realize a switched-capacitor biquad filter we therefore need a pair of complementary switched-capacitor integrators. Figure 14.16a shows a noninverting, or positive, integrator circuit. The reader is urged to follow the operation of this circuit during the two clock phases and thus show that it operates much the same way as the basic circuit of Fig. 14.15b, except for a sign reversal.

In addition to realizing a noninverting integrator function, the circuit in Fig. 14.16a is insensitive to stray capacitances; however, we shall not explore this point any further, and we refer the interested reader to reference 14.5. By reversal of the clock phases on two of the switches the circuit in Fig. 14.16b is obtained. This latter circuit realizes the inverting integrator function, like the circuit of Fig. 14.15b, but is insensitive to stray capacitances (which the original circuit of Fig. 14.15b is not). At the time of this writing the pair of complementary integrator circuits of Fig. 14.16 has become the standard building block in the design of switched-capacitor filters.

Let us now consider the realization of a complete biquad circuit. Figure 14.17a shows the active-RC two-integrator-loop circuit previously studied. By considering the cascade of integrator 2 and the inverter as a positive integrator and then simply replacing each resistor by its switched-capacitor equivalent we obtain the circuit in Fig. 14.17b. Ignore the damping around the first integrator (that is, the switched capacitor C_5) for the time being and note that the feedback loop indeed consists of one inverting and one noninverting integrator. Then note the phasing of the switched capacitor used for damping. Reversing the phases here would convert the feedback to positive and move the poles to the right half of the s plane. On the other hand, the phasing of the

---

[2] In the two-integrator loop of Fig. 14.14 the noninverting integrator is realized by the cascade of a Miller integrator and an inverter.

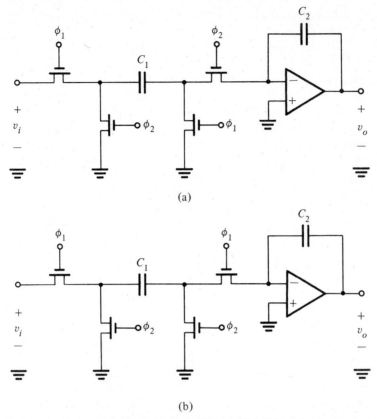

(a)

(b)

Fig. 14.16   A pair of complementary stray-insensitive switched-capacitor integrators. (a) Noninverting switched-capacitor integrator. (b) Inverting switched-capacitor integrator.

feed-in switched capacitor (C_6) is not that important; a reversal of phases would result only in an inversion in the sign of the function realized.

Having identified the correspondences between the active-RC biquad and the switched-capacitor biquad, we can now derive design equations. Analysis of the circuit in Fig. 14.17a yields

$$\omega_0 = \frac{1}{\sqrt{C_1 C_2 R_3 R_4}} \tag{14.32}$$

Substituting for R_3 and R_4 by their SC equivalent values, that is,

$$R_3 = \frac{T_c}{C_3} \quad \text{and} \quad R_4 = \frac{T_c}{C_4}$$

gives ω_0 of the SC biquad as

$$\omega_0 = \frac{1}{T_c} \sqrt{\frac{C_3}{C_2} \frac{C_4}{C_1}} \tag{14.33}$$

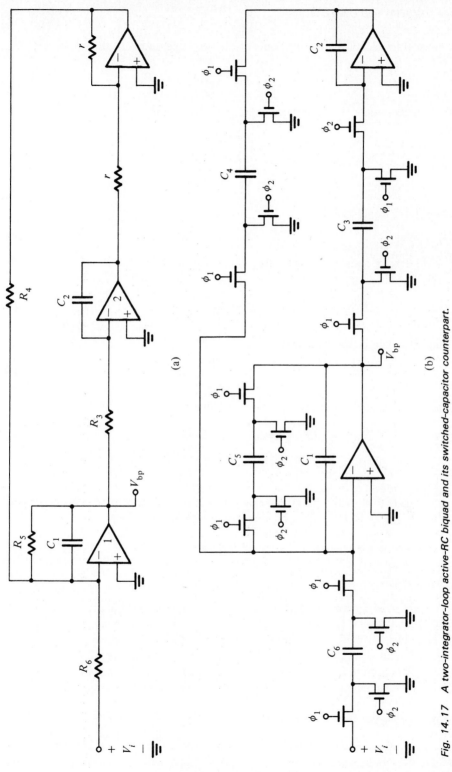

Fig. 14.17   A two-integrator-loop active-RC biquad and its switched-capacitor counterpart.

It is usual to select the time constants of the two integrators equal:

$$\frac{T_c}{C_3} C_2 = \frac{T_c}{C_4} C_1 \tag{14.34}$$

If we further select the two integrating capacitors C_1 and C_2 to be equal,

$$C_1 = C_2 = C \tag{14.35}$$

then

$$C_3 = C_4 = KC, \tag{14.36}$$

where from Eq. (14.33)

$$K = \omega_0 T_c \tag{14.37}$$

For the case of equal time-constants the Q factor of the circuit in Fig. 14.17a is given by R_5/R_4. Thus the Q factor of the corresponding SC circuit in Fig. 14.17b is given by

$$Q = \frac{T_c/C_5}{T_c/C_4} \tag{14.38}$$

Thus C_5 should be selected from

$$C_5 = \frac{C_4}{Q} = \frac{KC}{Q} = \omega_0 T_c \frac{C}{Q} \tag{14.39}$$

Finally, the center-frequency gain of the bandpass function is given by

$$\text{Center-frequency gain} = \frac{C_6}{C_5} = Q \frac{C_6}{\omega_0 T_c C} \tag{14.40}$$

EXERCISE

14.7 Use $C_1 = C_2 = 20$ pF and design the circuit in Fig. 14.17b to realize a bandpass function with $f_0 = 10$ kHz, $Q = 20$, and unity center-frequency gain. Use a clock frequency $f_c = 200$ kHz. Find the values of C_3, C_4, C_5, and C_6.
Ans. 6.283 pF; 6.283 pF; 0.314 pF; 0.314 pF.

A Final Remark

In the above we have attempted only to provide an introduction to this newly introduced (1977) and highly promising filter technology. We have made many simplifying assumptions, the most important being the switched-capacitor resistor equivalence [Eq. (14.30)]. This equivalence is correct only at $f_c = \infty$ and is approximated for $f_c \gg f_0$. Switched-capacitor filters, moreover, are sampled-data networks whose analysis and design is more appropriately carried out using z-transform techniques (Ref. 14.7).

14.6 BASIC PRINCIPLES OF SINUSOIDAL OSCILLATORS

The generation of sine waves is an important task that electronics engineers are often required to undertake. Basically there are two approaches to the design of sine-wave generators. The first is to design a *nonlinear oscillator*, which generates square and triangular waveforms, and apply the triangular wave to a *sine-wave shaper*, which usually consists of diodes and resistors. Nonlinear oscillators, or *function generators* as they are usually called, are studied in Chapter 5.

The second approach, which is the subject of the remainder of this chapter, employs a *positive-feedback loop* that contains a *frequency-selective network*. The loop is designed to have a gain of unity at a single frequency determined by the frequency-selective network. In this type of oscillator, called a *linear oscillator*, sine waves are generated essentially by a resonance phenomenon.

In spite of the name linear oscillator, some form of nonlinearity has to be employed to provide control of the amplitude of the output sine wave. In fact, all oscillators are essentially nonlinear circuits. This complicates the task of analysis and design of oscillators; no longer is one able to apply transform methods (*s* plane) directly. Nevertheless, techniques have been developed by which the design of sinusoidal oscillators can be performed in two steps. The first step is a linear one, and frequency-domain methods of feedback circuit analysis can be readily employed. Subsequently, a nonlinear mechanism for amplitude control can be provided.

In the following we shall be specifically concerned with the design of sinusoidal oscillators using op amps. The circuits we shall study can be employed directly only for the generation of sine waves whose frequencies cover the range 10 Hz to 100 kHz (or 1 MHz as a maximum). While the lower limit is dictated by the size of passive components required, the upper limit is governed by the frequency-response and slew-rate limitations of op amps. For lower frequencies the function-generator approach is preferred; for higher frequencies circuits that employ transistors and tuned LC circuits or crystals are frequently used.

The Oscillator Feedback Loop

The basic structure of a sinusoidal oscillator consists of an amplifier and a frequency-selective network connected in a positive-feedback loop, such as that shown in block-diagram form in Fig. 14.18. While in an actual oscillator circuit no input signal will be present, we include an input signal here to help explain the principle of operation. It is important to note that unlike the negative-feedback loop of Fig. 12.1, here the feedback signal x_f is summed with a *positive* sign. Thus the gain with feedback is given by

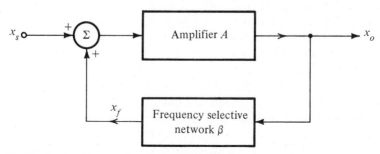

Fig. 14.18 *The basic structure of a sinusoidal oscillator.*

$$A_f(s) = \frac{A(s)}{1 - A(s)\beta(s)} \tag{14.41}$$

where we note the negative sign in the denominator.

According to the definition of loop gain in Chapter 12, the loop gain of the circuit in Fig. 14.18 is $-A(s)\beta(s)$. However, for our purposes here it is more convenient to drop the minus sign and define the loop gain $L(s)$ as

$$L(s) = A(s)\beta(s) \tag{14.42}$$

The characteristic equation thus becomes

$$1 - L(s) = 0 \tag{14.43}$$

Note that this new definition of loop gain corresponds directly to the actual gain of the feedback loop of Fig. 14.18.

The Oscillation Criterion

If at a specific frequency f_0 the loop gain $A\beta$ is equal to unity, it follows from Eq. (14.41) that A_f will be infinite. That is, at this frequency the circuit will have a finite output for zero input signal, which is by definition an oscillator. Thus the condition for the feedback loop of Fig. 14.18 to provide sinusoidal oscillations of frequency ω_0 is that

$$L(j\omega_0) \equiv A(j\omega_0)\beta(j\omega_0) = 1 \tag{14.44}$$

That is, at ω_0 the phase of the loop gain should be zero and the magnitude of the loop gain should be unity. This is known as the Barkhausen criterion. Note that for the circuit to oscillate at one frequency the oscillation criterion should be satisfied at one frequency only (that is, ω_0), otherwise the resulting waveform will not be sinusoidal.

An intuitive feeling for the Barkhausen criterion can be gained by considering the feedback loop of Fig. 14.18. For this loop to produce and sustain an output x_o with no input applied ($x_s = 0$), the feedback signal x_f,

$$x_f = \beta x_o,$$

should be sufficiently large so that when multiplied by A it produces x_o,

$$A x_f = x_o$$

that is,

$$A\beta x_o = x_o$$

which results in

$$A\beta = 1$$

It should be noted that the frequency of oscillation ω_0 is determined solely by the phase characteristics of the feedback loop; the loop oscillates at the frequency for which the phase is zero. It follows that the stability of the frequency of oscillation will be determined by the manner in which the phase $\phi(\omega)$ of the feedback loop varies with frequency. A "steep" function $\phi(\omega)$ will result in a more stable frequency. This can be

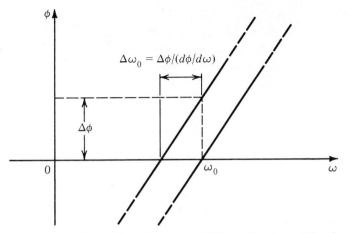

$$\Delta\omega_0 = \Delta\phi/(d\phi/d\omega)$$

Fig. 14.19 Dependence of the oscillator frequency stability on the slope of the phase response.

seen if one imagines a change in phase $\Delta\phi$ due to a change in one of the circuit components. If $d\phi/d\omega$ is large, the resulting change in ω_0 will be small, as illustrated in Fig. 14.19.

An alternative approach to the study of oscillator circuits consists of examining the circuit poles, which are the zeros of the characteristic equation [Eq. (14.43)]. For the circuit to produce sustained oscillations at a frequency ω_0 the characteristic equation has to have zeros at $s = \pm j\omega_0$. Thus $1 - A(s)\beta(s)$ should have a factor of the form $s^2 + \omega_0^2$.

Nonlinear Amplitude Control

The oscillation condition discussed above guarantees sustained oscillations in a mathematical sense. It is well known, however, that the parameters of any physical system cannot be maintained constant for any length of time. In other words, suppose we work hard to make $A\beta = 1$ at $\omega = \omega_0$; then the temperature changes and $A\beta$ becomes slightly less than unity. Obviously, oscillations will cease in this case. Conversely, if $A\beta$ exceeds unity, oscillations will grow in amplitude. We therefore need a mechanism for forcing $A\beta$ to remain equal to unity at the desired value of output amplitude. This task is accomplished by providing a nonlinear circuit for gain control.

Basically, the function of the gain-control mechanism is as follows: First, to ensure that oscillations will start, one designs the circuit such that $A\beta$ is slightly greater than unity. This corresponds to designing the circuit so that the poles are in the right half of the s plane. Thus as the power supply is turned on, oscillations will grow in amplitude. When the amplitude reaches the desired level, the nonlinear network comes into action and causes the loop gain to be reduced to exactly unity. In other words, the poles will be "pulled back" to the $j\omega$ axis. This action will cause the circuit to sustain oscillations at this desired amplitude. If, for some reason, the loop gain is reduced below unity, the amplitude of the sine wave will diminish. This will be detected by the nonlinear network, which will cause the loop gain to increase to exactly unity.

As will be seen, there are two basic approaches to the implementation of the non-linear amplitude-stabilization mechanism. The first approach makes use of a limiter circuit (see Chapter 5). Oscillations are allowed to grow until the amplitude reaches the level to which the limiter is set. Once the limiter comes into operation, the amplitude remains constant. Obviously, the limiter should be "soft" in order to minimize nonlinear distortion. Such distortion, however, is reduced by the filtering action of the frequency-selective network in the feedback loop. In fact, in one of the oscillator circuits studied in Section 14.7 the sine waves are hard limited, and the resulting square waves are applied to a bandpass filter present in the feedback loop. The "purity" of the output sine waves will be a function of the selectivity of this filter. That is, the higher the Q of the filter, the less the harmonic content of the sine wave output.

The other mechanism for amplitude stabilization is more elaborate. The amplitude of the sine wave output is detected and converted to a dc level, which is compared with a preset value. The output of the comparator is then used to adjust the value of a resistance that determines the loop gain. This latter resistance, a voltage-controlled one, may be implemented for example by a JFET operated in the triode region (see Chapter 7).

14.7 OSCILLATOR CIRCUITS

In this section we shall study a number of practical oscillator circuits.

The Wien-Bridge Oscillator

One of the simplest oscillator circuits is based on the Wien bridge. Figure 14.20 shows a Wien-bridge oscillator without the nonlinear gain-control network. The circuit consists of an op amp connected in the noninverting configuration with a closed-loop gain of $1 + R_2/R_1$. In the feedback path of this positive-gain amplifier an RC network is connected. The loop gain can be easily obtained by multiplying the transfer function $V_a(s)/V_o(s)$ of the feedback network by the amplifier gain,

$$L(s) = \left[1 + \frac{R_2}{R_1}\right] \frac{Z_p}{Z_p + Z_s}$$

Thus

$$L(s) = \frac{1 + R_2/R_1}{3 + sCR + 1/sCR} \tag{14.45}$$

Substituting $s = j\omega$ results in

$$L(j\omega) = \frac{1 + R_2/R_1}{3 + j(\omega CR - 1/\omega CR)} \tag{14.46}$$

The loop gain will be a real number (that is, the phase will be zero) at one frequency given by

$$\omega_0 CR = \frac{1}{\omega_0 CR}$$

That is,

$$\omega_0 = 1/CR \tag{14.47}$$

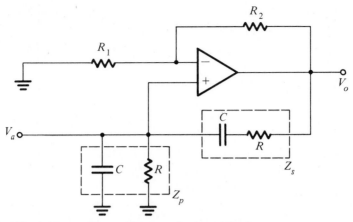

Fig. 14.20 Wien-bridge oscillator without amplitude stabilization.

To obtain sustained oscillations at this frequency one should set the magnitude of the loop gain to unity. This can be achieved by selecting

$$\frac{R_2}{R_1} = 2 \tag{14.48}$$

To ensure that oscillations will start, one chooses R_2/R_1 slightly greater than 2. The reader can easily verify that if $R_2/R_1 = 2 + \delta$, where δ is a small number, the roots of the characteristic equation $1 - L(s) = 0$ will be in the right half of the s plane.

The amplitude of oscillation can be determined and stabilized by using a nonlinear control network. Two different implementations of the amplitude control are shown in Figs. 14.21 and 14.22. The circuit in Fig. 14.21 employs a symmetrical feedback limiter (see Section 5.8) formed by diodes D_1 and D_2 together with resistors R_3, R_4, R_5, and R_6. The limiter operates in the following manner: At the positive peak of the output voltage v_O the voltage at node b will exceed the voltage v_1 (which is about $\frac{1}{3}v_O$), and diode D_2 conducts. This will clamp the positive peak to a value determined by R_5, R_6, and the negative power supply. The value of the positive output peak can be calculated by writing a node equation at node b and neglecting the current through D_2. Similarly, the negative peak of the output sine wave will be clamped to the value that causes diode D_1 to conduct. The value of the negative peak can be determined from the node equation for node a while neglecting the current through D_1. Finally, note that in order to obtain a symmetrical output waveform, R_3 is chosen equal to R_6 and R_4 equal to R_5.

EXERCISE

14.8 For the circuit in Fig. 14.21,
 (a) Disregard the limiter circuit and find the location of the closed-loop poles.
 (b) Find the frequency of oscillation.
 (c) Find the amplitude of the output sine wave (assume that the diode drop is 0.7 V).
 Ans. $(10^5/16)$ $(0.015 \pm j)$; 1 kHz; 21.36 V (peak-to-peak)

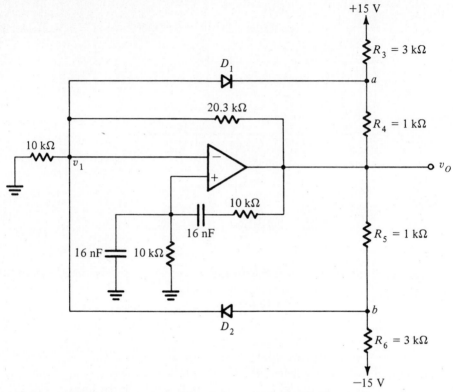

Fig. 14.21 A Wien-bridge oscillator with a limiter used for amplitude control.

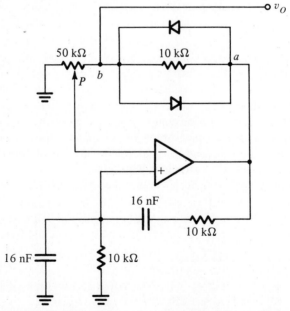

Fig. 14.22 A Wien-bridge oscillator with an alternative method for amplitude stabilization.

The circuit of Fig. 14.22 employs an inexpensive implementation of the parameter-variation mechanism of amplitude control. Potentiometer P is adjusted until oscillations just start to grow. As the oscillations grow, the diodes start to conduct, causing the effective resistance between a and b to decrease. Equilibrium will be reached at the output amplitude that causes the loop-gain to be exactly unity. The output amplitude can be varied by adjusting potentiometer P.

As indicated in Fig. 14.22, the output is taken at point b rather than at the op-amp output terminal because the signal at b has lower distortion than that at a (why?). Node b, however, is a high-impedance node, and a buffer will be needed if a load is to be connected.

EXERCISE

14.9 For the circuit in Fig. 14.22 find the following:

    (a) The setting of potentiometer P at which oscillations start;

    (b) The frequency of oscillation.

    Ans. (a) 20 kΩ to ground; (b) 1 kHz

The Phase-Shift Oscillator

The basic structure of the phase-shift oscillator is shown in Fig. 14.23. It consists of a negative-gain amplifier $(-K)$ with a three-section (third-order) RC ladder network in

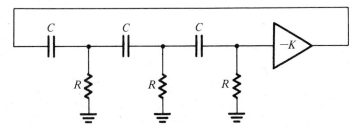

Fig. 14.23 Phase-shift oscillator.

the feedback. The circuit will oscillate at the frequency for which the phase shift of the RC network is 180°. Only at this frequency will the total phase shift around the loop be 0 or 360°. Here we should note that the reason for using a three-section RC network is that three is the minimum number of sections (that is, lowest order) that is capable of producing 180° phase shift at a finite frequency.

For oscillations to be sustained the value of K should be equal to the inverse of the magnitude of the RC network transfer function at the frequency of oscillation. However, in order to ensure that oscillations start, the value of K has to be chosen slightly higher than the value that satisfies the unity-loop-gain condition. Oscillations will then grow in magnitude until limited by some nonlinear control mechanism.

Figure 14.24 shows a practical phase-shift oscillator with a feedback limiter, consisting of diodes D_1 and D_2 and resistors R_1, R_2, R_3, and R_4 for amplitude stabilization. To start oscillations, R_f has to be made slightly greater than the minimum required value. Although the circuit stabilizes more rapidly and provides sine waves with more

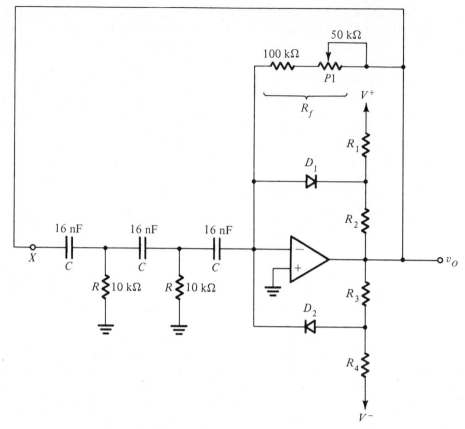

Fig. 14.24   Practical phase-shift oscillator with a limiter for amplitude stabilization.

stable amplitude if R_f is made much larger than this minimum, the price paid is an increased output distortion.

EXERCISES

14.10   Consider the circuit of Fig. 14.24 *without* the limiter. Break the feedback loop at X and find the loop gain $A\beta \equiv V_o(j\omega)/V_x(j\omega)$. To do this it is easier to start at the output and work backward, finding the various currents and voltages, and eventually V_x, in terms of V_o.

Ans. $\dfrac{\omega^2 C^2 R R_f}{4 + j(3\omega CR - 1/\omega CR)}$

14.11   Use the expression derived in Exercise 14.10 to find the frequency of oscillation f_o and the minimum required value of R_f for the oscillator circuit in Fig. 14.24.
Ans.   574.3 Hz; 120 kΩ

The Quadrature Oscillator

The *quadrature oscillator* is based on the two-integrator loop studied in Section 14.5. As an active filter the loop is damped so as to locate the poles in the left half of the s

plane. Here no such damping will be used, since we wish to locate the poles on the $j\omega$ axis in order to provide sustained oscillations. In fact, to ensure that oscillations start, the poles are initially located in the right half-plane and then "pulled back" by the nonlinear gain control.

Figure 14.25 shows a practical quadrature oscillator. Amplifier 1 is connected as an inverting Miller integrator with a limiter in the feedback for amplitude control.

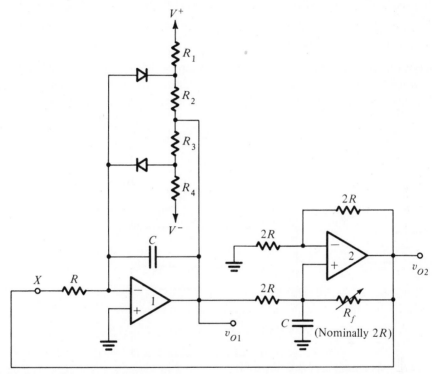

Fig. 14.25 *A quadrature oscillator circuit.*

Amplifier 2 is connected as a noninverting integrator (thus replacing the cascade connection of the Miller integrator and the inverter in the two-integrator loop of Fig. 14.14b). This noninverting integrator circuit is studied in Example 3.3.

The resistance R_f in the positive-feedback path of op-amp 2 is made variable, with a nominal value of $2R$. Decreasing the value of R_f moves the poles to the right half-plane (Problem 14.37) and ensures that the oscillations start. Too much positive feedback, although it results in better amplitude stability, also results in higher output distortion (because the limiter has to operate "harder"). In this regard, note that the output v_{o2} will be "purer" than v_{o1} because of the filtering action provided by the second integrator on the limited output of the first integrator.

If we disregard the limiter and break the loop at X, the loop gain can be obtained as

$$L(s) \equiv \frac{V_{o2}}{V_x} = -\frac{1}{s^2 C^2 R^2}$$

Thus the loop will oscillate at frequency ω_0, given by

$$\omega_0 = 1/CR$$

Finally, it should be pointed out that the name *quadrature oscillator* is used because the circuit provides two sinusoids with 90° phase difference. This should be obvious, since v_{O2} is the integral of v_{O1}. There are many applications for which quadrature sinusoids are required.

The Active-Filter Tuned Oscillator

The last oscillator circuit that we shall discuss is quite simple in principle and in design. Nevertheless, the approach is general and versatile and can result in high-quality (that is, low-distortion) output sine waves. The basic principle is illustrated in Fig. 14.26. The

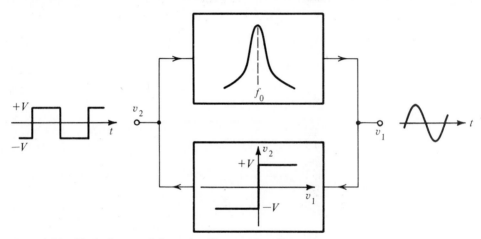

Fig. 14.26   Block diagram of the active-filter tuned oscillator.

circuit consists of a high-Q bandpass filter connected in a positive-feedback loop with a hard limiter. To understand how this circuit works, assume that oscillations have already started. The output of the bandpass filter will be a sine wave whose frequency is equal to the center frequency of the filter, f_0. The sine wave signal v_1 is fed to the limiter, which produces at its output a square wave whose levels are determined by the limiting levels and whose frequency is f_0. The square wave is in turn fed to the bandpass filter, which filters out the harmonics and provides a sinusoidal output v_1 at the fundamental frequency f_0. Obviously, the purity of the output sine wave will be a direct function of the selectivity (or Q factor) of the bandpass filter.

The simplicity of this approach to oscillator design should be apparent. We have independent control of frequency and amplitude as well as distortion of the output sinusoid. Any filter circuit with positive gain can be used to implement the bandpass filter. The frequency stability of the oscillator will be directly determined by the frequency stability of the bandpass-filter circuit. Also, a variety of limiter circuits (see

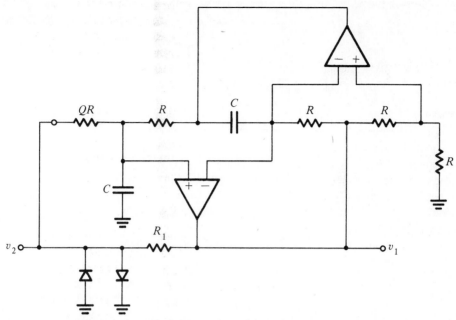

Fig. 14.27 Practical implementation of the active-filter tuned oscillator.

Chapter 5) with different degrees of sophistication can be used to implement the limiter block.

Figure 14.27 shows one possible implementation of the active-filter tuned oscillator. This circuit uses the GIC-based bandpass filter studied in Section 14.5. The limiter used is a very simple one consisting of a resistance R_1 and two diodes.

EXERCISE

14.12 Use $C = 16$ nF and find the value of R such that the circuit of Fig. 14.27 produces 1-kHz sine waves. If the diode drop is 0.7 V, find the peak-to-peak amplitude of the output sine wave. (*Hint:* A square wave with peak-to-peak amplitude of V volts has a fundamental component with $4V/\pi$ volts peak-to-peak amplitude.)
Ans. 10 kΩ; 1.8 V

14.8 CONCLUDING REMARKS

The material studied in this chapter should serve as an introduction to the topics of filter and oscillator design. For a more complete study of filter design the interested reader may consult reference 14.1. References 14.8 to 14.14 contain additional material on oscillator design.

LOGIC-CIRCUIT 15 FAMILIES

The fundamentals of digital circuits and systems were studied in Chapter 6. Various logic and memory functions were introduced, and their implementations were described in black-box form. This black-box approach, although useful at an introductory level, is not sufficient for the electronics design engineer. Only through knowledge of the circuit inside an IC package will an electronics designer be able to intelligently and effectively apply such a package. Therefore it is our purpose in this chapter and the next to provide the reader with some familiarity with digital integrated circuits.

The material studied is divided into two parts. The first part, which is the topic of this chapter, deals with the various circuit types used in implementing logic functions. These circuit types are known as *logic circuit families*. The second part, studied in Chapter 16, deals with digital memory circuits. Memory chips represent the state-of-the-art in VLSI circuit technology.

In each of the logic families considered we shall study the basic gate circuit. This will enable us to outline the various design tradeoffs and develop the characteristics and features that distinguish the given family. Circuits belonging to a given family share the basic features and limitations such as noise immunity, power dissipation, speed of operation, maximum fan-out, and so on. One of our objectives will be to relate circuit operation to manufacturer's specifications.

At the time of this writing (1981) the most popular logic-circuit families are transistor-transistor logic (TTL or T^2L), emitter-coupled logic (ECL), and complementary-symmetry MOS (CMOS or COSMOS). In addition, a recently introduced digital circuit technology called integrated injection logic (I^2L) will be studied. Although not commercially available as a conventional logic family, I^2L is used in LSI and VLSI circuits. It is also available for semicustom IC design, as will be described in Chapter 16.

In addition to the popular IC logic families mentioned above we shall study resistor-transistor logic (RTL) and diode-transistor logic (DTL). RTL circuits are simple and can

660

therefore be used as a vehicle for introducing basic concepts. DTL is considered briefly because it is the precursor of TTL.

15.1   THE BJT AS A DIGITAL CIRCUIT ELEMENT

We shall begin our study of BJT logic circuits with a summary of pertinent BJT characteristics. In addition, a popular large-signal model for the BJT will be introduced. Before proceeding with this material the reader is advised to review Chapter 9.

Saturating and Nonsaturating Logic

The most common usage of the BJT in digital circuits is to employ its two extreme modes of operation: cutoff and saturation. The resulting logic circuits are called *saturated* (or *saturating*) logic. The advantages of this mode of application are: relatively large, well defined logic swings and reasonably low power dissipation. The main disadvantage is the relatively slow response due to the long turnoff times of saturated transistors.

To obtain faster logic, one has to arrange the design such that the BJT does not saturate. We shall study two forms of nonsaturating BJT logic: emitter-coupled logic (ECL), which is based on the differential pair studied in Section 10.5, and Schottky TTL, which is based on the use of special low-voltage-drop diodes called Schottky diodes.

The Ebers-Moll (EM) Model

Although the simple large-signal transistor model developed in Chapter 9 is usually quite adequate for the approximate analysis of BJT digital circuits, more insight can be obtained from a more formal approach using a popular large-signal model of the BJT known as the *Ebers-Moll* (EM) model.

The EM model is a low-frequency (static) model based on the fact that the BJT is composed of two *pn* junctions, the emitter–base junction and the collector–base junction. One can therefore express the terminal currents of the BJT as the superposition of the currents due to the two *pn* junctions, as follows:

Figure 15.1 shows an *npn* transistor together with its EM model. The model consists of two diodes and two controlled sources. The diodes are D_E, the emitter–base junction diode, and D_C, the collector–base junction diode. The diode currents i_{DE} and i_{DC} are given by the diode equation:

$$i_{DE} = I_{SE}(e^{v_{BE}/V_T} - 1) \tag{15.1}$$

$$i_{DC} = I_{SC}(e^{v_{BC}/V_T} - 1) \tag{15.2}$$

where I_{SE} and I_{SC} are the saturation or scale currents of the two diodes. Since the collector–base junction is usually of larger area than the emitter–base junction, I_{SC} is usually larger than I_{SE} (by factors of 2 to 50).

As explained in Chapter 9, part of the emitter–base junction current i_{DE} reaches the collector and registers as collector current. It is this component that gives rise to the

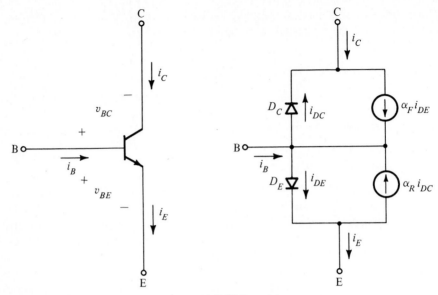

Fig. 15.1   An npn transistor and its Ebers-Moll (EM) model.

current source $\alpha_F i_{DE}$ in the model of Fig. 15.1. Here α_F denotes the *forward* α of the transistor (which is the parameter we simply called α previously). The value of α_F is usually very close to unity. Similarly, part of the collector–base junction current i_{DC} is transported across the base region and reaches the emitter. This component is represented in the EM model by the current source $\alpha_R i_{DC}$, where α_R denotes the *reverse* α of the transistor. Since the transistor structure is not physically symmetric but rather is optimized to have a large forward α, α_R is usually small (0.02 to 0.5).

A relationship exists (Ref. 15.1) between the four parameters of the EM model and the transistor current scale I_S (see Chapter 9):

$$\alpha_F I_{SE} = \alpha_R I_{SC} = I_S \tag{15.3}$$

Since $\alpha_F \cong 1$, we see that

$$I_{SE} \cong I_S \tag{15.4}$$

Recall that I_S is of the order of 10^{-14} to 10^{-15} A and is proportional to the area of the emitter–base junction.

The Transistor Terminal Currents

Having provided a qualitative physical justification for the EM model, we shall now use it to express the BJT terminal currents in terms of the junction voltages. From Fig. 15.1 we can write

$$i_E = i_{DE} - \alpha_R i_{DC} \tag{15.5}$$
$$i_C = -i_{DC} + \alpha_F i_{DE} \tag{15.6}$$
$$i_B = (1 - \alpha_F)i_{DE} + (1 - \alpha_R)i_{DC} \tag{15.7}$$

Substituting for i_{DE} and i_{DC} from Eqs. (15.1) and (15.2) and using the relationship in Eq. (15.3) gives

$$i_E = \frac{I_S}{\alpha_F}(e^{v_{BE}/V_T} - 1) - I_S(e^{v_{BC}/V_T} - 1) \qquad (15.8)$$

$$i_C = I_S(e^{v_{BE}/V_T} - 1) - \frac{I_S}{\alpha_R}(e^{v_{BC}/V_T} - 1) \qquad (15.9)$$

$$i_B = \frac{I_S}{\beta_F}(e^{v_{BE}/V_T} - 1) + \frac{I_S}{\beta_R}(e^{v_{BC}/V_T} - 1) \qquad (15.10)$$

where β_F is the forward β and β_R is the reverse β,

$$\beta_F = \frac{\alpha_F}{1 - \alpha_F} \qquad (15.11)$$

$$\beta_R = \frac{\alpha_R}{1 - \alpha_R} \qquad (15.12)$$

While β_F is usually large, β_R is very small.

Application of the EM Model

We shall now consider the application of the EM model to characterize transistor operation in various modes.

The Normal Active Mode. Here the emitter–base junction is forward-biased and the collector–base junction is reverse-biased. The word normal is used to distinguish this mode from that in which the roles of the two junctions are interchanged (the reverse active mode). Since v_{BC} is negative and its magnitude is usually much greater than V_T, Eqs. (15.8) through (15.10) can be approximated as

$$i_E \cong \frac{I_S}{\alpha_F}e^{v_{BE}/V_T} + I_S\left(1 - \frac{1}{\alpha_F}\right) \qquad (15.13)$$

$$i_C \cong I_Se^{v_{BE}/V_T} + I_S\left(\frac{1}{\alpha_R} - 1\right) \qquad (15.14)$$

$$i_B \cong \frac{I_S}{\beta_F}e^{v_{BE}/V_T} - I_S\left(\frac{1}{\beta_F} + \frac{1}{\beta_R}\right) \qquad (15.15)$$

In each of these three equations one can normally neglect the second term on the right-hand side. This results in the familiar current–voltage relationships that characterize the active mode of operation.

EXERCISES

15.1 Use Eq. (15.8) to show that the i-v characteristic of the diode-connected transistor of Fig. E15.1 is given by

$$i = \frac{I_S}{\alpha_F}(e^{v/V_T} - 1) \cong I_Se^{v/V_T}$$

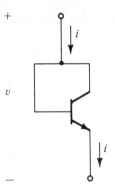

Fig. E15.1

The Saturation Mode. Consider first the normal (as opposed to reverse) saturation mode, as can be obtained in the circuit of Fig. 15.2. Assume that a current I_B is pushed into the base and that its value is sufficient to drive the transistor into saturation. Thus the collector current will be $\beta_{\text{forced}} I_B$, where $\beta_{\text{forced}} < \beta_F$. We wish to use the EM equations to derive an expression for $V_{CE\text{sat}}$.

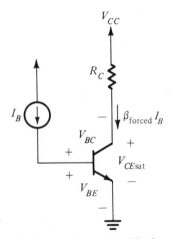

Fig. 15.2   Circuit in which the transistor can be operated in the normal saturation mode.

In saturation both junctions are forward-biased. Thus V_{BE} and V_{BC} are both positive, and their values are much greater than V_T. Thus in Eqs. (15.9) and (15.10) we can assume that $e^{V_{BE}/V_T} \gg 1$ and $e^{V_{BC}/V_T} \gg 1$. Making these approximations and substituting $i_B = I_B$ and $i_C = \beta_{\text{forced}} I_B$ results in two equations that can be solved to obtain V_{BE} and V_{BC}. The saturation voltage $V_{CE\text{sat}}$ can be then obtained as the difference between these two voltage drops:

$$V_{CE\text{sat}} = V_T \ln \frac{1 + (\beta_{\text{forced}} + 1)/\beta_R}{1 - \beta_{\text{forced}}/\beta_F} \tag{15.16}$$

It is instructive to use Eq. (15.16) to find V_{CEsat} in a typical case. Table 15.1 provides numerical values for the case $\beta_F = 50$, $\beta_R = 0.1$, and various values of β_{forced}. Also, Fig. 15.3 shows a sketch of V_{CEsat} versus β_{forced}. This curve is simply the v_{CE}-i_C characteristic for a constant base current I_B. From Table 15.1 and Fig. 15.3 we note that the infinite value of V_{CEsat} obtained at $\beta_{forced} = \beta_F$ is an indication that the transistor

TABLE 15.1   Numerical Values

β_{forced}	50	48	45	40	30	20	10	1	0
V_{CEsat} (mV)	∞	235	211	191	166	147	123	76	60

is at the boundary between saturation and active mode. Figure 15.3 illustrates this further by showing the independence of v_{CE} on β_{forced} (or i_C) in the active mode. As β_{forced} is reduced, the transistor is driven deeper into saturation, V_{BC} increases, and V_{CEsat} is reduced. Finally, for $\beta_{forced} = 0$, which corresponds to the collector being open-circuited, we obtain a small value of V_{CEsat}. This small value is almost equal to the *offset voltage* of the BJT switch, as defined in Fig. 9.49.

The numerical values of Table 15.1 suggest that for a saturated transistor $V_{CEsat} \cong 0.1$–0.3 V. For approximate calculations we shall henceforth assume that for a transistor on the verge of saturation $V_{CEsat} = 0.3$ V; for a transistor "comfortably" saturated $V_{CEsat} = 0.2$ V; and for a transistor deep into saturation $V_{CEsat} = 0.1$ V.

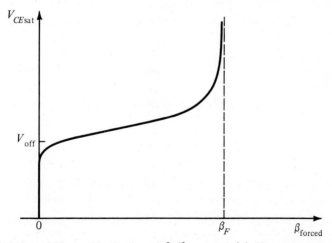

Fig. 15.3   Variation of V_{CEsat} with the forced β ($\beta_{forced} = I_C/I_B$). The vertical line obtained for $\beta_{forced} = \beta_F$ indicates that the transistor has left saturation and entered the active mode.

EXERCISE

15.2 Use the entries for $\beta_{forced} = 1$ and 10 in Table 15.1 to calculate an approximate value for the collector–emitter saturation resistance of a transistor having $I_B = 1$ mA.
Ans. 5.2 Ω

The Inverse Mode. We shall next consider the operation of the BJT in the inverse or reverse mode. Figure 15.4 shows a simple circuit in which the transistor is used with its collector and emitter interchanged. Note that the currents indicated—namely, I_B, I_1, and I_2—have positive values. Thus since $i_C = -I_2$ and $i_E = -I_1$, both i_C and i_E will be negative.

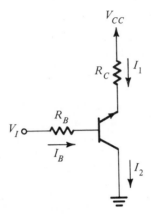

Fig. 15.4 *Circuit in which the transistor is used in the reverse (or inverse) mode.*

Since the roles of the emitter and collector are interchanged, the transistor in the circuit of Fig. 15.4 will operate in the active mode (called *reverse active mode* in this case) when the emitter-base junction is reverse-biased. In this case

$$I_1 = \beta_R I_B$$

Since β_R is usually very low, it makes little sense to operate the BJT in the reverse active mode.

The transistor in the circuit of Fig. 15.4 will saturate (that is, operate in the reverse saturation mode) when the emitter-base junction becomes forward-biased. In this case

$$\frac{I_1}{I_B} < \beta_R$$

We can use the EM equations to find an expression for V_{ECsat} in this case. Such an expression can be directly obtained from Eq. (15.16) as follows: Replace β_{forced} by $-I_2/I_B$ and then replace I_2 by $I_1 + I_B$. The result is

$$V_{ECsat} = V_T \ln \frac{1 + \dfrac{1}{\beta_F} + \left(\dfrac{I_1}{I_B}\right)\left(\dfrac{1}{\beta_F}\right)}{1 - \left(\dfrac{I_1}{I_B}\right)\left(\dfrac{1}{\beta_R}\right)} \qquad (15.17)$$

From this equation it can be seen that the minimum V_{ECsat} is obtained when $I_1 = 0$. This minimum is very close to zero. Furthermore, we observe that the condition $I_1/I_B < \beta_R$ has to be satisfied in order that the denominator remain positive. This, of course, is the condition for the transistor to operate in the reverse saturation mode. Finally, note that since β_R is usually very low, I_1 has to be much smaller than I_B, with the result that V_{ECsat} will be very small. This indeed is the reason for operating the BJT in the reverse saturation mode. Saturation voltages as low as a fraction of a millivolt have been reported. The disadvantage of the reverse saturation mode of operation is a relatively long turnoff time.

EXERCISE

15.3 For the circuit in Fig. 15.4 let $R_B = 1$ kΩ and $V_{CC} = V_I = +5$ V. Assume that $V_{BC} = 0.6$ V, $\beta_R = 0.1$, and $\beta_F = 50$. Calculate approximate values for the emitter voltage in the following cases: $R_C = 1$ kΩ; $R_C = 10$ kΩ; and $R_C = 100$ kΩ.
    Ans. $+4.56$ V; $+0.6$ V; $+3.5$ mV

15.2 RESISTOR-TRANSISTOR LOGIC (RTL)

RTL is an early form of bipolar transistor logic that has become obsolete because of the availability of the higher-performance logic-circuit families studied in subsequent sections. Nevertheless, we shall study RTL and use its simple circuit structure to explain some of the fundamental concepts of digital-circuit design.

Basic Gate Circuit

The basic RTL gate circuit is shown in Fig. 15.5. As will be verified shortly, this gate implements the NOR function. The circuit can be thought of as three transistor inverters whose outputs are connected in parallel. Thus R_C represents the parallel equivalent of the load resistances of the individual inverters. The values shown for R_B, R_C, and V_{CC} are typical of IC RTL gates. Addition of more inputs is straightforward.

Consider now the operation of the circuit, assuming a positive-logic system. If any one of the input voltages is low, that is lower than the transistor *cut-in voltage,* which

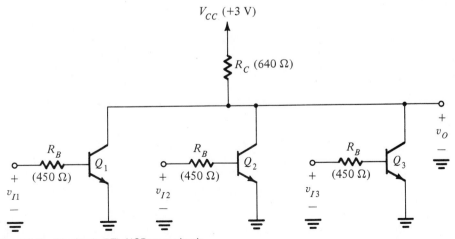

Fig. 15.5 The basic RTL NOR gate circuit.

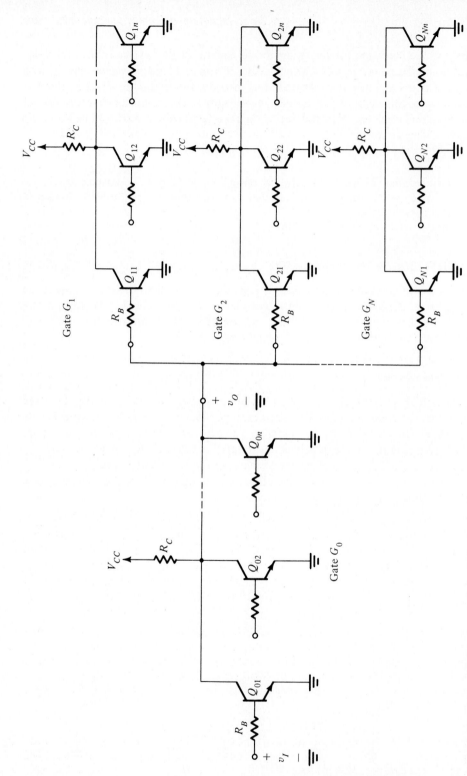

Fig. 15.6 RTL gate G_0 driving N identical gates G_1 to G_N.

is about 0.6 V, the corresponding transistor will be off. If all inputs are low, all transistors will be off and the output will be high ($v_O = V_{CC}$). If any of the inputs is high, the corresponding transistor will be on and (assuming that component values are appropriately chosen) saturated. In this case the output will be equal to V_{CEsat} ($\cong 0.2$ V), that is, a logic 0. If more than one input is high, this situation persists.

From the above we see that the output will be high only in one case: when all inputs are low. Thus the logic function can be expressed as

$$v_O = \overline{v_{I1}} \cdot \overline{v_{I2}} \cdot \overline{v_{I3}}$$

or using DeMorgan's law,

$$v_O = \overline{v_{I1} + v_{I2} + v_{I3}}$$

which is the NOR function. (Note that we have used the voltage symbols to denote also the logic variables.)

Transfer Characteristic

The static transfer characteristic (v_O versus v_I) of a logic gate provides an effective means for displaying many of the gate properties. To derive the transfer characteristic of the RTL gate consider the situation depicted in Fig. 15.6. The RTL gate under study, G_0, is shown driving N identical gates. Gate G_0 has a *fan-in* of n and a *fan-out* of N. In deriving v_O versus v_I we shall assume, for simplicity, that all other inputs to G_0 are low. Thus transistors $Q_{02}, \ldots, Q_{0n}$ will be off, and their existence can be ignored altogether. Furthermore, we shall assume that in each of the driven gates all transistors except the one connected to G_0 are off. Also, all driven transistors will be assumed identical. These simplifying assumptions enable us to eliminate all the off transistors and thus obtain the simplified circuit shown in Fig. 15.7.

Figure 15.8 shows a sketch of the transfer characteristic v_O versus v_I. The characteristic has three distinct regions: the low-input region ($v_I < V_{IL}$); the transition region ($V_{IL} \leq v_I \leq V_{IH}$), and the high-input region ($v_I > V_{IH}$). Input voltages less than V_{IL} are acknowledged by the gate as representing logic 0. Thus V_{IL} is the maximum allowable logic 0 value. For RTL, V_{IL} is equal to the cut-in voltage of the transistor ($\cong 0.6$ V). For $v_I < V_{IL}$ transistor Q_{01} (see Fig. 15.7) is off. The output voltage, however, is not equal to V_{CC}, since current flows through R_C and into the bases of the N driven transistors $Q_{11}, Q_{21}, \ldots, Q_{N1}$. The current supplied to the base of each of these fan-out transistors should be sufficient to drive the transistor into saturation. As the number of fan-out gates (N) increases, this current will decrease, and eventually a value of N will be reached for which the driven transistors do not saturate. This is the mechanism that limits the maximum fan-out of an RTL gate. An expression for the maximum fan-out will be derived at a later stage.

At the moment we are interested in finding the output high voltage V_{OH} assuming a fan-out of N. A simplified equivalent circuit for finding V_{OH} is shown in Fig. 15.9. This equivalent circuit is obtained by replacing the N load transistors by a resistance

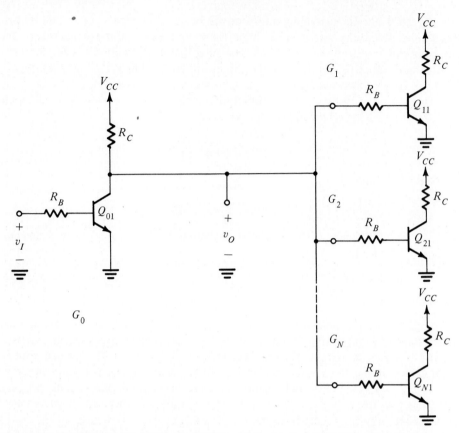

Fig. 15.7   Simplified version of the circuit in Fig. 15.6 with the off transistors not shown.

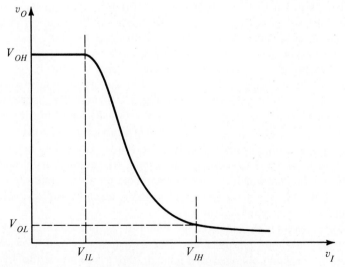

Fig. 15.8   Sketch of the transfer characteristic of gate G_o in Fig. 15.7.

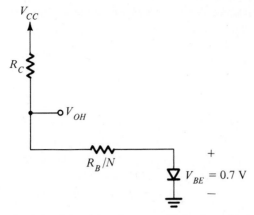

Fig. 15.9 Equivalent circuit for determining the output high voltage (V_{OH}) of gate G_o in Fig. 15.7.

R_B/N in series with an ideal diode having a voltage drop equal to V_{BE} ($\cong 0.7$ V). From this equivalent circuit we obtain

$$V_{OH} = V_{CC} - R_C \frac{V_{CC} - V_{BE}}{R_C + R_B/N}$$

It can readily be seen that as N is increased, V_{OH} decreases. Commercially available RTL is specified for a maximum fan-out of 5. Substituting $N = 5$, $R_B = 450$ Ω, $R_C = 640$ Ω, $V_{BE} \cong 0.7$ V, and $V_{CC} = 3$ V, we obtain $V_{OH} \cong 1$ V.

As the input voltage v_I exceeds V_{IL}, transistor Q_{01} turns on and operates in the active mode. This mode of operation gives rise to the transition region of the gate transfer characteristic. If we continue to increase v_I, a value V_{IH} will be reached at which Q_{01} enters saturation. Input voltages greater than V_{IH} drive the transistor deeper into saturation, and the output voltage decreases slightly below the value V_{OL} ($\cong 0.2$ V). Thus V_{IH} is the minimum input voltage which is interpreted by the gate as logic 1. For the RTL gate V_{IH} is simply the input voltage required to saturate transistor Q_{01} so that $V_{CEsat} = 0.2$ V.

Example 15.1
For the RTL gate find the value of V_{IH} assuming that $\beta_F = 50$ and $\beta_R = 0.1$.

Solution
We first use Eq. (15.16) to find the value of β_{forced} required to obtain $V_{CEsat} = 0.2$ V. The result is $\beta_{forced} = 42.7$. Now, since the collector current is given by

$$I_C = \frac{V_{CC} - V_{CEsat}}{R_C} = \frac{3 - 0.2}{0.64 \text{ k}\Omega} = 4.375 \text{ mA}$$

it follows that the base current should be

$$I_B = \frac{I_C}{\beta_{forced}} = \frac{4.375}{42.7} = 102.5 \text{ }\mu\text{A}$$

Corresponding to this value of base current we obtain the value of input voltage V_{IH} as

$$V_{IH} = V_{BE} + I_B R_B$$
$$= 0.7 + 0.1025 \times 0.45 \cong 0.75 \text{ V}$$

· · ·

Noise Margins

Figure 15.10 shows a piecewise-linear representation of the RTL gate transfer characteristic with the values of the four parameters V_{IL}, V_{IH}, V_{OL}, and V_{OH} indicated. These values were calculated above assuming the gate to be loaded with five identical gates, five being the maximum fan-out typically specified by manufacturers of RTL.

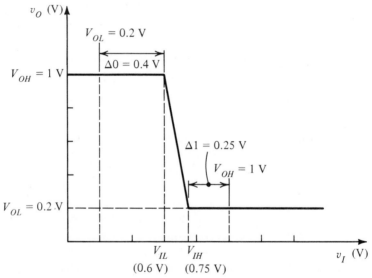

Fig. 15.10   Piecewise-linear approximation of the RTL gate transfer characteristic, illustrating the definition of noise margins.

In a logic system one gate usually drives another. Thus a gate whose output is high at $V_{OH} = 1$ V drives an identical gate whose specified minimum input logic 1 level is $V_{IH} = 0.75$ V. It follows that in this situation there is a margin of safety of $1 - 0.75 = 0.25$ V. This margin implies that if noise were superimposed on the 1-V output signal of the driving gate the driven gate would not be bothered as long as the amplitude of noise voltage is below 0.25 V; for any value of input voltage greater than 0.75 V is properly interpreted as logic 1. This margin is called the *logic 1 noise margin* and is denoted $\Delta 1$,

$$\Delta 1 = V_{OH} - V_{IH}$$

The *logic 0 noise margin*, $\Delta 0$, is similarly defined,

$$\Delta 0 = V_{IL} - V_{OL}$$

For the RTL gate $\Delta 0 = 0.4$ V.

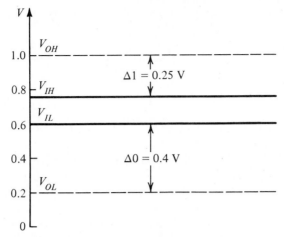

Fig. 15.11   *Logic level band diagram for RTL, indicating noise margins.*

Because of unavoidable variability in the values of circuit components and power supply voltage, the manufacturer usually specifies worst-case values for the four parameters V_{OH}, V_{IH}, V_{OL}, and V_{IL}. These values apply at a given temperature, for a given ΔV_{CC}, and at the specified maximum fan-out. For the definitions of these worst-case values, refer to Section 6.4. As mentioned in Section 6.4, rather than drawing the complete gate transfer characteristic, one is usually satisfied with the *logic band diagram* of Fig. 15.11.

EXERCISES

15.4   IC gate manufacturers usually specify the parameters of the gate transfer characteristic over a range of temperature. For instance, "military grade" ICs are specified over the temperature range $-55\,°C$ to $125\,°C$. The primary reason for the changes in V_{OL}, V_{OH}, etc. with temperature is the change in V_{BE} by -2 mV/$°C$. Assuming that the values calculated in the above apply at room temperature $(25\,°C)$, find the values of V_{IL}, V_{OH}, V_{OL}, V_{IH}, $\Delta1$, and $\Delta0$ at $125\,°C$. Neglect the temperature dependence of V_{CEsat}, α, and β.
Ans.   0.4 V; 0.81 V; 0.2 V; 0.55 V; 0.26 V; 0.2 V

15.5   Repeat Exercise 15.4 at $-55\,°C$.
Ans.   0.76 V; 1.12 V; 0.2 V; 0.91 V; 0.21 V; 0.56 V

Maximum Fan-Out

We have already explained the mechanism that limits the fan-out of RTL gates. We shall now calculate the maximum fan-out of one such gate.

Consider the situation illustrated in Fig. 15.6. Since the fan-out limitation occurs when the driving gate G_0 is off, we shall assume this to be the case and draw the simplified circuit shown in Fig. 15.12. Here we have also assumed that each of the driven gates has a fan-in of 2.

Since we wish to calculate the maximum allowable fan-out, we have to consider the worst-case situation. This happens when the driven gates draw maximum current through resistance R_C of G_0. It can be shown (see Problem 15.15) that this current is maximum when the other transistors (those not driven by G_0) in all but one of the driven

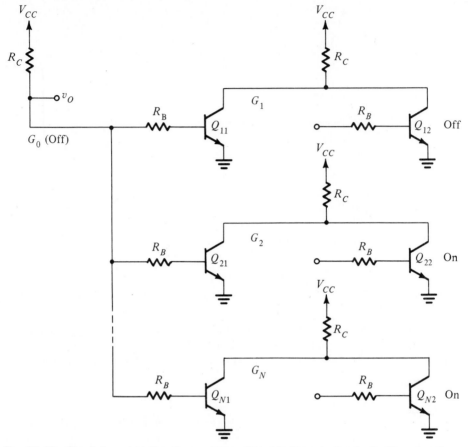

Fig. 15.12 Circuit for calculating the maximum allowable fan-out of gate G_0. The case shown results in maximum current being drawn through R_C of G_0 and hence the worst case.

gates are already saturated. To be specific, assume that in G_1 transistor Q_{12} is off but in G_2 to G_N transistors Q_{22}, Q_{32}, ..., Q_{N2} are already saturated. It can be shown (Ref. 15.3) that while Q_{11} needs a base current of about 100 μA to bring it into saturation, each of Q_{21}, Q_{31}, ..., Q_{N1} will draw (at the corresponding input voltage) a current of about 300 μA. Using these numbers we can now calculate the maximum current drawn through R_C of G_0 as

$$I = 100 + (N - 1)300 \quad \mu\text{A}$$

The output voltage of G_0 can then be obtained as $V_{CC} - IR_C$. For proper operation, this voltage should be at least equal to V_{IH} (which we have found to be 0.75 V). Performing this simple calculation results in a maximum value of N of 11.4. Thus the maximum allowable fan-out of the RTL gate is 11. Commercially available gates are conservatively specified to have a maximum allowable fan-out of 5.

Before leaving the discussion of fan-out we wish to point out the relationship between the gate fan-out and the rise time of its output waveform. Once again consider

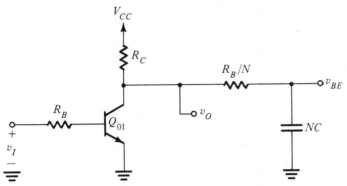

Fig. 15.13 Circuit for finding the relationship between the rise time of v_o (as v_I goes low) and the number of fan-out gates, N.

the situation depicted in Fig. 15.6. Let gate G_0 be initially on. Thus its output voltage is initially low (approx. 0.2 V), and all the driven transistors in the N fan-out gates are initially off. The base–emitter junction of each of the driven transistors can be represented by a capacitance C. This gives rise to the equivalent circuit shown in Fig. 15.13, where Q_{01} is initially on and saturated.

Now let v_I go to the low level. Transistor Q_{01} will turn off and the output voltage v_o will rise to logic 1 level with a time constant of

$$NC\left(R_C + \frac{R_B}{N}\right) = NCR_C + CR_B$$

It follows that the rise time increases with the number of fan-out gates, N.

EXERCISE

15.6 Consider the circuit of Fig. 15.13. If initially $v_o = V_{CEsat} = 0.2$ V, find the time for v_{BE} to reach the cut-in voltage (0.6 V) of the fan-out transistors. Use the component values given in Fig. 15.5 and assume that $N = 5$ and $C = 5$ pF and is constant (independent of the value of v_{BE}).
    Ans. 2.8 ns

The Wired-AND Connection

The output terminals of two or more RTL gates can be connected together. The logic function realized is simply the AND of the functions initially available at the individual output terminals. Since each gate realizes the NOR function of its inputs, this *wired-AND* connection results in the NOR function of all the input variables. It thus effectively increases fan-in. The wired-AND connection of two two-input gates is illustrated in Fig. 15.14.

Propagation Delay

An indicated in Fig. 15.15 the RTL gate propagation delay is measured from the point on the waveform 0.5 V above the logic 0 level. This point is approximately at the middle

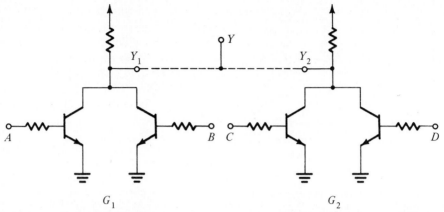

Fig. 15.14   The wired-AND connection of a pair of two-input RTL NOR gates: $Y_1 = \overline{AB}$, $Y_2 = \overline{CD}$ connecting the outputs together results in $Y = Y_1 Y_2 = \overline{ABCD}$, which is in effect a four-input NOR.

of the transition region of the transfer characteristic (see Fig. 15.10). Generally t_{PLH} and t_{PHL} are not equal; their average is of the order of 10 ns. (Note that the subscripts are referred to the output waveform.)

Delay-Power Product

The finite gate propagation delay is due to the time required to charge and discharge the transistor junction capacitances and the various parasitic capacitances as well as the time required to remove the excess minority-carrier charge stored in the base of the saturated transistor. The gate delay is therefore a function of both fan-in and fan-out. One can reduce the gate delay by reducing the values of the resistances R_B and R_C, thereby increasing the current available to charge and discharge the various capacitances. The price paid is an increase in the power dissipated in the gate circuit. This is undesirable not only because of the additional current drawn from the power supply but more importantly because the extra power has to be dissipated in the IC chip and could result in an unacceptable increase in chip temperature. The problem becomes especially acute in LSI and VLSI chips, which contain thousands of gates.

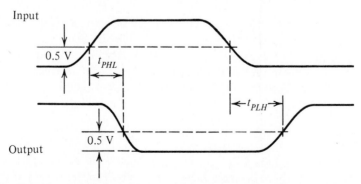

Fig. 15.15   Definition of the propagation delay times t_{PHL} and t_{PLH} for RTL.

The tradeoff between speed of operation and power dissipation is a property that exists in all digital logic circuits. For this reason a figure of merit that combines speed and power and hence can be used for comparing various logic-circuit families has been devised. It is simply the product of the gate propagation delay and its power dissipation. The figure of merit is appropriately called the *delay-power product* (DP). It has the units of energy and is usually expressed in picojoules (pJ).

Like many of the other logic families, RTL is available in two versions: medium-power RTL and low-power RTL. The two versions differ only in resistor values. Medium-power RTL has a delay-power product of the order of 140 pJ.

A Final Remark

Although it is simple and reasonably high speed, RTL has lost favor as a logic family mainly because of its limited fan-out and its rather poor noise margins.

EXERCISE

15.7 Figure E15.7 shows a special RTL circuit. Convince yourself that the function realized is the Exclusive OR, that is,

$$Y = A\bar{B} + \bar{A}B$$

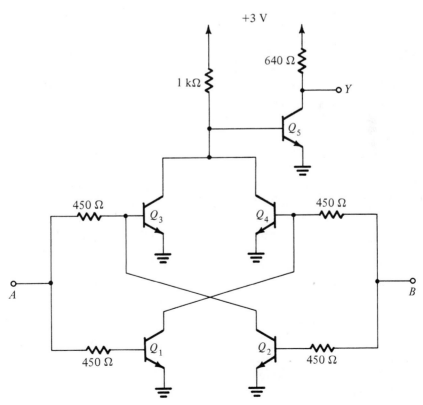

Fig. E15.7

15.3   INTEGRATED INJECTION LOGIC (I^2L)

Integrated injection logic (I^2L) is the most modern form of bipolar transistor logic. It is not commercially available as a set of general-purpose chips but rather is used in LSI and VLSI circuits. I^2L offers a very low (fraction of a pJ) delay-power product coupled with the unique ability to trade power dissipation for increased speed of operation in a given IC chip by simply controlling the dc current supplied to the chip. In addition, it will be shown that I^2L circuits are simple in design and fabrication, making it feasible to fabricate a large number of gates per unit of silicon area.

The Basic Gate Circuit

The basic I^2L gate circuit is shown in Fig. 15.16a. It consists of a number of *npn* transistors (three are shown) having their emitter–base junctions connected in parallel. A constant current I is *injected* to the common-base terminal of the parallel transistors.

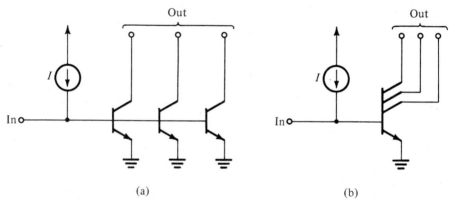

Fig. 15.16   The basic I^2L gate circuit.

From a circuit point of view, and as will be seen shortly from a fabrication point of view as well, the parallel transistors are equivalent to a multiple-collector transistor. We may therefore draw the I^2L gate as indicated in Fig. 15.16b. The gate has one input terminal, the base terminal of the *npn* transistor, and a number of output terminals (three are indicated in Fig. 15.16) corresponding to the collectors of the *npn* transistor. This structure should be contrasted with that of other logic families, such as RTL, whose basic gate has a number of input terminals and one output terminal. In operation the collectors of the I^2L gate will not be left open, of course, but will be connected to the input terminals of other I^2L gates, as explained below.

Circuit Operation

To understand the circuit operation of I^2L, consider the situation shown in Fig. 15.17. Here we show a cascade of three I^2L gates, or inverters, since we use single-output devices. The gate under study is G_2; it is driven by an identical gate G_1 and is driving another gate G_3.

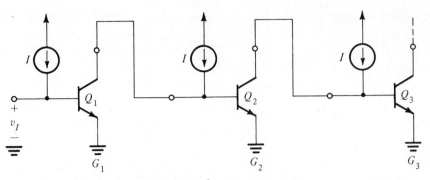

Fig. 15.17 A cascade of three single-output I^2L gates (inverters).

Consider first the case when the input voltage at the base of Q_1, v_I, is low (say, zero volts). Transistor Q_1 will be off and the current I from its associated current source will flow through the source that provides the low input voltage v_I. Since Q_1 is off, the current I of the second gate will flow through the base of Q_2. Thus transistor Q_2 will be on and its collector will conduct the current I of the third gate. Thus Q_2 will be saturated and operating at a forced β of unity. Its collector-to-emitter voltage V_{CEsat} will be small (about 0.1 V) and this will keep Q_3 off.

Next consider the case when v_I is high or the input terminal of G_1 is simply left open. We see that Q_1 will be on and saturated. Its collector will sink the current I of the second gate and its collector-to-emitter voltage will be small ($\cong 0.1$ V). Since this is the input voltage to the second gate, Q_2 will be off. Thus the current I of the third gate will flow into the base of Q_3. This establishes the high-voltage level at the collector of Q_2 at about 0.7 V.

From the above description we see that in I^2L the logic 0 level is V_{CEsat} ($\cong 0.1$ V) and the logic 1 level is V_{BE} ($\cong 0.7$ V). The inverter transfer characteristic is sketched in Fig. 15.18. We note the rather small logic swing and noise margins. Nevertheless,

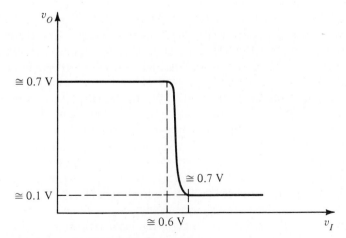

Fig. 15.18 Transfer characteristic of the I^2L inverter.

these values are quite sufficient for proper logic operation within a large array of I^2L gates on a VLSI chip. To interface to off-chip circuitry, however, large swings are usually required at the chip output. This can be obtained by connecting the collector of the output transistor to a suitable power supply (say, $+5$ V if interfacing to TTL circuits is required) through a resistance R_C. In this case the output logic levels will be V_{CEsat} and V_{CC}.

Logic Operation of I^2L

We shall now consider the implementation of logic functions using I^2L circuits. To start, consider the circuit in Fig. 15.19. Here we have tied together the outputs (collectors) of two I^2L inverters, Q_1 and Q_2. The common-collector terminal is then connected to

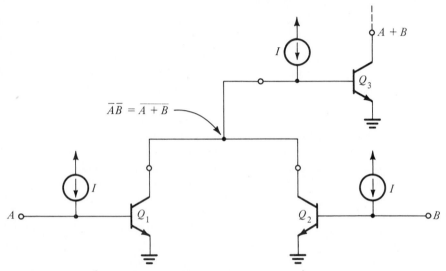

Fig. 15.19   *Use of I^2L to realize the NOR function.*

the input (base) of a third inverter, Q_3. At the outputs of Q_1 and Q_2 we initially (that is, before tying them together) have $\overline{A}$ and $\overline{B}$, respectively. Tying the two collectors together implements a wired-AND function; thus at the common collector we obtain $\overline{A}\overline{B}$. Since $\overline{A}\overline{B} = \overline{A + B}$, we thus see that I^2L gates can be used to implement the NOR function.

Another view of the logic operation of I^2L can be seen by considering Q_3 in Fig. 15.19. The base of Q_3 is connected to two collectors. Thus at the base of Q_3 we have the AND function of the variables initially available at each of these collectors. Q_3 then inverts this function, thus providing at its output the NAND function of the variables initially available at each of the separate collectors connected to its base.

This alternative point of view of the logic capability of I^2L can be generalized as follows: Consider an I^2L structure in which the logic variables $X_1, X_2, \ldots, X_n$ are available at n collector terminals (of n separate gates). Wiring these collectors together and connecting the common collector terminal to the input terminal (base) of an I^2L gate results in the function $X_1 X_2 \cdots X_n$ at the base terminal and the function $\overline{X_1 X_2 \cdots X_n}$ at each of the collectors of the output gate. This process is illustrated in Fig. 15.20.

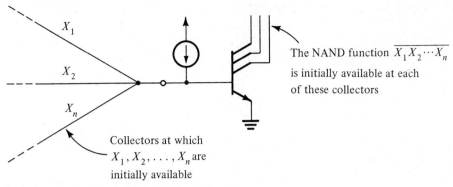

The NAND function $\overline{X_1 X_2 \cdots X_n}$ is initially available at each of these collectors

Collectors at which $X_1, X_2, \ldots, X_n$ are initially available

Fig. 15.20   Use of I^2L to realize the NAND function.

Logic Design With I^2L

Whatever view one takes of the logic operation of I^2L, it should be evident that the two fundamental logic operations performed are the inversion provided by the transistor and the wired-AND function obtained by joining collector terminals from different gates together.

To further illustrate the logic operations of I^2L, consider the circuit in Fig. 15.21. Here we have two input variables A and B and we generate the four minterms $\overline{A}B$, $\overline{A}\overline{B}$, $A\overline{B}$, and AB. To generate these AND functions we need the four variables A, $\overline{A}$, B, and $\overline{B}$ to be available at collector terminals. At the collectors of Q_1 initially (that is,

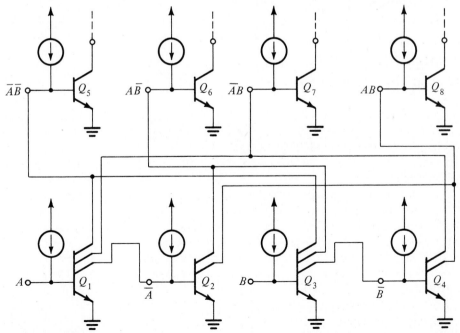

Fig. 15.21   Logic operation of I^2L. This circuit generates the four minterms of the two logic variables A and B.

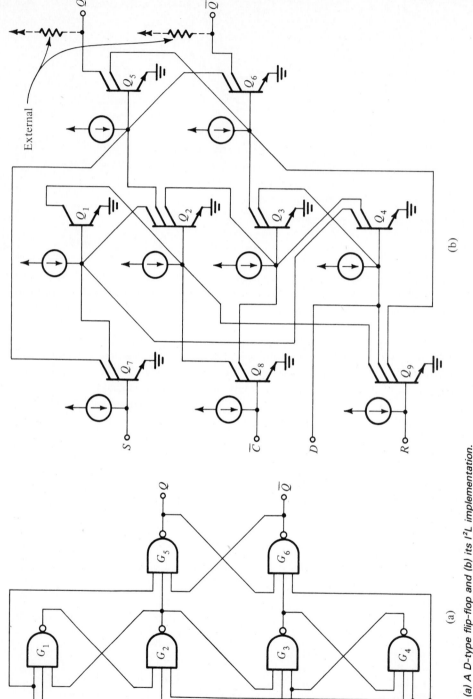

(a)

(b)

Fig. 15.22 (a) A D-type flip-flop and (b) its I²L implementation.

before connecting these collectors to other collectors) we have $\overline{A}$ and at the collectors of Q_3 we have $\overline{B}$. Connecting one of the collectors of Q_1 to the base of another I^2L gate, Q_2, results in A becoming available at the collectors of this gate, Q_2. Similarly, connecting one of the collectors of Q_3 to the base of another I^2L gate, Q_4, results in B becoming available at the collectors of this gate, Q_4.

Now that we have A, $\overline{A}$, B, and $\overline{B}$ available at the collectors of separate I^2L gates, all we need to do to generate a particular minterm is join the two appropriate collectors together and connect the resulting common terminal to the base of another I^2L gate. For instance, $\overline{A}\overline{B}$ is obtained by joining together one of the Q_1 collectors and one of the Q_3 collectors and connecting the common terminal to the base of Q_5. At the base of Q_5 we then have $\overline{A}\overline{B}$, and so on for the other minterms.

From the discussion above we can formulate a simple procedure for the realization of arbitrary logic functions using I^2L. The procedure consists of first obtaining a realization using NAND gates (see Section 6.3). Then each NAND gate is replaced by an I^2L gate. The input terminal of the I^2L gate should be tied to the collectors at which the input variables to the NAND are available. The number of collectors of the I^2L gate should be equal to the number of different nodes to which the output terminal of the NAND gate is connected. To illustrate this procedure we shall consider an example:

Figure 15.22a shows the realization of a D-type flip-flop,[1] with direct set and reset inputs, using NAND gates. The I^2L implementation of this logic circuit is given in Fig. 15.22b. It is simply obtained by replacing each NAND gate with the correspondingly numbered I^2L gate. In addition, Q_7, Q_8, and Q_9 are included in order to make the variables $\overline{S}$, C, and $\overline{R}$ available at collectors. Since the variable D is fed to one input terminal only, no additional gate is included.

The reader is urged to attempt to generate the I^2L implementation on his or her own and use the circuit given as a check.

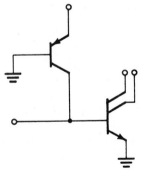

Fig. 15.23 The actual I^2L gate circuit showing a pnp transistor realizing the injection current source I.

Actual Gate Circuit

Up until this point we have shown the I^2L gate circuit with the injection current I supplied by an ideal current source. The actual I^2L gate circuit is shown in Fig. 15.23,

---

[1]D-type flip-flops are discussed in Section 6.5. For the purpose of the present discussion detailed knowledge of the nature of D-type flip-flops is not necessary.

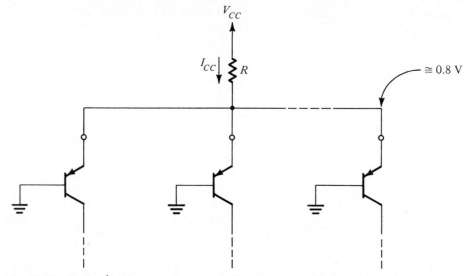

Fig. 15.24 On the I²L chip the emitters of the pnp injectors are usually tied together and fed with the chip current I_{CC}.

from which we see that the *injector* current source is realized by a *pnp* transistor. On an I²L chip containing many gates the emitters of the injector *pnp* transistors are connected together and fed through a common resistor R connected to the power supply V_{CC}, as shown in Fig. 15.24. Usually the voltage at the common-emitter terminal is about 0.8 V, which sets the total dc current supplied to the chip, I_{CC}, to the value

$$I_{CC} \cong \frac{V_{CC} - 0.8}{R}$$

The dc power dissipated in the chip is then given by

$$P \cong I_{CC} \times 0.8$$

from which we note that the power dissipation is directly proportional to I_{CC} and thus can be controlled by controlling the value of I_{CC}.

The *pnp* devices on a given chip are usually reasonably matched, so that one can assume that the current I_{CC} divides more or less equally between them. Thus for a chip with *n* pnp injectors the emitter current of each of the injectors is given by

$$I_E = \frac{I_{CC}}{n}$$

Let us return to the gate circuit in Fig. 15.23. We see that when the input is low ($\cong 0.1$ V) the *pnp* transistor operates in the active mode and its collector current is almost equal to I_E. This current is sunk by the collector of the feeding transistor, and the *npn* transistor is off. On the other hand, when the input is high ($\cong 0.7$ V) the *pnp* injector will be saturated and its collector current, which is fed to the base of the *npn* transistor, will be somewhat smaller than I_E. A more detailed investigation of the operation of I²L is beyond the scope of this text and can be pursued in reference 15.5.

Physical Structure

The success of I²L in LSI and VLSI is due in part to the simplicity of its fabrication. To help explain this point we show in Fig. 15.25 the structure of integrated-circuit *npn* transistors in a somewhat simplified form. (For more details refer to Appendix A.) The transistors are fabricated on a *p*-type substrate with the *p*-type material extending upward between transistors to electrically isolate one transistor from another. This iso-

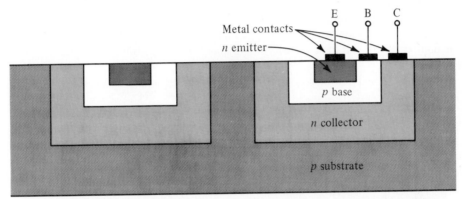

Fig. 15.25   *Cross section of an IC chip illustrating the structure of a standard npn transistor.*

lation is achieved because the *p*-type substrate is connected to the most negative voltage in the circuit. Since current in the *npn* transistor flows vertically, the device is called a *vertical transistor*.

As indicated in Fig. 15.25 the collector region virtually surrounds the emitter, making it difficult for the electrons injected into the thin base to escape being collected by the collector. This physical structure together with the relative doping concentrations used (see Section 9.2) ensure that the forward α (α_F) is close to unity and hence that the forward β (β_F) is large.

The standard IC fabrication process can be used to construct *pnp* transistors as shown in Fig. 15.26. Into the *n*-type material used for the *npn* collector region two *p*-

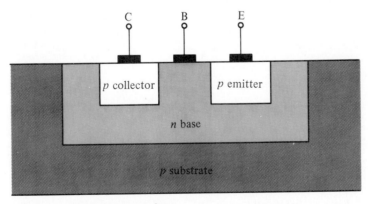

Fig. 15.26   *Structure of lateral pnp transistor.*

type wells are diffused. These p-type wells together with the n-type material form a *pnp* transistor in which current is conducted laterally (rather than vertically as in the *npn* device). This gives this *pnp* transistor the name *lateral pnp*. Because of the physical structure of the lateral *pnp* transistor, its forward β is rather low (5 to 10, typically).

Next consider the physical structure of I²L shown in Fig. 15.27a. The I²L integrated circuit is formed in an n-type material that is connected to ground and serves both as the emitter of the *npn* transistor and as the base of the *pnp* injector (refer also to the circuit shown in Fig. 15.27b). The p-type region, shown at the left, acts both as the base of the *npn* transistor and as the collector of the *pnp* injector. n-type diffusions (of which three are shown) into this p-type region act as collectors for the *npn* transistor. The emitter of the *pnp* injector is formed by another p-type diffusion, indicated at the right side.

An important point to note is that the *npn* transistor is a vertical device but the

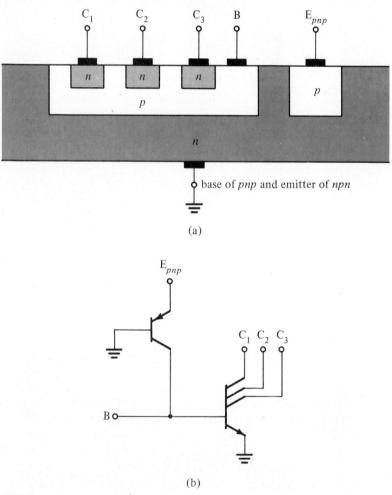

(a)

(b)

Fig. 15.27   Physical structure of I²L gate.

pnp transistor is of the lateral type. The vertical *npn* device, however, differs from the standard structure of Fig. 15.25 in that in the I²L case the collector has a small area and is surrounded by the emitter. We should therefore expect that for this inverted transistor β_R is large and β_F is small. This is indeed the case, with β_R typically around 50 and β_F about 5. It should be noted that β_F of the multiple-collector transistor is defined as the ratio of the sum of collector currents to the base current.

Another interesting point worth observing is the multiple usage of the various regions in the IC. In fact, a total of four separate regions (excluding the multiplicity of collectors used) are used to form two transistors. One can therefore say that the *npn* and *pnp* transistors are *merged,* which gives rise to the name *merged transistor logic* (MTL) originally used for I²L.

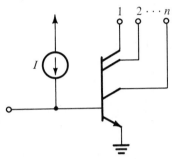

Fig. 15.28   An I²L gate with a fan-out of n.

EXERCISES

15.8 Consider the I²L gate shown in Fig. 15.28: If the *npn* transistor has a forward β of 5, what is the maximum value for the fan-out *n* such that the transistor saturates with $\beta_{\text{forced}} \leq 0.8 \beta$?
Ans. $n = 4$

15.9 Consider an I²L gate having a fan-out of 3 and an injection current *I*. Let the *npn* transistor have $\beta_F = 5$ and $\beta_R = 50$. Calculate the saturation voltage when each collector is sinking a load current equal to *I*.
Ans. 25 mV

Speed of Operation

The speed of operation or the propagation delay of the I²L gate is determined by the wiring (stray) capacitance, the transistor depletion capacitance, the transistor diffusion capacitance, the storage time of the saturated transistor, and the value of the injection current *I*. The dependence on the current *I* is twofold. First, since it is the current *I* that charges and discharges the various capacitances, increasing the value of *I* shortens the charging and discharging times and hence shortens the gate propagation delay. Second, the value of the transistor diffusion capacitance and the amount of charge stored in the saturated transistor are both proportional to the value of *I*. What, then, is the overall dependence of gate delay on the injection current *I*?

At low levels of injection current the diffusion capacitance and the charge storage effects are negligible, and wiring and depletion capacitances dominate. Since these latter capacitances are independent of current, increasing the value of I reduces the gate delay t_P; thus at low injection t_P is inversely proportional to I. Since the power dissipation is directly proportional to I, it follows that at low injection currents the delay-power product is constant. One can therefore increase the speed of operation by simply increasing the injection current (that is, by increasing the chip supply current I_{CC}) at the expense of a proportionate increase in power dissipation. This convenient trade-off applies only up to a point, beyond which high-injection effects occur.

At high levels of injection current I the diffusion capacitance dominates over the wiring and depletion capacitances. Since the value of the diffusion capacitance is proportional to the current I, it follows that the charging and discharging times and hence the gate delay will be constant independent of the value of I. That is, at high injection t_P is constant. Thus increasing the current I in this range of operation does not result in any increase in speed but only in increased power dissipation.

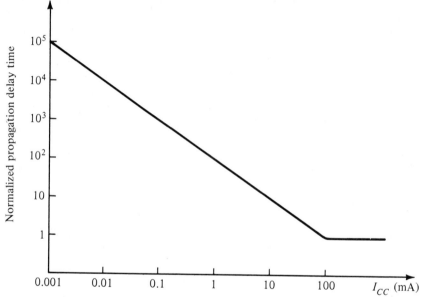

Fig. 15.29 *Variation of gate propagation delay with the current injected to an I²L chip.*

In conclusion, we show in Fig. 15.29 the variation of gate propagation delay (normalized) versus the dc current supplied to a commercially available I²L chip. Note that up to $I_{CC} = 100$ mA one can trade off dissipation for speed.

Finally it should be mentioned that in the low-injection region I²L achieves delay-power products less than 1 pJ.

EXERCISE

15.10 Assume that the *pnp* emitter is at a voltage approximately equal to 0.8 V. Find the delay time of an I²L gate having $I = 10$ μA provided that the delay-power product is 0.8 pJ.

Ans. 100 ns

15.4   *TRANSISTOR-TRANSISTOR LOGIC (TTL OR T²L)*

Over the past decade TTL has enjoyed immense popularity. For the bulk of digital-systems applications employing SSI and MSI packages, TTL is rivaled only by CMOS.

We shall begin this section with a study of the evolution of TTL from an earlier form of BJT logic known as diode-transistor logic (DTL). In this way we shall explain the function of each of the stages of the complete TTL gate circuit. Characteristics of standard TTL gates will be studied in Section 15.5. This will be followed by a brief exposition of special forms of TTL that feature improved performance in one or more characteristics.

Evolution of TTL From DTL

The basic DTL gate circuit in discrete form was discussed in Chapter 9 (see Fig. 9.44). The integrated-circuit form of the DTL gate is shown in Fig. 15.30 with only one input indicated. As a prelude to introducing TTL, we have drawn the input diode as a diode-connected transistor (Q_1), which corresponds to how diodes are made in IC form.

This circuit differs from the discrete DTL circuit of Fig. 9.44 in two important aspects. First, one of the "steering" diodes is replaced by the base–emitter junction of a transistor (Q_2) that is either cut off (when the input is low) or in the active mode (when the input is high). This is done to increase the fan-out capability of the gate. A detailed explanation of this point, however, is not relevant to our study of TTL. Second,

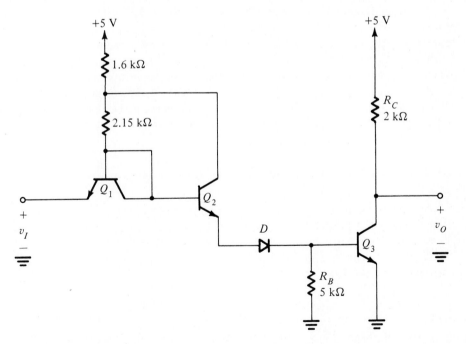

Fig. 15.30   *IC form of DTL gate with the input diode shown as a diode-connected transistor (Q_1). Only one input terminal is shown.*

the resistance R_B is returned to ground rather than to a negative supply, as was done in the earlier discrete circuit. An obvious advantage of this is the elimination of the additional power supply. The disadvantage, however, is that the reverse base current available to remove the excess charge stored in the base of Q_3 is rather small. We shall elaborate on this point below.

EXERCISE

15.11 Consider the DTL gate circuit shown in Fig. 15.30 and assume that $\beta(Q_2) = \beta(Q_3) = 50$.
(a) When $v_I = 0.2$ V, find the input current.
(b) When $v_I = +5$ V, find the base current of Q_3.
Ans. (a) 1.1 mA; (b) 1.76 mA

Reasons for the Slow Response of DTL

The DTL gate has relatively good noise margins and reasonably good fan-out capability. Its response, however, is rather slow. This is due to two problems. First, when the input goes low and Q_2 and D turn off, the charge stored in the base of Q_3 has to leak through R_B to ground. The initial value of the reverse base current that accomplishes this "base discharging" process is approximately 0.7 V/R_B, which is about 0.14 mA. Because this current is quite small in comparison to the forward base current, the time required for the removal of base charge is rather long, which contributes to lengthening the gate delay.

The second reason for the relatively slow response of DTL derives from the nature of the output circuit of the gate, which is simply a common-emitter transistor. Figure 15.31 shows the output transistor of a DTL gate driving a capacitive load C_L. The capacitance C_L represents the input capacitance of another gate and/or the wiring and parasitic capacitances that are inevitably present in any circuit. When Q_3 is turned on, its collector voltage cannot instantaneously fall because of the existence of C_L. Thus Q_3 will not immediately saturate but rather will operate in the active region. The collector of Q_3 will therefore act as a constant-current source and will sink a relatively large

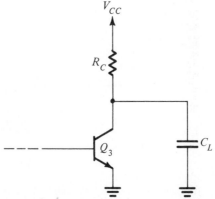

Fig. 15.31 The output circuit of a DTL gate driving a capacitive load C_L.

current (βI_B). This large current will rapidly discharge C_L. We thus see that the common-emitter output stage features a short turn-on time. However, turn-off is another matter.

Consider next the operation of the common-emitter output stage when Q_3 is turned off. The output voltage will not rise immediately to the high level (V_{CC}). Rather, C_L will charge up to V_{CC} through R_C. This is a rather slow process, and it results in lengthening the DTL gate delay (and similarly the RTL gate delay).

Having identified the two reasons for the slow response of DTL, we shall see in the following how these problems are remedied in TTL.

Input Circuit of the TTL Gate

Figure 15.32 shows a conceptual TTL gate with only one input terminal indicated. The most important feature to note is that the input diode has been replaced by a transistor.

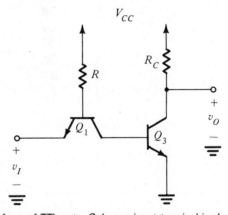

Fig. 15.32   Conceptual form of TTL gate. Only one input terminal is shown.

One can think of this simply as if the short circuit between base and collector of Q_1 in Fig. 15.30 has been removed.

To see how the conceptual TTL circuit of Fig. 15.32 works, let the input v_I be high (say, $v_I = V_{CC}$). In this case current will flow from V_{CC} through R, thus forward-biasing the base–collector junction of Q_1. Meanwhile, the base–emitter junction of Q_1 will be reverse-biased. Therefore Q_1 will be operating in the *inverse active mode*—that is, in the active mode but with the roles of emitter and collector interchanged. The voltages and currents will be as indicated in Fig. 15.33, where the current I can be calculated from

$$I = \frac{V_{CC} - 1.4}{R}$$

In actual TTL circuits Q_1 is designed to have a very low reverse β ($\beta_R \cong 0.02$). Thus the gate input current will be very small and the base current of Q_3 will be approximately equal to I. This current will be sufficient to drive Q_3 into saturation, and the output voltage will be low (0.1 to 0.2 V).

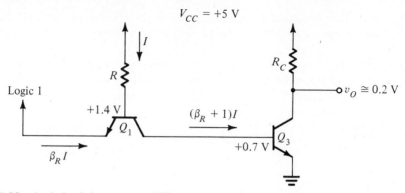

Fig. 15.33 *Analysis of the conceptual TTL gate when the input is high.*

Next let the gate input voltage be brought down to the logic 0 level (say, $v_I \cong 0.2$ V). The current I will then be diverted to the emitter of Q_1. The base–emitter junction of Q_1 will become forward-biased, and the base voltage of Q_1 will therefore drop to 0.9 V. Since Q_3 *was* in saturation, its base voltage will remain at $+0.7$ V pending the removal of the excess charge stored in the base region. Figure 15.34 indicates the various voltage and current values immediately after the input is lowered. We see that Q_1

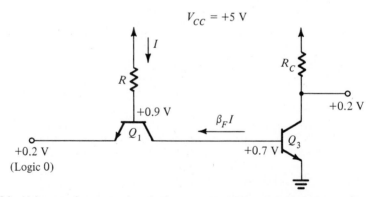

Fig. 15.34 *Voltage and current values in the conceptual TTL circuit immediately after the input is lowered.*

will be operating in the normal active mode and its collector will carry a large current ($\beta_F I$). This large current rapidly discharges the base of Q_3 and drives it into cutoff. We thus see the action of Q_1 in speeding up the turnoff process.

As Q_3 turns off, the voltage at its base is reduced and Q_1 enters the saturation mode. Eventually the collector current of Q_1 will become negligibly small, which implies that its V_{CEsat} will be approximately 0.1 V and the base of Q_3 will be at about 0.3 V, which keeps Q_3 in cutoff.

Output Circuit of the TTL Gate

The above discussion illustrates how one of the two problems that slow down the operation of DTL is solved in TTL. The second problem, the long rise time of the output waveform, is solved by modifying the output stage, as we shall now explain.

First, recall that the common-emitter output stage provides fast discharging of load capacitance but rather slow charging. The opposite is obtained in the emitter-follower output stage shown in Fig. 15.35. Here as v_I goes high the transistor turns on and pro-

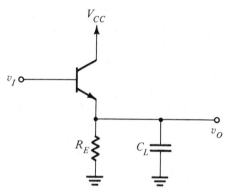

Fig. 15.35   An emitter-follower output stage with capacitive load.

vides a low output resistance (characteristic of emitter followers), which results in fast charging of C_L. On the other hand, when v_I goes low the transistor turns off and C_L is then left to discharge slowly through R_E.

It follows that an optimum output stage would be a combination of the common-emitter and the emitter-follower configurations. Such an output stage is shown in Fig. 15.36. The stage has to be driven by two *complementary* signals v_{I1} and v_{I2}. When v_{I1} is high v_{I2} will be low, and in this case Q_3 will be on and saturated and Q_4 will be off. The common-emitter transistor Q_3 will then provide the fast discharging of load capacitance and in steady state provide a low resistance (R_{CEsat}) to ground. Thus when the output is low the gate can *sink* substantial amounts of current through the saturated transistor Q_3.

When v_{I1} is low and v_{I2} is high, Q_3 will be off and Q_4 will be conducting. The emitter-follower Q_4 will then provide fast charging of load capacitance. It also provides the gate with a low output resistance in the high state and hence with the ability to *source* a substantial amount of load current.

Because of the appearance of the circuit in Fig. 15.36 with Q_4 stacked on top of Q_3, the circuit has been given the name *totem-pole output stage*. Also, because of the action of Q_4 in *pulling up* the output voltage to the high level, Q_4 is referred to as the *pull-up transistor*. Since the pulling up is achieved here by an active element (Q_4), the circuit is said to have an *active pull-up*. This is in contrast to the *passive pull-up* of RTL and DTL gates. Finally, note that a special *driver circuit* is needed to generate the two complementary signals v_{I1} and v_{I2}.

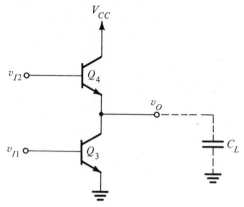

Fig. 15.36 The totem-pole output stage.

Example 15.2

We wish to analyze the circuit shown together with its driving waveforms in Fig. 15.37 to determine the waveform of the output signal v_O. Assume that Q_3 and Q_4 have $\beta = 50$.

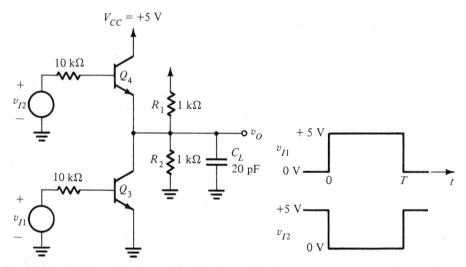

Fig. 15.37 Circuit and input waveforms for Example 15.2.

Solution

Consider first the situation before v_{I1} goes high—that is, at time $t < 0$. In this case Q_3 is off and Q_4 is on, and the circuit can be simplified to that shown in Fig. 15.38. In this simplified circuit we have replaced the voltage divider (R_1, R_2) by its Thévenin equivalent. In the steady state C_L will be charged to the output voltage v_O, whose value can be obtained as follows

$$5 = 10 \times I_B + V_{BE} + I_E \times 0.5 + 2.5$$

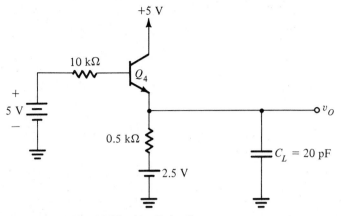

Fig. 15.38   *The circuit of Fig. 15.37 when Q_3 is off.*

Substituting $V_{BE} \cong 0.7$ V and $I_B = I_E/(\beta + 1) = I_E/51$ gives $I_E = 2.59$ mA. Thus the output voltage v_O is given by

$$v_O = 2.5 + I_E \times 0.5 = 3.79 \text{ V}$$

We next consider the circuit as v_{I1} goes high and v_{I2} goes low. Transistor Q_3 turns on and transistor Q_4 turns off, and the circuit simplifies to that shown in Fig. 15.39. Again we have used the Thévenin equivalent of the divider (R_1, R_2). We shall also

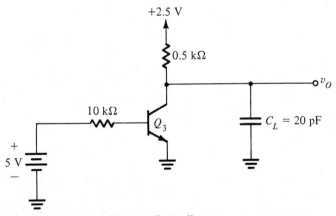

Fig. 15.39   *The circuit of Fig. 15.37 when Q_4 is off.*

assume that the switching times of the transistors are negligibly small. Thus at $t = 0+$ the base current of Q_3 becomes

$$I_B = \frac{5 - 0.7}{10} = 0.43 \text{ mA}$$

Since at $t = 0$ the collector voltage of Q_3 is 3.79 V and since this value cannot change instantaneously because of C_L, we see that at $t = 0+$ transistor Q_3 will be in the active

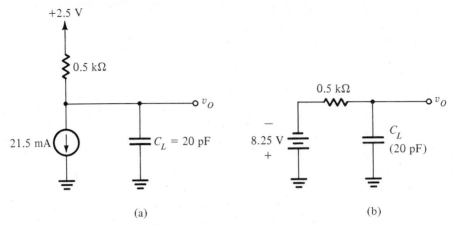

(a)   (b)

Fig. 15.40 (a) Equivalent circuit for the circuit in Fig. 15.39 when Q_3 is in the active mode. (b) Simpler version of the circuit in (a) obtained using Thévenin's theorem.

mode. The collector current of Q_3 will be βI_B, which is 21.5 mA, and the circuit will have the equivalent shown in Fig. 15.40a. A simpler version of this equivalent circuit, obtained using Thévenin's theorem, is shown in Fig. 15.40b.

The equivalent circuit of Fig. 15.40 applies as long as Q_3 remains in the active mode. This condition persists while C_L is being discharged and until v_O reaches about +0.3 V, at which time Q_3 enters saturation. This is illustrated by the waveform in Fig. 15.41. The time for the output voltage to fall from +3.79 V to +0.3 V, which can be considered the *fall time* t_f, can be obtained from

$$-8.25 - (-8.25 - 3.79)e^{-t_f/\tau} = 0.3$$

which results in

$$t_f \cong 0.34\tau$$

where

$$\tau = C_L \times 0.5 \text{ K} = 10 \text{ ns}$$

Thus $t_f = 3.4$ ns.

After Q_3 enters saturation the capacitor discharges further to the final steady-state value of V_{CEsat} ($\cong 0.2$ V). The transistor model that applies during this interval is more complex; since the interval in question is quite short, we shall not pursue the matter further.

Consider next the situation as v_{I1} goes low and v_{I2} goes high at $t = T$. Transistor Q_3 turns off as Q_4 turns on. We shall assume that this occurs immediately and thus at $t = T+$ the circuit simplifies to that in Fig. 15.38. We have already analyzed this circuit in steady state and thus know that eventually v_O will reach +3.79 V. Thus v_O rises exponentially from +0.2 V toward +3.79 V with a time constant of $C_L\{0.5 \text{ K} \| [10 \text{ K}/(\beta + 1)]\}$, where we have neglected the emitter resistance r_e. Denoting this time constant τ_1 we obtain $\tau_1 = 2.8$ ns. Defining the rise time t_r as the time for v_O to reach 90% of the final value we obtain $3.79 - (3.79 - 0.2)e^{-t_r/\tau_1} = 0.9 \times 3.79$

which results in $t_r = 6.4$ ns. Figure 15.41 illustrates the details of the output voltage waveform.

• • •

The Complete Circuit of the TTL Gate

Figure 15.42 shows the complete TTL gate circuit. It consists of three stages: the input transistor Q_1, whose operation has already been explained, the driver stage Q_2, whose

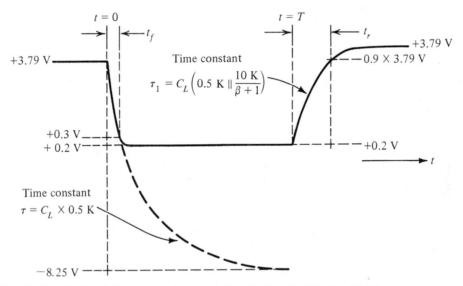

Fig. 15.41 Details of the output voltage waveform for the circuit in Fig. 15.37.

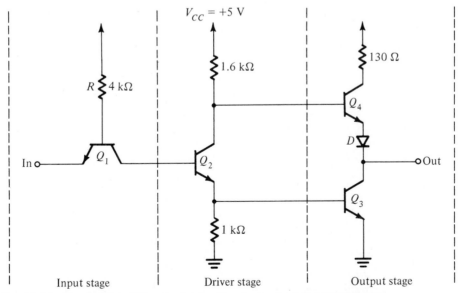

Fig. 15.42 The complete TTL gate circuit with only one input terminal indicated.

function is to generate the two complementary voltage signals required to drive the totem-pole circuit, which is the third and output stage of the gate. The totem-pole circuit in the TTL gate has two additional components: the 130-Ω resistance in the collector circuit of Q_4 and the diode D in the emitter circuit of Q_4. The function of these two additional components will be explained shortly. Note that the TTL gate is shown with only one input terminal indicated. Inclusion of additional input terminals will be considered in Section 15.5.

Because the driver stage Q_2 provides two complementary (that is, out of phase) signals, it is known as a *phase splitter*.

We shall now provide a detailed analysis of the TTL gate circuit in its two extreme states: with the input high and with the input low.

Analysis When the Input Is High

When the input is high (say, +5 V), the various voltages and currents of the TTL circuit will have the values indicated in Fig. 15.43. The analysis illustrated in Fig. 15.43 is quite straightforward. As expected, the input transistor is operating in the inverse active mode and the input current, called the *input high current I_{IH}*, is small:

$$I_{IH} = \beta_R I \cong 15 \ \mu A$$

where we assume that $\beta_R \cong 0.02$.

The collector current of Q_1 flows into the base of Q_2, and its value is sufficient to saturate the phase-splitter transistor Q_2. The latter supplies the base of Q_3 with suffi-

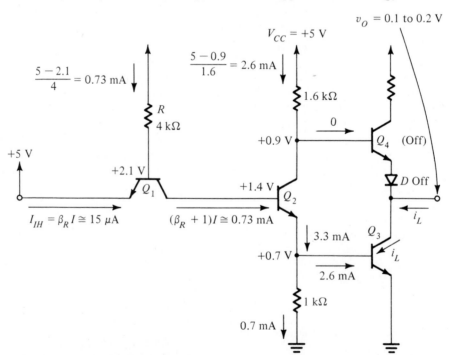

Fig. 15.43 Analysis of the TTL gate with the input high.

cient current to drive it into saturation and lower its output voltage to V_{CEsat} (0.1 to 0.2 V). The voltage at the collector of Q_2 is $V_{BE3} + V_{CEsat}$ (Q_2), which is approximately +0.9 V. If diode D were not included, this voltage would be sufficient to turn Q_4 on, which is contrary to the proper operation of the totem-pole circuit. Including diode D ensures that both Q_4 and D remain off. The saturated transistor Q_3 then establishes the low output voltage of the gate (V_{CEsat}) and provides a low impedance to ground.

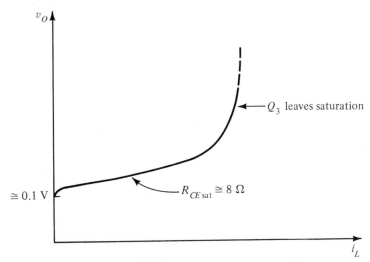

Fig. 15.44 *The v_O-i_L characteristic of the TTL gate when the output is low.*

In the low-output state the gate can sink a load current i_L provided that the value of i_L does not exceed $\beta \times 2.6$ mA, which is the maximum collector current that Q_3 can sustain while remaining in saturation. Obviously the greater the value of i_L, the greater the output voltage will be. To maintain the logic 0 level below a certain specified limit, a corresponding limit has to be placed on the load current i_L. As will be seen shortly, it is this limit that determines the maximum fan-out of the TTL gate.

Figure 15.44 shows a sketch of the output voltage v_O versus the load current i_L of the TTL gate when the output is low. This is simply the i_C-v_{CE} characteristic curve of Q_3 measured with a base current of 2.6 mA. Note that at $i_L = 0$, v_O is the offset voltage, which is about 100 mV.

EXERCISE

15.12 Assume that the saturation portion of the v_O-i_L characteristic shown in Fig. 15.44 can be approximated by a straight line (of slope $= 8\ \Omega$) that intersects the v_O axis at 0.1 V. Find the maximum load current that the gate is allowed to sink if the logic 0 level is specified to be ≤ 0.3 V.
Ans. 25 mA

Analysis When the Input Is Low

Consider next the operation of the TTL gate when the input is at the logic 0 level ($\cong 0.2$ V). The analysis is illustrated in Fig. 15.45, from which we see that the base–

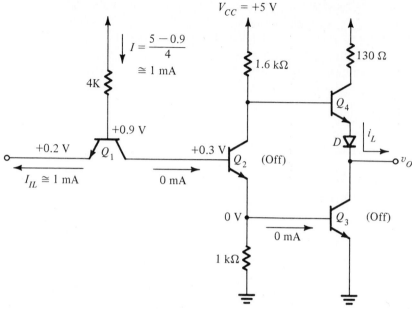

Fig. 15.45   *Analysis of the TTL gate when the input is low.*

emitter junction of Q_1 will be forward-biased and the base voltage will be approximately $+0.9$ V. Thus the current I can be found to be approximately 1 mA. Since 0.9 V is insufficient to forward-bias the series combination of the collector–base junction of Q_1 and the base–emitter junction of Q_2 (at least 1.2 V would be required), the latter will be off. Therefore the collector current of Q_1 will be almost zero and Q_1 will be saturated, with $V_{CEsat} \cong 0.1$ V. Thus the base of Q_2 will be at approximately $+0.3$ V, which is indeed insufficient to turn Q_2 on.

The gate input current in the low state, called *input low current* I_{IL}, is approximately equal to the current I ($\cong 1$ mA) and flows out of the emitter of Q_1. If the TTL gate is driven by another TTL gate, the output transistor Q_3 of the driving gate should sink this current I_{IL}. Since the output sink current of a TTL gate is limited to a certain maximum value, the maximum fan-out of the gate is directly determined by the value of I_{IL}.

EXERCISES

15.13   Consider the TTL gate analyzed in Exercise 15.12. Find its maximum allowable fan-out using the value of I_{IL} calculated above.
Ans.   25

15.14   Use Eq. (15.16) to find V_{CEsat} of transistor Q_1 when the input of the gate is low (0.2 V). Assume that $\beta_F = 50$ and $\beta_R = 0.02$.
Ans.   98 mV

Let us continue with our analysis of the TTL gate. When the input is low we see that both Q_2 and Q_3 will be off. Transistor Q_4 will be on and will supply (source) the load current i_L. Depending on the value of i_L, Q_4 will be either in the active mode or in the saturation mode.

With the gate output terminal open, the current i_L will be very small (mostly leakage) and the two junctions (base–emitter junction of Q_4 and diode D) will be barely conducting. Assuming that each junction has a 0.65-V drop and neglecting the voltage drop across the 1.6-kΩ resistance, we find that the output voltage will be

$$v_O \cong 5 - 0.65 - 0.65 = 3.7 \text{ V}$$

As i_L is increased, Q_4 and D conduct more heavily, but for a range of i_L, Q_4 remains in the active mode, and v_O is given by

$$v_O = V_{CC} - \frac{i_L}{\beta + 1} \times 1.6 \text{ K} - V_{BE4} - V_D$$

If we keep increasing i_L, a value will be reached at which Q_4 saturates. Then the output voltage becomes determined by the 130-Ω resistance according to the approximate relationship

$$v_O \cong V_{CC} - i_L \times 130 - V_{CEsat}(Q_4) - V_D$$

Function of the 130-Ω Resistance

At this point the reason for including the 130-Ω resistance should be evident: simply to limit the current that flows through Q_4, especially in the event that the output terminal is accidentally short-circuited to ground. This resistance also limits the supply current in another circumstance, namely, when Q_4 turns on while Q_3 is still in saturation. To see how this occurs, consider the case where the gate input was high and then is suddenly brought down to the low level. Transistor Q_2 will turn off relatively fast because of the availability of large reverse current supplied to its base terminal by the collector of Q_1. On the other hand, the base of Q_3 will have to discharge through the 1-kΩ resistance, and thus Q_3 will take some time to turn off. Meanwhile Q_4 will turn on, and a large current pulse will flow through the series combination of Q_4 and Q_3. Part of this current will serve the useful purpose of charging up any load capacitance to the logic 1 level. The magnitude of the current pulse will be limited by the 130-Ω resistance to about 30 mA.

The occurrence of these current pulses or spikes raises another important issue. The current spikes have to be supplied by the V_{CC} source and because of its finite source resistance will result in voltage spikes (or "glitches") superimposed on V_{CC}. These voltage spikes could be coupled to other gates and flip-flops in the digital system and thus might produce false switching in other parts of the system. This effect, which might loosely be called *crosstalk,* is a problem in TTL systems. To reduce the size of the voltage spikes, capacitors (called bypass capacitors) should be connected to ground at frequent locations on the supply rail. These capacitors lower the impedance of the supply voltage source and hence reduce the magnitude of the voltage spikes. Alternatively, one can think of the bypass capacitors as supplying the impulsive current spikes.

EXERCISES

15.15 Assuming that Q_4 has $\beta = 50$ and that at the verge of saturation $V_{CEsat} = 0.3$ V, find the value of i_L at which Q_4 saturates.
Ans. 4.16 mA

15.16 Assuming that at a current of 1 mA the voltage drops across the emitter–base junction of Q_4 and the diode D are each 0.7 V, find v_O when $i_L = 1$ mA and 10 mA.

Ans. 3.6 V; 2.74 V

15.17 Find the maximum current that can be sourced by a TTL gate while the output high level (V_{OH}) remains greater than the minimum guaranteed value of 2.4 V.

Ans. 12.3 mA

15.5 CHARACTERISTICS OF STANDARD TTL

Because of its popularity and importance, TTL will be studied further in this and the next sections. In this section we shall consider some of the important characteristics of standard TTL gates. Special improved forms of TTL will be dealt with in Section 15.6.

Transfer Characteristic

Figure 15.46 shows the TTL gate together with a sketch of its transfer characteristic drawn in a piecewise-linear fashion. The actual characteristic is, of course, a smooth curve. We shall now explain the transfer characteristic and calculate the various break-points and slopes. It will be assumed that the output terminal of the gate is open.

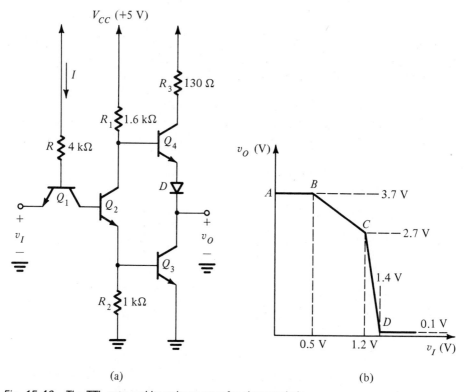

(a)   (b)

Fig. 15.46 *The TTL gate and its voltage transfer characteristic.*

Segment *AB* is obtained when transistor Q_1 is saturated, Q_2 and Q_3 are off, and Q_4 is on. The output voltage is approximately two diode drops below V_{CC}. At point *B* the phase splitter (Q_2) begins to turn on because the voltage at its base reaches 0.6 V (0.5 V + V_{CEsat} of Q_1).

Over segment *BC* transistor Q_1 remains saturated, but more and more of its base current *I* gets diverted to its base–collector junction and into the base of Q_2, which operates as a linear amplifier. Transistor Q_4 and diode *D* remain on, with Q_4 acting as an emitter follower. Meanwhile the voltage at the base of Q_3, although increasing, remains insufficient to turn Q_3 on (less than 0.6 V).

Let us now find the slope of segment *BC* of the transfer characteristic. Let the input v_I increase by an increment Δv_I. This increment appears at the collector of Q_1, since the saturated Q_1 behaves (approximately) as a three-terminal short circuit as far as signals are concerned. Thus at the base of Q_2 we have a signal Δv_I. Neglecting the loading of emitter-follower Q_4 on the collector of Q_2, we can find the gain of the phase-splitter from

$$\frac{v_{c2}}{v_{b2}} = -\frac{\alpha_2 R_1}{r_{e2} + R_2} \tag{15.18}$$

The value of r_{e2} will obviously depend on the current in Q_2. This current will range from zero (as Q_2 begins to turn on) to the value that results in a voltage of about 0.6 V at the emitter of Q_2 (the base of Q_3). This latter value is about 0.6 mA and corresponds to point *C* on the transfer characteristic. Assuming an average current in Q_2 of 0.3 mA, we obtain $r_{e2} \cong 83\ \Omega$. For $\alpha = 0.98$, Eq. (15.18) results in a gain value of 1.45. Since the gain of the output-follower Q_4 is close to unity, the overall gain of the gate, which is the slope of the *BC* segment, is about -1.45.

As already implied, breakpoint *C* is determined by Q_3 starting to conduct. The corresponding input voltage can be found from

$$v_I(C) = V_{BE3} + V_{BE2} - V_{CEsat}(Q_1)$$
$$= 0.6 + 0.7 - 0.1 = 1.2\ \text{V}$$

At this point the emitter current of Q_2 is approximately 0.6 mA. The collector current of Q_2 is also approximately 0.6 mA; neglecting the base current of Q_4, the voltage at the collector of Q_2 is

$$v_{c2}(C) = 5 - 0.6 \times 1.6 \cong 4\ \text{V}$$

Thus Q_2 is still in the active mode. The corresponding output voltage is

$$v_O(C) = 4 - 0.65 - 0.65 = 2.7\ \text{V}$$

As v_I is increased past the value of $v_I(C) = 1.2$ V, Q_3 begins to conduct and operates in the active mode. Meanwhile, Q_1 remains saturated, and Q_2 and Q_4 remain in the active mode. The circuit behaves as an amplifier until Q_2 and Q_3 saturate and Q_4 cuts off. This occurs at point *D* on the transfer characteristic, which corresponds to an input voltage $v_I(D)$ obtained from

$$v_I(D) = V_{BE3} + V_{BE2} + V_{BC1} - V_{BE1}$$
$$= 0.7 + 0.7 + 0.7 - 0.7 = 1.4\ \text{V}$$

Note that we have in effect assumed that at point D transistor Q_1 is still saturated but with $V_{CEsat} \cong 0$. To see how this comes about, note that from point B on, more and more of the base current of Q_1 is diverted to its base–collector junction. Thus while the drop across the base–collector junction increases, that across the base–emitter junction decreases. At point D these drops become almost equal. For $v_I > v_I(D)$ the base–emitter junction of Q_1 cuts off; thus Q_1 leaves saturation and enters the inverse active mode.

Calculation of gain over the segment CD is a relatively complicated task. This is due to the fact that there are two paths from input to output: one through Q_3 and one through Q_4. A simple but gross approximation for the gain over this segment can be obtained from the coordinates of points C and D in Fig. 15.46b as follows:

$$\text{Gain} = -\frac{v_O(C) - v_O(D)}{v_I(D) - v_I(C)}$$

$$= -\frac{2.7 - 0.1}{1.4 - 1.2} = -13$$

EXERCISE

15.18 Taking into account the fact that the voltage across a forward-biased *pn* junction changes by about -2 mV/$^\circ$C, find the coordinates of points A, B, C, and D of the gate transfer characteristic at -55°C and at $+125^\circ$C. Assume that the characteristic in Fig. 15.46b applies at 25°C, and neglect the small temperature coefficient of V_{CEsat}.
Ans. at -55°C: (0,3.38), (0.66,3.38), (1.52,2.38), (1.72,0.1); at $+125^\circ$C: (0,4.1), (0.3,4.1), (0.8,3.1), (1.0,0.1)

Manufacturer's Specifications

Manufacturers of TTL usually provide curves for the gate transfer characteristics, the input *i-v* characteristics, and the output *i-v* characteristics measured at the limits of the specified operating temperature range. In addition, guaranteed values are usually given for the parameters V_{OL}, V_{OH}, V_{IL}, and V_{IH}. For standard TTL these values are $V_{OL} = 0.4$ V, $V_{OH} = 2.4$ V, $V_{IL} = 0.8$ V, and $V_{IH} = 2$ V. These limit values are guaranteed for a specified tolerance in power-supply voltage and for a maximum fan-out of 10. From our discussion in Section 15.4 we know that the maximum fan-out is determined by the maximum current that Q_3 can sink while remaining in saturation and while maintaining a saturation voltage lower than a guaranteed maximum ($V_{OL} = 0.4$ V). Calculations performed in Section 15.4 indicate the possibility of a maximum fan-out of 20 to 30. Thus the figure specified by the manufacturer is appropriately conservative.

As we have done in the case of RTL, the parameters V_{OL}, V_{OH}, V_{IL}, and V_{IH} can be used to compute the noise margins. Thus we have

$$\Delta 1 = V_{OH} - V_{IH} = 0.4 \text{ V}$$
$$\Delta 0 = V_{IL} - V_{OL} = 0.4 \text{ V}$$

EXERCISES

15.19 In Section 15.4 we found that when the gate input is high the base current of Q_3 is approximately 2.6 mA. Assume that this value applies at 25°C and that at this temperature $V_{BE} \cong 0.7$ V. Taking into account the -2 mV/°C temperature coefficient of V_{BE} and neglecting all other changes, find the base current of Q_3 at -55°C and at $+125$°C.

Ans.   2.2 mA; 3 mA

15.20   Figure E15.20 shows sketches of the i_L-v_o characteristics of a TTL gate when the output is low. Use these characteristics together with the results of Exercise 15.19 to calculate the value of β of transistor Q_3 at -55°C, $+25$°C, and $+125$°C.

Ans.   16; 25; 28

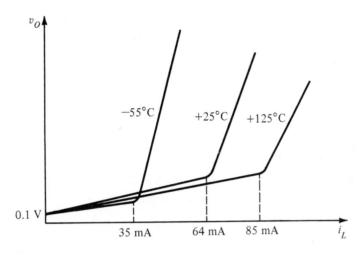

Fig. E15.20

Propagation Delay

The propagation delay of TTL gates is defined as the time between the 1.5 V points of corresponding edges of the input and output waveforms. For standard TTL (also known as *medium-speed* TTL) t_p is typically about 18 ns.

As far as power dissipation is concerned it can be shown (see Exercise 15.21 below) that when the gate output is high the gate dissipates 5 mW and when the output is low the dissipation is 16.7 mW. Thus the average power dissipation is 11 mW, resulting in a delay-power product of about 200 pJ. This is more than two orders of magnitude greater than the corresponding figure for I²L.

EXERCISE

15.21   Calculate the value of the supply current (I_{CC}) and hence the power dissipated in the TTL gate when the output terminal is open and the input is (a) low at 0.2 V (see Fig. 15.45) and (b) high at +5 V (see Fig. 15.43).

Ans.   (a) 1 mA, 5 mW; (b) 3.33 mA, 16.65 mW

Dynamic Power Dissipation

In Section 15.4 the occurrence of supply current spikes was explained. These spikes give rise to additional power drain from the V_{CC} supply. This *dynamic power* is also dissipated in the gate circuit. It can be evaluated by multiplying the average current due to the spikes by V_{CC}, as illustrated by the solution of Exercise 15.22.

EXERCISE

15.22   Consider a TTL gate that is switched on and off at the rate of 1 MHz. Assume that each time the gate is turned off (that is, the output goes high) a supply current pulse of 30 mA amplitude and 2 ns width occurs. Also assume that no current spikes occur when the gate is turned on. Calculate the dynamic power dissipation.
Ans.   0.3 mW

The TTL NAND Gate

Figure 15.47 shows the basic TTL gate. Its most important feature is the *multiemitter transistor Q_1* used at the input. Figure 15.48 shows the structure of the multiemitter transistor.

It can be easily verified that the gate of Fig. 15.47 performs the NAND function. The output will be high if one (or both) of the inputs is (are) low. The output will be

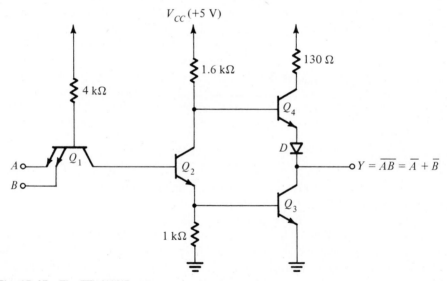

Fig. 15.47   *The TTL NAND gate.*

low in only one case: when both inputs are high. Extension to more than two inputs is straightforward and is achieved by diffusing additional emitter regions.

Although theoretically an unused input terminal may be left open-circuited, this is generally not a good practice. An open-circuit input terminal could act as an "antenna" that "picks up" interfering signals and thus could cause erroneous gate switching. An unused input terminal should therefore be connected to the positive power supply *through a resistance* (of, say, 1 kΩ). In this way the corresponding base–emitter junction of Q_1 will be reverse-biased and thus will have no effect on the operation of the gate. The series resistance is included in order to limit the current in case of breakdown of the base–emitter junction due to transients on the power supply.

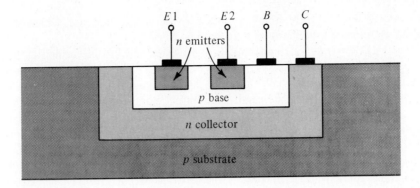

Other TTL Logic Circuits

On a TTL MSI chip there are many cases in which logic functions are implemented using "stripped-down" versions of the basic TTL gate. As an example we show in Fig. 15.49 the TTL implementation of the AND-OR-INVERT function. As shown, the phase-splitter transistors of two gates are connected in parallel and a single output stage is used. The reader is urged to verify that the logic function realized is as indicated.

At this point it should be noted that the totem-pole output stage of TTL does *not* allow the wired-AND connection. To see the reason for this, consider two gates whose outputs are connected together and let one gate have a high output while the other have a low output. Current will flow from Q_4 of the first gate through Q_3 of the second gate.

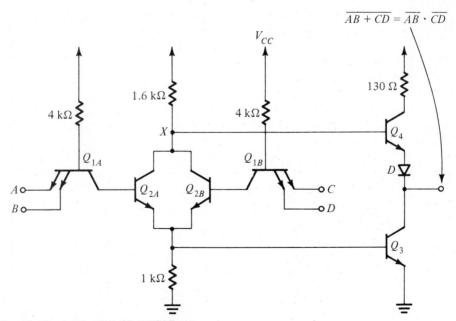

$$\overline{AB + CD} = \overline{AB} \cdot \overline{CD}$$

Fig. 15.49 A TTL AND-OR-INVERT gate.

The current value will fortunately be limited by the 130-Ω resistance. Obviously, however, no useful logic function is realized by this connection.

The lack of wired-AND capability is a drawback of TTL. Nevertheless, the problem is solved in a number of ways, including doing the paralleling at the phase-splitter stage, as illustrated in Fig. 15.49. Another solution consists of deleting the emitter-follower transistor altogether. The result is an output stage consisting solely of the common-emitter transistor Q_3 without even a collector resistance. Obviously, one can connect the outputs of such gates together to a common collector resistance and achieve a wired-AND capability. TTL gates of this type are known as *open-collector TTL*. The obvious disadvantage is the slow rise time of the output waveform.

Another useful variant of TTL is the *tri-state* output arrangement explored in Exercise 15.23.

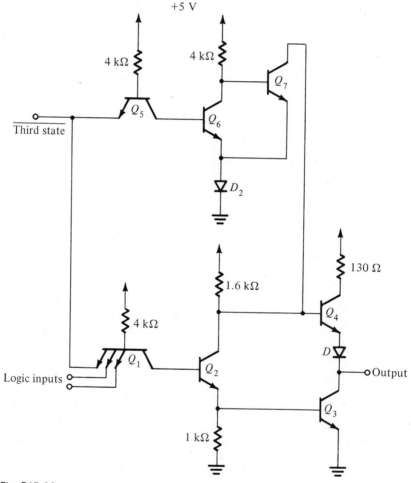

Fig. E15.23

15.23   The circuit shown in Fig. E15.23 is called Tristate TTL. Verify that when the terminal labeled $\overline{\text{Third}}$ state is high, the gate functions normally and that when this terminal is low, transistor Q_4 is cut off and the output of the gate is an open-circuit. This latter state is the third state, or the high-output-impedance state.

Tristate TTL enables the connection of a number of TTL gates to a common output line (or *bus*). At any particular time the signal on the bus will be determined by the one TTL gate which is *enabled* (by raising its third-state input terminal). All other gates will be in the third state and will thus have no control of the bus.

15.6   SPECIAL FORMS OF TTL

One of the reasons for the popularity of the TTL logic family is its versatility. Many different logic functions are available in SSI, MSI, and some LSI packages. These diverse functions are available from a large number of manufacturers at quite moderate prices.

Another reason for the success and longevity of TTL is the variety of improved versions introduced over the years. In this section we shall provide a brief introduction to these special forms of TTL.

High-Speed TTL

The speed of the standard TTL gate is limited by two mechanisms. First, transistors Q_1, Q_2, and Q_3 saturate, and hence we have to contend with their finite storage time. Although Q_2 is discharged reasonably quickly because of the active mode of operation of Q_1, as already explained, this is not true for Q_3, whose base charge has to leak out through the 1-kΩ resistance in its base circuit. Second, the resistances in the circuit together with the various transistor capacitances form relatively long time constants, which contribute to the gate delay.

It follows that there are two approaches to speeding up the operation of TTL. The first is to prevent transistor saturation. We shall discuss this approach below. The second approach, which can be used alone or in conjunction with the saturation-prevention approach, is to reduce the values of all resistances. This second approach alone results in some increase in speed. The circuit topology remains the same as that of standard TTL, but all resistance values are decreased. The price paid, of course, is a corresponding increase in power dissipation.

Before discussing this form of high-speed TTL, it is appropriate to mention another type of TTL called *low-power TTL*. In this form the circuit resistances are increased (from the values used in standard TTL) in order to achieve reduced power dissipation. As expected, low-power TTL is slower than standard TTL.

The high-speed TTL that we shall now discuss has a number of additional features other than the simple reduction in resistor values. These are illustrated in Fig. 15.50, which shows the high-speed TTL gate. The added special features are as follows:

1. A Darlington stage is used in place of Q_4 in order to provide increased current gain and hence increased current sourcing capability. This together with the lower output resistance of the gate (in the output-high state) yields a reduction in the time required to charge load capacitances to the high level. Note that the use of the

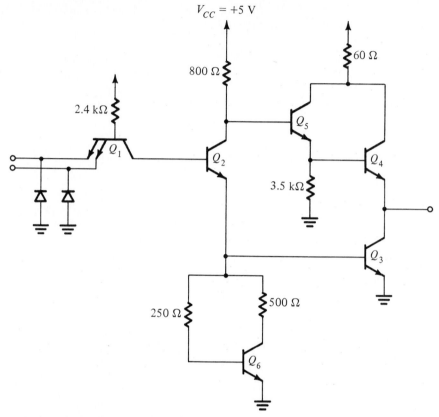

Fig. 15.50 Typical high-speed TTL gate circuit

additional transistor in the Darlington pair obviates the need for the diode D (used in the standard gate).

2. Input clamping diodes are included. These diodes conduct only when the input voltages go below ground level. This could happen due to "ringing" on the wires connecting the inputs of the gate to the output of another gate. Ringing occurs because such connecting wires behave as *transmission lines* that are not properly terminated. Without the clamping diodes ringing can cause the input voltage transiently to go sufficiently negative to cause the substrate to become forward-biased and hence result in improper operation of the gate. Also, ringing can cause the input voltage to go sufficiently positive to result in false gate switching. The input diodes clamp the negative excursions of the input ringing signal. Their conduction also provides a power loss in the transmission line, which results in *damping* of the ringing waveform and thus a reduction in its positive-going part.

3. The resistance between the base of Q_3 and ground has been replaced by a *nonlinear resistance* realized by transistor Q_6 and two resistances. This nonlinear resistance is known as *active pull-down;* in analogy to the active pull-up provided by the emitter-follower part of the totem pole. This feature is an ingenious one, and we shall discuss it in more detail.

Active Pull-Down

Figure 15.51 shows a sketch of the *i-v* characteristic of the nonlinear network composed of Q_6 and its two associated resistors. Also shown for comparison is the linear *i-v* characteristic of a resistor that would be connected between the base of Q_3 and ground if the active pull-down were not used.

The first characteristic to note of the active pull-down is that it conducts negligible current and thus behaves as a high resistance until the voltage across it reaches a V_{BE}

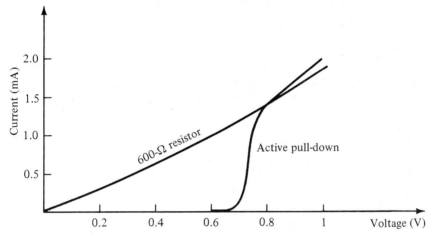

Fig. 15.51 *Comparison of the i-v characteristic of the active pull-down with that of a 600-Ω resistor.*

drop. Thus the gain of the phase splitter (as a linear amplifier) will remain negligibly small until a V_{BE} drop develops between its emitter and ground, which is the onslaught of conduction of Q_3. Thus segment *BC* of the transfer characteristic (see Fig. 15.46b) will be absent and the gate transfer characteristic will take the form shown in Fig. 15.52. This is an important improvement from the point of view of noise immunity.

The fact that the active pull-down draws a negligible current over a good part of its characteristics means that the current supplied by the phase splitter will initially be diverted into the base of Q_3. This speeds up the turn-on of Q_3.

Next note that above a certain voltage, the active pull-down draws more current than the corresponding passive pull-down (resistor) does. Again this is fortuitous; it implies that extra current supplied by the phase splitter will not go into the base of Q_3 but rather will be diverted into the active pull-down. Thus Q_3 will not be heavily saturated, and its storage time will thus be reduced.

As Q_3 begins to turn off, the active pull-down will initially draw a larger current than that drawn by a resistor. Furthermore, because of the internal dynamics of the active pull-down transistor this large current persists for a time, with the result that Q_3 will be quickly discharged.

There are further advantageous properties of the active pull-down that improve the performance of the gate in an environment in which temperature changes. As an exam-

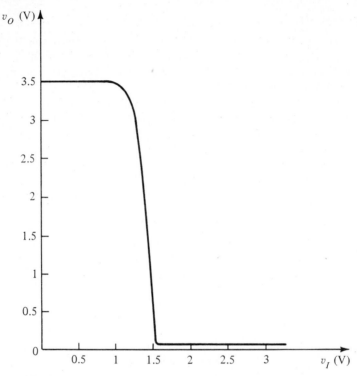

Fig. 15.52   *Transfer characteristic of the high-speed TTL gate.*

ple of these additional advantages, consider the situation as temperature is increased. Since β increases with temperature, transistor Q_3 would (in the absence of the active pull-down) operate deeper into saturation, and its storage time would increase. However, as temperature increases, V_{BE} of Q_6 decreases and the active pull-down conducts a larger current. It therefore diverts more current away from the base of Q_3 and thus protects it from heavy saturation. The large current conducted by the active pull-down again helps discharge Q_3 quickly.

Finally, it should be noted that the active pull-down is employed in all modern forms of TTL.

High-speed TTL features a delay time of about 10 ns. The gate dissipation, however, is increased to about 22 mW. It is currently not very popular because of the superior performance obtained in the Schottky TTL family (or subfamily), discussed next.

Schottky TTL

Another special and very important form of TTL that achieves a very high speed of operation is *Schottky* TTL. Low propagation delay is obtained by preventing the transistors from operating in the saturation mode. Schottky TTL is one of two nonsaturating bipolar transistor logic families; the other is emitter-coupled logic (ECL), studied in Section 15.7.

In Schottky TTL, transistors are prevented from saturation by connecting a low-voltage-drop diode between base and collector, as shown in Fig. 15.53. These diodes, formed as a metal-to-semiconductor junction (see Appendix A), are called Schottky diodes and have forward voltage drops of about 0.4 V. Schottky diodes are easily fabricated and do not increase chip area. In fact, the Schottky TTL process has been designed to yield transistors with smaller areas than those produced by the standard TTL process. Thus Schottky transistors have higher β and f_T. Figure 15.53 also shows the symbol used to denote a Schottky transistor.

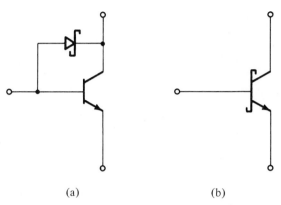

(a)   (b)

Fig. 15.53   *(a) A transistor with a Schottky diode clamp. (b) Circuit symbol for Schottky transistor.*

The Schottky transistor does not saturate, since some of its base drive is shunted by the Schottky diode. The latter then conducts and *clamps* the base-collecto· junction to a voltage smaller than that required for saturation. In this way the storage time is eliminated and the transistor turn-off time is decreased.

Figure 15.54 shows the circuit for a Schottky TTL NAND gate. It is quite similar to the high-speed TTL circuit except that *Schottky clamps* are added to all transistors except Q_4. Transistor Q_4 never saturates and therefore does not need a Schottky clamp. Note that the input clamping diodes are also of the Schottky type.

The circuit in Fig. 15.54 operates in much the same way as the high-speed TTL circuit except that here the transistors do not saturate. It is interesting to note that when the output is high, transistor Q_4 does not conduct until the output current exceeds about 0.6 mA. Currents less than 0.6 mA are simply supplied by Q_5 through the 1-kΩ resistor in its emitter. The 50-Ω resistance in the collector of Q_5 limits the current sourced by the gate during transients. The mechanism here is that the voltage at the collector of Q_5 is lowered and its Schottky diode conducts, thus limiting the current through the base of Q_5. Correspondingly, the emitter current of Q_5 and hence the output current of the gate are limited.

Schottky TTL features gate delays of about 5 ns. Thus it is about twice as fast as high-speed TTL. Nevertheless, the gate dissipation remains almost equal to that of high-speed TTL.

The most recent and very important development in the TTL family has been the advent of *low-power Schottky TTL.* Low-power Schottky circuits use Schottky clamps,

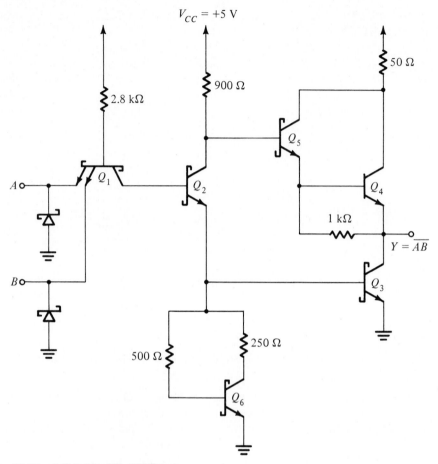

Fig. 15.54 *A Schottky TTL NAND gate.*

increased resistance values, and a number of other circuit techniques the details of which are beyond the scope of this book. It achieves a gate power dissipation of about a tenth of that of Schottky TTL at the expense of increasing the gate delay to about twice that of Schottky TTL. Its power-delay product is therefore about 20 pJ. A comparison of the various logic circuit families from a power-delay-product point of view is provided in Section 15.9.

15.7 EMITTER-COUPLED LOGIC (ECL)

Emitter-coupled logic (ECL) is currently (and likely to remain) the fastest form of logic, since the active devices within it are arranged to operate out of saturation. In addition to avoiding stored charge by this means, ECL is made even faster by arranging that the logic signal swings are relatively small (about 0.8 V). Thus the time required for charging and discharging the various load and parasitic capacitances is quite short.

Unlike Schottky TTL, where saturation is prevented actively by feedback, it is avoided in an open loop (or passive) way in ECL. ECL is a current-switching logic form whose basic switch is the BJT differential pair studied in detail in Chapter 10. We urge the reader to review Sections 10.5 and 10.6 before proceeding with the remainder of this section.

ECL Families

The first integrated version of ECL was introduced by Motorola in 1962 and given the name MECL. Another version, called MECL II, was introduced in 1966. Both of these forms are now obsolete. Currently available are MECL III (introduced in 1968) and MECL 10,000 (introduced in 1971). MECL III has the highest speed; its gate propagation delay and its edge speeds (rise and fall time) are of the order of 1 ns. Flip-flops made with MECL III technology can be toggled at speeds up to 500 MHz. MECL III, however, has a relatively high gate power dissipation of about 60 mW, giving it a delay-power product of 60 pJ. MECL III finds applications in high-speed test and communications systems.

For more general-purpose applications MECL 10,000 offers a slightly slower but easier-to-use logic family. It features a gate propagation delay of 2 ns and an edge speed of 3.5 ns. It is this slower edge speed that reduces crosstalk between adjacent signal lines and makes the family easier to use. The power dissipation in a MECL 10,000 gate is about 25 mW, and the delay-power product is 50 pJ.

MECL 10,000 has been extended to the design of VLSI products such as microprocessors. In such applications a lower-power version is used with a typical 2 mW dissipation per gate.

The Basic MECL Gate

The basic circuit found in the current MECL families is shown in Fig. 15.55. Resistor values are those that apply to MECL 10,000.

As indicated, the gate consists of three parts. The network composed of Q_1, D_1, D_2, and R_1 through R_3 generates a reference voltage V_{BB} whose value at room temperature is -1.29 V. As will be shown below, the value of this reference voltage is made to change with temperature in a predetermined manner so as to keep the noise margins almost constant. Also, the reference voltage V_{BB} is made relatively insensitive to variations in the power supply voltage V_{EE}.

The second part, and the heart of the gate, is the differential amplifier formed by Q_R and either Q_A or Q_B. This differential amplifier is biased not by a current source, as was done in the circuits of Chapter 10, but with a resistance R_E connected to the negative supply V_{EE}. One side of the differential amplifier consists of the reference transistor Q_R whose base is connected to the reference voltage V_{BB}. The other side consists of a number of transistors (two in the case shown) connected in parallel. If the voltages applied to A and B are low (logic 0), both Q_A and Q_B will be off and the current I_E in R_E will flow through the reference transistor Q_R. Here it should be noted that the resistance connecting each input terminal to the negative supply enables the user to leave

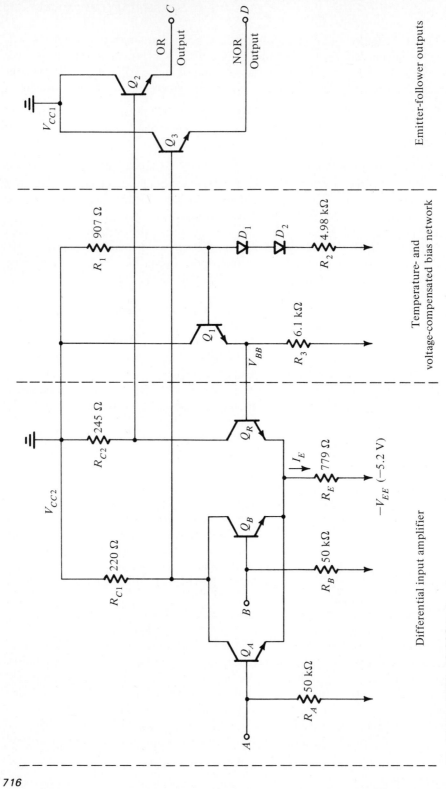

Fig. 15.55 MECL 10,000 basic gate.

an unused input terminal open. An open input terminal will then be *pulled down* to the negative supply voltage, and its associated transistor will be off.

When A or B is high, the corresponding transistor will be on and the current I_E will flow through it. In this case Q_R will be off. It follows that the collector of Q_R will be high when Q_R is off, which occurs when A or B is high. Thus at the collector of Q_R we obtain the OR function $A + B$. Similarly we see that at the common collector of Q_A and Q_B we obtain the NOR function $\overline{A + B}$.

From our previous analysis of the differential amplifier we know that a small input signal (approximately 0.2 V) is sufficient to steer the bias current from one side of the pair to the other. Thus the logic swing required at the input of an ECL gate is very small.

The third part of the ECL gate circuit is composed of the two emitter followers, Q_2 and Q_3. The emitter followers do not have on-chip loads, since in most applications of high-speed logic circuits the gate output drives a transmission line terminated at the other end, as indicated in Fig. 15.56. The arrangement shown in Fig. 15.56 is a common one, and in fact MECL 10,000 is specified for the termination values indicated.

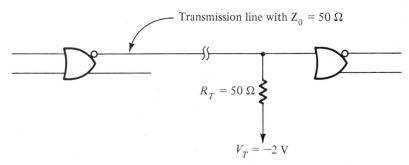

Fig. 15.56   *Proper way for connecting high-speed logic gates such as ECL. Properly terminating the transmission line connecting the two gates eliminates "ringing" that would otherwise corrupt the logic signals.*

The emitter followers have two purposes. First, they shift the level of the output signals by one V_{BE} drop; this, as will be seen, makes the output logic levels compatible with the input logic levels. Second, the emitter followers provide the gate with low output resistances and with the large output currents required for charging load capacitances. Since such large transient currents flow mostly through the output emitter followers, one can avoid spikes on the power-supply line of the reference and switching parts of the gate circuit by connecting the collectors of the followers to a separate supply line. Although normally both terminals V_{CC1} and V_{CC2} are connected to ground, separate paths are used.

We conclude our introduction of the ECL circuit by noting that the gate provides two complementary outputs: the OR output at C and the NOR output at D. The availability of complementary outputs is another advantage of ECL; it simplifies logic design and avoids use of an additional inverter with its associated time delay.

Transfer Characteristics

We shall now derive the transfer characteristics of the ECL gate in Fig. 15.55 under the condition that its output terminals are terminated in the manner indicated in Fig. 15.56. Assuming that the B input (Fig. 15.55) is low, which implies that Q_B is off, we obtain the simplified circuit in Fig. 15.57, which we shall analyze to determine v_C versus v_A and v_D versus v_A.

In the analysis to follow we shall make use of the exponential i_C-v_{BE} characteristics of the BJTs. Since the BJTs used in MECL circuits have small areas (in order to have

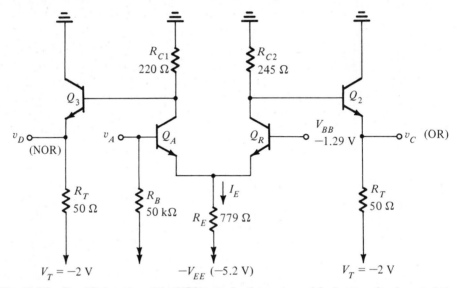

Fig. 15.57   *Simplified version of the MECL gate for the purpose of finding transfer characteristics.*

small capacitances and hence high f_T), their scale currents I_S are small. We will therefore assume that at an emitter current of 1 mA a MECL transistor has a V_{BE} drop of 0.75 V.

The OR Transfer Curve

Figure 15.58 shows a sketch of the v_C-versus-v_A characteristic, which is the OR transfer characteristic. Indicated on this sketch are the parameters V_{OL}, V_{OH}, V_{IL}, and V_{IH}. These parameters are defined in the usual manner. Note specifically that V_{IL} and V_{IH} define the limits of the transition region, points x and y. From our study of differential amplifiers (Section 10.5) we know that the transition region will be centered around the reference voltage V_{BB}. That is,

$$V_{IH} - V_{BB} = V_{BB} - V_{IL}$$

which follows directly by assuming that in the circuit of Fig. 15.57 Q_A and Q_R are identical devices.

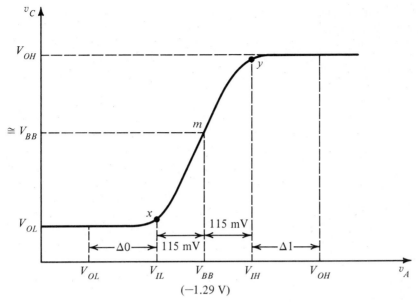

Fig. 15.58 The OR transfer characteristic, v_C versus v_A for the circuit in Fig. 15.57.

For $v_A < V_{IL}$, Q_A is off and Q_R conducts all the bias current I_E. Conversely, for $v_A > V_{IH}$, Q_R is off and Q_A conducts all the current I_E. Over the transition region $V_{IL} \leq v_A \leq V_{IH}$, both Q_A and Q_R will be on and will thus share the bias current I_E. Of course, I_E will not remain constant. Nevertheless, it will be shown that the change in I_E is indeed quite small.

We shall arbitrarily define the threshold points x and y as those points at which one transistor carries 99% of I_E and the other carries 1% of I_E. Thus at x the emitter currents of Q_A and Q_R will have the ratio

$$\frac{I_{ER}}{I_{EA}} = 99$$

Using the exponential i_E-v_{BE} relationship we obtain

$$V_{BER} - V_{BEA} = V_T \ln 99 = 115 \text{ mV}$$

which gives $V_{IL} = -1.405$ V. Similarly, at point y we have $I_{EA}/I_{ER} = 99$, which gives

$$V_{IH} - V_{BB} = 115 \text{ mV}$$

Thus $V_{IH} = -1.175$ V. Thus the transition region is 0.23 V in width and is centered around the reference voltage V_{BB}.

We next determine the output voltages V_{OL} and V_{OH} as well as the voltage corresponding to the midpoint of the transition region. To obtain V_{OL} we note that Q_A is off and Q_R carries the entire current I_E, given by

$$I_E = \frac{V_{BB} - V_{BER} + V_{EE}}{R_E}$$

Using the values given in the circuit diagram, and iterating two times to determine a reasonably accurate value for V_{BER}, we find that $I_E \cong 4$ mA and $V_{BER} \cong 0.785$ V. Since β of Q_R is large ($\cong 100$), the collector current of Q_R will be approximately 4 mA. Neglecting the base current of Q_2, we obtain for the collector voltage of Q_R

$$V_{CR} \cong -4 \times 0.245 = -0.98 \text{ V}$$

Thus a first approximation for the value of v_C is

$$v_C = -0.98 - 0.75 = -1.73 \text{ V}$$

We can use this value to find the emitter current of Q_2 and then iterate to determine a better estimate of its base–emitter voltage. The result is $V_{BE2} \cong 0.79$ V, and correspondingly the value of v_C, which is V_{OL}, is $V_{OL} \cong -1.77$ V. At this value of output voltage Q_2 supplies a load current of about 4.6 mA.

To find the value of V_{OH} we assume that Q_R is completely off (because $v_A > V_{IH}$). Thus the circuit for determining V_{OH} simplifies to that in Fig. 15.59. Solving this circuit and assuming $\beta_2 = 100$ results in $V_{BE2} \cong 0.83$ V, $I_{E2} \cong 22.4$ mA; $V_{OH} \cong -0.88$ V.

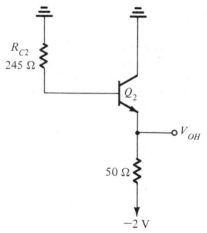

Fig. 15.59 *Circuit for determining V_{OH}.*

EXERCISES

15.24 Determine the values of I_E obtained when $v_A = V_{IL}$; V_{BB}; and V_{IH}. Also, find the value of v_C corresponding to $v_A = V_{BB}$.
Ans. 4.01 mA; 4.03 mA; 4.16 mA; -1.31 V

15.25 Determine the slope of the transfer characteristic in Fig. 15.58 at the midpoint, m. Note that the slope is simply the gain of the differential pair.
Ans. -9.4 V/V

Noise Margins

The results of Exercise 15.24 indicate that the bias current I_E remains approximately constant. Also, the output voltage corresponding to $v_A = V_{BB}$ is approximately equal to

V_{BB}. Notice further that this is also approximately the midpoint of the logic swing; specifically,

$$\frac{V_{OL} + V_{OH}}{2} = -1.325 \cong V_{BB}$$

Thus the output logic levels are centered around the midpoint of the input transition band. This is an ideal situation from the point of view of noise margins, and it is one of the reasons for selecting the rather arbitrary looking numbers $V_{BB} = -1.29$ V and $V_{EE} = 5.2$ V.

The noise margins can be now evaluated as follows:

$$\Delta 1 = V_{OH} - V_{IH}$$
$$= -0.88 - (-1.175) = 0.295 \text{ V}$$
$$\Delta 0 = V_{IL} - V_{OL}$$
$$= -1.405 - (-1.77) = 0.365 \text{ V}$$

These values are approximately equal.

Finally, it should be emphasized that all our calculations are approximate.

The NOR Transfer Curve

The NOR transfer characteristic, which is v_D versus v_A for the circuit in Fig. 15.57, is sketched in Fig. 15.60. The values of V_{IL} and V_{IH} are identical to those found above for the OR characteristic. To emphasize this we have labeled the threshold points x and y, the same letters used in Fig. 15.58.

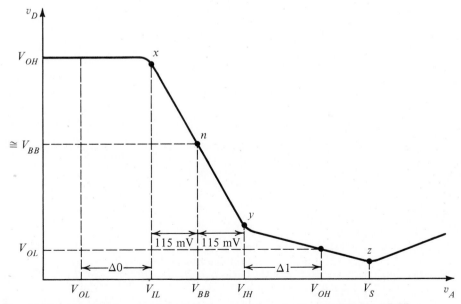

Fig. 15.60   The NOR transfer characteristic, v_D versus v_A for the circuit in Fig. 15.57.

For $v_A < V_{IL}$, Q_A is off and the output voltage at D can be found by analyzing the circuit composed of R_{C1}, Q_3, and its 50-Ω termination. Except that R_{C1} is slightly smaller than R_{C2}, this circuit is identical to that in Fig. 15.59. Thus the output voltage will be only slightly greater than the value V_{OH} found previously. In the sketch of Fig. 15.60 we have assumed that the output voltage is approximately equal to V_{OH}.

For $v_A > V_{IH}$, Q_A is on and is conducting the entire bias current. The circuit then simplifies to that in Fig. 15.61. This circuit can be easily analyzed to obtain v_D versus v_A for the range $v_A \geq V_{IH}$. A number of observations are in order. First, note that $v_A = V_{IH}$ results in an output voltage slightly higher than V_{OL}. This is because R_{C1} is smaller than R_{C2}. In fact, R_{C1} is chosen lower in value than R_{C2} so that with v_A equal to the normal logic 1 value (that is, V_{OH}, which is approximately -0.88 V) the output will be equal to the V_{OL} value previously found for the OR output.

Second, note that as v_A exceeds V_{IH}, transistor Q_A operates in the active mode and the circuit of Fig. 15.61 can be analyzed to find the gain of this amplifier, which is the slope of the segment yz of the transfer characteristic. At point z transistor Q_A saturates. Further increments in v_A (beyond the point $v_A = V_S$) cause the collector voltage and hence v_D to increase. The slope of the segment of the transfer characteristic beyond point z, however, is not unity but is about 0.5 because as Q_A is driven deeper into saturation a portion of the increment in v_A appears as an increment in the base–collector forward-bias voltage. The reader is urged to solve Exercise 15.26, which is concerned with the details of the NOR transfer characteristic.

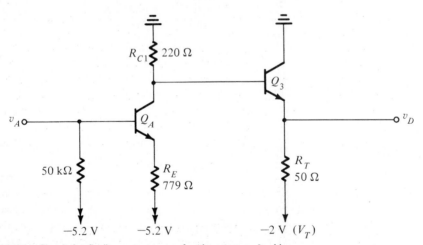

Fig. 15.61 Circuit for finding v_D versus v_A for the range $v_A > V_{IH}$.

EXERCISE

15.26 Consider the circuit in Fig. 15.61.
(a) For $v_A = V_{IH} = -1.175$ V, find v_D.
(b) For $v_A = V_{OH} = -0.88$ V, find v_D.
(c) Find the slope of the transfer characteristic at the point $v_A = V_{OH} = -0.88$ V.
(d) Find the value of v_A at which Q_A saturates (that is, V_S). Assume that on the verge of saturation $V_{CE} \cong 0.3$ V.
Assume $\beta = 100$.
Ans. (a) -1.71 V; (b) -1.78 V; (c) -0.24 V/V; (d) -0.68 V

Manufacturer's Specifications

ECL manufacturers supply gate transfer characteristics of the form shown in Figs. 15.58 and 15.60. The manufacturer usually provides such curves measured at a number of temperatures. In addition, at each relevant temperature worst-case values for the parameters V_{IL}, V_{IH}, V_{OL}, and V_{OH} are given. These worst-case values are specified with the inevitable component tolerances taken into account. As an example, Motorola specifies that for MECL 10,000 at 25°C the following worst-case values apply:

$$V_{ILmax} = -1.475 \text{ V}, \qquad V_{IHmin} = -1.105 \text{ V}$$
$$V_{OLmax} = -1.630 \text{ V}, \qquad V_{OHmin} = -0.980 \text{ V}$$

These values can be used to determine worst-case noise margins:

$$\Delta 0 = 0.155 \text{ V}, \qquad \Delta 1 = 0.125 \text{ V}$$

which are about half the *typical* values previously calculated.

For additional information on MECL specifications the interested reader is referred to references. 15.9 and 15.10.

Fan-Out

When the input signal to an ECL gate is low, the input current is equal to the current that flows in the 50-kΩ pull-down resistor. Thus

$$I_{IL} = \frac{-1.77 + 5.2}{50} \cong 69 \ \mu A$$

When the input is high the input current is greater because of the base current of the input transistor. Thus assuming transistor β of 100 we obtain

$$I_{IH} = \frac{-0.88 + 5.2}{50} + \frac{4}{101} \cong 126 \ \mu A$$

Both of these current values are quite small, which, coupled with the fact that the output resistance of the ECL gate is very small, ensures that little degradation of logic signal levels results from the input currents of fan-out gates. It follows that the fan-out of ECL gates is not limited by logic level considerations but rather by the degradation of the circuit speed (rise and fall times). This latter effect is due to the capacitance that each fan-out gate presents to the driving gate (approximately 3 pF). Thus while the *dc fan-out* can be as high as 90 and thus does not represent a design problem, the *ac fan-out* is limited by considerations of circuit speed to 5 or so.

Speed

The speed of operation of a logic family is measured by the delay of its basic gate and by the rise and fall times of the output waveforms. Typical values of these parameters for MECL have already been given. Here we should note that because the output circuit is an emitter follower the rise time of the output signal is shorter than its fall time, since on the rising edge of the output pulse the emitter follower functions and provides the output current required to charge up the load and parasitic capacitances. On the other hand, as the signal at the base of the emitter follower falls, the emitter follower cuts off and the load capacitance discharges through the combination of load and pull-down resistances. This point was explained in detail in Section 15.4.

Signal Transmission

In order to take full advantage of the very high speed of operation possible with ECL, special attention should be paid to the method of interconnecting the various logic gates in a system. To appreciate this point we shall briefly discuss the problem of signal transmission.

ECL deals with signals whose rise times may be 1 ns or even less, the time it takes for light to travel only 30 cm or so. For such signals a wire and its environment becomes a relatively complex circuit element along which signals propagate with finite speed (perhaps half the speed of light, which is 30 cm/ns). Unless special care is taken, energy that reaches the end of such a wire is not absorbed but rather returns as a *reflection* to the transmitting end, where (without special care) it may be re-reflected. The result of this process of reflection is what can be observed as *ringing,* a damped oscillatory excursion of the signal about its final value.

Unfortunately ECL is particularly sensitive to ringing because the signal levels are so small. Thus it is important that transmission of signals be well controlled and surplus energy absorbed to prevent reflections. The accepted technique is to limit the nature of connecting wires in some way. One way is to insist that they be very "short," where short is taken with respect to the signal rise time. The reason for this is that if the wire connection is so short that reflections return while the input is still rising, the result becomes only a somewhat slowed and "bumpy" rising edge.

If, however, the reflection returns *after* the rising edge, it produces not simply a modification of the initiating edge but an *independent second event*. This is clearly bad! The restriction is thus made that the time taken for a signal to go from one end of line and back should be less than the rise time of the driving signal by some factor— say, 5. Thus for a signal with 1 ns rise time and for propagation at the speed of light (30 cm/ns), a double path of only 0.2 ns equivalent length, or 6 cm, would be allowed, representing in the limit a wire only 3 cm from end to end.

Such is the restriction on MECL III. However, MECL 10,000 has intentionally slower rise time of about 3.5 ns. Using the same rules, wires can accordingly be as long as about 10 cm for MECL 10,000.

If greater lengths are needed, then transmission lines must be used. These are simply wires in a controlled environment in which the distance to a ground reference plane or second wire is highly controlled. Thus they might simply be twisted pairs of wires, one of which is grounded, or parallel ribbon wires every second of which is grounded, or so-called microstrip lines on a printed-circuit (PC) board. The last are simply copper strips of controlled geometry on one side of a printed circuit board the other side of which consists of a grounded plane.

Such transmission lines have a *characteristic impedance* R_0 which ranges from a few tens of ohms to 1 kΩ or so. Signals propagate on such lines somewhat slower than the speed of light, perhaps half as fast. When a transmission line is terminated at its receiving end in a resistance equal to its characteristic impedance R_0, all the energy sent on the line is absorbed at the receiving end, and no reflections occur. Thus signal integrity is maintained. Such transmission lines are said to be *properly terminated*. A properly terminated line appears at its sending end as a resistor of value R_0. The followers of MECL 10,000 with their open emitters and low output resistances (7 Ω maximum) are ideally suited for driving transmission lines. MECL is also good as a line receiver. The simple gate with its high (50 kΩ) pull-down input resistor represents a very high resistance to the line. Thus a few such gates can be connected to a terminated line with little difficulty.

Much more on the subject of logic signal transmission in ECL can be found in references 15.3 and 15.10.

Power Dissipation

Because of the differential-amplifier nature of ECL, the gate current remains approximately constant and is simply steered from one side of the gate to the other depending on the input logic signals. Thus unlike TTL the supply current and hence the gate power dissipation of unterminated ECL remain relatively constant independent of the logic state of the gate. It follows that no voltage spikes are introduced on the supply line. Such spikes are a dangerous source of noise in a digital system, as explained in connection with TTL. It follows that in ECL the need for supply-line bypassing is not as great as in TTL. This is another advantage of ECL.

At this point we should mention that although the ECL gate would operate with $V_{EE} = 0$ and $V_{CC} = +5.2$ V, the selection of $V_{EE} = -5.2$ V and $V_{CC} = 0$ V is recommended because it results in much less susceptibility to noise or fluctuations that inevitably appear on the power supply line. For further details consult Ref. 15.3.

EXERCISE

15.27 For the ECL gate in Fig. 15.55 calculate an approximate value for the power dissipated in the circuit under the condition that all inputs are low and that the emitters of the output followers are left open.
Ans. 27.4 mW

Thermal Effects

In our analysis of the MECL gate of Fig. 15.55 it has been assumed that at room temperature the reference voltage V_{BB} is -1.29 V, which is the figure supplied by the manufacturer. We have shown that the midpoint of the output logic swing is approximately equal to this voltage, which is an ideal situation in that it results in equal 1 and 0 noise margins. In the following we shall (or, more appropriately, the reader shall) first show in Exercise 15.28 that V_{BB} is indeed equal to -1.29 V. Next, in Example 15.3 we shall derive expressions for the temperature coefficients of the reference voltage and of the output low and high voltages. In this way it will be shown that the midpoint of the output logic swing varies with temperature with the same magnitude as the changes in the reference voltage. As a result, although the magnitudes of the 1 and 0 noise margins change with temperature, their values remain equal. This is an added advantage of ECL and is a demonstration of the design optimization of this gate circuit.

15.28 Figure E15.28 shows the circuit that generates the reference voltage V_{BB}. Assume that D_1, D_2, and the base–emitter junction of Q_1 each have drops of 0.75 V at a current of 1 mA, and neglect the base current of Q_1. Find the value of V_{BB}.
Ans. -1.31 V

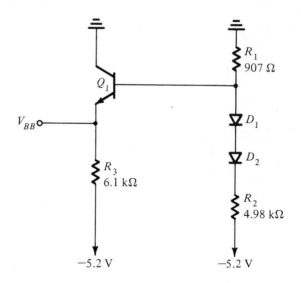

Fig. E15.28

The value of V_{BB} obtained in Exercise 15.28 is a good approximation to the -1.29 V value given by the manufacturer. The difference results from not knowing the exact characteristics of D_1, D_2, and Q_1.

Example 15.3
We wish to determine the temperature coefficient of the reference voltage V_{BB} and of the midpoint between V_{OL} and V_{OH}.

Solution
To determine the temperature coefficient of V_{BB}, consider the circuit in Fig. E15.28 and assume that the temperature changes by $+1\,°\text{C}$. Denoting the temperature coefficient of the diode and transistor voltage drops by δ, where $\delta \cong -2$ mV/$°$C, we obtain the equivalent circuit shown in Fig. 15.62. In this latter circuit the changes in device voltage drops are considered as signals, and hence the power supply is shown as a signal ground.

In the circuit of Fig. 15.62 we have two signal generators, and we wish to analyze the circuit to determine the change in V_{BB}, ΔV_{BB}. We shall do so using the principle of superposition. Consider first the branch R_1, D_1, D_2, 2δ, and R_2 and neglect the signal base current of Q_1. The voltage signal at the base of Q_1 can be easily obtained from

$$v_{b1} = \frac{2\delta \times R_1}{R_1 + r_{D1} + r_{D2} + R_2}$$

where r_{D1} and r_{D2} denote the incremental resistances of diodes D_1 and D_2, respectively. The dc bias current through D_1 and D_2 is approximately 0.64 mA, and thus $r_{D1} = r_{D2}$

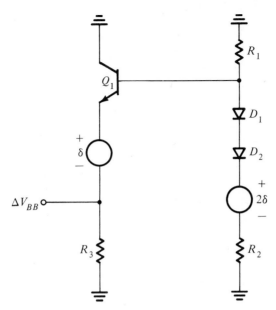

Fig. 15.62 Equivalent circuit for determining the temperature coefficient of the reference voltage V_{BB}.

$= 39.5 \, \Omega$. Hence $v_{b1} \cong 0.3\delta$. Since the gain of the emitter-follower Q_1 is approximately unity, it follows that the component of ΔV_{BB} due to the generator 2δ is approximately equal to v_{b1}, that is, $\Delta V_{BB1} = 0.3\delta$.

Consider next the component of ΔV_{BB} due to the generator δ. Reflection of the total resistance of the base circuit, $[R_1 \| (r_{D1} + r_{D2} + R_2)]$, into the emitter circuit by dividing it by $\beta + 1$ ($\beta \cong 100$) results in the following component of ΔV_{BB}:

$$\Delta V_{BB2} = - \frac{\delta \times R_3}{[R_B/(\beta + 1)] + r_{e1} + R_3}$$

where R_B denotes the total resistance in the base circuit and r_{e1} denotes the emitter resistance of Q_1 ($\cong 40 \, \Omega$). This calculation yields $\Delta V_{BB2} \cong -\delta$. Adding this value to that due to the generator 2δ gives $\Delta V_{BB} \cong -0.7\delta$. Thus for $\delta = -2 \, \text{mV}/^\circ\text{C}$ the temperature coefficient of V_{BB} is $+1.4 \, \text{mV}/^\circ\text{C}$.

We next consider determination of the temperature coefficient of V_{OL}. The circuit for accomplishing this analysis is shown in Fig. 15.63. Here we have three generators whose contributions can be considered independently and the resulting components of ΔV_{OL} summed. The result is

$$\Delta V_{OL} \cong \Delta V_{BB} \frac{-R_{C2}}{r_{eR} + R_E} \frac{R_T}{R_T + r_{e2}}$$

$$- \delta \frac{-R_{C2}}{r_{eR} + R_E} \frac{R_T}{R_T + r_{e2}}$$

$$- \delta \frac{R_T}{R_T + r_{e2} + R_{C2}/(\beta + 1)}$$

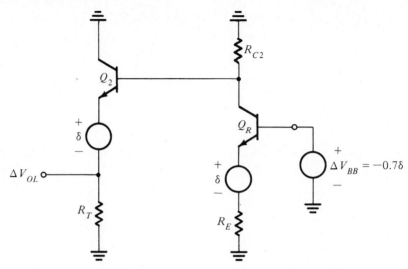

Fig. 15.63 Equivalent circuit for determining the temperature coefficient of V_{OL}.

Substituting the values given and obtained throughout the analysis of this section, we find $\Delta V_{OL} \cong -0.43\delta$.

The circuit for determining the temperature coefficient of V_{OH} is shown in Fig. 15.64, from which we obtain

$$\Delta V_{OH} = -\delta \frac{R_T}{R_T + r_{e2} + R_{C2}/(\beta + 1)} \cong -0.93\delta$$

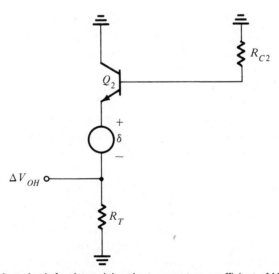

Fig. 15.64 Equivalent circuit for determining the temperature coefficient of V_{OH}.

We now can obtain the variation of the midpoint of the logic swing as

$$\frac{\Delta V_{OL} + \Delta V_{OH}}{2} = -0.68\delta$$

which is approximately equal to that of the reference voltage V_{BB} (-0.7δ).

$$\cdot \quad \cdot \quad \cdot$$

The Wired-OR Capability

The emitter-follower output stage of the ECL families allows an additional level of logic to be performed at very low cost by simply wiring the outputs of several gates in parallel. This is illustrated in Fig. 15.65, where the outputs of two gates are wired together. Note

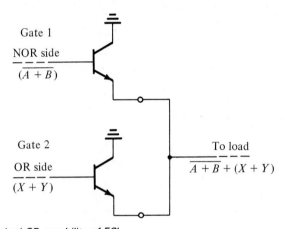

Fig. 15.65   *The wired-OR capability of ECL.*

that the base–emitter diodes of the output followers provide a positive OR function: this *wired-OR* connection may be used to provide gates with high fan-in as well as to increase the flexibility of ECL in logic design.

A Final Remark

ECL is an important logic family that has been successfully applied in the design of high-speed digital communications systems as well as computer systems. It is also currently applied in LSI and VLSI circuit design.

15.8   COMPLEMENTARY-SYMMETRY MOS LOGIC (CMOS)

The CMOS logic family is one of the most popular in the design of medium-speed logic systems (up to 25 MHz). It offers excellent noise immunity, simplicity, versatility, low power dissipation, operation over a wide range of supply voltages, and the ability to perform a variety of analog functions. For these reasons CMOS has penetrated newer markets such as consumer and industrial applications: it is being used in watches, clocks, automotive safety and control devices, appliance controls, electronic toys, and so on. Furthermore, CMOS technology is currently undergoing a number of improvements that make it a strong contender for the VLSI systems area.

The fundamentals of CMOS were introduced in Sections 8.11 and 8.12, which we urge the reader to review before proceeding with the following material. Building on these fundamentals, we shall study the characteristics and important features of the standard CMOS logic family.

Basic Building Blocks

CMOS integrated circuits are formed of two basic circuits: the logic inverter, which we studied in Section 8.11, and the bilateral switch or transmission gate, which was studied in Section 8.12. Figure 15.66a shows the inverter circuit; a simplified circuit diagram is shown in Fig. 15.66b.

The CMOS inverter consists of one n-channel and one p-channel enhancement-mode MOSFET. The source of the p device is connected to the positive terminal of the power supply (V_{DD}), and the source of the n device is connected to the negative terminal (usually ground) of the power supply (V_{SS}). Only a single power supply is required, and the circuit operates satisfactorily for power supply voltages in the range 3 to 18 V. Increasing the power-supply voltage (within this range) provides greater noise immunity and faster response at the expense of increased power dissipation.

When the inverter input is low (0 V) the n-channel device will be off and the p-channel device will be on. In the absence of an external load, the current conducted by the p device will be negligibly small (in the nanoamp range) and the power dissipation will be correspondingly small. The voltage drop across the p device will be very small (a few millivolts) and the output high level (V_{OH}) will be almost equal to V_{DD}. In this

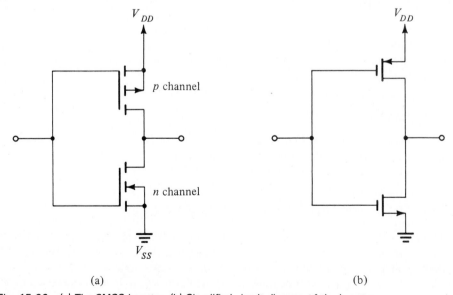

(a)   (b)

Fig. 15.66   (a) The CMOS inverter. (b) Simplified circuit diagram of the inverter.

state the p device (called the pull-up device) provides a low impedance between the output terminal and the positive supply V_{DD}. The current-sourcing capability of the gate is directly determined by the i-v characteristics of the p device.

When the inverter input is high (V_{DD} volts) the n-channel device will be on and the p-channel device will be off. The operation in this state is exactly the complement of that described above, and the output low voltage (V_{OL}) will be within a few millivolts of V_{SS}. In this state the current-sinking capability of the gate is determined by the i-v characteristics of the n-channel pull-down transistor.

The ability of the CMOS gate to operate over a supply voltage range of 3 to 18 V should be contrasted to the typical allowed range for TTL of 4.75 to 5.25 V. Also, the static power dissipation of the CMOS inverter is two to three orders of magnitude lower than that of low-power TTL.

The other basic building block of CMOS is the transmission gate, shown in basic form in Fig. 15.67 together with its circuit symbol. (There exist augmented versions of

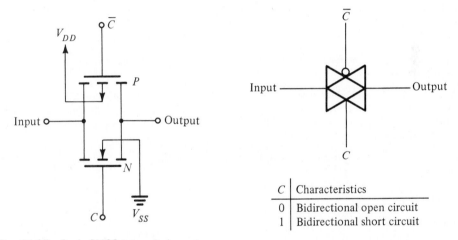

C	Characteristics
0	Bidirectional open circuit
1	Bidirectional short circuit

Fig. 15.67 Basic CMOS transmission gate.

this basic transmission gate circuit with improved performance.) The CMOS bilateral switch includes an inverter to produce the complement ($\overline{C}$) of the control signal. Operation of the transmission gate was explained in detail in Section 8.12.

Transfer Characteristic

The transfer characteristic of the CMOS inverter can be derived using the expressions for the MOSFET i-v characteristics, as was done in Exercise 8.11. The transfer characteristic is given in Fig. 15.68. Assuming that the n and p devices are matched, the switching threshold occurs exactly at half the power supply voltage. Furthermore, the gain at the switching point is usually high. These features, together with the fact that the logic 0 level is very close to 0 V and the logic 1 level is very close to V_{DD} volts, make the CMOS transfer characteristic very nearly ideal from the point of view of noise immunity.

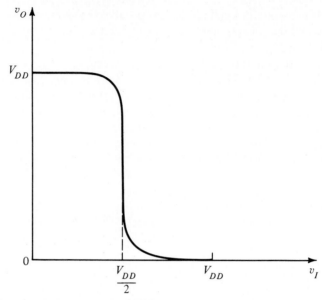

Fig. 15.68   Transfer characteristic of a CMOS inverter with matched n and p devices.

In actual CMOS circuits mismatches exist between the p and n devices and hence switching does not occur at exactly $V_{DD}/2$. A second effect that contributes to the variability of the switching voltage will be illustrated shortly. Manufacturers of CMOS circuits therefore specify minimum and maximum transfer characteristics, as indicated in Fig. 15.69. From these characteristics one can find the noise immunity, as explained next.

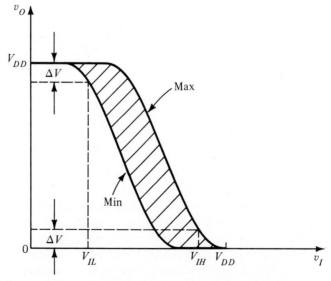

Fig. 15.69   The two limit transfer characteristics specified for a CMOS gate.

Noise Margins

The definitions of the worst-case values of V_{IL} and V_{IH} are illustrated in Fig. 15.69. Normally the deviation in output voltage ΔV is taken as 0.1 V_{DD}. Correspondingly, the following values of V_{IL} and V_{IH} are specified by Motorola for standard CMOS logic:

$$V_{DD} = 5 \text{ V}, \qquad V_{IL} = 1 \text{ V}, \qquad V_{IH} = 4 \text{ V}$$
$$V_{DD} = 10 \text{ V}, \qquad V_{IL} = 2 \text{ V}, \qquad V_{IH} = 8 \text{ V}$$
$$V_{DD} = 15 \text{ V}, \qquad V_{IL} = 2.5 \text{ V}, \qquad V_{IH} = 12.5 \text{ V}$$

These values can be used to find the noise margins $\Delta0$ and $\Delta1$:

$$V_{DD} = 5 \text{ V}, \qquad \Delta0 = \Delta1 = 0.5 \text{ V}$$
$$V_{DD} = 10 \text{ V}, \qquad \Delta0 = \Delta1 = 1 \text{ V}$$
$$V_{DD} = 15 \text{ V}, \qquad \Delta0 = \Delta1 = 1 \text{ V}$$

It should be noted that these are worst-case values. Furthermore, as will be discussed shortly, there exists an improved version of CMOS (called *buffered CMOS* or *B series*) that has increased noise immunity.

EXERCISE

15.29   The buffered CMOS family is specified to have
   (a)   At $V_{DD} = 5$ V, $V_{IL} = 1.5$ V and $V_{IH} = 3.5$ V;
   (b)   At $V_{DD} = 10$ V, $V_{IL} = 3$ V and $V_{IH} = 7$ V;
   (c)   At $V_{DD} = 15$ V, $V_{IL} = 4$ V and $V_{IH} = 11$ V.
   Calculate the corresponding noise margins.
   Ans.   (a) 1 V; (b) 2 V; (c) 2.5 V

NOR and NAND Gates

Figure 15.70 shows a two-input NOR gate and a two-input NAND gate of the standard CMOS family. It is easy to verify that these circuits indeed implement the logic functions indicated. For the NOR gate note that the output will be high only when both inputs are simultaneously low:

$$C = \overline{AB} = \overline{A + B}$$

For the NAND gate the output will be high when either A or B is low:

$$C = \overline{A} + \overline{B} = \overline{AB}$$

It is interesting to observe that the NOR gate can be converted into a NAND gate (in positive logic), and vice versa, by replacing the n-channel devices by p-channel devices, and vice versa, and by interchanging the power-supply terminals.

Increasing the fan-in of CMOS gates is straightforward. A practical limit on fan-in exists, however, due to the resulting shift in the gate threshold voltage. This problem is illustrated by Example 15.4.

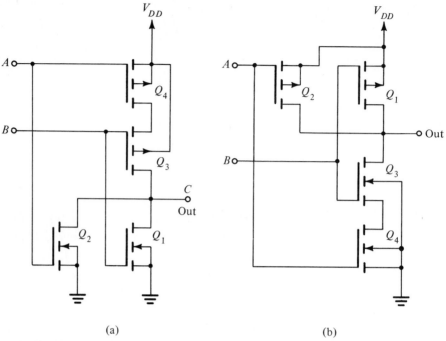

(a) (b)

Fig. 15.70 CMOS two-input (a) NOR and (b) NAND gates.

Example 15.4

Consider the two-input NOR gate of Fig. 15.70a. Assume that all devices have equal values of β and V_T. Calculate the value of the switching threshold of the gate, V_t, when input terminal A is connected to ground and when input terminal A is connected to input terminal B. For numerical calculations use $V_{DD} = 10$ V and $V_T = 2$ V.

Solution

Figure 15.71a shows a simplified circuit diagram of the NOR circuit with terminal A connected to ground. Since Q_2 will be off, we have eliminated it from the circuit altogether. The switching threshold V_t is the value of input voltage v_I at which both Q_1 and Q_3 are in the active (pinch-off) mode. For the no-load condition, the currents in Q_1 and Q_3 will be equal and given by

$$I = \tfrac{1}{2}\beta(V_t - V_T)^2 = \tfrac{1}{2}\beta(V_1 - V_t - V_T)^2 \qquad (15.19)$$

where V_1 denotes the voltage at the source of Q_3. Since the value of V_t will be around $V_{DD}/2$, V_1 will be higher than this value by at least V_T. Hence it is reasonable to expect that Q_4 will be operating in the triode region. We shall make this assumption and later verify its validity. Thus, using Eq. (8.3), which describes the MOSFET operations in the triode region, we can write for Q_4

$$I = \beta[(V_{DD} - V_T)(V_{DD} - V_1) - \tfrac{1}{2}(V_{DD} - V_1)^2] \qquad (15.20)$$

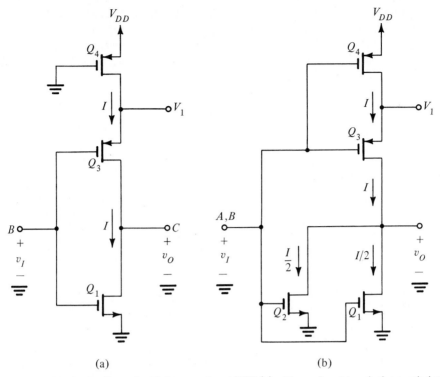

Fig. 15.71   *Circuits for Example 15.4: a two-input NOR (a) with one input terminal grounded and with (b) the input terminals tied together.*

Equations (15.19) and (15.20) are two equations in the two unknowns V_1 and V_t. The solution for $V_{DD} = 10$ V and $V_T = 2$ V is $V_1 = 9.5$ V and $V_t = 4.75$ V. Thus Q_4 is indeed in the triode region as assumed. Note that the value of V_t is lower than that obtained for the CMOS inverter, a result of the the additional series device Q_4. At the switching threshold voltage V_t of 4.75 V the output voltage abruptly changes from 4.75 $+ V_T$ to 4.75 $- V_T$—that is, from 6.75 to 2.75 V (see Fig. 8.42).

Next we consider the case where the two input terminals are tied together as indicated in the circuit of Fig. 15.71b. Again we shall assume that with $v_I = V_t$ the value of V_1 will be such that Q_4 will be operating in the triode region. The other three transistors will be assumed to operate in the active (pinch-off) region. Straightforward analysis gives $V_1 \cong 9$ V and $V_t = 4$ V. Thus we see that the gate threshold has been further reduced from the ideal value of $V_{DD}/2 = 5$ V.

· · ·

Gate Delay

The dynamic response of a standard CMOS inverter with a capacitive load was analyzed in Section 8.11. It was shown that the rise and fall times of the output waveform are almost equal (assuming that the p and n devices are matched), and their value is proportional to the value of load capacitance.

Figure 15.72 illustrates the definition of the switching times of CMOS gates as they are specified on the device data sheets. Motorola provides the following formulas for their standard CMOS logic when operated with $V_{DD} = 10$ V:

$$t_{TLH} = (1.5 \text{ ns/pF}) C_L + 15 \text{ ns}$$
$$t_{THL} = (0.75 \text{ ns/pF}) C_L + 12.5 \text{ ns}$$
$$t_{PLH}, t_{PHL} = (0.66 \text{ ns/pF}) C_L + 22 \text{ ns}$$

Note that the rise time is longer than the fall time because of the lower current drive capability of the p device (as compared to the n device).

The above formulas can be used to determine the switching times corresponding to a given value of load capacitance C_L. Typically, a CMOS gate has an input capacitance of 5 pF. Maximum fan-out is therefore limited by the dynamic response rather than by the degradation in noise immunity.

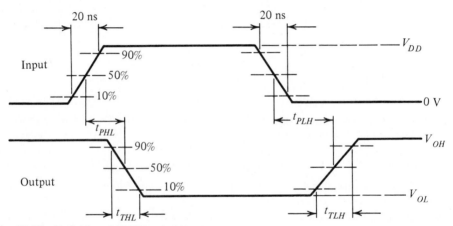

Fig. 15.72   Definition of CMOS switching times.

Power Dissipation

The quiescent supply current of CMOS gates is very small, as was explained in Section 8.11. For instance, a standard CMOS gate operated at $V_{DD} = 10$ V is specified to have a dc supply current of 1 nA (typical). Thus the static power dissipation of CMOS is the nanowatt range.

However, when CMOS gates are switched, their drain current increases and peaks at the switching threshold because at the switching threshold both the n and p devices of a CMOS inverter operate in the active mode. This phenomenon gives rise to power dissipation, called *dynamic power dissipation* (see Section 6.4), that is proportional to the frequency at which the gate is switched.

The dynamic power dissipation of a CMOS gate increases considerably when the gate is driving a capacitive load, which is the case in practice. As explained in Section 8.11, the current to charge and discharge the load capacitance gives rise to power dissipation in the p and n devices. This component of dynamic power dissipation is the dominant one. As an example, Motorola specifies that a standard CMOS gate operating with $V_{DD} = 10$ V and loaded with $C_L = 50$ pF draws an average dc supply current of 0.6 μA/kHz. Thus at a frequency of 1 MHz such a gate dissipates 6 mW, which is much greater than the static power dissipation.

Finally, it should be noted that the higher operating speeds possible at high supply voltages are obtained at the expense of increased dynamic power dissipation.

Buffered or B-Series CMOS

The buffered or B-series CMOS includes additional inverter stages on each gate output. These additional stages provide for increased gain in the transition region and hence result in a more ideal transfer characteristic. This in turn implies an increase in the gate noise margins. The B-series gates also have increased output drive capability. However, the additional stages result in a slight increase in propagation delay.

For applications in which CMOS logic gates are employed as amplifiers or oscillators (that is, in analog applications) standard CMOS (also referred to as UB-series CMOS) is generally preferred: the lower number of poles results in better stability and a "cleaner" output waveform.

CMOS Flip-Flops

In Chapter 6 we studied the terminal characteristics of a variety of flip-flop types. We also considered their implementation using logic gates. Such implementations are followed (with some variations) in logic families such as TTL and ECL. For this reason we have not looked at TTL and ECL flip-flop circuits. CMOS, however, presents an exception. The availability of transmission gates in the CMOS technology provides for very interesting schemes for the realization of clocked flip-flops. In the following we shall present one example to give the reader a flavor of this approach.

Figure 15.73 shows a logic diagram of a D-type flip-flop of the edge-triggered master-slave type. As indicated, this flip-flop also has direct set and reset inputs. The operation of the flip-flop is best explained by considering the transmission gates. When the clock is low the transmission gates are in the following states:

$$TG1 \rightarrow ON, \quad TG2 \rightarrow OFF, \quad TG3 \rightarrow OFF, \quad TG4 \rightarrow ON$$

Thus the slave is isolated from the master, and the feedback loop of the slave flip-flop is closed, making it retain its previous state. Meanwhile the feedback loop of the master flip-flop is open and the output of the master, Q', simply follows the complement of the input D.

When the clock input goes high the transmission gates change to the following states:

$$TG1 \rightarrow OFF, \quad TG2 \rightarrow ON, \quad TG3 \rightarrow ON, \quad TG4 \rightarrow OFF$$

The result is that the master is disconnected from input D and its feedback loop is closed. Thus its output Q' adopts the complement of the signal that existed on input line D just prior to the positive transition of the clock. Meanwhile, the feedback loop of

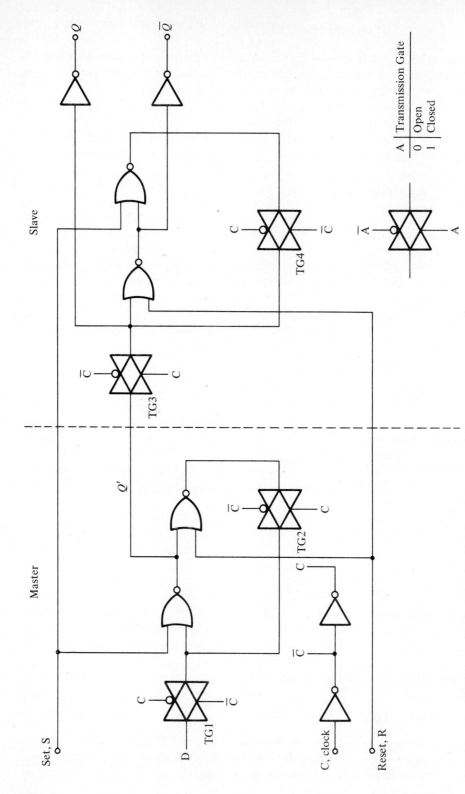

Fig. 15.73 Logic diagram of a CMOS D-type flip-flop.

the slave is opened and its output Q adopts the complement of Q', which is the input D. The overall effect is that on the positive transition of the clock the output Q adopts the value of input D that existed just prior to the transition.

Input Circuit Protection

Because of the very high input resistance of MOS devices, static electricity can cause charge to accumulate on the input capacitances during handling. Such charge can cause large voltages, which in turn can cause device breakdown. CMOS circuits include input diode networks, together with appropriate series resistors, for the purpose of limiting the voltages that may appear at the MOS device inputs. For details on the input protection circuitry the reader may consult Ref. 15.11.

Attention to Unused Inputs

Since the input resistances of a CMOS gate are very high, a gate input left unconnected will float at an unknown voltage. Usually, however, leakage currents are such that the input devices enter the active mode, allowing large currents to flow and causing overheating to result. Accordingly it is important that spare gate inputs be connected to an appropriate local power supply pin or paralleled with another input (keeping in mind the effect of this on the gate-switching threshold).

Tristate Outputs

Like TTL gates, the output terminals of CMOS gates cannot be connected together to create wired-AND or wired-OR logic (why not?). The solution to this problem is the use of tristate output devices. We have explained the tristate concept in connection with TTL. Here the approach is similar and will not be considered further (see Ref. 15.11 for details).

A Final Remark

CMOS is very close to living up to its initial description as the "ideal logic family." Currently it exists in SSI, MSI, LSI and VLSI packages. The last includes memory and microprocessors. Furthermore, CMOS technology is undergoing many improvements, allowing it to make deep inroads into the VLSI systems area.

15.9 CONCLUDING REMARKS

The emphasis in this chapter has been for the most part on logic families for which a broad range of integrated circuits can be purchased and for which small-scale integrated (SSI) packages as well as medium-scale (MSI) and large-scale (LSI) components are available. The TTL, ECL, and CMOS families all fall into this class. The I^{2}L family, on the other hand, does not appear (and is unlikely to appear) in SSI and general-purpose forms; however, it is of great importance in systems of LSI and VLSI scale, particularly where both analog and digital technologies merge.

As a consequence of the emphasis on availability of general-purpose parts we have not discussed PMOS and NMOS as logic families. These logic-circuit technologies, which were introduced in Chapter 8, are specifically useful in LSI and VLSI products, such as memory chips. This topic will be studied in Chapter 16.

Finally, to place all of these developments in context, Fig. 15.74 provides a chart in which various digital circuit technologies are compared from the point of view of gate delay and power dissipation. Here a distinct feature is the broad range of capabilities of I^2L and its maintenance of a constant delay-power product over a power range of about two decades.

Note also the features of the MOS-family speed hierarchy. We see that PMOS is basically slower than NMOS by a factor of 10 but that the combination of both in CMOS provides the best of both worlds.

Note that there is some justification for extending the CMOS region much more toward the low-power end of the scale to represent its very low static power consumption. The chart, however, has been prepared to represent dynamic activity consistent with the characteristic propagation delay. The extended power range shown represents the behavior of a CMOS gate as its power supply is varied over the broad range of 5 or

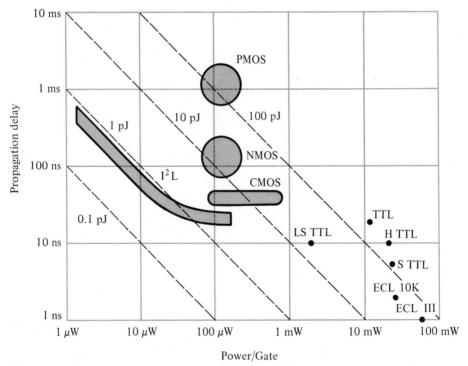

Fig. 15.74 *Comparison of speed and power capabilities of various logic families. NOTE: (1) The broken lines represent contours of delay-power product. (2) S TTL denotes Schottky TTL; LS TTL denotes low-power Schottky TTL; H TTL denotes high-speed TTL. (Adapted from Ref. 15.13.)*

6 to 1, which is unique to the CMOS family (as compared to TTL or ECL). The reason that the gate delay remains essentially constant is that as the supply voltage is increased the current available to charge and discharge load capacitances is increased, but the output voltage swing is also increased.

It is worthwhile to note that while speed can be bought with the currency of power, other limitations exact a tax on the transaction. Thus we see that a representative trend line drawn through the high-power/high-speed part of the chart has a slope that is somewhat less than that of the constant delay-power-product lines. Furthermore, we may sense a turnup of this line similar to that seen in the I^2L case.

Finally, it should be mentioned that each of the current digital circuit technologies (that is, TTL, ECL, CMOS, NMOS, and I^2L) is still undergoing development and improvement and is making inroads into the VLSI circuits area. In Chapter 16 we shall study an example of the area of VLSI circuits applications, namely digital memory.

MEMORY CIRCUITS

16

Introduction

In this, the last chapter of the book, we shall study an important application of the current Very Large Scale Integrated (VLSI) circuit technology, namely, digital computer memory circuits. Digital computer systems are composed of logic gates and memory circuits. The former were studied in the last chapter and the latter is the topic of the present chapter.

VLSI circuit technology permits the fabrication, on a single silicon chip, of circuitry with complexity greater than the equivalent of 1,000 logic gates. Since the development of the MOS transistor in the late 1950s, and through to the late 1970s, complexity of IC chips has doubled every year, reaching the current level of about 500,000 transistors per chip. An important problem posed by such a high level of integration is finding high-volume applications for the resulting circuits. One solution to this problem has been the invention of the microprocessor (μP)—a highly complex logic circuit that is organized so that it can be field-tailored to a variety of applications by software.

Microprocessors provided another impetus to the development of VLSI: They require memory in which the software can reside, and because memory can be implemented with chips having a highly regular structure it is the most suitable functional block for VLSI implementation. Current estimates indicate that by 1983 the market value of microprocessor and memory chips will reach about $1.5 billion per year.

To give the reader a flavor of the state-of-the-art in VLSI circuit technology, we shall study a variety of memory circuits. To reflect current trends, only MOS circuits are discussed. Nevertheless, in the concluding section of this chapter we will briefly consider other VLSI circuit technologies. We shall also mention an important application of VLSI, namely, system design using *gate array* chips.

16.1 MEMORY TYPES AND TERMINOLOGY

A computer system, whether a large machine or a microcomputer, requires memory for storing data and program instructions. Furthermore, within a given computer system, there usually are various types of memory utilizing a variety of technologies and having different *access times*. Specifically, we first wish to distinguish between the computer *main memory* and other mass storage devices.

Random Access Memory (RAM)

The main memory is usually the most rapidly accessible memory and the one from which most, often all, instructions in programs are executed. The main memory is usually of the *random access* type. A random access memory (RAM) is one in which the time required for storing (writing) information and for retrieving (reading) information is independent of the physical location (within the memory) in which the information is stored.

Serial Memory

Random access memories should be contrasted with *serial* or *sequential* memories, such as disks and tapes, from which data is available only in the same sequence in which it is originally stored. Thus, in a serial memory the time to access given information depends on the memory location in which the required information is stored, and the average access time (also known as *latency time*) is longer than the access time of random access memories.

In a computer system, serial memory is used for mass storage. Items not frequently accessed, such as the computer operating system, are usually stored in a *moving-surface memory* such as magnetic disk, drum, or tape. In this chapter we shall not study moving-surface memories. The only serial memory we will consider is that made of *charge-coupled devices* (CCDs). At the time of this writing, CCDs are starting to find application as modest-sized higher-speed (compared to moving-surface devices) serial memories.

Read/Write and Read-Only Memory

Another important classification of memory is whether it is a *read/write* or a *read-only* memory. Read/write (R/W) memory permits data to be stored and retrieved at comparable speeds. Computer systems require random access, read/write memory for data and program storage.

Read-only memories (ROM) permit reading at the same high speeds of R/W memories, but restrict the writing operation. Strictly speaking, ROMs are intended to be written only once—at the time of manufacture. However, there exists a type of ROM that can be written (programmed) in the field by the user. These *programmable ROMs* (or *PROMs*) can be programmed only once. Still there exists a type of ROM that can be erased and reprogrammed as many times as desired. It is known as *erasable programmable ROM* (*EPROM*) and is clearly the most versatile type of ROM. EPROMs are currently very popular in microprocessor systems where they are used to store the μP operating system program. ROMs are also used in a variety of operations that require table lookup, such as finding the values of trigonometric functions. We shall study read-only memory circuits in Section 16.6.

Memory Organization

Consider a random-access read/write memory. Such a memory is formed by a large number of basic units referred to as *memory cells*. Each memory cell is a device or an electronic circuit that has two states and is thus capable of storing a binary digit or *bit*. These cells or bits are physically grouped into chunks such as *bytes* and *words*. All the digits of a byte or a word in a memory are simultaneously accessed for a read or a write operation.

Viewed from a system standpoint, main memory can be considered as W words, each of B bits for a total storage capacity of $W \times B$ bits. As mentioned before, the B bits of a word are simultaneously available for reading or writing, as schematically indicated in Fig. 16.1.

Figure 16.1 also indicates that to select 1 out of W words (for reading or writing) one requires an address of n bits where $2^n = W$. Thus a 10-bit address is required for a memory having $2^{10} = 1,024$ words. Such a memory is said to have a capacity of 1K words. A 2K-word memory has 2,048 words, a 16K-word memory has $16 \times 1024 = 16,384$ words, and so on. A 16K-word memory requires a 14-bit address, and so on.

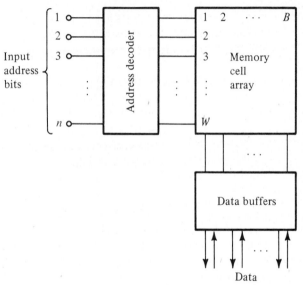

Fig. 16.1   Conceptual organization of a memory consisting of W words of B bits each, requiring an address of n bits ($2^n = W$).

Memory Timing

In using the memory we speak of *access time* and *cycle time*. Access time is the time required to read out any randomly accessed word from memory. It is the time between the initiation of a read operation (accomplished by applying the address of the word to be read and then applying a pulse commonly called a *read strobe* signal) and the time at which the bits of the addressed word become available at the output terminals. The cycle time is the minimum time interval required between the initiation of two successive independent memory operations. To be general, each of the operations is usually of the *read-modify-write* type. That is, a word is read out and then new information is written into the addressed location.

16.2 AN OVERVIEW OF MEMORY TECHNOLOGIES[1]

Magnetic Core Memory

The oldest, and still existing, memory technology is that of ferrite cores (see Ref. 16.2). The idea behind the use of toroidal ferrite cores as memory devices is that the core magnetization can take one of two possible directions. Thus a core can be used to store one bit of information with one of the two magnetization directions representing a stored 0 and the other a stored 1. The direction of magnetization is established by the polarity of the electric current passing through a wire that threads the toroidal core. Furthermore, once the core is magnetized in a certain direction, this condition is maintained even if the magnetizing current is eliminated. Thus magnetic core memories retain their content when the system power supply is switched off. This property is known as *nonvolatility*. In contrast to magnetic core memories, most commercially available semiconductor memories are *volatile*. Nonvolatility, however, is no longer considered of paramount importance in memory systems. This is especially true for low-power MOS memories that can be made to retain their contents using a modest-sized battery as a backup power supply, in what is known as the *standby* or *power-down* mode.

Reading of magnetic core memories is a destructive process. Therefore each read operation has to be followed by a write operation in order to restore the data to its original value. Magnetic core memory is thus said to be of the *destructive readout* (*DRO*) type. As will be seen, modern MOS RAMs are also of the DRO type.

The output signals obtained from reading magnetic core memories are quite small and, hence, very sensitive amplifiers, called *sense amplifiers,* are required. Also, relatively large current signals are required in the writing process, implying the need for high-power drivers. In summary, relatively sophisticated and expensive circuitry is required for interfacing magnetic core memories to the computer logic circuits. This makes such memories economically viable only in applications that require large storage capacity. In such applications the overhead cost is shared by a large number of bits, reducing the cost per bit. Even then, their cost per bit is still higher than that currently achieved in semiconductor memories.

However, the major disadvantage of magnetic core memory is the incompatibility of its technology with modern IC technology. This together with the bulkiness of magnetic core memories is contributing to the rapid demise of this memory technology in favor of semiconductor memory which is the main topic of this chapter.

Semiconductor Memory

Semiconductor memory is known by various names: monolithic memory, microelectronic memory, integrated-circuit memory, LSI memory, and active memory. All these

---

[1]For a more-detailed, highly readable overview of memory technologies see Ref. 16.1

names refer to binary digital memories that employ an electronic circuit for each memory cell.

Traditionally, semiconductor memories have been used in computer systems for applications requiring high operating speed (short access time) such as *scratch pad* memories, which are used for storing intermediate results in arithmetic and logic operations. These were low-density, high-cost units employing bipolar junction transistors as flip-flops. During the last decade, semiconductor memories have undergone tremendous development and have virtually taken over the main memory market. We shall now present an overview of semiconductor RAMs placing emphasis on the terminology that has grown around this relatively new area. This will be followed in the succeeding sections with a study of representative semiconductor memory circuits.

Semiconductor RAMs can be divided into two major types: bipolar RAMs and MOS RAMs. Bipolar RAMs utilize BJTs and are manufactured using TTL, ECL, or I²L technology. Each memory cell is a flip-flop and the access time and cycle time are quite short, typically in the 10–100 ns range. Bipolar memories are therefore especially useful in applications that require high operating speeds. This is achieved at the expense of the relatively large power dissipation characteristic of TTL and ECL circuits. As mentioned in Chapter 15, I²L has the unique feature of a continuous speed/power trade-off. Nevertheless, the combination of high packing density, low power dissipation, reasonable operating speed, and, very importantly, low cost per bit, make the MOS RAM the device of choice for the bulk of main memory applications.

At the time of this writing, 64K-bit NMOS dynamic RAM chips (we will explain the meaning of "dynamic" shortly) that operate from a singe 5-V power supply and that are housed in convenient 16-pin dual-in-line packages have just become available commercially. Such chips feature access times as low as 150 ns and cycle times as low as 280 ns. Furthermore, these memory chips interface directly with TTL logic circuits. In short, the availability of such VLSI memory chips enables one to design large-capacity memory systems in a relatively small physical space and without the need for elaborate peripheral circuitry. We will discuss a slightly less complex but quite similar and very popular memory chip, the 16K-bit dynamic RAM, in Section 16.3.

MOS RAMs

Modern MOS RAMs are manufactured using NMOS or CMOS technology. Although NMOS technology is currently the dominant one, CMOS RAMs are especially useful in applications that require very low power dissipation.

While bipolar RAMs are almost all of the static type, MOS RAMs can be either *static* or *dynamic*. Static RAMs utilize flip-flops for storage and each memory cell requires 6–8 MOS transistors. Thus each cell requires a relatively large silicon area, which limits the chip packing density or level of integration. Also in the *fully static RAMs,* those in which the sense amplifiers are turned on all the time, the power dissipation is relatively high. In return, very high operating speeds, rivaling those characteristic of bipolar RAMs, have been achieved. The state-of-the-art in static MOS RAMs is represented by a 16K-bit NMOS chip that features 45-ns access and cycle times and requires a single 5-V power supply. We will study static MOS RAMs in Section 16.5.

Fully static MOS RAMs do not require clocks for their operation. Clocks may be, and usually are, used for gating and synchronization purposes but are not essential for the operation of the memory chip. To reduce the power dissipation, some static RAMs use sense amplifiers that are normally off and are activated only when needed. Also, static MOS RAMs usually have a power down or standby mode of operation in which the chip retains its contents but cannot be accessed for reading and writing. The power dissipation in the standby mode is usually a small fraction of that in the *active mode*.

In summary, static MOS RAMs, although useful in special applications, have not at this time achieved the chip packing density and cost per bit obtained in dynamic MOS RAMs. In fact there is little doubt that the breakthrough in semiconductor memories came with the development of dynamic MOS RAMs.

Dynamic MOS RAMs do not utilize flip-flops as storage cells. Rather, the memory cells are very simple, each consisting of a small capacitor (typically 0.05 pF) and one MOS transistor. A logic 0 or 1 is stored as a low or high voltage on the capacitor. The obvious problem with this dynamic storage is that charge eventually leaks off the capacitor, leading to loss of the stored information. This is unlike static storage where the cell flip-flop retains its information as long as the power supply remains on.

If data are to be retained in the dynamic memory for any length of time (typically for times greater than 2 ms in the older units and 4 ms in the new 64K units), then the stored data must be sensed before being lost and the voltages restored to their original good levels. The operation of restoring the cell voltages to good levels is known as *refresh*.

Because dynamic RAMs require refreshing, a clock becomes an essential part of the memory system. The refresh requirement makes dynamic RAMs less convenient to use than the static units. In modern devices, however, this inconvenience is a very slight one and is more than offset by the many advantages of dynamic memories.

In summary, at the present time dynamic NMOS RAMs are rapidly replacing magnetic core memories in main memory applications. We will study dynamic RAMs in some detail in Sections 16.3 and 16.4.

Organization of Semiconductor Memory Chips

Although the various RAM chips differ in their internal organization there are common features that we will now discuss. As an example, Fig. 16.2 shows the organization of a 1K-bit memory chip. This chip is said to be organized as 1,024 words $\times$ 1 bit (or simply 1,024 $\times$ 1) which implies that each bit is individually addressable. Thus, a 10-bit address is required ($2^{10} = 1,024$).

As indicated, the cells are physically organized in a square 32×32 array. The cell array has 32 rows and 32 columns. Each cell is connected to one of the 32 row lines, known as *word lines,* and to one of the 32 column lines, known as *digit lines* or *bit lines.* A particular cell is selected by activating its word line and its digit line. This in turn is achieved by the row address decoder and the column address decoder. As indicated, five of the ten address bits form the *row address* and the other five address bits form the *column address.* The row address bits, labeled A_0 to A_4, are fed to the row address decoder which selects one out of the 32 word lines. Similarly, the column address bits, A_5 to A_9, are fed to the column address decoder which selects one out of the 32 digit

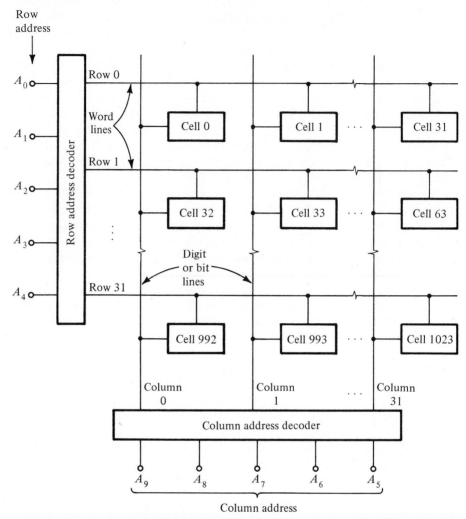

Fig. 16.2 *Organization of a 1,024-bit chip. Note that from a system point of view this chip is organized as 1,024 words × 1 bit.*

lines. Each address decoder is a combinational logic circuit that causes one of the output lines to be activated (i.e., raised or lowered in potential) in correspondence to the particular combination of input values. We will have more to say about memory chip organization in the succeeding sections.

16.3 DYNAMIC MOS RAM

As mentioned in the previous section, dynamic MOS RAMs are rapidly replacing magnetic cores in main memory applications. Therefore, it appears that of the many RAM technologies available, dynamic NMOS is the most successful. For this reason we shall study in some detail a popular dynamic MOS RAM—the 16K-bit chip, which is currently available from a large number of suppliers. As a measure of its popularity, we note that in excess of 50 million units were produced in 1979. Although at the time of this writing 64K-bit chips have become commercially available, many of the circuit techniques are similar to those used in the simpler 16K units. Some of the differences between the 64K and the 16K chips will be pointed out. Finally, it should be mentioned that some of the more specific comments that we will make on the 16K chip apply to the MK4116 manufactured by Mostek.

The One-Transistor Memory Cell

A major breakthrough in the development of dynamic MOS RAMs was the invention of the *one-transistor memory cell* shown in Fig. 16.3a. The cell consists of an enhance-

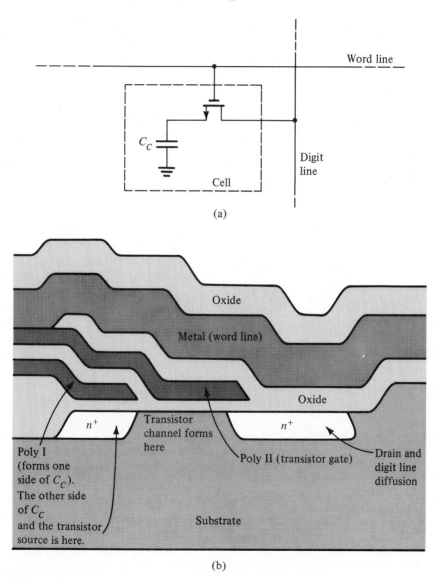

Fig. 16.3   The one-transistor cell and its cross section.

ment-mode *n*-channel MOS transistor, known as the *access transistor,* and a storage capacitor C_C (subscript C denotes cell). The gate of the transistor is connected to the word (or row) line and its drain is connected to the digit (bit or column) line. Data is stored in this cell as charge on the cell capacitor C_C. Thus a logic 1 is stored by charging C_C to a high voltage (V_{DD}) and a logic 0 is stored by discharging C_C to a low voltage (ground).

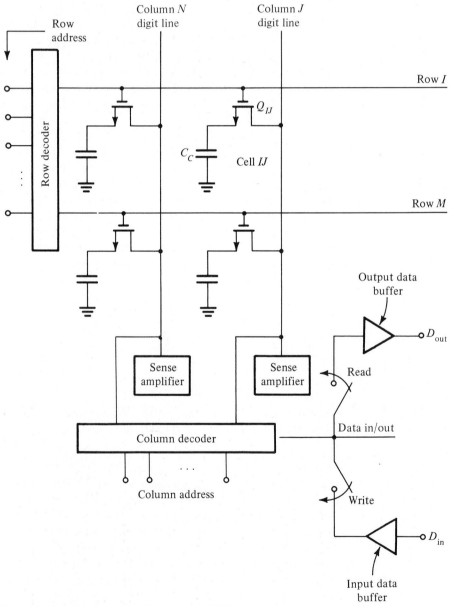

Fig. 16.4   Simplified structure of a dynamic MOS RAM chip.

A cross section of the one-transistor cell is shown in Fig. 16.3b. The cell is made using the *double-poly process* (see Appendix A). As indicated, the gate of the transistor is formed by a second layer of polysilicon (Poly II). The storage capacitor is formed between the first layer of polysilicon and the source diffusion. While the drain n^+ diffusion forms the digit line, the word line is metal. Note that the use of double poly results in a cell that extends vertically and is quite economical in its use of chip area.

To help explain the operation of the one-transistor memory cell we show in Fig. 16.4 a very simplified schematic of a dynamic RAM chip. In response to the row address, the row decoder selects a particular row by raising the voltage of its word line. This causes all the transistors in this particular row to become conductive, thus connecting the storage capacitors of all the cells in the selected row to their respective digit lines. We shall explain in the following how reading, writing, and refreshing are accomplished. For the purpose of this explanation, it is important to note that each digit line is connected to a sense amplifier.

The Read Operation

The read operation follows the following sequence. Before any particular cell is selected all word lines are charged (or said to be *precharged*) to 0 V and all digit lines are precharged to V_{DD} ($\cong 12$ V). Then to select a particular cell, the appropriate word line is raised to a high voltage. This causes all the transistors in this row to become conductive. The storage capacitor of each of the cells in the selected row is thus shorted to the corresponding digit line. If the digit line capacitance (to ground) is denoted by C_L the situation of Fig. 16.5 results. Here we should note that C_L is usually about 10–30 times greater than C_C.

Fig. 16.5   When the voltage of the selected word line is raised, the transistor conducts, thus connecting the storage capacitor C_C to the digit line capacitance C_L.

Now if a cell in the selected row has a stored 0, then its storage capacitor C_C initially has 0 V. Shorting this C_C to the digit line capacitance C_L, which was precharged to V_{DD}, will cause C_C to charge and C_L to discharge (note here the destructive nature of the readout process). The voltage of the corresponding digit line will therefore drop and this drop will be detected by the sense amplifier to which the digit line is connected (see Fig. 16.4). The sense amplifier, in turn, amplifies this voltage drop and produces a logic-0 value of 0 V. The amplifier then impresses the resulting 0 V on the digit line, thus discharging the storage capacitor C_C to 0 V. In this way the voltage of the cell is returned to its original low value.

On the other hand, if a cell in the selected row had a stored 1 then its capacitor C_C

was originally charged to a high voltage. This high voltage is usually lower than the "good" logic 1 level (V_{DD}) because C_C discharges due to leakage. When the word line is selected and all the transistors in the row become conductive, capacitor C_C charges up (only slightly) and thus the drop in the voltage of the corresponding digit line is very small. This small voltage drop is interpreted by the sense amplifier as corresponding to logic 1 and the output of the sense amplifier goes high. This high output is impressed on the digit line and serves to charge C_C all the way to V_{DD}, thus restoring the stored 1 to a good level.

From the above we note that during a read operation all the cells in the selected word (row) are refreshed. Furthermore, the data stored in all the cells of the selected row appear on the corresponding digit lines. The column address decoder then connects one of the digit lines (the one corresponding to the cell being addressed) to the data in/out bus. Since the operation is a read, this data bus gets connected to the output data buffer whose output is brought out of the memory package as the D_{out} terminal. The end result of a read cycle is that the data bit stored in the selected cell appears on the D_{out} terminal of the memory package and that all the cells in the selected row are refreshed.

EXERCISE

16.1 In a particular one-transistor-per-cell dynamic MOS memory utilizing a 64 × 64 array, the storage cell capacitance is 0.05pF, while the capacitance per cell on the digit line is 0.04 pF and the input capacitance of the sense amplifier and associated circuitry is 0.5 pF. If, after degeneration, the smallest signal allowed on the cell capacitance is 6 V, what is the corresponding signal available at the input of the sense amplifier when the access switch is closed?
Ans.   96.5 mV

The Write Operation

The write operation proceeds similarly to the read operation except that the data to be written (0 or 1) is impressed on the data in/out bus. The column address decoder then connects this bus to the appropriate digit line. Now depending on whether the data to be written is 1 or 0, capacitor C_C of the selected cell will charge to V_{DD} or discharge to 0 V, respectively.

The Refresh Operation

As already mentioned, dynamic RAMs require periodic refreshing in order to restore the stored data to good levels. While the 16K units have to be refreshed at least once every 2 ms, some of the newer 64K units need refreshing only every 4 ms. The refresh operation can be accomplished in a variety of ways, perhaps the simplest is a "burst mode" in which the memory chip is declared to be unavailable for reading or writing for an interval of time. During this interval the various rows are sequentially selected. The process for a given row is just like that which occurs during the read operation except that for refresh no column is selected (i.e., the column decoder remains inactive). Usually the interval required to refresh the entire chip is less than about 2% of the time (2% of 2 or 4 ms). In other words, the memory chip remains available for normal operation for more than 98% of the time.

16.2 A dynamic memory cell using a 0.05-pF capacitor and an MOS selection transistor employs 0- and 5-V levels for information storage. Sensing circuitry is adequate to permit the stored charge to decay to $1/e$ of its original value before refresh is required. The maximum allowed refresh interval for this design is 2 ms. What is the smallest equivalent resistance that can be allowed to shunt the storage capacitor? If the leakage phenomenon is best characterized as a current, what is the largest such current that can be tolerated? *Ans.* $40 \times 10^9 \Omega$; 79 pA

The 16K Memory Chip

We shall now consider the actual organization and operation of the 16K memory chip. Figure 16.6 shows an outline of the 16-pin package that houses the RAM chip. Indicated on the diagram are the signals and supply voltages to be connected to the 16

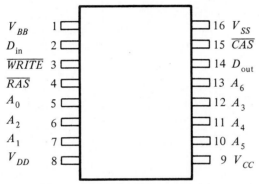

Fig. 16.6 *Outline of the 16K RAM package with the pin assignment indicated. This package has become the industry standard for 4K, 16K, and 64K RAMs.*

terminals. We will define these external signals in the course of explaining the internals of the chip, indicated in the block diagram of Fig. 16.7. In the following discussion, we shall make frequent reference to this block diagram.

The 16K RAM requires three power supplies: $V_{BB} = -5$ V (pin 1), $V_{DD} = +12$ V (pin 8), and $V_{CC} = +5$ V (pin 9). The reference terminal (ground) of each of the three supplies must be connected to the V_{SS} terminal (pin 16). It is worth mentioning at this point that the inconvenience of requiring three power supplies has been eliminated in the new 64K RAM chips. These latter units require one $(+5$ V) power supply only and generate other required voltages using on-chip circuitry.

In order to address 1 out of the 16,384 memory cells of the 16K RAM chip, 14 address bits are required $(2^{14} = 16,384)$. To conserve the number of pins and thus enable housing the chip in the convenient, standard 16-pin package, the 16K chip employs *address multiplexing*, which works as follows. To the seven address terminals, labeled A_0 to A_6, one first presents the 7-bit row address. After the input address bits have "settled" (said to have become valid), the signal $\overline{RAS}$ (*Row Address Strobe*) is activated by presenting a logic 0 to pin 4. This signal initiates a memory cycle and strobes in (latches) the row address. Furthermore, $\overline{RAS}$ should be maintained active

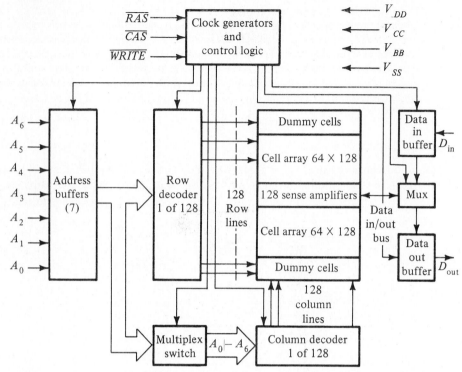

Fig. 16.7 Simplified block diagram of 16K RAM chip. (Adapted from the Mostek 1980 Memory Data Book.)

throughout the entire memory cycle, because it controls the operation of the row address decoder and other internal circuits.

The row address decoder selects 1 out of the 128 rows, and the 128 cells in this row are then refreshed as explained before. After a certain time interval, the contents of the 128 cells of the selected row become available at the output terminals of the respective sense amplifiers (there are 128 sense amplifiers). At this point it should be noted that, as indicated in Fig. 16.7, the memory cell array is divided into two halves (64 rows in the top half and 64 rows in the bottom half) with the 128 sense amplifiers placed in the middle. The reason for this structure and for the inclusion of "dummy cells" will be explained later.

The column address (7 bits) is then applied to the same address terminals (labeled A_0–A_6) and the Column Address Strobe ($\overline{CAS}$) is activated by lowering the voltage of terminal 15 to the logic-0 level. $\overline{CAS}$ strobes-in and latches the column address which is fed to the column address decoder. The latter selects 1 out of the 128 sense amplifiers and connects its output to the data in/out bus. The control logic in turn connects this bus, through a multiplexer, to the data output buffer. The retrieved data thus becomes available at the D_{out} terminal. Here it should be mentioned that when inactive the D_{out} terminal is in the open-circuit state (that is, the output is tristate).

A write cycle proceeds in the same sequence as the read cycle except that the

$\overline{WRITE}$ signal is activated by applying a logic 0 to terminal 3. This signal causes the input data applied to the D_{in} terminal to be multiplexed onto the data in/out bus which in turn is connected to the appropriate column line (as determined by the column address), and thus the input data is written in the selected cell.

Timing of the Read Cycle

The events of a read cycle as seen from the external chip terminals, are as follows:

1. Apply the 7-bit row address to the A_0–A_6 terminals.
2. Apply the $\overline{RAS}$ signal. The row address bits should remain valid for a minimum time interval, called *row address hold time,* after the $\overline{RAS}$ signal is applied. For the MK4116 this interval is 20 ns.
3. Apply the 7-bit column address to the A_0–A_6 terminals.
4. Apply the $\overline{CAS}$ signal. The column address bits should be maintained valid for a minimum time interval, called *column address hold time,* after the $\overline{CAS}$ signal is applied. For the MK4116 this interval is 45 ns. The $\overline{CAS}$ signal can be activated at any time past 20 ns after $\overline{RAS}$ is applied. $\overline{CAS}$, however, is not needed until 50 ns after $\overline{RAS}$. Nevertheless, if it occurs earlier than this 50-ns instant, it will be automatically delayed on the chip. In other words, there is no critical timing relationship between $\overline{CAS}$ and $\overline{RAS}$ from the user's point of view.
5. The output data appears on the D_{out} terminal after a certain specified time delay. This time interval is the access time and is measured relative to the activation of $\overline{RAS}$. For the MK4116 the access time is 150 ns, under the condition that $\overline{CAS}$ occurs no later than 50 ns after $\overline{RAS}$. If $\overline{CAS}$ is activated later than this point then the access time is 100 ns from the point $\overline{CAS}$ is applied.

Since the above description pertains to a read cycle, the $\overline{WRITE}$ signal should be maintained inactive. It should also be mentioned that although the access time is 150 ns (for the MK4116), the minimum interval for a complete read cycle is specified as 320 ns. The difference between the two intervals is required for the purpose of "setting up" the various dynamic circuits on the chip. For more details on timing and other specifications the reader may consult the data sheets for the particular chip. For the MK4116 a convenient reference is 16.4. However, before we leave this topic it is important to mention the impact of the active and standby operating modes: the MK4116 consumes 462 mW when active at 3 MHz and 20 mW when in the standby mode. The large difference between the two figures is a result of the dynamic nature of the circuit. The active power dissipation can be reduced by operating the chip at a lower cycle rate.

The 64K Chip

We shall now briefly mention the important differences between the 64K and the 16K chips. The 64K chip is housed in the same standard 16-pin package whose outline is shown in Fig. 16.6. However, because the 64K unit requires only one +5-V power supply which is applied to pin 8, pins 1 and 9 are freed-up for other functions. While one of these two (pin 9) is used for the additional address bit (A_7) needed (address multiplexing is of course still applied), the other (pin 1) is used for a new control signal that simplifies the refresh operation.

The MK4164 (manufactured by Mostek) dissipates 330 mW in the active mode and 22 mW in the standby mode. It features an access time of 120 ns and a cycle time of 265 ns. As with the 16K chip, the 64K chip accepts TTL-level inputs. For further details consult Ref. 16.4.

16.4   SENSE AMPLIFIERS AND OTHER DYNAMIC CIRCUITS

The Sense Amplifier

The sense amplifier is perhaps the most important circuit in a dynamic RAM. To appreciate the reason for the need of sensitive, well-balanced sense amplifiers, note that to integrate many cells on the same chip the cell capacitors have to be small in area and, hence, in capacitance. In fact, for the MK4116 the cell capacitors are 0.04 pF each. Such a small capacitor is capable of storing only a very small amount of charge. The problem is further compounded by the way in which sensing is accomplished: Recall that when a cell is selected its storage capacitor is shorted to the stray capacitance of the respective digit line. Since the digit line connects to a large number of cells, it is physically long and its stray capacitance is therefore relatively large. As implied by the block diagram of Fig. 16.7 the digit lines in the MK4116 are each split in half, thus reducing the digit-line capacitance by a factor of 2. Nevertheless, the capacitance of this half digit line is still high, being approximately 1 pF. Thus, the charge transferred from the 0.04-pF cell capacitor to the 1-pF digit-line capacitance causes a voltage signal on the digit line that is only ($\frac{1}{25}$) that of the cell. The sense amplifier is therefore called upon to detect this severely attenuated signal. Furthermore since such signals can be much smaller than those picked up by crosstalk from other parts of the chip, the sense amplifier should be capable of rejecting noise. This leads to the conclusion that the sense amplifier must be a differential amplifier with a high common-mode rejection (picked-up interference is usually a common-mode signal).

Another consideration in the design of sense amplifiers is power dissipation. Since the 16K chip requires 128 sense amplifiers, the power dissipated in each amplifier should be kept as low as possible. This requirement implies a dynamic sense amplifier that is activated only when required (to do the sensing).

We will now discuss possible designs of the sense amplifer. The design discussed first is that of a static sense amplifier which, although it works well, dissipates too much power and slows down the chip operation. An alternative design known as a dynamic sense amplifier is used in the MK4116 and will be discussed also.

Static Sense Amplifier

Figure 16.8 shows the static sense amplifier together with some of the cells of the column to which it is connected. The sense amplifier is basically a flip-flop formed by transistors Q_1 and Q_2 with resistors R_1 and R_2 serving as loads. To conserve power, the flip-flop is gated by transistor Q_5. This transistor is normally off and is turned on only when sensing is to be performed.

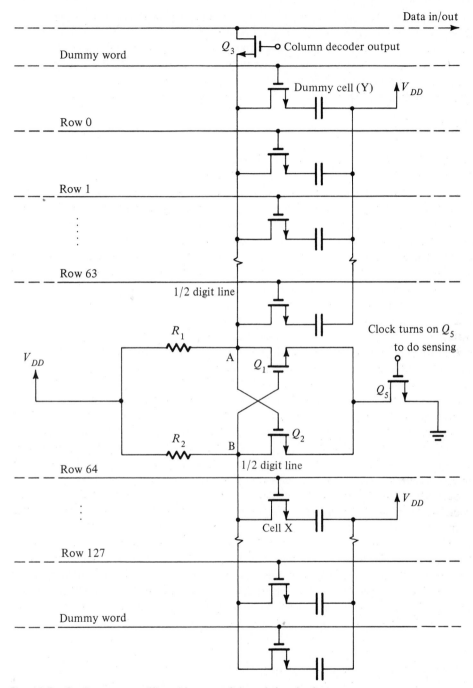

Fig. 16.8   Static sense amplifier with some of the column circuitry.

As indicated in Fig. 16.8, the digit line is split into two halves with each half connected to 64 memory cells and an additional dummy cell. The top half digit line is connected to the data in/out bus through transistor Q_3 which is turned on by the output of the column decoder (when this particular column is selected). Also note that Fig. 16.8 shows one side of the cell capacitors connected to V_{DD} rather than ground as we have been assuming. This, however, does not change the principles of operation already studied.

Let us now see how the sense amplifier of Fig. 16.8 operates. Before any particular row is selected, both halves of the digit line are precharged to precisely the same voltage. When a memory cycle is initiated by $\overline{RAS}$ going low, the two half digit lines are allowed to momentarily float. Then as a particular row is selected, charge will be transferred from the cells of this row to their respective digit lines (or more appropriately, half digit lines). Assume, for example, that row 64 is the one selected. If cell X initially contained a low value, the lower half digit line will experience a voltage drop and node B will go low. On the other hand, if cell X initially contained a high value, node B will remain high. For reasons already explained, the difference between this low and high voltage, however, is no larger than about 0.5 V. It is this small voltage signal that the sense amplifier is required to detect and amplify.

Let us proceed now with our example where we assume that row 64 is the one selected. In this case, the upper dummy word is also selected and thus the upper dummy cell (labeled Y) is connected to the upper half digit line. This cell is designed in such a way as to cause the voltage of the upper half digit line to assume a value in the middle of the low to high range, that is, 0.25 V higher than the low value. Thus, if cell X contained a low value then node A will be higher than node B by about 0.25 V. This imbalance gets amplified by the positive feedback of the flip-flop with the result that Q_1 turns off and Q_2 turns fully on. Transistor Q_2 thus pulls node B to ground and the low value stored in cell X is restored to a good level. Meanwhile, node A is pulled through R_1 to V_{DD}. If this particular column is selected, then Q_3 will be turned on and the high voltage (V_{DD}) at node A will be impressed on the data in/out bus. Note that the value appearing on the bus is the complement of that read from the selected cell. This inversion is due to the fact that the selected row (64) is in the lower half of the array. The control logic of the chip will introduce an inversion before the data is presented to the output terminal.

Let us proceed further with our example and assume that we wish to write a high value in cell X. Since cell X is in the lower half of the array, the complement signal (in this case ground voltage) should be presented to the data in/out bus. Through transistor Q_3 node A will be pulled to a low voltage close to ground. Transistor Q_2 then turns off and the positive feedback of the flip-flop causes Q_1 to turn on. As a result, node B is pulled through R_2 to V_{DD} and a high value is written in cell X, as required.

The disadvantage of this sense amplifier is the relatively large amount of power dissipated in the resistors R_1 and R_2. This power can be reduced by using high resistor values. However, this results in lengthening the write cycle time. To see how this comes about consider the case of writing a high level in cell X. We have seen that Q_2 turns off and node B is pulled to V_{DD} through R_2. A high-valued resistance R_2 would, together with the stray capacitance of the half-digit line, form a long time constant that slows down the writing operation.

These disadvantages have been overcome in the sense amplifier design used in the MK4116 as explained next.

Dynamic Sense Amplifier

The dynamic sense amplifier used in the MK4116 is shown in Fig. 16.9. When we compare this circuit to that of the static amplifier in Fig. 16.8 we note that in the dynamic circuit the flip-flop load resistors are replaced by the two MOS transistors Q_6 and Q_7 whose gates are connected to an internally generated large positive voltage ($+16$ V). Furthermore, in the dynamic circuit both the data in/out line and its complement data in/out are used. When a particular column is selected, its top half digit line is connected to the data line (via transistor Q_3) and its bottom half digit line is connected to the $\overline{\text{data}}$ line (via transistor Q_4).

To see the improved performance of the dynamic amplifier consider, as we did before, the process of writing a high value in cell X. Since cell X is in the bottom half of the array, the data line will be driven low and the $\overline{\text{data}}$ line will be driven high. Transistor Q_4 connects the $\overline{\text{data}}$ line to the bottom half digit line and thus causes node B to be driven high. A high level is thus written in cell X. Now since node B is driven high directly by the $\overline{\text{data}}$ line, the write-delay encountered in the static amplifier is considerably reduced.

Clock Input Buffer and Driver

A critical aspect of the design of any logic device, particularly those using LSI and VLSI circuit technologies, is the provision of clock signals. Difficulties arise because the clock signal (whether generated on or off the chip) must usually feed many parts of the circuit. This need for large fan-out can be met by applying the clock signal to a string of cascaded inverters and feeding the output of each inverter to a different part of the circuit, a structure known as a *branching fan-out tree*. However, problems may arise with this approach as a result of the differing time delays that the clock signal experiences on the paths to the various parts of the system. To be specific, if two physically remote segments of a logic network must intercommunicate, it is essential that the relative time variation, or *skew,* of their clocks be controlled and limited to ensure reliable operation.

The problem in VLSI circuit design is further compounded because large clock fan-out and long clock lines imply large capacitive loads and two conflicting dangers: very slow rise and fall times if drive current is inadequate or, alternatively, enormous charging and discharging currents if the driver is too capable and fast changing. Thus, the design of clock systems in VLSI circuits is a challenging problem, involving control of signals that are large both in amplitude and in rate of rise.

Figure 16.10 shows a schematic of an interesting circuit, illustrating a means to provide large reliable clock signals that make the greatest possible use of the power supply. This circuit is a *bootstrap clock buffer.* It takes TTL-level signals and produces 12-V output signals with fast but controlled rise time in the output transition from 0 to 12 V. The input is a clock signal such as $\overline{RAS}$ which is active when low. As will be seen, the circuit relies for correct operation on capacitor charge storage. Thus it must be operated at a sufficiently high rate to ensure that leakage does not impair operation.

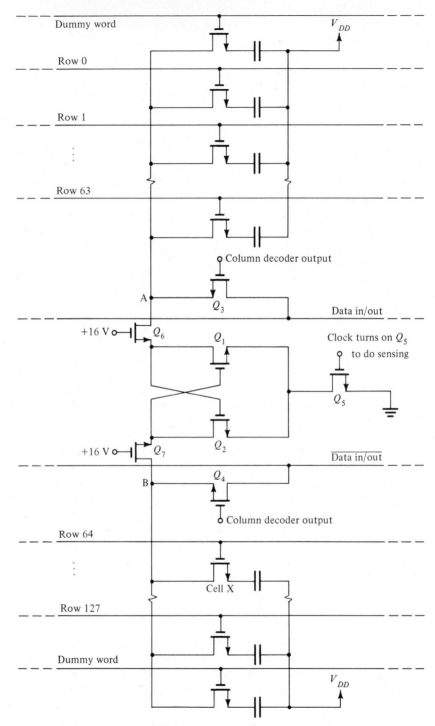

Fig. 16.9   The dynamic sense amplifier with some of the column circuitry.

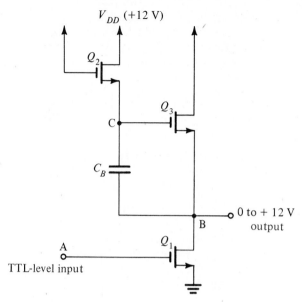

Fig. 16.10   Clock input buffer and driver.

With the clock input at rest (high), the high TTL level (≥ 2 V) exceeds the 1 V or so threshold (V_T) of Q_1, turning it on. When Q_1 turns on, it causes the Q_2, Q_3 cascade to be forward-biased, with node B at a voltage close to ground and node C at a voltage lower than V_{DD} by V_T (about 11 V). In this steady state, Q_2 conducts a very small, mainly leakage, current. On the other hand, Q_3 would be heavily biased, conducting about 1 mA (in one particular design).

Now, when the clock input is activated (that is, goes low), Q_1 is turned off while C_B, the bootstrap capacitor, keeps Q_3 conducting. In this way, Q_3 provides charging current to the capacitive load. As the load voltage rises, so does the upper end of C_B, maintaining a constant V_{GS} for Q_3 but turning Q_2 off. Ultimately the load voltage will reach nearly V_{DD} and node C will rise above the power supply to a voltage of about $2V_{DD} - V_T$. Of course, C_B will loose some charge due to stray capacitance at node C and due to leakage.

After a relatively short time, the clock input will return to the rest state (high), turning on Q_1. When Q_1 turns on it takes away all of the current from Q_3 and lowers the voltage at both nodes B and C. At the bottom end of the swing, Q_2 conducts, thus replacing the charge lost on C_B and readying the circuit for a new cycle.

It should be noted that the rise time of the output waveform is well controlled, being determined by the current of Q_3 (which is operating at a constant V_{GS}) and the load capacitance which is well known in a VLSI circuit.

EXERCISE

16.3 Consider the circuit of Fig. 16.10. All devices have $V_T = 1$ V. With node C at +11 V, Q_3 conducts 1 mA while node B is at +0.5 V and the input is at +3 V. Q_2 is a small device whose β is $\frac{1}{10}$ that of Q_3.

(a) Find β for each of the three MOS devices.

(b) Find the 0 to 90% rise time when the load capacitance is 5 pF. Note here that Q_3 operates in the active region for only part of the rise time.

   Hint: You will need the integral $\int dx/(ax^2 - x) = \ln(1 - 1/ax)$

   Ans. (a) $\beta_1 = 1.14$ mA/V^2, $\beta_2 = 2.2$ µA/V^2, $\beta_3 = 22.2$ µA/V^2; (b) $t_r \simeq 75$ ns

Address Buffer and Latch

To make the address multiplexing scheme discussed in Section 16.3 possible, it is necessary to capture the address inputs both rapidly and precisely and to store them for the remainder of the cycle. A circuit for this purpose is shown in Fig. 16.11.

Operation of the clocks ϕ_P, ϕ_1, and ϕ_2 shown, is as follows. While the external (TTL-level) clock (or strobe, say $\overline{RAS}$) is inactive (high) ϕ_P is high while ϕ_1 and ϕ_2 are low. Immediately upon the TTL clock going active (low), ϕ_P goes low, and ϕ_1 is energized (high) for a short period, equal to the address hold time (t_{AH}). Shortly after ϕ_1 goes high, ϕ_2 goes high and stays high for the entire part of the cycle during which stable address data are needed internally.

Initially, because of ϕ_P, both nodes B and C are charged to within 1 V or so (V_T) of the supply (V_{DD}) by Q_7 and Q_8. Furthermore, Q_{10} ensures that nodes B and C are at the same voltage. Also, node A is held close to ground by Q_{12}. Now, when the external TTL clock goes high, ϕ_P goes low and ϕ_1 rises turning Q_1 on and capturing the input on the capacitor C. The sampling action occupies the short interval t_{AH} while the holding action on C lasts essentially until the external clock goes inactive. Meanwhile, slightly

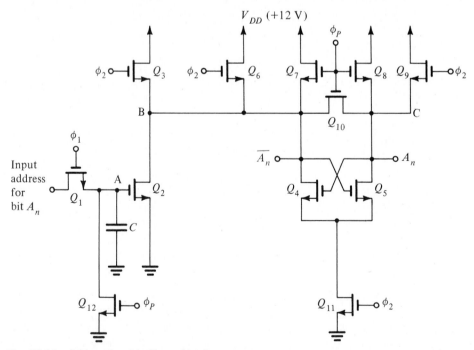

Fig. 16.11   Address input-buffer and latch.

after the rise of ϕ_1, ϕ_2 goes high turning on Q_{11}, thus activating the regenerative pair Q_4 and Q_5, whose drains are initially equally high and now are connected to the load devices Q_6 and Q_9. The direction in which the voltages at the drains of Q_4 and Q_5 move, and hence the value of the bit stored in the flip-flop, depends on the current flow through Q_2 and Q_3. If the strobed input address bit is low, Q_2 is cut off and Q_3 conducts, raising the voltage at node B. Node C then falls as Q_5 turns on. On the other hand, if the input bit is high, Q_2 conducts (more current than supplied by Q_3) and node B falls, biasing the action of Q_4 and Q_5 to cause node C to remain high. The brief interval between ϕ_1 and ϕ_2 allows Q_2 to act prior to Q_3, making the competition between Q_2 and Q_3 some- what less critical.

Finally, we should note that in the MK4116 the input address buffers are multi- plexed for both the row and column address. Each address, of course, is latched in a different set of flip-flops.

A Final Remark

The objective of this section has been to introduce the reader to some of the design techniques employed in dynamic MOS circuits. The intricacy of such circuits and the challenge that their design poses should be appreciated.

16.5   STATIC MOS RAM

For a while the development of static MOS RAMs was limited by the success of the dynamic MOS RAM in achieving extremely high packing density and spectacular chip capacity. As a result, static MOS RAM development stopped for a period at the 1K level, with the dominant units being 1K × 1 or 256 × 4 chips. More recently, however, considerable progress has been made in the development of high-speed, low-power, high-density static MOS RAMs. The current state-of-the-art in commercially available static MOS RAM chips is represented by the IMS 1400 manufactured by Inmos. This is a 16K chip featuring 45-ns access time and a maximum power dissipation of 660 mW when in operation and 110 mW when in the standby mode.

Both CMOS and NMOS technologies are being used in the static MOS RAM area. CMOS provides the advantage of extremely low power dissipation and thus allows "nonvolatile" operation by powering the memory via small rechargeable batteries. Nevertheless, recently available devices, such as the IMS 1400 mentioned above, use NMOS technology.

In this section we will study some of the design techniques employed in static MOS RAMs.

Static RAM Cells

Figure 16.12 shows typical static RAM cells in NMOS and CMOS. Each of the cells shown consists of a flip-flop formed by cross-coupling two inverters, and two *access transistors,* Q_5 and Q_6. The access transistors are turned on when the word line is selected (raised in voltage) and they connect the flip-flop to the column (digit or D) line and the $\overline{\text{column}}$ ($\overline{\text{digit}}$ or $\overline{D}$) line. The access transistors act as transmission gates, allow-

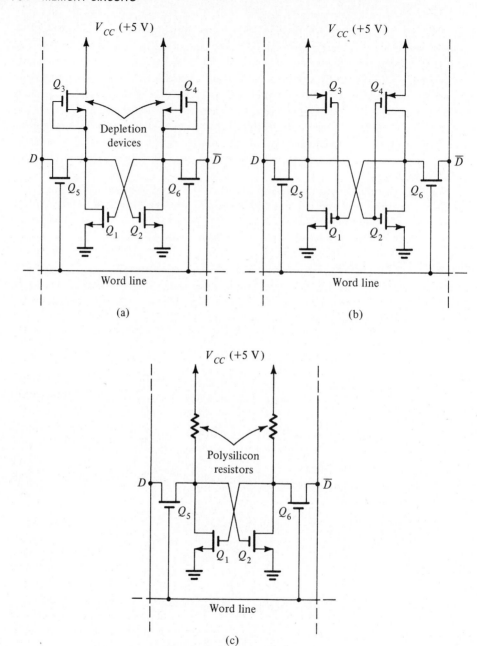

Fig. 16.12 Static RAM cells: (a) depletion load; (b) CMOS; and (c) polysilicon load.

ing bidirectional current flow between the flip-flop and the D and $\overline{D}$ lines. (To emphasize this point, their drains and sources are not distinguished.)

The cells of Fig. 16.12 differ only in the type of device used as load. In Fig. 16.12a
depletion MOSFETs are employed as loads. The advantages of depletion-load technol-

ogy (over the all-enhancement circuits) were discussed in Chapter 8. The cell of Fig. 16.12b utilizes a CMOS flip-flop formed by cross-coupling two CMOS inverters of the type studied in Section 8.11.

The cell of Fig. 16.12c utilizes two ion-implanted polysilicon resistors[2] as inverter loads. Such resistors can be made to have very large values without requiring additional chip area. Thus the memory cell of Fig. 16.12c requires a smaller chip area and dissipates less power than the depletion-load cell of Fig. 16.12a. The polysilicon-load cell is employed in the IMS 1400 16K chip.

Accessing the Cell

To access a memory cell, for reading or writing, the voltage of its word line is raised, thus turning the access transistors Q_5 and Q_6 on. In this way, one side of the cell flip-flop is connected to the D line and the other side is connected to the $\overline{D}$ line. Consider as an example the read operation of the cell in Fig. 16.12a and assume that the cell is storing a 0. In this case Q_1 is on and Q_2 is off. When Q_5 and Q_6 are turned on, current flows from the D line through Q_5 and Q_1 to ground. This causes the voltage of the D line to be pulled down to ground. Simultaneously, current flows from V_{CC} through Q_4 and Q_6 and onto the $\overline{D}$ line. The voltage of the $\overline{D}$ lines thus rises toward V_{CC}. The resulting voltage difference between the D and $\overline{D}$ lines is detected by the column sense amplifier, as will be discussed shortly.

From the above description we note that the finite currents available in Q_1 and Q_4 together with the capacitances of the D and $\overline{D}$ lines determine the fall and rise times of the signals on the D and $\overline{D}$ lines. These times, in turn, contribute to the access time of the RAM. Another component of the access time is contributed by the finite rise time of the signal on the word line. This finite rise time is a result of the finite capacitance of the word line and the limited current drive available from the output of the row decoder.

In order to speed up RAM operation, chip designers attempt to reduce both components of time delay mentioned above. In addition, an ingenious approach which results in a further reduction in access time is employed in the IMS 1400 chip: While the signal on the word line is rising, the voltages of the D and $\overline{D}$ lines are brought from the values acquired in the previous cycle to a value midway between 0 and V_{CC}. Thus, as Q_5 and Q_6 turn on, the D and $\overline{D}$ lines will have to charge and discharge by voltages smaller than those required if D and $\overline{D}$ were to start from the extreme values of 0 and V_{CC}. In this way, the time taken by the column sense amplifier to reliably detect a voltage difference between D and $\overline{D}$ will be reduced. Figure 16.13 illustrates the technique. The process is known as *equilibration* and *precharge* and is controlled by a pulse automatically generated whenever a change in the row address bits is detected (see Ref. 16.11).

To complete our discussion of cell access consider the write operation. The data bit to be written and its complement are transferred to the D and $\overline{D}$ lines, respectively. Thus, if a 1 is to be written, the D line is raised to V_{CC} and the $\overline{D}$ line is lowered to

---

[2]See Appendix A.

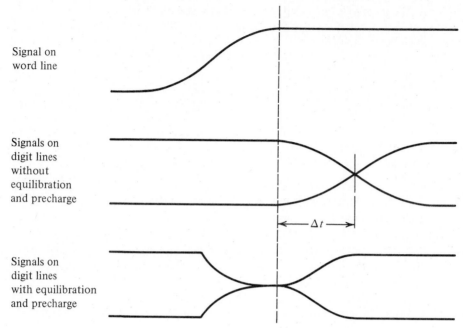

Signal on
word line

Signals on
digit lines
without
equilibration
and precharge

Signals on
digit lines
with equilibration
and precharge

Fig. 16.13   Equilibrating and precharging the digit lines reduces the access time by the interval Δt.

ground. The conducting Q_5 and Q_6 (see Fig. 16.12a) then cause the high voltage to appear at the gate of Q_2 and the low voltage to appear at the gate of Q_1. The flip-flop is then forced into the state in which the drain of Q_1 is high and that of Q_2 is low. This state, which denotes a stored 1, will be maintained indefinitely unless changed by another write operation.

EXERCISES

16.4   A large static RAM (16K $\times$ 1) utilizing the cell shown in Fig. 16.12c, requires 150 mW in standby. If the standby current is nearly all required by the storage cells, what value (approximately) must the polysilicon resistors have if V_{CC} is +5 V? If V_{CC} of the cell is reduced to +2 V during standby, what must the resistance be? In each case calculate the maximum current that can be drawn from the flip-flop at full voltage (5 V) without the loss of information that can occur if the drain of Q_1 or Q_2 is pulled below 1.5 V (V_T plus a safety margin). If the device β is designed to provide a low level $\leq$ 10 mV at the largest standing current and full voltage applied, what is the greatest current the flip-flop can sink while retaining an output $\leq$ 0.5 V (V_T less a safety margin).
Ans.   2.73 MΩ; 1.09 MΩ; 3.2 μA and 1.28 μA; 0.23 mA

16.5   Estimate the time saving to be made in a system for which the 10 to 90% rise time of the digit lines is 100 ns by the addition of the equilibration technique. If an amplifier is added to the equilibrated system to allow digit detection to be made at the point where the digit lines are different by 20% of full signal, what further time saving is achieved? For simplicity assume the rising and falling edges to be linear.
Ans.   56 ns; 31 ns

The Sense Amplifier

Because the output voltage available from a static RAM cell is generally higher than that obtained from a dynamic RAM cell, the design of sense amplifiers for static RAMs is an easier problem than that for dynamic RAMs. Nevertheless, in the static RAM case the challenge is to perform the sensing in a very short time.

The simplest arrangement for performing sensing is illustrated in Fig. 16.14. Here we show the Mth column of a static RAM chip. The D and $\overline{D}$ lines of this column are connected to the N cells of the column through the access transistors, which are turned on when a particular cell is selected. Two enhancement transistors, Q_{L1} and Q_{L2}, serve as loads for the D and $\overline{D}$ lines. These devices establish a dc voltage of $V_{CC} - V_T$ on each digit line when no cell is being selected.

As shown in Fig. 16.14, the D_M line is connected to the data line via transistor Q_{C1}, and the $\overline{D}_M$ line is connected to the $\overline{\text{data}}$ line via transistor Q_{C2}. It should be clear that each column has two such transistors which connect its digit lines to the pair of data

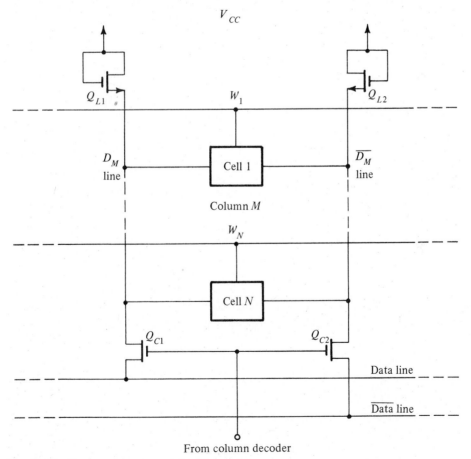

Fig. 16.14   *Common arrangement for data sensing.*

lines. However, only in the selected column (let that be column M) will transistors Q_{C1} and Q_{C2} be turned on (by the output of the column decoder) and thus the D_M line will be connected to the data line and the $\overline{D}_M$ lines to the $\overline{data}$ line. If the operation is a read then signals are transferred from the D_M and $\overline{D}_M$ lines to the data and $\overline{data}$ lines; if it is a write, reverse transfer occurs. In a read operation, the data lines are connected to the data output buffer of the chip; in a write operation they are driven by the data input buffer.

Because of the finite capacitance of the data and $\overline{data}$ lines and the limited current available from Q_{C1} and Q_{C2}, the transfer of data from D and $\overline{D}$ to data and $\overline{data}$ exhibits a finite time delay which represents an additional component of the chip access time. To shorten this time delay, some designs employ bootstrapping by which the signal at the gate of Q_{C1} and Q_{C2} is made greater than V_{CC}. This causes Q_{C1} and Q_{C2} to conduct more heavily than otherwise possible and thus charge the load capacitances more rapidly.

There is a fundamental problem, however, with the sensing arrangement of Fig. 16.14: There is no actual sense amplifier; Q_{C1} and Q_{C2} merely act as transfer gates that are controlled by the column select signal. Including an actual sense amplifier can conceivably reduce the detection time and hence the chip access time. This is achieved in a recently reported alternative sensing scheme that is illustrated in Fig. 16.15 (see Ref. 16.13). Here the D and $\overline{D}$ lines are connected to the gates of transistors Q_7 and Q_8. This pair forms a differential amplifier which is activated only when Q_9 is turned on. This in turn happens when the particular column is selected for reading or writing. Then, the Q_7, Q_8 pair amplifies the difference signal generated between the D and $\overline{D}$ lines. The amplifier output appears between the data out and $\overline{data}$ out lines.

In a write operation, the input data bit and its complement are applied to the gates of transistors Q_{10} and Q_{11}. This pair forms another differential amplifier and provides its output between the D and $\overline{D}$ lines.

Note that in the sensing arrangement of Fig. 16.14 the total capacitance of the data lines is increased due to the capacitances of the large devices Q_{C1} and Q_{C2} (of which there are two per column). On the other hand, in the circuit of Fig. 16.15, the data lines see the much lower capacitances of the small-sized amplifying transistors Q_7, Q_8, Q_{10}, and Q_{11}. The reduction of the capacitance of the data lines is an important factor in the faster operation of the circuit in Fig. 16.15.

Before leaving this part we should mention that the load devices Q_{L1} and Q_{L2} (see Fig. 16.14) usually dissipate a considerable amount of power. To reduce this component of power dissipation, the IMS 1400 chip (because it employs equilibration and precharge) eliminates the load devices altogether.

Address Buffers and Decoders

The address input terminals of a static RAM usually are buffered using relatively simple circuits such as that in Fig. 16.16. Unlike dynamic RAMs, here the input address bits are not latched. As indicated in Fig. 16.16, the input buffers provide the true and the complement of the input address bits. Availability of the complements simplifies decoding.

Figure 16.17 shows a simple decoder circuit formed of a depletion-load NOR gate

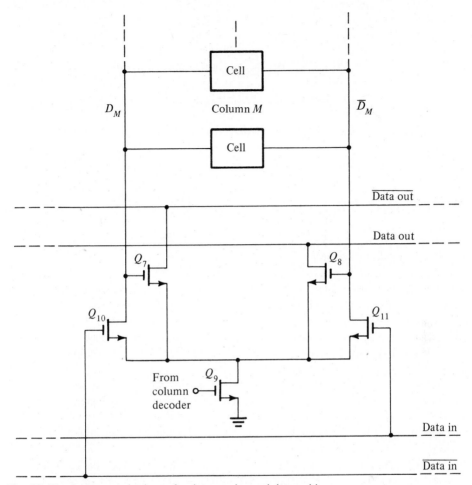

Fig. 16.15   *An improved scheme for data sensing and data writing.*

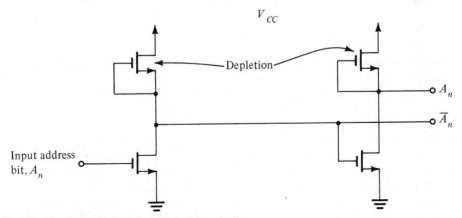

Fig. 16.16   *A simple depletion-load address buffer.*

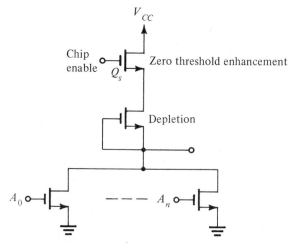

Fig. 16.17   Simple NOR *decoder.*

with an additional enhancement-mode transistor Q_s (subscript s denotes switch) connected in series with the load. Q_s is usually a specially designed device that has zero threshold voltage (known as a *natural device*). It acts as a switch that turns on, and thus connects the power supply V_{CC} to the circuit, when the chip is selected (by raising the *chip enable* signal). The chip enable signal determines whether the chip is to operate in the active mode or in the standby mode.

When the chip is selected and Q_s is turned on, the output of NOR decoder will be high (indicating that the particular row or column to which this decoder is connected is selected) when all inputs are low. It is not difficult to figure out the appropriate combination of true and complement address bits that should be connected to the input terminals of each NOR decoder.

The decoder output terminal must drive a word or a column line. To reduce the rise time of the signal on the word or column line, and thus reduce the access time, the decoder output may be buffered. Buffering also reduces the load capacitance seen by the NOR circuit and hence reduces its dynamic power dissipation. One such buffer circuit, that is used in the Intel 2147 which is a 4K static RAM chip, is shown in Fig. 16.18. This buffer circuit utilizes a variant of the bootstrap clock buffer discussed in Section 16.4. Both circuits employ a capacitor (C_B in Fig. 16.18) to provide a constant V_{GS} for the output transistor (Q_6 in Fig. 16.18) of the buffer, and hence a large current sourcing capability.

We will now explain the operation of the buffered NOR decoder of Fig. 16.18. When any of the input address bits is high, node A will be low, node B will be high, and the output node C will be low. If all address inputs go low, node A rises to V_{CC} and thus a voltage equal to V_{CC} develops across the bootstrap capacitor C_B. The Q_4, Q_5 inverter then operates and node B goes low, turning Q_7 off. Here we should note that the delay through the Q_4, Q_5 inverter must be sufficiently long to allow for the precharging of C_B before Q_7 turns off. As Q_7 turns off, node C is pulled toward V_{CC} and node A moves toward $2V_{CC}$ (because of the V_{CC} volts across C_B). Transistor Q_6 therefore operates with a large, constant V_{GS} and thus provides a strong drive for rapidly charging up the load

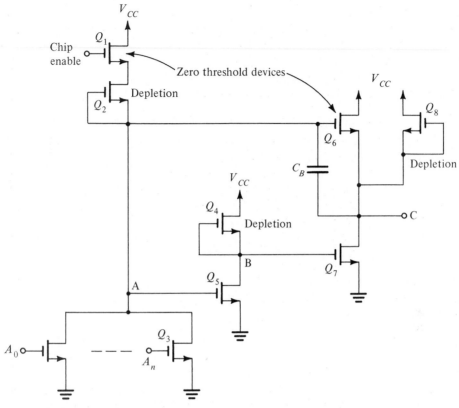

Fig. 16.18 Buffered NOR decoder.

capacitance at node C. Transistor Q_8 provides the dc pull-up that maintains node C at V_{CC}.

Although the circuit of Fig. 16.18 has been successfully employed, it suffers form certain problems which we will not discuss (see Ref. 16.10). An improved version of this buffer is employed in the IMS 1400 chip.

The 16K Chip

We will conclude this section with a brief discussion of the important features of the IMS 1400 16K chip. Figure 16.19 shows a block diagram of the chip. The chip is housed in a 20-pin package. It operates from a single +5-V supply. To individually address each of 16K bits, a 14-bit address (A_0 to A_{13}) is required. The chip has two control signals: chip enable (*CE*) which determines whether the chip is selected or not and write enable (*WE*) which determines whether the operation is a read or a write. A chip that is not selected (said to be *deselected*) operates in the standby mode with a considerable reduction in its power dissipation (110 mW compared to 660 mW for active mode operation).

The function of the precharge and equilibration circuitry and its associated control

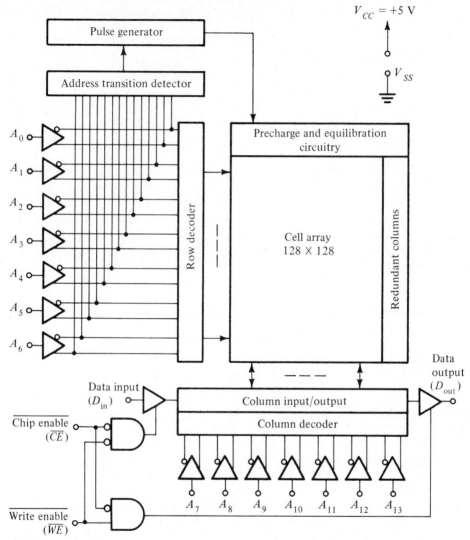

Fig. 16.19 Block diagram of the IMS 1400 16K static RAM. (Courtesy Inmos Corp.)

blocks, namely, the row address transition detector and the pulse generator, has already been explained. We have not discussed, however, the reason for including the *redundant columns,* shown in Fig. 16.19. It is an interesting feature of this memory chip that increases its production yield. The redundant columns (two) are included to be used as substitutes for any column that contains a defective cell. At the time of initial testing (by the manufacturer) such a defective column is detected, its address is stored on the chip, and it is electrically isolated and replaced with one of the two redundant columns. Now if the user addresses a bit in the defective column, this condition is detected by the column selection circuitry, and the corresponding bit in the substitute column is selected automatically. This is all done with on-chip circuits and the process is transparent to the user.

The input and output signal levels of the IMS 1400 are TTL compatible. For a read operation, the write enable input should be inactive (high) and the chip enable should be activated. The access time is defined as the time from the instant $\overline{CE}$ goes low to the instant at which the output data becomes valid. It is specified to be a maximum of 45 ns. Furthermore, since no set-up time is needed in a static RAM, the cycle time is equal to the read access time. It should be also mentioned that the output terminal is of the tristate type, allowing a simple increase in memory size by OR-tying a number of memory chips.

For a write operation, $\overline{WE}$ is activated (lowered in voltage). The write cycle time, that is, the minimum time allowed between successive writing cycles, is specified to be 40 ns. As indicated by the logic in the block diagram of Fig. 16.19, the data output buffer is disabled during a write operation. The output terminal thus goes to the high-impedance state. This also occurs if the device is deselected.

For further information on the IMS 1400 chip the reader may consult its data sheet.

As has already been mentioned, fully static memories, such as the one described above, require no external clocks and need no refreshing. Their application is therefore quite simple. This, combined with the very high speed of operation, makes static MOS RAMs very attractive devices.

16.6 READ-ONLY MEMORIES

As mentioned in Section 16.1, read-only memories (ROMs) are memories that contain fixed data patterns. They are used in a variety of digital system applications. Currently, a very popular application of ROMs is in microprocessor systems where they are used to store the instructions of the system operating program. ROMs are particularly suited for such an application because they are nonvolatile.

A ROM can be viewed as a combinational logic circuit for which the input is the collection of address bits of the ROM and the output is the set of data bits retrieved from the addressed location. This viewpoint leads to the application of ROMs in code conversion, that is, to change the code of the signal from one code (say, binary) to another. Code conversion is employed, for instance, in secure communication systems where the process is known as *scrambling*. It consists of feeding the code of the data to be transmitted to a ROM that provides corresponding bits in a supposedly secret code. The reverse process, which also uses a ROM, is applied at the receiving end.

In this section we will study the various types of read-only memory. These are: fixed ROM, which we refer to simply as ROM, programmable ROM (PROM), and erasable programmable ROM (EPROM).

Basic ROM Structure

Figure 16.20 shows a greatly simplified 32-bit (or 8 × 4 bit) ROM. As indicated, the memory consists of an array of diodes joining the word lines to the digit lines. A diode

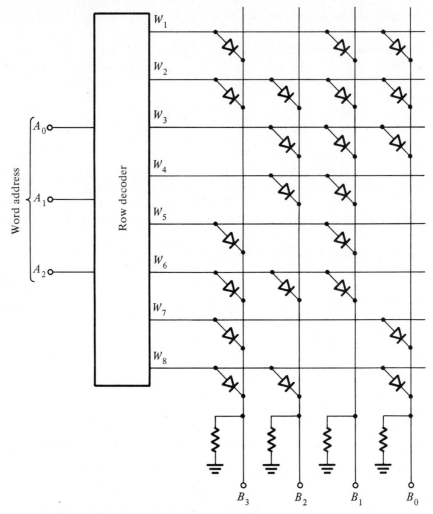

Fig. 16.20 *Simple ROM organized as 8 words* $\times$ *4 bits.*

exists in a particular cell if the cell is storing a 1; a cell storing a 0 has no diode. This ROM can be thought of as 8 words of 4 bits each. The row decoder selects one of the 8 words by raising the voltage of the corresponding word line. The cell diodes connected to this word line will then conduct, thus shorting the digit lines to the word lines and causing a high voltage (logic 1) to appear on these digit lines. The digit lines that are connected to cells (of the selected word) without diodes (that is, those cells that are storing 0s) will remain at ground voltage (logic 0). In this way, the bits of the addressed word are read.

In the simplified ROM structure discussed, the row decoder drives both the word line and the digit lines. This is obviously not a desirable situation, especially from the viewpoint of speed of operation. The problem can easily be solved by replacing the diodes with amplifying devices, that is, bipolar or MOS transistors. Figure 16.21 shows a BJT ROM cell. This cell is storing a 1; a cell storing a 0 will have no transistor or, more commonly, will have a transistor that is not connected to the digit line. A MOS ROM can be formed by replacing the BJTs with MOSFETs. Note that when BJTs or MOSFETs are used, the row decoder drives only the bases of the BJTs (or the gates of the MOSFETs). The transistor then turns on and drives the digit line with current provided by the dc supply V_{CC}.

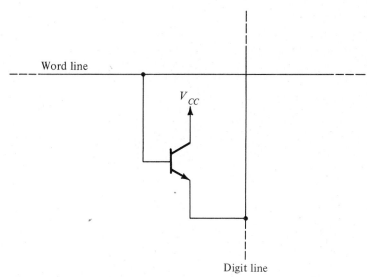

Fig. 16.21   Typical BJT ROM cell.

Mask Programmable ROMs

The data stored in the ROMs discussed above is determined at the time of fabrication, according to the user's specifications. However, in order to avoid having to custom design each ROM from scratch (which would be an extremely costly process), ROMs are manufactured using a process known as *mask programming*. As explained in Appendix A, integrated circuits are fabricated on a wafer of silicon using a sequence of processing steps that include photomasking, etching, and diffusion. In this way, a pattern of junctions and interconnections is created on the surface of the wafer. One of the final steps in the fabrication process consists of coating the surface of the wafer with a layer of aluminum and then selectively (using a mask) etching away portions of the aluminum, leaving aluminum only where interconnections are desired. This last step can be used to program (that is, store a desired pattern in) a ROM. For instance, if the ROM is made of enhancement MOS transistors, then the gates of those transistors where 1s are to be stored are connected to the word lines; the gates of transistors where 0s are to be stored are not connected. This pattern is determined by the mask which is produced according to the user's specifications.

The economic advantages of the mask programming process should be obvious: All ROMs are fabricated similarly; customization occurs only during one of the final steps in fabrication.

Programmable ROMs (PROMs)

PROMs are ROMs that can be programmed by the user, only once. Figure 16.22 shows a typical cell used in bipolar PROM. The cell is identical to that in Fig. 16.21 except that here the emitter is connected to the digit line through a fuse. This fuse can be made in a number of ways but the basic idea is the same: In a cell where a 0 is to be stored the fuse is blown, thus opening the connection between the emitter and the digit line.

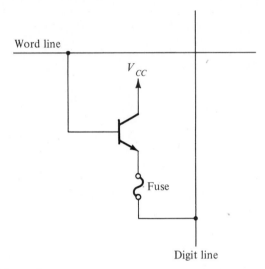

Fig. 16.22 *Typical fuse cell.*

The fuse is left intact if the cell is to store a 1. Blowing up the fuses can be done by applying large currents and is obviously an irreversible process.

In early PROM designs the fuses were made of nichrome, an alloy of nickel and chrome. Nichrome is deposited as a very thin film link to the digit lines. Heavy currents cause such a film to "blow." Technological problems, however, developed with nichrome fuses and, as an alternative to nichrome, polysilicon fuses have been successfully employed (see Ref. 16.6).

Erasable Programmable ROMs (EPROMs)

An EPROM is a ROM that can be erased and reprogrammed as many times as the user wishes. It is therefore the most versatile type of read-only memory. It should be noted, however, that the process of erasure and reprogramming is time consuming and is intended to be performed only infrequently.

The EPROM Cell

State-of-the-art EPROMs use variants of the memory cell whose cross section is shown in Fig. 16.23a. The cell is basically an enhancement-type n-channel MOSFET with two gates made of polysilicon material[3]. One of the gates is not electrically connected to any

---

[3]See Appendix A for a description of silicon-gate technology.

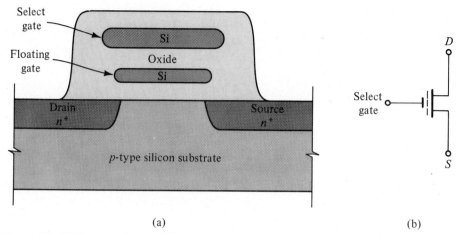

(a)   (b)

Fig. 16.23 (a) Cross section and (b) circuit symbol of the floating-gate transistor which is used as EPROM cell.

other part of the circuit; rather, it is left floating and is appropriately called a *floating gate*. The other gate, called a *select gate*, functions in the same manner as the gate of a regular enhancement MOSFET.

The MOS transistor of Fig. 16.23a is known as a *floating-gate transistor* and is given the circuit symbol shown in Fig. 16.23b. In this symbol the broken line denotes the floating gate. The memory cell is known as the *stacked-gate cell*.

Let us now examine the operation of the floating-gate transistor. Before the cell is programmed (we will shortly explain what this means), no charge exists on the floating gate and the device operates as a regular n-channel enhancement MOSFET. It thus exhibits the i_D-v_{GS} characteristic shown as curve (a) in Fig. 16.24. Note that in this case

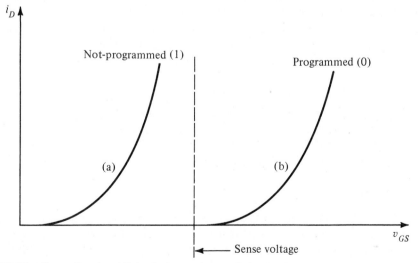

Fig. 16.24 Illustrating the shift in the i_D-v_{GS} characteristic of a floating-gate transistor as a result of programming.

the threshold voltage (V_T) is rather low. This state of the transistor is known as the *not-programmed state*. It is one of two states in which the floating-gate transistor can exist. Let us arbitrarily take the not-programmed state to represent a stored 1. That is, a floating-gate transistor whose i_D-v_{GS} characteristic is that shown as curve (a) in Fig. 16.24 will be said to be storing a 1.

To program the floating-gate transistor, a large voltage (16–20 V) is applied between its drain and source. Simultaneously, a large voltage (about 25 V) is applied to its select gate. Figure 16.25 shows the floating-gate MOSFET during programming.

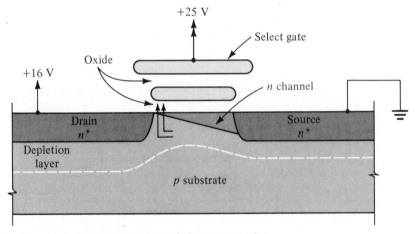

Fig. 16.25 *The floating-gate transistor during programming.*

In the absence of any charge on the floating gate the device behaves as a regular *n*-channel enhancement MOSFET. An *n*-type inversion layer (channel) is created at the wafer surface as a result of the large voltage applied to the select gate. Because of the large positive voltage at the drain, the channel has a tapered shape.

The drain-to-source voltage accelerates electrons through the channel. As these electrons reach the drain-end of the channel they acquire sufficiently large kinetic energy and are referred to as *hot electrons*. The large positive voltage on the select gate (greater than the drain voltage) establishes an electric field in the insulating oxide. This electric field attracts the hot electrons and accelerates them toward the floating gate. In this way the floating gate is charged, and the charge that accumulates on it becomes trapped.

Fortunately, the process of charging the floating gate is self-limiting: The negative charge that accumulates on the floating gate reduces the strength of the electric field in the oxide to the point that it eventually becomes incapable of accelerating any more of the hot electrons.

Let us now inquire about the effect of the floating gate's negative charge on the operation of the transistor. The negative charge trapped on the floating gate will cause electrons to be repelled from the surface of the substrate. This implies that to form a channel, the positive voltage that has to be applied to the select gate will have to be greater than that required when the floating gate is not charged. In other words, the threshold voltage V_T of the programmed transistor will be higher than that of the not-programmed device. In fact, programming causes the i_D-v_{GS} characteristic to shift to that labeled (b) in Fig. 16.24. In this state, known as the *programmed state,* the cell is said to be storing a 0.

Once programmed, the floating-gate device retains its shifted *i-v* characteristic [curve (b)] even when the power supply is turned off. In fact, extrapolated experimental results indicate that the device can remain in the programmed state for as long as 100 years!

Reading the content of the stacked-gate cell is easy: A voltage V_{GS} somewhere between the low and high threshold values (see Fig. 16.24) is applied to the select gate. While a programmed device (one that is storing a 0) will not conduct, a not-programmed device (one that is storing a 1) will conduct heavily.

To return the floating-gate MOSFET to its not-programmed state, the charge stored on the floating gate has to be returned to the substrate. This *erasure* process can be accomplished by illuminating the cell with ultraviolet light of the correct wavelength (2,537 Å) for a specified duration. The ultraviolet light imparts sufficient photon energy to the trapped electrons, allowing them to overcome the inherent energy barrier and thus to be transported through the oxide, back to the substrate. To allow this erasure process, the EPROM package contains a quartz window. Finally, it should be mentioned that the device is extremely durable and can be erased and programmed many times.

The EPROM Cell Array

Figure 16.26 shows a portion of an EPROM cell array, indicating how the stacked-gate cells are connected to the word and digit lines. As shown, the select gates of each row of cells are connected to the word line of that row, the drains of the transistors in each column are connected to the digit line of the column, and all the sources are grounded.

In the read operation, the word line of the selected row is raised to a voltage midway between the V_T values of the not-programmed and the programmed cells. The transistor of a not-programmed cell will conduct heavily, thus lowering the voltage of its digit line. On the other hand, a programmed cell will not conduct and its digit line remains at a high voltage (the digit lines are precharged to high voltages). The column decoder selects one of the digit lines and connects it to the sense amplifier which, in turn, detects the change in voltage of the digit line and thus determines whether the stored bit is a 1 or a 0.

A 64K-Bit EPROM Chip

The state-of-the-art EPROM is a 64K-bit chip organized (from an external viewpoint) as 8K words of 8 bits (one byte) each. Such an organization is known as a *byte-wide organization.*

To address one of the 8K bytes, a 13-bit address is used. This chip is internally organized as 256 rows × 256 columns. Thus, 8 of the 13 address bits are fed to the row decoder to select one of the 256 rows. The remaining 5 address bits are fed to the column decoder which selects 1 of 32 column select lines. Each of these column select lines

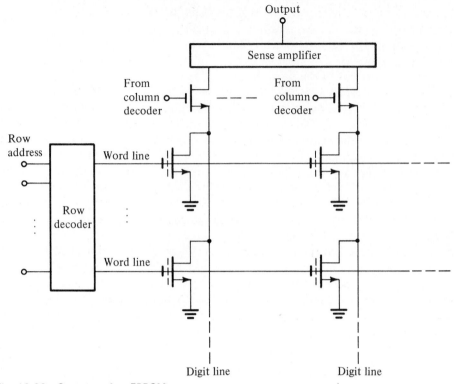

Fig. 16.26   Structure of an EPROM.

controls the connection of 8 digit lines (those of the byte being addressed) to the 8 output buffers. Thus, the result of a read operation is the retrieval of the 8 bits of the addressed byte.

The chip operates from a single $+5$-V supply and dissipates 500 mW of power when active. It automatically reverts to the low-power (50 mW) standby mode when it is deselected. The chip features a read access time of 200 ns which is about half the figure of the previous generation of EPROMs (those with 32K-bit capacity). The higher operating speed is a result of reducing device dimensions (a process known as *device scaling*) as well as innovations in circuit design.

Programming the chip requires 20-V pulses. We will not discuss here the programming procedure and refer the interested reader to Ref. 16.14.

16.7   CHARGE-COUPLED DEVICE (CCD) MEMORIES

The memories studied thus far are of the random access type. As mentioned in Section 16.1, there usually is a need in a computer system for a bulk-storage memory in which speed of operation is traded for a much lower price per bit. Traditionally, this need has been fulfilled by magnetic disks, drums, and tapes. All these devices provided sequential or serial memory. Data can be retrieved only in the sequence in which it is originally stored. As a result, the access time is variable and depends on the position of the data in the memory.

Magnetic disks, tapes, and drums all utilize physical motion to bring the data to the output device. The need for moving parts results in reduced reliability and in a relatively slow performance.

In this section we will discuss another approach to serial memory that utilizes an electronic device called a *charge-coupled device* (CCD). It will be shown that the CCD is, in effect, a long shift register. Thus data is moved to the readout device by shifting it through the register using a multiphase clock with no need for physical motion. CCDs therefore are capable of speeds of operation much higher than those of other serial memories. At the present time, however, it does not appear likely that CCDs will replace moving-surface memories, such as disks, drums, and tapes. In fact improvements continue to be made in the design of moving-surface memories, and they currently provide enormous storage capacities at a very low price per bit. The challenge such memories face is from another type of serial memory, namely, *magnetic bubble memory*. For reasons of space limitations, we will not discuss bubble memories in this book.

Therefore, it appears that CCD memories will be useful in applications that require modest-sized serial storage with faster access than movable-surface memories.

Charge-Coupled Devices (CCDs)

Figure 16.27 shows a cross section of a CCD. The device is built on a p substrate using the MOS process. In fact, the entire device can be thought of as one single n-channel enhancement MOSFET where the source and drain diffusions are at the two ends (not shown in the figure) and where many separate gates are employed. This view of the CCD, however, is of limited use in understanding the details of its operation.

Like other dynamic MOS memory devices, the CCD stores data in the form of charge. When a positive voltage is applied to a gate electrode, the majority substrate carriers (holes) are repelled from the region underneath this gate electrode. Thus a

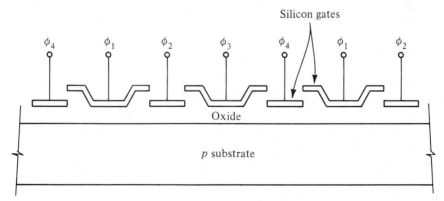

Fig. 16.27 *Cross section of a CCD operated from a four-phase clock. (Reprinted by permission of Intel Corp. All mnemonics copyright Intel Corporation.)*

depletion region is formed under the gate electrode. This depletion region, known as a *potential well,* is capable of attracting and storing electrons, if such electrons can somehow be supplied, and provided of course that the positive voltage on the gate electrode is maintained. Thus in a CCD, data are stored as negative charge packets in potential wells formed under the gate electrodes. In a binary digital system, data is stored by the presence or absence of such charge. In an analog application, the amount of charge in the well is proportional to the signal sample stored.

Let us now proceed to see how data are shifted through the CCD. Figure 16.28 illustrates the operation of a CCD that uses a four-phase clock (ϕ_1, ϕ_2, ϕ_3, and ϕ_4). Figure 16.28a shows the position of potential wells, along the CCD, at the different time intervals A, B, C, D, and E. These time intervals are indicated in Fig. 16.28b, which shows the clock waveforms. Operation is as follows: At time A only clock ϕ_2 is high and a storage well is formed under its gate electrode (and of course under every gate electrode connected to ϕ_2). Assume that this well is filled with a negative charge packet that has been externally injected. The method of generating and injecting such charge is not relevant to our current discussion.

At time B, both ϕ_2 and ϕ_4 are high and an additional storage well is formed under the ϕ_4 gate. This storage well is at this moment empty. At time C, ϕ_2, ϕ_3, and ϕ_4 are all high, causing a storage well to form under the ϕ_3 gate also. This newly formed well overlaps the storage wells under the ϕ_2 and ϕ_4 gates, thus allowing charge packets under the ϕ_2 gate to disperse throughout the charge wells of all three gates.

At time D, ϕ_2 goes low, causing the well under its electrode to disappear. This forces the charge packet to move into the combined well under the ϕ_3 and ϕ_4 electrodes. However, at time E the ϕ_3 clock goes low and its corresponding well also disappears, forcing the charge packet to be transferred to the well under the ϕ_4 gate. This completes a shift cycle: charge has been transferred from under the ϕ_2 gates to under the ϕ_4 gates. This transfer is effected by the ϕ_3 clock and is referred to as a ϕ_3 shift. Note that the ϕ_1 clock has not been utilized in this shift. The ϕ_1 clock comes into effect in the next shift cycle which causes data to be transferred from the ϕ_4 locations to the ϕ_2 locations. The ϕ_1 shift cycle follows a sequence similar to that described above for the ϕ_3 shift cycle.

From the above it can be seen that if somehow we can inject into the CCD, charges that represent a sequence of data bits, these data bits can be shifted through the CCD using a multiphase clock. CCD memory, however, is dynamic and thus needs periodic refreshing. The dynamic nature is due to charge leakage, on the one hand, and to thermally generated carriers that act to fill uncharged potential wells, on the other hand. This latter phenomenon is known as the *dark current effect*. Thus if we take the presence of charge to represent a stored 1 and the absence of charge to represent a stored 0, we see that the 1 level will deteriorate due to leakage and the 0 level will deteriorate due to the dark current effect. This points to the need for refreshing.

In a CCD, refreshing is achieved by including one or two refresh amplifiers in cascade with the CCD shift register. Figure 16.29 illustrates the technique by showing a circular 256-stage register that is split into two 128-stage registers, connected at both ends by refresh amplifiers. Note that the need for refreshing imposes a minimum frequency for circulating the data in the CCD.

Figure 16.29 also shows how data can be read from and written into the CCD.

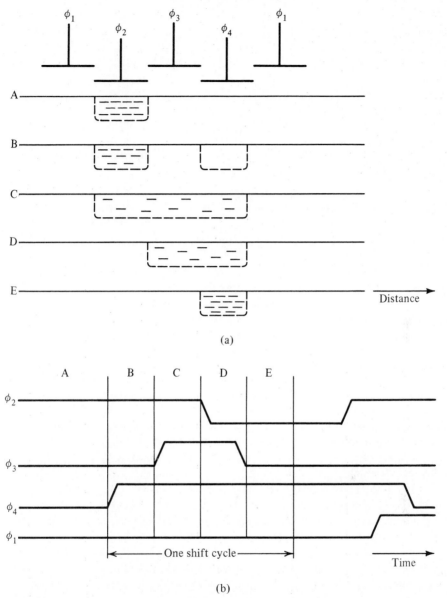

Fig. 16.28 *Illustrating the shifting process in a CCD. The clock waveforms are shown in (b). (Reprinted by permission of Intel Corp. All mnemonics copyright Intel Corp.)*

Reading is done sequentially as the bits reach amplifier 1 which is gated to the data out line. To write data, the output of the data in amplifier is connected to the CCD. The output of the amplifier causes charge to be injected to, or removed from, the n^+ region. Data is then clocked in by the shifting clock. Details of the refresh amplifier, read and write circuitry, and timing can be found in Ref. 16.6.

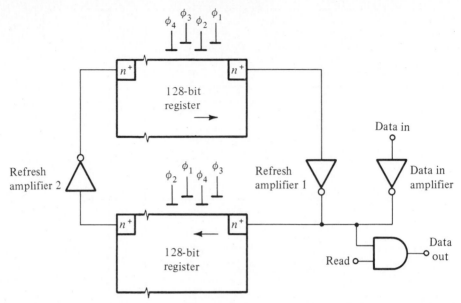

Fig. 16.29 A circulating 256-stage CCD shift register is split into two halves connected at the two ends with refresh amplifiers. Also shown are the write and read circuits. (Reprinted by permission of Intel Corp. All mnemonics copyright Intel Corp.)

CCD Memory Chips

A number of CCD memory chips are available having a variety of organizations and capacities. 64K-bit chips have been available for a number of years and higher-capacity units are under development. In order to reduce the access time, most CCD chips are organized as a number of independent individually addressed registers. As an example, Fig. 16.30 shows a block diagram of the Intel 2416 which is a 16K-bit chip organized as 64 registers each of 256 bits length.

Each of the 64 registers of the Intel 2416 chip is of the type shown in Fig. 16.29. Data are continuously circulated in each of the registers under the control of the externally supplied four-phase clock. This circulation is performed synchronously in all registers. As mentioned before, there is a minimum rate at which this data shifting is to be performed. This minimum frequency is in the range of 1–100 kHz, depending on the device ambient temperature and other operating conditions. The device also has a maximum specified shift frequency of about 1.3 MHz. At this frequency, the average access time is about 100 μs.

To perform a read or a write operation, the chip must first be selected by raising the CS input. The chip enable (CE) input is then pulsed high to initiate a read or a write cycle. For a write operation the write enable (WE) input also should be raised. The decoder then selects one of the 64 registers and connects it to the input/output circuitry.

Read and write operations are performed between consecutive shift cycles. At the option of the user, from 1 to 64 read or write cycles can be performed at any one time. That is, after one shift cycle is completed, one data bit from each of the 64 registers can

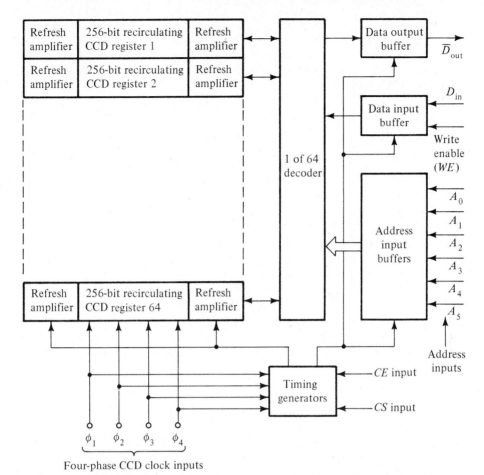

Fig. 16.30 *Block diagram of the Intel 2416 16K-bit CCD memory chip. (Reprinted by permission of Intel Corp. All mnemonics copyright Intel Corp.)*

be read or written. Thus we can see that the data rate of the chip can be much greater than the data shift frequency. The Intel 2416 has been operated with data rates as high as 4M bits/s.

A Final Remark

We have already mentioned that CCDs are not likely to replace moving-surface serial memories. They are eminently suited, however, for applications that demand speeds of operation higher than those available in moving-surface devices. In such applications, CCDs have to compete with dynamic MOS RAMs, both in packing density and in cost per bit. Since, in both of these respects, dynamic MOS RAMs have reached very impressive levels of performance, the future role of CCDs in the digital memory area is not very clear at the moment. However, CCDs are interesting devices that find important applications in analog signal processing and other areas (see Ref. 16.15).

16.6 Leakage in a CCD is such that stored charge is lost at the rate of $l\%$ per microsecond. At this rate how long does it take to lose half the charge? if $d\%$ of the initial charge is barely detectable, what is the minimum clock rate f required for an s stage register?
Ans. $(69.3/l)\mu s$; $f = sl/100 \ln (100/d)$

16.8   OTHER VLSI CIRCUIT APPLICATIONS: PLAs AND GATE ARRAYS

The continual development of VLSI circuit technology allows the integration of more and more functions on the same silicon chip. As a result, the chips become more and more specialized. This in turn implies lower production volumes and higher cost per unit. Memory, of course, presents an exception to this rule, a fact that, together with the essentially regular structure of memory chips, has made memory the proving ground for VLSI circuit technology. Another exception to the rule is found in microprocessors which are in effect logic networks of very general structure that can be tailored to a variety of applications via software programming.

The problem, however, remains of finding ways for the circuit designer to make use of the VLSI circuit technology. Although in general one can replace some logic circuits by a microprocessor, this approach is not always the best. Specifically, because μPs have a rather general structure they tend to be slow as logic replacements. In this section we shall discuss two related approaches for what may be termed *semicustom VLSI* circuit design.

Programmable Logic Array (PLAs)

The programmable logic array (PLA) is a device that is quite like a ROM. In fact in our discussion of ROM in Section 16.6 it was mentioned that a ROM may be thought of as a combinatorial logic network whose input is the address bits and whose output is the bits stored in the addressed word.

To see how PLAs resemble ROMs consider the simple PLA shown in Fig. 16.31. This PLA has four input variables, A, B, C, and D, and two output variables, X and Y. The four input variables are first inverted to obtain the complements $\overline{A}$, $\overline{B}$, $\overline{C}$, and $\overline{D}$. The eight lines of the input variables and their complements together with the five vertical lines labeled L_1 to L_5 form a matrix of 40 intersection points. These intersection points correspond to the memory cells of a ROM, and in fact we see that like the ROM of Fig. 16.20 some of the cells contain diodes and some do not. It is easy to see that at the intersections where diodes exist, the logic AND of the variables connected to the cathodes of the diodes are realized. For instance, on the vertical line labeled L_1 we obtain the logic function $\overline{A}\,\overline{B}\,C\overline{D}$; on line L_2 we obtain the function $AB\overline{C}$, and so on. Note that these functions are *products* of the input variables and their complements. However, they are *partial products* and not necessarily minterms.

From the above, we see that the diode matrix allows the realization of any five arbitrary partial products of the input variables and their complements. Which partial product terms are realized is determined by where diodes are connected. Any terms can be realized by placing diodes at the appropriate intersections—a process identical to mask programming a ROM. Note also that the number of product terms realized (five

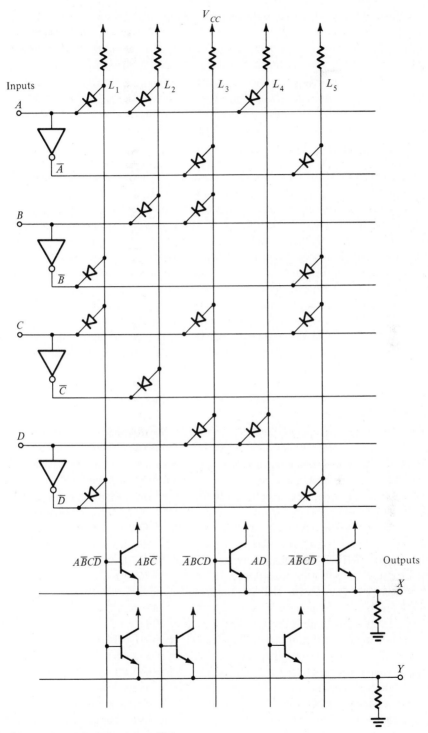

Fig. 16.31   A four-input two-output PLA.

in this example) is arbitrary; we can have any number desired up to the maximum possible of 16.

Consider next the second matrix formed by the intersection of the five vertical lines, L_1 to L_5, and the two horizontal lines, X and Y. Here we have 10 intersection points at some of which transistors are connected—again similar to a ROM array. It is easy to see that the emitter-follower transistors in each row realize at their emitters the OR function of the variables at their bases, which are the product terms available on the vertical lines. As an example the output variable X is given by

$$X = A\overline{B}C\overline{D} + \overline{A}BCD + \overline{A}\,\overline{B}C\overline{D}$$

The particular combination of product terms in this expression is determined by the location of the various transistors. This again is similar to programming a ROM.

From the above we see that the PLA consists of two ROM arrays, the upper ROM array forms product terms and the lower ROM array forms the sum of products. A given PLA can be mask programmed to realize any arbitrary logic function specified by the user. Furthermore, like a PROM one can have a *field programmable PLA* (called FPLA). Such a device contains fuse links that can be selectively blown by the user.

FPLAs are versatile devices that enable easy realization of complex logic functions. They are available from a number of manufacturers in a variety of organizations. A typical PLA can be used to replace 4–12 SSI/MSI chips. Some PLAs include flip-flops also, thus enabling the realization of sequential circuits for timing and control applications. Finally, we should mention that PLAs are also known as *programmable array logic* (PAL) (see Ref. 16.17).

Gate Arrays

While PLAs have been available for some time, *gate arrays* are recent innovations that promise to have great impact on the electronics industry. A gate array, which is also known by the names *uncommitted logic array* and *master slice,* is basically a VLSI circuit that contains as many as 4000 gates that are *not* connected to form a logic network. Rather, the interconnections are formed by metallizing the chip using a mask specified by the user.

Gate arrays offer the logic designer the opportunity to replace parts or all of his logic circuit (perhaps 10–40 MSI packages) with one gate-array chip. This not only saves board space but it also increases system reliability, reduces the power dissipation, and increases the speed of operation.

To employ gate arrays, the potential user has to first decide on a particular technology and a specific gate array chip. Currently, gate arrays are available in CMOS, ECL, and I²L, and some in NMOS and Schottky TTL. ECL is by far the fastest, featuring gate delays in the subnanosecond range. Its power dissipation, however, is relatively high. CMOS on the other hand offers very low power dissipation. Also, recent improvements in the CMOS technology have increased its operating speed considerably. CMOS also offers the very attractive possibility of allowing the fabrication of both analog and digital circuits on the same silicon chip. In fact some of the available CMOS arrays allow the design of such analog circuits as op amps, comparators, and switched-capacitor filters (see Section 14.5) on the same silicon chip that contains logic gates.

Among the technologies for gate array chips, I^2L currently ranks third (after CMOS and ECL). Nevertheless, it is a technology that is undergoing continual development. It also provides the ability to integrate analog and digital circuits on the same silicon chip.

Before concluding this section, we should point out that using gate arrays is not a simple process. For one thing, the potential user has to become thoroughly familiar with the gate-array chip and with its technology. Then there is the problem of generating a *layout* for the chip in order to realize the desired function. This step can be accomplished manually only for chips that contain a maximum of a few hundred gates. Layouts for more complex chips have to be generated with the aid of a computer. Finally, there is the problem of devising test procedures for such a complex chip. Here again, computer aids are advised.

It is interesting to note that gate-array chips usually have a large number of external terminals (up to 225 terminals have been used) and are therefore housed in new and interesting packages.

16.9 CONCLUDING REMARKS

The objective of this chapter has been to introduce the topic of VLSI circuits. Through a study of memory circuits and an introduction to the new area of gate arrays, the challenges and opportunities that this technology offers should have become apparent.

Apart from the brief introduction to microprocessors in Section 6.9 and some frequent comments on them in this chapter, we have not studied the μP in detail. Such a study requires knowledge of computer structures and is beyond the scope of a circuits book.

In conclusion, we live in an era in which microelectronics is pervading almost every aspect of our daily lives. Needless to say this presents tremendous opportunities to those who have chosen to pursue this area of study. We hope that we have conveyed to the reader a sense of excitement and enthusiasm for microelectronics.

INTEGRATED-CIRCUIT TECHNOLOGY

The purpose of this appendix is to familiarize the reader with integrated-circuit terminology. An explanation is given of standard integrated-circuit processes. The characteristics of the devices available in integrated-circuit form are also presented. This is done to aid the reader in understanding those aspects of integrated-circuit design, which are distinct from discrete-circuit design. In order to take proper advantage of the economics of integrated circuits, designers have had to overcome some serious device limitations (such as poor resistor tolerances), while exploiting device advantages (such as good resistor matching). An understanding of device characteristics is therefore essential in designing good integrated circuits, and also helps when applying commercial integrated circuits to system designs.

This appendix will consider only silicon technology. Although germanium and gallium arsenide are also used to make semiconducting devices, silicon is still the most popular material and will remain so for some time. The physical properties of silicon make it suitable for fabricating active devices with good electrical characteristics. In addition, silicon can easily be oxidized to form an excellent insulating layer (glass). This insulator is used to make capacitor structures and allows the construction of field-controlled devices. It also serves as a good mask against foreign impurities which could diffuse into the high-purity silicon material. This masking property allows the formation of integrated circuits; active and passive circuit elements can be built together on the same piece of material (substrate), at the same time, and interconnected to form a circuit function.

INTEGRATED-CIRCUIT PROCESSES

The basic processes involved in the fabrication of integrated circuits will be described in the following sections.

Wafer Preparation

The starting material for modern integrated circuits is very high purity silicon. The material is grown as a single crystal. It takes the shape of a solid cylinder, up to 10 cm in diameter and 1 m in length, and has a steel-grey color. This crystal is then sawed (like a loaf of bread) to produce wafers 10 cm in diameter and 250 microns thick (a micron is a millionth of a meter). The surface of the wafer is then polished to a mirror finish.

The basic electrical and mechanical properties of the wafer depend on the direction in which the crystal was grown (crystal orientation) and the number and type of impurities present. Both variables are strictly controlled during crystal growth. Impurities can be added on purpose to the pure silicon in a process known as *doping*. Doping allows controlled alteration of the electrical properties of the silicon, in particular the resistivity. It is also possible to control the type of carrier used to produce electrical conduction, being either holes (in *p*-type silicon) or electrons (in *n*-type silicon). If a large number of impurity atoms is added, then the silicon is said to be heavily doped. When designating relative doping concentrations on diagrams of devices, it is common to use $+$ and $-$ symbols. Thus a heavily doped (low resistivity) *n*-type silicon wafer would be referred to as n^+ material. This ability to control the doping of silicon permits the formation of diodes, transistors, and resistors in integrated circuits.

Oxidation

Oxidation refers to the chemical process of silicon reacting with oxygen to form silicon dioxide. To speed up the reaction, it is necessary to heat up the wafers to the 1000 to 1200°C range. The heating is performed in special high-temperature furnaces. To avoid the introduction of even small quantities of contaminants (which could significantly alter the electrical properties of the silicon), it is necessary to maintain an ultra clean environment for the processing. This is true for all processing steps involved in the fabrication of an integrated circuit. Specially filtered air is circulated in the processing area, and all personnel must wear special lint-free clothing.

The oxygen used in the reaction can be introduced either as a high-purity gas (referred to as a "dry oxide") or as water vapor (forming a "wet oxide"). In general, a wet oxide has a faster growth rate, but a dry oxide has better electrical characteristics. The oxide layer grown has excellent electrical insulation properties. It has a dielectric constant of about 3.5, and it can be used to form excellent capacitors. It also serves as a good mask against many impurities. It can therefore be used to protect the silicon surface from contaminants. It can also be used as a masking layer, allowing the introduction of dopants into the silicon only in regions which are not covered with oxide. This masking property is what permits the convenient fabrication of integrated circuits.

The silicon dioxide layer is a thin transparent film and the silicon surface is highly reflective. If white light is incident on the oxidized wafer, constructive and destructive interference effects occur in the oxide, causing certain colors to be absorbed strongly. The wavelengths absorbed depend on the thickness of the oxide layer. This absorption produces different colors in the different regions of a processed wafer. The colors can be quite vivid, ranging from blue to greens and reds, and are immediately obvious when a finished chip is viewed under the microscope. It should be remembered though that the color is due to an optical effect. The oxide layer is transparent and the silicon underneath it is a steel-grey color.

Diffusion

Diffusion is the process by which atoms move through the crystal lattice. In this case, it relates to the introduction of impurity atoms (dopants) into silicon to change its doping. The rate at which dopants diffuse in silicon is a strong function of temperature. This allows us to introduce the impurities at a high temperature (1000 to 1200°C) to obtain the desired doping. The slice is then cooled to room temperature and the impurities are essentially "frozen" in position. This diffusion process is performed in furnaces similar to those used for oxidation. The depth to which the impurities diffuse depends both on the temperature and time allowed.

The two most common impurities used as dopants are boron and phosphorus. Boron is a *p*-type dopant and phosphorus is an *n*-type dopant. Both dopants are effectively masked by thin silicon dioxide layers. By diffusing boron into an *n*-type substrate, a *pn* junction is formed (diode). A subsequent phosphorus diffusion will produce an *npn* structure (transistor). In addition, if the doping concentration is heavy, the diffused layer can be used as a conductor.

Ion Implantation

In addition to diffusion, impurities can be introduced into silicon by the use of an ion implanter. An ion implanter produces ions of the desired impurity, accelerates them by an electric field, and allows them to strike the silicon surface. The ions become imbedded in the silicon. The depth of penetration is related to the energy of the ion beam, which can be controlled by the accelerating field voltage. The quantity of ions implanted can be controlled by varying the beam current (flow of ions). Since both voltage and current can be accurately measured and controlled, ion implantation results in much more accurate and reproducible impurity profiles than can be obtained by diffusion. In addition, ion implantation can be performed at room temperature. Ion implantation normally is used when accurate control of the dopant is essential for device operation.

Chemical Vapor Deposition

Chemical vapor deposition (CVD) is a process by which gases or vapors are chemically reacted, leading to the formation of a solid on a substrate. In our case, CVD can be used to deposit silicon dioxide or silicon on a silicon substrate. For instance, if silane gas and oxygen are mixed above a silicon substrate, silicon dioxide deposits as a solid on the silicon. The oxide layer formed is not as good as a thermally grown oxide, but it is good enough to act as an electrical insulator. The advantage of a CVD layer is that the oxide deposits at a faster rate and a lower temperature (below 500°C).

If silane gas alone is used, then a silicon layer deposits on the wafer. If the reaction temperature is high enough (above 1000°C), then the layer is deposited as a crystalline layer (assuming the substrate is crystalline silicon). This is because the atoms have enough energy to align themselves in the proper crystal directions. Such a layer is said to be an epitaxial layer, and the deposition process is referred to as *epitaxy* instead of CVD. At lower temperatures, or if the substrate is not single-crystal silicon, the atoms are not all aligned along the same crystal direction. Such a layer is called *polycrystalline silicon,* since it consists of many small crystals of silicon aligned in various directions. Such layers are normally doped very heavily to form a high conductivity region which can be used for interconnecting devices.

Metallization

The purpose of metallization is to interconnect the various components of the integrated circuit (transistors, resistors, etc.) to form the desired circuit. Metallization involves the deposition of a metal (aluminum) over the entire surface of the silicon. The required interconnection pattern is then selectively etched. The aluminum is deposited by heating it in vacuum until it vaporizes. The vapors then contact the silicon surface and condense to form a solid aluminum layer.

Packaging

A finished silicon wafer may contain from 100 to 1000 finished circuits. The circuits are first tested electrically (while still in wafer form) using an automatic probing station. Bad devices are marked for later identification. The devices are then separated from each other (dicing) and the good devices (dies) are mounted in packages (headers). Fine gold wires are then used to interconnect the pins of the package to the metallization pattern on the die. Finally, the package is sealed under vacuum or in an inert atmosphere.

Photolithography

The surface geometry of the various integrated-circuit components is defined photographically: The silicon surface is coated with a photosensitive layer and then exposed to light through a master pattern on a photographic plate. The layer is then developed to reproduce the pattern on the wafer. Very fine surface geometries can be reproduced accurately by this technique. The resulting layer is not attacked by the chemical etchants used for silicon dioxide or aluminum and so forms an effective mask. This allows "windows" to be etched in the oxide layer in preparation for subsequent diffusion processes. This process is used to define transistor regions and isolate one transistor from another.

INTEGRATED-CIRCUIT STRUCTURES

The structure of the basic integrated-circuit components available to the circuit designer will be presented in this section. Only a standard bipolar junction-isolated process will be considered initially. The processing steps involved are outlined in Figure A.1 along with the corresponding transistor cross section. The process requires six masking steps, four diffusions, and one epitaxial layer growth. The possible components produced include *npn* and *pnp* bipolar transistors, *n*-channel junction field-effect transistors, resistors, and capacitors. Each of these components will be described in the following sections.

Resistors

Resistors can be made from the collector layer (*n*-type), base region (*p*-type), emitter region (n^+-type), or the "pinched-base" region (base region between the n^+ diffused region and the *n*-type epitaxial region). The value of resistance obtained is related to the surface geometry and the region resistivity (doping). Figure A.2 gives the surface and cross-sectional views of the possible resistor structures. Table A.1 gives typical characteristics of the resulting resistors. Note that it is difficult to obtain high resistor values. Also note that while the tolerance of the resistor value is very poor (20 to 50%), matching of two similar resistor values is quite good (5%). Thus circuit designers should design circuits which exploit resistor matching, and should avoid all circuits which require a specific resistor value. Also note that diffused resistors have a significant temperature coefficient.

Capacitors

Two types of capacitor structure are available in integrated circuits. Figure A.3 shows cross-sections of those structures. The junction capacitor makes use of one of the junctions available by diffusion. Either the collector-base or emitter-base junction can be used, but the low breakdown voltage of the emitter-base junction (6 to 7 V) makes it less popular. The capacitance is determined by geometry and doping levels, but normally is restricted to values less than 20 pF.

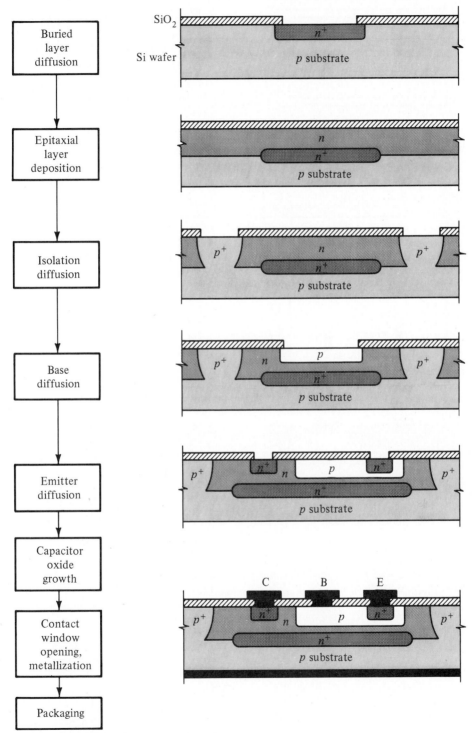

Fig. A.1 Standard bipolar integrated-circuit process.

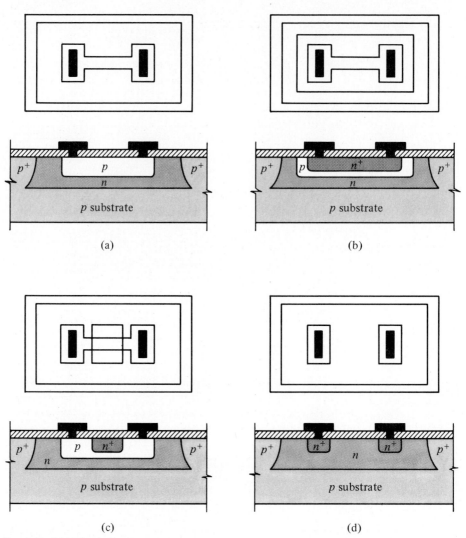

Fig. A.2   Integrated-circuit resistor structures. (a) Base resistor. (b) Emitter resistor. (c) Pinched base resistor. (d) Collector resistor.

TABLE A.1   Diffused Resistor Characteristics

	Base	Emitter	Pinched Base	Collector
			Resistor Type	
Range (ohms)	50–50K	5–100	10K–500K	1K–10K
Tolerance	20%	20%	50%	50%
Matching	5%	5%	5%	5%
Temperature coefficient	0.1 %/C	0.2 %/C	0.5 %/C	0.8 %/C
Breakdown voltage (V)	40	6	6	70

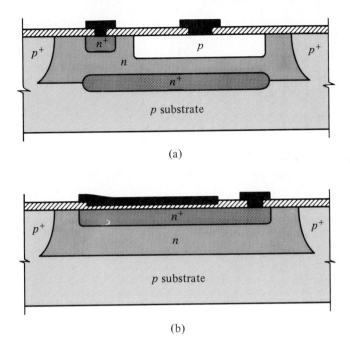

Fig. A.3   Integrated-circuit capacitor structures. (a) Collector-base junction capacitor. (b) Oxide capacitor.

The tolerance is 20%. Junction capacitors must always be reverse-biased and have a large voltage dependence.

A capacitor can also be formed using the oxide as a dielectric. The aluminum metallization and the emitter region form a parallel plate structure. This structure makes an excellent capacitor with very low voltage and temperature coefficients, low leakage, and high breakdown voltage. Although the tolerance is 10%, capacitors on the same chip can be matched to better than 1%. Again the values are restricted to a few ten's of picofarads.

Bipolar Junction Transistors

Three basic types of bipolar transistors are available: *npn, lateral pnp,* and *substrate pnp.* The *npn* structure is repeated in cross-section in Figure A.4. It has a beta typically of 100 to 500, a collector breakdown voltage of 40 V, and a cutoff frequency of 500 MHz. The normal operating current range is from microamps up to 10 milliamps, but higher operating currents can be obtained by increasing the surface area of the device.

The lateral *pnp* is shown in Figure A.5. The base region in this case is the lateral area between the two *p* diffusions, hence the name (the other transistor structures are vertical devices). This is the most commonly used *pnp* structure. It suffers from very low beta, being typically about 20 at low collector currents (a few microamps) and diminishing rapidly as current levels are increased. At milliamp current levels, its gain may be only slightly greater than unity. It also has a low cutoff frequency around 1 MHz. Its breakdown voltage is over 40 V. Despite its poor characteristics, the lateral *pnp* is the best *pnp* transistor available in integrated circuits.

The substrate *pnp* structure is shown in Figure A.6. In this case, the collector is electrically connected to substrate, which is normally a signal ground. This limits application of the device to common collector configurations. It does, however, have a slightly better current gain than

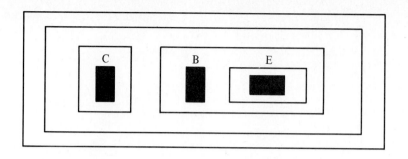

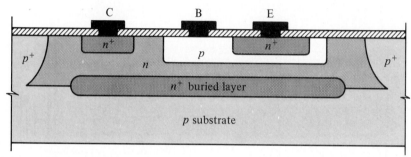

Fig. A.4   npn bipolar transistor.

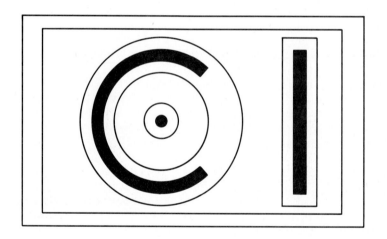

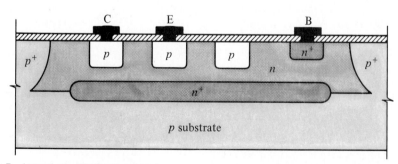

Fig. A.5   Lateral pnp bipolar transistor.

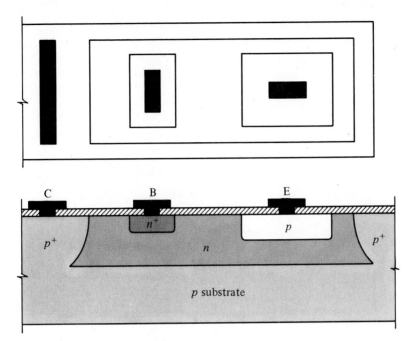

Fig. A.6   Substrate pnp bipolar transistor.

the lateral *pnp* and exhibits somewhat better performance at higher currents. The characteristics are only slightly improved, so integrated-circuit designers avoid using *pnp* transistors, if at all possible. If needed, they are best used in unity gain configurations to avoid serious degradation of bandwidth due to their low cutoff frequency.

For digital integrated circuits, the storage time of the *npn* transistor is most important if the device is operated in the saturation region. One technique for reducing the storage time is to allow gold to diffuse into the silicon. Gold acts as an impurity in the silicon and serves to reduce the carrier lifetime, hence the storage time. It also reduces the current gain and breakdown voltage, but these reductions are not important in digital circuits. It is also not possible to make *pnp* transistor structures with gains greater than unity due to the low lifetime, but again the *pnp* is not needed in digital circuits. An alternative to gold doping is to avoid saturation region operation by Schottky clamping the collector-base junction. This will not effect any transistor parameters, and yet the storage time can be eliminated. Fortunately, a suitable Schottky diode can be formed by using aluminum as the metal and the collector *n* region as the semiconductor. This leads to the very simple structure shown in Figure A.7.

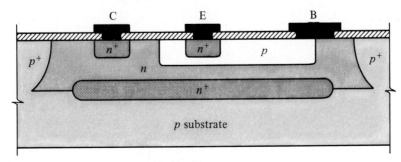

Fig. A.7   Schottky clamped npn bipolar transistor.

Junction Field-Effect Transistors

A *p*-channel JFET can be formed using the bipolar process. The corresponding cross-section is shown in Figure A.8. This structure suffers from a low gate-drain breakdown voltage of 6 V, and poor matching of the pinch-off voltages. The use of ion implantation as an extra processing step, however, permits much improvement in pinch-off matching and control. The structure can be used to functionally replace *pnp* bipolar devices in circuits. It is capable of providing

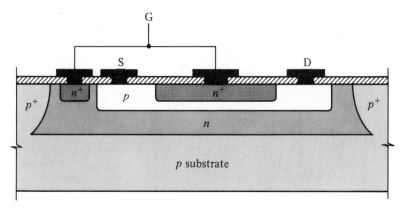

Fig. A.8   p-channel JFET.

increased gain and cutoff frequency, with values similar to those of the *npn* device. Thus the *p*-channel JFET is very useful in analog integrated-circuit design, and is to be preferred over the use of lateral *pnp* devices.

MOS Technology

Aside from the standard bipolar technology presented above, a technology based on MOSFET transistors also exists. It is used primarily for the production of digital circuits, where only transistors and small capacitors are required. MOS processing is simpler than bipolar processing and consumes less surface area. Thus MOS technology is best suited for producing complex digital circuits in integrated-circuit form. Three possible MOS processes will be considered here: *p*-channel metal gate (PMOS), *n*-channel silicon gate (NMOS), and complementary MOS (CMOS). The difficulty of the processing is in the order presented, that is, CMOS is the most difficult and hence the most costly process. On the other hand, PMOS offers the poorest performance (slow speed). Thus if cost is the primary consideration, PMOS is normally

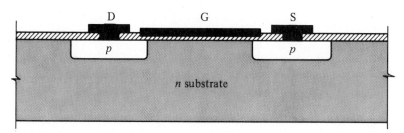

Fig. A.9   PMOS transistor (aluminum gate).

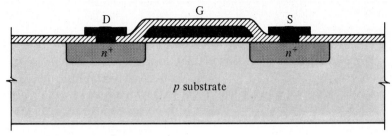

Fig. A.10   NMOS transistor (silicon gate).

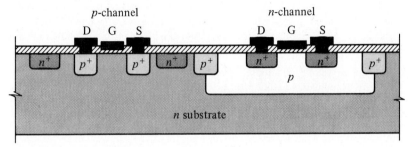

Fig. A.11   CMOS transistors (aluminum gate).

selected. If performance is important, then NMOS or CMOS is selected. Cross-sections of the transistor structures produced by the three technologies are presented in Figures A.9 to A.11. Note that the NMOS process uses polysilicon as the gate electrode (it is deposited by a CVD process, and then covered with a CVD oxide). The CMOS structure can be made with either an aluminum or a polysilicon gate electrode. Ion implantation is used extensively in modern MOS processing to accurately control the threshold voltage of the MOSFET, as well as to control certain other transistor parameters.

When comparing the capabilities of the various technologies, it is common to consider how many transistors can be fabricated per die (or circuit). The bipolar technology can produce circuits with a couple of thousand transistors in digital designs; significantly fewer are possible in analog designs. The PMOS process can produce a few thousand transistors, as can the CMOS process. The NMOS process, however, is capable of producing chips with many ten's of thousands of transistors.

PROBLEMS

CHAPTER 1

1.1 A symmetrical square wave of frequency 1 kHz has a peak-to-peak amplitude of 20 V. What is the angular frequency of the wave? What is the frequency of the fundamental and its amplitude? What is the amplitude of the second harmonic? What is the ratio of the amplitude of the third harmonic to the amplitude of the fundamental? What are the relative phases of the third harmonic and the fundamental?

1.2 For a square-wave audio signal of 10 kHz, what fraction of the available energy is perceived by an average adult listener of age 40 whose hearing extends only to 16 kHz?

1.3 A sampled-data signal is quantized to one of ten levels (0 through 9) by a process of "rounding," that is, selection of the nearest whole value. If this channel is used to transmit information, what is the level of uncertainty of the received result? Express this as a percentage of full scale and of the smallest nonzero signal.

1.4 Sketch the output of an amplifier such as that in Exercise 1.3 (which has a gain of 1,000 and limits the output to ± 10 V) when supplied with a 1-V, 1-kHz sine-wave signal. Convince yourself that the output waveform is a good approximation to a square wave. Hence, with the aid of Eq. (1.1), find the "effective gain" of the amplifier for the component at 1 kHz.

1.5 In Problem 1.4, the components of the output at frequencies other than the fundamental constitute distortion whose rms value is the square root of the sum of the squares of the rms values of the components. For the previous conditions, what is the rms value of (a) the signal component at 1 kHz? (b) all components at other frequencies (up to 10 kHz)? What is the ratio of (b) to (a) expressed as a percentage? It is called the percent harmonic distortion.

1.6 A 1-V rms signal from a phonograph cartridge is supplied to an amplifier with a voltage gain of unity which in turn drives a 4-Ω loudspeaker? What is the power available at the loudspeaker?

1.7 A square-wave audio signal of 1 kHz is transmitted through a telephone channel which acts as a filter, cutting off all signal components outside the 300- to 3,400-Hz band. What fraction of the signal energy is received at the output of the channel?

1.8 An AM broadcast station operating at an assigned frequency of 1,010 kHz broadcasts a pure tone of 2,000 Hz at a level of one-tenth full modulation. Use Eq. (1.6) to draw the spectrum of the resulting transmission, labeling both frequency and relative amplitude.

1.9 A microprocessor is to be used to demultiplex a ten-channel TDM signal. For each sample the program requires 1.5 μs to read a signal at the input and 2.0 μs to distribute it to one of ten output channels. It further requires 2.5 μs for other tasks related to synchronization. What is the maximum possible TDM sample rate that can be handled?

CHAPTER 2

2.1   A two-terminal, one-port device A is characterized by the i-v characteristic shown in Fig. P2.1. What incremental conductances pertain in the segments WO, OX, and XY? This device is biased to operate at point P, midway between O and X, and a triangular signal is superimposed having a 1-V peak-to-peak amplitude. Sketch and label the waveform of the current flowing in the device. What is its peak-to-peak amplitude? What is the average current conducted by the device? If a 1-kΩ resistor is inserted in series with the device and the series combination of bias and signal sources, what do the average and peak currents become?

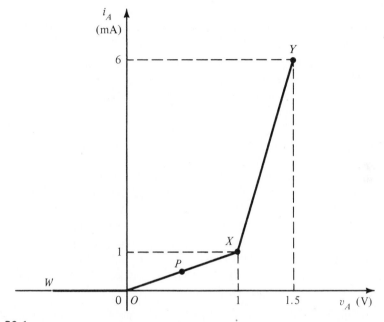

Fig. P2.1

2.2   The device in Problem 2.1 is biased at the point O. What is the average current with no signal? A triangular voltage signal of 1-V peak-to-peak amplitude is applied. Sketch the current waveform. What peak current flows? What average current flows?

2.3   In Problem 2.1, the operating point is shifted to X. What average bias current is required for no signal? What are the (positive and negative) peak currents which flow if a 1-V peak-to-peak triangular voltage source is added in series with the bias supply? What is the average current which flows? Sketch and label the current waveform.

2.4   For the one-port device described in Exercise 2.1, find the voltage required to establish a 10-mA operating current. What is the incremental resistance of the device at this bias point?

2.5   An amplifier characterized as a linear two-port as shown in Fig. 2.2b has an incremental input resistance of 1 kΩ. It is supplied by a sine-wave source whose open-circuit voltage is 10 mV peak-to-peak and whose source resistance is 9 kΩ. What peak-to-peak voltage will appear across the input of the two-port when the source is connected?

2.6   An amplifier operating off a single 15-V supply provides a 10-V rms signal to a 1-kΩ load. The dc current taken from the 15-V supply is 8 mA. What is the power lost in the amplifier?

2.7   An amplifier operating between positive and negative 15-V power supplies and driven by a 1-V rms source draws a current of 1 mA from each supply when unloaded and a current of 10 mA from each supply when providing a 10-V rms output to a 1-kΩ load. In both cases the current drawn from the signal source is negligibly small. What is the power dissipation in the amplifier under unloaded and loaded conditions?

2.8   An ideal voltage amplifier of gain $\mu = 100$ is connected between a 10-mV rms source and a 1-kΩ load. What is the power delivered to the load? What is the power gain of the circuit?

2.9   If in Problem 2.8 the input and output resistances of the amplifier are each 1 kΩ, what is the power delivered to the 1-kΩ load? What is the power lost in the amplifier? What is the power gain of the amplifier? What is the minimum power delivered by the amplifier power supply?

2.10   A phonograph cartridge characterized by a voltage of 1-V rms and a resistance of 1 MΩ is available to drive a 10-Ω loudspeaker. If connected directly, what voltage and power levels result at the loudspeaker? What is the power gain from source to load? If an ideal unity-gain buffer amplifier is interposed between source and load, what do the output voltage and power levels become? What are the voltage and power gains of the new arrangement?

2.11   If in Problem 2.10 the buffer amplifier has an input resistance of 1 MΩ and an output resistance of 10 Ω, calculate the levels of output voltage and power. What are the voltage and power gains of this arrangement?

2.12   An amplifier has a voltage gain of 100 and a current gain of 1,000. Express its voltage, current, and power gains in dB.

2.13   An amplifier with a voltage gain of +40 dB, an input resistance of 10 kΩ, and an output resistance of 1 kΩ is used to drive a 1-kΩ load. Express the power gain of this arrangement in dB.

2.14   An amplifier consists of three cascaded stages each having a voltage gain of 10, an input resistance of 100 kΩ, and an output resistance of 10 kΩ. The cascade is driven from a 1-MΩ source and supplies a 1-kΩ load. Express the ratio of output to source voltage in dB. What is the gain in dB of the amplifier as measured from its input to its output terminals? Why do these differ? Calculate the loss in dB of the input network.

2.15   Consider a unity-gain buffer amplifier with a 1-MΩ input resistance and a 100-Ω output resistance connected to a 1-MΩ source and a 100-Ω load. Calculate the voltage, current, and power gains in dB.

2.16   A current amplifier with gain $\beta = 100$, an input resistance of 100 Ω, and an output resistance of 100 kΩ is connected between a source and load whose resistances are 100 kΩ and 100 Ω, respectively. Calculate the resulting current gain from source to load in dB. What is the voltage gain of the amplifier when loaded? When unloaded?

2.17 Consider a transconductance amplifier with gain of 10 mA/V, an input resistance of 10 kΩ, and an output resistance of 100 kΩ connected between a 100-kΩ source and a 10-kΩ load. What is the voltage gain of the resulting system?

2.18 A transistor modeled by the techniques displayed in Example 2.2 is biased to operate as a linear amplifier for small signals. A small current variation of 1 μA peak-to-peak is introduced into the base and an emitter current variation of 100 μA peak-to-peak is measured. Furthermore, a base-to-emitter voltage variation of 2 mV peak-to-peak is noted. For this situation, calculate α, β, r_e, g_m, and r_x.

2.19 If port 2 of the gyrator of Example 2.3 is terminated by a capacitance C shunted by a resistance R, what is the equivalent circuit seen looking into port 1? (*Hint:* Use frequency-domain analysis to find the input impedance looking into port 1.)

2.20 Consider a transistor modeled as in Fig. 2.11b connected between a voltage source having a 10-kΩ source resistance and a load of 10 kΩ. Let $r_x = 1$ kΩ and $\beta = 100$. Express the voltage, current, and power gains of this arrangement as ratios and in dB.

2.21 A transistor modeled as in Fig. 2.11b is operated with c grounded and b grounded through a resistor R. What is the input resistance of this circuit as seen from terminal e? That is, consider a test current i supplied to terminal e, and calculate the ratio of the voltage v which results at e to the current supplied.

2.22 For a 60-Hz sine wave, calculate the period T and the radian frequency ω. What is the time taken by the wave to transit from a positive-going zero crossing to a level corresponding to its rms value? For what fraction of a cycle does the absolute value of a sine wave exceed its rms value?

2.23 Find the frequencies f and ω of a sine wave of period 1 μs.

2.24 Derive an expression for the transfer function of the two-port network shown in Fig. P2.24. Sketch the magnitude and phase response on a logarithmic frequency axis using a dB scale for magnitude. What do you call this filter?

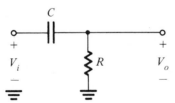

Fig. P2.24

2.25 For the low-pass network whose response is shown in Fig. 2.16a and b, calculate the magnitude in dB and the phase in degrees at one-tenth and ten times the corner frequency ω_0.

2.26 Consider the RC circuit analyzed in Example 2.4 to which is supplied a composite signal whose components are 10-V-rms sine waves at frequencies of 100, 1,000 and 10,000 Hz. Find an expression (a sum of sinusoids) for the composite output waveform. Use $C = 0.1$ μF and $R = 1.59$ kΩ.

2.27 A source is measured to have a 10-V open-circuit voltage and to provide 1 mA into a short circuit. Calculate its Thévenin and Norton equivalent source parameters.

2.28 A voltage source is found to provide 10 V to a 1-kΩ load and a voltage of 11 V to a 2-kΩ load. Calculate the equivalent Thévenin voltage and resistance.

2.29 Repeat Example 2.6 under the condition that the grounded base and collector connections incorporate series resistors R_B and R_C, respectively. Find an expression for the resistance looking into the emitter.

2.30 Consider the circuit of Fig. 2.25 as the gain is varied. What load impedances are presented to the source and to the amplifier if the amplifier gain is changed to $-2, -1, 0$, or $+1$?

2.31 In Fig. E2.9, if the supply voltage is increased to 20 V, what does the diode current become? Assume $V_D = 0.7$ V independent of current level.

2.32 In Fig. E2.10, M is a controlled current source whose current I_M depends on the voltage V_M; I_M mA $= 0.1$ V_M V. What voltage V_M results?

2.33 A voltage amplifier with gain of 1.1 and input resistance of 100 kΩ has a 1-MΩ resistor connected from output back to input. What is the equivalent input resistance of the resulting circuit?

2.34 A network is characterized by the following h parameters: $h_{11} = 2.6$ kΩ, $h_{12} = 2.5 \times 10^{-4}$, $h_{21} = 100$, $h_{22} = 10^{-5}$ ℧. Sketch and label the equivalent circuit corresponding. Consider an alternate representation for the h_{12} element. Use the describing equations to calculate the gain v_0/v_s of this network when connected between a source v_s having a 10-kΩ internal resistance, and a load of 10 kΩ across which v_0 is measured.

2.35 A network can be described in many different ways. Find expressions for each y parameter in terms of h parameters and each h parameter in terms of y parameters.

2.36 The terminal properties of a two-port network are measured with the following results: With the output short-circuited and an input current of 0.01 mA, the output current is 1.0 mA and the input voltage is 26 mV. With the input open-circuited and a voltage of 10 V applied to the output, the current in the output is 0.1 mA and the voltage measured at the input is 2.5 mV. Find values for the h parameters of this network.

2.37 A source having a source resistance of 2 kΩ is coupled to a load of 2 kΩ via a capacitor of 1.0 μF. What is the time constant of the resulting circuit? The source undergoes a 10-V positive step change. What output signal results? How long does it take for the output to recover from all but 5% of the initial change?

2.38 Reduce this circuit to an equivalent STC network; then use the rules of Table 2.1 to deduce whether it is low pass or high pass. What is the time constant of the STC network?

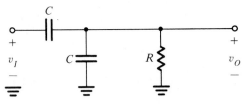

Fig. P2.38

2.39   For the phase response curve of Fig. 2.37b verify that the error made by the asymptotic approximation at $\omega = \omega_0/10$ and $\omega = 10\,\omega_0$ is 5.7°.

2.40   An ideal voltage amplifier with gain μ of $-1,000$ has a small capacitance of 1.9 pF connected from output to input. The amplifier has in addition an input capacitance of 100 pF. It is driven by a 50-kΩ source. Show that the response of this arrangement is equivalent to that of an STC network. What is the time constant? What is the transfer function from input to output?

2.41   The circuit of Fig. E2.13 is connected to a load consisting of a 100-pF capacitor and a 10-kΩ resistor in parallel. For the combination find the dc transmission, the corner frequency f_0, and the transmission at $f = 2$ MHz.

2.42   The ideal amplifier of Fig. E2.14 is shunted from output to input by a resistance of 20 MΩ. Show that the resulting circuit can be modeled by an STC network. Find the high-frequency gain, the 3-dB frequency f_0, and the gain at $f = 1$ Hz.

2.43   The ideal amplifier of Fig. E.2.14 is shunted from output to input by a capacitance of 500 pF. Show that the resulting circuit is STC. Find the high-frequency gain, the 3-dB frequency f_0, and the gain at $f = 1$ Hz.

2.44   Imagine Fig. 2.45b to represent the output current of an STC high-pass network for which $P = 10$ mA, $T = 1$ ms, and $\tau = 100$ ms. For these conditions what is ΔP? If the pulse represents forward charge flow, what is the amount of charge that flows backwards after the pulse? Express this in absolute terms and as a percentage of that in the forward direction.

2.45   The equivalent circuit of the input attenuator and amplifier of an oscilloscope consists of a 1-MΩ resistance in parallel with a 30-pF capacitance. To reduce the sensitivity of the oscilloscope by a factor of 10, an input network (called "an X10 probe") is to be designed. It should reduce the magnitude of input signals at very low and very high frequencies to one-tenth of their original value. Use the results of Examples 2.9, 2.10, and 2.12 to design the two components of the probe. Characterize the input impedance of the probe and oscilloscope combination.

2.46   An experimenter having only an "X10 probe" as described in Problem 2.45 wishes to raise the input resistance of his oscilloscope still further at the expense of a loss in signal. He decides that an "X100 probe" would be a possible solution. However, although he has a resistor of suitable value, the minimum capacitance available is limited by the body of the resistor to 1 pF. To retain a flat frequency response and the highest possible input impedance, he conceives of using a capacitor in his probe network in another way. What is his design? What is its input resistance and capacitance?

2.47   In the circuit of Fig. 2.36d, find an expression for i_o if i_I is a 3-mA step, $R = 1$ kΩ, and $C = 100$ pF.

2.48   In the circuit of Fig. 2.36e, find an expression for i_o if i_I is a 2-mA step, $R = 2$ kΩ, and $L = 10\ \mu$H.

2.49   In the circuit of Fig. 2.36f, a current i_I or 20 mA that has been flowing for a long time is suddenly cut off. Find v_o if $R = 2$ kΩ, and $L = 10\ \mu$H.

2.50   A particular oscilloscope with an upper cutoff of 50 MHz is intended for use in the measurement of digital signals. On the assumption (not a bad one) that the oscilloscope can be modeled as an STC network, what is the fastest rise time which can be observed?

2.51   When a pulse of finite rise time t_p is propagated through a network whose step response is characterized by rise time t_n, the rise time of the output is approximately $t_r = \sqrt{t_n^2 + t_p^2}$. What is the ratio K of t_n to t_p or t_p to t_n at which the rise time is dominated (to within 90%) by the longer of t_p or t_n. If an amplifier of digital signals is to reproduce rise times at 100 ns or less, what must the bandwidth of the amplifier (assumed STC) be?

2.52   An RC differentiator is to be used to convert a step voltage change V to a single pulse for a digital logic application. The logic which it drives distinguishes signals above $V/2$ as "high" and below $V/2$ as "low." What must the time constant of the network be to convert a step input into a pulse which will be interpreted as "high" for 10 μs?

2.53   An ac amplifier is required for a digital application in which pulses of duration ranging from 0.1 to 10 ms must be transmitted relatively faithfully. The greatest sag allowed is one-quarter of the pulse amplitude. Using exact expressions, calculate the low-frequency 3-dB cutoff of the amplifier considered (at the low-frequency end) as an STC network. What time constant would you have chosen by means of the percentage sag formula given? What would the percentage sag actually be if this time constant were used?

2.54   An STC low-pass RC network is to be used as an integrator to distinguish the duration of rectangular pulses of amplitude V to within 10% over the range 10 μs to 10 ms. The circuit should have the largest possible width-to-amplitude conversion factor consistent with this linearity requirement. What is the time constant to be used? What is the resulting output voltage range?

2.55   A coupling capacitor between two stages of an amplifier connects a 2-kΩ source to a 3-kΩ load. Find the value of C so that a 1-ms pulse exhibits less than 1% sag. What is the associated 3-dB frequency?

2.56   Percent sag is a measure in the time domain of the low-frequency cutoff of an amplifier and may be used, for example, in audio amplifier testing by means of square waves. For an amplifier displaying an STC cutoff at 20 Hz, what is the frequency of the test square wave for which a droop of 10% is observed? For a square wave whose frequency is equal to that of the low cutoff, what droop is observed?

2.57   A high-pass STC network with a time constant of 1 ms is excited by a pulse of 10 V height and 1 ms width. Calculate the value of the undershoot in the output waveform. If an undershoot of 1 V or less is required, what is the time constant necessary?

CHAPTER 3
3.1   Design a simple op-amp circuit with a gain of -10 and an input resistance of 100 kΩ.

3.2   Using the idea included in Exercise 3.2, design an op-amp circuit with a gain of -100 and an input resistance of 1 MΩ using resistors no larger than 1 MΩ.

3.3   Design a current to voltage converter having an input impedance of 0 Ω and a transresistance (the ratio of output voltage to input current) of -10 V per mA or -10 kΩ.

3.4   Design an op-amp circuit with a dc gain of -10 and an input resistance of 10 kΩ that incorporates an STC low-pass transfer function with 3-dB cutoff at 10^5 rad/s.

3.5   Design an op-amp circuit with 10-kΩ input resistance which converts a symmetrical square wave at 1 kHz having 2 V peak-to-peak amplitude and zero average value into a triangle wave of 2 V peak-to-peak amplitude.

3.6 Design a two op-amp circuit with inputs v_1 and v_2 and input resistances of 100 kΩ whose output is $v_O = v_1 - 10v_2$.

3.7 Design an op-amp differentiator circuit which provides a positive output pulse of 1 V amplitude and 1 ms duration when presented with an input which falls at a constant rate from $+3$ to $+2$ V in 1 ms. The design should limit the input current under the stated conditions to a maximum of 1 μA.

3.8 Design a low-pass STC op-amp circuit having a 3-dB cutoff at 100 rad/s, a unity-gain frequency of 10^4 rad/s, and an input resistance of 1 kΩ. What is the gain of your circuit at 10 rad/s?

3.9 Design an STC high-pass inverting op-amp circuit with 3-dB cutoff at 10^2 rad/s, unity gain at 10 rad/s, and resistive component of the input impedance of 10 kΩ. What is the gain of this circuit at 10^3 rad/s?

3.10 Design op-amp circuits with gains of 1, 2, and 1.5.

3.11 Design a circuit utilizing one op-amp and having three inputs v_1, v_2, and v_3 and output $v_O = v_1 + v_2 - 2v_3$. The input resistance at each port should be 10 kΩ.

3.12 Design a high-impedance analog voltmeter with a full-scale reading of 1 V using an op amp and a meter movement having 1-mA full-scale calibration and a 2-kΩ internal resistance. What must be done to this circuit if it is provided with 1-kΩ meter movements at some later time?

3.13 Design a differential amplifier with a gain of 10 and a differential input resistance of 20 kΩ. What is the resistance (to ground) seen at the positive input? At the negative input? Revise your circuit to make these equal while maintaining other specifications.

3.14 Design a differential amplifier with a gain of 10, a differential input resistance of 2 MΩ, but which uses a small number of additional resistors to allow a design in which no individual resistor is larger than 1 MΩ. (*Hint:* Study Fig. E3.2.)

3.15 Design a negative impedance converter having an input resistance of -1 kΩ. This circuit is connected to the output terminal of a source whose open-circuit voltage is 1 mV and whose output resistance is 900 Ω. What voltage is then measured at the output of the source?

3.16 Using a simple modification of the noninverting integrator of Fig. 3.22, design a circuit having an STC low-pass transfer function, a low-frequency gain of $+20$, an upper 3-dB frequency of 1,591 Hz, and an input resistance at high frequencies of 100 kΩ.

3.17 Using the idea incorporated in Fig. E3.8, design an all-pass circuit having a gain magnitude of unity for all frequencies but a phase angle which ranges from 0° at high frequencies to 180° at low frequencies, with 90° phase at 1,000 rad/s. The input resistance of the circuit at very high and very low frequencies should be 50 kΩ. What is the input resistance at 1,000 rad/s?

3.18 What is the gain $A = v_2/v_1$ as a function of x, the fraction of full rotation of the potentiometer R for each of these circuits?

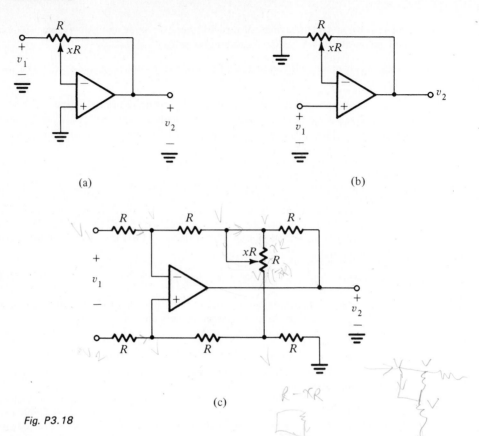

Fig. P3.18

3.19   Express i_o as a function of v.

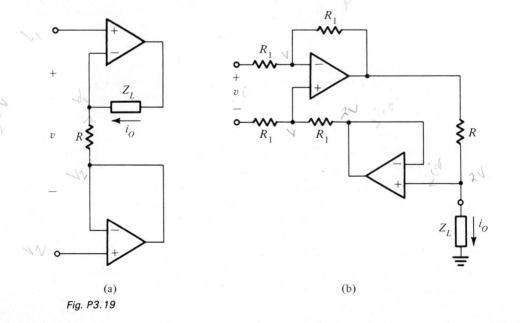

(a)   (b)

Fig. P3.19

3.20 An internally compensated op amp has a low-frequency open-loop gain of 10^6 and a unity-gain bandwidth of 10^6 Hz. What is its 3-dB cutoff frequency? What is the upper 3-dB cutoff frequency of an inverting amplifier of gain -100 using this op amp? What phase shift is associated with the amplifier of gain -100 at frequencies of 10^3, 10^4, 10^5, and 10^6 Hz?

3.21 Contrast the bandwidths of two designs of an amplifier with gain of $+100$. The first uses a single op amp, while the second uses two op amps. Each op amp has a gain of 10^5 and a gain bandwidth product of 10^5 Hz.

3.22 Derive an expression for the bandwidth of a differential amplifier of gain K using an op amp with open-loop gain A and unity-gain frequency f_t.

3.23 An amplifier consisting of an op amp connected in a gain of 10 inverting configuration has a dc closed-loop gain of 9.5 and a step response with rise time (10 to 90%) of 10 μs. Assuming the op amp to be dominant-pole compensated (i.e., the usual case), estimate its open-loop gain and 3-dB cutoff frequency.

3.24 An op amp whose output-limiting voltages are ± 10 V, whose open-loop gain is 10^5, and whose slew rate is 10 V/μs, is driven by a sine-wave signal at 10 kHz. For what amplitude of input signal does the amplifier output begin to limit? What is the rise time of the output signal when the input is highly overdriven with signal amplitudes well beyond the limiting value? What is the smallest input signal at 10 kHz for which the best possible square-wave output results?

3.25 Consider an op amp having a rated output voltage of ± 10 V, $f_t = 1$ MHz, and $SR = 1$ V/μs. What is the largest gain available from an inverting amplifier that uses this op amp, in which small-signal and full-power bandwidths are the same?

3.26 An op amp with a CMRR of 60 dB is connected in a differential amplifier configuration with closed-loop gain of 100. Calculate the common-mode gain of the complete amplifier.

3.27 An op amp with CMRR of 100 dB is connected in a unity-gain differential configuration to whose input a 10-V common-mode signal at 60 Hz and a 0.1-mV signal at 1 kHz is applied. Calculate the ratio of 60 Hz to 1 kHz signals at the output. Express this as a signal-to-noise ratio in decibels.

3.28 Contrast the input impedance of noninverting amplifiers of gain 1 and of gain 10, using an op amp for which $f_t = 10^6$ Hz, $R_{id} = 10^6$ Ω, and $R_{icm} = 10^8$ Ω. Consider the relative effects of the use of each to measure the voltage across an LC circuit resonant at 10^4 Hz with a Q of 1,000 and using a 100-pF tuning capacitor. (*Note:* For a parallel RLC circuit, $\omega_0 = 1\sqrt{LC}$ and $Q = \omega_0 CR$.)

3.29 Calculate the output impedance of an inverting amplifier of gain 100 utilizing an op amp having an open-loop gain of 10^4, an output resistance of 10^3 Ω and a unity-gain frequency of 10^6 Hz.

3.30 An inverting amplifier with gain of 10 utilizes an op amp with a voltage offset of 1 mV to amplify ac signals of 0.1-mV amplitude and 0.0-mV average value. What are the ac and dc components of the output? Consider the effect of capacitor coupling, that is, the connection of the signal to R_1 via a capacitor. Now what are the two components of the output signal?

3.31 An inverting amplifier utilizes R_1 at the input, R_2 for feedback, and R_3 connected between the positive input terminal and ground. It is designed to have an input resistance of 100 kΩ and a gain of −10. The op-amp input offset voltage is 2 mV, input bias current is 20 nA directed inward, and offset current is 3 nA. Choose a value for R_3 to minimize the output dc offset. What is the remaining worst-case output offset voltage? What does the offset become if R_3 is made zero?

3.32 A Miller integrator utilizes R_1 as an input resistor and C_2 as a feedback capacitor. In addition, C_2 is shunted by R_2 to reduce the effects of op-amp imperfections, notably an input offset voltage of 1 mV and a bias current of 10 nA. Design the integrator to have an input resistance of 100 kΩ, an integration time constant of 10^{-3} s and a maximum static output offset of 200 mV.

3.33 Find the output of the nonideal integrator designed in Problem 3.32 when fed with a 25-Hz, 2-V peak-to-peak, zero-average symmetrical square wave. Contrast this response to that of an ideal integrator having an equal integrating time constant.

3.34 An inverting bandpass audio amplifier is constructed from an ideal op-amp using an input network consisting of R_1 and C_1 in series and a feedback network of R_2 and C_2 in parallel. Design R_1, R_2, C_1, C_2 to meet the following specifications: an input resistance at high frequencies of 10 kΩ; a midband gain of −100; a lower 3-dB cutoff frequency of 100 Hz; an upper 3-dB cutoff frequency of 10 kHz. What is the minimum f_t of the op amp to ensure that the upper cutoff is relatively independent of the amplifier rolloff? Use a factor of 10 in separation of critical frequencies.

3.35 In a differential amplifier of the standard topology designed for a differential gain of 10, all resistors are as specified except one of the larger ones which is 0.1% too large. Assuming the op amp to be ideal, what is the CMRR of the resulting unbalanced amplifier?

3.36 The circuit shown in Fig. P3.36 is that of an "instrumentation amplifier," having high-impedance differential inputs and gain control by means of a single resistor (R_4). Assuming the op amps to be ideal, show that

$$v_O = -\frac{R_2}{R_1}\left(1 + \frac{2R_3}{R_4}\right)(v_2 - v_1)$$

Thus the circuit amplifies differential input signals only.

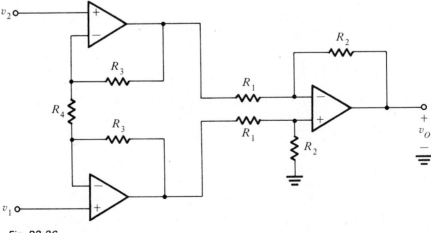

Fig. P3.36

3.37 This problem is a challenging one. The objective is to find the effect of the finite CMRR of the op amps on the common-mode rejection of the instrumentation amplifier of Fig. P3.36. Use the technique discussed and illustrated in Fig. 3.33 to find an expression for v_O of the instrumentation amplifier in terms of v_1 and v_2 assuming that each op amp has a finite common-mode rejection ratio, denoted CMRR. Then let $v_1 = v_{icm}$ and $v_2 = v_{icm} + v_{id}$ and thus find the differential gain and the common-mode gain. Show that the common-mode rejection ratio of the instrumentation amplifier is approximately equal to $\text{CMRR} \times [1 + (2R_3/R_4)]$.

3.38 The bias current of the amplifier in Fig. P3.38 is I_B, while the offset current is zero. Express v_O in terms of v_I, I_B, and R. What is the value of the input current i_I?

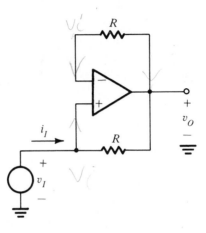

Fig. P3.38

3.39 The op amp in Fig. P3.39 limits at ± 13 V and is otherwise ideal. What is the voltage at A with switch S closed? Sketch the waveform at A after S opens at time zero. How long after S opens does it take for A to reach $+11$ V?

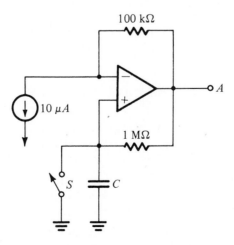

Fig. P3.39

3.40 The op amp in Fig. P3.40 has an open-loop gain of 10^3 and an open-loop output resistance of 1 kΩ but is otherwise ideal. What is the output voltage at A with no load? If a current of 10 mA is extracted from the output, what does the output voltage become? What is the equivalent closed-loop output resistance?

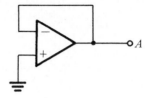

Fig. P3.40

3.41 In the circuit of Fig. P3.41, when S is opened, the voltage at A falls at the rate of 10 mV/s. What is the input bias current of the op amp and in which direction does it flow?

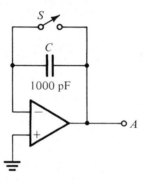

Fig. P3.41

3.42 An operational summing amplifier utilizes input resistors R_1 and R_3 and feedback resistor R_2. Using the methods of Section 3.8, calculate the effects on both gain and bandwidth, as viewed from the R_1 input, of the independent input via R_3. For this purpose consider the latter input grounded. Note that the effects of R_3 can be used in bench testing the consequences of bandwidth reduction of any inverting amplifier.

CHAPTER 4

4.1 In the circuits of Fig. P4.1, find the voltages V_n and the currents I_n. Assume all diodes to be ideal.

4.2 Two ideal diodes, A and B, having cathode terminals identified, are connected in series. How many possible connections are there? How many retain the nonlinearity of a single diode? What do the remainder appear to be?

4.3 Two ideal diodes, A and B, with cathodes marked, are connected in parallel. How many possible connections are there? How many retain the nonlinearity of a single diode? What do the remainder appear to be?

4.4 A meter circuit consisting of an ideal diode connected in series with a 10-V full-scale moving-coil instrument is used to measure the amplitude of a symmetrical square-wave

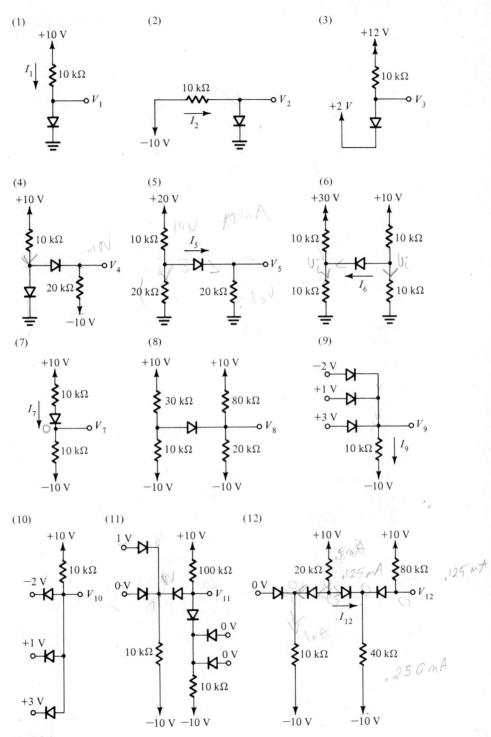

Fig. P4.1

signal. When the meter reading is 50 V, what is the peak-to-peak value of the square wave? For a triangular wave of the same peak-to-peak value, what would the reading be?

4.5 Design an ac voltmeter utilizing a moving-coil meter having 1-mA full-scale deflection and 1-kΩ series resistance, a series resistor R, and a diode, acting as a clipper, connected between the terminals of the meter movement. The ac voltmeter should read 10-V rms full scale for sine-wave voltages. What value is R? What terminal of the diode is connected to the positive terminal of the movement?

4.6 Design a charger for a 12-V automobile battery using an ideal diode and a 12.6-V rms source. Use a resistor to limit the charging current supplied to a fully charged battery (whose terminal voltage rises to 14 V) to a "trickle charging" current of 1 A. What is the charging current when the battery voltage is discharged to 12 V?

4.7 Calculate values for n and I_S for two junction diodes. The first, "a 1-mA diode," has a forward voltage of 0.7 V at 1 mA and of 0.8 V at 10 mA. The second, "a 10-A diode," has a forward voltage of 0.7 V at 10 A and of 0.6 V at 1 A.

4.8 For a diode which conducts 10 mA at a forward voltage of 0.7 V, calculate the "cut-in current," the current at a forward voltage of 0.5 V, if the diode follows (a) the 0.1 V per decade model or (b) the $n = 2$ model.

4.9 Use the observation that junction-diode leakage doubles for every 10°C rise in temperature to estimate the leakage of a rectifier diode in equipment left in an automobile in the hot sun. The diode is known to leak 1 μA at 25°C. The automobile interior temperature may rise to 125°C.

4.10 A shunt voltage regulator consists of a 6.8-V zener diode supplied by a constant current of 20 mA. Near this operating current the zener resistance is specified to be no greater than 5 Ω. Estimate the drop in voltage if the regulator is loaded (a) lightly by a resistor of 2 kΩ or (b) very heavily by a resistor of 200 Ω.

4.11 It is possible to model the breakdown operating region of a zener diode as a series combination of a battery and a small number of forward-conducting junctions. For example, a zener diode rated at 6.8 V at 20 mA can be modeled by a 6.1-V battery and a junction providing 0.7 V at 20 mA (a "20-mA diode" by our convention). Using this idea and the "tenth-volt-per-decade-of current" approximation, what is the effect of reducing the zener current to a 1% level? Contrast this with a regulator formed by a series combination of 10 forward-conducting diodes. Calculate the voltage regulation (in percent) for both regulators.

4.12 A series circuit consisting of a junction diode and a 1-kΩ resistor is supplied by a voltage which varies from 1.0 to 10.0 V. What are the extreme values of diode current and voltage which result? The diode exhibits 0.7 V at 1.0 mA and $n = 1.8$.

4.13 A junction diode is operated in a circuit in which it is supplied with a constant current of I. What is the effect on forward voltage of shunting it by a second identical diode? Assume $n = 2$.

4.14 Two diodes connected in parallel are supplied from a 1-V supply via a 100-Ω resistor. One is a "1-mA diode," the other a "1-A diode." What is the forward voltage across the pair if the tenth-volt-per-decade characterization applies?

4.15 A regulator supplied by a fixed current of 10 mA consists of a single diode to ground across which a 100-Ω resistor is connected to a second grounded diode. Calculate the voltage across the second diode if the diodes are identical "1-mA diodes" with $n = 2$.

4.16 Calculate the small-signal incremental resistance at a junction voltage of 0.7 V for a 1-mA and a 1-A diode, both having $n = 2$. What is the incremental resistance of the 1-A diode at 1 mA?

4.17 A remotely controlled attenuator is formed by a junction diode with cathode grounded to which a dc current I is fed. The anode of the diode is connected via a capacitor to a 1-mV radio frequency (RF) voltage source with a source resistance of 1 kΩ. For $I = 10\ \mu$A, what is the RF voltage across the diode? For $I = 10$ mA, what does this voltage become?

4.18 A shunt regulator utilizing a single forward-conducting junction is to be designed for an application in which no output current is required, using a "1-mA diode" to provide a nominal output of 0.7 V. Two nonregulated power supplies are available. One is nominally 1 V with a $\pm 10\%$ variability. The other is 5 V with a $\pm 50\%$ variability. In each case, what resistor is required? What is the percent variability of the regulator output? Which is the better design? Assume $n = 2$.

4.19 Design a shunt regulator to provide an output which is nominally $+1.5$ V under the following conditions: The unregulated supply varies from $+4$ to $+7$ V. The maximum current available from it is 15 mA. The load on the shunt regulator is either a 1-kΩ or a 10-kΩ resistor. Use "1-mA diodes." The intent is to minimize output voltage variation. What variation of output voltage did you achieve?

4.20 In the circuits of Fig. P4.1, find the voltages V_n and the currents I_n. Assume all diodes to be 1-mA diodes, that is, conducting a current of 1 mA at a voltage of 0.7 V. Also assume that the value of n is such that the diode voltage changes by 0.1 V per decade of current change.

4.21 Using the simple constant-voltage-drop (0.7 V) model for each of the diodes, find the transfer characteristic of the circuit shown in Fig. P4.21.

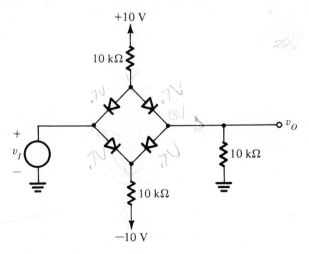

Fig. P4.21

CHAPTER 5

5.1 Derive an analytic expression for the forward conduction part of the transfer characteristic of a half-wave rectifier, that is, of the lower curve in Fig. 5.2b, assuming an exponential diode model.

5.2 Consider the operational rectifier or superdiode circuit of Fig. 5.3, with $R = 1$ kΩ. For input signals of 1 mV and 1 V, what are the voltages which result at the rectifier output and at the op-amp output? Assume the op amp to be ideal and the diode to be a 1-mA unit with $n = 2$.

5.3 Augment the circuit of Fig. 5.3 with a resistor of value $3R$ in series with the diode, connected from its cathode to the top of the resistor R. What is the transfer characteristic of the revised circuit as measured from the input to the cathode of the diode? For an input of $+1$ V, what is the output?

5.4 Sketch the transfer characteristic of the circuit shown in Fig. P5.4. Also sketch the output waveform when the input is a sinusoid of 5 V peak. Find the value of v_O and of the op-amp output voltage when $v_I = +5$ V and when $v_I = -5$ V. Assume that the output saturation voltages of the op amp are ± 10 V.

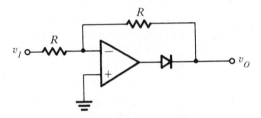

Fig. P5.4

5.5 In Fig. 5.4 consider the consequence of shunting R_2 by a capacitor $C_2 = 0.1$ μF. If $R_1 = R_2 = 1$ MΩ and the input is a symmetrical square wave of 100-Hz frequency and 5 V peak amplitude, sketch and quantify the resulting output. What is its peak value? To what value does it fall? What is its average value?

5.6 As mentioned in the text, the op amp in the circuit of Fig. 5.3 saturates when v_I goes negative. In an attempt to avoid amplifier saturation one might think of adding a "catching diode," as is done in the circuit of Fig. 5.4. The resulting circuit is shown in Fig. P5.6 where D_2 is the catching diode. Unfortunately the circuit no longer functions

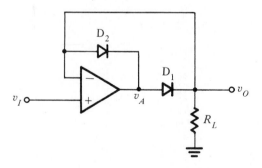

Fig. P5.6

as a half-wave rectifier. Convince yourself of this fact by finding the transfer characteristic v_O versus v_I. Also sketch the transfer characteristic v_A versus v_I. (The latter transfer characteristic has some applications.)

5.7 In the full-wave rectifier circuit of Fig. 5.7c show that op-amp saturation can be avoided by the addition of: a catching diode to A_1 (in the manner of Fig. 5.4), a catching diode to A_2 (in the manner of Fig. P5.6), and a resistor between node C and the inverting input of A_2.

5.8 (a) Sketch the transfer characteristic of the circuit in Fig. E5.3.

(b) Suppose that in this circuit both diodes are reversed and the +15-V source is replaced by a −15-V source. Sketch the transfer characteristic of the resulting circuit.

(c) Consider the circuit formed by combining the circuit of Fig. P5.3 with the circuit of (b) above as follows: Connect the two input terminals together and connect the two output terminals to an op-amp summer circuit having input resistors R and R, and a feedback resistor R. Sketch the transfer characteristic of the resulting composite circuit, i.e., from the input to the output terminal of the summing amplifier.

5.9 In a particular application of the circuit of Fig. 5.9, the average current in the load is 1.0 A. What is the peak current in each diode?

5.10 In Fig. 5.11c what is the output voltage produced across R_L if the input is a 10-V peak-to-peak square wave and each diode is modeled by a 0.7-V forward voltage for all currents?

5.11 Consider the metering circuit of Fig. 5.12b. For a sinusoidal input of 0.1-V rms and $R = 100\ \Omega$, what is the average current in the meter? What happens if, during manufacture of an instrument using this circuit, the supplier of the meter raises the meter resistance by 1 kΩ?

5.12 In the circuit shown in Fig. P5.12, inputs A and B are connected to signals having only two possible values: +1 V or −1 V. Prepare a table (called a truth table) which lists all possible combinations of A and B (there are 4) and the corresponding output at C. This circuit is often called a MAX (for maximum) circuit.

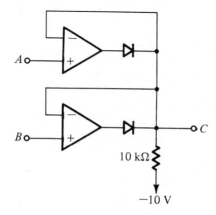

Fig. P5.12

5.13 It is required to analyze the circuit of Fig. 5.14a when excited with the triangular signal shown in Fig. 5.14b. Using the exponential diode model, show that shortly after time t_0 and for $t \leq t_1$ the output is given by

$$v_O \simeq nV_T \ln \left(\frac{I_S T}{4CV_p} \right) + \left(\frac{V_p}{T/4} \right) t$$

Use this relationship to find the value reached by v_O at time t_1 for the case $T = 40$ ms, $C = 10 \ \mu\text{F}$, and $V_p = 10$ V. Assume the diode to be a 1-mA device whose voltage drop changes by 0.1 V per decade of current change. Continue the analysis to determine the value that v_O reaches when the diode cuts off beyond time t_1.

5.14 A peak rectifier connected to a 10-V rms sine-wave source at 60 Hz drives a 1-kΩ load. It utilizes an ideal diode and a 100-μF capacitor. What is the peak output voltage? What is the peak-to-peak ripple? What is the average output voltage? Estimate the peak diode current.

5.15 As a result of the finite output ripple voltage, the average (dc) output voltage of a peak rectifier is a function of the load resistance R. Thus, as a dc source, the peak rectifier can be modeled by a dc source and a series resistance. Derive an expression for this "output resistance" assuming that the diode is ideal.

5.16 Consider an AM detector circuit that is designed to operate up to 5 kHz with a 10-kΩ resistive load (such as the circuit in Example 5.2). Let it be driven by a 1-kΩ source providing an RF signal that is modulated to a depth $m = 0.5$ by a 100-Hz square wave. What are the 10 to 90% rise and fall times of the demodulated wave? Assume the diode to be ideal.

5.17 A diode-clamped capacitor is used to couple a signal which varies from +90 to +105 V to an amplifier operating near ground. The clamping diode, considered ideal, has a grounded cathode. If the average value of the high-voltage signal is +95 V, what will the average value of the clamped output be?

5.18 A clamped capacitor circuit driven by a 1-kΩ source utilizes a 1-μF capacitor, a 100-kΩ load, and a grounded-anode diode. The diode conducts 1 mA at 0.7 V with 0.1-V variation for each decade of current change. Calculate the positive and negative limits of the output signal when the input signal is a series of 1-ms negative pulses of 10-V amplitude occurring at 10-ms intervals. (*Hint:* The 1-kΩ source resistance plays a dominant role in determining the negative peak of the output waveform.)

5.19 Design limiter circuits using only diodes, and 10-kΩ resistors to provide an output signal limited to the range:
(a) −0.7 V and above
(b) −2.1 V and above
(c) +0.7 V and half the instantaneous value of the input
(d) ±1.4 V

5.20 Augment the circuit designed in Example 5.4 with a resistance $R_f = 60$ kΩ connected in the negative-feedback path of the op amp. Find the transfer characteristic.

5.21 Consider the circuit of Fig. 5.33 in which the amplifier output limits at ±10 V, $R_2 = 2$ kΩ and $R_1 = 10$ Ω. Calculate the two threshold voltages. Now consider the addition of a resistor R connected to the positive op-amp input and +10 V. What do the thresholds become if $R = 10$ kΩ?

5.22   Assume that the saturation voltages of the op amp in Fig. P5.22 are ± 9 V. Show that the circuit operates as a bistable. Sketch v_2 versus v_1. Give the threshold voltages. Assume that a conducting diode has 0.7-V drop.

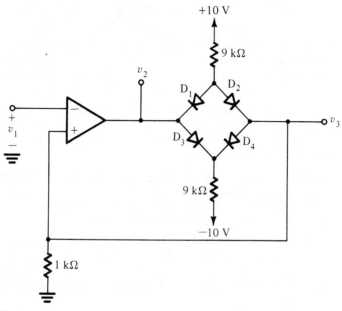

Fig. P5.22

5.23   The circuit shown in Fig. P5.23 can function as a frequency multiplier for some purposes. Illustrate this by considering a 10-V peak-to-peak triangular wave applied at the

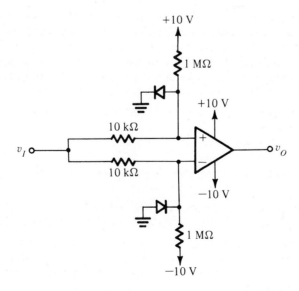

Fig. P5.23

input. Sketch and label the input and output waveforms. Assume that the diodes are 1-mA units and that the saturation voltages of the op amp are ± 10 V.

5.24 This is a challenging problem. Assuming that the diodes in Fig. P5.24 operate at 0.7 V, plot the transfer characteristic v_O versus v_I. Note that the input thresholds and output voltage levels are independent of the op-amp characteristics and the power supplies. What happens as R_1 is reduced toward 10 kΩ? When $R_1 = 10$ kΩ? When $R_1 < 10$ kΩ? (You will benefit by observing that part of this circuit is the negative impedance converter of Example 3.3.)

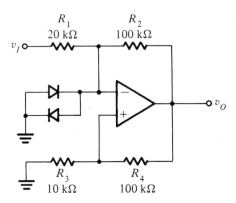

Fig. P5.24

5.25 Consider the circuit of Fig. 5.34. For $R_1 \ll R_2$ (that is, $\beta \ll 1$), the waveform at C becomes almost triangular, consisting of the initial nearly linear parts of the exponential charging waveform. Assuming the $L_+ = -L_-$, use the geometry of the resulting waveform to show that the period $T = 4CR_3\beta/(1 + \beta)$. Design the circuit to produce oscillation at 1 kHz. Use $R_2 = 10R_1$ and select appropriate component values.

5.26 Design a two op-amp circuit suggested by Fig. 5.35 for operation at 1 kHz, assuming op-amp limiting voltages of ± 10 V. Arrange that the triangular-wave output has a 5-V amplitude and that a capacitor of 0.1 μF is used.

5.27 Design a two-segment sine-wave shaper using a 10-kΩ input resistor, two diodes, and two clamping voltages. The circuit, fed by a 10-V peak-to-peak triangular wave, should limit the amplitude of the output signal via a 0.7-V diode to a value corresponding to that of a sine wave whose zero crossing slope matches that of the triangle. What are the clamping voltages you have chosen?

CHAPTER 6

6.1 To function, a desk lamp must be plugged in ($P = 1$) and be switched on ($S = 1$). Prepare a four-line truth table to enumerate the conditions necessary for the occurrence of light ($L = 1$) and all of the reasons for which it may not appear. Emphasize the latter difficulty by tabulating the variable darkness ($D = \overline{L}$). Express L as a function of P and S.

6.2 Using two op amps and two diodes, design a two-input OR (positive logic) which operates with no signal degradation due to diode drops. Arrange that at the output the most negative logic 0 is -5 V, and that for a logic 1 output of $+5$ V, the diode current is 1 mA. (*Hint:* Use superdiode circuits.)

6.3 An op amp which saturates at $\pm V$ can be used in a digital system in which logic 1 is $+V$ and logic 0 is $-V$. For this purpose one op-amp input is grounded while the other is connected to the common node of three resistors of value R, two of which connect to inputs, A and B, while the third connects either to logic 0 or 1 (that is, $+V$ or $-V$). Label this third input C. Prepare four truth tables listing output D and inputs A and B under the conditions: (a) negative input grounded with $C = 0$, and (b) with $C = 1$; (c) positive input grounded with $C = 0$, and (d) with $C = 1$. Identify the logic functional relationship of D with A and B in the four cases. Combine the tables in pairs to identify the relation between D and A, B, C, that is, considering C to be acting as a third input.

6.4 Minimize the following, both by means of truth tables and by Boolean algebraic manipulation:

$$f_1 = (A + B)(B + \overline{A})$$
$$f_2 = ABC + A\overline{C}B + BA$$
$$f_3 = (A + B + C)(A + \overline{B})(\overline{B} + C)$$

6.5 See Fig. P6.5. For signal-voltage levels of ± 10 V and the positive logic convention, express D as a function of A, B, and C.

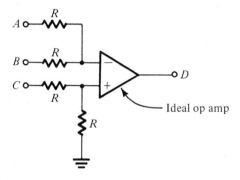

Fig. P6.5

6.6 See Fig. P6.6. For signal-voltage levels of ± 10 V, the positive logic convention, and $X = Y = 0$, find $D = f_1(A, B, C)$. For the negative logic convention and $X = Y = 0$, find $D = f_2(A, B, C)$. Now in the positive logic convention, if $X = Y = 1$, find

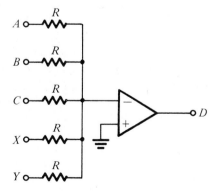

Fig. P6.6

$D = f_3(A, B, C)$. Finally in the positive logic convention find $D = f_4(X, Y, A, B, C)$. It is called a majority logic function. Why?

6.7   The op amp has a gain of 10^4 and limits at ± 10 V output. Sketch a transfer characteristic of v_O versus v_I. For application as a logic inverter, what are V_{IL} and V_{IH}? and V_{OL} and V_{OH}? Calculate the noise immunities or noise margins.

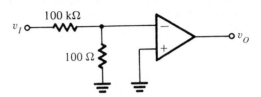

Fig. P6.7

6.8   A logic inverter utilizing an op amp that limits at ± 10 V, drives a twisted pair 30 m in length having a capacitance of 30 pF per meter. What is the power dissipated in this inverter operating at a 10-kHz rate. If the op-amp output is current-limited to 10 mA and if the op-amp slew rate is 2 V/μs, estimate the rise and fall times of the output.

6.9   Using three NAND gates, design an SR flip-flop which is set by a logic 1 but reset by a logic 0. Label the inputs and the outputs.

6.10   Using NOT and OR gates, design a gated D flip-flop which accepts the data input when $G = 1$. Label inputs and outputs.

6.11   Design a clocked master-slave D flip-flop using ten two-input NAND gates. Label inputs and outputs. Indicate the additional connection required to make the circuit count, that is, reverse state at the rising edge of every clock pulse. Why is the circuit of Fig. 6.37 not usable for this task?

6.12   Three JK flip-flops are interconnected in a ring. Output Q and $\overline{Q}$ of each flip-flop is connected to J and K, respectively, of the next, the third being connected back to the first. Initially the first flip-flop is set to 1 and the others reset to 0. Prepare a table to indicate the states of Q_1, Q_2, and Q_3 for a sequence of seven clock pulses.

6.13   Connect three JK flip-flops as in Problem 6.12 except join Q_3 to K_1 and $\overline{Q}_3$ to J_1. Initially all flip-flops are cleared to zero. Prepare a table to indicate the sequence of states of Q_1, Q_2, and Q_3 produced by a string of seven clock pulses.

6.14   Using the idea provided in Fig. E6.13, design a modulo 5 counter using three JK flip-flops and one AND gate. (*Hint:* In a parallel clocked counter an AND of the outputs of all prior stages is required to signal the time of a reversal of a high-order stage.)

6.15   A useful logic-circuit device applicable to most logic families (particularly CMOS) is a delay element consisting of an *RC* low-pass circuit (series *R*, shunt *C* grounded) connected between two logic gates. Draw the signal waveforms for such a circuit consisting of two two-input NANDs connected as inverters with switching threshold at halfway between the two logic levels. Let $RC = 10$ μs and let the driving square wave have a 10-kHz frequency. Indicate with a sketch what happens as the frequency is raised. Indicate the result if the resistor is shunted by a diode whose cathode is at the capacitor end.

6.16 In the circuit shown in Fig. P6.16, τ is a delay element, such as that investigated in Problem 6.15. Find the signal waveform at C when the input terminal A is driven by a logic 1 pulse whose duration is ten times that of the delay.

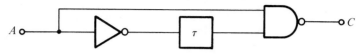

Fig. P6.16

6.17 Consider the circuit of Fig. P6.16 driven by a two-input NAND gate. One of the inputs of the NAND is connected to the output of the circuit (node C), while the other input is driven by a logic 0 pulse of duration which is variable from very small to very large. Find the output waveform. What would you call this circuit? If each logic gate has a propagation delay of 50 ns, what is the shortest input pulse that can be used?

6.18 Three NAND gates, each with inputs paralleled, are connected in a ring. Specifications for this family of gates indicates a typical propagation delay of 30 ns for high-to-low output transitions and 70 ns for low-to-high. Assuming that for some reason the input of one of the gates undergoes a low-to-high transition, by sketching timing diagrams for the three gate outputs, show that the circuit functions as an oscillator. What is the frequency of oscillation of this ring oscillator? What is the average propagation delay of a typical gate in a ring of 3 if an oscillation frequency of 12.5 MHz is measured?

6.19 Design a bistable circuit with two inputs A, B, and output C using one op amp and three 10-kΩ resistors, such that C becomes the same as A and B if they agree and otherwise retains its past value. This device could be called a "conservative voter." Write a logic equation for the new state C' in terms of A, B, C. Assume in your design that logic signal levels are determined by amplifier limiting levels of ± 10 V. (*Hint:* Base your design on the bistable circuit of Fig. 5.33.)

6.20 For logic signals defined by amplifier-limiting levels of ± 10 V, indicate the function of the circuit in Fig. P6.20 and its inputs by means of a truth table using a positive logic convention. Identify the roles of A and B. Is the state of C known when $A = B = 1$? (*Hint:* The circuit embodies positive feedback and hence memory.)

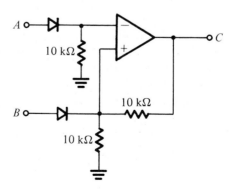

Fig. P6.20

6.21   What is the effect on node C in Fig. P6.21 of successive operations of the push button? Quantify the interval required between successive closures of the push button contact. Suggest an application.

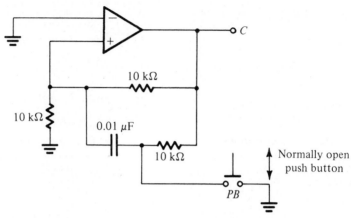

Fig. P6.21

6.22   In Fig. P6.22 the diodes in the quad are reasonably matched, and similar to the other two. Control signals X and $\overline{X}$ are complementary with levels of ± 5 V. Plot the transfer characteristic A to C for the cases (in positive logic): $X = 1$, $X = 0$. Suggest the connection of a second op amp which will correct for mismatch of the diodes in the quad. What function does this circuit perform?

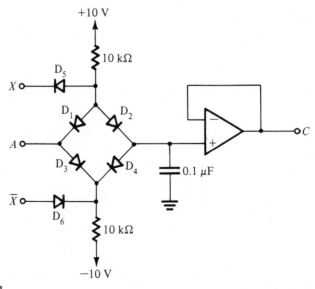

Fig. P6.22

6.23 Using a symbolic representation, show how to connect a D flip-flop of the kind shown in Fig. 6.38 as a toggle or counter to convert a string of equally spaced short pulses to a square wave.

6.24 Two CMOS D flip-flops utilizing a zero and +6-V supply, individually connected to operate in the counting mode to reverse state upon each positive input transition, are interconnected to form a 4-state counter. Resistors are connected from the true outputs of the flip-flops to a common node to which a follower op amp is connected. From the MSB the resistor is 10 kΩ; from the LSB it is 20 kΩ. Sketch on a timing diagram an input waveform at 1 kHz, the digital outputs of each stage, and the output signal from the follower. Label the voltage levels of the latter. Allow your sketch to extend over a 10-ms interval.

6.25 Show the circuit of a 1-bit A/D converter operating in the analog range from 0 to 10 V, utilizing two resistors, a 10-V power supply, and an op amp which limits at ± 10 V.

6.26 A 2-bit A/D converter operating over a 0- to 10-V range utilizes three comparators referenced to the three intermediate nodes of a voltage divider consisting of four 10-kΩ resistors connected between 0 and +10 V. The positive terminals of the comparators are connected in common to the analog input. The comparator outputs, which are digital signals labeled from the least to most significant as A, B, C, are connected to a combinatorial logic network from which two outputs X_1 and X_2 (X_1 least significant) provide a binary-coded version of the analog input. Prepare a truth table, with quantized ranges of the input voltage included on the right, which lists A, B, C, and X_1, X_2, suitably coded. Identify the functional relationships among X_1, X_2 and A, B, C. Utilizing NOR logic gates, draw the circuit that produces X_1 and X_2 from A, B, and C. Now minimize the number of gates by utilizing the fact that interchanging the inputs of a comparator produces a signal inversion.

6.27 A very useful circuit in many general-purpose signal-processing applications is the Exclusive OR, $C = A \oplus B = A\overline{B} + \overline{A}B$, and its inverse, the *Coincidence* circuit,

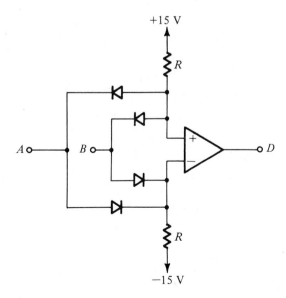

Fig. P6.27

whose output $D = \overline{C} = \overline{A \oplus B}$, is 1 when its inputs agree, and is otherwise 0. Show by means of a truth table that, for a positive logic system in which the logic levels are ± 10 V, the circuit in Fig. P6.27 implements the Coincidence function. Modify the circuit to implement the Exclusive OR function.

6.28 In a ± 10-V logic system, what does the circuit in Fig. P6.28 do when driven by the "short" positive input pulse shown? (Note that the input normally rests at -10 V.) What happens when the input pulse is too long? What is the length of the input pulse at which the operating mode of the circuit changes?

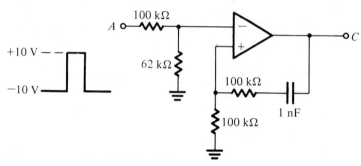

Fig. P6.28

CHAPTER 7

7.1 The drain of an n-channel JFET is connected to the positive terminal of a 9-V battery. Its gate is connected directly to the negative terminal, to which the source is also connected through a digital voltmeter (DVM) having a very large series resistance R (typically 10 MΩ). What JFET parameter does the DVM reading closely represent? Find an approximate analytic expression for the reading in terms of $|V_P|$, I_{DSS}, and R, where R is very large.

7.2 A JFET with gate joined to source by a 1-MΩ resistance is connected in series with a milliammeter and a 9-V battery. For an n-channel FET, the drain is connected to the positive end; for a p-channel FET the drain is made negative. Identify the FET parameter which is read on the meter. What is the reading if the source and drain terminal connections are accidentally reversed?

7.3 For a JFET with $I_{DSS} = 10$ mA and $V_P = -2$ V to which a small v_{DS} is applied, what is r_{DS} for (a) $V_{GS} = 0$ and (b) $V_{GS} = -1$ V?

7.4 Use the relationships expressed in Eqs. (7.4), (7.9), and (7.11) to prepare a sketch of the static characteristics of an n-channel JFET for which $I_{DSS} = 10$ mA and $V_P = -2$ V. Include the characteristic for $v_{GS} = 0$, $v_{GS} = V_P/2$, and $v_{GS} = V_P$. Indicate also the effect of a (low) 10-V gate-to-channel breakdown voltage.

7.5 For a JFET with $I_{DSS} = 10$ mA and $V_P = -2$ V, what is the minimum value of v_{DS} for which the device can operate in pinch-off with $v_{GS} = -1.5$ V.

7.6 For a JFET with $I_{DSS} = 10$ mA and $V_P = -2$ V, utilize Eq. (7.11), the control relation, to find the change in i_D corresponding to a change in v_{GS} from -1.0 to -1.1 V and to a change in v_{GS} from -1.0 to -0.9 V. Now from Eq. (7.11) calculate $\partial i_D / \partial v_{GS}$ and use the result to verify the effect of a 0.1-V change in v_{GS} around an operating point at $V_{GS} = -1$ V.

7.7 For a JFET with $I_{DSS} = 10$ mA and $V_P = -2$ V, calculate r_{DS} at $v_{GS} = 0$, at $v_{GS} = -1$ V and at $v_{GS} = -1.9$ V. Assume v_{DS} is small.

7.8 A p-channel JFET with $I_{DSS} = 3$ mA and $V_P = 3$ V is connected in series with a resistor R to a 10-V supply polarized to place the FET source more positive than the drain. For v_{GS} held at zero, what is the voltage across the FET when: (a) $R = 1$ kΩ, (b) $R = 10$ kΩ, (c) $R = 100$ kΩ.

7.9 A p-channel FET with $I_{DSS} = 10$ mA and $V_P = 3$ V is operated with gate grounded, drain connected via a 1-kΩ resistor to -10 V, and a 2.5-mA constant current injected into the source. What are the voltages, v_{GS}, v_{GD}, and v_{DS}?

7.10 For the circuits in Fig. P7.10 find the values of the labeled voltages and current variables. All FETs have $I_{DSS} = 4$ mA and $|V_P| = 2$ V.

7.11 Using a copy of the characteristic curves of Fig. 7.7 relabeled to correspond to a complementary p-channel device, draw the load line corresponding to a 12-V supply and a 667-Ω load. Indicate the operating current and voltages for fixed-bias operation at $V_{GS} = 1$ V.

7.12 A self-bias design utilizing a source resistor is required to establish $V_{GS} = -1$ V at a current level of 8 mA. On a copy of the characteristic of Fig. 7.8 draw the load line that corresponds. What value of source resistor is needed? Is the graphical approach to this design essential?

7.13 Using the transfer characteristics of Fig 7.8, illustrate the effect on a self-bias design using a 125-Ω resistor of a reduction in the device I_{DSS} by 50% to 8 mA. What is the change in bias current expressed as a percentage of the original value? What is the effect of simultaneous reduction of I_{DSS} and V_P to half their original values?

7.14 Repeat Problem 7.13 with a combination of fixed and self-bias using a 3-V gate-bias supply and a 500-Ω resistor in the source lead. What are the percentage changes in bias current for the reduction in I_{DSS} to half value while V_P remains the same, and for a reduction of both I_{DSS} and V_P to half their original values?

7.15 Without resorting to characteristic curves, provide a bias design for an n-channel JFET for which $I_{DSS} = 8$ mA, $V_P = -4$ V. Design for operation of the FET at 4 mA with the 18-V supply utilized to provide equal voltages across the load, across the FET (that is, V_{DS}), and across the bias resistor. Use a 10-MΩ resistor in the upper end of the gate-bias divider. What do the voltages across the three parts of the circuit become if a device with I_{DSS} of 16 mA is used? What if an 8-mA, -2 V FET is used?

7.16 The output resistance r_o of the model shown in Fig. 7.25 is proportional to the reciprocal of the transconductance. That is $r_o = \mu / g_m$ where μ is called the amplification factor. In a JFET in which $\mu = 100$, what is the largest possible gain of a common source amplifier (obtained as the load resistor is made larger and larger)? What value must the load resistance be to achieve a gain of 50 if $\mu = 100$ and $g_m = 1$ mA/V?

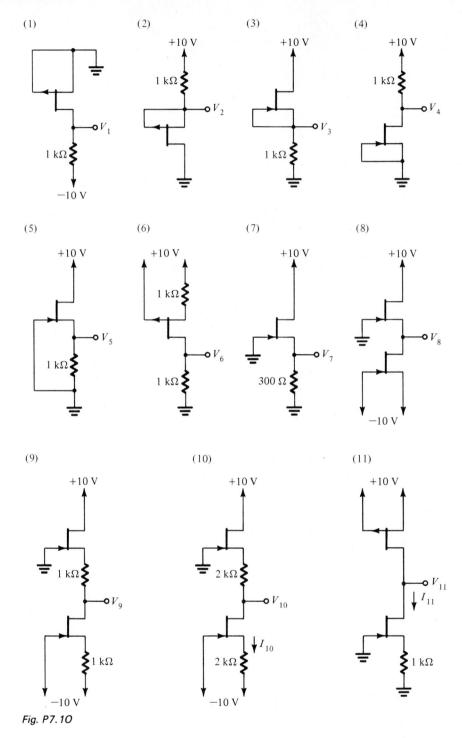

Fig. P7.10

7.17 A JFET amplifier for which $I_{DSS} = 8$ mA and $|V_P| = 4$ V is operated at $I_D = 2$ mA using a constant-current source feeding the source terminal. The bias design allows a 10-V drop across the load resistor. The FET source is bypassed to ground via a large capacitor. What is the gain of the resulting amplifier?

7.18 In the circuit of Fig. 7.20 the FET has $V_P = -4$ V and $I_{DSS} = 16$ mA. For $V_{GS} = -3$ V, find the value of g_m. Calculate the value of voltage gain when $R_D = 10$ kΩ. For $V_{DD} = 20$ V and the input signal a 0.1 V-amplitude square wave, what is the output signal amplitude at the drain? What is the lowest drain-to-source voltage that occurs?

7.19 A JFET amplifier is to be designed with the topology given in Fig. 7.28. A power supply of 15 V is provided. The JFET to be used has $I_{DSS} = 10$ mA and $|V_P| = 3$ V. Design the bias network to provide $I_D = 2.5$ mA and equal voltages across the load, across the FET (that is V_{DS}), and across R_S. While the input resistance to the amplifier is to be maximized, the largest available resistor value is 1 MΩ. Assuming all capacitors to be very large and $R = 100$ kΩ, $R_L = 10$ kΩ, what is the gain v_o/v_i of the resulting amplifier?

7.20 A source follower resembling that of Fig. 7.30 can be constructed in which advantage is taken of a second FET matched to the first that is used for bias. This FET, which replaces R_S, is connected with its gate and source joined to a negative supply of value $-V_{DD}$. Its drain is connected to the source of the follower FET. For this connection what is the dc voltage at the source of the source follower? What is the voltage offset of the follower? What is the output resistance of the follower? Is C_{C2} needed?

7.21 A JFET source follower is biased to operate with a g_m of 2 mA/V utilizing a source resistor of 10 kΩ. What is the output resistance of the follower? What is the gain of the follower when it is capacitively coupled to a 10-kΩ load?

7.22 A JFET amplifier, biased to have a g_m of 2 mA/V has a drain resistor of 10 kΩ and a capacitively coupled load resistance of 10 kΩ. What is the gain of the amplifier if the source resistor is totally bypassed? What source resistance must remain unbypassed to reduce the gain by a factor of 2?

7.23 In the circuits shown in Fig. P7.23, assume that all FETs have $I_{DSS} = 8$ mA and $|V_P| = 2$ V. Find the voltage gain v_o/v_i.

7.24 In the circuit of Fig. P7.24, all the FETs are matched and have $I_{DSS} = 4$ mA and $V_P = -2$ V. (a) Find the dc voltages at D_1, D_2, and D_3. (b) If a small, capacitively coupled signal is applied between G_1 and G_2, and the output signal voltage is taken between D_1 and D_2, find the voltage gain. (*Hint:* Note that Q_3 acts as a constant-current source and use the small-signal model of Fig. 7.24 for each of Q_1 and Q_2.)

7.25 Repeat the analysis of the amplifier of Fig. 7.31 under the conditions that $I_{DSS} = 4$ mA and $|V_P| = 2$ V.

7.26 Repeat the analysis of the amplifier of Fig. 7.31 under the conditions that $I_{DSS} = 16$ mA and $|V_P| = 4$ V.

7.27 A circuit like that in Fig. 7.34 utilizing a FET with I_{DSS} of 8 mA, $|V_P|$ of 2 V, a supply of 50 V, $R_D = 50$ kΩ, and a capacitor of 0.1 μF is driven by a pulse which falls from 0 to -5 V for 0.1 ms. Sketch and label the waveform of the voltage across the capacitor.

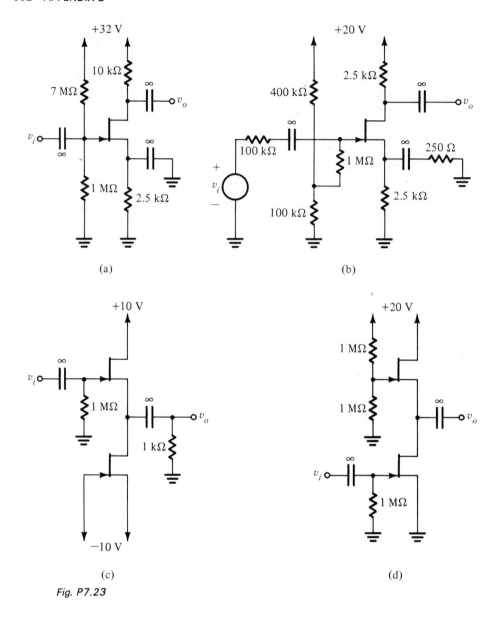

(a)

(b)

+10 V

+20 V

(c)

(d)

Fig. P7.23

7.28 The circuit shown in Fig. P7.28 provides at its output a sawtooth waveform. The purpose of the JFET is to rapidly discharge the capacitor at the end of every period. To
accomplish this, the control signal v_C is a pulse waveform which stays negative for most
of the period and goes positive for the relatively short interval required for capacitor
discharge. Assume that JFET has $I_{DSS} = 10$ mA and $V_P = -2$ V.

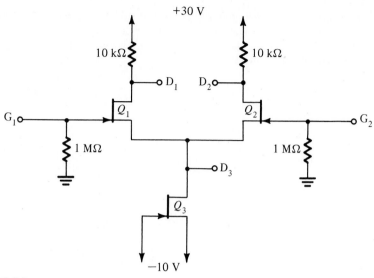

Fig. P7.24

(a) Find the time interval during which v_C must remain negative, so that the sawtooth output has a 5-V amplitude.

(b) Find the time interval during which v_C must go positive, so that 99% of the capacitor charge is removed.

Hint: You will need the following integral: $\int dx/(ax^2 - x) = \ln(1 - 1/ax)$.

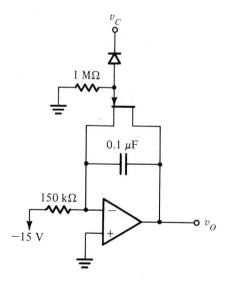

Fig. P7.28

7.29 A *complementary source follower* can be formed by interconnecting *n*- and *p*-channel FETs with similar I_{DSS} and $|V_P|$ parameters, as shown in Fig. P7.29. For an input voltage of 0 V, what is the standby current used by the follower? What is the output resistance of the follower?

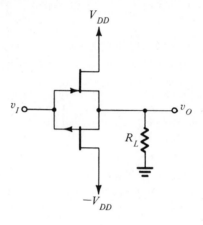

Fig. P7.29

7.30 For positive v_I, and assuming the op amp to be ideal, find the relationship between I_O and v_I in Fig. P7.30. Assume that the value of R_D is selected such that the FET remains in pinch-off for the range of v_I of interest. For $I_{DSS} = 8$ mA, $|V_P| = 3$ V and $R = 1$ kΩ, find I_O and V_A when $v_I = +4$ V.

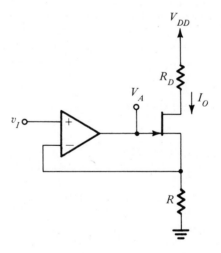

Fig. P7.30

CHAPTER 8

8.1 A depletion-mode NMOS transistor with $V_P = -1$ V and $I_{DSS} = 1$ mA operates at $V_{GS} = +1$ V. What is the minimum v_{DS} required for the device to operate in pinch-off? What is the corresponding value of i_D?

8.2 A PMOS transistor with $V_P = +2$ V and $I_{DSS} = 8$ mA is operated with $V_{GS} = -1$ V. What is the value of v_{DS} required to ensure pinch-off? What is the corresponding value of i_D?

8.3 An NMOS transistor with $V_P = -0.5$ V and $I_{DSS} = 0.5$ mA is operated at $V_{GS} = +10$ V. What is the current that results if operation is in pinch-off? What drain-to-source voltage is required?

8.4 Equations (7.16) and (8.4) can be viewed as conveying the same information, with Eq. (8.4) being somewhat more general since I_{DSS} is defined only for depletion or mixed-mode (enhancement-depletion) devices. By considering V_P and V_T equivalent and equal to V_0, calculate β in terms of I_{DSS}. Note that for a depletion device I_{DSS} is that current which flows when the gate is raised above the conduction threshold by an amount equal to the threshold voltage. Discuss for an enhancement device its equivalence to the current which flows when $v_{GS} = 2V_T$.

8.5 An enhancement n-channel device with $V_T = 2$ V is to be operated in the pinch-off region with a gate-to-source voltage of 4 V. What is the lowest drain-to-source voltage for which this is possible?

8.6 Consider an enhancement n-channel device with $V_T = 2$ V and for which $i_D = 2$ mA when $v_{GS} = 4$ V. What value of i_D is found when this NMOS device is operated in pinch-off at $v_{GS} = 8$ V?

8.7 An n-channel enhancement MOSFET with $V_T = 1$ V conducts 2 mA at $v_{GS} = v_{DS} = 4$ V. What is the linear incremental resistance of the FET operating as a switch in the triode region with $V_{DS} = 0$ and $V_{GS} = 4$ V.

8.8 A p-channel enhancement MOSFET with $V_T = 1$ V is operated as a "diode" with gate and drain joined. For an applied voltage of 2 V it conducts 10 mA. What is the voltage drop of the diode at 1 mA and 0.1 mA?

8.9 The diode-connected FET described above is connected in series with a 5-kΩ resistor across a 10-V supply. What voltage is measured across the diode?

8.10 The diode-connected PMOS FET of Problem 8.8 is to be operated as a shunt voltage regulator to which a current of 10 mA is supplied. What is the output voltage of the regulator? What is the output resistance of the regulator? By what amount will the output voltage drop if 1 mA of load current is taken from the regulator?

8.11 An n-channel enhancement MOSFET for which $i_D = 2$ mA at $v_{GS} = 2V_T$ is to be used as a constant-current output of 1 mA. If the FET operates in pinch-off, what gate-to-source voltage is required? What is the minimum voltage from drain to source to ensure constant-current operation?

8.12 Two closely matched n-channel FETs having $V_T = 1$ V and I_D at $2V_T$ of 1 mA are connected as shown in Fig. P8.12. What is the output voltage v_O? If R is reduced to 4 kΩ, what does v_O become? With $R = 8$ kΩ and a third matched FET connected in parallel with Q_2, what does v_O become?

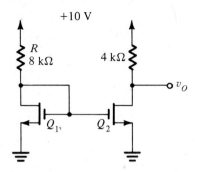

+10 V

Fig. P8.12

8.13 A diode-connected enhancement MOSFET (with gate and drain joined) is wired in series with a high-resistance digital voltmeter across a 10-V power supply polarized to allow the FET to conduct. Interpret the DVM reading V_M in terms of appropriate FET parameters.

8.14 A diode-connected enhancement MOSFET is supplied by a constant current I polarized to cause the FET to conduct. The voltage across the FET is measured with a high-resistance DVM. For 4 mA supplied, a voltage of 4 V is measured. For a current of 1 mA, the voltage is 3 V. What are V_T and β of the FET? What is the value of i_D for $v_{GS} = 2V_T$ and operation in pinch-off?

8.15 In the circuit of Fig. P8.15, the drain-to-source voltage is 2 V with the switch open. What is it with the switch closed?

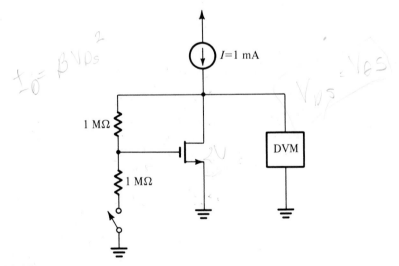

Fig. P8.15

8.16 In the circuit of Fig. P8.16 the current meter reads 6.5 mA with the switch open and 4 mA with the switch closed. Characterize the FET with two (i_D, v_{GS}) coordinates. What is V_T? What is β?

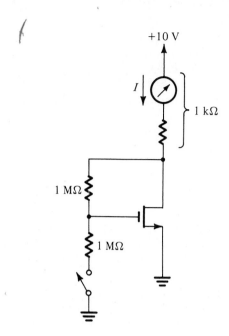

Fig. P8.16

8.17 In the circuit shown in Fig. P8.17, the current reading is 4 mA with the switch open. With the switch closed the reading is 3 mA. What is V_T? What is β?

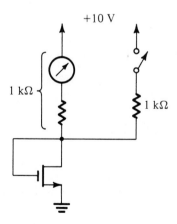

Fig. P8.17

8.18 An *n*-channel MOSFET amplifier is biased using a variation of Fig. 8.16. Here the gate is grounded via a 1-MΩ resistor; the drain is returned to a positive supply of 20 V via a resistor of 10 kΩ, the source is connected to a current source of value *I*. The FET has $V_T = +2$ V and $\beta = 0.5$ mA/V^2. It is desired to operate the drain at $+10$ V. What is the required value of *I*. What is the corresponding voltage at the source? What is the largest available signal swing at the drain? What is the effect of substituting an FET with $V_T = 4$ V in the circuit?

8.19 Two matched FETs are connected as shown in Fig. P8.19: $V_T = 2$ V, $\beta = 0.5$ mA/V^2, $V_{DD} = 20$ V. Design for a current of 1 mA in each device, and the largest possible input resistance at the gate of Q_1. The largest available resistor is 10 MΩ. For a processing change which reduces the V_T of both devices to 1.5 V, what does the current I_D become? If a resistor had been used in place of Q_2, what would be the change in current as V_T of Q_1 is changed from 2 to 1.5 V?

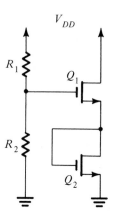

Fig. P8.19

8.20 It is possible to fabricate very high current, high-voltage MOSFETs having a V-shaped cross section and called VFETs. Some such units are capable of operation at several hundred volts and tens of amperes. One such device has a threshold which is 0.8 V minimum and 1.2 V typically. Its pinched-off drain current at $v_{GS} = 10$ V is 1.0 A minimum and 2.0 A typically. If the device is operated at low drain voltages as a power switch with $v_{GS} = 5$ V, what is the typical series resistance it represents? From the data given, what is the largest resistance you may expect? For typical values, what is the voltage drop of the switch at 0.1 A?

8.21 Design an amplifier resembling the circuit of Fig. 8.19 utilizing a 12-V power supply and a power NFET having $V_T = 1$ V and conducting 4 A at $v_{GS} = 10$ V in which the quiescent voltage across the load is 5 V and the quiescent power dissipation in the FET is 7 W. Use a feedback resistor of 1 MΩ and an additional resistor from gate to source chosen to establish the conditions stated. What does the FET power dissipation become if a 2-V peak-to-peak sinusoidal signal is produced across the load? What is the power level of the signal component of the load voltage? What peak-to-peak input voltage is required at the gate?

8.22 An n-channel MOSFET with $V_T = 2$ V and conducting 3 mA at $v_{GS} = 4$ V is biased in a simple feedback configuration (Fig. 8.19) using a 10-MΩ feedback resistor, a 10-kΩ load, and a 12-V supply. What value of V_{DS} is produced? A negative input pulse of 0.1 V amplitude is introduced at the gate. What signal is produced at the output? What pulse input current is required from the signal source?

8.23 In Problem 8.22 what is the effect on the dc output voltage of adding a 22-MΩ resistor between gate and source of the FET? On the pulse output signal? On the equivalent input resistance?

8.24 An *n*-channel enhancement MOSFET with $V_T = 1$ V and $\beta = 0.5$ mA/V^2 is to be biased at $V_{GS} = 2$ V or $V_{GS} = 3$ V. What is the operating current and transconductance which corresponds to each of these cases?

8.25 An *n*-channel enhancement MOSFET with $V_T = 1$ V and $\beta = 0.5$ mA/V^2 is "feedback" biased (Fig. 8.19) using $V_{DD} = 15$ V, $R_D = 5$ kΩ, and $R_G = 1$ MΩ. What is the voltage gain of the resulting amplifier and the output resistance under the following conditions:

(a) no load and r_o ignored

(b) a capacitor-coupled 5-kΩ load, and r_o ignored

(c) no load and r_o modeled as μ/g_m for μ a constant equal to 100.

8.26 For conditions as in Problem 8.25 and with r_o infinite and no load, calculate the voltage gain of the circuit for $R_D = 500$ Ω, 5 kΩ, and 50 kΩ.

8.27 Review the conditions under which Eq. (8.23) has been derived to determine the boundaries of the linear gain region for which it applies.

8.28 Consider a circuit whose topology is that of Fig. 8.31a in which both transistors are enhancement devices with $V_T = 1$ V, $\beta_1 = 0.36$ mA/V^2, $\beta_2 = 0.04$ mA/V^2, and the power supply voltage equal to 15 V. Calculate the gain of the resulting amplifier.

8.29 A voltage divider resembling that in Fig. E8.6 consists of four transistors, the upper and lower of which are identical with $\beta = 100$ μA/V^2 and the middle two of which both have $\beta = 1$ μA/V^2. For all devices, $V_T = 2$ V. Calculate the voltage at the three internal nodes and the standing current.

8.30 An NMOS inverter operating from a 10-V power supply utilizes two enhancement devices with $V_T = 1$ V. The switching device is constructed with $\beta_1 = 49$ μA/V^2 while the load has $\beta_2 = 1$ μA/V^2. Calculate the output voltage levels reached with signals of $+10$ and 0 V applied at the input. At what input voltage does the output (a) fall by 10% of the full output voltage range; (b) rise by 10% of the full range. What is the slope of the transfer characteristic in the middle of the switching range? What is the equivalent resistance of the load at the midpoint of the output swing? Use this value to estimate the (10 to 90%) rise time of the output loaded by 1 pF. Use Eq. (8.26) to calculate a more precise value.

8.31 An NMOS inverter circuit utilizes a depletion load for which $V_{T2} = -1$ V and $\beta_2 = 1$ μA/V^2. The switch is characterized by $V_{T1} = +1$ V and $\beta_1 = 10\mu$A/V^2. A 5-V supply is used. Calculate the limiting output voltages. At what input does the output fall to $+4.5$ V? At what input is the output at $+2.5$ V?

8.32 In the circuit described in Problem 8.31, each FET includes an output resistance $r_o = \mu/g_m$ where the factor μ is about 100. Calculate the gain of this inverter when feedback biased as a linear amplifier (for which the dc input and output voltages are equal, that is, using an arrangement similar to that depicted in Fig. 8.31a for the enhancement-load case).

8.33 A CMOS inverter utilizes matched devices with $|V_T| = 1$ V and $\beta = 60$ μA/V^2. For power supply voltages of (a) 5 V and (b) 15 V: What are the expected output voltages? What is the maximum output current that the inverter can tolerate if the resulting change in output voltage is limited to $0.1V_{DD}$?

8.34 It is required to derive an expression for the rise and fall times of the output waveform of a CMOS inverter loaded with a capacitance C. (The rise and fall times are equal if the two transistors are matched.) For this purpose refer to Fig. 8.43 and note that C initially discharges linearly until the point at which $v_O = V_{DD} - V_T$ where Q_1 enters the triode region. The 90 to 10% fall time t_f thus consists of two intervals: t_{f1}, during which v_O drops from $0.9V_{DD}$ to $V_{DD} - V_T$, and t_{f2}, during which v_O drops from $V_{DD} - V_T$ to $0.1V_{DD}$. Find the expression for t_f and calculate its value for the parameter values given in Problem 8.33 and for $C = 10$ pF. [*Hint:* $\int dx/(ax^2 - x) = \ln(1 - 1/ax)$.]

8.35 Draw the circuit of a two input CMOS logic NAND.

8.36 Draw the circuit of an SR flip-flop using two CMOS NOR gates. What polarity of input signal causes a change of state? Leaves the state unchanged?

8.37 Repeat Exercise 8.13 for $|V_T| = 1$ V and $\beta = 30$ μA/V^2.

8.38 Calculate V_O in Fig. P8.38 if:
(a) $\beta_1 = \beta_2$, $V_{T1} = V_{T2}$.
(b) $\beta_1 = \beta_2/4$, $V_{T1} = V_{T2}$.
(c) $\beta_1 = \beta_2$, $V_{T1} = 4V_{T2}$.

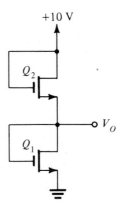

Fig. P8.38

8.39 What are I_2 and I_3 in Fig. P8.39 if: $V_{T1} = V_{T2} = V_{T3}$, $\beta_1 = \beta_2 = \beta_3/2$, and $I_1 = 1$ mA?

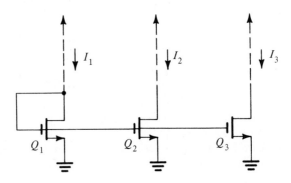

Fig. P8.39

8.40 What is V_O for all devices matched in Fig. P8.40?

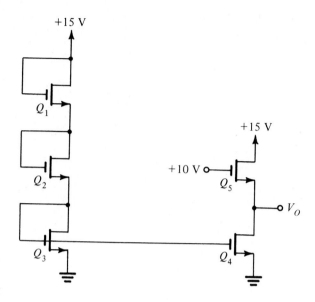

Fig. P8.40

8.41 For $V_{T1} = V_{T2} = V_{T3} = 2$ V, $\beta_1, = \beta_2 = \beta_3 = 2$ mA/V^2, and $I_1 = 1$ mA in Fig. P8.41, calculate I_2, V_1, and V_2.

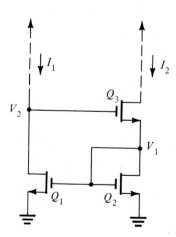

Fig. P8.41

8.42   In Fig. P8.42, $V_T = 2$ V and $\beta = 0.5$ mA/V². What are V_O and R_o?

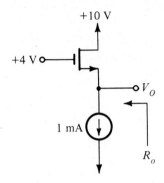

Fig. P8.42

8.43   In Fig. P8.43, $V_T = 2$ V and $\beta = 0.5$ mA/V². What are V_O and R_o?

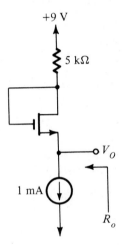

Fig. P8.43

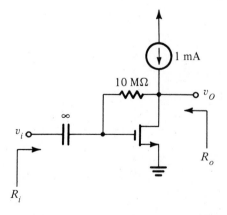

Fig. P8.44

8.44   In Fig. P8.44, $V_T = 2$ V and $\beta = 0.5$ mA/V^2. What is the dc voltage at the output? What is the voltage gain? What is the input resistance, R_i? What is the output resistance R_o? (*Hint:* Apply a small voltage and calculate the resulting current.)

8.45   For $V_{T1} = V_{T2} = 2$ V and $\beta_1 = 0.5$ mA/V$^2 = 36\beta_2$, in Fig. P8.45, calculate the dc output voltage, the voltage gain, the input resistance, the output resistance, and the low cutoff frequency. (*Hint:* The low cutoff frequency is f_{3dB} of the STC network at the input.)

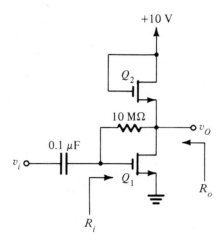

Fig. P8.45

8.46   For $V_{T1} = V_{T2} = 2$ V, $\beta_1 = 0.5$ mA/V$^2 = 4\beta_2$, and C_1, C_2 large (Fig. P8.46), calculate the dc voltage at the output, the voltage gain, and the input resistance. For $C_2 = 0$, what does the input resistance become?

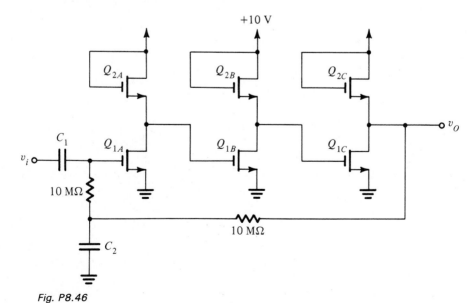

Fig. P8.46

8.47   For $V_T = 1$ V and $\beta = 0.5$ mA/V², in Fig. P8.47, identify the positive logic function this circuit can perform. If one input is at $+5$ V and the other is at 0 V, what is the output voltage?

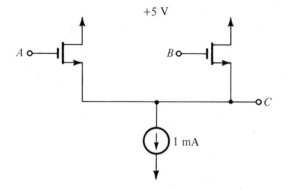

Fig. P8.47

8.48   For the circuit shown in Fig. P8.48, express C as a logic function of A and B. Calculate the three values of output voltage which are possible if $V_T = 1$ V and $\beta = 0.5$ mA/V². Assume that the logic levels at the input are 0 and $+5$ V.

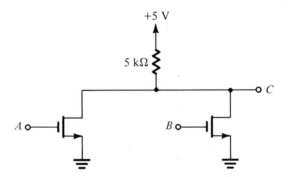

Fig. P8.48

CHAPTER 9

9.1   An *npn* silicon BJT exhibits v_{BE} of 0.64 V at $i_C = 0.1$ mA. Find v_{BE} at $i_C = 1$ mA, 10 mA and 100 mA.

9.2   A particular transistor is found to be operating with a base current of 2.7 μA and a collector current of 0.37 mA. What is β for this device?

9.3   A particular transistor is said to have α ranging from 0.900 to 0.999. What range of β does this represent? To what α does a β of 100 correspond?

9.4   Measurement of an *npn* BJT in a particular circuit shows the base current to be 14.46 μA, the emitter current to be 1.460 mA, and the base-emitter voltage to be 0.7 V. For these conditions calculate α, β, and I_S.

9.5   An *npn* transistor intended for operation at low currents exhibits a base-emitter voltage of 0.8 V at i_c = 10 mA and 0.9 V at i_c = 100 mA. What are the values of I_s and n which would apply to this situation?

9.6 Measurements on three *pnp* transistors in a particular circuit provide the following terminal voltages:

	Transistor 1	Transistor 2	Transistor 3
Emitter	5.3	7.3	4.9
Base	4.6	7.1	4.2
Collector	3.9	2.1	4.7

Indicate the mode in which each of these transistors appears to be operating.

9.7 In the circuit shown in Fig. P9.7, the voltage on the emitter is measured to be +0.7 V. If β is 10, find I_E, I_B, I_C, and the collector voltage. If a collector voltage of -10.7 V were measured, what would you conclude about β?

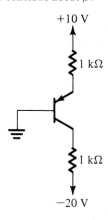

Fig. P9.7

9.8 In the circuit shown in Fig. P9.8, measurement indicates the base voltage to be +0.40 V and the emitter voltage to be +1.10 V. What is β of the BJT? What voltage would

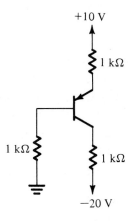

Fig. P9.8

you measure on the collector? Approximately what voltage would you expect on the base and on the emitter if the collector lead were opened?

9.9 An *npn* BJT that is operated in the active mode at a constant collector current $i_C = 1$ mA has $v_{BE} = 0.710$ V at 25°C. It is to be used as a temperature sensor. At location A the v_{BE} is measured to be 0.760 V while at location B it becomes 0.560 V. Estimate the temperatures at A and B.

9.10 Consider a *pnp* BJT with $v_{EB} = 0.700$ V at 1 mA and high β, with base grounded, emitter connected via 300 Ω to a +1-V supply, and collector connected via a 3-kΩ resistor to a −10-V supply. If the temperature is raised by 50°C, what change in the emitter and collector voltages results?

9.11 Calculate the emitter and collector voltages for the circuit of Fig. P9.11 for $\beta = 20$ and v_{BE} approximately constant at 0.70 V.

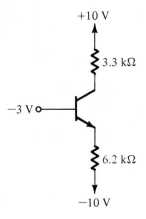

+10 V

3.3 kΩ

−3 V

6.2 kΩ

−10 V

Fig. P9.11

9.12 In the circuit shown in Fig. P9.11, what is the highest voltage to which the base can be raised while maintaining the transistor in the active mode? If a very-high-beta transistor is substituted, what does this critical voltage become?

9.13 In the circuit shown in Fig. P9.13, the BJT is a high-beta device with $v_{BE} = 0.7$ V at $i_C = 1$ mA. Calculate the collector current for an input voltage $V = 10.76, 1.70, 0.74$, and 0.59 V.

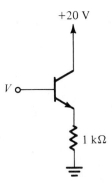

+20 V

V

1 kΩ

Fig. P9.13

9.14 In the circuit shown in Fig. P9.14, the transistor operates with $V_{BE} = 0.70$ V and $\beta = 150$. What is the collector current and collector voltage? In what mode does the device operate?

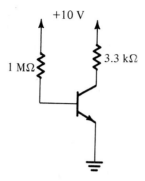

Fig. P9.14

9.15 In the circuit of Fig. P9.14, a transistor with $\beta = 400$ is substituted. What collector current is possible if the transistor remains in the active mode? What collector voltage would that imply? What is the lowest possible collector voltage (which occurs when the base-collector and base-emitter junctions conduct with equal voltages)? What mode of operation is this an extreme case of?

9.16 In the circuit shown in Fig. P9.16, the BJT operates with $V_{BE} = 0.7$ V and $\beta = \infty$. What are the emitter, base, and collector voltages which result?

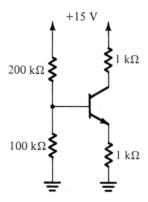

Fig. P9.16

9.17 In Problem 9.16 calculate the effect of using a BJT with $\beta = 50$.

9.18 Reconsider Problem 9.16 with $\beta = 50$ and the base divider resistances reduced by a factor of 10.

9.19 In the circuit shown in Fig. P9.19, all junctions are assumed to operate at 0.7 V. Calculate the voltages at nodes A, B, C, and D if β of each device is (a) infinite, (b) 10.

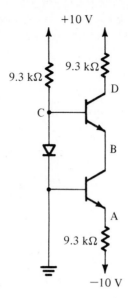

Fig. P9.19

9.20 All junctions in the circuit shown in Fig. P9.20 are assumed to operate at 0.7 V and β is assumed infinite. What are the voltages at nodes A, B, C, D, and E?

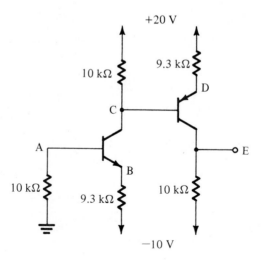

Fig. P9.20

9.21 In Problem 9.20, what do the node voltages become if $\beta = 100$? If $\beta = 10$?

9.22 If the two transistors in Fig. P9.22 are perfectly matched with $V_{BE} = 0.7$ V, what are the voltages at the collectors of Q_1 and Q_2 if (a) $\beta_1 = \beta_2 = \infty$, (b) $\beta_1 = \beta_2 = 100$.

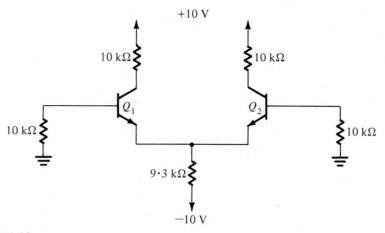

Fig. P9.22

9.23 Repeat Example 9.7 with R_{B2} = 33 kΩ.

9.24 In Fig. P9.24 with $|V_{BE}|$ = 0.7 V and β = ∞, what are the voltages at nodes A, B, C, and D?

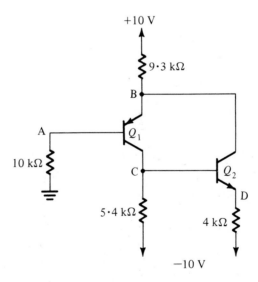

Fig. P9.24

9.25 Calculate the node voltages in Problem 9.24 if β is reduced to 10.

9.26 Utilize the idea expressed in Eq. (9.19) and the basic relationship in Eq. (9.1) to derive an expression for g_m in terms of the collector bias current.

9.27 Beginning with the basic definition of r_e incorporated in Eq. (9.25), and that for r_π in Eq. (9.23), derive the relationship between r_π and r_e.

9.28 Consider the effect of doubling the collector and emitter resistances in the circuit of Fig. 9.26. What is the voltage gain of the resulting circuit? What is the peak value of input signal which can cause the transistor to cut off? What is the input resistance of this circuit? What is the peak output signal available if the voltage change across the junction is limited to 10 mV?

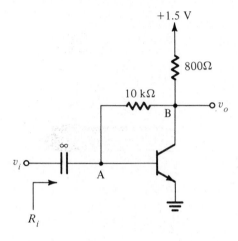

Fig. P9.29

9.29 The simple bias arrangement shown in Fig. P9.29 is useful provided only small output signals are required. In practice, operation remains linear until the base collector junction becomes forward-biased by as much as 0.3 V. What are the dc voltages at A and B, the voltage gain from A to B, the input resistance, the largest peak-to-peak output at B, and the corresponding largest peak-to-peak input voltage at A for $V_{BE} = 0.7$ V and (a) $\beta = \infty$, (b) 100? (*Hint:* For gain calculations replace the transistor by its (g_m, r_π) equivalent circuit model.)

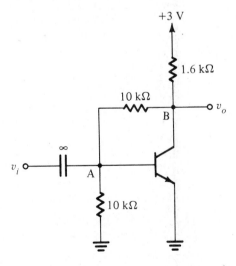

Fig. P9.30

9.30 For $\beta = 100$ and $V_{BE} = 0.7$ V in Fig. P9.30, what is the voltage at B, the gain from A to B, the input resistance, and the largest sine-wave input for which distortion is not great?

9.31 For the circuit of Fig. 9.18a and very large β, calculate I_E, r_e, and the voltage gains from base to emitter and from base to collector.

9.32 In Fig. 9.36 the base voltage is raised from 6 to 8 V. What do the emitter and collector voltages become? Use these and the corresponding 6-V values to estimate "voltage gains" from base to emitter and base to collector. Note the signs of these "gains." Interpret this result.

9.33 The emitter of a *pnp* BJT is connected to ground via a 10-kΩ resistor while its base and collector are each connected to -10 V via individual 10-kΩ resistors. Transistor $\beta = 100$. What collector, base, and emitter voltages result? What is the value of forced β for the circuit? For what range of β does the BJT remain saturated?

9.34 In Fig. 9.39 the input is raised to $+7$ V. Calculate all node voltages and branch currents.

9.35 In Fig. 9.39 the input is lowered to 0 V. Calculate all node voltages and branch currents.

9.36 For the circuit of Fig. 9.39, draw the transfer characteristic for the input varying over a ± 10-V range. Quantify critical breakpoints.

9.37 For the transfer characteristic of Fig. 9.42, calculate the slope at which the output crosses $V_{CC}/2$ in terms of β, R_B, R_C, and V_{CC}.

9.38 For the transfer characteristic of Fig. 9.42, calculate the voltage V_2 in terms of β, R_B, R_C, and V_{CC}.

9.39 Design an RTL inverter with a minimum fan-out of 5 to operate with a 5-V supply on a maximum of 1-mA supply current. Available transistors have β as low as 20. (*Hint:* The inverter should be capable of supplying each of the five driven transistors with sufficient base current to cause them to saturate.)

9.40 Consider the RTL inverter of Fig. 9.42. It utilizes a transistor which, in saturation at the current levels expected, has a base-to-emitter voltage of 0.7 V which reduces by 100 mV for each factor of 10 reduction in current. Assuming β to be large, what is the turnon threshold of this gate (that is, the input voltage at which the collector current is 1% of maximum). Indicate how to augment the circuit with a single resistor which serves to raise the turnon threshold. What value should this resistor have to raise the threshold to 1.50 V, if $R_B = 10$ kΩ?

9.41 Design a two-input positive-logic NAND following the principle illustrated in Fig. 9.43 using (a) only *pnp* BJTs, (b) only *npn* BJTs.

9.42 Augment the circuit of Fig. 9.44 with three diodes and one resistor to implement a circuit with four inputs to provide the logic function $Y = \overline{X_1 \cdot X_2 + X_3 \cdot X_4}$.

9.43 In the circuit shown in Fig. P9.43, each transistor exhibits $|V_{EB}| = 0.7$ V at 10 mA reducing by about 60 mV for each factor of 10 reduction in current, and a β of 100 at relatively low currents. Sketch the transfer characteristic, v_O versus v_I, and the input characteristic, i_I versus v_I, for v_I ranging from 0 to $+1$ V and again to 0 to -3 V and back to 0.

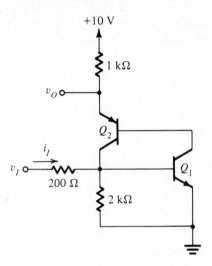

Fig. P9.43

9.44 If $v_A = +3$ V and $v_B = -2$ V in Fig. P9.44, what is v_C? Transistors are matched; V_{BE} = 0.7 V. What is the function of Q_3?

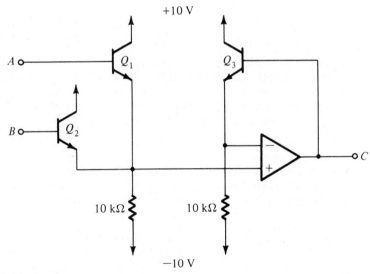

Fig. P9.44

CHAPTER 10

10.1   Prepare a design for an amplifier using the topology of Fig. 10.2 and a 9-V power supply. Arrange that the voltages across R_C, R_E, and the transistor are equalized and that the transistor operates at an emitter current of 0.5 mA and V_{BE} of 0.7 V. Utilize onetenth of the emitter current in the input voltage divider. Produce a design which meets these specifications when a very high β transistor is used. What is the effect on your design if a BJT with $\beta = 100$ is substituted? What percentage change from the prototype design does this represent? Reflect on the relationship of this value to the ratio of currents in R_2 and R_E?

10.2   A BJT for which $\beta = 100$ is biased at 0.5-mA emitter current utilizing a 1-kΩ unbypassed emitter resistance and a base bias equivalent resistance of 10 kΩ. Calculate the small-signal input resistance at the base terminal. What does this resistance become if the emitter is bypassed to ground with a large capacitor?

10.3   In Problem 10.2, the equivalent resistance of the collector resistor and coupled load is 5 kΩ. Calculate the voltage gain of the amplifier from base to load (v_o/v_b), with the emitter resistor unbypassed. What does the gain become if the emitter resistance is completely bypassed? If a source v_s having a 1-kΩ source resistance is connected to the base, what is the gain v_o/v_s in each case?

10.4   A signal v of 100-mV amplitude is to be amplified by a BJT in the grounded-emitter topology with part of the emitter resistance, R_e, unbypassed. To minimize distortion, a signal of at most 10 mV is to be allowed across the base-emitter junction. Derive an expression for R_e in terms of the emitter current I_E.

10.5   In this problem we investigate the possible design optimization mentioned in Section 10.2. Calculate the fractional change (use an increase) in emitter current which a 10-mV peak signal v_{be} causes. For an amplifier whose sole load is the collector resistor, R_C, and whose bias design equalizes the voltages across R_C and between transistor collector and base, what is the possible increase in gain available if the collector resistor is raised so that the "largest undistorted" output is obtained with a 10-mV input signal?

10.6   An *npn* BJT with base connected directly to ground has its collector connected to +10 V via a 10-kΩ resistor, and a current of 0.5 mA extracted from its emitter. An input signal from a 50-Ω source is connected via a large capacitor to the emitter. What is the input resistance at the emitter seen by the source? What is the voltage gain for small signals from the source to a 10-kΩ load resistor capacitively coupled to the BJT collector and the 10-kΩ collector resistor? For reasonably small distortion, what is the largest usable source signal amplitude and the output signal which results?

10.7   Consider a capacitively coupled emitter follower utilizing the topology of Fig. 10.5 for which $V_{CC} = 12$ V, $R_{B1} = R_{B2} = 1$ MΩ, $R_E = 10$ kΩ, $R_s = 10$ kΩ, $R_L = 1$ kΩ, and $\beta = 100$. Calculate the dc base and emitter voltages. What is the gain of the follower as measured from source to load? What is the input resistance of the loaded follower? What is the output resistance of the follower when driven by the source specified? Verify the consistency of these three results by using them to calculate the voltage gain of the unloaded follower.

10.8   For $\beta = 100$ and $V_{BE} = 0.7$ V in Fig. P10.8, what are the dc voltages at nodes A, B, and C? What is the output resistance R_o?

10.9   For $\beta = 100$ and $V_{BE} = 0.7$ V in Fig. P10.9, what are the dc voltages at nodes A, B, and C? What is the output resistance R_o?

10.10 For $\beta = 100$ and $V_{BE} = 0.7$ V in Fig. P10.10, calculate the dc collector voltage. For $V_{CEsat} = 0.3$ V, what is the largest unclipped sine wave that can be accommodated at the output? What is the voltage gain v_o/v_i, the output resistance R_o, and the input resistance R_i?

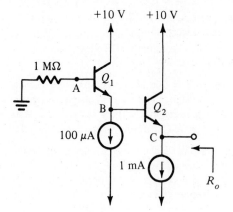

Fig. P10.8

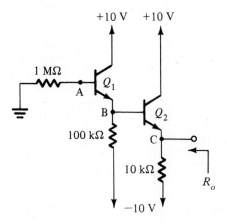

Fig. P10.9

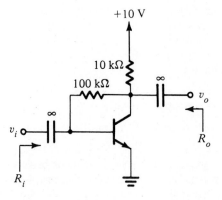

Fig. P10.10

10.11 For the four circuits shown in Fig. P10.11, let $V_{BE} = 0.7$ V and $\beta = 50$. Calculate the voltage gains indicated.

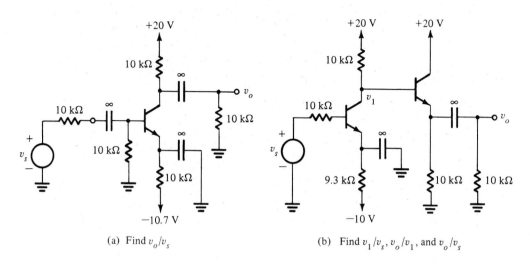

(a) Find v_o/v_s

(b) Find v_1/v_s, v_o/v_1, and v_o/v_s

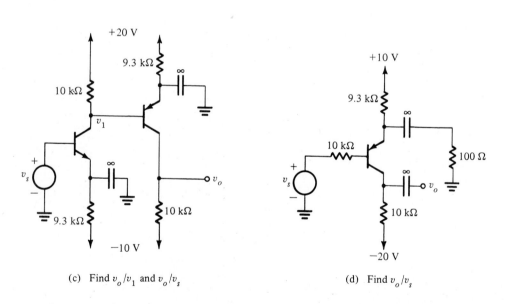

(c) Find v_o/v_1 and v_o/v_s

(d) Find v_o/v_s

Fig. P10.11

10.12 For β very large and $V_{BE} = 0.7$ V, what are the dc voltages at nodes A and B? Calculate the voltage gain and input and output resistances with the circuit: (a) as shown in Fig. P10.12, and (b) with capacitor C removed. What is the largest unclipped sine wave that can be accommodated at the ouptut if $V_{CEsat} = 0.3$ V?

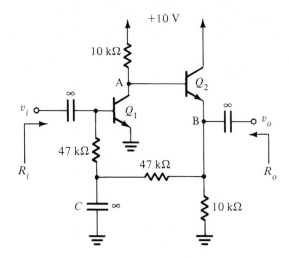

Fig. P10.12

10.13 Consider the circuit in Fig. P10.13 in which transistors are matched with emitter-base voltages of 0.7 V, $|V_{CEsat}| = 0.3$ V, and $\beta = 100$. What are the dc voltages at nodes A, B, and C? What is the consequence of substituting a transistor of $\beta = \infty$ for Q_1 on these voltages? For each of these cases what is the largest unclipped sine-wave output voltage that can be accommodated? What is the voltage gain and the input resistance of the circuit for (a) no load and (b) a 1-kΩ load?

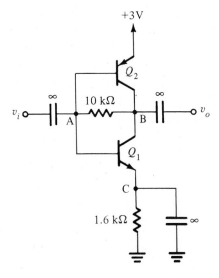

Fig. P10.13

10.14 Consider the circuit in Fig. P10.14 in which all devices are matched with $\beta = 100$ and $|V_{EB}| = 0.7$ V. What are the dc voltages at nodes A, B, C, D, and E? What are the gain and input resistance of the circuit as shown with (a) no load, (b) 200-Ω load? What does the gain become if C_1 and C_2 are removed at (a) no load, (b) 200-Ω load? For no

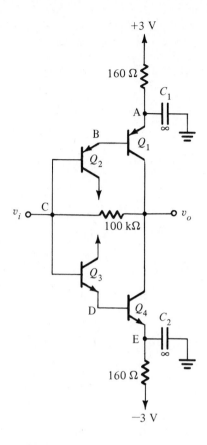

Fig. P10.14

load and C_1 and C_2 connected, what is the largest unclipped sine-wave output if $V_{CEsat} = 0.3$ V?

10.15 In Fig. P10.15, what is the peak-to-peak value of the largest sine-wave signal applied as v_s for which (a) the output negative peak begins to clip, (b) the output positive peak begins to clip? Assume $V_{BE} = 0.7$ V, $\beta = \infty$, and $V_{CEsat} = 0.3$ V.

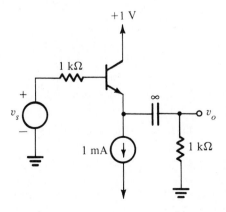

Fig. P10.15

10.16   For the two-terminal device in Fig. P10.16 and I sufficiently large, calculate the dc voltage drop and the incremental resistance as functions of R_1, R_2, V_{BE}, and r_e for β very high. What is the voltage and resistance when $\beta = 10$, $R_1 = R_2 = 1$ kΩ, $I = 10$ mA, and $V_{BE} = 0.7$ V. In general, what value of I is "sufficiently large" to meet the implied conditions?

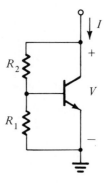

Fig. P10.16

10.17   For the amplifier in Fig. P10.17, what are the largest possible positive and negative output voltages? $V_{BE} = 0.7$ V and $V_{CEsat} = 0.3$ V.

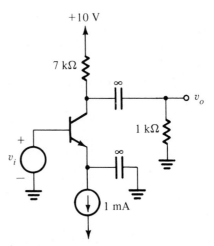

Fig. P10.17

10.18   For the circuit of Fig. 10.9 find the condition for which the output current is independent of variation in $V_{BE} = V_{BE1} = V_{BE2} = V_{BE3}$.

10.19   In this current-sink circuit (Fig. P10.19), the transistors are matched and have high β. To increase the share of the power supply available to the load resistor, R_2 is reduced to one-half of R_1. Design R_3 and R_4 so that the dependence of I_O on variation of $V_{BE} = V_{BE1} = V_{BE2}$ is minimized. Arrange that the current in R_4 is one-tenth that in Q_1 and Q_2 and that I_O is 10 mA.

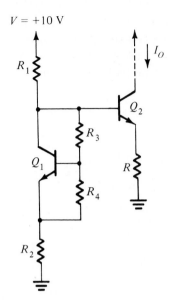

Fig. P10.19

10.20 In Fig. P10.19, consider the effect of reducing R_2 (and R) with R_1 held fixed. What change must be made in R_3? As R_2 (and R) approach zero, relate the resulting circuit to that in Fig. 10.12.

10.21 A current mirror resembling that in Fig. 10.12 is constructed with three transistors in which two, Q_1 and Q_3, are paralleled. The output is taken from Q_2. What would you expect the mirror transfer ratio to be for very high β? What is the precise ratio for finite β?

10.22 For the circuit of Fig. 10.13, calculate the current transfer ratio from input I to output 1 and output 2. Assume β very large.

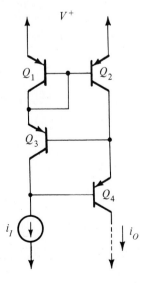

Fig. P10.23

10.23 For the current mirror in Fig. P10.23, calculate the current transfer ratio i_o/i_I. Compare the performance with the circuit in Fig. E10.12.

10.24 For the circuit shown in Fig. P10.24, find the node voltages and branch currents assuming that $V_{EB} = 0.7$ V, for (a) $\beta = \infty$ and (b) $\beta = 10$.

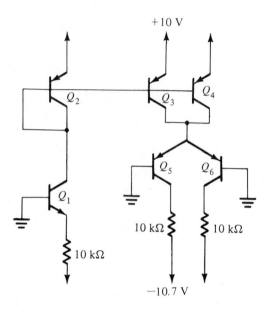

+10 V

Q_2 Q_3 Q_4

Q_5 Q_6

Q_1

10 kΩ

10 kΩ   10 kΩ

10 kΩ

−10.7 V

Fig. P10.24

10.25 Consider the circuit of Fig. E10.13. If each transistor has a V_{EB} of 0.7 V at 1 mA, what voltage at the base of Q_1 reduces the current level in Q_1 to 1% of that of the maximum in Q_2?

10.26 The input offset voltage of a differential amplifier is defined as the dc voltage that has to be applied between the input terminals in order to reduce the output to zero. In the circuit of Fig. 10.17, a manufacturing defect causes one of the collector resistors of Q_1 to increase to 22 kΩ. Assuming the amplifier to be ideal otherwise, what input offset voltage is required to compensate? The devices are specified to have a V_{BE} of 0.7 V at 1 mA.

10.27 Calculate the input resistance R_i and the voltage gain v_o/v_i of the amplifier in Fig. P10.27. $\beta = 100$.

10.28 A differential amplifier with a 1.0-V peak-to-peak signal connected to both inputs is found to have signals of 10.0 mV and 11.0 mV at its two outputs. When a 1-mV signal is applied between its inputs, a 105-mV signal is measured between its outputs. Calculate the worst-case common-mode gain for single-ended and differential outputs. What is the CMRR as a ratio and in dB for single-ended and differential outputs?

10.29 The circuit of Fig. 10.24a operates with a constant-current source of 1 mA having an output resistance of 1 MΩ and with 10-kΩ collector resistors. What is the common-mode gain of the circuit for (a) single-ended output, (b) differential output, if one of the collector resistors is lower than nominal by 10%?

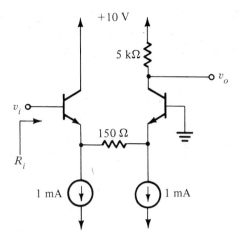

Fig. P10.27

10.30 Repeat the analysis of Exercise 10.15 under the condition that the power supplies are reduced to ± 7.5 V. Compare the ac results with those for ± 15 V supplies.

10.31 Assuming that $|V_{BE}| = 0.7$ V and $\beta = 100$, find the dc voltages in Fig. P10.31 at B_1 and B_2, the dc voltage at the output, and the voltage gain v_o/v_i.

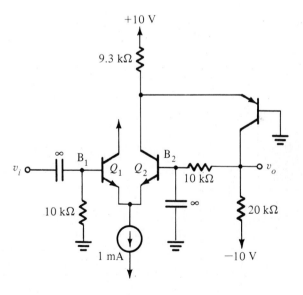

Fig. P10.31

10.32 Consider a differential amplifier in which the output is taken single-endedly. Let the input signals be $v_1 = v_{icm} + v_{id}/2$ and $v_2 = v_{icm} - v_{id}/2$ and let the biasing current source be of value I and having an output resistance R. Express v_o as a function of v_1, v_2, I, R, R_C, and V_T. Assume $R \gg r_e$.

10.33 In Problem 10.32 let $I = 1$ mA. What is the minimum value of R required to obtain a CMRR of at least 80 dB?

10.34 Consider the circuit of Fig. 10.26 with $V_{DD} = 20$ V, $I = I_{DSS} = 2$ mA, $R_D = 12$ kΩ and $V_P = -3$ V. What value of v_{id} will switch the current entirely to Q_2? What is the small-signal voltage gain, if the output is taken differentially?

10.35 In the circuit shown in Fig. P10.35, all FETs have $I_{DSS} = 4$ mA, $V_P = -2$ V, and an output resistance r_o related to g_m by $g_m r_o = 100$. Calculate: (a) The differential gain, (b) the CMRR in dB, and (c) the common-mode range (that is, the range of input common-mode signals over which the amplifier remains in the linear range).

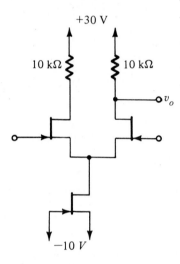

Fig. P10.35

10.36 In the amplifier of Fig. 10.28, consider the effect of adding 100-Ω resistors in the emitter of Q_4 and of Q_5 on (a) the dc bias, (b) the amplifier gain.

10.37 Consider the amplifier of Fig. 10.28: Calculate the effect on the output dc voltage of changing the power supplies (separately) by 1 V under the condition that $|V_{EB}|$ of each transistor remains constant at 0.7 V. These output changes, when referenced to the input by dividing by the amplifier gain, are a measure of power-supply noise rejection in V/V. Use a nominal gain of 8500 to calculate the two supply-voltage rejection ratios.

10.38 Consider the enhancement MOS differential amplifier shown in Fig. P10.38. Assume the two devices to be matched with identical values for β and V_T. Derive expressions for: (a) the value of $v_{id} \equiv v_{G1} - v_{G2}$ at which Q_2 just cuts off, (b) g_m of each device, (c) the total currents i_{D1} and i_{D2} as functions of v_{id}, I, and β. Show that the expressions in (c) reduce to those based on the small-signal model. What is the small-signal condition for this case? Find numerical values for all the results above for the case $I = 2$mA, $\beta = 0.5$ mA/V^2, and $V_T = 2$ V.

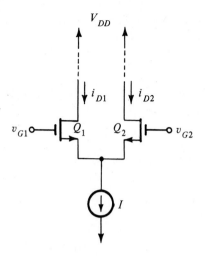

V_{DD}

i_{D1} i_{D2}

v_{G1} Q_1 Q_2 v_{G2}

I

Fig. P10.38

CHAPTER 11

11.1 The circuit of Fig. E11.1 is augmented by a capacitor C_1 shunting R_1. Let capacitor C be relabeled C_2. Find the voltage transfer function $T(s) = V_o(s)/V_i(s)$ for the resulting network. Under what condition is $T(s)$ frequency independent?

11.2 An operational amplifier has the voltage transfer function

$$T(s) = \frac{-10^4}{(1 + s/10)(1 + s/10^6)}$$

What are the poles and zeros of this function? Sketch the magnitude of the gain versus frequency. Estimate the gain at $\omega = 0, 10, 100, 10^5,$ and 10^6 rad/s. What is the net phase shift at unity gain?

11.3 A grounded-emitter BJT amplifier having an input-coupling capacitor and an emitter bypass capacitor shows the following relationship between its collector current (in milliamperes) and its input voltage (in volts):

$$T(s) = \frac{5s(1 + s/10)}{(1 + s/1)(1 + s/100)}$$

Find the poles and zeros of this transconductance function. Sketch the magnitude of the transconductance versus frequency. What is the transconductance magnitude (in mA/V) at 0.1, 1, 10, 100, 1,000 rad/s? What is the phase shift at 100 rad/s?

11.4 Consider an amplifier whose high-frequency response is characterized by three real poles with frequencies of $10^5, 10^6,$ and 10^6 rad/s. Find the exact value of the upper 3-dB cutoff, ω_H. Also determine the approximate value of ω_H using the superposition method.

11.5 Consider an amplifier whose low-frequency response is characterized by two real poles with frequencies of 1 and 2 rad/s, and two zeros, 0 and -0.5 rad/s. Find the exact value of ω_L. Also find an approximation to ω_L using the superposition method.

11.6   An FET common-source amplifier biased to operate with $g_m = 5$ mA/V and having $C_{gd} = C_{gs} = C_{ds} = 1$ pF, is driven by a 100-kΩ source and direct coupled to a load consisting of a 10-kΩ resistor shunted by a 50-pF capacitance. (a) Using the Miller approximation, find the frequency of the input dominant pole. Also find the frequency of the nondominant pole at the output. Use these pole frequencies to determine ω_H. (b) Use the superposition method to determine ω_H. (c) Analyze the circuit to determine its transfer function and thus find the exact value of ω_H.

11.7   A JFET amplifier with the topology shown in Fig. 11.6 utilizes $R = 1$ MΩ, $R_{G1} = 22$ MΩ, $R_{G2} = 10$ MΩ, $R_S = 5$ kΩ, $R_D = 5$ kΩ, $R_L = 10$ kΩ, $C_{C1} = 0.01$ μF, $C_S = 10$ μF, $C_{C2} = 1.0$ μF, with the FET operating at a g_m of 10 mA/V and $C_{gs} = C_{gd} = C_{ds} = 1$ pF. Calculate the complete frequency response of the amplifier. What are its midband gain and upper and lower 3-dB cutoff frequencies? Sketch a Bode plot of the response.

11.8   If the amplifier described in Problem 11.7 is excited by a positive pulse of 0.1 V height and 0.1 ms width, find the height, rise and fall times, and sag of the resulting output pulse.

11.9   Calculate the voltage gain of a common-emitter amplifier whose collector load resistance R'_L is very large, $R'_L \gg r_o$. Utilize the model of Fig. 11.11 for which $g_m r_o = \mu$ and $r_\mu = k\beta r_o$. What is the input resistance of such an amplifier for $k = 1$?

11.10   Find a value for the output resistance of a current source similar to that shown in Fig. 11.12a under the condition that $\beta_0 = 100$, $\mu = 1,000$, $r_\mu = \beta_0 r_o$, $R_E = 10$ kΩ, and the source current $= 1$ mA. What does the output resistance become if R_E is replaced by a grounded-emitter BJT whose output resistance is essentially r_o?

11.11   The following parameters were measured on a BJT biased at $I_C = 10$ mA: $h_{ie} = 350$ Ω, $h_{fe} = 100$, $h_{re} = 0.5 \times 10^{-4}$, and $h_{oe} = 1.2 \times 10^{-4}$ A/V. Determine values for g_m, r_π, r_x, r_μ, and r_o.

11.12   For the transistor in Problem 11.11 operating at the same bias point, determine C_π if C_μ is measured to be 2 pF, and $|h_{fe}| = 10$ at 50 MHz. What is f_T under these conditions?

11.13   A sine-wave source at 1,000 Hz is applied at A in this circuit (Fig. P11.13). Ac rms voltage measurements are taken at A, B, C with the following results: $V_a = 1$ V, $V_b = 3 \times 10^{-3}$ V, and $V_c = 12 \times 10^{-3}$ V. Assuming $r_x = 0$, calculate h_{ie}, h_{fe}, g_m, r_x, r_e, β, and I.

11.14   Consider an amplifer using the topology of that in Fig. 11.21 with $R_s = 10$ kΩ, $R_1 \parallel R_2 = 10$ kΩ, $R_E = 3$ kΩ, $R_C = 3$ kΩ, $R_L = 3$ kΩ, $C_{C1} = 1$ μF, $C_{C2} = 1$ μF, and $C_E = 10$ μF, biased to operate at 1 mA at which the BJT has $\beta_0 = 100$, $C_\pi = 13.9$ pF, $C_\mu = 2$ pF, $r_o = 100$ kΩ, and $r_x = 50$ Ω. Find A_M, f_H, f_L, and the frequency of the zero due to C_E.

11.15   Repeat the analysis of Problem 11.14 for a circuit modified to include a 100-Ω resistor in series with C_E. (*Hint:* Use the superposition method for the determination of f_H and f_L.)

11.16   A BJT with grounded base, biased by a constant emitter current of 0.5 mA, is driven by a 50-Ω source capacitively coupled to the emitter. The collector resistor of 10 kΩ is capacitively coupled to a load consisting of 1 kΩ shunted by 10 pF. The BJT has $h_{fe} = 100$, $C_\mu = 2$ pF, and f_T of 400 MHz. Determine the voltage gain of the amplifier and the two high-frequency poles. Estimate the upper 3-dB cutoff frequency.

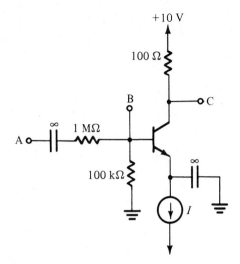

+10 V

100 Ω

B

C

∞ 1 MΩ

A

100 kΩ

∞

I

Fig. P11.13

11.17 A cascode amplifier having the basic configuration shown in Fig. 11.25 operates with BJT collector currents of 0.5 mA and a source resistance of 1 kΩ. The base bias network has a resistance high enough to be ignored while the parallel combination of collector and load resistances is 1 kΩ. The load also includes a 5-pF wiring capacitance. For each BJT, $h_{fe} = 100$, $r_x = 100$ Ω, $C_\mu = 1$ pF, and $f_T = 400$ MHz. Identify and calculate three high-frequency poles. What are the midband gain and 3-dB cutoff frequency? What are the gain and cutoff if the source resistance is reduced to 100 Ω?

11.18 An emitter follower biased at $I_C = 0.5$ mA and utilizing $R_E = 1$ kΩ and a BJT with $f_T = 400$ MHz, $C_\mu = 2$ pF, $r_x = 100$ Ω, and $\beta_0 = 100$ is driven by a 100-Ω source. Evaluate the midband gain and the dominant upper pole.

11.19 In order to save power and space (for capacitors), the amplifier discussed in Example 11.7 and shown in Fig. 11.29 is to have all of its impedances raised by a factor of 2. The load and source resistances remain the same at 4 kΩ. Calculate the gain of the amplifier and the upper 3-dB cutoff frequency.

11.20 In the circuit for Problem 11.19 identify and calculate four low-frequency singularities: three poles and a zero. What is the lower 3-dB frequency?

11.21 In the circuit of Fig. 11.29 which has been analyzed in Example 11.7, Q_1 and Q_2 are modified to the Darlington configuration, that is, the collector of Q_1 is connected to that of Q_2. Calculate the effect on the high-frequency performance of the amplifier. What does the upper 3-dB frequency become?

11.22 A common-base amplifier is directly driven by an emitter follower which is biased to share a common 1-mA bias current. The common-base amplifier has a 100-kΩ collector resistor capacitively coupled to a load consisting of 1 kΩ shunted by 100 pF. The BJTs each have $h_{fe} = 100$, $C_\mu = 2$ pF, and $f_T = 400$ MHz. Determine the voltage gain of the amplifier and the upper 3-dB frequency when the follower is driven by a 1-kΩ source.

11.23 Repeat Exercise 11.13 for the case in which $\beta_1 = 100$ and $\beta_2 = 10$. Find R_{in}, the gain and R_{out}.

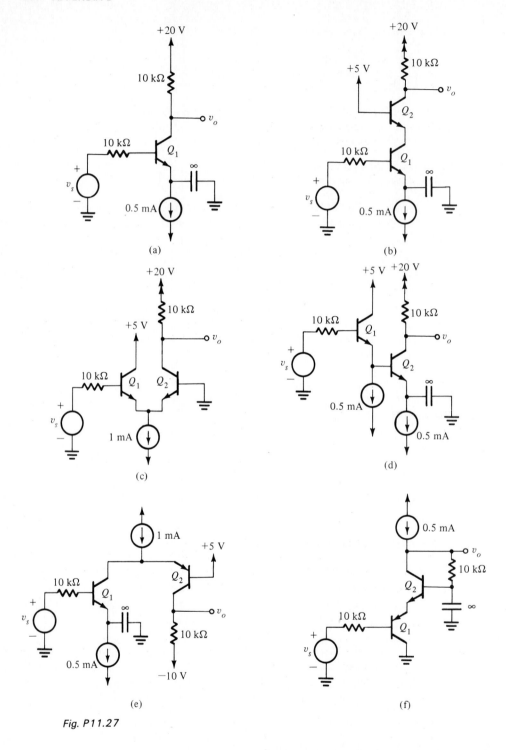

Fig. P11.27

11.24 Repeat Exercise 11.14 for the case in which $I = 2$ mA.

11.25 Repeat Exercise 11.15 with resistors of 450 Ω included in each emitter. Calculate A_0, f_H, and the gain bandwidth product. Compare the latter with the results of Exercises 11.14 and 11.15.

11.26 Consider the circuit of Fig. 11.37 in which $I = 0.5$ mA and $R_C = 10$ kΩ, when fed with a source having a resistance of 20 kΩ. Let $f_T = 400$ MHz and $C_\mu = 2$ pF. Evaluate the low-frequency gain and the high 3-dB cutoff.

11.27 In each of the six circuits in Fig. P11.27, let $h_{fe} = 100$, $C_\mu = 2$ pF, and $f_T = 400$ MHz. Calculate the midband gain and the upper 3-dB cutoff.

11.28 For the circuit in Fig. P11.28 $I_{DSS} = 2$ mA, $V_P = -2$ V, $C_{gd} = C_{gs} = C_{ds} = 2$ pF. Find the midband gain, all poles and zeros, f_L, and f_H.

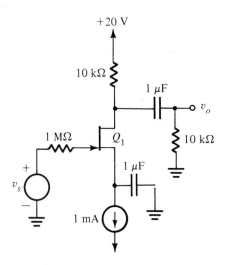

Fig. P11.28

11.29 Repeat Problem 11.28 for the circuit in Fig. P11.29.

11.30 For the circuit in Fig. P11.30, $V_T = 2$ V, $\beta = 0.5$ mA/V^2, $C_{dg} = C_{gs} = C_{ds} = 1$ pF. Find A_M, f_L, and f_H.

11.31 For the circuit in Fig. P11.31, $|V_T| = 2$ V, $\beta = 0.5$ mA/V^2, $C_{dg} = C_{gs} = C_{ds} = 1$ pF, $\mu = g_m r_o = 100$. Find A_M, f_L, and f_H.

11.32 For the circuit in Fig. P11.32, $|V_T| = 2$ V, $\beta = 0.5$ mA/V^2, $C_{dg} = C_{gs} = C_{ds} = 1$ pF. Find A_M and f_H. What is the effect of inserting a 1-MΩ resistor in the path between the gate of Q_2 and ground?

11.33 For the circuit in Fig. P11.33, $\beta_0 = 100$, $r_x = 100$ Ω, $f_T = 400$ MHz, $C_\mu = 2$ pF. Find A_M, f_L, and f_H with $R_s = 0$ and 1 kΩ.

11.34 Repeat Problem 11.33 for the circuit in Fig. P11.34.

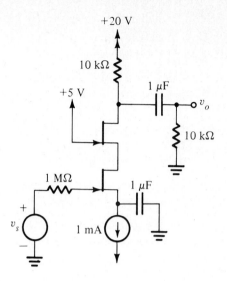

Fig. P11.29

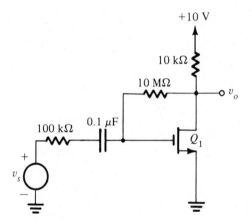

Fig. P11.30

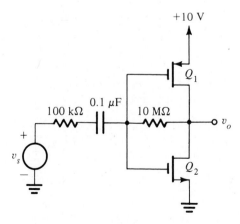

Fig. P11.31

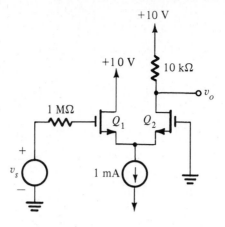

Fig. P11.32

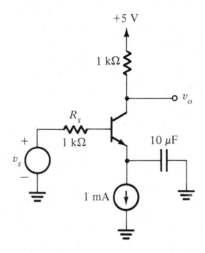

Fig. P11.33

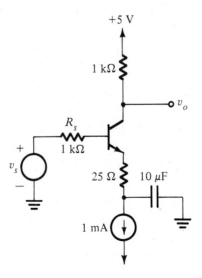

Fig. P11.34

11.35 Repeat Problem 11.33 for the circuit in Fig. P11.35.

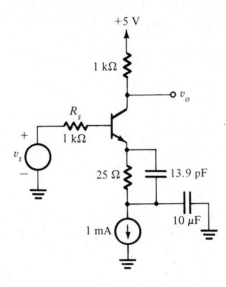

Fig. P11.35

11.36 Repeat Problem 11.33 for the circuit in Fig. P11.36.

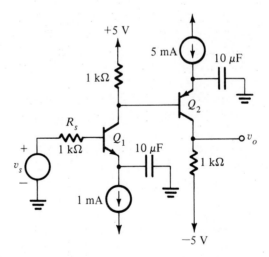

Fig. P11.36

CHAPTER 12

12.1 Consider the noninverting op-amp circuit of Exercise 12.1. For $A = 10^2$ and $R_2/R_1 = 9$, what is the closed-loop gain? What value of (R_2/R_1) is required to obtain a closed-loop gain $A_f = 10$? In the latter case what is the loop gain and the amount of feedback in decibels?

12.2 Consider the noninverting op-amp circuit of Exercise 12.1. The open-loop gain A has a low-frequency value of 10^2 with a single-pole rolloff at 10 kHz. Find the closed-loop gain and the upper 3-dB frequency of the closed-loop amplifier with $R_1 = 1$ kΩ and $R_2 = 9$ kΩ.

12.3 A power amplifier with a voltage gain $A_1 = 1$ operates with an output containing two components—1 volt of signal and 1 volt of power supply hum. Assume that this power amplifier is preceded by a small-signal stage with gain $A_1 = 10$ V/V and that overall feedback with $\beta = 1$ is utilized. If the applied signal and the amplifier noise (hum) remain unchanged, find the signal and hum voltages at the output and hence the improvement in signal-to-noise ratio.

12.4 Without reference to the solution provided, repeat Example 12.1 for the circuit of Fig. 12.12 under the conditions that $\mu = 10^4$, $R_{id} = 100$ kΩ, $R_{icm} = 10$ MΩ, $r_o = 10$ kΩ, $R_L = 2$ kΩ, $R_1 = 100$ kΩ, $R_2 = 10$ MΩ and $R_s = 100$ kΩ. Then follow the derivation provided using the new values to check your result.

12.5 If the frequency response of the op amp in Problem 12.4 is characterized by a single-pole rolloff with 3-dB frequency at 1 kHz, find the 3-dB frequency of the closed-loop gain V_o/V_s.

12.6 The circuit of Fig. E12.5 is modified by replacing R_2 by a 4-kΩ resistor. Find values of A, β, A_f, R'_{if}, and R'_{of}.

12.7 The circuit of Fig. E12.5 is modified by replacing R_2 by a short circuit. Find values of A, β, A_f, R'_{if}, and R'_{of}.

12.8 The circuit of Fig. E12.5 is modified by incorporating a capacitance of 1 μF in series with R_1. Find A, β, A_f, R'_{if}, and R'_{of} at: (a) relatively high frequencies where the capacitor can be considered a short circuit (while the transistor dynamics can still be ignored); and (b) relatively low frequencies where the capacitor can be considered an open circuit. By examining the circuit, convince yourself that the frequency response exhibits a zero at $s \simeq -1/C(R_1 + R_2)$. (You may do this by considering the amplifier to approximate an ideal op amp.) Using the calculated frequency of the zero, give a sketch of the amplifier frequency response and calculate the pole frequency.

12.9 Consider the circuit of Fig. E12.6 with $R_F = 0$ Ω. Assuming as in Exercise 12.6 that $I_{C1} = 0.6$ mA, $I_{C2} = 1.0$ mA, and $I_{C3} = 4.0$ mA and that $h_{fe} = 100$ and $r_o = \infty$, find A, β, V_o/V_s, and R_{if}.

12.10 In the circuit of Fig. 12.21 the resistor R_f is shunted by a series combination of 0.1 μF and 47 kΩ. Using the fact that the device is biased at 1.5 mA with $h_{fe} = 100$, find the voltage gain V_o/V_s and the input and output resistances R'_{if} and R_{of} at: (a) low frequencies where the capacitor effectively acts as an open-circuit; and (b) high frequencies where the capacitor effectively acts as a short circuit (while the transistor dynamics can still be neglected). Sketch the frequency response of the amplifier. From the equivalent circuit of the closed-loop amplifier, find the frequency of the real zero. (Note that this can be done by finding the value of s at which $V_o = 0$.) Use this frequency together with the sketch of frequency response to determine the frequency of the pole.

12.11 Consider the amplifier of Fig. 12.25 to have its output at the emitter of the right-most transistor Q_2. Use the technique for a shunt-shunt amplifier to calculate (V_{out}/I_{in}) and R_{in}. Using this result, calculate I_{out}/I_{in} directly. Compare this with the results obtained in Example 12.4.

12.12 In Example 12.4, the capacitor coupling R_f is $C_f = 0.1\ \mu F$. Sketch the low-frequency response of the amplifier. Calculate I_{out}/I_{in} at frequencies above and below the effects caused by C_f. Calculate the frequencies of the associated pole and zero. (*Hint:* Use the approach suggested in Problem 12.10.)

12.13 Repeat Exercise 12.7 assuming that the open-loop gain of the op amp drops to 1,000.

12.14 Consider a feedback amplifier for which the open-loop transfer function $A(s)$ is given by

$$A(s) = \frac{1{,}000}{(1 + s/10^4)(1 + s/10^5)^2}$$

If the feedback factor β is independent of frequency, find the frequency at which the phase shift is 180° and the critical amount of feedback at which oscillation will commence. For the amount of feedback reduced from this critical value by 20 dB, what is the largest value of β which can be used.

12.15 Consider a modification of the circuit of Fig. 12.34a in which the low-pass filter section connected to the amplifier input is raised in impedance by a factor of 10, using a capacitance $C/10$ and a resistance $10R$. Find the characteristic equation in terms of amplifier gain K. What are the pole frequency and Q factor? At what value of K does the circuit become unstable (and oscillate)?

12.16 Consider a feedback amplifier for which the open-loop transfer function $A(s)$ is given by

$$A(s) = \frac{1{,}000}{(1 + s/10^4)^2}$$

and the feedback factor $\beta(s)$ is given by

$$\beta(s) = \frac{\beta}{(1 + s/10^4)}$$

Sketch the root locus as a function of β. Find the value of β for which the loop is unstable.

12.17 In Problem 12.16, an alternative feedback network is being considered whose pole can be either at 10^3 or 10^5 rad/s. Calculate the value of β for which instability results in each case, the corresponding ω_0 and β for which Q of the complex poles reaches 0.707, as well as the corresponding low-frequency gain.

12.18 Consider an op amp with a two-pole open-loop response with $A_0 = 10^5$, $f_{P1} = 10$ Hz, $f_{P2} = 10^4$ Hz. If this amplifier is connected in a noninverting configuration with a nominal low-frequency closed-loop gain of 100, find the frequency at which $|A\beta| = 1$ and the corresponding phase margin.

12.19 An op amp with open-loop gain of 80 dB and poles at 10^5, 10^6, and 2×10^6 Hz is connected as a differentiator. On the basis of the "rate of closure" rule, what is the smallest time constant that should be used for stable performance?

12.20 An amplifier with open-loop gain of 10^4 and poles at 10^5, 10^6, and 10^7 Hz is to be compensated by the addition of a fourth dominant pole to operate stably with unity feedback ($\beta = 1$). What is the frequency of the required dominant pole? The compensation network is to consist of an RC low-pass network interposed between the feedback network and the amplifier negative input terminal. Dc bias conditions are such that a 1-

MΩ series resistor can be tolerated in series with each of the negative and positive inputs. What capacitor is required between the negative input and ground to implement the required fourth pole?

12.21 An op amp with open-loop gain of 80 dB and poles at 10^5, 10^6, and 2×10^6 Hz is to be compensated to be stable for unity β. Assume that the op amp incorporates an amplifier equivalent to that in Fig. 12.40 with $C_1 = 150$ pF, $C_2 = 5$ pF, $g_m = 40$ mA/V and that f_{P1} is caused by the input circuit and f_{P2} by the output circuit of this amplifier. Find the value of the compensating Miller capacitance if (a) the effect of pole splitting is ignored, (b) the effect of pole splitting is utilized.

12.22 For $h_{fe} = 100$, find the gain (v_o/v_s) and input and output resistances in Fig. P12.22 using feedback techniques. Verify by direct calculation.

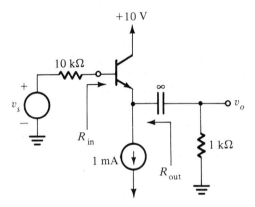

Fig. P12.22

12.23 For $h_{fe} = 100$, find the gain (v_o/v_s) and input and output resistances in Fig. P12.23 using feedback techniques. Verify by direct calculation.

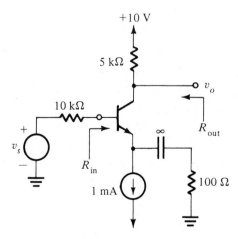

Fig. P12.23

12.24 For $h_{fe} = 100$, find the gain (v_o/v_s) and input and output resistances in Fig. P12.24 using feedback analysis. Verify by other means.

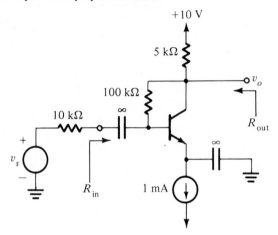

Fig. P12.24

12.25 For $V_T = 2$ V and $\beta = 0.5$ mA/V^2 in Fig. P12.25, calculate the voltage gain (v_o/v_s) and input and output resistances using feedback analysis. Verify by other means.

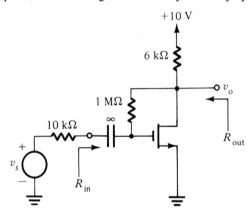

Fig. P12.25

12.26 Consider Problem 12.23 in which the emitter capacitance is reduced to 1 μF. Calculate the low-frequency cutoff using feedback analysis. Verify by direct computation.

12.27 For $V_T = 2$ V, $\beta = 0.5$ mA/V^2 in Fig. P12.27, calculate the voltage gain (v_o/v_s) and input resistance at low and high frequencies. Sketch the frequency response and find the frequencies of the zero and pole. Use feedback analysis techniques.

12.28 In the circuit in Fig. P12.28 the op amp has a gain of 10^6 at low frequencies with a single-pole rolloff and $f_t = 10$ MHz, and $r_o = 0$ and $r_i = \infty$. Prepare a Bode plot for the transfer function (V_o/V_s). What function can the circuit perform? For this purpose what is the relevant time constant? (*Hint:* The transmission zero can be found most easily as the value of s at which $V_o = 0$.)

12.29 For $\beta = 100$ and $V_{BE} = 0.7$ V in Fig. P12.29, calculate the input resistance R_{in}, output resistance R_{out}, and the voltage gain at relatively high frequencies, first by approximate

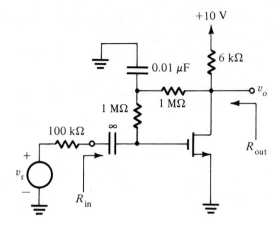

Fig. P12.27

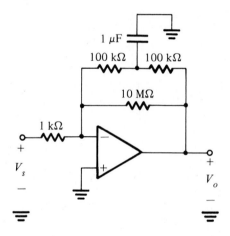

Fig. P12.28

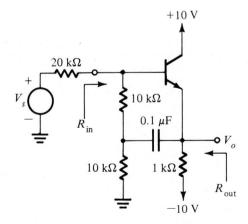

Fig. P12.29

methods and then by feedback analysis. Plot a Bode diagram for gain, identifying critical frequencies, etc.

12.30 For $|V_P| = 2$ V, $I_{DSS} = 8$ mA in Fig. P12.30, find the dc output voltage and voltage gain. What is R_{in} at very low and very high frequencies? For a 2-MΩ source, provide a Bode plot of (V_o/V_s).

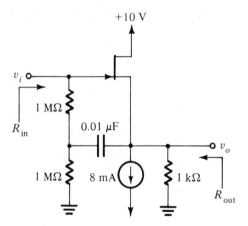

Fig. P12.30

12.31 For $\beta = 100$ and $|V_{BE}| = 0.7$ V in Fig. P12.31, calculate the input resistance R_{in}, output resistance R_{out}, and the voltage gain V_o/V_s.

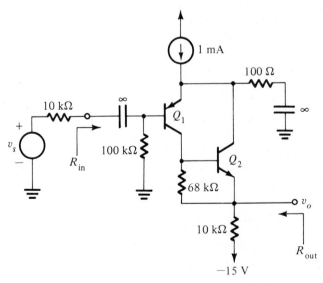

Fig. P12.31

12.32 For $\beta = 100$, $V_{BE} = 0.7$ V in Fig. P12.32, calculate the dc output voltage, R_{in}, R_{out}, and V_o/V_i for $C_1 = \infty$ and $C_1 = 0$. For $C_1 = 1$ μF, sketch a low-frequency Bode plot for V_o/V_i.

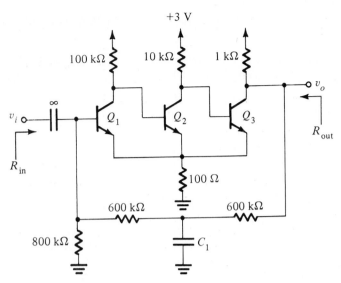

Fig. P12.32

12.33   Solve Problems 10.10, 10.12, 10.13, and 10.14 using feedback analysis techniques.

CHAPTER 13

13.1   Transistor Q_{13} in Fig. 13.1 consists in effect of two transistors in parallel for which $I_{SA} = 0.25 \times 10^{-14}$ A, $I_{SB} = 0.75 \times 10^{-14}$ A while $\beta = 50$ and $\mu = 2,000$ for each. For operation at $I_E = 1.0$ mA, find values for the parameters V_{BE}, g_m, r_e, r_π, r_o, r_μ, assuming $r_\mu = 10\beta r_o$.

13.2   A two-transistor mirror supplied by current I_1 utilizes transistor Q_1 diode-connected to bias Q_2 whose collector provides the output current I_2. Use the exponential relationship between i_C and v_{BE} to derive the coupling between I_2 and I_1 in terms of device saturation currents and current gains. For Q_1 having twice as great a junction area as Q_2 and $\beta_1 = \beta_2 = 100$, what ratio of I_2 to I_1 results? Repeat for Q_1 having half as great a junction area as Q_2.

13.3   Provide designs of the Widlar current source of Fig. 13.2 for output currents of 1 and 100 μA utilizing a 1-mA reference current and devices for which $V_{BE} = 0.70$ V at a collector current of 1 mA. What is the resistor required and base-emitter voltage in each case?

13.4   If, as suggested in Exercise 13.5, two diode-connected transistors were to be used in place of the Q_{18}-Q_{19} network in Fig. 13.1, what size devices would be needed? Express this in terms of the required saturation current.

13.5   In the circuit of Fig. 13.5, what value of R_{10} is required to reduce the current in Q_{14} and Q_{20} to 77 μA?

13.6   Repeat Exercise 13.6 for devices with $I_S = 2 \times 10^{-14}$ A and resistors R_1, R_2, R_3 halved in value. Note that you first will need to find the new value of dc bias currents.

13.7   Through a processing imperfection, the β of Q_4 in Fig. 13.1 is reduced to 25, while the

β of Q_3 remains at its regular value of 50. Find the input offset voltage that this mismatch introduces.

13.8   What is the maximum input offset which can be compensated by completely short-circuiting either R_1 or R_2 in Fig. 13.1.

13.9   In Exercise 13.9, what is the effect on CMRR of using external components to reduce $\Delta R / R$ by a factor of 10?

13.10   Use the circuit of Fig. E13.8 to determine the CMRR, defined as in Exercise 13.9, under the condition that $\Delta R = 0$ but β of $Q_4 = 60$ while that of Q_3 is 50. Assume $R_o = 2.4$ MΩ.

13.11   What is the effect on amplifier gain of short-circuiting one, or the other, or both of R_1 and R_2 in Fig. 13.1 (Refer to Fig. 13.7.)

13.12   We wish to determine the loop gain of the common-mode feedback loop consisting of Q_8-Q_{10} and Q_1-Q_4. This can be done by breaking the loop at the connection between the collectors of Q_1 and Q_2 and the current mirror. Apply an input current signal to the Q_8-Q_9 mirror and find the returned current signal in the common collector terminal of Q_1 and Q_2, then determine the loop gain. For this purpose the constant-current source transistor Q_{10} can be omitted and a resistance $R_o = 2.4$ MΩ placed between node Y and ground (to model the output resistances of Q_9 and Q_{10}). Using the result of Problem 13.10, find the CMRR taking into account the effect of negative feedback.

13.13   Consider the situation in which an emitter-follower class A output stage with constant-current bias is driven by a 5-V peak-to-peak square wave with zero average. A ± 5-V power supply is used with a bias current of 20 mA. What is the lowest resistance load that can be driven? For this load, what is the signal output power. What is the total power-supply power consumption? What is the average power lost in the follower? What is the efficiency under these conditions?

13.14   Consider a complementary follower class B output amplifier driving a 100-Ω load and utilizing ± 5 V supplies when driven by a 4-V peak-to-peak square wave having zero average value. Find the power drawn from the two power supplies and the efficiency of the amplifier under these signal conditions. What does the efficiency become for a 6-V peak-to-peak signal?

13.15   In the class AB complementary output stage shown in Fig. 13.20, D_1 and D_2 are diode-connected transistors having one-tenth the saturation current of Q_N and Q_P. Choose the bias current supply so as to limit the variation in emitter-base voltage of either output transistor to 0.1 V under the conditions of a ± 5-V signal provided across a 100-Ω load. Assume $\beta = \infty$.

13.16   A circuit which can be used in place of simple diodes to bias the complementary followers of a class AB output stage is shown. Compare and contrast this circuit with one formed using two similar transistors diode-connected in series. Assume that a bias current of 1 mA is available, that $I_S = 1 \times 10^{-14}$ A and that β is 100 or more. Further, assume that the base current required by the output devices varies from 0 to 0.9 mA. In your design utilize 0.05 mA in the resistor network. In each case calculate the extreme variation in the voltage A to B, and the equivalent series resistance of the device at each extreme.

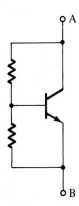

Fig. P13.16

13.17 Consider the two-terminal circuit of Fig. E13.12, in which the dc bias current flowing from A to A' is 180 μA, in the event that R_{10} is (a) 10 kΩ or (b) 80 kΩ. For $I_{S18} = I_{S19} = 10^{-14}$ A and $\beta = 100$, find the equivalent voltage and the incremental resistance of the two-terminal network.

13.18 The amplifier of Fig. 13.1 is to be operated from ± 9-V supplies. What does the open-loop gain of the amplifier become under these conditions?

13.19 As indicated in Section 13.7, the compensated amplifier has a phase margin of 80° for unity gain. At the unity-gain frequency, the pole at 4.1 Hz provides about 90° of phase. The excess phase can be considered to be due to a second pole beyond f_t. Find the frequency of this second pole. Now find a new value for the compensating capacitor C_C that would reduce the phase margin at the new f_t to 45°. Assume that changing C_C does not affect the location of the second pole. What is the 3-dB frequency of the modified op amp? Compare its open-loop gain at 1 kHz to that of the original op amp.

13.20 What is the bandwidth of a 741 op amp connected as a unity-gain follower for input sine-wave signals of ± 1 V peak amplitude. To what frequency will ± 1 V symmetric triangle waves be faithfully reproduced?

CHAPTER 14

14.1 Calculate the values of ω_0 and Q if the poles of a second-order transfer function are:
(a) $s = -0.5$ and $s = -2$
(b) $s = -1$ and $s = -1$
(c) $s = -1/\sqrt{2} \pm j\,1/\sqrt{2}$
(d) $s = -0.1 \pm j0.995$
(e) $s = \pm j1$

14.2 Consider the low-pass transfer function of Eq. (14.2) evaluated for physical frequencies $s = j\omega$. Show that $|T(j\omega)|$ exhibits a peak at $\omega = \omega_0 \sqrt{1 - (1/2Q^2)}$ provided that $Q > 1/\sqrt{2}$. Also show that at the peak,

$$|T(j\omega)| = \frac{n_0 Q}{\omega_0^2 \sqrt{1 - (1/4Q^2)}}$$

14.3   Give the transfer function $T(s)$ for a second-order low-pass filter having: dc gain = 10 and maximum gain = 20 at $\omega = 1$ rad/s. Specify the value of ω_0 and Q. (*Hint:* Use the results of Problem 14.2.)

14.4   For the high-pass transfer function of Eq. (14.3), find the condition for $|T(j\omega)|$ to exhibit a peak, the frequency of the peak, and $|T(j\omega)|$ at the peak.

14.5   Show that for the second-order bandpass function of Eq. (14.4), $|T(j\omega)|$ is greatest at $\omega = \omega_0$. Find $|T(j\omega_0)|$ and the frequencies ω_1 and ω_2 for which $|T(j\omega_1)| = |T(j\omega_2)| = |T(j\omega_0)|/\sqrt{2}$. Hence find the 3-dB bandwidth ($\omega_2 - \omega_1$).

14.6   Give the transfer function of a second-order bandpass filter having a center frequency of 1,000 rad/s, a center-frequency gain of 10, and a 3-dB bandwidth of 50 rad/s. For this filter calculate the magnitude of transmission at $\omega = 100$ rad/s and at $\omega = 10^4$ rad/s.

14.7   Plot the phase response of a second-order all-pass filter having $\omega_0 = 1$ rad/s and $Q = 1$.

14.8   For the bandpass circuit of Fig. 14.3c, let $C = 10$ nF. Find the values of L and R that will result in a center-frequency of 10^4 rad/s and a 3-dB bandwidth of 10^3 rad/s.

14.9   Derive the transfer functions of the bridged-T networks of Fig. 14.5.

14.10   Consider the connection of the bridged-T network of Fig. 14.5b in the negative feedback path of an infinite-gain op amp. Let $R_1 = R_2 = R$, $C_4 = C$, and $C_3 = C/m$. Find m and (CR) so that the closed loop has a pair of complex conjugate poles characterized by ω_0 and Q.

14.11   Design the circuit of Fig. 14.6 to realize a pair of poles with $\omega_0 = 10^4$ rad/s and $Q = 1/\sqrt{2}$. Use $C_1 = C_2 = 1$ nF.

14.12   Design the circuit of Problem 14.10 to realize a pair of poles with $\omega_0 = 10^4$ rad/s and $Q = 1$. Select $R_1 = R_2 = 10$ kΩ.

14.13   Consider the bridged-T network of Fig. 14.5a with $C_1 = C_2 = 1$ F, $R_3 = 1$ Ω, and $R_4 = 0.5$ Ω. Find the zeros and poles of this network. If the network is placed in the negative-feedback path of an infinite-gain op amp (as in Fig. 14.6), find ω_0 and Q of the poles of the closed-loop circuit.

14.14   Consider the bandpass circuit shown in Fig. 14.7. Let $C_1 = C_2 = C$, $R_3 = R$, $R_4 = R/4Q^2$, $CR = 2Q/\omega_0$, and $\alpha = 1$. Let the input signal V_i also be applied to the positive input terminal of the op amp through a voltage divider consisting of a series resistance R_1 and a resistance R_2 between the positive terminal and ground. Find the voltage divider ratio $R_2/(R_1 + R_2)$ so that the circuit realizes a second-order all-pass function. What is the magnitude of transmission of the all-pass filter?

14.15   Repeat Problem 14.14 to realize a notch filter. What is the frequency of the notch?

14.16   Consider the feedback loop of Problem 14.10. Split the capacitance C_4 into two parts: αC_4, connected to an input voltage V_i; and $(1 - \alpha)C_4$, connected to ground. Derive an expression for the transfer function V_o/V_i.

14.17   Consider the circuit of Fig. 14.6 with $C_1 = C_2 = C$, $R_3 = R$, $R_4 = R/4Q^2$, and $CR = 2Q/\omega_0$. Unground the positive input terminal of the op amp and connect it to terminal d of the network shown in Fig. P14.17. Find the transfer function $V_o(s)/V_i(s)$.

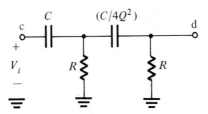

Fig. P14.17

14.18 Using the circuit of Fig. 14.7, design a bandpass filter having a center frequency of 10^4 rad/s, a Q of 5 and a center-frequency gain of 10. Use $C_1 = C_2 = 1$ nF.

14.19 Derive the transfer function of the circuit in Fig. 14.10b. Thus show that the circuit realizes a high-pass function. What is the high-frequency gain of the circuit? Design the circuit for a maximally flat response with a 3-dB frequency of 10^4 rad/s. Use $C_1 = C_2 = 1$ nF. (*Hint:* For a maximally flat response, $Q = 1/\sqrt{2}$ and $\omega_{3dB} = \omega_0$.)

14.20 This is a continuation of Problem 14.10 and Exercise 14.4. Design the resulting low-pass circuit to obtain a dc gain of unity and maximally flat response with a 3-dB frequency of 1,000 rad/s. Use $R_1 = R_2 = 10$ kΩ. (*Hint:* For a maximally flat response, $Q = 1/\sqrt{2}$ and $\omega_{3dB} = \omega_0$.)

14.21 Evaluate the sensitivities of ω_0 and Q relative to R, L, and C of the bandpass circuit in Fig. 14.3c.

14.22 Verify the following sensitivity identities:
(a) If $y = uv$, then $S_x^y = S_x^u + S_x^v$.
(b) If $y = u/v$, then $S_x^y = S_x^u - S_x^v$.
(c) If $y = ku$, where k is a constant, then $S_x^y = S_x^u$.
(d) If $y = u^n$, where n is a constant, then $S_x^y = nS_x^u$.
(e) If $y = f_1(u)$ and $u = f_2(x)$, then $S_x^y = S_u^y \cdot S_x^u$.

14.23 For the feedback loop generated by placing the bridged-T of Fig. 14.5b in the negative feedback of an op amp, find the sensitivities of ω_0 and Q relative to all passive components, for the design in which $R_1 = R_2$.

14.24 Use the fact illustrated in Fig. 14.8 to show that if $t(s)$ is a bandpass function with a center-frequency gain of 2, then an all-pass function can be obtained by interchanging the input and ground terminals. Apply this process to the two-amplifier bandpass of Fig. 14.12a (with $R_1 = R_3 = R_4 = R_5 = R$, $C_2 = C_6 = C$, and $R_7 = QR$). Sketch the circuit of the resulting all-pass.

14.25 What filter function is realized by interchanging R_7 and C_6 in the circuit of Fig. 14.12a?

14.26 Use the circuit of Fig. 14.13b to realize a bandpass filter with $f_0 = 10$ kHz, $Q = 20$, and unity center-frequency gain. For $R = 10$ kΩ, find suitable values for C, R_1, R_2, and R_3.

14.27 Use the circuit of Fig. 14.14b to realize a low-pass filter with $f_0 = 10$ kHz, $Q = 1/\sqrt{2}$, and unity low-frequency gain. If $R = 10$ kΩ, give the values of C, R_d, and R_g.

14.28   Consider the circuit of Fig. 14.13b augmented as follows. The outputs V_{1p}, V_{bp}, and V_{hp} are fed to a summing op amp through resistors R_4, R_5 and R_6, respectively. The feedback resistance of the summing op amp is R_7. Derive an expression for the transfer function from the circuit input to the output of the summing op amp. Show that to obtain a realization for the notch function of Eq. (14.5) the following resistance values may be used: R_6 = arbitrary, $R_5 = \infty$, $R_4 = R_6(\omega_0/\omega_n)^2$, and $R_7 = R_6[n_2/(2 - 1/Q)]$.

14.29   Repeat Exercise 14.7 for $f_0 = 10$ kHz, $Q = 10$ and for $f_0 = 5$ kHz, $Q = 20$.

14.30   Design the circuit of Fig. 14.17b to realize at the output of the second (noninverting) integrator a maximally flat low-pass function with $\omega_{3dB} = 10^4$ rad/s and unity dc gain. Use a clock frequency $f_c = 100$ kHz and select $C_1 = C_2 = 10$ pF. Give the values of C_3, C_4, C_5 and C_6. (*Hint:* For a maximally flat response, $Q = 1/\sqrt{2}$ and $\omega_{3dB} = \omega_0$.)

14.31   The circuit shown in Fig. P14.31 is known as a *Colpitts oscillator*. Assuming that the input impedance of the FET amplifier is very large, break the loop at the node marked X and determine the loop gain (V_{c2}/V_{gs}). Hence, find the frequency of oscillation ω_0 and the minimum gain ($g_m R_d$) required for the circuit to oscillate. Assume that the FET is adequately characterized by an ideal voltage-controlled current-source model.

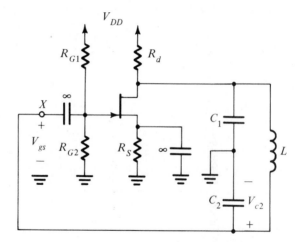

Fig. P14.31

14.32   If the circuit of Fig. 14.20 is modified to raise the impedance of the Z_s network by a factor of 10, while retaining its time constant equal to CR, what is the value of R_2/R_1 required for oscillation?

14.33   In the vicinity of the frequency ω_0, compare the phase sensitivity of the basic oscillator in Fig. 14.20 with the modified one suggested in Problem 14.32.

14.34   The frequency-selective circuit of Fig. 14.21 is modified by reducing the capacitors to 1,600 pF. What is the expected frequency of oscillation? In addition, R_4 and R_5 are reduced to 500 Ω. What is the amplitude of the resulting sine wave for a diode drop of 0.7 V?

14.35   Repeat Exercise 14.10 with a resistor $R = 10$ kΩ connected in series with the rightmost capacitor C. What is the frequency of oscillation and what minimum R_F is required?

14.36   Calculate the loop gain of two related circuits. The first is that in Fig. 14.23 and the second includes a fourth RC section. In each case find an expression for the frequency of oscillation and the required gain. Evaluate for each circuit the phase sensitivity as a function of frequency around ω_0 (that is, find $d\phi/d\omega$ at $\omega = \omega_0$).

14.37   Consider the quadrature oscillator circuit of Fig. 14.25 without the limiter. Let the resistance R_f be equal to $2R/(1 + \Delta)$ when $\Delta \ll 1$. Show that the poles of the characteristic equation are in the right-half s-plane and given by $s \simeq (1/CR)[(\Delta/4) \pm j]$.

14.38   Design the circuit of Fig. 14.27 to operate at 10 kHz with a 3.6-V peak-to-peak amplitude using capacitors of 16 nF. Utilize a Q of 20.

14.39   Assuming that the diode-clipped waveform in Problem 14.38 is nearly an ideal square wave and that the resonator Q is 20, provide an estimate of the distortion in the output sine wave by calculating the magnitude (relative to the fundamental) of (a) the second harmonic, (b) the third harmonic, (c) the fifth harmonic, (d) the root mean square of harmonics to the tenth. Note that a square wave of amplitude V and frequency ω is represented by the series

$$\frac{4}{\pi}V(\cos \omega t - 1/3 \cos 3\omega t + 1/5 \cos 5\omega t - 1/7 \cos 7\omega t + \ldots)$$

CHAPTER 15

15.1   Consider a grounded-emitter *npn* transistor supplied with a small base current under the condition that the collector is:
  (a)   connected to a large positive voltage
  (b)   connected to the base
  (c)   connected to the emitter
  (d)   left open
  In each case use the EM equations and appropriate approximations to find expressions for the base-to-emitter voltage that results. For a device with $I_S = 10^{-14}$ A, $\beta_F = 100$, $\beta_R = 1$, and $I_B = 100$ μA, find numerical values for V_{BE}.

15.2   Calculate the collector-emitter saturation voltage of a transistor for which $\beta_F = 100$ and $\beta_R = 20$ operating with a forced β of 10. Repeat for $\beta_R = 2$.

15.3   Reconsider Problem 15.2 when the transistor is operated with its collector and emitter leads interchanged. What is the voltage drop between collector and emitter which results?

15.4   Calculate V_{CEsat} of a symmetric junction device for which $\beta_F = \beta_R = 25$ operating at a forced β of 5.

15.5   For a grounded-emitter transistor for which $\beta_F = 100$ and $\beta_R = 1$ in a circuit in which $I_B = 1$ mA and $\beta_{forced} = 10$, label all currents in the branches of the EM model shown in Fig. 15.1b.

15.6   Repeat Problem 15.5 for the situation in which the collector lead is cut while the base connection remains. Estimate the expected change in base-emitter voltage which results.

15.7   Use the EM equations to find the i-v characteristics of the base-emitter junction of a BJT with (a) the collector open and (b) the collector connected to the emitter.

15.8 Use Table 15.1 to calculate the incremental saturation resistance of the tabulated transistor for collector currents in the range of 40 to 60 mA and a base current of 2 mA.

15.9 For the circuit of Fig. 15.4, let $R_B = 10$ kΩ and $V_{CC} = V_I = +5$ V. Assuming $V_{BC} = 0.6$ V, $β_R = 1$ and $β_F = 100$, calculate approximate values for the emitter voltage in the case where (a) $R_C = 10$ kΩ and (b) $R_C = 100$ kΩ.

15.10 If in Example 15.1 a transistor with $β_F = 200$ and $β_R = 0.05$ is substituted, what value of V_{IH} would result if $V_{CEsat} = 0.2$ V is assured?

15.11 For the RTL gate of Fig. 15.5, what is the lowest $β_F$ that can be tolerated while maintaining $V_{IH} \leq 1$ V for devices for which the $β_F/β_R$ ratio is 500 and $V_{OL} \leq 0.2$ V?

15.12 In an RTL logic system, ground currents cause a voltage difference of 0.1 V to appear between the ground references of interconnected logic gates. What do the two noise margins for the interconnected gates become under these conditions?

15.13 With reference to Fig. 15.10, what change is required in the power-supply voltage to equalize the noise margins of an RTL gate with fan-out of 5?

15.14 Consider the circuit shown in Fig. P15.14 as the basis of an RTL-like logic with increased noise margins. For the BJT, $β_F = 50$, $β_R = 0.1$ and $V_{BE} = 0.7$ V. What voltage V_{CC} is required to equalize the noise margins for a fan-out of 5? What is the resulting noise margin?

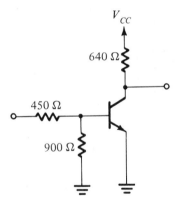

Fig. P15.14

15.15 Note that Eq. (15.15) implies that the base input of a transistor in the *active* region acts as a junction of reduced area whose *scale* current is $1/β_F$ times that of the actual base-emitter diode. However, when the transistor is saturated (consider the limiting case of the collector open), the junction itself is available. On the basis of this observation, calculate the reduction in base-emitter voltage at a fixed value of I_B, which will be observed when a transistor is allowed to (or made to) saturate. Assume $β_F = 50$.

NOTE: Before attempting Problems 15.16 to 15.20, the reader is advised to review the material on transistor switching times given in Section 9.8.

15.16 To reduce storage delay in RTL, a variant called resistor-capacitor transistor logic (RCTL) has been introduced. Here the capacitor serves to remove stored charge

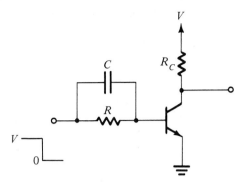

Fig. P15.16

directly from the base of the BJT. However, if the capacitor is too large, it seriously loads the driving gate. In the circuit shown in Fig. P15.16, first find the amount of charge that C is capable of instantaneously removing from the transistor base when the negative step input is applied. Then by equating this to the excess charge stored in the base, Q_x, find the required value of the "speed up" capacitor C.

15.17   When well-saturated at a low forced β, a particular class of BJT is known to have a storage delay time of 10 ns when its forward base current (I_{B2} in Fig. 9.40) is exchanged for a reverse current (I_{B1} in Fig. 9.40) of equal magnitude. It is to be used in a circuit in which the forward base current is 1 mA while the reverse current is limited to 0.25 mA. What is the storage delay time to be expected?

15.18   The transistor described in the previous problem, for which V_{BE} is 0.7 V, is used to switch a small collector load current. Its base is controlled via a 1-kΩ base resistor from a source which falls from $+1$ to $+0.2$ V. Estimate the storage delay time to be observed.

15.19   Someone unaware of the subtleties of BJT storage delay proposes to use the switch described in Problem 15.18 under control of a signal that he believes to be more than adequate. This signal falls from $+10$ to 0 V. Estimate the storage delay he will experience.

15.20   If the BJT switch described in Problems 15.17 and 15.18 is required to switch a collector current of 20 mA, estimate the resulting storage delay. $\beta_F = 100$.

15.21   Commercial-grade ICs generally are specified over a temperature range of 0 to 70°C. On the basis of the data on thresholds provided in the text for operation of RTL at 25°C and the fact that junction voltages fall 2 mV for each °C rise in temperature, calculate V_{OL}, V_{IL}, V_{IH}, V_{OH}, $\Delta 0$, and $\Delta 1$ at the extreme commercial temperatures.

15.22   A low-powered circuit resembling that in Fig. 15.5 is being considered for a battery-powered application. It is proposed to raise the impedance level of the circuit by a factor of 4. For a fan-out N = 5 and equivalent base input capacitance of 5 pF (independent of v_{BE}), calculate the time taken in the new system for a driven gate to reach cutin (0.6 V) from the moment the output of the driving gate rises from $+0.2$ V.

15.23   Five RTL inverters with component values as shown in Fig. 5.5 are connected in a ring to form an oscillator. The oscillator operates at 15 MHz. What is the average propagation delay of each gate? Describe the waveform at each gate output. Sketch two cycles of the waveforms of the five stages in correct relative positions. (*Note:* The idea of the ring oscillator is explored in Problem 6.18.)

15.24 For the ring oscillator in Problem 15.23, calculate the power provided by the 3-V supply: (a) ignoring dynamic effects, and (b) assuming that the need for switching power can be modeled by a 5-pF capacitor connected to the output of each inverter.

15.25 If in the ring oscillator of Problem 15.23, the values of all resistances used are raised by a factor of 4 over the original, what do the "static" and dynamic power consumptions become? Note for the latter part of the question that the oscillation frequency will decrease. Assuming that the dominant capacitive attributes of the gate are current-level independent, estimate the new frequency.

15.26 Figure P15.26 depicts a special RTL-derived circuit which can be used under special conditions, for example, internal to an IC where the terminal conditions are well known. For normal RTL signal levels, what is the logic function performed? Using the threshold values for RTL shown in Fig. 15.10, sketch and label the transfer characteristic of the circuit from A to C with (a) β high; (b) β low.

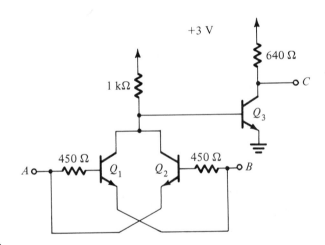

Fig. P15.26

15.27 Sketch an I²L circuit that realizes the Exclusive OR function, $C = A\bar{B} + \bar{A}B$.

15.28 The Equivalence function $D = AB + \bar{A}\bar{B}$ can be realized by adding an inverter to the I²L circuit of Problem 15.27. The additional delay of the inverter, however, slows down the operation. Show an alternative, faster realization of the Equivalence function. Now suppose that we wish to realize both the Exclusive OR and the Equivalence functions simultaneously. Suggest two different approaches for accomplishing this task. Compare the two approaches using the following criteria:
(a)   speed, as measured by the number of cascaded stages
(b)   cost, as measured by the number of collectors and/or the number of injectors.

15.29 An I²L gate which must operate with a fan-out of 4 has a minimum forward β of 6. What is the highest level of forced β at which the device must work, measured as a fraction of β?

15.30 Consider an I²L gate with fan-out of 4 and an injection current I. Let the *npn* transistor have $\beta_F = 5$ and $\beta_R = 50$. Calculate the saturation voltage of each collector when each is loaded by a current I.

15.31 An I²L chip incorporating 1,000 *pnp* injectors uses a gate geometry characterized by a delay-power product of 1 pJ. Its injector operates at a voltage of about 0.8 V. A system is to be designed in which a cascade of five stages of logic must settle adequately during one cycle of a 3-MHz clock. What total injector current should be provided to the chip if a factor of safety of 1.5 is allowed for gate-to-gate speed variation?

15.32 Consider the D flip-flop of Fig. 15.22b implemented in 0.5-pJ I²L operated at an injection current I of 50 μA and an injection voltage of 0.8 V. For proper operation of this edge-triggered data flip-flop, the data input on the D line must be stable in the vicinity of the triggering edge (the negative going edge) of the clock input $\overline{C}$, for a time interval called the *hold time*. Hold time is positive if it extends past the clock edge, and negative when it precedes it. What is the hold time of the circuit shown for the conditions stated? Assume that the variability of gate delay is $\pm 10\%$. (*Hint:* For this purpose the critical part of the circuit consists of Q_4 and Q_8.) Also, how long after the clock edge do both outputs stabilize?

15.33 Consider the DTL gate shown in Fig. 15.30. Let $V_{BE} = 0.7$ V, $\beta = 50$, and let the smallest collector current of Q_3 (for a fan-out of 2) be 4.5 mA. Find a new value for R_B that results in the base overdrive current (that is, the base current in excess of the value required to saturate Q_3) being equal to the reverse turn-off current of Q_3.

15.34 Consider the input circuit of the DTL gate in Fig. 15.30 with $V_I = +5$ V, and a junction voltage of 0.7 V for Q_1, Q_2, D, and Q_3. Calculate the base current of Q_3 if β of Q_2 is (a) 25, (b) 100.

15.35 Consider the output circuit of the DTL gate of Fig. 15.30 when loaded by a capacitance of 20 pF and a second gate which can be modeled as a 3.8-kΩ resistor connected to a $+4.3$-V supply.
(a) Calculate the time taken by the output voltage to rise from $+0.2$ V to the threshold voltage of the driven gate (about $+1.2$ V).
(b) Calculate the 0 to 90% rise time of the output waveform. (Be careful to note that the input of the driven gate turns off beyond the 1.2-V threshold.)

15.36 Repeat the analysis done in Fig. 15.43 for standard TTL on a low-power TTL gate of the same topology for which the following resistance exchanges have been made: 4 to 40 kΩ, 1 to 12 kΩ; 1.6 to 20 kΩ; 130 to 500 Ω. What is the total current drawn from the 5-V supply?

15.37 For the low-power TTL gate of Problem 15.36, find the input current when the input voltage is low at $+0.2$ V. What is the total current drawn from the supply in this state, assuming that the gate is not loaded?

15.38 For the low-power TTL gate of Problem 15.36 find:
(a) the output short-circuit current to ground for output high
(b) the output short-circuit current to $+5$ V for output low
Assume $V_{BE} = 0.7$ V and $\beta = 30$, and use the results of Problem 15.36.

15.39 In the low-output state, the output switch of a low-power TTL gate is characterized as follows: When sinking 1 mA, the output is 0.13 V; when sinking 12 mA, the output is 0.49 V. Estimate the saturation resistance and the offset voltage of the grounded-emitter output device.

15.40 In Fig. 15.45, assuming for Q_4 that $\beta = 50$ and the edge of saturation occurs at $V_{CEsat} = 0.3$ V, find the value of v_O at which Q_4 just saturates. Assume junction voltages to be 0.7 V.

15.41 For the low-power TTL gate described in Problem 15.36, find the maximum current that can be sourced by the output terminal while V_{OH} remains greater than 2.4 V. The β of Q_4 is 50.

15.42 Proceed to find a second estimate of the slope of the segment CD of the transfer characteristic in Fig. 15.46 as follows: Find the dc current in each of Q_2, Q_3, and Q_4 so that the output voltage v_O is somewhere in the middle of the range, say at $+2.0$ V. Use these dc bias currents to determine the small-signal parameters of the devices. Then find the gain from the base of Q_2 to the output through the Q_2, Q_3 path; and the gain from the base of Q_2 to the output through the Q_2, Q_4 path. Sum these two gain values to obtain the overall gain from the base of Q_2 to the output. Assuming no loss of gain in the saturated transistor Q_1, the gain value obtained gives a good estimate of the slope of CD. Assume the junction voltages to be 0.7 V and $\beta = 30$.

15.43 For a low-power TTL gate as described in Problems 15.36 and 15.37, calculate the total supply current and power consumption with the input (a) low, (b) high. What is the average power dissipation of the gate? If the delay-power product of low-power TTL is said to be 40 pJ, what propagation delay would you expect the gate to show?

15.44 A TTL gate characterized statically by 5-mW dissipation in the input low state and 17 mW in the input-high state is found to show an increase of 0.6 mW when operating at 10 MHz. What is the total device power dissipation at 10 MHz? What is the dissipation likely to be at 20 MHz? If the additional (dynamic) loss at 20 MHz results from a cyclic short circuit in the output totem pole, how long must it last during each cycle? Recall that the totem pole consists of a 130-Ω resistor, two forward-biased junctions (0.7 V), and a saturated BJT (0.3 V) across a 5-V supply.

15.45 Two conventional TTL gates with totem pole outputs are wire ORed by accident. What is the short-circuit current that can flow if all transistors have a β of 30? What voltage results on the connection? If the β of the device on the up side is 100, what does the voltage become?

15.46 In the circuit of Fig. E15.23, what is the maximum input current required to enable the tristate function?

15.47 (a) Sketch the circuit of an open-collector TTL inverter.
(b) Use two inverters of the type in (a) above with an output resistor to realize the function $\overline{AB}$.
(c) Use the circuit in (b) with part of the circuit in Fig. 15.49 to obtain an economical realization of the Exclusive OR function $Y = A\overline{B} + \overline{A}B$.

15.48 For the circuit of Fig. 15.50, what are the input and supply currents with input (a) low ($+0.2$ V) and (b) high. $\beta_F = 50$, $\beta_R = 0.02$, $V_{BE} = 0.7$ V, and $V_{CEsat} = 0.2$ V.

15.49 Consider the circuit in Fig. 15.50. The output when high is accidentally grounded. What is the current that flows if β of all devices is quite high? What is the lowest β (of all devices at once) at which the current reaches 90% of this value?

15.50 Sketch the i-v characteristic of the two-terminal active pull-down formed by Q_6 and its two associated resistors (see Fig. 15.50). Assume that for Q_6, $I_S = 10^{-15}$ A. Calculate the voltage values at currents of 0.1, 0.5, 1.0, 1.5, and 2.0 mA. Assume $\beta_F = 20$ and $\beta_R = 5$.

15.51 Consider Fig. 15.54. Sketch the output characteristic, v_O versus i_L, of the gate with the output high, as the current i_L extracted from the output terminal increases from zero. Devices have $V_{BE} = 0.7$ V, $V_D = 0.4$ V, and $\beta = 30$. (Note that the Schottky transistor Q_5 will saturate when its v_{CB} reaches 0.4 V.) What is the output current with the output shorted to ground?

15.52 Using the logic/circuit flexibility of ECL indicated by Figs. 15.55 and 15.65, draw the circuit of an Exclusive OR ECL gate. *Hint:* $\overline{A}B + A\overline{B} = \overline{A + B} + \overline{A} + B$.

15.53 For the circuit of Fig. 15.57 with $V_A = V_{BB}$ and a small-signal applied to A, find the gain from A to C and the gain from A to D. Take into account the loading effects of the emitter followers. Use $\beta = 100$.

15.54 For signals whose rise and fall times are 3.5 ns, what length of unterminated gate-to-gate wire interconnect can be used if a rise-time to return-time ratio of 5 to 1 is required? Assume the environment of the wire to be such that the signal propagates at two-thirds the speed of light (which is 30 cm/ns).

15.55 For the symmetrically loaded circuit of Fig. 15.57 and for typical output signal levels ($V_{OH} = -0.88$ V and $V_{OL} = -1.77$ V), calculate the power lost in both load resistors R_T and both output followers. What then is the total power dissipation of a single ECL gate including symmetrical output terminations?

15.56 The bias circuit of Fig. E15.28 serves to generate a voltage $V_{BB} = -1.29$ V at 25°C. Assuming D_1 and D_2 to be diode-connected transistors identical to Q_1, what must the junction voltage at 1 mA at 25°C actually be? (Note that Exercise 15.28 indicated it is not exactly 0.75 V.)

15.57 A particular CMOS gate operated at 15 V will provide 9.8 mA when its output is shorted to the opposite supply but only 0.8 mA when operated at 5 V. Calculate values of V_T and β for its MOS devices.

15.58 Consider a three-input CMOS NOR resembling that shown in Fig. 15.70a. Calculate the switching threshold with 1, 2, or 3 inputs connected in parallel while the remainder are grounded. Use $V_{DD} = 10$ V and $V_T = 2$ V. All devices are assumed matched.

15.59 At what frequency will a ring oscillator formed of five CMOS inverters powered by a 10-V supply oscillate? To calculate the gate delay, use the formula given in the text and assume that the load of each gate on the preceding gate in the ring is 5 pF.

15.60 In 1 ms a CMOS gate is cycled from ON to OFF to ON again. The gate operates from a 10-V supply and has a 50-pF load. Calculate the amount of charge supplied by the 10-V supply in 1 ms and, hence, the average current required from the supply for periodic operation, expressed in μA/kHz.

15.61 In a CMOS inverter operated from a 10-V supply, both devices have 2-V thresholds. The n-channel device has $\beta = 0.25$ mA/V^2 while the p device has $\beta = 0.125$ mA/V^2. For an input signal which ramps linearly from 0 to $+10$ V in 10 ns, sketch the current which flows between devices. If the inverter is operated at 10 MHz with an input square-wave signal having approximately linear ramp edges with 10-ns rise and fall times, estimate a value for the average current drawn from the supply. Ignore capacitive effects.

15.62 For CMOS gates formed of devices as described in Problem 15.61, find the short-circuit current that would result if the output terminals of two gates in different states were mistakenly joined. What output voltage results?

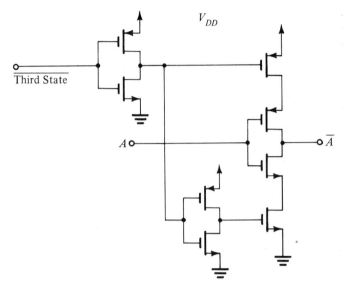

Fig. P15.63

15.63   Show that the CMOS circuit in Fig. P15.63 functions as a tristate inverter.

CHAPTER 16

16.1   Calculate the number of binary-coded lines required to address each of the following memories: 256×4, $1K \times 1$, $1K \times 4$, $4K \times 1$, $4K \times 4$, $16K \times 1$, $64K \times 1$.

16.2   A design for a digital memory array of 2^N cells is being considered. Various arrangements of row and column drivers (which are the circuits that connect the outputs of the decoders to the rows and columns) are available. In system A each driver has unit cost. In system B, the cost of row drivers is 1.5 units, while the corresponding column drivers cost 0.75 units. Considering the cost of driving alone, what are the dimensions and total cost of the lowest cost array in each system? (For N odd and N even.)

16.3   Further consideration of the situation in Problem 16.2 causes the designer to realize that he must examine a larger part of the system in his optimization. Thus he realizes that the cost of a sense amplifier for each column must be included. In system A, a sense amplifier (per column) costs 1 unit, while in system B, it costs 1.25 units. If the total cost of drivers and amplifiers is to be minimized, what is the minimum cost and what are the dimensions of the corresponding arrays for systems A and B?

16.4   In a dynamic MOS memory, the chain of circuitry beginning at the storage capacitor and ending at the digit output pin of the IC package may be thought of as a power amplifier. If the input is thought to be an 8-V change on the 0.05-pF storage capacitor and the output to be a 5-V change on a 50-pF output line, what is the power gain corresponding.

16.5   Consider the operation of the 16K RAM package shown in Fig. 16.6 during one cycle of a READ operation. For each signal lead of the package, note the number of signal level transitions during each cycle. For the entire package, what is the total number of external signal changes per cycle? If the external levels are 0 and 5 V and the external line capacitance is 50 pF, what is the total energy required for each cycle? What is the total power dissipated in a continuous sequence of READ cycles initiated at 320 ns intervals?

16.6   A memory chip with a 200-ns cycle and operating from a single 5-V supply, consumes 100 mW at low speeds and 500 mW at full speed. Assuming each cycle consists of four signal transitions on all active internal lines, what must the effective capacitance of the network within the chip be? If voltage multipliers are used internally to raise internal signals to 10 V, what do you estimate the equivalent capacitance to be? What might you conclude about memory cell packing density and chip capacity in each case?

16.7   A 16K × 1 dynamic MOS RAM with a READ cycle of 320 ns, refreshes 128 cells for each row address applied. If the entire memory must be refreshed every 2 ms and refresh is done all at once "in burst mode," what percentage of the time is the memory available for regular operation?

16.8   An add-to-register operation in a particular microprocessor requires four memory cycles each of which retrieves or stores one byte: The first retrieves the instruction byte from ROM. The second and third retrieve the (two byte) address of the operand from ROM. The fourth retrieves the operand from RAM. If the ROM cycle time is 100 ns and the RAM cycle is 400 ns, what is the shortest time in which the add operation could be done by an infinitely fast microprocessor? However, in parallel with each access the microprocessor must complete a corresponding cycle of activity before the next step can be taken. If each such μP cycle is 250 ns, what time is taken by the add operation. What changes in memory cycle time may permit a faster more economical system?

16.9   A memory chip dissipates 600 mW at an operating frequency of 4 MHz and 400 mW at 2 MHz. What is the power dissipation of the chip when operating at 100 kHz? Express power dissipation P as a function of operating frequency f for this chip.

16.10   While following the description of the timing of the READ cycle for signals applied to the pins of the Mostek MK4116 chip, prepare a timing diagram in which critical delays are labeled and causes and effects are indicated by arrowed connectives. Note that barred signals are active when low. (The answer can be found in the data sheet of the MK 4116, Ref. 16.4.)

16.11   As indicated in the description of the READ cycle of the MK4116, one aspect of the relative timing of the $\overline{RAS}$ and $\overline{CAS}$ signals is handled internally. Show that the simple circuit shown in Fig. P16.11 implements the desired timing relationship and produces

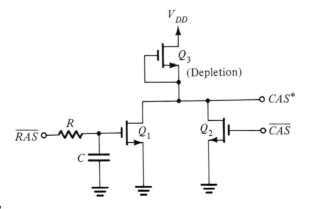

Fig. P16.11

a version of $\overline{CAS}$ called CAS^* which follows $\overline{RAS}$ by at least 50 ns. Assume signal transitions of 5 V and device thresholds of 1 V. Find the required value for R if $C = 0.5$ pF.

16.12 Consider the circuit of Fig. 16.10 with V_T of Q_1, Q_2, and Q_3 being 1 V. What is the voltage at node C if A has been high for a time? If, with A at $+5$ V, the voltage at B is 0.5 V and the power supply current is 1.0 mA, what is β_3 of Q_3 and β_1 of Q_1? To what voltage does C rise when A falls to ground if there is no load on C or B and C_B is relatively large? To what voltage does C rise if (for large C_B) there is a 0.1-mA load to ground at node B? If B is unloaded and C is loaded only by a capacitance of 1 pF, what capacitance must C_B have to ensure at least $+18$ V transiently at C when A goes low?

16.13 Consider the input section of the address buffer of Fig. 16.11 for the case where $C = 0.1$ pF, $V_T = 1$ V, and the input high level is $+5$ V. If the address hold time is as little as 20 ns, what is the minimum β of Q_1 necessary to ensure a signal of at least 3 V at node A following the return of ϕ_1 to ground from its upper level of 12 V? [*Hint:* $\int dx/(ax^2 - x) = \ln(1 - 1/ax)$].

16.14 In the circuit of Fig. 16.11, if leakage currents from C can be as large as 10 nA at elevated temperatures and a voltage change of at most 0.5 V in 1 μs can be tolerated, what value of C is necessary?

16.15 Consider the regenerative part of Fig. 16.11, with the flip-flop consisting of Q_4, Q_5, Q_6, Q_9, and Q_{11} modeled as a positive-feedback amplifier that presents a negative input resistance of $-R$ ohms to the Q_2, Q_3 circuit. Now consider the operation as Q_2 turns on and assume that the combination of Q_2 and Q_3 can be modeled as a current source having a step waveform with amplitude I and having an output resistance R_o. As well, let the total capacitance between node B and ground be denoted C. For $R = 100$ kΩ, $R_o = 10$ MΩ, $C = 0.2$ pF, and $I = 10$ μA, calculate the time taken for the voltage at node B to fall by 10 V.

16.16 Consider the operation of the flip-flop in the circuit of Fig. 16.11 just as ϕ_2 goes high. Assume that Q_{11} turns on instantaneously, thus lowering the voltage at the common-source terminal of Q_4 and Q_5 to $+0.5$ V. Simultaneously, assume that the drains of Q_4 and Q_5 are still at equal voltages, denoted V. At this time, Q_7 and Q_8 are off while Q_6 and Q_9 act as the load devices of the flip-flop. The flip-flop behaves instantaneously as a positive-feedback amplifier that presents a negative input resistance between node B and ground. Assuming that $V_T = 1$ V, $\beta_6 = \beta_9 = \frac{1}{2}\beta_4 = \frac{1}{2}\beta_5$, and the high level of ϕ_2 is $+12$ V, find the value of V. Also find the value of β_4 so that the incremental input resistance (seen by node B) is -10 kΩ. Also find the value of β_{11}.

16.17 A 1K $\times$ 1 static RAM utilizing a 5-V supply has a power dissipation of 800 mW when in operation and 200 mW in the standby mode. If standby current is assumed to be predominantly that required by the storage cells, what is the current used by each cell? If the cell design is that shown in Fig. 16.12a and the load devices have the same β as the switches, calculate β for all devices. The magnitude of the threshold voltage of the enhancement and of the depletion devices is 1 V. What are the logic levels achieved by this design?

16.18 In Fig. 16.12a calculate the maximum possible currents which can be made available to the digit lines without loss of state of flip-flop. For the flip-flop to remain stable, its drain voltages must remain outside the range ($V_T \pm 0.5$ V). Assume that all four transistors have $|V_T| = 1.0$ V and that β of Q_1 and Q_2 is 20 μA/V^2 and that of Q_3 and Q_4

is 2 $\mu A/V^2$. For a digit-line capacitance of 3 pF, what are the best rise and fall times to be expected on the digit lines?

16.19   For the memory cell in Fig. 16.12a, what is the ratio of β for switch and load that ensures that the maximum drain-source voltage with the switch on is 0.2 V. Assume $|V_T| = 1$ V.

16.20   For the memory cell in Fig. 16.12c with polysilicon resistors of 1 MΩ, what is the β of Q_1 and Q_2 required to ensure that $v_{DSon} \leq 0.1$ V? 0.2 V? What is the standby current and power in a collection of 16K such cells for a standby supply voltage of 5 V? Assume $V_T = 1$ V.

16.21   Consider the access arrangement shown in Fig. 16.14. When the column decoder input rises, the D_M and $\overline{D_M}$ lines are lowered by the currents through Q_{C1} and Q_{C2} required to charge the capacitance of the Data and $\overline{\text{Data}}$ lines. This will cause a high level on a D_M line to fall. If the resulting temporary low level on the D_M line is not to affect the state of the cell, it must remain above a critical level (say, $V_T + 0.5$ V or 1.5 V). Under these conditions, what is the greatest allowed ratio of capacitances of the Data line to the D_M line? Note that the problem arises only during the initial 1.5-V rise in the Data line since thereafter D_M and Data, connected through Q_C, both rise. $V_{CC} = +5$ V.

16.22   In Fig. 16.15 the Data out lines are pulled positive by a depletion load whose I_{DSS} is 1 mA. Transistors Q_7, Q_8, and Q_9 are identical with $V_T = 1$ V and a β large enough to guarantee a low level ≤ 0.5 V. What difference between the D_M and $\overline{D_M}$ lines centered around 2.5 V is sufficient to cause the currents in Q_7 and Q_8 to differ by a factor of 9? Assume that $V_{CC} = 5$ V and that the logic 1 levels are approximately $+5$ V.

16.23   In Fig. 16.16 the load devices are depletion and the switches enhancement with $|V_T| = 1$ V in each case and the power supply is 5 V. Select β values in the circuit to ensure that the low-level output is 0.1 V and the 10 to 90% rise and fall times with a 1-pF load are 20 ns or less. (*Hint:* The rise time is longer than the fall time.)

16.24   Using the gated NOR decoder circuit of Fig. 16.17 as a building block, show the circuit of a 2-bit decoder with a common chip-enable input.

16.25   In the circuit shown in Fig. 16.17 consider the voltages at each FET to FET connection with a single logic input at a high level. The input transistors have $V_T = +1$ V, the load has $V_T = -1$ V and the enable transistor has $V_T = 0$. What are the node voltages if the supply voltage and control voltages are 5 V and the β of all devices is the same? What change is necessary if the output signal must fall to $+0.1$ V?

16.26   In the circuit of Fig. 16.17, suppose that the depletion device is eliminated. What should the β of Q_s be so that the current when the input devices are on remains the same as in the original circuit? Give β of Q_s in terms of the β of the devices (all having the same β) of the original circuit. What is the disadvantage of this modified circuit? Let $V_{CC} = +5$ V and $|V_T|$ of the depletion device be 1 V.

16.27   It is desired to use one enable transistor to drive four NOR decoders of the type shown in Fig. 16.17. To keep the logic output levels relatively independent, find the β of the enable device (relative to that of the load device) which allows only a 1-V variation in the upper logic level as 0 to 4 of the decoders are activated. $|V_T|$ of all devices (except the enable transistor) is 1 V. The enable device has a zero threshold.

16.28   A (256 $\times$ 1) ROM can be used as a *parity generator* for single byte (8 bit) signals in a communications system. Sketch an arrangement which generates both even parity

(1 indicating that an even number of the 8 input bits is 1) and odd parity. Show the content of the bottom five and the top five addressed locations in ROM and of the location whose address is 117. Which of the two outputs required is the Exclusive OR of the input variables? What is the other output called?

16.29   Sketch the circuit of a Row Decoder suitable for driving a simple diode ROM (Fig. 16.20), using inverters and a resistor-loaded diode array.

16.30   Design the bit pattern to be stored in a (16 × 4) ROM which provides the 4-bit product of two 2-bit variables.

16.31   How large a ROM is required to accommodate a decimal table multiplier capable of providing the two-digit product of two BCD-coded decimal digits? (Be careful of unused states.)

16.32   A CCD memory chip consists of 128 registers each of 128 stages capable of operating at 2 MHz. What is the average access time for this memory?

16.33   The minimum operating frequency of the Intel 2416 CCD chip varies from 1 to 100 kHz over the normal operating temperature range. If this requirement can be attributed to the usual effect of temperature on junction leakage, what is the range of temperature (assumed centered at $25°C$) over which operation is intended?

16.34   Using a four-input two-output PLA of the type shown in Fig. 16.31, find the pattern of diodes and transistors to implement voting circuits which could be called "exactly 2 of 4" and "exactly 3 of 4." How many diodes and transistors are needed in each case?

16.35   Using a four-input, two-output PLA of the type shown in Fig. 16.31, find the pattern of diodes and transistors to implement voting circuits which could be called "2 or more of 4" and "3 or more of 4." How many diodes and transistors are required in each case? Is there a simpler version if a single inverter is available?

16.36   Using a four-input two-output PLA of the type shown in Fig. 16.31, show the device connections for a two-, three-, and four-input Exclusive OR. (*Hint:* The Exclusive OR of three variables is the EXOR of one with the result of the EXOR of the other two.)

16.37   A CMOS master-slice array contains 500 *p*- and *n*-channel device pairs having a common gate connection and separate source and drain connections. If a particular class of application uses building blocks in the following proportion, how large a system can one expect to put on the array assuming 72% overall device utilization: three NOT, five 2-input NAND, one 4-input NAND, one 3-input NOR, $\frac{1}{2}$ SR flip-flop, 1 D flip-flop with direct set and clear. State your answer in terms of the number of D flip-flops used. For each of the building blocks, use topologies introduced earlier in this text.

ANSWERS TO SELECTED PROBLEMS

CHAPTER 1

1.1 6.28 krad/s; 6.28 krad/s, 12.73 V; 0 V; 1/3; in phase.

1.3 5.5%; 50%.

1.4 12.73

1.6 0.25 W

1.9 16.7 kHz

CHAPTER 2

2.1 *WO:* 0; *OX:* 1 mA/V; *XY:* 2.5 mA/V; *p-p* amplitude = 1mA; average = 0.5 mA; With 1 kΩ in series: average = 0.25 mA and *p-p* = 0.5 mA.

2.2 0 mA; 0.5 mA; 0.125 mA

2.5 1 mV

2.6 20 mW

2.8 1 mW; ∞

2.10 10 μV, 10^{-11} W; 10^{-5} W/W; 1 V, 0.1 W; 1 V/V, ∞ W/W.

2.12 40 dB; 60 dB; 50 dB

2.13 44 dB

2.16 40 dB; 40 dB; 100 dB

2.18 0.99; 99; 20 Ω; 49.5 mA/V; 2 kΩ

2.21 $(R + r_x)/(\beta + 1)$

2.28 12.2 V; 222 Ω

2.31 2.21 mA

2.33 101 kΩ

2.37 $\tau = 4$ ms; $v_o = 5e^{-t/\tau}$; 12 ms

2.40 $\tau = 0.1$ ms; $\dfrac{V_o}{V_s} = -\dfrac{1000}{1 + j\omega\tau}$

2.42 100 V/V; 8 Hz; 12.5 V/V

2.44 0.1 mA; 10 μC; 100%

2.48 $i_o = 2e^{-2\times10^8 t}$

2.51 $K \simeq 2$; $f_0 \simeq 7$ MHz

2.54 $\tau \simeq 100$ ms; $10^{-4} \times V$ to $0.9 \times V$

CHAPTER 3

3.1 Circuit in Fig. 3.4 with $R_1 = 100$ kΩ and $R_2 = 1$ MΩ.

3.3 Circuit is an op amp with positive input grounded and a 10-kΩ resistor connected from the output to the negative input.

3.5   Miller integrator with $R = 10$ kΩ and $C = 25$ nF.

3.7 $C = 1$ nF and $R = 1$ MΩ.

3.9   Circuit is that in Fig. 3.44 with $C = 1$ μF, $R_1 = 10$ kΩ, and $R_2 = 100$ kΩ.

3.13   Circuit in Fig. 3.17 with $R_1 = 55/3$ kΩ, $R_2 = 550/3$ kΩ, $R_3 = 5/3$ kΩ, and $R_4 = 50/3$ kΩ.

3.15   Voltage measured $= 10$ mV.

3.16   Circuit as in Fig. 3.22 but with the feedback resistance changed to $R/0.9$ and with $R = 100$ kΩ and $C = 0.1$ μF.

3.20   1 Hz; 9900 Hz; $-5.8°$, $-45.3°$, $-84.3°$, $-89.4°$.

3.23 $A_0 = 209$ V/V; $f_{3\text{ dB}} = 1.84$ kHz.

3.24   61.8 V/V

3.25   Rise time (from 0 to 10 V) $= 15.9$ μs; 1.59-mV peak.

3.26   0.1 V/V.

3.30   With direct coupling: dc output $= 11$ mV and ac output $= 1$ mV; with capacitor coupling: dc output $= 1$ mV and ac output $= 1$ mV.

3.32 $R_1 = 100$ kΩ; $C_2 = 10$ nF; $R_2 = 10$ MΩ.

3.34 $R_1 = 10$ kΩ; $R_2 = 1$ MΩ; $C_1 = 0.159$ μF; $C_2 = 15.9$ pF; $f_t = 10$ MHz.

3.39   1 V; Waveform is a linear ramp starting at $t = 0$ at 1 V and reaching $+13$ V at $t = 12$ ns; 10s.

3.40   0 V; -10 mV; 1 Ω.

3.41   10 pA; outward.

3.42   For $A_0 \gg 1$, the low-frequency gain remains $(-R_2/R_1)$; R_3, however, reduces the 3-dB bandwidth of the closed-loop amplifier, $f_{3\text{ dB}} = f_t / \left[1 + \dfrac{R_2}{(R_1 \| R_3)} \right]$.

CHAPTER 4

4.1 $I_1 = 1$ mA, $V_1 = 0$ V; $I_2 = 0$ mA, $V_2 = -10$ V; $V_3 = +2$ V; $V_4 = 0$ V; $V_5 = 10$ V, $I_5 = 0.5$ mA; $I_6 = 0$ mA; $I_7 = 1$ mA, $V_7 = 0$ V; $V_8 = -5.3$ V; $V_9 = +3$ V, $I_9 = 1.3$ mA; $V_{10} = -2$ V; $V_{11} = 0$ V; $I_{12} = 0.125$ mA.

4.5   3.5 kΩ; the cathode.

4.8   (a) 0.1 mA; (b) 0.18 mA.

4.9   1.024 mA

4.10   (a) 17 mV; (b) 2.8 V.

4.14   472 mV.

4.17   0.83 mV; 5 μV.

4.18   300 Ω; 4.3 kΩ; $\pm 2\%$; $\pm 4.1\%$; the first.

4.21   For -4.65 V $\le v_I \le +4.65$ V: $v_O = v_I$;
for $v_I \ge +4.65$ V: $v_O = +4.65$ V;
for $v_I \le -4.65$ V: $v_O = -4.65$ V.

CHAPTER 5

5.1 $v_I = v_O + nV_T \ln(v_O/I_S R)$

5.2 $v_I = 1$ mV: $v_O = 1$ mV and $v_A = 0.356$ V; $v_I = 1$ V: $v_O = 1$ V and $v_A = 1.7$ V

5.5   2.564 V; 2.436 V; 2.5 V

5.9   1.57 A

5.10   3.6 V dc

5.11   0.09 mA; nothing happens

5.13   9.2 V; 9.23 V

5.14   14.14 V; 2.36 V; 12.96 V; 0.3 A

5.17 -5 V

5.22 ± 0.93 V

5.23   The output goes positive (to $+10$ V) for the interval during which the input is in the range ± 0.6 V (approximately); for the remainder of the time the output is negative (-10 V). Thus the output is a pulse waveform having a frequency twice that of the input triangular waveform.

CHAPTER 6

6.1

P	S	L	D
0	0	0	1
0	1	0	1
1	0	0	1
1	1	1	0

$L = PS$

6.2   Use the circuit of Fig. P5.12 with the negative power supply changed to -5 V.

6.3   (a) $D = AB$; (b) $D = A + B$; (c) $D = \overline{AB}$; (d) $D = \overline{A + B}$; Negative input grounded: $D = AB\overline{C} + \overline{A}BC + A\overline{B}C + ABC$; Positive input grounded: $D = $ NOT of previous expression.

6.4 $f_1 = B; f_2 = AB; f_3 = AC + \overline{B}(A + C)$

6.7 $V_{OL} = -10$ V; $V_{OH} = +10$ V; $V_{IL} = -1$ V; $V_{IH} = +1$ V, $\Delta 1 = 9$ V; $\Delta 0 = 9$ V.

6.8   3.6 mW; 10 μs; 10 μs

6.16   The output is a negative pulse (logic 0) of duration equal to the delay τ.

6.17   One-shot; 100 ns.

6.18   3.3 MHz; 13.3 ns

6.21   Circuit is bistable; when PB is pressed, output at C is reversed ($+10$ V to -10 V and vice versa); can be used as Push on/Push off.

6.26 $X_1 = A\overline{B} + C; X_2 = B$

CHAPTER 7

7.1 $|V_P|; |V_P|(1 - \sqrt{|V_P|/I_{DSS}R})$

7.3   100 Ω; 200 Ω

7.5 $+0.5$ V

7.9 $+1.5$ V; $+7.5$ V; -6 V

7.10 -6 V; $+6$ V; $+4$ V; $+6$ V; $+1$ V; $+ 1$ V; $+0.6$ V, 0 V; 0 V; 0 V; 0.61 mA; $+9.73$ V; 1 mA

7.16   100; 100 kΩ

7.17 -10

7.18   2 mA/V; -20; square wave of 2 V amplitude; 8 V

7.20   0 V; 0 V; $1/g_m$; no.

7.21   476 Ω; 0.91

7.22 -10; 500 Ω

7.23   (a) -40; (b) -3.84; (c) 0.89; (d) -1

7.28   (a) 5 ms; 73.5 μs

7.29 $I_{DSS}; |V_P|/4I_{DSS}$

CHAPTER 8

8.1 $+2$ V; 4 mA

8.2 -3 V; 18 mA

8.5 2 V

8.7 750 Ω

8.8 1.3 V; 1.1 V

8.10 2 V; 50 Ω; 1.95 V

8.11 1.7 V_T; 0.7 V_T

8.13 $V_M = 10 - V_T$

8.15 $\simeq 4$ V

8.18 1 mA; -4 V; -2 V to $+20$ V; V_S becomes -6 V.

8.20 5.1 Ω; 10.2 Ω; 0.5 V

8.22 $+3.09$ V; positive pulse of 1.635 V amplitude; 0.17 μA.

8.29 $+12$ V, 0 V, -12 V; 50 μA

8.33 Approximately 0 and V_{DD} volts; (a) 0.11 mA and (b) 0.41 mA.

8.34 $t_{f1} = 2C(V_T - 0.1V_{DD})/\beta(V_{DD} - V_T)^2$;

$$t_{f2} = \frac{C}{\beta(V_{DD} - V_T)} \ln\left(20\frac{V_{DD} - V_T}{V_{DD}} - 1\right)$$

8.35 See Fig. 15.70b

8.40 $+5$ V

8.46 $+4$ V; 8 V/V; 10 MΩ; 2.22 MΩ

8.48 $C = \overline{A + B}$; $+5$ V, 0.48 V, and 0.25 V

CHAPTER 9

9.4 0.99; 100; 10^{-15} A

9.6 Active; cutoff; saturation

9.8 21.25; -11.5 V; $+4.65$ V, and $+5.35$ V

9.10 Emitter voltage drops from $+0.7$ V to approximately $+0.5$ V, and collector voltage rises from -7 V to approximately -5 V.

9.19 (a) -0.7, 0, $+0.7$, $+0.7$ V; (b) No change

9.20 0, -0.7, $+10$, $+10.7$, 0 V

9.21 $\beta = 100$: -0.1, -0.8, $+10.3$, $+11.0$, -0.4;
$\beta = 10$: -0.83, -1.53, $+12.39$, $+13.09$, -3.25.

9.24 0, $+0.7$, -7.3, -8 V

9.29 (a) $V_A = V_B = +0.7$ V; $\dfrac{v_o}{v_i} = -29.6$; $R_i = 327$ Ω; peak-to-peak voltage at output = 0.6 V and at input = 20 mV.

9.30 $+1.49$ V; -47.4 V/V; 189 Ω; 20 mV peak-to-peak.

9.39 $R_B = 62.5$ kΩ; $R_C = 4.7$ kΩ.

CHAPTER 10

10.2 9.14 kΩ; 3.36 kΩ

10.3 -4.7 V/V; -99 V/V; -4.2 V/V; -76.3 V/V

10.5 40%; 71.4%

10.8 -1.1, -1.8, and -2.5 V; 125 Ω

10.10 $+1.54$ V; 1.24-V peak; -305; 9.1 kΩ; 295 Ω

10.12 $+1.4$ V and $+0.7$ V; (a) -344 V/V, 47 kΩ, and 23.3 Ω; (b) -344 V/V, 272 Ω, and 23.3 Ω; 1.1-V peak

10.13 $+2.3$ V, $+2.3$ V, $+1.6$ V; V_B changes to $+2.2$ V; 0.4 V and 0.3 V peak; (a) -800 V/ V, 12.5 Ω; (b) -72.6 V/V, 123 Ω

10.15 (a) 2 V; (b) 2.8 V

10.21 $0.5; 0.5/\left(1 + \dfrac{1.5}{\beta}\right)$

10.25 $+0.12$ V

10.27 20.2 kΩ; 25 V/V

10.28 Differential output: CMG $= 10^{-3}$, CMRR $= 1.05 \times 10^5$ or 100.4 dB; Single-ended output: CMG $= 11 \times 10^{-3}$, CMRR $= 5 \times 10^3$ or 74 dB

10.33 1 MΩ

10.34 -3 V; 11.3 V/V

10.38 (a) $v_{id} = \sqrt{2I/\beta}$

(b) $g_m = \sqrt{\beta I}$

(c) $i_{D1} = \dfrac{I}{2} + \sqrt{\beta I}\left(\dfrac{v_{id}}{2}\right)\sqrt{1 - \left(\dfrac{\beta v_{id}^2}{4I}\right)},$

$i_{D2} = \dfrac{I}{2} - \sqrt{\beta I}\left(\dfrac{v_{id}}{2}\right)\sqrt{1 - \left(\dfrac{\beta v_{id}^2}{4I}\right)};$

small-signal condition: $v_{id} \ll 2\sqrt{I/\beta}$

CHAPTER 11

11.2 Two zeros at $s = \infty$; poles at $s = -10$ and $s = -10^6$ rad/s; gain values are 80, 77, 60, 0 and -23 dB; $-95.7°$

11.5 2.3 rad/s; 3 rad/s

11.10 1.11 MΩ; 1.19 MΩ

11.11 0.4 A/V; 250 Ω; 100 Ω; 5 MΩ; 10 kΩ

11.12 125.3 pF; 500 MHz

11.16 9.1 V/V; 1061 MHz and 14.6 MHz; 14.6 MHz

11.18 0.49 V/V; 1.35 MHz

11.22 8.9 V/V; 30.3 MHz

11.30 -7.3 V/V; 1.3 Hz; 173.6 kHz

11.31 -50 V/V; 31.8 Hz; 15.6 kHz

CHAPTER 12

12.1 9.09 V/V; 10.11; 19.1 dB; 20 dB

12.3 Signal-to-noise ratio increases by a factor of 10.

12.5 6.49 kHz

12.7 76.3 V/V; 1 V/V; 0.987; 1.54 MΩ; 2 Ω

12.9 40.25 A/V; 50 Ω; -12; 18.65 MΩ

12.10 (a) -4.16 V/V, 165 Ω, 495 Ω; (b) -2.19 V/V, 94 Ω, 268.5 Ω; zero frequency $= 212.8$ rad/s; pole frequency $= 112$ rad/s

12.14 1.1×10^5 rad/s; 28 dB; 1.5×10^{-3}

12.17 For the case in which the pole of the β network is at 10^3 rad/s: $\beta_{cr} = 0.0242$; β for $Q = 0.707$ is 2.33×10^{-3} and corresponding $\omega_0 = 4.85 \times 10^3$ rad/s and low-frequency gain $\simeq 300$ V/V

12.18 10^4 Hz; 45°

12.21 118 pF; 59 pF

12.22   0.89 V/V; 103.5 kΩ; 124 Ω

12.25   −5.58 V/V; 145.6 kΩ; 5.63 kΩ

12.28   Differentiator with time constant of 0.05 s

CHAPTER 13

13.2 $\dfrac{I_2}{I_1} = \dfrac{I_{S2}/I_{S1}}{1 + 1/\beta_1 + \left(\dfrac{I_{S2}}{I_{S1}}\right)\left(\dfrac{1}{\beta_2}\right)}$

13.4 $I_S = 3.5 \times 10^{-14}$ A, that is, 3.5 times as large as the standard *npn* devices used.

13.7   0.94 mV

13.10   40.2 dB

13.12   Loop gain = 913 V/V; CMRR = 99.4 dB

13.14   115 mW; 26%; 46%

13.15 $I_{bias} \simeq 0.09$ mA

13.19   5.67 MHz; 5.3 pF; 23.3 Hz; gain increases by 15 dB

CHAPTER 14

14.3 $T(s) = 11.55/[s^2 + 0.557s + 1.156]$; $\omega_0 = 1.075$ rad/s; $Q = 1.93$

14.4 $Q > \dfrac{1}{\sqrt{2}}$; $\omega_0 \bigg/ \sqrt{1 - \dfrac{1}{2Q^2}}$; $n_2 Q \bigg/ \sqrt{1 - \dfrac{1}{4Q^2}}$

14.6 $T(s) = 500s/[s^2 + 50s + 10^6]$; 0.05; 0.05

14.8   1 H; 100 kΩ

14.10 $m = 4Q^2$; $CR = 2Q/\omega_0$

14.13   Zeros: $-1 \pm j$; poles: -0.586 and -3.414; $\omega_0 = \sqrt{2}$ rad/s; $Q = 1/\sqrt{2}$

14.15 $\dfrac{2Q^2}{2Q^2 + 1}$; ω_0

14.18 $R_3 = 1$ MΩ; $R_4/\alpha = 50$ kΩ; $R_4/(1 - \alpha) = 12.5$ kΩ

14.20 $C_3 = 0.0707$ μF; $C_4 = 0.1414$ μF

14.21 $S_R^{\omega_0} = 0$; $S_L^{\omega_0} = -0.5$; $S_C^{\omega_0} = -0.5$; $S_R^Q = 1$; $S_L^Q = -0.5$; $S_C^Q = +0.5$

14.25   High-pass

14.27   1.59 nF; 7.07 kΩ; 10 kΩ

14.30   1 pF; 1 pF; 1.414 pF; 1 pF

14.31 $\omega_0 = 1/\sqrt{LC_1 C_2/(C_1 + C_2)}$; $g_m R_d = C_2/C_1$

14.32   20

14.33   Original circuit: $\dfrac{d\phi}{d\omega} = \dfrac{2}{3}$; Modified circuit: $\dfrac{d\phi}{d\omega} = \dfrac{20}{21}$

14.38   To double the output amplitude, add a diode in series with each of the diodes in Fig. 14.27; $R = 1$ kΩ; $QR = 20$ kΩ

CHAPTER 15

15.1   0.691 V; 0.575 V; 0.575 V; 0.592 V

15.2   13.6 mV; 49.4 mV

15.3   19.9 mV

15.8   0.95 Ω

15.10   0.72 V

15.12 $\Delta 0 = 0.3$ V; $\Delta 1 = 0.15$ V

15.13 Ignoring the effect of changing V_{CC} on the value of V_{IH}, we find that by raising V_{CC} to $+4.35$ V the two noise margins become equal to 0.4 V.

15.15 $\Delta V_{BE} = V_T \ln \beta_F$; 98 mV for $\beta_F = 50$

15.17 40 ns

15.19 133 ns

15.22 11.3 ns

15.23 6.7 ns

15.24 (a) 48.3 mW; (b) 50 mW

15.29 0.67β

15.30 42.6 mV

15.31 28.125 mA

15.33 835 Ω

15.35 (a) 6.5 ns; (b) 87.6 ns

15.37 0.1025 mA; 0.1025 mA

15.40 3.46 V

15.44 11.6 mW; 12.2 mW; 0.47 ns

15.48 (a) $I_{IL} = 1.71$ mA, $I_{CC} = 2.93$ mA; (b) $I_{IH} = 24$ μA, $I_{CC} = 6.4$ mA

15.55 26.15 mW; 27.8 mW; 74.8 mW

15.59 3.95 MHz

CHAPTER 16

16.1 8; 10; 10; 12; 12; 14; 16

16.4 391

16.6 1600 pF; 400 pF; In the higher voltage case, could use smaller capacitances, smaller connections, and greater packing density

16.9 210 mW; $P = 200 + 100f$, mW where f is in MHz

16.13 0.6 μA/V^2

16.15 48.3 ns

16.16 $+3.8$ V; 15.1 μA/V^2; 14.9 μA/V^2

16.18 1 μA (source) and 36.5 μA (sink); 15 μs (rise) and 0.41 μs (fall)

16.21 2.3 to 1

16.23 $\beta_{load} = 0.4$ mA/V^2; $\beta_{switch} = 0.5$ mA/V^2

16.26 β of Q_s is $1/25$ of original β; slower rise time

16.32 32 μs

16.33 $-8.2°$C to $+58.2°$C

16.37 System can have 9 D flip-flops for a total of 360 device pairs.

REFERENCES AND SOURCES FOR FURTHER READING

GENERAL

G.1 E. J. Angelo, Jr., *Electronics: BJTs, FETs and Microcircuits,* New York: McGraw-Hill, 1969.

G.2 D. J. Comer, *Modern Electronic Circuit Design,* Reading, Mass.: Addison-Wesley, 1976.

G.3 *Data Books* and *Application Notes* from the various semiconductor manufacturers such as Motorola, Fairchild, Texas Instruments, Intel, Mostek, RCA, etc.

G.4 *Digest of Technical Papers of the International Solid-State Circuits Conference,* held annually, sponsored by the IEEE.

G.5 *Electronics,* a magazine published biweekly by McGraw-Hill.

G.6 M. S. Ghausi, *Electronic Circuits,* New York: Van Nostrand Reinhold, 1971.

G.7 P. R. Gray and R. G. Meyer, *Analysis and Design of Analog Integrated Circuits,* New York: Wiley, 1977.

G.8 P. E. Gray and C. L. Searle, *Electronic Principles,* New York: Wiley, 1969.

G.9 A. B. Grebene, *Analog Integrated Circuit Design,* New York: Van Nostrand Reinhold, 1972.

G.10 V. H. Grinich and H. G. Jackson, *Introduction to Integrated Circuits,* New York: McGraw-Hill, 1975.

G.11 D. J. Hamilton and W. G. Howard, *Basic Integrated Circuit Engineering,* New York: McGraw-Hill, 1975.

G.12 W. H. Hayt and G. W. Neudeck, *Electronic Circuit Analysis and Design,* Boston: Houghton Mifflin, 1976.

G.13 C. A. Holt, *Electronic Circuits,* New York: Wiley, 1978.

G.14 *IEEE Journal of Solid State Circuits,* published bimonthly by the Institute of Electrical and Electronics Engineers.

G.15 J. Millman, *Microelectronics,* New York: McGraw-Hill, 1979.

G.16 J. Millman and C. Halkias, *Integrated Electronics,* New York: McGraw-Hill, 1972.

G.17 S. D. Senturia and B. D. Wedlock, *Electronic Circuits and Applications,* New York: Wiley, 1975.

G.18 H. Taub and D. Schilling, *Digital Integrated Electronics,* New York: McGraw-Hill, 1977.

CHAPTER 1

1.1 M. E. Van Valkenburg, *Network Analysis,* 3rd ed., Englewood Cliffs, N.J.: Prentice-Hall, 1974. Fourier Series and Fourier Transforms are studied in Chaps. 15 and 16.

1.2 P. R. Gray and R. G. Meyer, *Analysis and Design of Analog Integrated Circuits,* Chap. 11, New York: Wiley, 1977.

1.3 *Scientific American,* special issue on microelectronics, Sept. 1977.

1.4 B. M. Oliver, "The role of microelectronics in instrumentation and control," *Scientific American,* vol. 237, no. 3, pp. 180–190, Sept. 1977.

CHAPTER 2

2.1 M. E. Van Valkenburg, *Network Analysis,* 3rd ed., Englewood Cliffs, N.J.: Prentice-Hall, 1974.

2.2 W. H. Hayt and J. E. Kemmerly, *Engineering Circuit Analysis,* 3rd ed., New York: McGraw-Hill, 1978.

2.3 R. Littauer, *Pulse Electronics,* New York: McGraw-Hill, 1965.

2.4 J. Millman and H. Taub, *Pulse, Digital, and Switching Waveforms,* New York: McGraw-Hill, 1965.

CHAPTER 3

3.1 G. B. Clayton, *Operational Amplifiers,* 2nd ed., London: Newnes-Butterworths, 1979.

3.2 G. B. Clayton, *Experimenting with Operational Amplifiers,* London: MacMillan, 1975.

3.3 J. G. Graeme, G. E. Tobey, and L. P. Huelsman, *Operational Amplifiers: Design and Applications,* New York: McGraw-Hill, 1971.

3.4 W. Jung, *IC Op Amp Cookbook,* Indianapolis, Ind.: Howard Sams, 1974.

3.5 J. K. Roberge, *Operational Amplifiers: Theory and Practice,* New York: Wiley, 1975.

3.6 J. I. Smith, *Modern Operational Circuit Design,* New York: Wiley-Interscience, 1971.

3.7 J. V. Wait, L. P. Huelsman, and G. A. Korn, *Introduction to Operational Amplifier Theory and Applications,* New York: McGraw-Hill, 1975.

CHAPTER 4

4.1 See the general references, in particular G.1 and G.8.

CHAPTER 5

5.1 See the references listed for Chapter 3.

5.2 *Nonlinear Circuits Handbook,* Norwood, Mass.: Analog Devices, Inc., 1976.

CHAPTER 6

6.1 H. Taub and D. Schilling, *Digital Integrated Electronics,* New York: McGraw-Hill, 1977.

6.2 P. M. Chirlian, *Digital Circuits,* Portland, Ore.: Matrix, 1976.

6.3 J. B. Peatman, *The Design of Digital Systems,* McGraw-Hill, 1972.

6.4 J. B. Peatman, *Digital Hardware Design,* New York: McGraw-Hill, 1980.

6.5 B. P. Lathi, *Signals, Systems and Communication,* Chap. 11, New York: Wiley, 1965.

6.6 D. F. Hoeschele, *Analog-to-Digital/Digital-to-Analog Conversion Techniques,* New York: Wiley, 1968.

6.7 A. B. Grebene, *Analog Integrated Circuit Design,* Chap. 10, New York: Van Nostrand Reinhold, 1972.

6.8   *Analog-Digital Conversion Notes,* Norwood, Mass: Analog Devices, Inc., 1977.

6.9   V. C. Hamacher, Z. G. Vranesic, and S. G. Zaky, *Computer Organization,* New York: McGraw-Hill, 1978.

6.10   J. B. Peatman, *Microcomputer-Based Design,* 2nd ed., New York: McGraw-Hill, 1980.

6.11   D. P. Burton and A. L. Dexter, *Microprocessor Systems Handbook,* Norwood, Mass.: Analog Devices, Inc., 1977.

CHAPTER 7

7.1   See the general references, in particular G.1, G.12, and G.17.

7.2   R. S. C. Cobbold, *Theory and Applications of Field-Effect Transistors,* New York: Wiley, 1969.

7.3   A. D. Evans (Ed.), *Designing with Field-Effect Transistors,* New York: McGraw-Hill, 1981.

7.4   *Analog Switches and Their Applications,* Santa Clara, Calif: Siliconix.

CHAPTER 8

8.1   See the general references, in particular G.1, G.11, and G.12.

8.2   R. S. C. Cobbold, *Theory and Applications of Field-Effect Transistors,* New York: Wiley, 1969.

8.3   P. R. Gray, D. A. Hodges, and R. W. Brodersen, *Analog MOS Integrated Circuits,* New York: IEEE Press, 1980.

8.4   Y. P. Tsividis and P. R. Gray, "An integrated NMOS operational amplifier with internal compensation," *IEEE Journal of Solid-State Circuits,* vol. SC-11, pp. 748–754, Dec. 1976. Reprinted in Ref. 8.3.

8.5   W. N. Carr and J. P. Mize, *MOS/LSI Design and Application,* New York: McGraw-Hill, 1972.

8.6   W. M. Penny and L. Lau (Eds.), *MOS Integrated Circuits,* New York: Van Nostrand Reinhold, 1972.

8.7   H. Taub and D. Schilling, *Digital Integrated Electronics,* New York: McGraw-Hill, 1977.

8.8   *McMOS Handbook,* Phoenix, Ariz.: Motorola Inc., 1974.

CHAPTER 9

9.1   See the general references, in particular G.1, G.7, G.8, and G.18.

9.2   C. L. Searle, A. R. Boothroyd, E. J. Angelo, Jr., P. E. Gray, and D. O. Pederson, *Elementary Circuit Properties of Transistors,* Vol. 3 of the SEEC Series, New York: Wiley, 1964.

9.3   I. Getreu, *Modeling the Bipolar Transistor,* Beaverton, Ore.: Tektronix, Inc., 1976.

9.4   J. N. Harris, P. E. Gray, and C. L. Searle, *Digital Transistor Circuits,* Vol. 6 of the SEEC Series, New York: Wiley, 1966.

9.5   J. Millman and H. Taub, *Pulse, Digital, and Switching Waveforms,* Chap. 20, New York: McGraw-Hill, 1965.

CHAPTER 10

10.1   See the general references, in particular G.7, G.8, and G.11.

10.2   F. C. Fitchen, *Electronic Integrated Circuits and Systems,* New York: Van Nostrand Reinhold, 1970.

10.3 J. K. Roberge, *Operational Amplifiers: Theory and Practice,* New York: Wiley, 1975.

10.4 J. N. Giles, *Linear Integrated Circuits Applications Handbook,* Mountain View, Calif.: Fairchild Semiconductors, 1967.

10.5 *Linear Integrated Circuits and MOS Devices: Application Notes,* Somerville, N.J.: RCA, 1974.

CHAPTER 11

11.1 M. E. Van Valkenburg, *Network Analysis,* 3rd ed., Englewood Cliffs, N.J.: Prentice-Hall, 1974.

11.2 W. H. Hayt and J. E. Kemmerly, *Engineering Circuit Analysis,* 3rd ed., New York: McGraw-Hill, 1978.

11.3 See the general references, in particular G.7, G.8, and G.9.

CHAPTER 12

12.1 See the general references, in particular G.7, G.8 (perhaps the most thorough treatment of the subject), and G.9.

12.2 S. S. Haykin, *Active Network Theory,* Reading, Mass.: Addison-Wesley, 1970.

12.3 W.-K. Chen, *Active Network and Feedback Amplifier Theory,* New York: McGraw-Hill, 1980.

12.4 E. S. Kuh and R. A. Rohrer, *Theory of Linear Active Networks,* San Francisco: Holden-Day, Inc., 1967. (This is an advanced-level text.)

12.5 G. S. Moschytz, *Linear Integrated Networks: Design,* New York: Van Nostrand Reinhold, 1974.

12.6 *Linear Integrated Circuits,* Harrison, N.J.: RCA, 1967.

12.7 E. Renschler, *The MC1539 Operational Amplifier and Its Applications,* Application Note AN-439, Phoenix, Ariz: Motorola Semiconductor Products.

12.8 J. K. Roberge, *Operational Amplifiers: Theory and Practice,* New York: Wiley, 1975.

CHAPTER 13

13.1 P. R. Gray and R. G. Meyer, *Analysis and Design of Analog Integrated Circuits,* New York: Wiley, 1977.

13.2 A. B. Grebene, *Analog Integrated Circuit Design,* New York: Van Nostrand Reinhold, 1972.

13.3 J. A. Connely (Ed.), *Analog Integrated Circuits,* New York: Wiley-Interscience, 1975.

13.4 R. G. Meyer (Ed.), *Integrated-Circuit Operational Amplifiers,* New York: IEEE Press, 1978.

13.5 A. B. Grebene (Ed.), *Analog Integrated Circuits,* New York: IEEE Press, 1978.

13.6 *IEEE Journal of Solid-State Circuits.* The December issue of every year has been devoted to analog ICs.

13.7 (a) R. J. Widlar, "Some circuit design techniques for linear integrated circuits," *IEEE Transactions on Circuit Theory,* vol. CT-12, pp. 586–590, Dec. 1965.

     (b) —— "Design techniques for monolithic operational amplifiers," *IEEE Journal of Solid-State Circuits,* vol. SC-4, pp. 184–191, August 1969.

13.8 J. E. Solomon, "The monolithic op amp: A tutorial study," *IEEE Journal of Solid-State Circuits,* vol. SC-9, no. 6, pp. 314–332, Dec. 1974.

13.9 P. R. Gray, D. A. Hodges, and R. W. Brodersen, *Analog MOS Integrated Circuits,* New York: IEEE Press, 1980.

CHAPTER 14

14.1 A. S. Sedra and P. O. Brackett, *Filter Theory and Design: Active and Passive,* Portland, Ore.: Matrix, 1978.

14.2 R. Schaumann, M. Soderstrand, and K. Laker (Eds.), *Modern Active Filter Design,* New York: IEEE Press, 1981.

14.3 A. S. Sedra and L. Brown, "A refined classification of single-amplifier filters," *Int. Journal of Circuit Theory and Applications,* vol. 7, pp. 127–137, March 1979. (Reprinted in Ref. 14.2).

14.4 A. S. Sedra, M. Ghorab, and K. Martin, "Optimum configurations for single-amplifier biquadratic filters," *IEEE Transactions on Circuits and Systems,* vol. CAS-27, no. 12, pp. 1155–1163, Dec. 1980.

14.5 K. Martin, "Improved circuits for the realization of switched-capacitor filters," *IEEE Transactions on Circuits and Systems,* vol. CAS-27, no. 4, pp. 237–244, April 1980.

14.6 R. W. Brodersen, P. R. Gray, and D. A. Hodges, "MOS switched-capacitor filters," *Proceedings of the IEEE,* vol. 67, pp. 61–74, Jan. 1979.

14.7 M. S. Ghausi and K. Laker, *Modern Filter Design,* Chap. 6, Englewood Cliffs, N.J.: Prentice-Hall, 1981.

14.8 L. Strauss, *Wave Generation and Shaping,* 2nd ed., New York: McGraw-Hill, 1970.

14.9 J. G. Graeme, G. E. Tobey, and L. P. Huelsman, *Operational Amplifiers: Design and Applications,* New York: McGraw-Hill, 1971.

14.10 W. Jung, *IC Op Amp Cookbook,* Indianapolis, Ind.: Howard Sams, 1974.

14.11 J. K. Roberge, *Operational Amplifiers: Theory and Practice,* New York: Wiley, 1975.

14.12 J. I. Smith, *Modern Operational Circuit Design,* New York: Wiley-Interscience, 1971.

14.13 J. V. Wait, L. P. Huelsman, and G. A. Korn, *Introduction to Operational Amplifier Theory and Applications,* New York: McGraw-Hill, 1975.

14.14 M. E. Frerking, *Crystal Oscillator Design and Temperature Compensation,* New York: Van Nostrand Reinhold, 1978.

CHAPTER 15

15.1 J. N. Harris, P. E. Gray, and C. L. Searle, *Digital Transistor Circuits,* Vol. 6 of the SEEC Series, New York: Wiley, 1966.

15.2 I. Getreu, *Modeling the Bipolar Transistor,* Beaverton, Ore.: Tektronix Inc., 1976.

15.3 H. Taub and D. Schilling, *Digital Integrated Electronics,* New York: McGraw-Hill, 1977.

15.4 L. S. Garrett, "Integrated-circuit digital logic families," a three-part article published in *IEEE Spectrum,* Oct., Nov., and Dec. 1970.

15.5 J. E. Smith (Ed.), *Integrated Injection Logic,* New York: IEEE Press, 1980.

15.6 L. Strauss, *Wave Generation and Shaping,* 2nd ed., New York: McGraw-Hill, 1970.

15.7 Texas Instruments Staff, *Designing with TTL Integrated Circuits,* New York: McGraw-Hill, 1971.

15.8 *TTL Data Book,* Mountain View, Calif.: Fairchild Camera and Instruments Corp., Dec. 1978.

15.9 *MECL High-Speed Integrated Circuits,* Phoenix, Ariz.: Motorola Semiconductor Products, Inc., 1978.

15.10 *MECL System Design Handbook,* Phoenix, Ariz.: Motorola Semiconductor Products, Inc., 1972.

15.11 *McMOS Handbook,* Phoenix, Ariz.: Motorola Inc., 1974.

15.12 *COS/MOS Digital Integrated Circuits,* Publication No. SSD-203B, Somerville, N.J.: RCA Solid-State Division, 1974.

15.13 *Linear and Digital Semicustom IC Design Programs,* Sunnyvale, Calif.: Exar Integrated Systems, p. 25, Nov. 1979.

15.14 M. I. El-Masry (Ed.), *MOS Digital Circuits,* New York: IEEE Press, 1981.

15.15 *IEEE Journal of Solid-State Circuits.* The October issue of every year has been devoted to digital circuits.

CHAPTER 16

16.1 D. A. Hodges, "Microelectronic memories," *Scientific American,* vol. 237, no. 3, pp. 130–145, Sept. 1977.

16.2 K. C. Smith and A. S. Sedra, "Main Memory," in *Encyclopedia of Computer Science,* A. Ralston and C. L. Meek (Eds.), pp. 881–889, New York: Petrocelli/Charter, 1976.

16.3 D. A. Hodges (Ed.), *Semiconductor Memories,* New York: IEEE Press, 1972.

16.4 *1980 Memory Data Book and Design Guide,* Carrolton, Tex.: Mostek Corporation.

16.5 *Memory Design Handbook,* Santa Clara, Calif.: Intel Corporation, 1977.

16.6 Intel, *The Semiconductor Memory Book,* New York: Wiley-Interscience, 1978.

16.7 P. R. Schroeder and R. J. Proebsting, *Digest of Technical Papers, 1977 IEEE International Solid-State Circuits Conference,* pp. 12–13, Feb. 1977.

16.8 G. R. M. Rao and J. Hewkin, "64-K dynamic RAM needs only 5-volt supply to outstrip 16-K parts," *Electronics,* Sept. 28, 1978.

16.9 *65,536-BIT Dynamic Random Access Memory-TMS 4164 Data Sheet,* Dallas, Tex.: Texas Instruments Inc., July 1980.

16.10 R. Sud and K. C. Hardee, "16-K static RAM takes new route to high speed," *Electronics,* pp. 117–123, Sept. 11, 1980.

16.11 K. C. Hardee and R. Sud, "A fault-tolerant 30ns/375mw 16K $\times$ 1 NMOS static RAM," *IEEE Journal of Solid-State Circuits,* vol. SC-16, no. 5, pp. 435–443, Oct. 1981.

16.12 *IMS 1400 High Performance 16K Static RAM, Data Sheet,* Colorado Springs, Colo. INMOS Corporation, 1980.

16.13 S. D. Kang, J. D. Allan, B. Ashmore, T. H. Herndon, S. Wolpert, and W. C. Bruncke, "A 30-ns 16K $\times$ 1 fully static RAM," *IEEE Journal of Solid-State Circuits,* vol. SC-16, no. 5, pp. 444–448, Oct. 1981.

16.14 *EPROM Applications Manual,* Santa Clara, Calif.: Intel Corporation, Sept. 1980.

16.15 R. Melen and D. Buss (Eds.), *Charge Coupled Devices: Technology and Applications,* New York: IEEE Press, 1977.

16.16 M. Denzin, "The CCD answer to memory questions," *Progress,* vol. 8, no. 5/6, pp. 4–11, Oct. 1980.

16.17 S. Kazmi, "Design prototypes quickly with programmable arrays," *Electronic Design,* pp. 121–124, Feb. 19, 1981.

16.18 J. G. Posa, "Gate arrays," *Electronics,* pp. 145–158, Sept. 25, 1980.

16.19 *IEEE Spectrum, Special Issue on Technology '81,* Jan. 1981.

16.20 C. Mead and L. Conway, *Introduction to VLSI Systems,* Reading, Mass: Addison-Wesley, 1980.

16.21 *IEEE Journal of Solid-State Circuits.* The October issue of each year has been devoted to digital circuits.

APPENDIX A

A.1 A. B. Grebene, *Analog Integrated Circuit Design,* New York: Van Nostrand Reinhold, 1972.

A.2 D. J. Hamilton and W. G. Howard, *Basic Integrated Circuit Engineering,* New York: McGraw-Hill, 1975.

INDEX